Apoptotic Cell Biology: Clearance and Mechanisms

Apoptotic Cell Biology: Clearance and Mechanisms

Editor: Luke Foster

AMERICAN
MEDICAL PUBLISHERS
www.americanmedicalpublishers.com

AMERICAN
MEDICAL PUBLISHERS
www.americanmedicalpublishers.com

Cataloging-in-Publication Data

Apoptotic cell biology : clearance and mechanisms / edited by Luke Foster.
p. cm.
Includes bibliographical references and index.
ISBN 978-1-63927-041-5
1. Cytology. 2. Cells. 3. Biology. I. Foster, Luke.
QH581.2 .A66 2022
571.6--dc23

American Medical Publishers,
41 Flatbush Avenue,
1st Floor, New York,
NY 11217, USA

ISBN 978-1-63927-041-5 (Hardback)

Contents

Preface

The total number of cells in every multicellular organism is fundamentally fixed to a specific range. A type of programmed cell death happens in such organisms to effectively control the number of cells. This programmed cell death is called apoptosis. Characteristic cell changes and eventually death happens in cells due to certain biochemical events. It is a carefully controlled, energy-dependent process. An average adult human loses between 50 and 70 billion cells each day due to apoptosis, and an average human child loses about 20 to 30 billion cells. Apoptosis can be pathogenic when the death of healthy neurons leads to neurodegenerative diseases such as Parkinson's and Alzheimer's disease. This book contains some path-breaking studies in the field of apoptotic cell biology. It outlines the processes and mechanisms of apoptotic cells in detail. Researchers and students in this field will be greatly assisted by this book.

This book is a comprehensive compilation of works of different researchers from varied parts of the world. It includes valuable experiences of the researchers with the sole objective of providing the readers (learners) with a proper knowledge of the concerned field. This book will be beneficial in evoking inspiration and enhancing the knowledge of the interested readers.

In the end, I would like to extend my heartiest thanks to the authors who worked with great determination on their chapters. I also appreciate the publisher's support in the course of the book. I would also like to deeply acknowledge my family who stood by me as a source of inspiration during the project.

Editor

1

Apoptotic Cell Clearance in *Drosophila melanogaster*

Qian Zheng[1], AiYing Ma[1,2], Lei Yuan[1], Ning Gao[1], Qi Feng[1], Nathalie C. Franc[3*†] and Hui Xiao[1*†]

[1] Key Laboratory of the Ministry of Education for Medicinal Plant Resources and Natural Pharmaceutical Chemistry, National Engineering Laboratory for Resource Development of Endangered Crude Drugs in the Northwest of China, College of Life Sciences, Shaanxi Normal University Xi'an, Xi'an, China, [2] College of Biological Sciences and Engineering, Beifang University of Nationalities, Yinchuan, NingXia, China, [3] Department of Cell and Molecular Biology, The Scripps Research Institute, La Jolla, CA, United States

The swift clearance of apoptotic cells (ACs) (efferocytosis) by phagocytes is a critical event during development of all multicellular organisms. It is achieved through phagocytosis by professional or amateur phagocytes. Failure in this process can lead to the development of inflammatory autoimmune or neurodegenerative diseases. AC clearance has been conserved throughout evolution, although many details in its mechanisms remain to be explored. It has been studied in the context of mammalian macrophages, and in the nematode *Caenorhabditis elegans*, which lacks "professional" phagocytes such as macrophages, but in which other cell types can engulf apoptotic corpses. In *Drosophila melanogaster*, ACs are engulfed by macrophages, glial, and epithelial cells. *Drosophila* macrophages perform similar functions to those of mammalian macrophages. They are professional phagocytes that participate in phagocytosis of ACs and pathogens. Study of AC clearance in *Drosophila* has identified some key elements, like the receptors Croquemort and Draper, promoting *Drosophila* as a suitable model to genetically dissect this process. In this review, we survey recent works of AC clearance pathways in *Drosophila*, and discuss the physiological outcomes and consequences of this process.

Edited by:
Estee Kurant,
University of Haifa, Israel

Reviewed by:
Yoshinobu Nakanishi,
Kanazawa University, Japan
Kimberly McCall,
Boston University,
United States

***Correspondence:**
Nathalie C. Franc
nfranc@scripps.edu;
Hui Xiao
huixiao@snnu.edu.cn

† These authors have contributed equally to this work.

Keywords: phagocytosis, apoptosis, macrophages, signaling pathways, *Drosophila melanogaster*

INTRODUCTION

Programmed cell death is necessary for normal development and growth in multicellular organisms, which produce billions of apoptotic cells (ACs) daily (1, 2). Exogenous pathogenic microbes also threaten organisms' lives and development (3). Swift and efficient removal of ACs and pathogens is essential for maintaining tissue homeostasis. Failure in this process results in the release of potentially cytotoxic or antigenic molecules, causing inflammatory diseases or developmental autoimmune disorders (4–7). To clear ACs and pathogens, multicellular organisms have evolved a conserved cellular process named phagocytosis that is being carried out either by non-professional or professional phagocytes (8). The molecular mechanism of ACs clearance has been extensively studied in *Caenorhabditis elegans*, thus revealing relatively clear and detailed engulfment pathways (9). However, *C. elegans* lacks the professional phagocytes; instead ACs are engulfed by many neighboring cell types (10). Absence of a professional immune system in *C. elegans* may limit the extent to which these data can be applied to higher organisms. The fruitfly *Drosophila melanogaster* has also been used as a suitable model to study ACs clearance, in which ACs are engulfed by both non-professional phagocytes such as epithelial cells and professional phagocytes such as macrophages/hemocytes and glial cells (11), providing the advantages for studying phagocytosis in mammals. ACs clearance proceeds when ACs expose "eat me" signals, which are recognized by phagocytes, thereby triggering signaling cascades that lead to internalization of the apoptotic corpse and its degradation by the phagocytic vacuole known as phagosome matures by fusing with lysosomes (12, 13). In this review, we will summarize the current research on phagocytosis of ACs in *D. melanogaster*, and which signaling pathways regulate this process, thereby giving a systematic and general overview of this process.

SIGNALING BY ACs IN PHAGOCYTOSIS

Apoptotic cells generated by programmed cell death or physical wounds are quickly and silently removed, to maintain tissue homeostasis or prevent auto-inflammatory responses (14, 15). Once cells begin to undergo apoptosis, cell death pathway is activated, and they release multiple signaling to recruit phagocytes, which contains three steps: the release of "find me" signals, the presentation of "eat me" signals, and the removal of "don't eat me" signals (16).

At the beginning of cell death, "Find me" signals are released from ACs to promote the migration of phagocytes to ACs. Lauber firstly identified lysophosphatidylcholine (LPC) as a "find me" signal, which is released from ACs in a caspase-3-dependent manner. They furthermore showed that the activation of calcium-independent phospholipase A2 by caspase cleavage contributed to the release of LPC (17). Two other molecules, sphingosine-1-phosphate produced by sphingosine kinase in a caspase-dependent manner, and CX3CL1/fractalkine synthesized as a membrane-associated protein, have also been proposed to act as "find me" signals (18, 19). ATP and UTP that are released from ACs in a caspase-dependent manner have also recently been shown to act as "find me" signals for phagocytes (20). Whether these proposed "find me" signals are redundant or synergistic remains to be studied. Little evidence has shown that "find me" signals exist in *Drosophila*, but previous study revealed that H_2O_2 may be the immediate damage signal essential for the recruitment of hemocytes to wound regions in *Drosophila* embryos (21). Further research (22) found that Src42A–Draper–Shark signaling was important to recruitment of hemocytes by responding to wound-induced H_2O_2 in *Drosophila* embryos, which indicated that H_2O_2 may be *Drosophila* "find me" signal and Draper is responsible for the signal recognition. However, more evidence needs to be explored to verify this hypothesis.

"Don't eat me" signals (also known as self-associated molecular patterns) exist on healthy cells, playing inhibitory roles to prevent to be engulfed by phagocytes. Some examples of "don't eat me" signals include CD31, CD46, and CD47 in mammals (23).

"Eat me" signals are ligands, which can bind to engulfment receptors by moving to the surface of ACs. Engulfment receptors recognize and bind either directly to the apoptotic "eat me" signal, or through bridging molecules that bind the "eat me" signal. The best-studied and evolutionarily conserved "eat me" signal reported in human, *Drosophila*, and *C. elegans* is phosphatidylserine (PS) (24), a phospholipid exposed on the surface of ACs (25, 26). PS is a plasma membrane (PM) aminophospholipid maintained on the inner leaflet of live cells through aminophospholipid translocase activity (27, 28). After cell induced by apoptosis, aminophospholipid translocase is inactivated while a scramblase is activated to induce PS exposed to the cell surface in an ATP-independent manner (28). A recent study has shown that ACs can generate molecular memory in macrophages, priming them to recognize tissue wounds or microbes (29). This subsequently causes macrophages to produce pro-inflammatory signals and boost the innate response at sites associated with extensive AC death in *Drosophila* (29).

ENGULFMENT RECEPTORS AND RELATED SIGNAL PATHWAYS

In *Drosophila*, there are three cell types reported to function as phagocytic cells: professional phagocytes—macrophages/hemocytes, glial cells, and non-professional phagocytes—epithelial cells (30–32). Hemocytes are macrophage-like cells reported to engulf ACs or dendrite debris during pruning of *Drosophila* sensory dendrites (33) and embryogenesis (34). *Drosophila* glia act much similar role in engulfing dying cells or degenerating axons of the nervous system as their counterparts in mammals (35), degenerating dendrites are primarily cleared by the epidermal epithelia (36).

"Eat me" signals secreted by ACs are recognized by engulfment receptors, which are specifically expressed on the surface of phagocytic cells. In *C. elegans*, two seemingly independent engulfment signaling pathways have been genetically identified, which share similar functions both in fly and mouse, indicating that the process of ACs clearance is evolutionarily conserved. CED-1, a conserved transmembrane receptor protein, Draper in fly, MEGF10 in mouse, which have similar function in recognizing ACs, transducts the phagocytotic signal through its adaptor protein CED-6 (dCed-6 in fly, GULP in mouse) to regulate downstream effectors (37, 38). The CED-2, -5, -10, and -12 signaling pathway is believed to act downstream of the PS receptor PSR-1, a *C. elegans* homolog of mammalian PSR (39), which relates to ACs cytoskeletal rearrangements. Some of the abovementioned genes possess *Drosophila* counterparts, suggesting that fruitfly phagocytes share similar pathways to engulf ACs. Meanwhile, *Drosophila* has its own engulfment receptor, yet a more detailed mechanism remains to be unveiled in *Drosophila*.

Croquemort

In 1996, Franc and colleagues cloned the first *Drosophila* engulfment receptor on embryonic macrophages, Croquemort (Crq), which shares 23% identity with human CD36. In mammals, CD36 act as a scavenger receptor engulfing ACs (40) and regulates the host inflammatory responses (41, 42). Crq expresses specifically on *Drosophila* plasmatocytes, which become macrophages as they encounter ACs from late stage 11 of embryogenesis (43). Using AC-labeling and Crq immunostaining experiments, Crq was shown to be required for efficient phagocytosis of ACs, which was also confirmed *in vivo* (34). Crq is structurally unrelated to either CED-1 or PSR-1 (34), and how it promotes phagocytosis, including the identity of its ligand, is still unknown (44).

In addition to macrophages clearing ACs during embryogenesis, epithelial cells are responsible for prompt clearance of degenerating neurites to maintain tissue homeostasis and prevent inflammatory responses during development (36). Knocking out *crq* results in AC clearance defects by macrophages; however, it has no effect on engulfment of dendrites in epithelia. Further studies showed that *crq* was required for phagosome maturation during this process, while loss-of-function of *crq* leads to homotypic phagosome fusion defect, though it is not necessary for phagosomes to progress through the Rab7$^+$ positive stage (43). Besides, recent research revealed that *crq* mutant flies are susceptible to environmental microbes and infection, and that Crq is required for engulfment of bacteria in parallel to the Toll and Imd pathways, which play key roles in the innate immune system (45).

Draper

Freeman and colleagues first identified the homolog of CED-1 in *Drosophila*, named Draper (Drpr), which strongly expressed on glial and macrophage membranes, and found that it was required for the engulfment of apoptotic neurons and for larval locomotion (35). Similar to CED-1 in nematode and MEGF10 in human, Drpr encodes 15 extracellular atypical EGF repeats, a single transmembrane domain, and a novel intracellular domain (35). Manaka and colleagues confirmed the role of Drpr in glia and hemocytes/macrophages, showing that it plays a role in the phagocytosis of ACs (44), suggesting that the Drpr pathway plays similar role in *Drosophila* as the ced-1/6/7 pathway in *C. elegans*. Glial cells expressing Drpr are essential for the pruning of *Drosophila* mushroom body γ neurons, Awasaki et al. detected that *Drosophila ced-6* (mouse *gulp*) expressed in the same glial cells as *drpr* (46), genetic evidence showed that *drpr* and *ced-6* played role in engulfing γ neuron axon in the same pathway, meanwhile, the experiment *in vitro* confirmed that Ced-6 N-terminal might interact with the intracellular region of Drpr (47). Different from *C. elegans* Ced-7, an ABC transporter, which both expresses in ACs and engulfment cells for efficient phagocytosis, the homolog in *Drosophila*, has not yet been studied. *Drosophila* Shark, a non-receptor tyrosine kinase also plays an important role in removing cell corpses or debris mediated by Drpr through binding to its intracellular domain (48). The Src family kinase Src42A phosphorylates Drpr to allow its intracellular domain to interact with Dmel/Ced-6, thus activating the Drpr pathway and promoting phagocytosis of pruned axons and degenerating neurons by glial cells (47).

In addition to the Ced-1, -6, and -7 signaling pathway, Ced-2, -5, -10, and -12 were found to act in a parallel and yet partially redundant pathway that controls actin cytoskeleton rearrangement in cell corpse engulfment and cell migration (49). For *Drosophila*, although the homologs of CED-2, -5, and -10 correspond to CG1587, myoblast city, and Rac2, respectively, their function in ACs clearance has not been deeply studied. The *Drosophila* homolog of Ced-12, Dmel/ced-12, was found to be required for cell clearance in macrophages, function in a genetically distinct pathway compared with Drpr, which further indicated that the phagocytosis signal pathways are evolutionary conserved (50).

Integrin

Integrins are conserved heterodimeric transmembrane receptors, forming by two subunits called α and β (51, 52). The involvement of integrins in phagocytosis of ACs was first described in mammals (53). Ina-1, an α subunit of *C. elegans* integrin, was also reported to participate in cell corpse removal (54). In *Drosophila*, there are five α- and two β-subunits. Nagaosa and colleagues found that loss-of-function of *Drosophila* integrin *βv* results in reduced levels of AC clearance, while reexpressing *βv* in integrin *βv*-lacking fly hemocytes rescues their phagocytosis-defective phenotype (55). Flies lacking either integrin *βv* or Drpr showed almost the same level of phagocytosis, while loss of these two receptors further decreased phagocytosis, which indicated that integrin *βv* and Drpr act independently. As Drpr was shown to act upstream of CED-6 and CED-10, the integrin *βv* appears to act upstream of the other engulfment pathway CED-2–CED-5–CED-12. However, Crk and Mbc, the *Drosophila* homologs of *C. elegans* CED-2 and CED-5 have not been observed to participate in the phagocytosis of ACs at least by embryonic hemocytes, thus the molecular signaling downstream of *βv* remains unknown (55). Further research indicated that *Drosophila βv* acts as a phagocytic receptor to also promote clearance of *Staphylococcus aureus via* peptidoglycan binding on this bacterium (56). Another *Drosophila* integrin α-subunit, αPS3, also cooperates with *βv* in hemocytes and serves as an engulfment receptor for phagocytosis of ACs and *S. aureus* (57). In *Drosophila* ovary, highly polarized epithelial follicle cells (FCs) can engulf germline debris *via* their apical side. Meehan et al. (58) found that integrin heterodimer αPS3/βPS were apically enriched in engulfing FCs, which are required for engulfment of ACs by FCs. Thus, integrins are evolutionally conserved receptors that participate in AC clearance.

BRIDGING MOLECULE

Several engulfment receptors have been identified that mediate phagocytosis of ACs, yet little is known about their precise mechanism of action, or whether they cooperate or act alone. Several molecules have been characterized that function upstream of Drpr to recognize ACs that are considered as "bridging molecules."

Six-Microns-Under (Simu)

Kurant and colleagues characterized a transmembrane protein named Simu, which is highly expressed on the surface of glial cells in the nervous system and macrophages elsewhere (59). Simu acts upstream of Drpr promote the recognition and engulfment of ACs (59). It strongly binds to ACs, through its EMILIN-like domain without membrane anchoring. Furthermore, Kurant and colleagues demonstrated that SIMU recognizes and binds PS secreted on ACs through its N-terminal EMILIN (EMI)-like domain, while the C-terminal NIM3 and NIM4 repeats regulate Simu affinity to PS (60). In addition, caspase activity is required for clearance of ACs by glial cells (60). However, the interaction mechanism between Simu and Drpr during clearance of ACs remains unclear, as Kurant and colleagues were failed to detect a directly physical linkage between Simu and Drpr (59). Thus, it seems likely that other molecules are required to connect these proteins (61).

Calreticulin (Calr), Pretaporter (Prtp), and *Drosophila* Calcium-Binding Protein 1 (DmCaBP1)

Various proteins and lipids from the endoplasmic reticulum (ER) have also been found to be exposed at the surface of human

ACs (62). Nakanishi and colleagues identified three ER proteins acting upstream of Drpr to promote phagocytosis in *Drosophila* (63–65). They showed that *Drosophila* Calr existed at the surface of living cells and reassigned to form aggregates upon apoptosis without change of the amount and expression at the cell surface; and that in a *Drosophila* mutant strain with reduced level of Calr, the level of phagocytosis of ACs was about a half of that observed in wild-type embryos (63). Thus, like PS, Calr is considered as a marker for phagocytosis of ACs in *Drosophila*. Through protein pull-down analysis, Nakanishi isolated an ER protein binding to the extracellular region of Drpr, with a signal peptide at the N-terminal and an ER retention motif at the C terminal, named Prtp. They found that Prpt relocated from ER to cell surface during apoptosis in *Drosophila* S2 cells (64), and they further showed that loss-of-function of *prtp* leads to reduced level of AC clearance both by embryonic hemocytes and embryonic glia. Reexpression of *prtp* in hemocytes did not rescue this defect while the ubiquitous expression did, which indicated that Prtp functions in ACs to promote phagocytes' engulfment (64). The DmCaBP1 is released and externalized from ACs, to bind to the extracellular region of Drpr (65). Loss of either *prtp* or *DmCaBP1* led to a reduced level of AC clearance in *Drosophila* embryos, but the double mutant did not cause a further decreased in phagocytosis, which indicated that they act in the same pathway. As apoptosis induced, DmCaBP1 is externalized from ACs and serves as a bridging molecule to connect ACs and phagocytes, promoting efficient and timely phagocytosis to occur.

E3, UBIQUITIN PROTEASOME PATHWAY

By screening for genes required for efficient phagocytosis of ACs in *Drosophila* macrophages *in vivo*, Silva and colleagues identified *pallbearer* (*pall*), which encodes an F-box protein (66). F-box proteins are generally part of Skp/Cullin/F-box (SCF) complexes that act as E3 ligases targeting phosphorylated proteins to ubiquitylation and degradation *via* the 26S proteasome (67). In addition to F-box protein, the SCF complexes contain three constant polypeptides—Skp1, Cullin1 (Cul1), and Rbx1, which have their counterparts in *Drosophila*. In *Drosophila*, six Skp proteins have been identified; and only SkpA strongly expressed in the embryos (68), and Bocca reported that SkpA and Rbx1 interact with Lin19 (dCul1) respectively (69). Silva and colleagues showed that Pall physically interacts with SkpA *via* its F-box domain, the loss function of either Lin19 or SkpA resulted in phagocytosis-defective phenotype, which indicated that they constitute complexes to promote phagocytosis of ACs (66). Xiao and colleagues then identified one substrate of the Pall–SCF complex, namely, the ribosomal protein S6 (Rps6) (70). The F-box protein Pall interacts with phosphorylated Rps6, which induces its ubiquitination and degradation *via* the 26S proteasome pathway (70). As a consequence, Xiao and colleagues further showed that the Rac2 small GTPase was upregulated and activated, triggering actin cytoskeleton rearrangement and thus promoting the clearance of ACs (70). They also showed that Pall translocates from the nucleus to the cytoplasm upon AC exposure (70). However, the AC signal and molecular pathway that leads to Pall nuclear export has not yet been identified. Furthermore, the nature of the kinase that phosphorylates Rps6 upstream of its physical interaction with Pall and how the degradation of phosphorylated RpS6 results in higher levels and activation of Rac2 remain to be deciphered.

CALCIUM SIGNALING

Calcium signaling is a second messenger, which participates in a number of cellular processes (71). Studies have identified several Ca^{2+} signaling genes that are required for AC removal in *Drosophila*. Cuttell and colleagues identified Undertaker (Uta) (also known as retinophilin), a *Drosophila* protein with membrane occupational recognition nexus repeats related to Junctophilin-like proteins, as required for Drpr-mediated phagocytosis (72). Junctophilins form junctional complexes between the PM and the ER or sarcoplasmic reticulum (SR) Ca^{2+} storage compartments that allow for cross talk between Ca^{2+} channels at the PM and the ER/SR Ca^{2+} channels (73). Cuttell and colleagues showed that the *Drosophila* ryanodine receptor, Rya-r44F, a Ca^{2+} channel on the ER membrane, also plays role in phagocytosis of ACs mediated by the Drpr pathway (71). They found that *uta* genetically interacts with *rya-r44F* upstream of the Drpr and Dmel/Ced-6 pathway to activate their downstream signaling cascade for efficient phagocytosis of ACs (72). Thus presumably, Uta forms junctional complexes between the PM and the ER to trigger the release of Ca^{2+} from the ER/SR compartment *via* Rya-R44F. Conversely, they showed that *drpr* and *Dmel/ced-6* are required for store-operated calcium entry (SOCE) *via* Stim and Orai (71). Thus, signaling downstream of Drpr and Dmel/Ced-6 may promote and/or maintain Uta-mediated junctional complexes, consequently mediating ER Ca^{2+} release to SOCE *via* Stim and Orai. It appears that Ca^{2+} functions in Drpr signaling downstream during both recognition and internalization of ACs, and Uta plays a central role both in Ca^{2+} homeostasis and phagocytosis. A similar link between Ca^{2+} homeostasis and AC clearance has been found in mammalian systems and *C. elegans* (74). Interestingly, a novel mechanism has been found by Weavers that *Drosophila* embryonic macrophages generate a memory after the uptake of ACs, priming them to detect tissue damage or infections. Engulfment of ACs associates with calcium bursts, increasing Drpr expression, which is important for the macrophages to rapidly respond and migrate to subsequent injury or infections (29).

CROSS TALK WITH INNATE IMMUNE RESPONSE

As phagocytosis is crucial for the normal development, it also plays important role in the immune response for the removal of ACs and pathogens (3, 24). The mechanisms that mediate phagocytosis of bacteria and how it interacts with other innate immune responses defense remain elusive. Hashimoto and colleagues showed that Drpr promotes phagocytosis of *S. aureus*, and *drpr* mutant flies show reduced resistance to a septic infection with *S. aureus* (75). *ltaS* encodes an enzyme responsible for the synthesis of lipoteichoic acid in *S. aureus* that acts as a ligand for Drpr in phagocytosis of *S. aureus* by *Drosophila* hemocytes. The integrin βv subunit promotes phagocytosis of *S. aureus* by

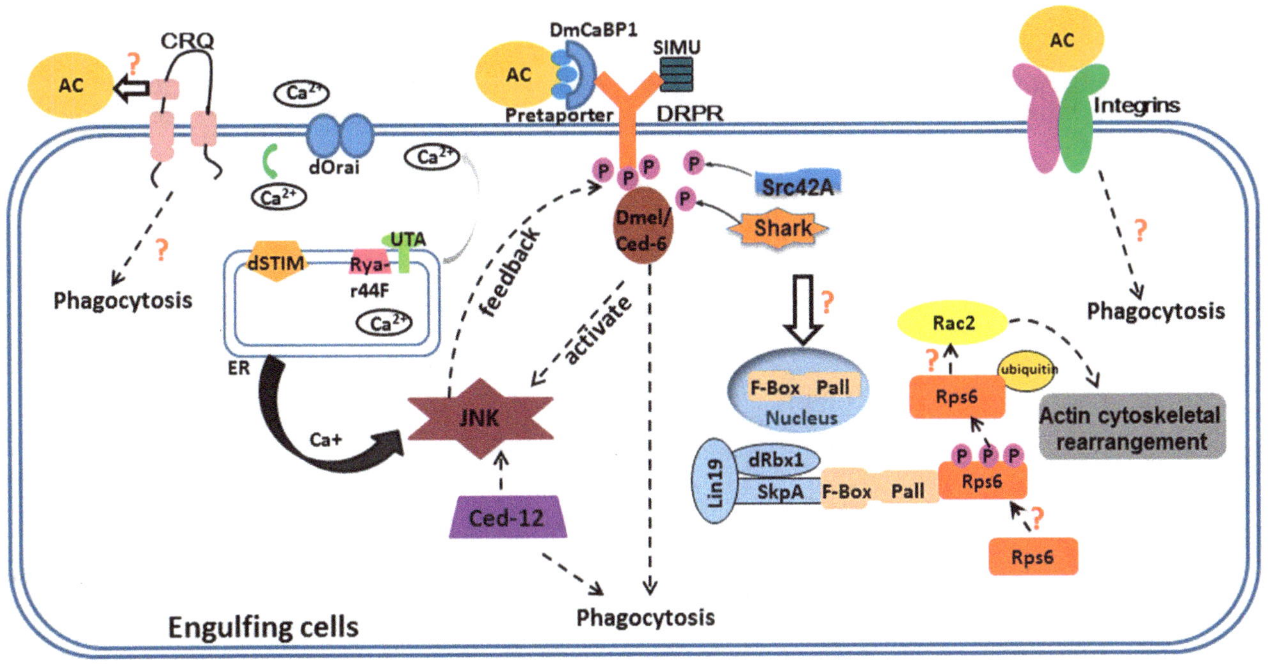

FIGURE 1 | Overview of corpse recognition and disposal in *Drosophila*. Key components positioned according to their roles in corpse recognition, internalization, and processing. Some important questions are also indicated. Three engulfment receptors located at the plasma membrane. CRQ (Croquemort), a membrane two-path protein, expressed mainly in the macrophages and its ligand and downstream have not studied clearly; Draper (Ced-1), a membrane single-path protein, expressed mainly in glia of CNS and hemocytes, with its ligands are shown in this figure, phosphorylated Draper interact with Ced-6 thus elevates Jun N-terminal kinase (JNK) signal and maintain Ca⁺ homeostasis; JNK promotes Drpr enrichment both in glia and follicular epithelia (under Ced-12 activation). Integrins, a heterodimer protein, function in epithelial follicle cells and hemocytes to engulf cell corpses. The solid arrows mean the relationship between two proteins; the dotted arrows mean the molecular mechanism linking two pathways or proteins unknown. The purple circles with "P" in them mean phosphate groups. The orange circles with "AC" in them mean apoptotic cells; the rectangle outlined by blue lines means engulfment cells; different proteins or molecules are represented by colored shapes as shown in this figure.

binding to peptidoglycan of this bacterium (56), and the integrin αPS3 subunit cooperates with *βv* in this process (57). Guillou and colleagues showed that *crq* defective mutant flies appeared to be more susceptible to environmental microbes both during development and at adulthood, they further demonstrated that *crq* is required for microbial phagocytosis (45). Interestingly, AC clearance by *Drosophila* macrophages appears essential in priming these cells to respond to subsequent microbial infections *in vivo* (29). Macrophages that have not engulfed ACs fail to take up *E. coli*, while those that have previously engulfed ACs can recognize and take up *E. coli*, ultimately mediating the bacterium phagosomal degradation (29).

JUN N-TERMINAL KINASE (JNK) PATHWAY

After recognition of ACs by macrophages and epithelial cells in mammals, the stress-activated MAP kinases JNK and p38 are activated at the early stage (76, 77). In *Drosophila* imaginal epithelia, normal imaginal cells exert an antitumor effect as oncogenic cells emerged to eliminate them (78). Ohsawa et al. revealed that the antitumor effect from surrounding cells was mediated by the activated JNK signaling, thus promoting the elimination of premalignant neighbors by engulfment (79). In *Drosophila* ovary, dying germline cells are cleared by neighbor follicular epithelia, which required Drpr signal pathway and activated JNK signal (80). During this process, Drpr acts upstream to activate JNK pathway, but another regulator exists to activate JNK pathway, which has not been studied. Their results suggested that the dying germline activates Drpr–JNK pathway, then JNK activity feeds back to increase Drpr expression in engulfing cells, which seem to be a circuit. Interestingly, although Ced-12 was showed to promote AC clearance in an independent pathway compared with Drpr, in *Drosophila* ovary, Timmons et al. (81) found that Ced-12 act upstream of JNK, which can to increase Drpr expression, similar to described earlier in *Drosophila* glia. As mentioned previously, glial cells play an important role in removing ACs during *Drosophila* embryonic development, neuronal pruning, and axonal degeneration (47, 59). Shklover showed that excess activation of JNK signaling in *Drosophila* embryonic glial cells does not affect the levels of Simu and Drpr expression but still promotes their apoptotic death and upregulates their phagocytic capacity by glial cells (82). As mentioned earlier, JNK signaling in follicular epithelia upregulates expression of Drpr, indicating that the phagocytosis induced by JNK signal may be tissue-specific. Recently, research showed that *Drosophila* glia upregulate their basal ability after neuronal injury, to phagocytosis through activation of the JNK pathway, which leads to the elevation of DRPR level (80, 83).

As mentioned previously, Weavers and colleagues proposed that ACs generate a molecular memory within macrophages, priming them to repair tissue damage and fight infection (29). They showed that JNK signaling is essential for macrophage detection of tissue damage and bacteria, as the uptake of ACs triggers calcium bursts in macrophages that induce JNK activation and signaling, ultimately leading to Drpr upregulation of expression (29).

CONCLUSION

Efficient and proper corpse clearance is important to maintain normal growth and prevent inappropriate inflammatory response, defective clearance of ACs often bring forth various diseases, such as autoimmune diseases, neurodegeneration, atherosclerosis, and Alzheimer's disease. As AC clearance pathways were conserved from invertebrate to mammals, the typical pathways—*ced-1*, *-6*, *-7*, and *ced-2*, *-5*, *-10*, and *-12*, most-studied in worms or in mammals, also exist in flies. Over the past few decades, researchers have unveiled some of the molecular mechanisms of AC clearance in *Drosophila*. The process is outlined in **Figure 1**. However, multiple important questions concerning clearance mechanisms remain to be answered, and more detailed mechanisms remain to be explored. How does Crq recognize ACs and what is the ligand of Crq? Are there other engulfment receptors or regulators required for phagocytosis of ACs in *Drosophila*? How is Crq expression regulated by ACs? How does the translocation of Pall from nucleus to cytoplasm happen and which pathway regulates this event? Are the same regulatory mechanisms involved? What are the molecular mechanisms that mediate phagocytosis of bacteria and how do they overlap or differ from that of ACs? How does phagocytosis interact with other innate immune responses defense? To answer these questions, further studies of engulfment signals and the phagocytic machinery is required. In conclusion, our understanding of AC engulfment mechanism in *Drosophila* will enhance our theoretical foundation in this area, and provide a powerful complement to the research in mammals that could be useful for the development of therapeutic strategies to control diseases related to defective cell clearance.

AUTHOR CONTRIBUTIONS

QZ, AM, LY, NG, QF, NF, and HX wrote and reviewed the manuscript.

FUNDING

This work was partially supported by the National Natural Science Foundation of China (Grant No. 31671439), Natural Science Foundation of Shaanxi Province, China (Grant No. 2017JM3008), the program of Innovative Research Team for the Central Universities (Grant No. GK201701005), the Fundamental Research Funds for the Central Universities (Grant No. 2017TS034), the Innovation Fund for undergraduate students (Grant No. KY2017ZD018), and Thousand Youth Talents Plan.

REFERENCES

1. Raff MC, Barres BA, Burne JF, Coles HS, Ishizaki Y, Jacobson MD. Programmed cell death and the control of cell survival. *Philos Trans R Soc Lond B Biol Sci* (1994) 345:265–8. doi:10.1098/rstb.1994.0104
2. Wyllie AH, Kerr JF, Currie AR. Cell death: the significance of apoptosis. *Int Rev Cytol* (1980) 68:251–306. doi:10.1016/S0074-7696(08)62312-8
3. Hoffmann JA, Kafatos FC, Janeway CA, Ezekowitz RA. Phylogenetic perspectives in innate immunity. *Science* (1999) 284:1313–8. doi:10.1126/science.284.5418.1313
4. Savill J, Dransfield I, Gregory C, Haslett C. A blast from the past: clearance of apoptotic cells regulates immune responses. *Nat Rev Immunol* (2002) 2:965–75. doi:10.1038/nri957
5. Nandrot EF, Finnemann SC. Lack of alphavbeta5 integrin receptor or its ligand MFG-E8: distinct effects on retinal function. *Ophthalmic Res* (2008) 40:120–3. doi:10.1159/000119861
6. Maderna P, Godson C. Phagocytosis of apoptotic cells and the resolution of inflammation. *Biochim Biophys Acta* (2003) 1639:141–51. doi:10.1016/j.bbadis.2003.09.004
7. Nagata S. Apoptosis and autoimmune diseases. *IUBMB Life* (2006) 58:358–62. doi:10.1080/15216540600746401
8. Krieser RJ, White K. Engulfment mechanism of apoptotic cells. *Curr Opin Cell Biol* (2002) 14:734–8. doi:10.1016/S0955-0674(02)00390-3
9. Wang X, Yang C. Programmed cell death and clearance of cell corpses in *Caenorhabditis elegans*. *Cell Mol Life Sci* (2016) 73:2221–36. doi:10.1007/s00018-016-2196-z
10. Ellis RE, Jacobson DM, Horvitz HR. Genes required for the engulfment of cell corpses during programmed cell death in *Caenorhabditis elegans*. *Genetics* (1991) 129:79–94.
11. Franc NC, White K, Ezekowitz RA. Phagocytosis and development: back to the future. *Curr Opin Immunol* (1999) 11:47–52. doi:10.1016/S0952-7915(99)80009-0
12. Shklover J, Levy-Adam F, Kurant E. Apoptotic cell clearance in development. *Curr Top Dev Biol* (2015) 114:297–334. doi:10.1016/bs.ctdb.2015.07.024
13. Fullard JF, Kale A, Baker NE. Clearance of apoptotic corpses. *Apoptosis* (2009) 14:1029–37. doi:10.1007/s10495-009-0335-9
14. Savill J, Fadok V. Corpse clearance defines the meaning of cell death. *Nature* (2000) 407:784–8. doi:10.1038/35037722
15. Danial NN, Korsmeyer SJ. Cell death: critical control points. *Cell* (2004) 116:205–19. doi:10.1016/S0092-8674(04)00046-7
16. Hochreiter-Hufford A, Ravichandran KS. Clearing the dead: apoptotic cell sensing, recognition, engulfment, and digestion. *Cold Spring Harb Perspect Biol* (2013) 5:a008748. doi:10.1101/cshperspect.a008748
17. Lauber K, Bohn E, Krober SM, Xiao YJ, Blumenthal SG, Lindemann RK, et al. Apoptotic cells induce migration of phagocytes via caspase-3-mediated release of a lipid attraction signal. *Cell* (2003) 113:717–30. doi:10.1016/S0092-8674(03)00422-7
18. Gude DR, Alvarez SE, Paugh SW, Mitra P, Yu J, Griffiths R, et al. Apoptosis induces expression of sphingosine kinase 1 to release sphingosine-1-phosphate as a "come-and-get-me" signal. *FASEB J* (2008) 22:2629–26381. doi:10.1096/fj.08-107169
19. Truman LA, Ford CA, Pasikowska M, Pound JD, Wilkinson SJ, Dumitriu IE, et al. CX3CL1/fractalkine is released from apoptotic lymphocytes to stimulate macrophage chemotaxis. *Blood* (2008) 112:5026–36. doi:10.1182/blood-2008-06-162404
20. Elliott MR, Chekeni FB, Trampont PC, Lazarowski ER, Kadl A, Walk SF, et al. Nucleotides released by apoptotic cells act as a find-me signal to promote phagocytic clearance. *Nature* (2009) 461:282–6. doi:10.1038/nature08296
21. Moreira S, Stramer B, Evans I, Wood W, Martin P. Prioritization of competing damage and developmental signals by migrating macrophages in the *Drosophila* embryo. *Curr Biol* (2010) 20:464–70. doi:10.1016/j.cub.2010.01.047
22. Evans IR, Rodrigues FS, Armitage EL, Wood W. Draper/CED-1 mediates an ancient damage response to control inflammatory blood cell migration in vivo. *Curr Biol* (2015) 25:1606–12. doi:10.1016/j.cub.2015.04.037

23. Poon IK, Lucas CD, Rossi AG, Ravichandran KS. Apoptotic cell clearance: basic biology and therapeutic potential. *Nat Rev Immunol* (2014) 14:166–80. doi:10.1038/nri3607
24. Nagata S, Hanayama R, Kawane K. Autoimmunity and the clearance of dead cells. *Cell* (2010) 140:619–30. doi:10.1016/j.cell.2010.02.014
25. Fadok VA, Voelker DR, Campbell PA, Cohen JJ, Bratton DL, Henson PM. Exposure of phosphatidylserine on the surface of apoptotic lymphocytes triggers specific recognition and removal by macrophages. *J Immunol* (1992) 148:2207–16.
26. Williamson P, Schlegel RA. Transbilayer phospholipid movement and the clearance of apoptotic cells. *Biochim Biophys Acta* (2002) 1585:53–63. doi:10.1016/S1388-1981(02)00324-4
27. Erwig LP, Henson PM. Clearance of apoptotic cells by phagocytes. *Cell Death Differ* (2008) 15:243–50. doi:10.1038/sj.cdd.4402184
28. Segawa K, Nagata S. An apoptotic 'eat me' signal: phosphatidylserine exposure. *Trends Cell Biol* (2015) 25:639–50. doi:10.1016/j.tcb.2015.08.003
29. Weavers H, Evans IR, Martin P, Wood W. Corpse engulfment generates a molecular memory that primes the macrophage inflammatory response. *Cell* (2016) 165:1658–71. doi:10.1016/j.cell.2016.04.049
30. Pazdera TM, Janardhan P, Minden JS. Patterned epidermal cell death in wild-type and segment polarity mutant *Drosophila* embryos. *Development* (1998) 125:3427–36.
31. Sonnenfeld MJ, Jacobs JR. Macrophages and glia participate in the removal of apoptotic neurons from the *Drosophila* embryonic nervous system. *J Comp Neurol* (1995) 359:644–52. doi:10.1002/cne.903590410
32. Tepass U, Fessler LI, Aziz A, Hartenstein V. Embryonic origin of hemocytes and their relationship to cell death in *Drosophila*. *Development* (1994) 120: 1829–37.
33. Williams DW, Truman JW. Cellular mechanisms of dendrite pruning in *Drosophila*: insights from in vivo time-lapse of remodeling dendritic arborizing sensory neurons. *Development* (2005) 132:3631–42. doi:10.1242/dev.01928
34. Franc NC, Heitzler P, Ezekowitz RA, White K. Requirement for croquemort in phagocytosis of apoptotic cells in *Drosophila*. *Science* (1999) 284:1991–4. doi:10.1126/science.284.5422.1991
35. Freeman MR, Delrow J, Kim J, Johnson E, Doe CQ. Unwrapping glial biology: Gcm target genes regulating glial development, diversification, and function. *Neuron* (2003) 38:567–80. doi:10.1016/S0896-6273(03)00289-7
36. Han C, Song Y, Xiao H, Wang D, Franc NC, Jan LY, et al. Epidermal cells are the primary phagocytes in the fragmentation and clearance of degenerating dendrites in *Drosophila*. *Neuron* (2014) 81:544–60. doi:10.1016/j.neuron.2013.11.021
37. Chimini G. Engulfing by lipids: a matter of taste? *Cell Death Differ* (2001) 8:545–8. doi:10.1038/sj.cdd.4400833
38. Schlegel RA, Williamson P. Phosphatidylserine, a death knell. *Cell Death Differ* (2001) 8:551–63. doi:10.1038/sj.cdd.4400817
39. Wang X, Wu YC, Fadok VA, Lee MC, Gengyo-Ando K, Cheng LC, et al. Cell corpse engulfment mediated by *C. elegans* phosphatidylserine receptor through CED-5 and CED-12. *Science* (2003) 302:1563–6. doi:10.1126/science.1087641
40. Savill J, Hogg N, Ren Y, Haslett C. Thrombospondin cooperates with CD36 and the vitronectin receptor in macrophage recognition of neutrophils undergoing apoptosis. *J Clin Invest* (1992) 90:1513–22. doi:10.1172/JCI116019
41. Stuart LM, Deng J, Silver JM, Takahashi K, Tseng AA, Hennessy EJ, et al. Response to *Staphylococcus aureus* requires CD36-mediated phagocytosis triggered by the COOH-terminal cytoplasmic domain. *J Cell Biol* (2005) 170:477–85. doi:10.1083/jcb.200501113
42. Gowda NM, Wu X, Kumar S, Febbraio M, Gowda DC. CD36 contributes to malaria parasite-induced pro-inflammatory cytokine production and NK and T cell activation by dendritic cells. *PLoS One* (2013) 8:e77604. doi:10.1371/journal.pone.0077604
43. Franc NC, Dimarcq JL, Lagueux M, Hoffmann J, Ezekowitz RA. Croquemort, a novel *Drosophila* hemocyte/macrophage receptor that recognizes apoptotic cells. *Immunity* (1996) 4:431–43. doi:10.1016/S1074-7613(00)80410-0
44. Manaka J, Kuraishi T, Shiratsuchi A, Nakai Y, Higashida H, Henson P, et al. Draper-mediated and phosphatidylserine-independent phagocytosis of apoptotic cells by *Drosophila* hemocytes/macrophages. *J Biol Chem* (2004) 279:48466–76. doi:10.1074/jbc.M408597200
45. Guillou A, Troha K, Wang H, Franc NC, Buchon N. The *Drosophila* CD36 homologue croquemort is required to maintain immune and gut homeostasis during development and aging. *PLoS Pathog* (2016) 12:e1005961. doi:10.1371/journal.ppat.1005961
46. Su HP, Nakada-Tsukui K, Tosello-Trampont AC, Li Y, Bu G, Henson PM, et al. Interaction of CED-6/GULP, an adapter protein involved in engulfment of apoptotic cells with CED-1 and CD91/low density lipoprotein receptor-related protein (LRP). *J Biol Chem* (2002) 277:11772–9. doi:10.1074/jbc.M109336200
47. Awasaki T, Tatsumi R, Takahashi K, Arai K, Nakanishi Y, Ueda R, et al. Essential role of the apoptotic cell engulfment genes draper and ced-6 in programmed axon pruning during *Drosophila* metamorphosis. *Neuron* (2006) 50:855–67. doi:10.1016/j.neuron.2006.04.027
48. Ziegenfuss JS, Biswas R, Avery MA, Hong K, Sheehan AE, Yeung YG, et al. Draper-dependent glial phagocytic activity is mediated by Src and Syk family kinase signalling. *Nature* (2008) 453:935–9. doi:10.1038/nature06901
49. Kinchen JM, Cabello J, Klingele D, Wong K, Feichtinger R, Schnabel H, et al. Two pathways converge at CED-10 to mediate actin rearrangement and corpse removal in *C. elegans*. *Nature* (2005) 434:93–9. doi:10.1038/nature03263
50. Van Goethem E, Silva EA, Xiao H, Franc NC. The *Drosophila* TRPP cation channel, PKD2 and Dmel/Ced-12 act in genetically distinct pathways during apoptotic cell clearance. *PLoS One* (2012) 7:e31488. doi:10.1371/journal.pone.0031488
51. Hynes RO. Integrins: bidirectional, allosteric signaling machines. *Cell* (2002) 110:673–87. doi:10.1016/S0092-8674(02)00971-6
52. Margadant C, Monsuur HN, Norman JC, Sonnenberg A. Mechanisms of integrin activation and trafficking. *Curr Opin Cell Biol* (2011) 23:607–14. doi:10.1016/j.ceb.2011.08.005
53. Hanayama R, Tanaka M, Miwa K, Shinohara A, Iwamatsu A, Nagata S. Identification of a factor that links apoptotic cells to phagocytes. *Nature* (2002) 417:182–7. doi:10.1038/417182a
54. Hsu TY, Wu YC. Engulfment of apoptotic cells in *C. elegans* is mediated by integrin alpha/SRC signaling. *Curr Biol* (2010) 20:477–86. doi:10.1016/j.cub.2010.01.062
55. Nagaosa K, Okada R, Nonaka S, Takeuchi K, Fujita Y, Miyasaka T, et al. Integrin betanu-mediated phagocytosis of apoptotic cells in *Drosophila* embryos. *J Biol Chem* (2011) 286:25770–7. doi:10.1074/jbc.M110.204503
56. Shiratsuchi A, Mori T, Sakurai K, Nagaosa K, Sekimizu K, Lee BL, et al. Independent recognition of *Staphylococcus aureus* by two receptors for phagocytosis in *Drosophila*. *J Biol Chem* (2012) 287:21663–72. doi:10.1074/jbc.M111.333807
57. Nonaka S, Nagaosa K, Mori T, Shiratsuchi A, Nakanishi Y. Integrin alphaPS3/betanu-mediated phagocytosis of apoptotic cells and bacteria in *Drosophila*. *J Biol Chem* (2013) 288:10374–80. doi:10.1074/jbc.M113.451427
58. Meehan TL, Kleinsorge SE, Timmons AK, Taylor JD, McCall K. Polarization of the epithelial layer and apical localization of integrins are required for engulfment of apoptotic cells in the *Drosophila* ovary. *Dis Model Mech* (2015) 8:1603–14. doi:10.1242/dmm.021998
59. Kurant E, Axelrod S, Leaman D, Gaul U. Six-microns-under acts upstream of Draper in the glial phagocytosis of apoptotic neurons. *Cell* (2008) 133: 498–509. doi:10.1016/j.cell.2008.02.052
60. Shklyar B, Levy-Adam F, Mishnaevski K, Kurant E. Caspase activity is required for engulfment of apoptotic cells. *Mol Cell Biol* (2013) 33:3191–201. doi:10.1128/MCB.00233-13
61. Elliott MR, Ravichandran KS. Death in the CNS: six-microns-under. *Cell* (2008) 133:393–5. doi:10.1016/j.cell.2008.04.014
62. Franz S, Herrmann K, Furnrohr BG, Sheriff A, Frey B, Gaipl US, et al. After shrinkage apoptotic cells expose internal membrane-derived epitopes on their plasma membranes. *Cell Death Differ* (2007) 14:733–42. doi:10.1038/sj.cdd.4402066
63. Kuraishi T, Manaka J, Kono M, Ishii H, Yamamoto N, Koizumi K, et al. Identification of calreticulin as a marker for phagocytosis of apoptotic cells in *Drosophila*. *Exp Cell Res* (2007) 313:500–10. doi:10.1016/j.yexcr.2006.10.027
64. Kuraishi T, Nakagawa Y, Nagaosa K, Hashimoto Y, Ishimoto T, Moki T, et al. Pretaporter, a *Drosophila* protein serving as a ligand for Draper in the phagocytosis of apoptotic cells. *EMBO J* (2009) 28:3868–78. doi:10.1038/emboj.2009.343
65. Okada R, Nagaosa K, Kuraishi T, Nakayama H, Yamamoto N, Nakagawa Y, et al. Apoptosis-dependent externalization and involvement in apoptotic cell

clearance of DmCaBP1, an endoplasmic reticulum protein of *Drosophila*. *J Biol Chem* (2012) 287:3138–46. doi:10.1074/jbc.M111.277921
66. Silva E, Au-Yeung HW, Van Goethem E, Burden J, Franc NC. Requirement for a *Drosophila* E3-ubiquitin ligase in phagocytosis of apoptotic cells. *Immunity* (2007) 27:585–96. doi:10.1016/j.immuni.2007.08.016
67. Ho MS, Tsai PI, Chien CT. F-box proteins: the key to protein degradation. *J Biomed Sci* (2006) 13:181–91. doi:10.1007/s11373-005-9058-2
68. Murphy TD. *Drosophila* skpA, a component of SCF ubiquitin ligases, regulates centrosome duplication independently of cyclin E accumulation. *J Cell Sci* (2003) 116:2321–32. doi:10.1242/jcs.00463
69. Bocca SN, Muzzopappa M, Silberstein S, Wappner P. Occurrence of a putative SCF ubiquitin ligase complex in *Drosophila*. *Biochem Biophys Res Commun* (2001) 286:357–64. doi:10.1006/bbrc.2001.5394
70. Xiao H, Wang H, Silva EA, Thompson J, Guillou A, Yates JR Jr, et al. The Pallbearer E3 ligase promotes actin remodeling via RAC in efferocytosis by degrading the ribosomal protein S6. *Dev Cell* (2015) 32:19–30. doi:10.1016/j.devcel.2014.11.015
71. Kinchen JM. A model to die for: signaling to apoptotic cell removal in worm, fly and mouse. *Apoptosis* (2010) 15:998–1006. doi:10.1007/s10495-010-0509-5
72. Cuttell L, Vaughan A, Silva E, Escaron CJ, Lavine M, Van Goethem E, et al. Undertaker, a *Drosophila* junctophilin, links Draper-mediated phagocytosis and calcium homeostasis. *Cell* (2008) 135:524–34. doi:10.1016/j.cell.2008.08.033
73. Takeshima H, Komazaki S, Nishi M, Iino M, Kangawa K. Junctophilins: a novel family of junctional membrane complex proteins. *Mol Cell* (2000) 6:11–22. doi:10.1016/S1097-2765(05)00005-5
74. Gronski MA, Kinchen JM, Juncadella IJ, Franc NC, Ravichandran KS. An essential role for calcium flux in phagocytes for apoptotic cell engulfment and the anti-inflammatory response. *Cell Death Differ* (2009) 16:1323–31. doi:10.1038/cdd.2009.55
75. Hashimoto Y, Tabuchi Y, Sakurai K, Kutsuna M, Kurokawa K, Awasaki T, et al. Identification of lipoteichoic acid as a ligand for draper in the phagocytosis of *Staphylococcus aureus* by *Drosophila* hemocytes. *J Immunol* (2009) 183:7451–60. doi:10.4049/jimmunol.0901032
76. Patel VA, Longacre A, Hsiao K, Fan H, Meng F, Mitchell JE, et al. Apoptotic cells, at all stages of the death process, trigger characteristic signaling events that are divergent from and dominant over those triggered by necrotic cells: implications for the delayed clearance model of autoimmunity. *J Biol Chem* (2006) 281:4663–70. doi:10.1074/jbc.M508342200
77. Patel VA, Lee DJ, Feng L, Antoni A, Lieberthal W, Schwartz JH, et al. Recognition of apoptotic cells by epithelial cells: conserved versus tissue-specific signaling responses. *J Biol Chem* (2010) 285:1829–40. doi:10.1074/jbc.M109.018440
78. Brumby AM, Richardson HE. Scribble mutants cooperate with oncogenic Ras or notch to cause neoplastic overgrowth in *Drosophila*. *EMBO J* (2003) 22:5769–79. doi:10.1093/emboj/cdg548
79. Ohsawa S, Sugimura K, Takino K, Xu T, Miyawaki A, Igaki T. Elimination of oncogenic neighbors by JNK-mediated engulfment in *Drosophila*. *Dev Cell* (2011) 20:315–28. doi:10.1016/j.devcel.2011.02.007
80. Etchegaray JI, Timmons AK, Klein AP, Pritchett TL, Welch E, Meehan TL, et al. Draper acts through the JNK pathway to control synchronous engulfment of dying germline cells by follicular epithelial cells. *Development* (2012) 139:4029–39. doi:10.1242/dev.082776
81. Timmons AK, Mondragon AA, Schenkel CE, Yalonetskaya A, Taylor JD, Moynihan KE, et al. Phagocytosis genes nonautonomously promote developmental cell death in the *Drosophila* ovary. *Proc Natl Acad Sci U S A* (2016) 113:E1246–55. doi:10.1073/pnas.1522830113
82. Shklover J, Mishnaevski K, Levy-Adam F, Kurant E. JNK pathway activation is able to synchronize neuronal death and glial phagocytosis in Drosophila. *Cell Death Dis* (2015) 6:e1649. doi:10.1038/cddis.2015.27
83. Macdonald JM, Doherty J, Hackett R, Freeman MR. The c-Jun kinase signaling cascade promotes glial engulfment activity through activation of *draper* and phagocytic function. *Cell Death Differ* (2013) 20:1140–8. doi:10.1038/cdd.2013.30

Conflict of Interest Statement: The authors declare that the research was conducted in the absence of any commercial or financial relationships that could be construed as a potential conflict of interest.

The reviewer KM declared a past coauthorship with the author NF to the handling editor.

2

Anti-inflammatory Mechanisms Triggered by Apoptotic Cells during their Clearance

Zsuzsa Szondy[1,2], Zsolt Sarang[1], Beáta Kiss[1], Éva Garabuczi[1] and Krisztina Köröskényi[1,2]*

[1] *Department of Biochemistry and Molecular Biology of Medical Faculty, University of Debrecen, Debrecen, Hungary,*
[2] *Department of Basic Medical Sciences of Dental Faculty, University of Debrecen, Debrecen, Hungary*

In the human body, billions of cells die by apoptosis every day. The subsequent clearance of apoptotic cells by phagocytosis is normally efficient enough to prevent secondary necrosis and the consequent release of cell contents that would induce inflammation and trigger autoimmunity. In addition, apoptotic cells generally induce an anti-inflammatory response, thus removal of apoptotic cells is usually immunologically silent. Since the first discovery that uptake of apoptotic cells leads to transforming growth factor (TGF)-β and interleukin (IL)-10 release by engulfing macrophages, numerous anti-inflammatory mechanisms triggered by apoptotic cells have been discovered, including release of anti-inflammatory molecules from the apoptotic cells, triggering immediate anti-inflammatory signaling pathways by apoptotic cell surface molecules *via* phagocyte receptors, activating phagocyte nuclear receptors following uptake and inducing the production of anti-inflammatory soluble mediators by phagocytes that may act *via* paracrine or autocrine mechanisms to amplify and preserve the anti-inflammatory state. Here, we summarize our present knowledge about how these anti-inflammatory mechanisms operate during the clearance of apoptotic cells.

Edited by:
Amiram Ariel, University of Haifa, Israel

Reviewed by:
Sylvain Perruche, Institut national de la santé et de la recherche médicale, France
Patrizia Rovere Querini, Vita-Salute San Raffaele University, Italy

***Correspondence:**
Zsuzsa Szondy
szondy@med.unideb.hu

Keywords: apoptotic cell, phagocytosis, anti-inflammatory, pro-resolving lipid mediators, phosphatidylserine

INTRODUCTION

Timed initiation of apoptotic type of cell death followed by prompt removal carried out by professional engulfers or by non-professional neighboring cells plays a central role in the maintenance of tissue homeostasis. Every day billions of our cells die and get removed without inducing inflammation and autoimmunity (1). And even when inflammation is induced, efficient clearance of apoptotic neutrophils by engulfing macrophages during the inflammatory response is a determining event in initiating resolution of inflammation and contributes to the tissue repair processes following injury (2–5).

To initiate effective clearance, apoptotic cells attract engulfing cells by secreting various chemotactic signals, such as CX3CL1/fractalkine (6), lysophosphatidylcholine (7), sphingosine-1-phosphate (8), thrombospondin-1 (TSP-1) (9), ATP and UTP (10), endothelial monocyte-activating polypeptide II (11), monocyte chemoattractant protein-1 (12), cleaved human tyrosyl-tRNA synthetase (13), or the S19 ribosomal protein cross-linked dimer (14). Upon approaching the dying cells, engulfing cells must make a distinction between dead and living cells, and they act so by recognizing apoptotic cell-associated molecular patterns (ACAMPs) displayed on the cell surface of apoptotic cells (15). Most of the ACAMPs are phagocytosis initiating "eat me" signals, which are able to trigger receptors on phagocytes either directly or *via* bridging molecules. Phosphatidylserine (PS) on the outer leaflet of the cell membrane is the most universally seen "eat me" signal appearing in apoptotic cells (16). Stabilin-2, the macrophage receptor T-cell immunoglobulin- and mucin-domain-containing molecule (Tim4), and brain-specific angiogenesis inhibitor 1 have the ability to directly engage PS on dying cells (17–19), other receptors, such as Mer tyrosine kinase (MerTk) (20), scavenger receptor SCARF1 (21), or integrin αv/β3/β5 either with the CD36 (22) or with the transglutaminase 2 (TG2) coreceptors (23) require bridging molecules for PS binding. While for MerTk Gas6 and Protein S serve as bridging molecules (24, 25), TSP-1 and milk-fat globulin-E8 play a similar role for the integrin αv/β3/CD36 and integrin αv/β3/TG2 receptor complexes, respectively (22, 26, 27). In the case of apoptotic neutrophils, however, the matricellular protein CCN1 bridges PS to the αv/β3 and αv/β5 integrins of macrophages (28).

Besides the PS-recognizing bridging molecules, additional bridging molecules also contribute to the phagocytosis of apoptotic cells. Thus, serum protein C1q links Annexin A2 and A5 on the apoptotic cells (29) to both the SCARF1 scavenger receptor and to the calreticulin-associated CD91 receptor on macrophages (20, 30, 31). Apoptotic neutrophils, T cells, and human mesangial cells release annexin I (32), and annexin I enhances phagocytosis of apoptotic cells *via* a mechanism that requires FPR2/ALX receptor and its internalization (33).

Interestingly, simultaneous triggering of all these phagocytic receptors seems to result in the activation of only two evolutionary conserved signaling pathways both leading to the activation of Rac1, a small GTPase that regulates cytoskeletal rearrangements required for the phagocytosis process (34).

Uptake of apoptotic cells delivers excess materials to the phagocytes, which is degraded after the newly formed phagosome fuses with the lysosomes (35, 36). However, for full protein degradation besides the lysosomal cathepsins (35), the proteosomal pathway also has to be activated (37). In addition, to successfully metabolize lipids originated from the apoptotic cells, phagocytes also require the lipid metabolism organizing function of their lipid sensing nuclear receptors, the liver X receptors (LXRs), and the perixosome proliferator-activated receptors (PPARs) (38). LXRs and PPARs are ligand-regulated transcription factors belonging to the nuclear receptor family. They function in the form of heterodimers with the retinoid X receptors (RXRs) (39). The unligated LXRs and PPARs are located on their respective DNA response elements and recruit co-repressor molecules that repress the transcription of their target genes. Binding of their ligand, however, induces such a conformational change in their structure that results in the exchange of co-repressors for co-activators, and in the consequent start of transcription. In addition, ligated LXR and PPAR heterodimers are also capable of transrepressing genes, the transcription of which would otherwise be initiated by other transcription factors (39).

The ligands of LXRs are sterol metabolites (40, 41), while for PPARs are unsaturated fatty acids, eicosanoids, and derivatives of linoleic acid (42). Metabolically, the main function of LXRs is to regulate whole body sterol metabolism (38). Since, in mammalians, sterols cannot be degraded; in macrophages, following the degradation of apoptotic cells, activated LXRs induce the expression of cholesterol efflux transporters ABCG1 and ABCA1, Apo lipoproteins ApoC and ApoE, and the lipoprotein remodeling enzyme PLTP (43) promoting this way the efflux of apoptotic cell-derived sterols onto serum apolipoproteins and their transport to the liver, from where sterols will be redistributed again. PPARs, on the other hand, are involved in nearly every facet of fatty acid metabolism (44). Thus, PPARγ controls the expression of genes involved in lipid metabolism (43), while PPARδ controls the macrophage energy homeostasis by regulating the expression of genes involved β-oxidation, mitochondrial respiration, and thermogenesis (45).

While engulfment of a number of pathogenic targets induces a pro-inflammatory program in macrophages, uptake of apoptotic cells initiates their transition into an anti-inflammatory phenotype. Furthermore, apoptotic cells are capable of actively inhibiting the inflammatory program. For example, the inflammatory response induced by lipopolysaccharide (LPS), a component of the cell wall of Gram-negative bacteria, is strongly attenuated by preincubation of macrophages with apoptotic cells (46–48). First, after the exposure to apoptotic cells, an immediate-early inhibition of macrophage pro-inflammatory cytokine gene transcription can be detected (46, 47), subsequently, both nuclear receptors (38) are activated and, as it was first recognized, soluble mediators, such as transforming growth factor (TGF)-β and interleukin (IL)-10, are released (48). These mediators act *via* paracrine or autocrine mechanisms to strengthen and maintain the anti-inflammatory state (47, 48). The aim of this review is to highlight how these efferocytotic processes are coupled to the anti-inflammatory mechanisms provoked simultaneously by apoptotic cells, since efficient apoptotic cell removal and induction of immunologic tolerance by apoptotic cells are the two crucial mechanisms that prevent chronic inflammation and autoimmunity (49).

APOPTOSIS IS AN IMMUNOLOGICALLY SILENT FORM OF CELL DEATH

Increasing evidence indicates that the caspase-dependent apoptosis is unique in a sense that it is an immunologically silent form of cell death (50). Whereas necrotic cells typically provoke inflammation, apoptotic cells generally do not. What is more, even if apoptotic cells enter secondary necrosis and are leaking their cellular contents, they retain this anti-inflammatory state, in contrast to cells that have entered necrosis directly (51, 52). This suggests that significant alterations to cellular composition must occur during apoptosis to diminish the activity of danger-associated molecules (DAMPs), and even if these are accidentally released, their pro-inflammatory activity is destroyed. There are several examples recently discovered that underline this assumption.

Genomic DNA is considered to be an important DAMP capable of initiating dendritic cell (DC) maturation and the initiation of immune responses to coadministered antigens (53). Conversely, hydrolysis of DNA with endonucleases strongly attenuates its immune-activating properties (54). In apoptotic cells, genomic DNA undergoes extensive hydrolysis to small ~200 bp fragments due to the actions of a caspase-activated DNAse (55). If DNA cannot be degraded by either apoptotic or phagocytic endonucleases, autoimmunity develops (55), which is related to incompletely digested DNA persisting in macrophages leading to the activation of the RIG-I/IRF-3 pathway which senses cytoplasmic DNA fragments (56). In addition to the caspase-dependent degradation of genomic DNA, activated caspase-8 itself also interferes with the RIG-I/IRF-3 pathway by proteolytically inactivating RIP kinase 1, a key signaling component of the RIG-I complex, and thereby attenuating expression of IRF-3-inducible genes that include interferons and other inflammatory factors (57). Apoptotic caspases, *via* interfering with the cGAS/STING pathway, prevent the induction of type I interferons also by mitochondrial DNA, which can be released into the cytosol during apoptosis following permeabilization of the mitochondrial outer membrane (58).

High mobility group 1 (HMGB1) is another well-known DAMP. It functions in the cell nucleus as an architectural chromatin-binding factor *via* bending DNA and facilitating this way the buildup of protein complexes on specific DNA sequences. But it associates only loosely with the DNA, and can be passively released by necrotic or damaged cells (59). In apoptotic cells, however, HMGB1 remains bound firmly to chromatin due to the generalized hypoacetylation of histones, another caspase-dependent event (59). Furthermore, caspases also inactivate HMGB1 indirectly *via* cleaving the mitochondrial protein p75NDUF. This event triggers a burst of reactive oxygen leading to oxidation of a critical cysteine residue on HMGB1 that neutralizes its pro-inflammatory activity (60). And finally, IL-33, a recently discovered alarmin, is also cleaved by caspases loosing this way its DAMP activity (61).

APOPTOTIC CELLS RELEASE ANTI-INFLAMMATORY MOLECULES

Apoptotic cells not only fail to be strongly immunogenic but were found to release various anti-inflammatory molecules as well. Thus, T cells express TGF-β, and release it during apoptosis (62). TGF-β is a potent anti-inflammatory cytokine (63). Its anti-inflammatory effects are reflected in the observation that loss of TGF-β1 in mice leads to wasting syndrome accompanied by a multifocal, mixed inflammatory cell response and tissue necrosis, resulting in organ failure and death (64, 65). Both the initiation and resolution of general inflammatory responses involve TGF-β1. Thus, TGF-β1 stimulates monocyte migration and growth factor production (66). But after initiation of an inflammatory response, it also exhibits potent anti-inflammatory effects, including inhibition of neutrophil and T-lymphocyte adhesion to endothelium (67), downregulation of macrophages (66, 68), and antagonism of tumor necrosis factor-α (TNF-α) function (68). TGF-β is crucial also in the initiation of differentiation of the regulatory T cells (69), which play a key role in preventing the development of autoimmunity (70). TGF-β plays such an essential role in the initiation of regulatory T cell differentiation that prior to the initiation of thymocyte apoptosis no Treg cells can be detected in the thymus (71).

Apoptotic cells can also release IL-10 (72). IL-10, together with TGF-β, induces an immune suppressive response by promoting regulatory T cell formation and by affecting their function (73).

Annexin I, which is released by apoptotic cells and promotes efferocytosis (32, 33), was also found to significantly attenuate IL-6 signaling and the release of TNF-α from endotoxin-challenged monocytes by activating annexin A1 receptors of the formyl peptide receptor family and the consequent JAK/STAT/SOCS signaling (74). And, finally, apoptotic cells were found to release lactoferrin that inhibits the migration of neutrophil granulocytes and eosinophils toward the chemotactic signals released by apoptotic cells (75, 76). As a result, apoptotic cells induce only migration of macrophages, but other classes of professional phagocytes are not recruited by the apoptotic cells (76).

Increasing evidence indicate that several chemotactic signals released from the apoptotic cells can also serve as anti-inflammatory molecules. Thus, TSP-1 functions as a major activator of TGF-β1 (77) which is secreted to the extracellular matrix in an inactive form by being in non-covalent association with the latency-associated peptide (78). First, TSP-1 releases TGF-β1 from its latent form *via* interacting with the N-terminal region of latency-associated peptide, then it binds the mature TGF-β1. This interaction induces such a conformational change in the structure of TGF-β1 (79), which allows its binding to its receptor. Furthermore, TSP-1 or its protease-cleaved derivative can also bind to immature DCs (iDCs) and induce their tolerogenic state (80). Fractalkine was also shown to act as an inhibitor of LPS-induced TNF-α production by microglia cells (81), while lysophosphatidylcholine was reported to interfere with the LPS-induced NO and pro-inflammatory cytokine production by macrophages (82).

Once released from the apoptotic cells *via* a caspase-regulated pannexin channel (83), ATP is fast degraded to AMP very often on the surface of apoptotic cells (84) and then to adenosine by the cell surface 5′ nucleotidase of engulfing macrophages (85, 86). Adenosine then triggers macrophage adenosine A_{2A} receptors (A2ARs) to suppress the NO-dependent formation of neutrophil migration factors, such as macrophage inflammatory protein-2, *via* activating the adenylate cyclase/protein kinase A pathway (86). Interestingly, both adenosine A_{2A} and A_3 (A3R) receptors are expressed by macrophages, and while A2ARs inhibit, A3Rs promote the release of neutrophil migration factors by engulfing macrophages (87). However, while A2AR expression increases (86), A3R is downregulated during the course of efferocytosis (87) potentiating and maintaining this way the anti-inflammatory effects of adenosine. In addition, adenosine, by triggering adenosine A3R, contributes also to the chemotactic navigation of macrophages toward the apoptotic cells driven by apoptotic cell-derived chemotactic signals, thus facilitates the fast clearance of apoptotic cells (88). As a result, downregulation of A3R during clearance prolongs their presence around the apoptotic cells.

In an inflammatory milieu, adenosine strongly suppresses also the LPS-induced pro-inflammatory cytokine formation of monocytes and macrophages by activating A2ARs (89). In monocytes adenosine increases the expression of the Nr4A orphan nuclear receptor which then inhibits the transcriptional activity of nuclear factor-κB (NF-κB) known to play a determining role in initiating the transcription of numerous pro-inflammatory cytokines (90). The adenosine-triggered adenylate cyclase pathway in LPS-activated macrophages, on the other hand, upregulates the expression of dual-specific phosphatase 1 that interferes with the activation of LPS-activated MAP kinases (91).

APOPTOTIC CELLS TRIGGER ANTI-INFLAMMATORY SIGNALING PATHWAYS BY ACTIVATING CELL SURFACE PHAGOCYTOSIS RECEPTORS ON THE PHAGOCYTE

Soon after it was discovered that PS recognition plays a central role in the uptake of apoptotic cells (92), it was also discovered that PS recognition in macrophages mediates also some of the anti-inflammatory responses observed during the uptake of apoptotic cells (93). This was proven by the observation that several anti-inflammatory responses provoked by apoptotic cells can be prevented by administration of annexin V, a naturally occurring PS binding protein (92), or are not induced by cells that do not express PS during apoptosis. In addition, in the induction of these anti-inflammatory effects, apoptotic cells can be replaced by PS liposomes (94). Some of the PS-induced anti-inflammatory responses are direct and can be detected as immediate inhibition of NF-κB transcriptional activity (46). Some others appear later, such as upregulation of the zinc finger nuclear factor, named GC binding protein (GC-BP) (95), or that of the Nr4a1 transcription factor (96). By binding to its promoter, GC-BP selectively inhibits IL-12 p35 gene transcription (95), while Nr4a1inhibits both NF-κB transcriptional activity and the induction of IL-12 (96). The induction of these transcription factors, however, might be macrophage type specific, because Nr4a1 induction is seen in peritoneal, but not in bone marrow-derived macrophages (96).

In addition to these intracellular anti-inflammatory effects, PS is also responsible for triggering TGF-β1 secretion from engulfing macrophages. Thus, apoptotic PLB-985 cells that are unable to express PS during apoptosis, fail to trigger TGF-β1 production, while PS directly transferred onto the PLB-985 cell surface membranes or PS liposomes can restore the secretion of TGF-β1 (94).

Since PS is recognized by a number of phagocytic receptors and opsonins that span a wide range of gene families, it is very likely that they induce immune suppression and tolerance *via* overlapping and non-overlapping mechanisms. Among the PS receptors, TIM-4 is not expected to transmit anti-inflammatory signals; since in the absence of a cytoplasmic tail, it alone cannot activate an intracellular signaling pathway (97). However, stabilin-2 was shown to be involved in inducing TGF-β1 release from engulfing macrophages (18), while MerTk was found to have a direct anti-inflammatory activity that suppresses NF-κB (98). The anti-inflammatory action of MerTk is independent of its effect on efferocytosis and is related to a signal transduction pathway that prevents the LPS-induced phosphorylation of IκB kinase and the consequent degradation of IκB (98). Whether other phagocytosis receptors also participate in the induction of the anti-inflammatory response of macrophages is still under investigation, but CD36 and $\alpha_v\beta_3$ receptors do not seem to participate in it (99). Interestingly, however, loss of both MerTk and TG2 leads to pro-inflammatory cytokine production during efferocytosis (100, 101). But the pro-inflammatory cytokine production in this latter case might be related not only to an improper integrin signaling for which TG2 is a cofactor but also to the fact that TG2 is required for proper TGF-β activation by macrophages (102).

Phosphatidylserine receptors are expressed not only by macrophages but by other immune cells as well. As a result, apoptotic cells can transmit further immune silencing signals *via* activating PS receptors on those cells as well. For example, it has been shown that in the presence of apoptotic cells, iDCs do not induce expression of DC maturation-markers, such as MHC class-II, CD40, CD80, CD83, and CD86, even after challenge with CD40-signaling, monocyte-conditioned medium, LPS, or TNF-α (103–108). Furthermore, activation of PS receptors in human DCs by PS liposomes reduces their IL-12p70 secretion and the capacity to stimulate allogeneic T cell proliferation and to activate IFN-γ-producing CD4$^+$ T cells (109). iDCs express MerTK, and activation of MerTK in iDCs triggers the phosphatidylinositol 3-kinase signaling pathway, which inhibits NF-κB activation and the consequent DC maturation (110). As a result, iDCs, which do not express MerTk or are treated with phosphatidylinositol 3-kinase inhibitors, do not respond to the immunosuppressing effect of apoptotic cells on LPS-induced pro-inflammatory cytokine formation (111). DCs also express Axl, another member of the TAM receptor family. The basal expression of Axl in DCs is very low, but it is significantly upregulated following TLR engagement. Axl, following activation, induces resolution of inflammation at the end of an inflammatory cycle (111). Ligation of the thrombospondin receptor CD36 also inhibits iDC maturation and function by suppressing the release of IL-12 and the secretion of high levels of IL-10 in response to DC-activation stimuli (105).

T cells, on the other hand, express the PS receptor TIM-3, another member of the TIM receptor family. In T cells, TIM-3, following PS exposure, transmits an immunosuppressive signal by sequestering lck, a critical tyrosine kinase participating in T cell receptor-mediated signal transduction (112).

In addition to PS, late apoptotic neutrophils were shown to express also pentraxin 3 (PTX3) in their membranes (113). PTX3 was shown not only to enhance their phagocytic removal during inflammation (113) but also to induce the expression of CD169 in macrophages (114), a molecule that interferes with the development of an autoimmune response (115).

NUCLEAR RECEPTORS ACTIVELY SUPPRESS PRO-INFLAMMATORY CYTOKINE FORMATION BY INHIBITING NF-κB

Once apoptotic cells are taken up, they have to be fast metabolized by macrophages in order to ingest further apoptotic cells. Uptake of high amount of extracellular material might induce a metabolic stress in the engulfing macrophages, and engulfing macrophages respond to it in many ways by altering their metabolism. Interestingly, these metabolic adaptors also seem to contribute to the immune silencing processes as well. Thus, the amino acid metabolic-stress sensing protein kinase GCN2 was implicated to participate in the signaling pathways that lead to tolerance (116).

Similarly, the lipid sensing nuclear receptors that function during engulfment as transcriptional regulators of lipid metabolic processes also interfere with inflammatory processes, such as those initiated by TLR signaling in macrophages. As it was reviewed by Kidani et al. (38), under non-inflammatory conditions, NF-κB target genes are kept in an inhibited state by co-repressor complexes associated with their promoters. As a response to inflammatory signaling, proteins of these co-repressor complexes become ubiquitinated and subsequently degraded by the 19S proteasome. At the same time, NF-κB activated simultaneously by the inflammatory signals translocates into the nucleus, binds to promoters of inflammatory genes, and induces their expression. Ligation of PPARγ during efferocytosis prevents NF-κB-regulated gene expression by sustaining co-repressor binding on the promoters of NF-κB target genes. This is related to the fact that ligand binding to PPARγ induces conformational changes that allows for SUMOylation of its ligand-binding domain. SUMOylated PPARγ, subsequently, binds to the co-repressor complex and prevents its degradation by the 19S proteasome, thereby sustaining the suppressed state. LXR-mediated transrepression of NF-κB target genes operates *via* similar mechanisms involving SUMOylation of LXR by HDAC4, SUMO2, or SUMO3 as the E3-ligase (38).

Ligated PPARδ also interferes with the NF-κB transcription, but it operates through a different mechanism. BCL-6 is an evolutionarily conserved zinc finger transcription factor that acts as a sequence-specific repressor of transcription. Unliganded PPARδ sequestrates BCL-6 away from the promoters of inflammatory genes; while following ligand binding PPARδ releases BCL-6 leading to suppression of inflammatory gene transcription (117).

Ligated nuclear receptors are also anti-inflammatory because during efferocytosis they upregulate the expression of various phagocytic receptors, thus, enhance the clearance capacity of macrophages (118–122). The upregulation of some phagocytic genes by these receptors, such as MerTk by LXR, is direct, while that of others is mediated *via* retinoic acid receptor (RAR)-α (121, 122). In engulfing macrophages, activation of all lipid sensing receptors leads to the upregulation of retinaldehyde dehydrogenase expression, an enzyme responsible for retinoic acid synthesis (123). Retinoic acids, ligands for the RAR and RXR receptors, then can promote both the transcriptional activity of lipid sensing nuclear receptors and trigger that of the RARs (122). In addition, LXRs are also involved in the induction of the synthesis of polyunsaturated fatty acids, which are precursors for the production of pro-resolving lipid mediators (124). And finally, ligated nuclear receptors are also responsible for the upregulation of A2ARs that mediate the anti-inflammatory effect of adenosine (86).

Besides the lipid sensing nuclear receptors, the orphan nuclear receptor Nr4a1 has also been implicated to act as an anti-inflammatory molecule during efferocytosis (96). Nr4a1 seems to interfere with the TLR signaling pathway at two levels: (1) it can interact with TRAF6, a central adaptor molecule in the TLR signaling pathway. The interaction affects TRAF6 auto-ubiquitination leading to the suppression of NF-κB activation and to that of the subsequent transcription of pro-inflammatory cytokines (125). (2) It directly associates with the p65 subunit of NF-κB and prevents its binding to the κB promoter. However, this latter effect of Nr4a1 might be suspended by its LPS-induced p38α phosphorylation (126).

MACROPHAGES RESPOND TO APOPTOTIC CELL UPTAKE BY RELEASING ANTI-INFLAMMATORY OR PRO-RESOLVING MOLECULES

Since the first discovery that macrophages engulfing apoptotic cells release TGF-β and IL-10 (47, 48), numerous other anti-inflammatory molecules have been described to be released by engulfing macrophages. Many of these anti-inflammatory molecules are lipid mediators, such as PGE_2, $PGF_{1\alpha}$, LXA_4, or PAF, the synthesis of which is interestingly dependent on TGF-β (127). TGF-β at the same time inhibits the synthesis of pro-inflammatory lipid mediators. While in the case of TGF-β, it was found that activation of PS receptors can regulate its production (94), the regulation of IL-10 synthesis by macrophages engulfing apoptotic cells was not studied in detail. Annexin 1, however, is a known inducer of it in macrophages (128).

Retinoids produced during engulfment to enhance efferocytosis (122, 123) can also be released into the surrounding environment and contribute to the TGF-β-induced development of regulatory T cells. They act so by stabilizing TGF-β-induced Foxp3 expression under inflammatory conditions *via* inhibiting DNA methylation, which otherwise would lead to the silencing of the Foxp3 gene (129). Since regulatory T cells play a central role in preventing autoimmunity, release of TGF-β and retinoids by engulfing macrophages strongly contributes to the prevention of an autoimmune response build up following engulfment of apoptotic cells (70).

In addition to producing anti-inflammatory mediators, engulfing macrophages converted from M1 pro-inflammatory macrophages to M2 $CD11b^{low}$ macrophages (130) during inflammation are also capable of producing pro-resolving lipid mediators, such as resolvins, protectins, and maresin (131–133). These molecules are synthetized from ω-3 fatty acids *via* the 12/15-lipoxygenase pathway. Pro-resolving mediators are stimulators of resolution of inflammation. Each facilitates cessation of neutrophil transmigration, microbial phagocytosis, and engulfment of apoptotic neutrophils. They also promote the formation of $CD11b^{low}$ macrophages, which in addition to producing pro-resolving lipid mediators, also contribute to the termination of efferocytosis and emigration to lymphoid organs (130) required for the proper termination of the inflammatory program. The conversion to pro-resolving macrophages is also facilitated by the atypical chemokine receptor D6 expressed on the surface of apoptotic neutrophils (134) and by the apoptotic cell uptake itself (135).

And, finally, apoptotic cells also modulate the function of immature iDCs in a way that they force them to secrete molecules which inhibit T cell function. For example, apoptotic cells induce in iDCs the secretion of interferon-γ that upregulates *via* autocrine and paracrine mechanisms their indoleamine 2,3-dioxygenase activity leading to the degradation of tryptophan into metabolites that inhibit T cell function (136). Apoptotic cells also induce the release of large amounts of nitric oxide from iDCs that impairs the T cell response (137). At the same time uptake of apoptotic cells by iDCs triggers TGF-β production, which also strongly contributes to the immunosuppressive effects of apoptotic cells *in vivo* (138). Interestingly, plasmacytoid DCs, which themselves cannot engulf apoptotic cells, can also initiate an apoptotic cell-induced anti-inflammatory response. However, their response is dependent on engulfing macrophage-derived soluble factors, including TGF-β (139).

CONCLUDING REMARKS

Apoptotic cell death is a determinant contributing factor to the cell turnover in most of the tissues. The silent removal of apoptotic cells maintains tissue integrity under healthy conditions and its anti-inflammatory nature contributes to the resolution of inflammation (**Figure 1**). Since the initial discovery that apoptotic cells possess potent immune suppressing effects on phagocytes of the immune system, numerous laboratories have initiated successful apoptotic cell-based therapies to induce donor-specific immunosuppression in transplantation (140). Though similar strategies were successful in mouse autoimmune experimental models as well, where apoptotic cells were given at the time of triggering of the disease (141), they would very likely fail in the treatment in chronic inflammatory diseases. This assumption is based on the observation that the establishment and/or progression of an increasing number of chronic inflammatory diseases seem to be coupled to improper phagocytosis of apoptotic cells (142). If the removal of apoptotic cells is delayed, their cell content is released and induces tissue damage leading to long-term inflammation or even to autoimmunity (143). During the past decade, we have understood that the development of chronic inflammation in these cases is related not only to the disturbed removal of dead cells but very likely also to the improper anti-inflammatory regulation that develops as a consequence of the improper clearance. Thus, enhancing efferocytosis and/or the related anti-inflammatory and pro-resolving mechanisms might provide a possibility in the treatment of those human chronic inflammatory diseases, in which impaired clearance of apoptotic cells appears as a driving or contributing force (49).

While the anti-inflammatory nature of apoptotic cell clearance is protective in healthy tissues, high PS expression of apoptotic tumor cells might promote tumor growth *via* suppressing immunity (93). Indeed, increasing evidence indicate that systemic administration of annexin A5 or other PS ligands that cover PS and block PS-mediated apoptotic tumor cell signaling may slow down tumor progression *via* interfering with the immunosuppressive properties of apoptotic tumor cells and tumor-derived microvesicles (144, 145). Thus, it was suggested that in combination with radio- or chemotherapy, annexin A5 could be used as a natural adjuvant to increase the immunogenicity of dying tumor cells thereby promoting the development of an anti-tumor immune response (93, 146).

And, finally, it is worth to note that though apoptotic cells are generally anti-inflammatory, cancer cells undergoing apoptosis in response to specific anticancer therapies can be immunogenic, if they emit precise DAMPs in a spatiotemporally defined fashion. Some of these DAMPs activate DCs during engulfment, which then can prime $CD4^+$ T cells, $CD8^+$ cytotoxic lymphocytes, and γδT lymphocytes against one or several tumor-associated antigens (147).

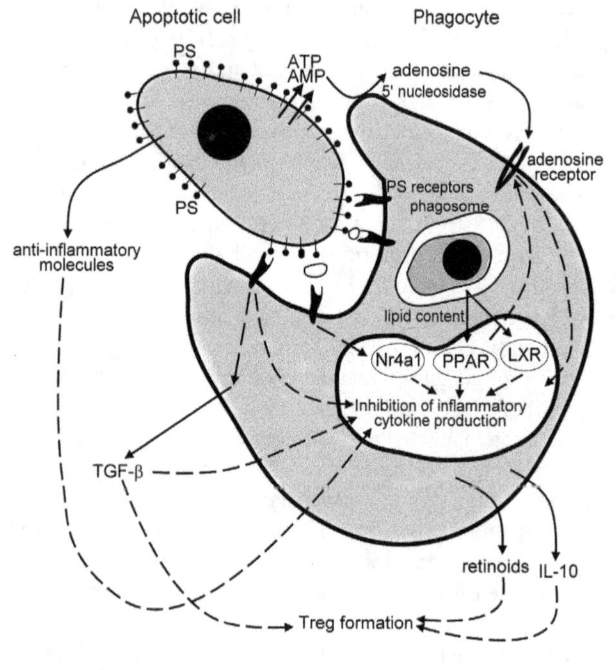

FIGURE 1 | Mechanisms of the anti-inflammatory effects of apoptotic cells. Apoptotic cells release various anti-inflammatory molecules, such as transforming growth factor (TGF)-β, interleukin (IL)-10, annexin I, thrombospondin-1 (TSP-1), fractalkine to inhibit pro-inflammatory cytokine formation of phagocytes. Once released from the apoptotic cells *via* a caspase-regulated pannexin channel, ATP is fast degraded to adenosine by the cell surface 5′ nucleotidase of engulfing macrophages. Adenosine then triggers macrophage adenosine A_{2A} receptors to suppress the NO-dependent formation of neutrophil migration factors, or in an inflammatory milieu the lipopolysaccharide (LPS)-induced pro-inflammatory cytokine formation by phagocytes. Apoptotic cells express phosphatidylserine (PS), which activates various PS sensing phagocytic receptors. Many of these receptors initiate such intracellular signaling pathways that interfere with the pro-inflammatory cytokine formation of phagocytes. Following engulfment, the lipid content of the apoptotic cells activate the nuclear lipid sensing receptors [liver X receptors (LXRs) and perixosome proliferator-activated receptors (PPARs)], which in ligated form can also interfere with the NF-κB-driven pro-inflammatory cytokine formation. And, finally, engulfing macrophages release TGF-β, IL-10, and retinoids, which act in a paracrine or autocrine fashion to amplify and sustain the anti-inflammatory response and strongly contribute to the formation of the regulatory T cells to prevent the development of autoimmunity.

AUTHOR CONTRIBUTIONS

ZSzondy, EG, BK, KK, and ZSarang contributed to the experiments related to the topic and write the paper together.

FUNDING

This study was supported by Hungarian National Research Fund (OTKA 104228) and by the GINOP-2.3.2-15-2016-00006 project (co-financed by the European Union and the European Regional Development Fund).

REFERENCES

1. Hart SP, Dransfield I, Rossi AG. Phagocytosis of apoptotic cells. *Methods* (2008) 44:280–5. doi:10.1016/j.ymeth.2007.11.009
2. Rigamonti E, Zordan P, Sciorati C, Rovere-Querini P, Brunelli S. Macrophage plasticity in skeletal muscle repair. *Biomed Res Int* (2014) 2014:560629. doi:10.1155/2014/560629
3. Thorp EB. Contrasting inflammation resolution during atherosclerosis and post myocardial infarction at the level of monocyte/macrophage phagocytic clearance. *Front Immunol* (2012) 3:39. doi:10.3389/fimmu.2012.00039
4. Wan E, Yeap XY, Dehn S, Terry R, Novak M, Zhang S, et al. Enhanced efferocytosis of apoptotic cardiomyocytes through myeloid-epithelial-reproductive tyrosine kinase links acute inflammation resolution to cardiac repair after infarction. *Circ Res* (2013) 113:1004–12. doi:10.1161/CIRCRESAHA.113.301198
5. Savill J, Dransfield I, Gregory C, Haslett C. A blast from the past: clearance of apoptotic cells regulates immune responses. *Nat Rev Immunol* (2002) 2:965–75. doi:10.1038/nri957
6. Truman LA, Ford CA, Pasikowska M, Pound JD, Wilkinson SJ, Dumitriu IE, et al. CX3CL1/fractalkine is released from apoptotic lymphocytes to stimulate macrophage chemotaxis. *Blood* (2008) 112:5026–36. doi:10.1182/blood-2008-06-162404
7. Lauber K, Bohn E, Kröber SM, Xiao YJ, Blumenthal SG, Lindemann RK, et al. Apoptotic cells induce migration of phagocytes via caspase-3-mediated release of a lipid attraction signal. *Cell* (2003) 113:717–30. doi:10.1016/S0092-8674(03)00422-7
8. Gude DR, Alvarez SE, Paugh SW, Mitra P, Yu J, Griffiths R, et al. Apoptosis induces expression of sphingosine kinase 1 to release sphingosine-1-phosphate as a "come-and-get-me" signal. *FASEB J* (2008) 22:2629–38. doi:10.1096/fj.08-107169
9. Moodley Y, Rigby P, Bundell C, Bunt S, Hayashi H, Misso N, et al. Macrophage recognition and phagocytosis of apoptotic fibroblasts is critically dependent on fibroblast-derived thrombospondin 1 and CD36. *Am J Pathol* (2003) 162:771–9. doi:10.1016/S0002-9440(10)63874-6
10. Elliott MR, Chekeni FB, Trampont PC, Lazarowski ER, Kadl A, Walk SF, et al. Nucleotides released by apoptotic cells act as a find-me signal to promote phagocytic clearance. *Nature* (2009) 461:282–6. doi:10.1038/nature08296
11. Knies UE, Behrensdorf HA, Mitchell CA, Deutsch U, Risau W, Drexler HC, et al. Regulation of endothelial monocyte-activating polypeptide II release by apoptosis. *Proc Natl Acad Sci U S A* (1998) 95:12322–7. doi:10.1073/pnas.95.21.12322
12. Cullen SP, Henry CM, Kearney CJ, Logue SE, Feoktistova M, Tynan GA, et al. Fas/CD95-induced chemokines can serve as "find-me" signals for apoptotic cells. *Mol Cell* (2013) 49:1034–48. doi:10.1016/j.molcel.2013.01.025
13. Wakasugi K, Schimmel P. Two distinct cytokines released from a human aminoacyl-tRNA synthetase. *Science* (1999) 284:147–51. doi:10.1126/science.284.5411.147
14. Nishiura H, Shibuya Y, Yamamoto T. S19 ribosomal protein cross-linked dimer causes monocyte-predominant infiltration by means of molecular mimicry to complement C5a. *Lab Invest* (1998) 78:1615–23.
15. Ravichandran KS. Find-me and eat-me signals in apoptotic cell clearance: progress and conundrums. *J Exp Med* (2010) 207:1807–17. doi:10.1084/jem.20101157
16. Fadok VA, Voelker DR, Campbell PA, Cohen JJ, Bratton DL, Henson PM. Exposure of phosphatidylserine on the surface of apoptotic lymphocytes triggers specific recognition and removal by macrophages. *J Immunol* (1992) 148:2207–16.
17. Miyanishi M, Tada K, Koike M, Uchiyama Y, Kitamura T, Nagata S. Identification of Tim4 as a phosphatidylserine receptor. *Nature* (2007) 450:435–9. doi:10.1038/nature06307
18. Park SY, Jung MY, Kim HJ, Lee SJ, Kim SY, Lee BH, et al. Rapid cell corpse clearance by stabilin-2, a membrane phosphatidylserine receptor. *Cell Death Differ* (2008) 15:192–201. doi:10.1038/sj.cdd.4402242
19. Park D, Tosello-Trampont AC, Elliott MR, Lu M, Haney LB, Ma Z, et al. BAI1 is an engulfment receptor for apoptotic cells upstream of the ELMO/Dock180/Rac module. *Nature* (2007) 450:430–4. doi:10.1038/nature06329
20. Scott RS, McMahon EJ, Pop SM, Reap EA, Caricchio R, Cohen PL, et al. Phagocytosis and clearance of apoptotic cells is mediated by MER. *Nature* (2001) 411:207–11. doi:10.1038/35079659
21. Ramirez-Ortiz ZG, Pendergraft WF III, Prasad A, Byrne MH, Iram T, Blanchette CJ, et al. The scavenger receptor SCARF1 mediates the clearance of apoptotic cells and prevents autoimmunity. *Nat Immunol* (2013) 14:917–26. doi:10.1038/ni.2670
22. Savill J, Hogg N, Ren Y, Haslett C. Thrombospondin cooperates with CD36 and the vitronectin receptor in macrophage recognition of neutrophils undergoing apoptosis. *J Clin Invest* (1992) 90:1513–22. doi:10.1172/JCI116019
23. Szondy Z, Sarang Z, Molnar P, Nemeth T, Piacentini M, Mastroberardino PG, et al. Transglutaminase 2-/- mice reveal a phagocytosis-associated crosstalk between macrophages and apoptotic cells. *Proc Natl Acad Sci U S A* (2003) 100:7812–7. doi:10.1073/pnas.0832466100
24. Chen J, Carey K, Godowski PJ. Identification of Gas6 as a ligand for Mer, a neural cell adhesion molecule related receptor tyrosine kinase implicated in cellular transformation. *Oncogene* (1997) 14:2033–9. doi:10.1038/sj.onc.1201039
25. Hall MO, Obin MS, Heeb MJ, Burgess BL, Abrams TA. Both protein S and Gas6 stimulate outer segment phagocytosis by cultured rat retinal pigment epithelial cells. *Exp Eye Res* (2005) 81:581–91. doi:10.1016/j.exer.2005.03.017
26. Hanayama R, Tanaka M, Miwa K, Shinohara A, Iwamatsu A, Nagata S. Identification of a factor that links apoptotic cells to phagocytes. *Nature* (2002) 417:182–7. doi:10.1038/417182a
27. Tóth B, Garabuczi E, Sarang Z, Vereb G, Vámosi G, Aeschlimann D, et al. Transglutaminase 2 is needed for the formation of an efficient phagocyte portal in macrophages engulfing apoptotic cells. *J Immunol* (2009) 182:2084–92. doi:10.4049/jimmunol.0803444
28. Jun JI, Kim KH, Lau LF. The matricellular protein CCN1 mediates neutrophil efferocytosis in cutaneous wound healing. *Nat Commun* (2015) 6:7386. doi:10.1038/ncomms8386
29. Martin M, Leffler J, Blom AM. Annexin A2 and A5 serve as new ligands for C1q on apoptotic cells. *J Biol Chem* (2012) 287:33733–44. doi:10.1074/jbc.M112.341339
30. Ogden CA, deCathelineau A, Hoffmann PR, Bratton D, Ghebrehiwet B, Fadok VA, et al. C1q and mannose binding lectin engagement of cell surface calreticulin and CD91 initiates macropinocytosis and uptake of apoptotic cells. *J Exp Med* (2001) 194:781–95. doi:10.1084/jem.194.6.781
31. Gardai SJ, McPhillips KA, Frasch SC, Janssen WJ, Starefeldt A, Murphy-Ullrich JE, et al. Cell-surface calreticulin initiates clearance of viable or apoptotic cells through trans-activation of LRP on the phagocyte. *Cell* (2005) 123:321–34. doi:10.1016/j.cell.2005.08.032
32. Scannell M, Flanagan MB, deStefani A, Wynne KJ, Cagney G, Godson C, et al. Annexin-1 and peptide derivatives are released by apoptotic cells and stimulate phagocytosis of apoptotic neutrophils by macrophages. *J Immunol* (2007) 178:4595–605. doi:10.4049/jimmunol.178.7.4595
33. Maderna P, Cottell DC, Toivonen T, Dufton N, Dalli J, Perretti M, et al. FPR2/ALX receptor expression and internalization are critical for lipoxin A4 and annexin-derived peptide-stimulated phagocytosis. *FASEB J* (2010) 24:4240–9. doi:10.1096/fj.10-159913
34. Kinchen JM, Cabello J, Klingele D, Wong K, Feichtinger R, Schnabel H, et al. Two pathways converge at CED-10 to mediate actin rearrangement and corpse removal in *C. elegans*. *Nature* (2005) 434:93–9. doi:10.1038/nature03263
35. Xu M, Liu Y, Zhao L, Gan Q, Wang X, Yang C. The lysosomal cathepsin protease CPL-1 plays a leading role in phagosomal degradation of apoptotic cells in *Caenorhabditis elegans*. *Mol Biol Cell* (2014) 25(13):2071–83. doi:10.1091/mbc.E14-01-0015
36. Kawane K, Ohtani M, Miwa K, Kizawa T, Kanbara Y, Yoshioka Y, et al. Chronic polyarthritis caused by mammalian DNA that escapes from degradation in macrophages. *Nature* (2006) 443:998–1002. doi:10.1038/nature05245
37. Silva E, Au-Yeung HW, Van Goethem E, Burden J, Franc NC. Requirement for a *Drosophila* E3-ubiquitin ligase in phagocytosis of apoptotic cells. *Immunity* (2007) 27:585–96. doi:10.1016/j.immuni.2007.08.016
38. Kidani Y, Bensinger SJ. Liver X receptor and peroxisome proliferator-activated receptor as integrators of lipid homeostasis and immunity. *Immunol Rev* (2012) 249:72–83. doi:10.1111/j.1600-065X.2012.01153.x
39. Evans RM, Mangelsdorf DJ. Nuclear receptors, RXR, and the Big Bang. *Cell* (2014) 157:255–66. doi:10.1016/j.cell.2014.03.012
40. Janowski BA, Willy PJ, Devi TR, Falck JR, Mangelsdorf DJ. An oxysterol signalling pathway mediated by the nuclear receptor LXR alpha. *Nature* (1996) 383:728–31. doi:10.1038/383728a0

41. Fu X, Menke JG, Chen Y, Zhou G, MacNaul KL, Wright SD, et al. 27-hydroxycholesterol is an endogenous ligand for liver X receptor in cholesterol-loaded cells. *J Biol Chem* (2001) 276:38378–87. doi:10.1074/jbc.M105805200
42. Castrillo A, Tontonoz P. Nuclear receptors in macrophage biology: at the crossroads of lipid metabolism and inflammation. *Annu Rev Cell Dev Biol* (2004) 20:455–80. doi:10.1146/annurev.cellbio.20.012103.134432
43. Calkin AC, Tontonoz P. Transcriptional integration of metabolism by the nuclear sterol-activated receptors LXR and FXR. *Nat Rev Mol Cell Biol* (2012) 13:213–24. doi:10.1038/nrm3312
44. Desvergne B, Wahli W. Peroxisome proliferator-activated receptors: nuclear control of metabolism. *Endocr Rev* (1999) 20:649–88. doi:10.1210/er.20.5.649
45. Wang YX, Lee CH, Tiep S, Yu RT, Ham J, Kang H, et al. Peroxisome-proliferator-activated receptor delta activates fat metabolism to prevent obesity. *Cell* (2003) 113:159–70. doi:10.1016/S0092-8674(03)00269-1
46. Cvetanovic M, Ucker DS. Innate immune discrimination of apoptotic cells: repression of proinflammatory macrophage transcription is coupled to specific recognition. *J Immunol* (2004) 172:880–9. doi:10.4049/jimmunol.172.2.880
47. Voll RE, Herrmann M, Roth EA, Stach C, Kalden JR, Girkontaite I. Immunosuppressive effects of apoptotic cells. *Nature* (1997) 390:350–1. doi:10.1038/37022
48. Fadok VA, Bratton DL, Konowal A, Freed PW, Westcott JY, Henson PM. Macrophages that have ingested apoptotic cells in vitro inhibit proinflammatory cytokine production through autocrine/paracrine mechanisms involving TGF-beta, PGE2, and PAF. *J Clin Invest* (1998) 101:890–8. doi:10.1172/JCI1112
49. Szondy Z, Garabuczi E, Joós G, Tsay GJ, Sarang Z. Impaired clearance of apoptotic cells in chronic inflammatory diseases: therapeutic implications. *Front Immunol* (2014) 5:354. doi:10.3389/fimmu.2014.00354
50. Martin SJ, Henry CM, Cullen SP. A perspective on mammalian caspases as positive and negative regulators of inflammation. *Mol Cell* (2012) 46:387–97. doi:10.1016/j.molcel.2012.04.026
51. Cocco R, Ucker DS. Distinct modes of macrophage recognition for apoptotic and necrotic cells are not specified exclusively by phosphatidylserine exposure. *Mol Biol Cell* (2001) 12:919–30. doi:10.1091/mbc.12.4.919
52. Birge RB, Ucker DS. Innate apoptotic immunity: the calming touch of death. *Cell Death Differ* (2008) 15:1096–102. doi:10.1038/cdd.2008.58
53. Marichal T, Ohata K, Bedoret D, Mesnil C, Sabatel C, Kobiyama K, et al. DNA released from dying host cells mediates aluminum adjuvant activity. *Nat Med* (2011) 17:996–1002. doi:10.1038/nm.2403
54. Enari M, Sakahira H, Yokoyama H, Okawa K, Iwamatsu A, Nagata S. A caspase-activated DNase that degrades DNA during apoptosis, and its inhibitor ICAD. *Nature* (1998) 391:43–50. doi:10.1038/34112
55. Kawane K, Fukuyama H, Yoshida H, Nagase H, Ohsawa Y, Uchiyama Y, et al. Impaired thymic development in mouse embryos deficient in apoptotic DNA degradation. *Nat Immunol* (2003) 4:138–44. doi:10.1038/ni881
56. Okabe Y, Sano T, Nagata S. Regulation of the innate immune response by threonine-phosphatase of eyes absent. *Nature* (2009) 460:520–4. doi:10.1038/nature08138
57. Rajput A, Kovalenko A, Bogdanov K, Yang SH, Kang TB, Kim JC, et al. RIG-I RNA helicase activation of IRF3 transcription factor is negatively regulated by caspase-8-mediated cleavage of the RIP1 protein. *Immunity* (2011) 34:340–51. doi:10.1016/j.immuni.2010.12.018
58. Rongvaux A, Jackson R, Harman CC, Li T, West AP, de Zoete MR, et al. Apoptotic caspases prevent the induction of type I interferons by mitochondrial DNA. *Cell* (2014) 159:1563–77. doi:10.1016/j.cell.2014.11.037
59. Scaffidi P, Misteli T, Bianchi ME. Release of chromatin protein HMGB1 by necrotic cells triggers inflammation. *Nature* (2002) 418:191–5. doi:10.1038/nature00858
60. Kazama H, Ricci JE, Herndon JM, Hoppe G, Green DR, Ferguson TA. Induction of immunological tolerance by apoptotic cells requires caspase-dependent oxidation of high-mobility group box-1 protein. *Immunity* (2008) 29:21–32. doi:10.1016/j.immuni.2008.05.013
61. Lüthi AU, Cullen SP, McNeela EA, Duriez PJ, Afonina IS, Sheridan C, et al. Suppression of interleukin-33 bioactivity through proteolysis by apoptotic caspases. *Immunity* (2009) 31:84–98. doi:10.1016/j.immuni.2009.05.007
62. Chen W, Frank ME, Jin W, Wahl SM. TGF-β released by apoptotic T cells contributes to an immunosuppressive milieu. *Immunity* (2001) 14:715–25. doi:10.1016/S1074-7613(01)00147-9
63. Li MO, Wan YY, Sanjabi S, Robertson AK, Flavell RA. Transforming growth factor-beta regulation of immune responses. *Annu Rev Immunol* (2006) 24:99–146. doi:10.1146/annurev.immunol.24.021605.090737
64. Shull MM, Ormsby I, Kier AB, Pawlowski S, Diebold RJ, Yin M, et al. Targeted disruption of the mouse transforming growth factor-beta 1 gene results in multifocal inflammatory disease. *Nature* (1992) 359:693–9. doi:10.1038/359693a0
65. Kulkarni AB, Huh CG, Becker D, Geiser A, Lyght M, Flanders KC, et al. Transforming growth factor beta 1 null mutation in mice causes excessive inflammatory response and early death. *Proc Natl Acad Sci U S A* (1993) 90:770–4. doi:10.1073/pnas.90.2.770
66. Wahl SM. Transforming growth factor beta (TGF-beta) in inflammation: a cause and a cure. *J Clin Immunol* (1992) 12:61–74. doi:10.1007/BF00918135
67. Gamble JR, Vadas MA. Endothelial adhesiveness for blood neutrophils is inhibited by transforming growth factor-beta. *Science* (1988) 242:97–9. doi:10.1126/science.3175638
68. Tsunawaki S, Sporn M, Ding A, Nathan C. Deactivation of macrophages by transforming growth factor-beta. *Nature* (1988) 334:260–2. doi:10.1038/334260a0
69. Chen W, Jin W, Hardegen N, Lei KJ, Li L, Marinos N, et al. Conversion of peripheral CD4+CD25- naive T cells to CD4+CD25+ regulatory T cells by TGF-beta induction of transcription factor Foxp3. *J Exp Med* (2003) 198:1875–86. doi:10.1084/jem.20030152
70. Sakaguchi S, Miyara M, Costantino CM, Hafler DA. FOXP3+ regulatory T cells in the human immune system. *Nat Rev Immunol* (2010) 10:490–500. doi:10.1038/nri2785
71. Konkel JE, Jin W, Abbatiello B, Grainger JR, Chen W. Thymocyte apoptosis drives the intrathymic generation of regulatory T cells. *Proc Natl Acad Sci U S A* (2014) 111:E465–73. doi:10.1073/pnas.1320319111
72. Gao Y, Herndon JM, Zhang H, Griffith TS, Ferguson TA. Antiinflammatory effects of CD95 ligand (FasL)-induced apoptosis. *J Exp Med* (1998) 188:887–96. doi:10.1084/jem.188.5.887
73. Levings MK, Bacchetta R, Schulz U, Roncarolo MG. The role of IL-10 and TGF-beta in the differentiation and effector function of T regulatory cells. *Int Arch Allergy Immunol* (2002) 129:263–76. doi:10.1159/000067596
74. Pupjalis D, Goetsch J, Kottas DJ, Gerke V, Rescher U. Annexin A1 released from apoptotic cells acts through formyl peptide receptors to dampen inflammatory monocyte activation via JAK/STAT/SOCS signalling. *EMBO Mol Med* (2011) 3:102–14. doi:10.1002/emmm.201000113
75. Bournazou I, Pound JD, Duffin R, Bournazos S, Melville LA, Brown SB, et al. Apoptotic human cells inhibit migration of granulocytes via release of lactoferrin. *J Clin Invest* (2009) 119:20–32. doi:10.1172/JCI36226
76. Bournazou I, Mackenzie KJ, Duffin R, Rossi AG, Gregory CD. Inhibition of eosinophil migration by lactoferrin. *Immunol Cell Biol* (2010) 88:220–3. doi:10.1038/icb.2009.86
77. Truman LA, Ogden CA, Howie SE, Gregory CD. Macrophage chemotaxis to apoptotic Burkitt's lymphoma cells in vitro: role of CD14 and CD36. *Immunobiology* (2004) 209:21–30. doi:10.1016/j.imbio.2004.02.001
78. Lopez-Dee Z, Pidcock K, Gutierrez LS. Thrombospondin-1: multiple paths to inflammation. *Mediators Inflamm* (2011) 2011:296069. doi:10.1155/2011/296069
79. Schultz-Cherry S, Chen H, Mosher DF, Misenheimer TM, Krutzsch HC, Roberts DD, et al. Regulation of transforming growth factor-beta activation by discrete sequences of thrombospondin 1. *J Biol Chem* (1995) 270:7304–10. doi:10.1074/jbc.270.13.7304
80. Krispin A, Bledi Y, Atallah M, Trahtemberg U, Verbovetski I, Nahari E, et al. Apoptotic cell thrombospondin-1 and heparin-binding domain lead to dendritic-cell phagocytic and tolerizing states. *Blood* (2006) 108:3580–9. doi:10.1182/blood-2006-03-013334
81. Zujovic V, Benavides J, Vigé X, Carter C, Taupin V. Fractalkine modulates TNF-alpha secretion and neurotoxicity induced by microglial activation. *Glia* (2000) 29:305–15. doi:10.1002/(SICI)1098-1136(20000215)29:4<305::AID-GLIA2>3.0.CO;2-V
82. Huang LS, Hung ND, Sok DE, Kim MR. Lysophosphatidylcholine containing docosahexaenoic acid at the sn-1 position is anti-inflammatory. *Lipids* (2010) 45:225–36. doi:10.1007/s11745-010-3392-5
83. Sandilos JK, Chiu YH, Chekeni FB, Armstrong AJ, Walk SF, Ravichandran KS, et al. Pannexin 1, an ATP release channel, is activated by caspase cleavage of its pore-associated C-terminal autoinhibitory region. *J Biol Chem* (2012) 287:11303–11. doi:10.1074/jbc.M111.323378

84. Yamaguchi H, Maruyama T, Urade Y, Nagata S. Immunosuppression via adenosine receptor activation by adenosine monophosphate released from apoptotic cells. *Elife* (2014) 3:e02172. doi:10.7554/eLife.02172
85. Sándor K, Pallai A, Duró E, Legendre P, Couillin I, Sághy T, et al. Adenosine produced from adenine nucleotides through an interaction between apoptotic cells and engulfing macrophages contributes to the appearance of transglutaminase 2 in dying thymocytes. *Amino Acids* (2017) 115:124–32. doi:10.1007/s00726-016-2257-5
86. Köröskényi K, Duró E, Pallai A, Sarang Z, Kloor D, Ucker DS, et al. Involvement of adenosine A2A receptors in engulfment-dependent apoptotic cell suppression of inflammation. *J Immunol* (2011) 186:7144–55. doi:10.4049/jimmunol.1002284
87. Duró E, Pallai A, Köröskényi K, Sarang Z, Szondy Z. Adenosine A3 receptors negatively regulate the engulfment-dependent apoptotic cell suppression of inflammation. *Immunol Lett* (2014) 162:292–301. doi:10.1016/j.imlet.2014.06.014
88. Joós G, Jákim J, Kiss B, Szamosi R, Papp T, Felszeghy S, et al. Involvement of adenosine A3 receptors in the chemotactic navigation of macrophages towards apoptotic cells. *Immunol Lett* (2017) 183:62–72. doi:10.1016/j.imlet.2017.02.002
89. Haskó G, Kuhe DG, Chen JF, Schwarzschild MA, Deitch EA, Mabley JG, et al. Adenosine inhibits IL-12 and TNF-[alpha] production via adenosine A2a receptor-dependent and independent mechanisms. *FASEB J* (2000) 14:2065–74. doi:10.1096/fj.99-0508com
90. Crean D, Cummins EP, Bahar B, Mohan H, McMorrow JP, Murphy EP. Adenosine modulates NR4A orphan nuclear receptors to attenuate hyperinflammatory responses in monocytic cells. *J Immunol* (2015) 195:1436–48. doi:10.4049/jimmunol.1402039
91. Köröskényi K, Kiss B, Szondy Z. Adenosine A2A receptor signaling attenuates LPS-induced pro-inflammatory cytokine formation of mouse macrophages by inducing the expression of DUSP1. *Biochim Biophys Acta* (2016) 1863:1461–71. doi:10.1016/j.bbamcr.2016.04.003
92. Krahling S, Callahan MK, Williamson P, Schlegel RA. Exposure of phosphatidylserine is a general feature in the phagocytosis of apoptotic lymphocytes by macrophages. *Cell Death Differ* (1999) 6:183–9. doi:10.1038/sj.cdd.4400473
93. Birge RB, Boeltz S, Kumar S, Carlson J, Wanderley J, Calianese D, et al. Phosphatidylserine is a global immunosuppressive signal in efferocytosis, infectious disease, and cancer. *Cell Death Differ* (2016) 23:962–78. doi:10.1038/cdd.2016.11
94. Huynh ML, Fadok VA, Henson PM. Phosphatidylserine-dependent ingestion of apoptotic cells promotes TGF-beta1 secretion and the resolution of inflammation. *J Clin Invest* (2002) 109:41–50. doi:10.1172/JCI0211638
95. Kim S, Elkon KB, Ma X. Transcriptional suppression of interleukin-12 gene expression following phagocytosis of apoptotic cells. *Immunity* (2004) 21:643–53. doi:10.1016/j.immuni.2004.09.009
96. Ipseiz N, Uderhardt S, Scholtysek C, Steffen M, Schabbauer G, Bozec A, et al. The nuclear receptor Nr4a1 mediates anti-inflammatory effects of apoptotic cells. *J Immunol* (2014) 192:4852–8. doi:10.4049/jimmunol.1303377
97. Park D, Hochreiter-Hufford A, Ravichandran KS. The phosphatidylserine receptor TIM-4 does not mediate direct signaling. *Curr Biol* (2009) 19:346–51. doi:10.1016/j.cub.2009.01.042
98. Tibrewal N, Wu Y, D'Mello V, Akakura R, George TC, Varnum B, et al. Autophosphorylation docking site Tyr-867 in Mer receptor tyrosine kinase allows for dissociation of multiple signaling pathways for phagocytosis of apoptotic cells and down-modulation of lipopolysaccharide-inducible NF-kappaB transcriptional activation. *J Biol Chem* (2008) 283:3618–27. doi:10.1074/jbc.M706906200
99. Lucas M, Stuart LM, Zhang A, Hodivala-Dilke K, Febbraio M, Silverstein R, et al. Requirements for apoptotic cell contact in regulation of macrophage responses. *J Immunol* (2006) 177:4047–54. doi:10.4049/jimmunol.177.6.4047
100. Lemke G. Biology of the TAM receptors. *Cold Spring Harb Perspect Biol* (2013) 5:a009076. doi:10.1101/cshperspect.a009076
101. Falasca L, Iadevaia V, Ciccosanti F, Melino G, Serafino A, Piacentini M. Tansglutaminase type II is a key element in the regulation of the anti-inflammatory response elicited by apoptotic cell engulfment. *J Immunol* (2005) 174:7330–40. doi:10.4049/jimmunol.174.11.7330
102. Nunes I, Shapiro RL, Rifkin DB. Characterization of latent TGF-beta activation by murine peritoneal macrophages. *J Immunol* (1995) 155:1450–9.
103. Gallucci S, Lolkema M, Matzinger P. Natural adjuvants: endogenous activators of dendritic cells. *Nat Med* (1999) 5:1249–55. doi:10.1038/15200
104. Sauter B, Albert M, Francisco L, Larsson M, Somersan S, Bhardwaj N. Consequences of cell death: exposure to necrotic tumor cells, but not primary tissue cells or apoptotic cells, induces the maturation of immunostimulatory dendritic cells. *J Exp Med* (2000) 191:423–35. doi:10.1084/jem.191.3.423
105. Urban BC, Willcox N, Roberts DJ. A role for CD36 in the regulation of dendritic cell function. *Proc Natl Acad Sci U S A* (2001) 98:8750–5. doi:10.1073/pnas.151028698
106. Stuart LM, Lucas M, Simpson C, Lamb J, Savill J, Lacy-Hulbert A. Inhibitory effects of apoptotic cell ingestion upon endotoxin-driven myeloid dendritic cell maturation. *J Immunol* (2002) 168:1627–35. doi:10.4049/jimmunol.168.4.1627
107. Verbovetski I, Bychkov H, Trahtemberg U, Shapira I, Hareuveni M, Ben-Tal O, et al. Opsonization of apoptotic cells by autologous iC3b facilitates clearance by immature dendritic cells, down-regulates DR and CD86, and up-regulates CC chemokine receptor 7. *J Exp Med* (2002) 196:1553–61. doi:10.1084/jem.20020263
108. Takahashi M, Kobayashi Y. Cytokine production in association with phagocytosis of apoptotic cells by immature dendritic cells. *Cell Immunol* (2003) 226:105–15. doi:10.1016/j.cellimm.2003.11.008
109. Chen X, Doffek K, Sugg SL, Shilyansky J. Phosphatidylserine regulates the maturation of human dendritic cells. *J Immunol* (2004) 173:2985–94. doi:10.4049/jimmunol.173.5.2985
110. Sen P, Wallet MA, Yi Z, Huang Y, Henderson M, Mathews CE, et al. Apoptotic cells induce Mer tyrosine kinase-dependent blockade of NF-κB activation in dendritic cells. *Blood* (2007) 109:653–60. doi:10.1182/blood-2006-04-017368
111. Rothlin CV, Ghosh S, Zuniga EI, Oldstone MB, Lemke G. TAM receptors are pleiotropic inhibitors of the innate immune response. *Cell* (2007) 131:1124–36. doi:10.1016/j.cell.2007.10.034
112. Rangachari M, Zhu C, Sakuishi K, Xiao S, Karman J, Chen A, et al. Bat3 promotes T cell responses and autoimmunity by repressing Tim-3–mediated cell death and exhaustion. *Nat Med* (2012) 18:1394–400. doi:10.1038/nm.2871
113. Jaillon S, Jeannin P, Hamon Y, Frémaux I, Doni A, Bottazzi B, et al. Endogenous PTX3 translocates at the membrane of late apoptotic human neutrophils and is involved in their engulfment by macrophages. *Cell Death Differ* (2009) 16:465–74. doi:10.1038/cdd.2008.173
114. Pilling D, Galvis-Carvajal E, Karhadkar TR, Cox N, Gomer RH. Monocyte differentiation and macrophage priming are regulated differentially by pentraxins and their ligands. *BMC Immunol* (2017) 18:30. doi:10.1186/s12865-017-0214-z
115. Klaas M, Crocker PR. Sialoadhesin in recognition of self and non-self. *Semin Immunopathol* (2012) 34:353–64. doi:10.1007/s00281-012-0310-3
116. Ravishankar B, Liu H, Shinde R, Chaudhary K, Xiao W, Bradley J, et al. The amino acid sensor GCN2 inhibits inflammatory responses to apoptotic cells promoting tolerance and suppressing systemic autoimmunity. *Proc Natl Acad Sci U S A* (2015) 112:10774–9. doi:10.1073/pnas.1504276112
117. Lee CH, Chawla A, Urbiztondo N, Liao D, Boisvert WA, Evans RM, et al. Transcriptional repression of atherogenic inflammation: modulation by PPARdelta. *Science* (2003) 302:453–7. doi:10.1126/science.1087344
118. A-Gonzalez N, Bensinger SJ, Hong C, Beceiro S, Bradley MN, Zelcer N, et al. Apoptotic cells promote their own clearance and immune tolerance through activation of the nuclear receptor LXR. *Immunity* (2009) 31:245–58. doi:10.1016/j.immuni.2009.06.018
119. Mukundan L, Odegaard JI, Morel CR, Heredia JE, Mwangi JW, Ricardo-Gonzalez RR, et al. PPAR-delta senses and orchestrates clearance of apoptotic cells to promote tolerance. *Nat Med* (2009) 15:1266–72. doi:10.1038/nm.2048
120. Rőszer T, Menéndez-Gutiérrez MP, Lefterova MI, Alameda D, Núñez V, Lazar MA, et al. Autoimmune kidney disease and impaired engulfment of apoptotic cells in mice with macrophage peroxisome proliferator-activated receptor gamma or retinoid X receptor alpha deficiency. *J Immunol* (2011) 186:621–31. doi:10.4049/jimmunol.1002230
121. Rébé C, Raveneau M, Chevriaux A, Lakomy D, Sberna AL, Costa A, et al. Induction of transglutaminase 2 by a liver X receptor/retinoic acid receptor alpha pathway increases the clearance of apoptotic cells by human macrophages. *Circ Res* (2009) 105:393–401. doi:10.1161/CIRCRESAHA.109.201855

122. Sarang Z, Joós G, Garabuczi É, Rühl R, Gregory CD, Szondy Z. Macrophages engulfing apoptotic cells produce nonclassical retinoids to enhance their phagocytic capacity. *J Immunol* (2014) 192:5730–8. doi:10.4049/jimmunol.1400284
123. Garabuczi É, Kiss B, Felszeghy S, Tsay GJ, Fésüs L, Szondy Z. Retinoids produced by macrophages engulfing apoptotic cells contribute to the appearance of transglutaminase 2 in apoptotic thymocytes. *Amino Acids* (2013) 44:235–44. doi:10.1007/s00726-011-1119-4
124. Serhan CN, Chiang N, Dalli J, Levy BD. Lipid mediators in the resolution of inflammation. Lipid mediators in the resolution of inflammation. *Cold Spring Harb Perspect Biol* (2014) 7:a016311. doi:10.1101/cshperspect.a016311
125. Li XM, Zhang S, He XS, Guo PD, Lu XX, Wang JR, et al. Nur77-mediated TRAF6 signalling protects against LPS-induced sepsis in mice. *J Inflamm (Lond)* (2016) 13:4. doi:10.1186/s12950-016-0112-9
126. Li L, Liu Y, Chen HZ, Li FW, Wu JF, Zhang HK, et al. Impeding the interaction between Nur77 and p38 reduces LPS-induced inflammation. *Nat Chem Biol* (2015) 11:339–46. doi:10.1038/nchembio.1788
127. Freire-de-Lima CG, Xiao YQ, Gardai SJ, Bratton DL, Schiemann WP, Henson PM. Apoptotic cells, through transforming growth factor-beta, coordinately induce anti-inflammatory and suppress pro-inflammatory eicosanoid and NO synthesis in murine macrophages. *J Biol Chem* (2006) 281:38376–84. doi:10.1074/jbc.M605146200
128. Parente L, Solito E. Annexin 1: more than an anti-phospholipase protein. *Inflamm Res* (2004) 53:125–32. doi:10.1007/s00011-003-1235-z
129. Liu ZM, Wang KP, Ma J, Guo Zheng S. The role of all-trans retinoic acid in the biology of Foxp3+ regulatory T cells. *Cell Mol Immunol* (2015) 12:553–7. doi:10.1038/cmi.2014.133
130. Schif-Zuck S, Gross N, Assi S, Rostoker R, Serhan CN, Ariel A. Saturated-efferocytosis generates pro-resolving CD11b low macrophages: modulation by resolvins and glucocorticoids. *Eur J Immunol* (2010) 41:366–79. doi:10.1002/eji.201040801
131. Godson C, Mitchell S, Harvey K, Petasis NA, Hogg N, Brady HR. Cutting edge: lipoxins rapidly stimulate nonphlogistic phagocytosis of apoptotic neutrophils by monocyte-derived macrophages. *J Immunol* (2000) 164:1663–7. doi:10.4049/jimmunol.164.4.1663
132. Serhan CN, Chiang N, Van Dyke TE. Resolving inflammation: dual anti-inflammatory and pro-resolution lipid mediators. *Nat Rev Immunol* (2008) 8:349–61. doi:10.1038/nri2294
133. Serhan CN, Chiang N. Resolution phase lipid mediators of inflammation: agonists of resolution. *Curr Opin Pharmacol* (2013) 4:632–40. doi:10.1016/j.coph.2013.05.012
134. Pashover-Schallinger E, Aswad M, Schif-Zuck S, Shapiro H, Singer P, Ariel A. The atypical chemokine receptor D6 controls macrophage efferocytosis and cytokine secretion during the resolution of inflammation. *FASEB J* (2012) 26:3891–900. doi:10.1096/fj.11-194894
135. Arnold L, Henry A, Poron F, Baba-Amer Y, van Rooijen N, Plonquet A, et al. Inflammatory monocytes recruited after skeletal muscle injury switch into antiinflammatory macrophages to support myogenesis. *J Exp Med* (2007) 204:1057–69. doi:10.1084/jem.20070075
136. Williams CA, Harry RA, McLeod JD. Apoptotic cells induce dendritic cell-mediated suppression via interferon-γ-induced IDO. *Immunology* (2007) 124:89–101. doi:10.1111/j.1365-2567.2007.02743.x
137. Ren G, Su J, Zhao X, Zhang L, Zhang J, Roberts AI, et al. Apoptotic cells induce immunosuppression through dendritic cells: critical roles of IFN-γ and nitric oxide. *J Immunol* (2008) 181:3277–84. doi:10.4049/jimmunol.181.5.3277
138. Perruche S, Zhang P, Liu Y, Saas P, Bluestone JA, Chen W. CD3-specific antibody-induced immune tolerance involves transforming growth factor-beta from phagocytes digesting apoptotic T cells. *Nat Med* (2008) 14:528–35. doi:10.1038/nm1749
139. Bonnefoy F, Perruche S, Couturier M, Sedrati A, Sun Y, Tiberghien P, et al. Plasmacytoid dendritic cells play a major role in apoptotic leukocyte-induced immune modulation. *J Immunol* (2011) 186:5696–705. doi:10.4049/jimmunol.1001523
140. Saas P, Daguindau E, Perruche S. Concise review: apoptotic cell-based therapies-rationale, preclinical results and future clinical developments. *Stem Cells* (2016) 34:1464–73. doi:10.1002/stem.2361
141. Saas P, Kaminski S, Perruche S. Prospects of apoptotic cell-based therapies for transplantation and inflammatory diseases. *Immunotherapy* (2013) 5:1055–73. doi:10.2217/imt.13.103
142. Elliott MR, Ravichandran KS. Clearance of apoptotic cells: implications in health and disease. *J Cell Biol* (2010) 189:1059–70. doi:10.1083/jcb.201004096
143. Muñoz LE, Lauber K, Schiller M, Manfredi AA, Herrmann M. The role of defective clearance of apoptotic cells in systemic autoimmunity. *Nat Rev Rheumatol* (2010) 6:280–9. doi:10.1038/nrrheum.2010.46
144. Lima LG, Chammas R, Monteiro RQ, Moreira ME, Barcinski MA. Tumor-derived microvesicles modulate the establishment of metastatic melanoma in a phosphatidylserine-dependent manner. *Cancer Lett* (2009) 283:168–75. doi:10.1016/j.canlet.2009.03.041
145. Munoz LE, Franz S, Pausch F, Furnrohr B, Sheriff A, Vogt B, et al. The influence on the immunomodulatory effects of dying and dead cells of Annexin V. *J Leukoc Biol* (2007) 81:6–14. doi:10.1189/jlb.0306166
146. Gaipl US, Multhoff G, Scheithauer H, Lauber K, Hehlgans S, Frey B, et al. Kill and spread the word: stimulation of antitumor immune responses in the context of radiotherapy. *Immunotherapy* (2014) 6:597–610. doi:10.2217/imt.14.38
147. Grag DA, Galuzzi L, Apetoh L, Baert T, Birge RB, Bravo-San Pedro SP, et al. Molecular and translational classifications of DAMPs in immunogenic cell death. *Front Immunol* (2015) 6:588. doi:10.3389/fimmu.2015.00588

Conflict of Interest Statement: The authors declare that the research was conducted in the absence of any commercial or financial relationships that could be construed as a potential conflict of interest.

3

Extracellular Vesicles Arising from Apoptotic Cells in Tumors: Roles in Cancer Pathogenesis and Potential Clinical Applications

*Catherine Lynch, Maria Panagopoulou and Christopher D. Gregory**

MRC Centre for Inflammation Research, Queen's Medical Research Institute, University of Edinburgh, Edinburgh, United Kingdom

It is known that apoptotic cells can have diverse effects on the tumor microenvironment. Emerging evidence indicates that, despite its renowned role in tumor suppression, apoptosis may also promote oncogenic evolution or posttherapeutic relapse through multiple mechanisms. These include immunomodulatory, anti-inflammatory, and trophic environmental responses to apoptosis, which drive tumor progression. Our group has introduced the term "onco-regenerative niche (ORN)" to describe a conceptual network of conserved cell death-driven tissue repair and regeneration mechanisms that are hijacked in cancer. We propose that, among the key elements of the ORN are extracellular vesicles (EVs), notably those derived from apoptotic tumor cells. EVs are membrane-delimited subcellular particles, which contain multiple classes of bioactive molecules including markers of the cell from which they are derived. EVs are implicated in an increasing number of physiological and pathological contexts as mediators of local and systemic intercellular communication and detection of specific EVs may be useful in monitoring disease progression. Here, we discuss the mechanisms by which EVs produced by apoptotic tumor cells—both constitutively and as a consequence of therapy—may mediate host responsiveness to cell death in cancer. We also consider how the monitoring of such EVs and their cargoes may in the future help to improve cancer diagnosis, staging, and therapeutic efficacy.

Edited by:
Kirsten Lauber, Ludwig-Maximilians-Universität München, Germany

Reviewed by:
Dmitri V. Krysko, Ghent University, Belgium
Angelo A. Manfredi, Vita-Salute San Raffaele University, Italy

***Correspondence:**
Christopher D. Gregory
chris.gregory@ed.ac.uk

Keywords: extracellular vesicle, apoptosis, exosome, ectosome, cancer pathogenesis, wound healing, regeneration

INTRODUCTION: APOPTOSIS AND ONCOGENESIS

Apoptosis plays important roles in regulating cell populations during ontogeny and in adult tissues. Emerging evidence indicates that, not only is apoptosis responsible for the well-established deletion of cells during organ sculpting (for example, removal of the cells of the interdigital webs during limb development of mammals) but also for stimulating the proliferation of cells in neighboring compartments, a process termed compensatory proliferation or apoptosis-induced proliferation (1–3). In adults, apoptosis is prominent in controlling responses to infection and other inflammatory stimuli and in cyclical turnover of tissues ranging from the lactating/post-lactating mammary gland to the turnover of epithelia. In the steady state in which cell gain is balanced by cell loss, apoptosis provides important tissue homeostatic signals, ensuring the safe, non-phlogistic removal of billions of cells daily.

In malignant disease, the homeostatic cell birth/cell death balance becomes dysregulated through acquisition of oncogenic mutations that lead to net expansion of transformed cell populations. In addition to the loss of normal constraints on proliferation, the capacity to evade apoptosis and survive inappropriately facilitates oncogenesis. Thus, mutations affecting genes that promote cell survival such as antiapoptotic Bcl-2 family members and molecules active in the PI3K/Akt pathway complement those that promote cell proliferation (4). In the face of frequent mitoses, tumors that grow slowly—a typical example being basal cell carcinoma—display relatively high rates of apoptosis [high apoptotic index (AI)] (5). Less intuitively, aggressive, rapidly growing tumors also display high AIs. Indeed, close association between high AI and high proliferative rate has been reported for multiple aggressive cancers, including colorectal carcinoma, bladder, lung and breast cancers, leukemia, and lymphoma (6). To place these observations in perspective, it has long been known that in Burkitt's lymphoma, for example, which displays evidence of both high mitotic and apoptotic indices in standard histological sections, a substantial proportion (around 70%) of the proliferating cells die (7). This accords with the principle that apoptosis is often prominent in association with proliferation in normal tissues.

Constitutive apoptosis in growing tumors is likely to be caused by multiple stresses characteristic of rapidly expanding tissues, cell growth outpacing (i) supply of nutrients and oxygen and (ii) removal of potentially toxic metabolites. Additional causes of constitutive apoptosis include survival pathway "insufficiency" (caused by genetic mutation), antitumor immunity, and cell competition effects. Apoptosis of tumor cells is also the basis for the effectiveness of anticancer chemotherapies and radiotherapy. Recent evidence indicates that microenvironmental tissue repair and regenerative responses to tumor cell apoptosis are critically important in promoting net tumor growth and posttherapeutic tumor repopulation/relapse (8–11). We propose that extracellular vesicles (EVs) play key roles in potentiating the microenvironmental effects of apoptosis in tumors.

EVs: CLASSES, CARGOES, AND FUNCTIONAL PROPERTIES IN CANCER

EVs are subcellular membrane-delimited particles, which can be released from cells both constitutively and in response to activation or stress. Although there is a lack of consensus on properties and nomenclature within the EV field, it is generally accepted that there are at least three different types of EVs, which have been classified according to size, biogenesis, or isolation technique (12–18) (**Figure 1**). Exosomes are the smallest category (30–150 nm) formed from endosomal membranes and released from the cell by exocytosis of multivesicular bodies. Ectosomes or microvesicles (100–1,000 nm) bud directly from the plasma membrane. Apoptotic bodies (typically described as 1,000–5,000 nm) are formed during apoptosis, and apoptotic cell-derived vesicles have very broad size ranges (our unpublished observations). Relatively large EVs (ectosome-like), >2 μm in diameter and not apparently associated with apoptosis, have also been described, showing intact organelles but not nuclear components [see, for example, Ref. (19)]. EV cargoes are diverse, ranging from nucleic acids such as miRNA, mRNA, and DNA, to lipids, and cytosolic and

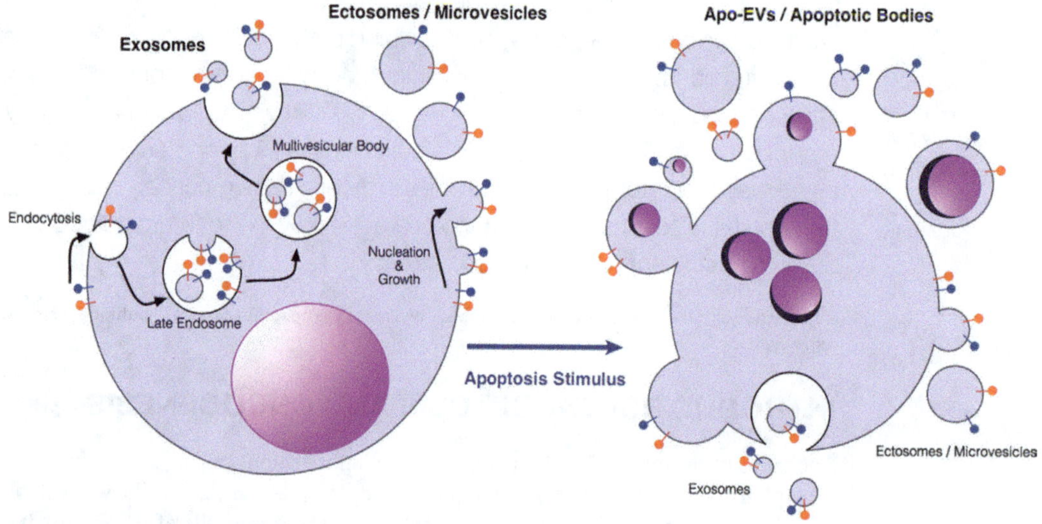

FIGURE 1 | Extracellular vesicles (EVs) produced by viable and apoptotic cells. Diagrammatic representation of the broad categories of EVs produced by viable cells (left) and cells undergoing apoptosis (right). Exosomes are produced through endosomal pathways involving the formation of multivesicular bodies that fuse with the plasma membrane. Ectosomes or microvesicles bud directly from the plasma membrane. Apo-EVs are produced in a diverse size range specifically by apoptotic cells—larger Apo-EVs are commonly known as apoptotic bodies. Apo-EVs can contain large cellular organelles including nuclear fragments as shown. Relatively little is yet known about the molecular mechanisms underlying the production of Apo-EVs/apoptotic bodies and the roots of their heterogeneity [see Ref. (18, 27), for reviews] or indeed their relationships with exosome or ectosome production. Preparations of Apo-EVs are likely to be contaminated by exosomes and ectosomes produced at least during pre-apoptosis stages of activation. Red and blue "lollipop" symbols serve to illustrate orientation of transmembrane proteins of the plasma membrane. Left-hand cartoon adapted from Ref. (17).

membrane proteins. It is thought that the sequestering of specific cargoes into EVs is directed, but the mechanisms by which this occurs have not yet been fully elucidated. EVs can be detected in most bodily fluids, including blood, urine, saliva, amniotic fluid, breast milk, and ascites. There has been a growing interest in using circulating EVs as carriers of biomarkers of disease, especially since EVs carry markers of their cell of origin and may represent the pathophysiological status of the cells (12–17, 20–22). EVs are involved in the regulation of tumor growth, progression, and antitumor immunity, in the latter context showing potential application in antitumor vaccination [see Ref. (23–26) and forthcoming Frontiers in Immunology Research Topic: *The Immunomodulatory Properties of Extracellular Vesicles from Pathogens, Immune Cells and Non-Immune Cells*].

EVs both from tumor cells and from stromal cell components of the tumor microenvironment can promote cancer growth and metastasis. Accruing evidence indicates that the pro-oncogenic properties of EVs are mediated *via* multiple mechanisms, reflecting the multitude of cargoes and the complex biological composition of EVs. For example, tumor cell-derived EVs can promote tumor growth and angiogenesis through transferral of mutant receptors, angiogenic proteins, and RNAs from tumor cells to neighboring cells, including endothelial cells and mutant growth receptor-deficient tumor cells in the tumor microenvironment (22, 28, 29). EVs can promote tumor metastasis through horizontal transfer of oncogenic molecules from cancer cells to bone marrow-derived stromal cells (30) or directly from malignant tumor cells to relatively benign counterparts, endowing the recipient cells with metastatic properties (31). EVs from tumor cells also endow stromal cells with switched metabolic pathways (32) and can alter the activation status of fibroblasts to resemble that of cancer-associated fibroblasts (CAFs) (33, 34). Furthermore, EVs derived from the stromal or immune cells of tumors have regulatory properties in cancer. For example, M2-polarized tumor-associated macrophages (TAMs) have been shown to transfer microRNA-21 to gastric cancer cells, suppressing apoptosis, and thereby causing cisplatin resistance (35).

APOPTOTIC TUMOR CELL-DERIVED EVs

Apoptotic cells in tumors communicate with neighboring cells not only by intercellular contact but also *via* soluble and EV-encapsulated signal mediators (36, 37). EVs from apoptotic cells display a broad size heterogeneity from around 50 nm to several microns and the term "apoptotic body" is often used to describe the larger varieties—commonly >1,000 nm—of apoptotic cell-derived EVs (Apo-EVs) (18). However, the terminology describing the different types of membrane-delimited subcellular-sized particles released from apoptotic cells is currently a matter of discussion, as a standardized nomenclature has not been established to date (38). We favor the concept that Apo-EVs represent a continuum (albeit heterogeneous) of vesicles released from apoptotic cells with wide variation in size, including those classed as apoptotic bodies (**Figure 1**). Although there seems little doubt that Apo-EVs will prove to be heterogeneous in other ways [e.g., some carrying genomic DNA and/or organelles such as mitochondria, together with heterogeneity in macromolecule content (39–42)], the critical definition of an Apo-EV is a vesicle that is apoptosis dependent. Clearly EVs may be released from apoptotic cells as a consequence of pre-apoptosis stress signals or as a result of post-apoptotic necrosis. The need for more information about the EVs released throughout the apoptotic process is reinforced by evidence of significant levels of proteins such as histones that are loaded into Apo-EVs prior to the loss of plasma membrane integrity (40). Since Apo-EVs encapsulate a wide variety of bioactive molecules and cellular organelles (39–43), they can be characterized as metabolically active structures that provide apoptotic cells with the ability to transduce signals over relatively long distances (6, 36, 37, 44).

Although several studies have been forthcoming in recent years, the structural characteristics, contents, and functional attributes of Apo-EVs in cancer remain poorly defined (6, 18, 44), particularly since apoptosis dependence of putative Apo-EVs has not been stringently investigated, for example, by comparing EVs from apoptotic and non-apoptotic cells subjected to identical stress signals. Notwithstanding this limitation, we highlight the potential for Apo-EVs and their cargoes to play important roles in the regulation of tumor growth and progression. Thus, it has been shown that cancer cells under stress transfer genetically active material to their neighbors and given that EVs are rich in nucleic acids—including both DNA and RNA [the former likely to be especially enriched specifically in Apo-EVs (our unpublished observations)]—they are likely to play a significant role in this communication (45). Furthermore, stromal cell-derived EVs (<100 nm) released as a consequence of cell stress may provide key signals supporting the neighboring tumor cells' capacity to metastasize, promoting proliferation and inhibiting apoptosis (46). Immunological functional heterogeneity of Apo-EVs is evident from studies indicating that, on the one hand they can be immunosuppressive (26), while on the other (albeit in a different context), immunostimulatory (41). Horizontal transfer of potentially oncogenic genomic DNA through phagocytosis of apoptotic bodies and its control through p53-mediated DNA damage responses has been demonstrated (47–49), raising the possibility that cells receiving Apo-EVs may, even transiently, express gene signatures from apoptotic tumor cells, with implications for tumor growth. It is also conceivable that oncogene transfer through Apo-EVs could lead to sustained transformation of recipient cells. Indeed, the production of EVs by tumor cells in response to anticancer therapies—which are well known to induce apoptosis in tumor cells—suggests that EV production is detrimental to therapeutic success (50–52). Mechanisms of EV-mediated drug resistance are beginning to emerge. For example, in pancreatic ductal adenocarcinoma (PDAC), EVs produced by CAFs as a consequence of chemotherapy transfer Snail and miR-146a to PDAC cells, enhancing both their survival and proliferation during treatment (53). Further work will clarify roles specifically for Apo-EV in tumor growth and progression pre- and posttherapy.

THE ONCO-REGENERATIVE NICHE (ORN)

Our group has proposed that Apo-EVs are key components in a conceptual tissue repair microenvironmental "module" we have termed the *ORN* (6, 44). This concept is based upon the well-rehearsed comparison of tumors to "wounds that fail to heal" or, perhaps more accurately, "wounds that fail to stop repairing." We propose that the ORN represents a microenvironmental signaling network driven by apoptosis and involving tumor cells, non-tumor stromal and immune cells, connective tissue, soluble factors, and EVs. The putative network engenders pro-repair and regenerative responses that promote tumor cell proliferation, angiogenesis, and invasiveness while at the same time suppressing antitumor immunity. Currently, the ORN remains conceptual, serving as a platform for rationalized experimentation to test the hypothesis that conserved tissue repair and regenerative responses to cell death in tumors provide fundamental pro-oncogenic signals that can facilitate tumor growth, metastasis, and posttherapeutic relapse.

Intercellular communication events concerning apoptotic cells, which lie at the heart of the ORN, involve TAMs, endothelial cells, and viable tumor cells among others including additional immune cells, CAFs, and mesenchymal stromal cells. Little is yet known about the mechanisms by which Apo-EVs interact with neighboring cells of the tumor microenvironment although EVs released by apoptotic cells harbor "find-me" signals that facilitate the directed migration of phagocytes to apoptotic cells. In some studies, active chemotactic molecules associated with EVs (CX_3CL1 and ICAM-3) have been described (54–56). In the context of phagocytosis, apoptotic cells are able to activate multiple lineages of cells in addition to macrophages, including dendritic cells, epithelial cells, bone marrow stromal cells, muscle cells, fibroblasts, and endothelial cells. Although direct cell-to-cell contact is an obviously essential prerequisite for phagocytosis of apoptotic cell bodies, information is currently limited on the potential roles of Apo-EVs in activating the phagocytes of apoptotic cells. Given the multiple modes by which EVs can interact with cellular targets, we speculate that both receptor-dependent and independent mechanisms operate in the ORN to provide communication pathways between apoptotic cell-derived EVs and other cells in the microenvironment—both phagocytes and non-phagocytes, non-transformed and transformed cells (**Figure 2**). First-line candidate receptors would be those already known to function in apoptotic cell clearance, especially the phosphatidylserine (PS)-binding glycoproteins, be they opsonins such as Gas6, Protein S, or MFG-E8, or cell surface PS receptors such as TIM-4, BAI-1, and Stabilin 2 [reviewed recently by Ref. (57)], especially since the majority of EVs, including those released from apoptotic cells tend to expose PS [(12) and our unpublished observations]. Additional, PS-independent mechanisms may also be involved, and mechanisms of activation of the cellular targets of Apo-EVs would be expected to include agonistic or antagonistic receptor ligation, localized release or transfer of highly labile, biologically active molecules and transfer of proteins, nucleic acids, and metabolites as has been described for EVs in other contexts.

Although limited mechanistic information is yet available, we propose that EVs released into the tumor microenvironment have multiple tumor-modulating activities. Apoptotic cells activate diverse responses in neighboring cells, producing mitogens, anti-inflammatory mediators, pro-angiogenic factors, and matrix-degrading enzymes in addition to chemoattractants (6, 10, 44). Given their biological potential as we have already discussed, it seems likely that Apo-EVs will prove to harbor key bioactive molecules that mediate at least some of the aforementioned tumor-modulating activities.

CLINICAL POTENTIAL OF APOPTOTIC TUMOR CELL-DERIVED EVs

It is widely accepted that EVs contain biomolecules indicative of the cell from which they derive, its state of activation, its metabolic activity, and in some cases its genotype. Since EVs are accessible in various body fluids, their cargoes may be useful as non-invasive biomarkers for diagnosis and prognosis of disease, including cancer. Recent investigations serve to highlight the association of EVs and their cargoes with diverse cancer types. For example, a recent study of EVs isolated from blood of colorectal cancer patients, based on $EpCAM^+$ selection, found increased levels of specific miRNAs in patients compared to healthy controls, which decreased after surgical removal of the tumor (59). EVs from patients with liver cancer (hepatocellular carcinoma and cholangiocarcinoma) are measurable and distinct from EVs derived from non-cancerous chronic liver disease such as cirrhosis (60). A study of multiple myeloma showed that the number of EVs expressing the plasma cell marker CD138 in the blood of patients corresponded to both disease state and therapeutic outcome, giving both diagnostic and prognostic relevance to the detection of EVs in this disease (61). In PDAC patients, mutant *KRAS* DNA can be detected with a higher frequency in circulating EVs than the more prevalent method of cell-free circulating DNA (62). Detection of miRNA in the EVs of Hodgkin lymphoma patients was followed along the course of treatment and was found to decrease concordantly with the FDG-PET scans that are routinely used to monitor this cancer type, and in patients with relapsing disease those miRNA levels were seen to rise (63). In breast cancer, the TRPC5 receptor channel, shown to be essential for chemoresistance, can be detected on circulating EVs and used to predict treatment response and relapse (64). In prostate cancer, two proteins—ADSV and TGM4—found in urinary EVs, can in combination predict not only the presence of the disease but also relapse of the patient after treatment (65).

These examples illustrate that there is conceptual concordance between the biological activities of EVs in mechanistic cancer models and their association with disease status in human patients. As yet, however, the particular significance of Apo-EVs in human malignant disease remains unclear. We speculate that a fraction of the Apo-EVs from cells of the tumor microenvironment, including tumor cells themselves, macrophages, and other immune cells, together with fibroblasts, mesenchymal stem cells, and endothelial cells, would gain access to the circulation [note the well-established circulation of tumor cells and nucleosomes in cancer patients (66, 67)] and as such would be readily amenable to analysis using appropriate technologies (*vide infra*) and could potentially provide useful diagnostic and prognostic

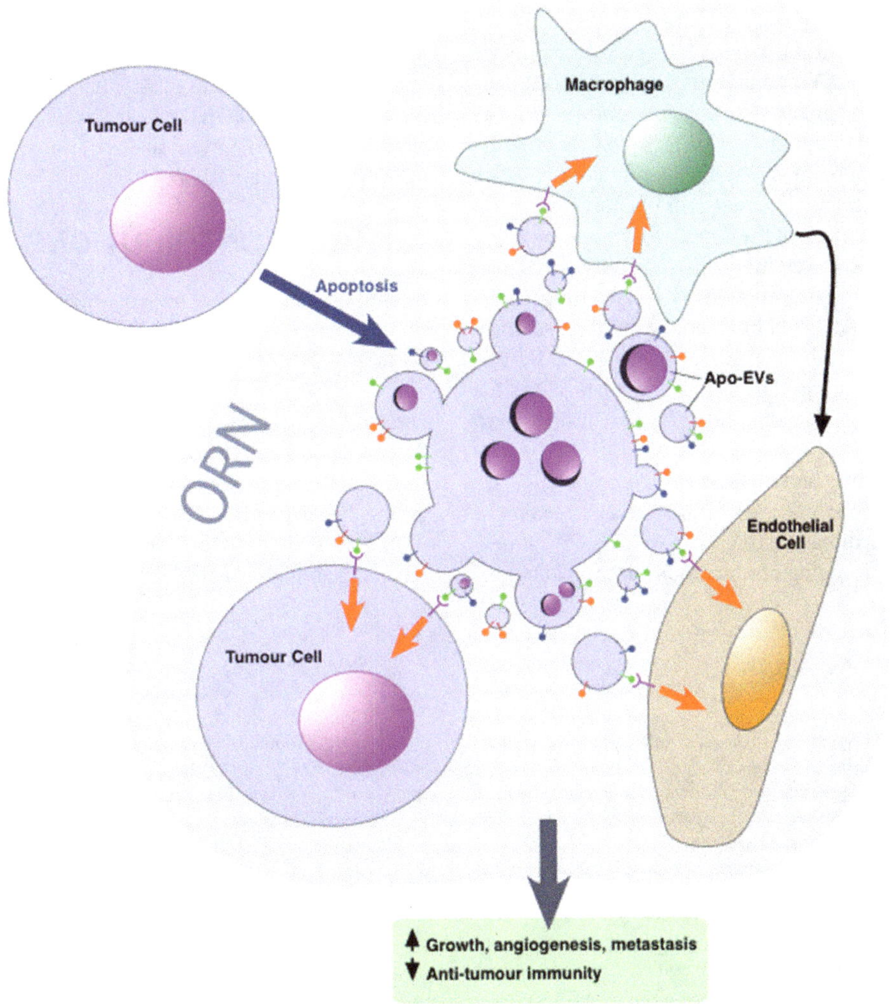

FIGURE 2 | Apoptotic tumor cell-derived extracellular vesicles (Apo-EVs) in the onco-regenerative niche (ORN). Cartoon depicting the potential involvement of Apo-EVs in providing key intercellular networking and activating elements of the ORN. Currently, the ORN is conceptualized as an apoptosis-driven oncogenic niche. Here, this is represented by Apo-EVs communicating with non-tumor cellular elements of the niche such as tumor-associated macrophages and endothelial cells as well as with viable tumor cells. Although little is yet known of the modes of interaction of Apo-EVs with recipient cells, it seems reasonable to assume that at least some of the molecules [see text and Ref. (57)] involved in clearance of apoptotic cells (and relatively large apoptotic bodies) will prove to be shared by smaller varieties of Apo-EVs. In this context, it is noteworthy that receptor-mediated cellular interaction of EVs involves the phosphatidylserine receptor TIM-4 and the integrins $\alpha v\beta 3$ and $\alpha v\beta 5$ (58), all of which are well defined as phagocyte receptors for apoptotic cells. Novel pathways also seem likely. Here, the surface of Apo-EVs is shown to display signals (green "lollipop" symbols), putatively involved in engagement with appropriate receptors on cells of the niche. We propose that, in addition to being targeted for degradation in phagolysosomes, Apo-EVs may transfer cargoes, including surface receptors and nucleic acids, to recipient cells, as has been described for other EV types, thereby altering cellular functional activities (red arrows) and intercellular communication (e.g., black arrow), culminating in the pro-oncogenic activities of the ORN such as the promotion of clonal tumor cell growth, angiogenesis, and metastasis, together with inhibition of antitumor immunity.

information of benefit to patient care. As well as providing valuable cargo information, EVs (and possibly Apo-EVs too) may also antagonize cancer therapies by acting as "decoy" vehicles for therapeutic antibodies, thereby diluting effective biopharmaceutical delivery to tumor cells. This has been shown to be the case for the anti-CD20 therapeutic antibody Rituximab, which binds EVs produced by lymphoma cells, effectively protecting the cells from immunotherapeutic destruction (68). Similarly, EVs from $HER2^+$ breast cancer carry HER2 molecules that can bind to the anti-HER2 therapeutic antibody, trastuzumab, compromising its effectiveness (69). Thus, EVs, possibly including Apo-EVs, are capable of inactivating therapy as well as promoting tumor growth, and so reducing EV production may provide a novel strategy to improve therapeutic responses.

CONCLUSION

Although EVs, including Apo-EVs, show much promise for use both in diagnosis and treatment monitoring, and potentially even as therapeutic targets, there are many hurdles to overcome before this can be translated effectively to the clinic. Perhaps the most significant challenge is EV heterogeneity and lack of standardization in terms of EV isolation and characterization in clinical samples (or indeed any samples). Much research effort is focused on purification of EV populations using advanced platforms such as immunomagnetic isolation and immunoaffinity arrays (70, 71), which also hold potential to become miniaturized, enabling the development and application of point-of-care technologies in the clinical setting. Methods of characterization are also a challenge, as the size of the smallest EVs pushes many techniques beyond their lower limits of reliability. However, progress is likely to be speedy, following the rapid progress in nanometrology that has occurred in recent years. The relatively recent explosion in EV interest and research has led to significant progress, not least in clinical oncology diagnostics (see, for example, products from Exosome Diagnostics http://www.exosomedx.com).

Because of their structure and highly specific cargoes, EVs are likely to prove to be most useful targets for biomarker screening in cancer, as well as in other disease settings. It is anticipated that new technologies will allow screening of tiny, readily accessible biopsy samples, most commonly blood, and that the vesicular nature of EVs will provide for concentration and protection of valuable biomarker cargoes, including labile molecules. Resolution of the phenotypic and functional heterogeneity of EVs is already under way, and we predict that substantial advances will continue to be made in EV research in the near future, not least in the functional attributes of Apo-EVs and their relevance to human cancer diagnosis, prognosis, and therapy.

AUTHOR CONTRIBUTIONS

All authors planned and cowrote the manuscript.

FUNDING

The authors' research is funded by Bloodwise, the Medical Research Council (MRC), and the Engineering and Physical Sciences Research Council (EPSRC). CL and MP are funded by the EPSRC and MRC under grant number EP/L019559/1 (OPTIMA) and by the University of Edinburgh College of Medicine and Veterinary Medicine.

REFERENCES

1. Bergmann A, Steller H. Apoptosis, stem cells, and tissue regeneration. *Sci Signal* (2010) 3:re8. doi:10.1126/scisignal.3145re8
2. Perez-Garijo A, Steller H. Spreading the word: non-autonomous effects of apoptosis during development, regeneration and disease. *Development* (2015) 142:3253–62. doi:10.1242/dev.127878
3. Dabrowska C, Li M, Fan Y. Apoptotic caspases in promoting cancer: implications from their roles in development and tissue homeostasis. *Adv Exp Med Biol* (2016) 930:89–112. doi:10.1007/978-3-319-39406-0_4
4. Westhoff MA, Marschall N, Debatin KM. Novel approaches to apoptosis-inducing therapies. *Adv Exp Med Biol* (2016) 930:173–204. doi:10.1007/978-3-319-39406-0_8
5. Kerr JF, Searle J. A suggested explanation for the paradoxically slow growth rate of basal-cell carcinomas that contain numerous mitotic figures. *J Pathol* (1972) 107:41–4. doi:10.1002/path.1711070107
6. Gregory CD, Ford CA, Voss JJ. Microenvironmental effects of cell death in malignant disease. *Adv Exp Med Biol* (2016) 930:51–88. doi:10.1007/978-3-319-39406-0_3
7. Iversen U, Iversen OH, Bluming AZ, Ziegler JL, Kyalwasi S. Cell kinetics of African cases of Burkitt lymphoma. A preliminary report. *Eur J Cancer* (1972) 8:305–8. doi:10.1016/0014-2964(72)90025-4
8. Li F, Huang Q, Chen J, Peng Y, Roop DR, Bedford JS, et al. Apoptotic cells activate the "phoenix rising" pathway to promote wound healing and tissue regeneration. *Sci Signal* (2010) 3:ra13. doi:10.1126/scisignal.2000634
9. Huang Q, Li F, Liu X, Li W, Shi W, Liu FF, et al. Caspase 3-mediated stimulation of tumor cell repopulation during cancer radiotherapy. *Nat Med* (2011) 17:860–6. doi:10.1038/nm.2385
10. Ford CA, Petrova S, Pound JD, Voss JJ, Melville L, Paterson M, et al. Oncogenic properties of apoptotic tumor cells in aggressive B cell lymphoma. *Curr Biol* (2015) 25:577–88. doi:10.1016/j.cub.2014.12.059
11. Lauber K, Herrmann M. Tumor biology: with a little help from my dying friends. *Curr Biol* (2015) 25:R198–201. doi:10.1016/j.cub.2015.01.040
12. Thery C, Zitvogel L, Amigorena S. Exosomes: composition, biogenesis and function. *Nat Rev Immunol* (2002) 2:569–79. doi:10.1038/nri855
13. Simons M, Raposo G. Exosomes – vesicular carriers for intercellular communication. *Curr Opin Cell Biol* (2009) 21:575–81. doi:10.1016/j.ceb.2009.03.007
14. Lee TH, D'Asti E, Magnus N, Al-Nedawi K, Meehan B, Rak J. Microvesicles as mediators of intercellular communication in cancer – the emerging science of cellular 'debris'. *Semin Immunopathol* (2011) 33:455–67. doi:10.1007/s00281-011-0250-3
15. Raposo G, Stoorvogel W. Extracellular vesicles: exosomes, microvesicles, and friends. *J Cell Biol* (2013) 200:373–83. doi:10.1083/jcb.201211138
16. Akers JC, Gonda D, Kim R, Carter BS, Chen CC. Biogenesis of extracellular vesicles (EV): exosomes, microvesicles, retrovirus-like vesicles, and apoptotic bodies. *J Neurooncol* (2013) 113:1–11. doi:10.1007/s11060-013-1084-8
17. Cocucci E, Meldolesi J. Ectosomes and exosomes: shedding the confusion between extracellular vesicles. *Trends Cell Biol* (2015) 25:364–72. doi:10.1016/j.tcb.2015.01.004
18. Atkin-Smith GK, Poon IK. Disassembly of the dying: mechanisms and functions. *Trends Cell Biol* (2017) 27:151–62. doi:10.1016/j.tcb.2016.08.011
19. Johnson SM, Dempsey C, Parker C, Mironov A, Bradley H, Saha V. Acute lymphoblastic leukaemia cells produce large extracellular vesicles containing organelles and an active cytoskeleton. *J Extracell Vesicles* (2017) 6:1294339. doi:10.1080/20013078.2017.1294339
20. Sheller S, Papaconstantinou J, Urrabaz-Garza R, Richardson L, Saade G, Salomon C, et al. Amnion-epithelial-cell-derived exosomes demonstrate physiologic state of cell under oxidative stress. *PLoS One* (2016) 11:e0157614. doi:10.1371/journal.pone.0157614
21. Roberson CD, Atay S, Gercel-Taylor C, Taylor DD. Tumor-derived exosomes as mediators of disease and potential diagnostic biomarkers. *Cancer Biomark* (2010) 8:281–91. doi:10.3233/CBM-2011-0211
22. Skog J, Wurdinger T, van Rijn S, Meijer DH, Gainche L, Sena-Esteves M, et al. Glioblastoma microvesicles transport RNA and proteins that promote tumour growth and provide diagnostic biomarkers. *Nat Cell Biol* (2008) 10:1470–6. doi:10.1038/ncb1800
23. Cho JA, Yeo DJ, Son HY, Kim HW, Jung DS, Ko JK, et al. Exosomes: a new delivery system for tumor antigens in cancer immunotherapy. *Int J Cancer* (2005) 114:613–22. doi:10.1002/ijc.20757
24. Iero M, Valenti R, Huber V, Filipazzi P, Parmiani G, Fais S, et al. Tumour-released exosomes and their implications in cancer immunity. *Cell Death Differ* (2008) 15:80–8. doi:10.1038/sj.cdd.4402237
25. Thery C, Ostrowski M, Segura E. Membrane vesicles as conveyors of immune responses. *Nat Rev Immunol* (2009) 9:581–93. doi:10.1038/nri2567
26. Xie Y, Bai O, Yuan J, Chibbar R, Slattery K, Wei Y, et al. Tumor apoptotic bodies inhibit CTL responses and antitumor immunity via membrane-bound transforming growth factor-beta1 inducing CD8+ T-cell anergy and CD4+ Tr1 cell responses. *Cancer Res* (2009) 69:7756–66. doi:10.1158/0008-5472.CAN-09-0496

27. Wickman G, Julian L, Olson MF. How apoptotic cells aid in the removal of their own cold dead bodies. *Cell Death Differ* (2012) 19:735–42. doi:10.1038/cdd.2012.25
28. Al-Nedawi K, Meehan B, Micallef J, Lhotak V, May L, Guha A, et al. Intercellular transfer of the oncogenic receptor EGFRvIII by microvesicles derived from tumour cells. *Nat Cell Biol* (2008) 10:619–24. doi:10.1038/ncb1725
29. Al-Nedawi K, Meehan B, Kerbel RS, Allison AC, Rak J. Endothelial expression of autocrine VEGF upon the uptake of tumor-derived microvesicles containing oncogenic EGFR. *Proc Natl Acad Sci U S A* (2009) 106:3794–9. doi:10.1073/pnas.0804543106
30. Peinado H, Aleckovic M, Lavotshkin S, Matei I, Costa-Silva B, Moreno-Bueno G, et al. Melanoma exosomes educate bone marrow progenitor cells toward a pro-metastatic phenotype through MET. *Nat Med* (2012) 18:883–91. doi:10.1038/nm.2753
31. Zomer A, Maynard C, Verweij FJ, Kamermans A, Schafer R, Beerling E, et al. In vivo imaging reveals extracellular vesicle-mediated phenocopying of metastatic behavior. *Cell* (2015) 161:1046–57. doi:10.1016/j.cell.2015.04.042
32. Johnson SM, Dempsey C, Chadwick A, Harrison S, Liu J, Di Y, et al. Metabolic reprogramming of bone marrow stromal cells by leukemic extracellular vesicles in acute lymphoblastic leukemia. *Blood* (2016) 128:453–6. doi:10.1182/blood-2015-12-688051
33. Webber J, Steadman R, Mason MD, Tabi Z, Clayton A. Cancer exosomes trigger fibroblast to myofibroblast differentiation. *Cancer Res* (2010) 70:9621–30. doi:10.1158/0008-5472.CAN-10-1722
34. Baroni S, Romero-Cordoba S, Plantamura I, Dugo M, D'Ippolito E, Cataldo A, et al. Exosome-mediated delivery of miR-9 induces cancer-associated fibroblast-like properties in human breast fibroblasts. *Cell Death Dis* (2016) 7:e2312. doi:10.1038/cddis.2016.224
35. Zheng P, Chen L, Yuan X, Luo Q, Liu Y, Xie G, et al. Exosomal transfer of tumor-associated macrophage-derived miR-21 confers cisplatin resistance in gastric cancer cells. *J Exp Clin Cancer Res* (2017) 36:53. doi:10.1186/s13046-017-0528-y
36. Gregory CD, Pound JD. Microenvironmental influences of apoptosis in vivo and in vitro. *Apoptosis* (2010) 15:1029–49. doi:10.1007/s10495-010-0485-9
37. Gregory CD, Pound JD. Cell death in the neighbourhood: direct microenvironmental effects of apoptosis in normal and neoplastic tissues. *J Pathol* (2011) 223:177–94. doi:10.1002/path.2792
38. Gould SJ, Raposo G. As we wait: coping with an imperfect nomenclature for extracellular vesicles. *J Extracell Vesicles* (2013) 2:20389. doi:10.3402/jev.v2i0.20389
39. Kerr JF, Wyllie AH, Currie AR. Apoptosis: a basic biological phenomenon with wide-ranging implications in tissue kinetics. *Br J Cancer* (1972) 26:239–57. doi:10.1038/bjc.1972.33
40. Wickman GR, Julian L, Mardilovich K, Schumacher S, Munro J, Rath N, et al. Blebs produced by actin-myosin contraction during apoptosis release damage-associated molecular pattern proteins before secondary necrosis occurs. *Cell Death Differ* (2013) 20:1293–305. doi:10.1038/cdd.2013.69
41. Dieude M, Bell C, Turgeon J, Beillevaire D, Pomerleau L, Yang B, et al. The 20S proteasome core, active within apoptotic exosome-like vesicles, induces autoantibody production and accelerates rejection. *Sci Transl Med* (2015) 7:318ra200. doi:10.1126/scitranslmed.aac9816
42. Atkin-Smith GK, Tixeira R, Paone S, Mathivanan S, Collins C, Liem M, et al. A novel mechanism of generating extracellular vesicles during apoptosis via a beads-on-a-string membrane structure. *Nat Commun* (2015) 6:7439. doi:10.1038/ncomms8439
43. Kowal J, Arras G, Colombo M, Jouve M, Morath JP, Primdal-Bengtson B, et al. Proteomic comparison defines novel markers to characterize heterogeneous populations of extracellular vesicle subtypes. *Proc Natl Acad Sci U S A* (2016) 113:E968–77. doi:10.1073/pnas.1521230113
44. Gregory CD, Paterson M. An apoptosis-driven "onco-regenerative niche": roles of tumour-associated macrophages and extracellular vesicles. *Philosoph Trans B* (2017). doi:10.1098/rstb.2017.0003
45. Ehnfors J, Kost-Alimova M, Persson NL, Bergsmedh A, Castro J, Levchenko-Tegnebratt T, et al. Horizontal transfer of tumor DNA to endothelial cells in vivo. *Cell Death Differ* (2009) 16:749–57. doi:10.1038/cdd.2009.7
46. Vallabhaneni KC, Hassler MY, Abraham A, Whitt J, Mo YY, Atfi A, et al. Mesenchymal stem/stromal cells under stress increase osteosarcoma migration and apoptosis resistance via extracellular vesicle mediated communication. *PLoS One* (2016) 11:e0166027. doi:10.1371/journal.pone.0166027
47. Holmgren L, Szeles A, Rajnavolgyi E, Folkman J, Klein G, Ernberg I, et al. Horizontal transfer of DNA by the uptake of apoptotic bodies. *Blood* (1999) 93:3956–63.
48. Bergsmedh A, Szeles A, Henriksson M, Bratt A, Folkman MJ, Spetz AL, et al. Horizontal transfer of oncogenes by uptake of apoptotic bodies. *Proc Natl Acad Sci U S A* (2001) 98:6407–11. doi:10.1073/pnas.101129998
49. Bergsmedh A, Ehnfors J, Kawane K, Motoyama N, Nagata S, Holmgren L. DNase II and the Chk2 DNA damage pathway form a genetic barrier blocking replication of horizontally transferred DNA. *Mol Cancer Res* (2006) 4:187–95. doi:10.1158/1541-7786.MCR-05-0262
50. Mitchell PJ, Welton J, Staffurth J, Court J, Mason MD, Tabi Z, et al. Can urinary exosomes act as treatment response markers in prostate cancer? *J Transl Med* (2009) 7:4. doi:10.1186/1479-5876-7-4
51. Shao H, Chung J, Balaj L, Charest A, Bigner DD, Carter BS, et al. Protein typing of circulating microvesicles allows real-time monitoring of glioblastoma therapy. *Nat Med* (2012) 18:1835–40. doi:10.1038/nm.2994
52. Aubertin K, Silva AK, Luciani N, Espinosa A, Djemat A, Charue D, et al. Massive release of extracellular vesicles from cancer cells after photodynamic treatment or chemotherapy. *Sci Rep* (2016) 6:35376. doi:10.1038/srep35376
53. Richards KE, Zeleniak AE, Fishel ML, Wu J, Littlepage LE, Hill R. Cancer-associated fibroblast exosomes regulate survival and proliferation of pancreatic cancer cells. *Oncogene* (2017) 36:1770–8. doi:10.1038/onc.2016.353
54. Segundo C, Medina F, Rodriguez C, Martinez-Palencia R, Leyva-Cobian F, Brieva JA. Surface molecule loss and bleb formation by human germinal center B cells undergoing apoptosis: role of apoptotic blebs in monocyte chemotaxis. *Blood* (1999) 94:1012–20.
55. Truman LA, Ford CA, Pasikowska M, Pound JD, Wilkinson SJ, Dumitriu IE, et al. CX3CL1/fractalkine is released from apoptotic lymphocytes to stimulate macrophage chemotaxis. *Blood* (2008) 112:5026–36. doi:10.1182/blood-2008-06-162404
56. Torr EE, Gardner DH, Thomas L, Goodall DM, Bielemeier A, Willetts R, et al. Apoptotic cell-derived ICAM-3 promotes both macrophage chemoattraction to and tethering of apoptotic cells. *Cell Death Differ* (2012) 19:671–9. doi:10.1038/cdd.2011.167
57. Fond AM, Ravichandran KS. Clearance of dying cells by phagocytes: mechanisms and implications for disease pathogenesis. *Adv Exp Med Biol* (2016) 930:25–49. doi:10.1007/978-3-319-39406-0_2
58. French KC, Antonyak MA, Cerione RA. Extracellular vesicle docking at the cellular port: extracellular vesicle binding and uptake. *Semin Cell Dev Biol* (2017) 67:48–55. doi:10.1016/j.semcdb.2017.01.002
59. Ostenfeld MS, Jensen SG, Jeppesen DK, Christensen LL, Thorsen SB, Stenvang J, et al. miRNA profiling of circulating EpCAM(+) extracellular vesicles: promising biomarkers of colorectal cancer. *J Extracell Vesicles* (2016) 5:31488. doi:10.3402/jev.v5.31488
60. Julich-Haertel H, Urban SK, Krawczyk M, Willms A, Jankowski K, Patkowski W, et al. Cancer-associated circulating large extracellular vesicles in cholangiocarcinoma and hepatocellular carcinoma. *J Hepatol* (2017) 67:282–92. doi:10.1016/j.jhep.2017.02.024
61. Krishnan SR, Luk F, Brown RD, Suen H, Kwan Y, Bebawy M. Isolation of human CD138(+) microparticles from the plasma of patients with multiple myeloma. *Neoplasia* (2016) 18:25–32. doi:10.1016/j.neo.2015.11.011
62. Allenson K, Castillo J, San Lucas FA, Scelo G, Kim DU, Bernard V, et al. High prevalence of mutant KRAS in circulating exosome-derived DNA from early-stage pancreatic cancer patients. *Ann Oncol* (2017) 28:741–7. doi:10.1093/annonc/mdx004
63. van Eijndhoven MA, Zijlstra JM, Groenewegen NJ, Drees EE, van Niele S, Baglio SR, et al. Plasma vesicle miRNAs for therapy response monitoring in Hodgkin lymphoma patients. *JCI Insight* (2016) 1:e89631. doi:10.1172/jci.insight.89631
64. Wang T, Ning K, Lu TX, Sun X, Jin L, Qi X, et al. Increasing circulating exosomes-carrying TRPC5 predicts chemoresistance in metastatic breast cancer patients. *Cancer Sci* (2017) 108:448–54. doi:10.1111/cas.13150
65. Sequeiros T, Rigau M, Chiva C, Montes M, Garcia-Grau I, Garcia M, et al. Targeted proteomics in urinary extracellular vesicles identifies biomarkers for diagnosis and prognosis of prostate cancer. *Oncotarget* (2017) 8:4960–76. doi:10.18632/oncotarget.13634

66. Balic M, Williams A, Lin H, Datar R, Cote RJ. Circulating tumor cells: from bench to bedside. *Annu Rev Med* (2013) 64:31–44. doi:10.1146/annurev-med-050311-163404
67. McAnena P, Brown JA, Kerin MJ. Circulating nucleosomes and nucleosome modifications as biomarkers in cancer. *Cancers (Basel)* (2017) 9:E5. doi:10.3390/cancers9010005
68. Aung T, Chapuy B, Vogel D, Wenzel D, Oppermann M, Lahmann M, et al. Exosomal evasion of humoral immunotherapy in aggressive B-cell lymphoma modulated by ATP-binding cassette transporter A3. *Proc Natl Acad Sci U S A* (2011) 108:15336–41. doi:10.1073/pnas.1102855108
69. Ciravolo V, Huber V, Ghedini GC, Venturelli E, Bianchi F, Campiglio M, et al. Potential role of HER2-overexpressing exosomes in countering trastuzumab-based therapy. *J Cell Physiol* (2012) 227:658–67. doi:10.1002/jcp.22773
70. Koliha N, Wiencek Y, Heider U, Jungst C, Kladt N, Krauthauser S, et al. A novel multiplex bead-based platform highlights the diversity of extracellular vesicles. *J Extracell Vesicles* (2016) 5:29975. doi:10.3402/jev.v5.29975
71. Baek R, Jorgensen MM. Multiplexed phenotyping of small extracellular vesicles using protein microarray (EV array). *Methods Mol Biol* (2017) 1545:117–27. doi:10.1007/978-1-4939-6728-5_8

Conflict of Interest Statement: The authors declare that the research was conducted in the absence of any commercial or financial relationships that could be construed as a potential conflict of interest.

4

Macrophage Clearance of Apoptotic Cells: A Critical Assessment

Siamon Gordon[1,2] and Annette Plüddemann[3]*

[1] College of Medicine, Graduate Institute of Biomedical Sciences, Chang Gung University, Taoyuan City, Taiwan, [2] Sir William Dunn School of Pathology, University of Oxford, Oxford, United Kingdom, [3] Nuffield Department of Primary Care Health Sciences, University of Oxford, Oxford, United Kingdom

As the body continues to grow and age, it becomes essential to maintain a balance between living and dying cells. Macrophages and dendritic cells play a central role in discriminating among viable, apoptotic, and necrotic cells, as selective and efficient phagocytes, without inducing inappropriate inflammation or immune responses. A great deal has been learnt concerning clearance receptors for modified and non-self-ligands on potential targets, mediating their eventual uptake, disposal, and replacement. In this essay, we assess current understanding of the phagocytic recognition of apoptotic cells within their tissue environment; we conclude that efferocytosis constitutes a more complex process than simply removal of corpses, with regulatory interactions between the target and effector cells, which determine the outcome of this homeostatic process.

Keywords: apoptosis, apoptotic cells, clearance of apoptotic cells, phagocytosis, macrophages

Edited by:
Amiram Ariel, University of Haifa, Israel

Reviewed by:
Dror Mevorach, Hadassah Medical Center, Israel
Sylvain Perruche, Institut National de la Santé et de la Recherche Médicale, France

Correspondence:
Siamon Gordon
siamon.gordon@path.ox.ac.uk

INTRODUCTION

Programmed cell death, apoptosis, is closely linked to recognition and clearance by phagocytosis, resulting in anti-inflammatory and compensatory growth responses during fetal and adult life. Prolonged or incomplete containment and destruction during the apoptotic process can merge into secondary necrosis to bring about pro-inflammatory and pathological responses by phagocytes that are still poorly understood. Many aspects of this process are covered in excellent reviews by leading investigators, including Henson (1), Arandjelovic and Ravichandran (2), Nagata and Tanaka (3), Lim et al. (4), and their colleagues, which should be consulted for detailed exposition. We have previously reviewed the process of phagocytosis in relation to infection and microbial recognition (5), emphasizing its cell biology in model systems. We focus here on clearance of apoptotic cells *in situ* by heterogeneous tissue mononuclear phagocytes, in health and disease.

A BRIEF HISTORY

Although current nomenclature settled around the early reports by Wyllie et al. (6), the observations of cell death in its many manifestations date back to the nineteenth century, from descriptions

Abbreviations: AIP, apoptosis-induced proliferation; DC, dendritic cell; Gas 6; growth arrest specific protein 6; GPCR, G protein-coupled receptor; IECs, intraepithelial cells; IGF-1, insulin-like growth factor-1; KLF, Kruppel-like factor; LPS, lipopolysaccharide; LXR, liver X receptor transcription factor; MPS, mononuclear phagocyte system; MerTK, proto-oncogene c-mer tyrosine kinase; Megf8, milk fat globule—EGF factor 8 (lactadherin); MNGC, multinucleated giant cell; NET, neutrophil extracellular trap; PGE2, prostaglandin E2; ProS1, gene for protein S(eattle); PRRs, pattern-recognition receptors; PS, phosphatidyl serine; PPAR, peroxisome proliferator activator receptor transcription factor; ROS, reactive oxygen species; SLE, systemic lupus erythematosus; TAM, tyrosine receptor kinases Tyro 3, Axl and MerTK; TLR, toll-like receptor; TNF, tumour necrosis factor; tgf beta, transforming growth factor beta.

by Virchow, Metchnikoff, and many pathologists. Wallach et al. have provided an instructive time line of concepts of tissue injury and cell death in inflammation (7). An historic perspective of macrophages, phagocytic mechanisms, and lysosomal digestion is provided elsewhere (5, 8). The pre-eminent role of tissue macrophages in clearance was emphasized in the twentieth century, as a primary function of the reticuloendothelial system, subsequently renamed the mononuclear phagocyte system (MPS) (9). Studies in *Caenorhabditis elegans* (10) stimulated genetic dissection of apoptosis and clearance by epithelial cells in organisms that lack "professional" phagocytes; important discoveries of macrophage clearance followed in *Drosophila* and other model organisms, such as zebra fish and mice. Uptake of dead cells by non-professional phagocytes in vertebrates became overshadowed by emphasis on macrophages and related dendritic cells (DCs), although recent studies (2) have to some extent redressed the balance; turnover of photoreceptors by retinal pigment epithelia and of aberrant sperm in the testis by Sertoli cells are highly active functions of non-hematopoietic phagocytic cells, and uptake of cell corpses has also been demonstrated in epithelia, fibroblasts, astrocytes, and cancer cells, the so-called non-professional phagocytes (11). Different terms have emerged for a range of distinct though related processes, in addition to efferocytosis (12); these include necroptosis (13), pyroptosis (14), phagoptosis (15), ferroptosis (16), trogocytosis (17), and entosis (18), depending on one or other characteristic feature. Mevorach and colleagues have introduced clarity into the terminology of this expanding topic, which will be defined as relevant, below (19). Henson and Bratton (20) provided early evidence that clearance of programmed apoptotic cell death by macrophages gave rise to anti-inflammatory effects, unlike the pro-inflammatory consequences of the uptake of necrotic cells, which could follow at a further stage of programmed cell death, during infection or as a result of accidental injury. Another time line of particular interest in this area is given by Nagata and Tanaka, who pioneered the role of phosphatidyl serine (PS) and membrane lipid reorganization in the recognition of apoptotic cells (3).

The physiological role of apoptotic cell clearance by macrophages has been documented in organ development, tissue remodeling, e.g., in the uterus and mammary gland, repair and potential cell replacement following injury and, in a few species, regeneration of complex organs. In pathology, monocytes, macrophages, and DCs are important contributors to inflammation and its resolution, following disposal of necrotic corpses and subcellular constituents, e.g., during infection, innate and autoimmunity, atherogenesis, and malignancy. Many authors have considered cell death, its recognition, disposal, and regulation as central homeostatic functions of the MPS; aspects of this topic are reviewed by the various contributors to this Frontiers of Immunology collection, cited as available at the time of writing (21, 22). We shall not deal in this review with the mechanisms of cell death itself.

MONONUCLEAR PHAGOCYTES ARE HIGHLY HETEROGENEOUS

The cells of the mammalian MPS constitute a widely dispersed population derived from common hematopoietic progenitors, which are distinct in the embryo and adult (23). Tissue-resident macrophage populations in the fetus are distributed from yolk sac and fetal liver precursors from mid gestation, and turn over locally to a variable extent throughout adult life (24). From birth, bone marrow-derived monocytes are recruited to replenish and supplement tissue macrophage populations in the steady state, and in response to inflammatory, metabolic, infectious, and malignant disease processes, as required. Circulating mononuclear cells contain precursors of macrophages, DCs, and osteoclasts, and subpopulations of monocytes that are characterized by distinct marker antigens and receptors (25). Recent single cell RNA analysis has revealed additional mononuclear cell subpopulations in blood without functional characterization (26). Trahtemberg and colleagues identified two human blood monocyte-derived DC subpopulations which differ in their expression of surface markers, phagocytic clearance, and responses to the uptake of apoptotic cells (27). The markedly heterogeneous mononuclear phagocytes in tissues display the ability to phagocytose particulates to a variable extent, depending on their differentiation, recruitment, and activation by tissue-specific and microbial stimuli. The ability of neutrophils which are highly phagocytic for opsonized microbes, and of eosinophils, basophils, and mast cells, to phagocytose apoptotic targets has been less studied. In addition, MPS cells variably display fluid phase, macropinocytic, and receptor-mediated endocytosis, which contribute to their clearance of solutes and smaller particulates. In response to local need, resident macrophages can migrate from adjacent reservoirs, as elegantly demonstrated through *in vivo* experiments by the groups of Wang and Kubes and Robbins, respectively, in liver (28) and heart (29). The prototypic and organ-specific phagocytic activity of the mixed populations of macrophages in different organs differ considerably, as shown by elegant clearance experiments *in vivo*, described in detail below (30, 31). Fusion of monocytes and macrophages gives rise to the formation of osteoclasts in bone and of multinucleated giant cells (MNGCs) in granulomatous diseases (32). We have demonstrated a striking difference in the phagocytosis of a range of particulates, including latex and complement-, but not antibody-opsonized sheep erythrocytes by IL4/13 cytokine-induced MNGC (33). Finally, activation of macrophages by innate and Th1- and Th2-dependent adaptive immune stimuli induces a spectrum of changes in cellular phenotype, including cytocidal and phagocytic capacity (34), hitherto mainly studied *in vitro*.

MYELOID CELL TURNOVER

Myeloid leukocytes themselves undergo apoptosis and necrosis, as well as inducing these processes in a range of target cells. We therefore briefly consider aspects of their own turnover, clearance, and mechanisms of death that may be relevant to subsequent consequences. Recent studies by Yona and colleagues have established the circulation time and precursor–product relationships of different human monocyte subsets (35). Macrophages can become long-lived, terminally differentiated tissue-resident cells which remain biosynthetically active while ceasing to proliferate, but recently recruited monocytes and activated macrophages and DC themselves can undergo apoptosis within days of responding to inflammatory and infectious stimuli. By comparison, neutrophils are short lived and readily proceed from apoptosis to secondary necrosis when activated during acute inflammation; eosinophils may survive longer in tissues than usually appreciated. The microbicidal and cytocidal activities of activated neutrophils and monocyte/macrophages depend on generation of reactive radicals derived from oxygen through a respiratory burst; in addition, degranulation by neutrophils and eosinophils releases potent neutral proteinases such as elastase, as well as myeloperoxidase, to initiate cytotoxicity, whereas interferon gamma activated macrophages contribute to target cell killing by release of nitrogen metabolites generated by inducible nitric oxide synthase. Other mechanisms of myeloid cell cytotoxicity include chromatin extracellular trap (neutrophil extracellular trap) production by neutrophils and probably activated macrophages and release of tumour necrosis factor (TNF)alpha. Inflammasome activation and IL-1 release depend on selected proteolysis by cytosolic caspases. Depletion of essential amino acids such as tryptophan and arginine in their local environment contribute to cell death.

INTERACTIONS BETWEEN MACROPHAGES AND APOPTOTIC CELLS

We have outlined the heterogeneity of the macrophage populations involved in clearance, and the complexity of their potential apoptotic/necrotic cargo. It has become customary to break the overall process down into distinct stages: "find me" (36, 37), by chemokine-receptor attractants and plexin–semaphorin guidance/repulsion (38, 39); "eat me" (40) and "don't eat" me (41) signals between apoptotic cells and phagocytes, involving diffusible products, plasma components such as thrombospondin, and intercellular interactions. We focus here on recognition (42) by macrophages of well-defined apoptotic cells. Apart from the well-documented chromatin and cytoplasmic changes, we consider PS (43) and other potential ligands expressed on the surface of senescent and apoptotic cells and the wide range of direct and indirect opsonic receptors expressed on the macrophage surface (**Figure 1**). This information is based mainly on *in vitro* studies, with some support obtained from mouse and other genetic models *in vivo*. Phagoptosis, the uptake of viable cells (15), resulting in their destruction, depends on exposure of "eat me" and/or loss of "don't eat me" signals; it mediates turnover of erythrocytes, neutrophils, and other cells, and also contributes to defense against infection.

Several aspects of our present understanding deserve further comment: there is a bewildering range of possible molecular interactions reported in this literature, depending to a large extent on the interests of the researcher rather than proven relative importance under different conditions *in vivo*. It is perhaps not surprising that so critical a housekeeping function should involve considerable redundancy; it is well known that inhibitors or ablation of favorite individual receptor–ligand pairs are at best only partial in their effect *in vitro*, and often undetectable *in vivo*. For example, this was found in our own studies of possible SR-A involvement in the thymus, a site of avid apoptotic thymocyte clearance by macrophages (44). Up to a point, one could argue that some of these candidates could act in a common pathway or be redundant. A more likely interpretation is that there is organ-specific involvement of particular receptor–ligand pairs, as indicated by new evidence, discussed in more detail below. Furthermore, most previous studies have failed to appreciate the modulation of phagocytic activity by physiological and pathological regulation of particular pathways in different tissue environments. We should expect wide variation from the experimental use of species as divergent as *Drosophila*, zebra fish, mouse, and man, with clearance mechanisms arising by convergent as well as divergent evolution. Finally, since different cell types can perform the same function, constitutively or after induction, it is essential to identify the particular cells involved by cell-specific deletion in any experimental model; for example, this was demonstrated by Ravichandran and colleagues in a receptor transgenic overexpression model which showed enhanced clearance was attributable to epithelial cells and not macrophages, as might have been assumed (45). The macrophage ablation studies during development, e.g., in PU-1 null mice (46) show remarkably little impact except for minor deficiencies in interdigital web regression; presumably, alternative mesenchymal cells can compensate for developmental clearance of apoptotic cells. Failure to clear erythrocyte nuclei, however, does create hemodynamic problems after birth and early death postnatally can also be ascribed to infection.

MECHANISMS OF CLEARANCE OF APOPTOTIC CELLS BY MACROPHAGES

We have previously outlined some of the broad features of opsonic, antibody, and complement-mediated phagocytosis (5), the phagocytic synapse model based on Dectin-1 mediated uptake of yeast particles (47), and a kinetic exclusion model of other plasma membrane molecules during uptake (4); its relevance to the mechanism of apoptotic cell uptake has not been determined. Intravital microscopy has revealed striking recruitment of different mononuclear cell populations during development (48) and following sterile and infectious injury and repair (28). Phosphatidylinositol kinases play an important role in signaling and ingestion of apoptotic cells (49). Various phosphoinositide mediators derive from preformed membrane phospholipids and reorganization of lipids within the plasma membrane bilayer with resultant PS exposure on the outer surface, as elucidated by the groups of Nagata and Tanaka (3) and Kay and Grinstein (50). BAI-1, an adhesion G protein-coupled receptor (51, 52), mediates uptake of apoptotic cells *via* PS recognition, followed by Dock and Elmo signaling. Effects of apoptotic cell uptake on membrane traffic, vesicular fusion and fission, the role of small GTPases, phagosome formation and function (53), as well as possible interactions with autophagy require further investigation. Similarly, there is need for more studies on the effects of apoptotic cell uptake on gene and protein expression, for example, by *in situ* hybridization and multiplex histochemistry, as well as of posttranslational and epigenetic modifications, in heterogeneous cell populations and individual macrophages, *in vivo* and *in vitro*.

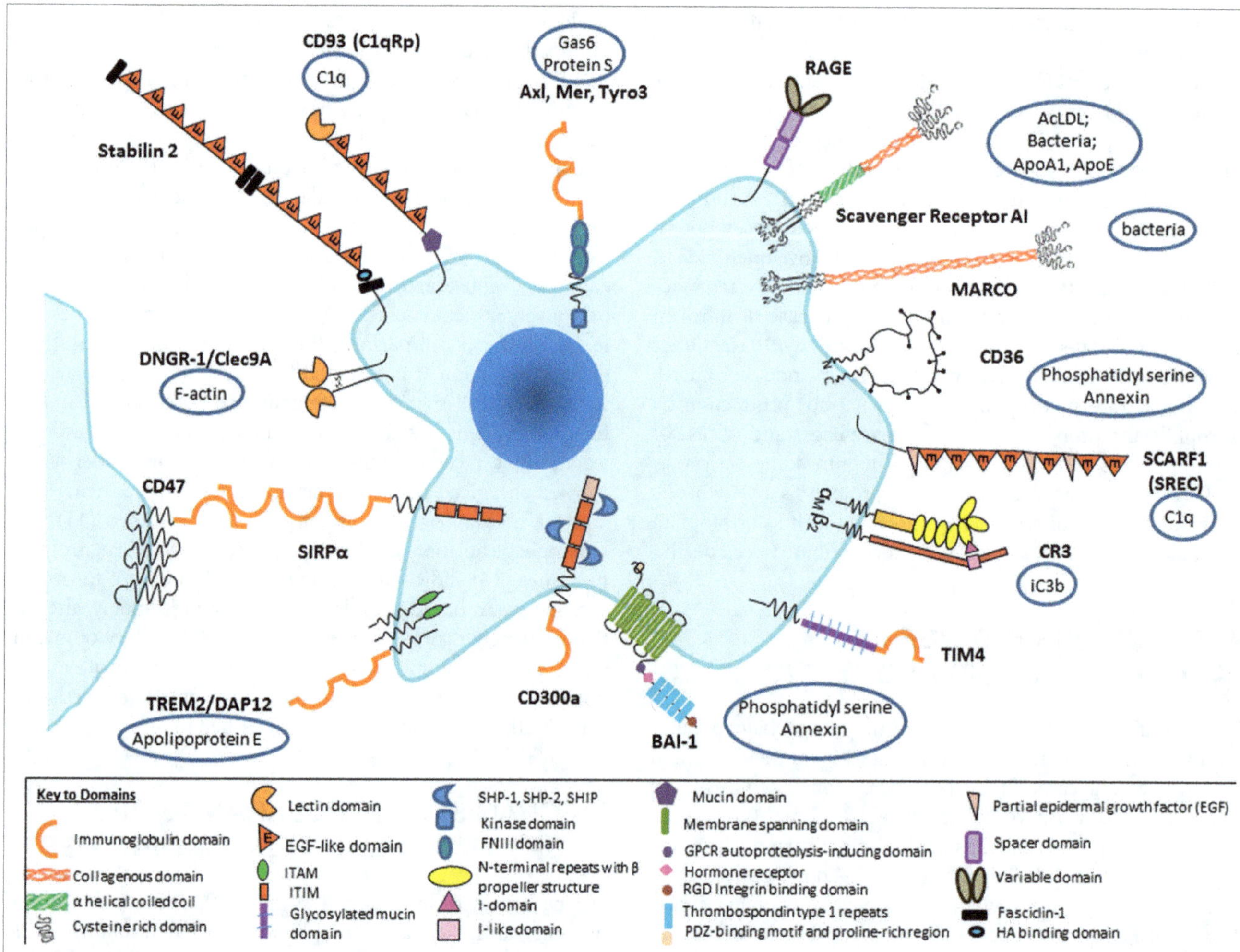

FIGURE 1 | Selected macrophage phagocytic receptors and ligands implicated in apoptotic cell clearance. Receptor–ligand interactions can be direct or indirect, *via* opsonic intermediates derived from plasma or tissue sources. Macrophage receptors such as CD11b/CD18 can bind *via* opsonic complement components, including secretion products of macrophages themselves, or promiscuously to other ligands, so that it becomes difficult to classify as direct or indirect.

CLEARANCE OF APOPTOTIC CELLS BY RESIDENT TISSUE MACROPHAGES *IN VIVO*

It has been known for some time that resident macrophage populations in different organs such as liver, gut, lung, bone marrow, spleen, and brain are heterogeneous in their morphology, turnover, and antigen expression profile. Gene expression studies by Lavin and colleagues have documented organ-specific gene and enhancer patterns, as well as canonical signatures for a range of tissue macrophage populations (54). Apoptotic cell clearance is mediated by distinct pathways in different organs and helps to shape the local phenotype, as shown by recent studies by the groups of Hidalgo (31) and Magarian Blander (30). A-Gonzalez and colleagues used a parabiotic model to study homeostatic leukocyte clearance *in vivo*. Distinct receptors, opsonins, and transcription factors were demonstrated in uptake by macrophages in the bone marrow, spleen, intestine, liver, and lung interstitium; alveolar macrophages in the lung were not accessed by blood delivery. Mutant mice deficient in milk fat globule—EGF factor 8 (lactadherin), liver X receptor transcription factor alpha/beta, peroxisome proliferator activator receptor transcription factor gamma, and proto-oncogene c-mer tyrosine kinase (MerTK) differed from wild-type controls and varied in their contribution to uptake in bone marrow, spleen, and liver. Annexin A1 and Timd4 were also utilized to determine tissue-specific dynamics and expression of different phagocytic mediators. Gene expression analysis demonstrated preservation of distinct core signatures, but phagocytosis imprinted a distinct anti-inflammatory profile, upregulation of CD163 and of CD206, the macrophage mannose receptor, and decrease of the pro-inflammatory expression of IL-1 beta. Expression of CD206 made it possible to identify phagocytic macrophages in the steady state in the absence of experimental manipulation. Clearance activity confirmed the diurnal rhythm of neutrophil turnover. Several interesting aspects of this study require further study: the role of variation in apoptotic rate in neutrophils, the impact of non-circulating apoptotic cells, e.g., generated in the thymus by dexamethasone treatment, and the absence of detectable annexin binding in some experiments, consistent with uptake and subsequent killing of viable cells by phagoptosis. Furthermore, the effects of concurrent inflammation and infection on the clearance of apoptotic as well as necrotic cells, add an additional layer of complexity.

The study by Cummings et al. used a different genetic strategy to study the effects of apoptosis in a single organ, the small intestine, on gene expression by macrophages and DC, after uptake of dying epithelial cells (30). Diphtheria receptor expression in mice was targeted to the intestine by a villin construct with an eGFP reporter; low dose diphtheria toxin induced synchronous apoptosis selectively to ileum epithelium without affecting barrier function. Five subpopulations of APC were isolated before and after apoptosis by FACS, and gene expression was analyzed. Two CD64$^+$ macrophages and one CD24$^+$ DC population were shown to display distinctive signatures after phagocytosis of apoptotic cells; the macrophage signatures included lipid metabolism and aromatic amino acid catabolism, whereas the DC subset program was consistent with activation of regulatory CD4$^+$ lymphocytes. A common signature of suppression of inflammation consisted of distinct genes in macrophages and DCs. Only the DCs were induced by apoptotic intraepithelial cell (IEC) uptake to migrate to draining lymph nodes and found able to induce immunological tolerance. Several of the differentially induced genes in phagocytes overlapped with susceptibility genes for inflammatory bowel disease. CD103$^+$CD11b$^+$ and distinct CD11b$^+$ only macrophages, which sampled the majority of apoptotic IEC$^+$, upregulated growth arrest specific protein 6, Cd 163, and MerTK genes, whereas CD103$^+$ DC expressed the gene for CD105, a member of the mannose receptor family. Other differentially expressed genes included those implicated in apoptotic cell clearance, enhanced anti-inflammatory, and reduced pro-inflammatory pathways.

A recent study by Roberts et al. (55) demonstrated that selected tissue-resident macrophages in the peritoneal and pleural cavities, and lung alveolar space, are locally programmed for silent clearance of apoptotic cells. Specific populations of macrophages clear apoptotic cells in different tissues; they share features that limit recognition of nucleic acids in response to signals from their local tissue microenvironment. The transcription factor Krüppel-like factor-2 (KLF2), and to a lesser extent KLF4, control the expression of many genes within the macrophage clearance program. These characteristics include high expression of receptors for apoptotic cells such as Tim4, but low expression of toll-like receptor (TLR)9 and reduced TLR responsiveness to nucleic acids. The environmental signals in different tissues, which are unknown, can override inflammatory stimulation and are lost when cells are isolated, *in vitro*. Other tissue-resident macrophages vary in their inhibition of inflammatory responses after apoptotic cell uptake, and it is unclear how homeostasis is maintained in the face of infection, or restored during resolution of inflammation.

Similarly, Baratin et al. reported that a population of CX3CR1$^+$CD64$^+$MerTK$^+$CD11c$^+$ resident macrophages, but not CD8alpha DC, were the main local efferocytic cells in the T zone of lymph nodes; however, these scavenger macrophages did not prime CD4 T cells efficiently, unlike DC (56). Tingible body macrophages, prominent in germinal centers in lymph nodes, spleen, and other lymphoid structures, clear apoptotic B lymphocytes. Their possible role in B lymphocyte affinity maturation and antibody diversification and interactions with follicular DCs and other mesenchymal stromal cells require further investigation (57).

ROLE OF PHAGOCYTIC CLEARANCE OF DYING CELLS IN INFLAMMATION AND ITS RESOLUTION, TISSUE GROWTH, AND IMMUNE REGULATION

The uptake of apoptotic cells by macrophages is not "silent," but anti-inflammatory. Secretory mediators include cytokines [transforming growth factor beta (tgf beta), IL-10] and arachidonate metabolites, e.g., prostaglandin E2, as well as a range of proresolving mediators. Addition of macrophages that have ingested early apoptotic cells, to lipopolysaccharide (LPS)-induced inflammatory cells *in vitro* or *in vivo*, reduced inflammatory pathways (20), whether through the above mediators or by undefined direct contact mechanisms. However, genome wide gene expression, proteomic and metabolomic studies may yield a more complex picture of pro- and anti-inflammatory signature markers which should be reinvestigated by newly developed population and single cell methods. *In vitro* studies need to take into account the tolerogenic and pro-inflammatory subtypes of macrophages and DC, the nature of the apoptotic cell targets, clearance receptor usage, and possible association with TLR. Previous exposure to priming agents, including apoptotic cells themselves, can convert the response to a secondary challenge (58). The *in vivo* studies described above have provided evidence for "imprinting," comparable to "trained immunity" (59), of an anti-inflammatory phenotype in several different tissues, but further studies are needed in the presence of infection or trauma, and associated cell death.

Studies in *Drosophila* (48) and a range of vertebrates have implicated macrophages in the replacement of epithelial and other cells during development and in tissue repair after wounding. Apoptosis and proliferation of microglia are coupled in the adult brain (60). Dying cells themselves are able to generate recruiting and mitogenic signals to attract and interact reciprocally with macrophages. These include a range of trophic factors such as insulin-like growth factor-1 (IGF-1) (45) and TNF-related cytokines (61), released after phagocytic clearance. Extracellular reactive oxygen species (ROS) produced by epithelial disk cells drive apoptosis-induced proliferation *via Drosophila* macrophages (61). The Ravichandran group has shown that IGF-1 released by macrophages in response to uptake of apoptotic cells, can inhibit apoptotic cell clearance by neighboring epithelial cells, while enhancing their uptake of smaller vesicles (45). Ueno et al. showed that CX3CR1$^+$ microglia were required for survival of selected cortical neurons during development, ascribed to IGF-1 production after uptake of apoptotic neurons (62). Conversely, Brown and Neher have postulated that microglial phagoptosis mechanisms contribute to neuronal loss, synaptic sculpting, and loss of neuronal processes during development and neurodegeneration (63).

A striking example of the link between phagocytic clearance by macrophages and cell growth and differentiation was demonstrated by Kawane et al. (64). DNAse II in macrophages is responsible for destruction of nuclear DNA from erythroid precursors after enucleation, and essential for definitive erythropoiesis in mouse fetal liver. Recent studies by Fearnhead et al. have shown that Ferroptosis, a regulated physiological cell death process which requires free ferrous iron, occurs through Fe(II)-dependent lipid peroxidation when the reduction capacity of a cell is insufficient (16). Most studies have not distinguished between apoptotic, anti-inflammatory and immunomodulatory, versus pro-inflammatory, necrotic targets, or the source of contact-dependent, vesicular, or soluble products such as ROS and lipid metabolites. The nature of the apoptotic stimulus, for example, programmed versus X-ray-induced cell death, the heterogeneity of mononuclear phagocyte populations, and the local tissue environment can all contribute to the outcome. Most striking is the regenerative capacity of species such as axolotls (65) after limb amputation, for example, which depends on early recruitment and engulfment by phagocytic macrophages (66). A recent comprehensive analysis of gene expression during regeneration of the axolotl limb provides a resource to identify critical genes (67).

Dying cells release a complex mixture of lipids, oxidized phosphoryl choline derivatives, which can display opposing activities, interfering with TLR signaling to prevent inflammatory responses, while hyperactivating DCs and macrophages to promote inflammation by IL-1. CD14, the microbial LPS receptor, expressed by monocytes and macrophages, and downregulated by DC, is an early central regulator of both activities (68). Zanoni et al. demonstrated further that isolated lipid constituents of the mixture induce inflammasome-dependent macrophage hyperactivation *via* CD14, without compromising macrophage viability, thus promoting inflammation in sepsis, but not lethality.

IL-4 and other Th2 cytokines, possibly of innate sources, induce growth and an alternative activation pathway of macrophages, which has been implicated in wound repair (32). The macrophage mannose receptor (CD206), one of the characteristic signature genes induced by IL-4, is upregulated in mouse models of intestinal apoptosis, in which there is regeneration of epithelium, although it has not been determined whether it is required. A recent study has shown that apoptotic cell phagocytosis selectively amplifies the gene signature induced in macrophages by IL-4/13 (69). The tyrosine receptor kinases Tyro 3, Axl, and MerTK (TAM) receptors and their ligands play distinct immunoregulatory roles in DCs and macrophages (70) (**Figure 2**). Tyro3 and gene for protein S(eattle) mediate a negative feedback loop between DC and activated Th2 lymphocytes to inhibit type 2 immunity (71); by contrast, Mer-TK and Axl on macrophages require a local apoptotic ligand to promote a

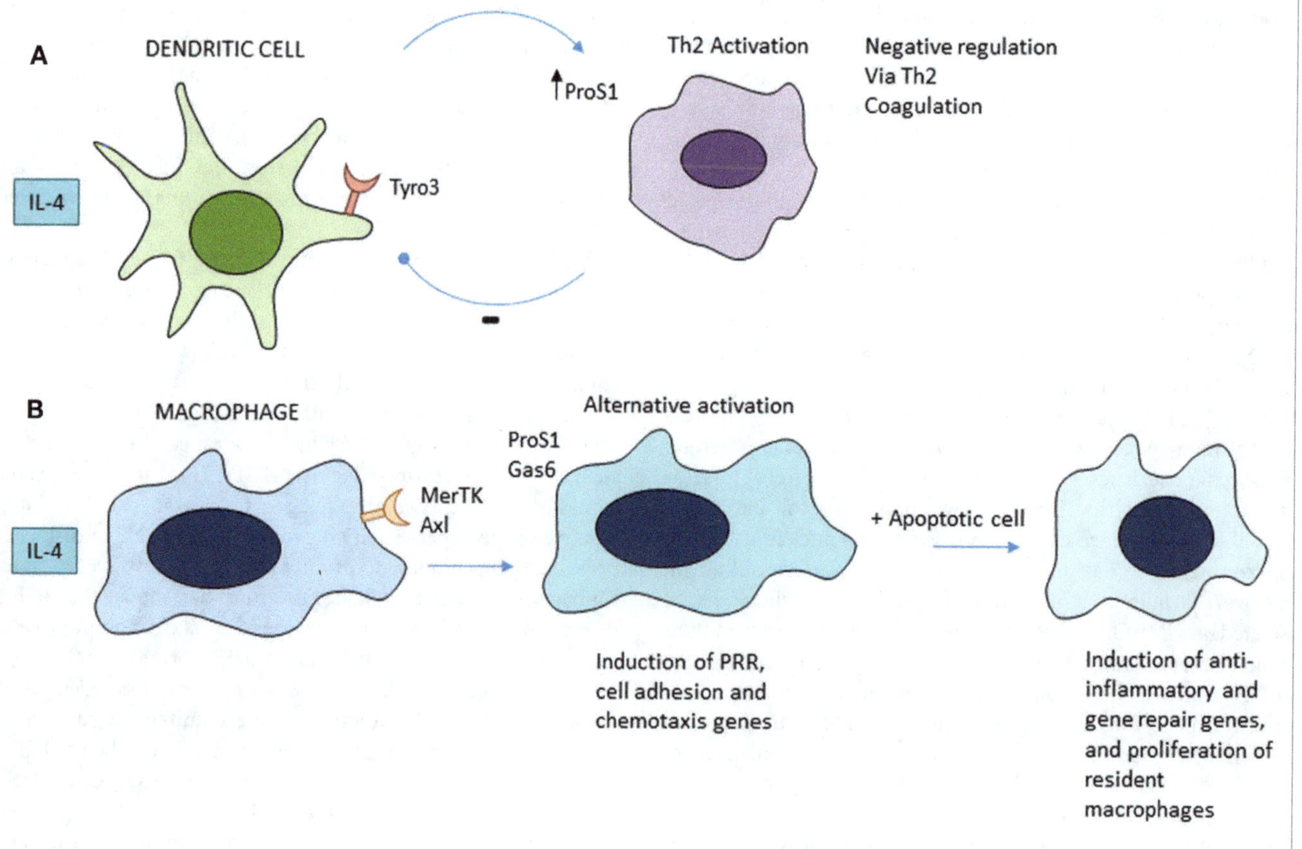

FIGURE 2 | Role of tyrosine receptor kinases Tyro 3, Axl, and MerTK family apoptotic receptors in regulation of Th2 immunity **(A)** and alternative activation of macrophages **(B)**. See text for further details and references.

full tissue repair program (69). If lysosomal peptides gain access to MHC or cytosol, cross-presentation can induce autoreactive CD8+ cytotoxic lymphocytes. Ly6+ monocytes are able to efferocytose apoptotic cells and cross-present cell-associated antigens to CD8+ T cells; this is enhanced by ligation of TLR7, but not TLR4 (72). Autoreactive Th17 lymphocytes can also be induced by apoptotic cell products in the context of microbial infection (73). Kleinclauss et al. reported that intravenous infusion of apoptotic spleen cells induced a tgf beta-dependent expansion of regulatory T lymphocytes, contributing to tolerance, induced by macrophages and not DCs (74). Saas and colleagues (21) have discussed the mechanism of early apoptotic cell infusion to prevent and limit ongoing autoimmune inflammation.

CONCLUDING REMARKS

Given the importance of the biological process of apoptotic cell clearance, it seems surprising that it might go wrong so infrequently; is there such redundancy and backup to keep it safe, or are we looking in the wrong place? Of course, we know of some autoimmune conditions, e.g., systemic lupus erythematosus (SLE), to which inefficient apoptotic cell clearance contributes; Munoz et al. (75) have discussed the role of nucleic acid and auto-antibody complexes in the pathogenesis of recurrent inflammation, resulting in production of type I interferon, a hallmark of SLE. A major effort is under way to use monoclonal antibodies directed against CD47, a "don't eat me" signal expressed by selected malignancies, to promote tumor cell ingestion by macrophages (40). A promising, safe protocol has been developed for adoptive transfer of mononuclear phagocytes that have started to engulf apoptotic cells, prophylactically, to facilitate allogeneic hematopoietic transplantation (76). It will be important to characterize the appropriate subpopulation of macrophages for such therapeutic applications. Conversely, we have good reason to believe that uptake of apoptotic HIV-infected lymphocytes by macrophages can provide a route to infection of healthy CD4+ T cells (77).

Although the research efforts in this area are mainly based on insights of several decades ago, it would be advisable to use the more advanced methods now available to revisit the fundamental aspects of the phagocytic process. It is exciting to see the recent adoption of intravital methods to study gene expression and growth responses *in vivo*. The holy grail might be to extend the limited regenerative capability of hematopoiesis, intestinal epithelium, and liver more widely to other tissues. In this regard, it would be wise to use a broad comparative approach to discover the factors that shut regeneration down during evolution. Or would that risk enhancing malignant diseases (78)?

AUTHOR CONTRIBUTIONS

SG conceived and wrote the manuscript. AP reviewed and edited the manuscript and designed the figures.

ACKNOWLEDGMENTS

We thank colleagues who have sent us preprints of their current work.

REFERENCES

1. Henson PM. Cell removal: efferocytosis. *Annu Rev Cell Dev Biol* (2017) 33:127–44. doi:10.1146/annurev-cellbio-111315-125315
2. Arandjelovic S, Ravichandran KS. Phagocytosis of apoptotic cells in homeostasis. *Nat Immunol* (2015) 16:907–17. doi:10.1038/ni.3253
3. Nagata S, Tanaka M. Programmed cell death and the immune system. *Nat Rev Immunol* (2017) 17:333–40. doi:10.1038/nri.2016.153
4. Lim JJ, Grinstein S, Roth Z. Diversity and versatility of phagocytosis: roles in innate immunity, tissue remodeling, and homeostasis. *Front Cell Infect Microbiol* (2017) 7:191. doi:10.3389/fcimb.2017.00191
5. Gordon S. Phagocytosis: an immunobiologic process. *Immunity* (2016) 44:463–75. doi:10.1016/j.immuni.2016.02.026
6. Wyllie AH, Kerr JF, Currie AR. Cell death: the significance of apoptosis. *Int Rev Cytol* (1980) 68:251–306. doi:10.1016/S0074-7696(08)62312-8
7. Wallach D, Kang TB, Kovalenko A. Concepts of tissue injury and cell death in inflammation: a historical perspective. *Nat Rev Immunol* (2014) 14:51–9. doi:10.1038/nri3561
8. Underhill DM, Gordon S, Imhof BA, Nunez G, Bousso P. Elie Metchnikoff (1845-1916): celebrating 100 years of cellular immunology and beyond. *Nat Rev Immunol* (2016) 16:651–6. doi:10.1038/nri.2016.89
9. Yona S, Gordon S. From the reticuloendothelial to mononuclear phagocyte system—the unaccounted years. *Front Immunol* (2015) 6:328. doi:10.3389/fimmu.2015.00328
10. Pinto SM, Hengartner MO. Cleaning up the mess: cell corpse clearance in *Caenorhabditis elegans*. *Curr Opin Cell Biol* (2012) 24:881–8. doi:10.1016/j.ceb.2012.11.002
11. Freeman SA, Grinstein S. Phagocytosis: how macrophages tune their non-professional counterparts. *Curr Biol* (2016) 26:R1279–82. doi:10.1016/j.cub.2016.10.059
12. deCathelineau AM, Henson PM. The final step in programmed cell death: phagocytes carry apoptotic cells to the grave. *Essays Biochem* (2003) 39:105–17. doi:10.1042/bse0390105
13. Vanden Berghe T, Linkermann A, Jouan-Lanhouet S, Walczak H, Vandenabeele P. Regulated necrosis: the expanding network of non-apoptotic cell death pathways. *Nat Rev Mol Cell Biol* (2014) 15:135–47. doi:10.1038/nrm3737
14. Cookson BT, Brennan MA. Pro-inflammatory programmed cell death. *Trends Microbiol* (2001) 9:113–4. doi:10.1016/S0966-842X(00)01936-3
15. Brown GC, Vilalta A, Fricker M. Phagoptosis-cell death by phagocytosis-plays central roles in physiology, host defense and pathology. *Curr Mol Med* (2015) 15:842–51. doi:10.2174/156652401509151105130628
16. Fearnhead HO, Vandenabeele P, Vanden Berghe T. How do we fit ferroptosis in the family of regulated cell death? *Cell Death Differ* (2017) 24:1991–8. doi:10.1038/cdd.2017.149
17. Joly E, Hudrisier D. What is trogocytosis and what is its purpose? *Nat Immunol* (2003) 4:815. doi:10.1038/ni0903-815
18. Overholtzer M, Mailleux AA, Mouneimne G, Normand G, Schnitt SJ, King RW, et al. A nonapoptotic cell death process, entosis, that occurs by cell-in-cell invasion. *Cell* (2007) 131:966–79. doi:10.1016/j.cell.2007.10.040
19. Mevorach D, Trahtemberg U, Krispin A, Attalah M, Zazoun J, Tabib A, et al. What do we mean when we write "senescence," "apoptosis," "necrosis," or "clearance of dying cells"? *Ann N Y Acad Sci* (2010) 1209:1–9. doi:10.1111/j.1749-6632.2010.05774.x
20. Henson PM, Bratton DL. Antiinflammatory effects of apoptotic cells. *J Clin Invest* (2013) 123:2773–4. doi:10.1172/JCI69344
21. Saas P, Bonnefoy F, Toussirot E, Perruche S. Harnessing apoptotic cell clearance to treat autoimmune arthritis. *Front Immunol* (2017) 8:1191. doi:10.3389/fimmu.2017.01191
22. Trahtemberg U, Mevorach D. Apoptotic cells induced signaling for immune homeostasis in macrophages and dendritic cells. *Front Immunol* (2017) 8:1356. doi:10.3389/fimmu.2017.01356

23. Gordon S, Pluddemann A. Tissue macrophages: heterogeneity and functions. *BMC Biol* (2017) 15:53. doi:10.1186/s12915-017-0392-4
24. Varol C, Mildner A, Jung S. Macrophages: development and tissue specialization. *Annu Rev Immunol* (2015) 33:643–75. doi:10.1146/annurev-immunol-032414-112220
25. Amit I, Winter DR, Jung S. The role of the local environment and epigenetics in shaping macrophage identity and their effect on tissue homeostasis. *Nat Immunol* (2016) 17:18–25. doi:10.1038/ni.3325
26. Villani AC, Satija R, Reynolds G, Sarkizova S, Shekhar K, Fletcher J, et al. Single-cell RNA-seq reveals new types of human blood dendritic cells, monocytes, and progenitors. *Science* (2017) 356:eaah4573. doi:10.1126/science.aah4573
27. Trahtemberg U, Grau A, Tabib A, Atallah M, Krispin A, Mevorach D. Identification and characterization of two human monocyte-derived dendritic cell subpopulations with different functions in dying cell clearance and different patterns of cell death. *PLoS One* (2016) 11:e0162984. doi:10.1371/journal.pone.0162984
28. Wang J, Kubes P. A reservoir of mature cavity macrophages that can rapidly invade visceral organs to affect tissue repair. *Cell* (2016) 165:668–78. doi:10.1016/j.cell.2016.03.009
29. Robbins CS, Chudnovskiy A, Rauch PJ, Figueiredo JL, Iwamoto Y, Gorbatov R, et al. Extramedullary hematopoiesis generates Ly-6C(high) monocytes that infiltrate atherosclerotic lesions. *Circulation* (2012) 125:364–74. doi:10.1161/CIRCULATIONAHA.111.061986
30. Cummings RJ, Barbet G, Bongers G, Hartmann BM, Gettler K, Muniz L, et al. Different tissue phagocytes sample apoptotic cells to direct distinct homeostasis programs. *Nature* (2016) 539:565–9. doi:10.1038/nature20138
31. A-Gonźalez N, Quintana JA, Garcia-Silva S, Mazariegos M, Gonzalez de la Aleja A, Nicolas-Avila JA, et al. Phagocytosis imprints heterogeneity in tissue-resident macrophages. *J Exp Med* (2017) 214:1281–96. doi:10.1084/jem.20161375
32. Helming L, Gordon S. Molecular mediators of macrophage fusion. *Trends Cell Biol* (2009) 19:514–22. doi:10.1016/j.tcb.2009.07.005
33. Milde R, Ritter J, Tennent GA, Loesch A, Martinez FO, Gordon S, et al. Multinucleated giant cells are specialized for complement-mediated phagocytosis and large target destruction. *Cell Rep* (2015) 13:1937–48. doi:10.1016/j.celrep.2015.10.065
34. Martinez FO, Gordon S. The M1 and M2 paradigm of macrophage activation: time for reassessment. *F1000Prime Rep* (2014) 6:13. doi:10.12703/P6-13
35. Patel AA, Zhang Y, Fullerton JN, Boelen L, Rongvaux A, Maini AA, et al. The fate and lifespan of human monocyte subsets in steady state and systemic inflammation. *J Exp Med* (2017) 214:1913–23. doi:10.1084/jem.20170355
36. Savill J, Dransfield I, Gregory C, Haslett C. A blast from the past: clearance of apoptotic cells regulates immune responses. *Nat Rev Immunol* (2002) 2:965–75. doi:10.1038/nri957
37. Medina CB, Ravichandran KS. Do not let death do us part: 'find-me' signals in communication between dying cells and the phagocytes. *Cell Death Differ* (2016) 23:979–89. doi:10.1038/cdd.2016.13
38. Wolf Y, Boura-Halfon S, Cortese N, Haimon Z, Sar Shalom H, Kuperman Y, et al. Brown-adipose-tissue macrophages control tissue innervation and homeostatic energy expenditure. *Nat Immunol* (2017) 18:665–74. doi:10.1038/ni.3746
39. Kong Y, Janssen BJ, Malinauskas T, Vangoor VR, Coles CH, Kaufmann R, et al. Structural basis for plexin activation and regulation. *Neuron* (2016) 91:548–60. doi:10.1016/j.neuron.2016.06.018
40. Weiskopf K, Schnorr PJ, Pang WW, Chao MP, Chhabra A, Seita J, et al. Myeloid cell origins, differentiation, and clinical implications. *Microbiol Spectr* (2016) 4. doi:10.1128/microbiolspec
41. Oldenborg PA. CD47: a cell surface glycoprotein which regulates multiple functions of hematopoietic cells in health and disease. *ISRN Hematol* (2013) 2013:614619. doi:10.1155/2013/614619
42. Penberthy KK, Ravichandran KS. Apoptotic cell recognition receptors and scavenger receptors. *Immunol Rev* (2016) 269:44–59. doi:10.1111/imr.12376
43. Segawa K, Nagata S. An apoptotic 'eat me' signal: phosphatidylserine exposure. *Trends Cell Biol* (2015) 25:639–50. doi:10.1016/j.tcb.2015.08.003
44. Platt N, Haworth R, Darley L, Gordon S. The many roles of the class A macrophage scavenger receptor. *Int Rev Cytol* (2002) 212:1–40. doi:10.1016/S0074-7696(01)12002-4
45. Han CZ, Juncadella IJ, Kinchen JM, Buckley MW, Klibanov AL, Dryden K, et al. Macrophages redirect phagocytosis by non-professional phagocytes and influence inflammation. *Nature* (2016) 539:570–4. doi:10.1038/nature20141
46. Wood W, Turmaine M, Weber R, Camp V, Maki RA, McKercher SR, et al. Mesenchymal cells engulf and clear apoptotic footplate cells in macrophageless PU.1 null mouse embryos. *Development* (2000) 127:5245–52.
47. Goodridge HS, Reyes CN, Becker CA, Katsumoto TR, Ma J, Wolf AJ, et al. Activation of the innate immune receptor Dectin-1 upon formation of a 'phagocytic synapse'. *Nature* (2011) 472:471–5. doi:10.1038/nature10071
48. Wood W, Martin P. Macrophage functions in tissue patterning and disease: new insights from the fly. *Dev Cell* (2017) 40:221–33. doi:10.1016/j.devcel.2017.01.001
49. Levin R, Hammond GR, Balla T, De Camilli P, Fairn GD, Grinstein S. Multiphasic dynamics of phosphatidylinositol 4-phosphate during phagocytosis. *Mol Biol Cell* (2017) 28:128–40. doi:10.1091/mbc.E16-06-0451
50. Kay JG, Grinstein S. Phosphatidylserine-mediated cellular signaling. *Adv Exp Med Biol* (2013) 991:177–93. doi:10.1007/978-94-007-6331-9_10
51. Billings EA, Lee CS, Owen KA, D'Souza RS, Ravichandran KS, Casanova JE. The adhesion GPCR BAI1 mediates macrophage ROS production and microbicidal activity against gram-negative bacteria. *Sci Signal* (2016) 9:ra14. doi:10.1126/scisignal.aac6250
52. Hamann J, Hsiao CC, Lee CS, Ravichandran KS, Lin HH. Adhesion GPCRs as modulators of immune cell function. *Handb Exp Pharmacol* (2016) 234:329–50. doi:10.1007/978-3-319-41523-9_15
53. Levin R, Grinstein S, Canton J. The life cycle of phagosomes: formation, maturation, and resolution. *Immunol Rev* (2016) 273:156–79. doi:10.1111/imr.12439
54. Lavin Y, Winter D, Blecher-Gonen R, David E, Keren-Shaul H, Merad M, et al. Tissue-resident macrophage enhancer landscapes are shaped by the local microenvironment. *Cell* (2014) 159:1312–26. doi:10.1016/j.cell.2014.11.018
55. Roberts AW, Lee BL, Deguine J, John S, Shlomchik MJ, Barton GM. Tissue-resident macrophages are locally programmed for silent clearance of apoptotic cells. *Immunity* (2017) 47:913.e–27.e. doi:10.1016/j.immuni.2017.10.006
56. Baratin M, Simon L, Jorquera A, Ghigo C, Dembele D, Nowak J, et al. T cell zone resident macrophages silently dispose of apoptotic cells in the lymph node. *Immunity* (2017) 47(349–62):e5. doi:10.1016/j.immuni.2017.07.019
57. Bannard O, Cyster JG. Germinal centers: programmed for affinity maturation and antibody diversification. *Curr Opin Immunol* (2017) 45:21–30. doi:10.1016/j.coi.2016.12.004
58. Weavers H, Evans IR, Martin P, Wood W. Corpse engulfment generates a molecular memory that primes the macrophage inflammatory response. *Cell* (2016) 165:1658–71. doi:10.1016/j.cell.2016.04.049
59. Moorlag S, Roring RJ, Joosten LAB, Netea MG. The role of the interleukin-1 family in trained immunity. *Immunol Rev* (2018) 281:28–39. doi:10.1111/imr.12617
60. Askew K, Li K, Olmos-Alonso A, Garcia-Moreno F, Liang Y, Richardson P, et al. Coupled proliferation and apoptosis maintain the rapid turnover of microglia in the adult brain. *Cell Rep* (2017) 18:391–405. doi:10.1016/j.celrep.2016.12.041
61. Fogarty CE, Diwanji N, Lindblad JL, Tare M, Amcheslavsky A, Makhijani K, et al. Extracellular reactive oxygen species drive apoptosis-induced proliferation via *Drosophila* macrophages. *Curr Biol* (2016) 26:575–84. doi:10.1016/j.cub.2015.12.064
62. Ueno M, Fujita Y, Tanaka T, Nakamura Y, Kikuta J, Ishii M, et al. Layer V cortical neurons require microglial support for survival during postnatal development. *Nat Neurosci* (2013) 16:543–51. doi:10.1038/nn.3358
63. Brown GC, Neher JJ. Microglial phagocytosis of live neurons. *Nat Rev Neurosci* (2014) 15:209–16. doi:10.1038/nrn3710
64. Kawane K, Fukuyama H, Kondoh G, Takeda J, Ohsawa Y, Uchiyama Y, et al. Requirement of DNase II for definitive erythropoiesis in the mouse fetal liver. *Science* (2001) 292:1546–9. doi:10.1126/science.292.5521.1546
65. Godwin JW, Pinto AR, Rosenthal NA. Macrophages are required for adult salamander limb regeneration. *Proc Natl Acad Sci U S A* (2013) 110:9415–20. doi:10.1073/pnas.1300290110
66. Yun MH, Davaapil H, Brockes JP. Recurrent turnover of senescent cells during regeneration of a complex structure. *Elife* (2015) 4. doi:10.7554/eLife.05505
67. Bryant DM, Johnson K, DiTommaso T, Tickle T, Couger MB, Payzin-Dogru D, et al. A tissue-mapped axolotl de novo transcriptome enables

identification of limb regeneration factors. *Cell Rep* (2017) 18:762–76. doi:10.1016/j.celrep.2016.12.063
68. Zanoni I, Tan Y, Di Gioia M, Springstead JR, Kagan JC. By capturing inflammatory lipids released from dying cells, the receptor CD14 induces inflammasome-dependent phagocyte hyperactivation. *Immunity* (2017) 47:697.e–709.e. doi:10.1016/j.immuni.2017.09.010
69. Bosurgi L, Cao YG, Cabeza-Cabrerizo M, Tucci A, Hughes LD, Kong Y, et al. Macrophage function in tissue repair and remodeling requires IL-4 or IL-13 with apoptotic cells. *Science* (2017) 356:1072–6. doi:10.1126/science.aai8132
70. Rothlin CV, Carrera-Silva EA, Bosurgi L, Ghosh S. TAM receptor signaling in immune homeostasis. *Annu Rev Immunol* (2015) 33:355–91. doi:10.1146/annurev-immunol-032414-112103
71. Chan PY, Carrera-Silva EA, De Kouchkovsky D, Joannas LD, Hao L, Hu D, et al. The TAM family receptor tyrosine kinase TYRO3 is a negative regulator of type 2 immunity. *Science* (2016) 352:99–103. doi:10.1126/science.aaf1358
72. Larson SR, Atif SM, Gibbings SL, Thomas SM, Prabagar MG, Danhorn T, et al. Ly6C(+) monocyte efferocytosis and cross-presentation of cell-associated antigens. *Cell Death Differ* (2016) 23:997–1003. doi:10.1038/cdd.2016.24
73. Campisi L, Barbet G, Ding Y, Esplugues E, Flavell RA, Blander JM. Apoptosis in response to microbial infection induces autoreactive TH17 cells. *Nat Immunol* (2016) 17:1084–92. doi:10.1038/ni.3512
74. Kleinclauss F, Perruche S, Masson E, de Carvalho Bittencourt M, Biichle S, Remy-Martin JP, et al. Intravenous apoptotic spleen cell infusion induces a TGF-beta-dependent regulatory T-cell expansion. *Cell Death Differ* (2006) 13:41–52. doi:10.1038/sj.cdd.4401699
75. Munoz LE, Lauber K, Schiller M, Manfredi AA, Herrmann M. The role of defective clearance of apoptotic cells in systemic autoimmunity. *Nat Rev Rheumatol* (2010) 6:280–9. doi:10.1038/nrrheum.2010.46
76. Mevorach D, Zuckerman T, Reiner I, Shimoni A, Samuel S, Nagler A, et al. Single infusion of donor mononuclear early apoptotic cells as prophylaxis for graft-versus-host disease in myeloablative HLA-matched allogeneic bone marrow transplantation: a phase I/IIa clinical trial. *Biol Blood Marrow Transplant* (2014) 20:58–65. doi:10.1016/j.bbmt.2013.10.010
77. Baxter AE, Russell RA, Duncan CJ, Moore MD, Willberg CB, Pablos JL, et al. Macrophage infection via selective capture of HIV-1-infected CD4+ T cells. *Cell Host Microbe* (2014) 16:711–21. doi:10.1016/j.chom.2014.10.010
78. Galluzzi L, Buque A, Kepp O, Zitvogel L, Kroemer G. Immunogenic cell death in cancer and infectious disease. *Nat Rev Immunol* (2017) 17:97–111. doi:10.1038/nri.2016.107

Conflict of Interest Statement: The authors declare that the research was conducted in the absence of any commercial or financial relationships that could be construed as a potential conflict of interest.

5

Apoptotic Cells Induced Signaling for Immune Homeostasis in Macrophages and Dendritic Cells

Uriel Trahtemberg[1] *and Dror Mevorach*[2]*

[1] *General Intensive Care Unit, Hadassah-Hebrew University Medical Center, Jerusalem, Israel,* [2] *Rheumatology Research Center, Department of Medicine, Hadassah-Hebrew University Medical Center, Jerusalem, Israel*

Inefficient and abnormal clearance of apoptotic cells (efferocytosis) contributes to systemic autoimmune disease in humans and mice, and inefficient chromosomal DNA degradation by DNAse II leads to systemic polyarthritis and a cytokine storm. By contrast, efficient clearance allows immune homeostasis, generally leads to a non-inflammatory state for both macrophages and dendritic cells (DCs), and contributes to maintenance of peripheral tolerance. As many as 3×10^8 cells undergo apoptosis every hour in our bodies, and one of the primary "eat me" signals expressed by apoptotic cells is phosphatidylserine (PtdSer). Apoptotic cells themselves are major contributors to the "anti-inflammatory" nature of the engulfment process, some by secreting thrombospondin-1 (TSP-1) or adenosine monophosphate and possibly other immune modulating "calm-down" signals that interact with macrophages and DCs. Apoptotic cells also produce "find me" and "tolerate me" signals to attract and immune modulate macrophages and DCs that express specific receptors for some of these signals. Neither macrophages nor DCs are uniform, and each cell type may variably express membrane proteins that function as receptors for PtdSer or for opsonins like complement or opsonins that bind to PtdSer, such as protein S and growth arrest-specific 6. Macrophages and DCs also express scavenger receptors, CD36, and integrins that function *via* bridging molecules such as TSP-1 or milk fat globule-EGF factor 8 protein and that differentially engage in various multi-ligand interactions between apoptotic cells and phagocytes. In this review, we describe the anti-inflammatory and pro-homeostatic nature of apoptotic cell interaction with the immune system. We do not review some forms of immunogenic cell death. We summarize the known apoptotic cell signaling events in macrophages and DCs that are related to toll-like receptors, nuclear factor kappa B, inflammasome, the lipid-activated nuclear receptors, Tyro3, Axl, and Mertk receptors, as well as induction of signal transducer and activator of transcription 1 and suppressor of cytokine signaling that lead to immune system silencing and DC tolerance. These properties of apoptotic cells are the mechanisms that enable their successful use as therapeutic modalities in mice and humans in various autoimmune diseases, organ transplantation, graft-versus-host disease, and sepsis.

Edited by:
Amiram Ariel,
University of Haifa, Israel

Reviewed by:
Philippe Saas,
Etablissement Français, du Sang BFC, France
Westley H. Reeves,
University of Florida, United States

***Correspondence:**
Dror Mevorach
mevorachd@hadassah.org.il

Keywords: clearance of apoptotic cells, toll-like receptor, nuclear factor kappa B, inflammasome, lipid-activated nuclear receptors, TAM receptors, signal transducer and activator of transcription 1, suppressor of cytokine signaling

INTRODUCTION

Inefficient clearance of apoptotic cells contributes to systemic autoimmune disease in humans and mice (1–6), and inefficient chromosomal DNA degradation by DNAse II leads to systemic polyarthritis and a cytokine storm (7, 8). By contrast, efficient clearance allows immune homeostasis, generally leads to a non-inflammatory state for both macrophages and dendritic cells (DCs), and contributes to maintenance of peripheral tolerance (9–12).

As many as 3×10^8 cells undergo apoptosis every hour in our bodies (13), where they are engulfed by macrophages and immature DCs and possibly by neighboring cells. What are the signaling patterns that induce these non-inflammatory responses?

Apoptotic cells characteristically expose "eat me" signals to macrophages (14, 15), and one of the primary "eat me" signals is phosphatidylserine (PtdSer). PtdSer is a phospholipid that localizes to the inner leaflet of plasma membranes in viable cells; however, when cells undergo apoptosis, it is exposed on the outer cell surface in a caspase-dependent manner (16–18). Other "eat me" signals exist and make important contributions to the process, including calreticulin from the dying cell endoplasmic reticulum and externally exposed chromatin and DNA, as well as alterations in surface charge and changes in glycosyl groups (19–21).

Apoptotic cells also produce "find me" signals to attract macrophages (22). Lysophosphatidylcholine (LPC) (22), fractalkine (23), sphingosine-1-phosphate (S1P) (24), and ATP/UTP (25) are released from apoptotic cells in a caspase-dependent manner. They also contribute to the "anti-inflammatory" nature of the engulfment process by secreting "tolerate me" signals *via* thrombospondin-1 (TSP-1) secretion (26) or "calm-down" signals *via* adenosine monophosphate (AMP) (27) and possibly other immune modulation signals yet to be discovered.

Another mechanism for immune modulation by apoptotic cells involves the caspase-dependent oxidation and deactivation of deactivation of high mobility group box 1 (HMGB1), a strong trigger of danger-associated-molecular-pattern (DAMP) that causes inflammatory responses (28). Similarly, in the context of viral infection, caspases can modify the mitochondria-initiated cell death process and inhibit the interferon (IFN) response, switching the result of the dying process from pro-inflammatory to immunologically silent (29, 30). Since the activation of caspases is not a necessary condition for apoptosis, it could be that caspase activation, which drives the apoptotic program toward tolerogenic consequences, is another way that apoptotic cells "instruct" the cells clearing them regarding the nature of their death (31).

Neither macrophage subpopulations or DCs are "uniform" and each cell type may variably express membrane proteins that function as receptors for PtdSer (Tim-4, stabilin 2, and BAI1), or for opsonins that bind to PtdSer, milk fat globule-EGF factor 8 protein (MFGE8), ProS, and growth arrest-specific 6 (GAS6) (14). Masking the PtdSer on apoptotic cells prevents their engulfment by macrophages and induces autoantibodies (4) and inflammation (32), supporting the idea that PtdSer is not only an important "eat me" signal but also a "tolerate me" signal. Macrophages also express integrins that function *via* bridging molecules such as TSP-1, MFGE8, and complement (2, 9, 33). These integrins can contribute to both phagocytosis and inhibition of a pro-inflammatory immune response, for example, by scavenger receptor (ScR) SCARF1 (34), the immunoglobulin superfamily member leukocyte-associated Ig-like receptor 1 (CD305) (35), CD11b or CD11c (2, 9, 36), other ScRs, CD36, and possibly additional receptors that are important in multi-ligand interactions between apoptotic cells and phagocytes (2, 19, 26, 37). In addition, "cross-talk" exists and, for example, C1q-dependent induction of opsonins Gas6 and Protein S has been described (38, 39).

Macrophages express specific receptors for some of these "find me" signals (CX3CR1 for fractalkine, aS1PR1 for S1P, and $P2Y_2$ for ATP and UTP), which may mediate migration to the dying cells (15). The "find me" signals are thought to prime macrophages for engulfment, as best exemplified by the enhanced expression of MFGE8 (40). On the other hand, some "find me" signals, for example, LPC, ATP/UTP, and S1P, may cause inflammation (41–43), contradicting the anti-inflammatory nature of the apoptotic process. How is the anti-inflammatory character of the apoptotic process maintained during cell death and engulfment? We will discuss several signaling patterns that have been identified.

Other modes of cell death that are immunogenic (44), including accidental cell death (necrosis), necroptosis, pyroptosis, and NETosis (45), will not be discussed here.

SIGNALING INHIBITION OF TOLL-LIKE RECEPTORS (TLRs), NUCLEAR FACTOR KAPPA B (NF-κB), AND THE INFLAMMASOME

Toll-Like Receptors

Toll-like receptors are membrane-associated innate immune sensors that recognize conserved microbial-associated molecular structures of invading pathogens. A classic example is lipopolysaccharide (LPS), which is expressed by Gram-negative bacteria that binds to TLR4 and induces pathogen-associated molecular patterns (PAMPs). What is mostly relevant to this review is that TLRs also detect host-derived, danger-associated molecular patterns (DAMPs) and alarmins that can be produced during immunogenic programmed cell death or cell necrosis, including HMGB1 and endogenous RNA and DNA that are normally hidden in TLR-inaccessible compartments but become exposed and are released during cell stress, inflammation, infection, or non-apoptotic death (46–48). TLRs are expressed by macrophages and DCs that are specifically important for interaction with apoptotic cells, but they are also expressed by natural killer (NK) cells, mast cells, and T- and B-lymphocytes, as well as by some non-immune cells, such as epithelial and endothelial cells (46–48).

Importantly, apoptotic cells downregulate the response to TLR receptors on both macrophages and DCs (9, 49–51). For ligand detection and co-receptor interactions, TLRs contain an ectodomain with multiple leucine-rich repeat domains involved in a portion of the transmembrane, and an intracellular toll/IL-1 receptor (IL-1R) homology domain (TIR) essential for signaling (46–48).

Toll-like receptors are expressed on the plasma membrane (e.g., TLR1, TLR5, TLR6, and TLR10), in intracellular endosomes (e.g., TLR3, TLR7–9, and TLR11), or in both compartments (e.g., TLR2 and TLR4) (52), and TLR localization critically regulates its signaling where the initial step following binding is recruitment of adaptor proteins. The two main known pathways are myeloid differentiation primary response gene 88 (MyD88)- and TIR-domain-containing adapter inducing interferon β (TRIF)-dependent. All TLRs except TLR3 use the MyD88-dependent pathway to initiate signaling, whereas TLR3 and TLR4 use the TRIF-dependent pathway to elicit induction of both pro-inflammatory cytokines and type I IFNs (52).

After recruitment to TLRs, MyD88 molecules cluster and recruit interleukin-1 (IL-1) receptor-associated kinases (IRAKs) through homotypic death–death domain interactions (53). An alternatively spliced form of MyD88 (MyD88s) lacks a short linker sequence between the death- and TIR domains. It binds to the TIR domain of TLRs but fails to recruit IRAK1, thereby inhibiting signaling (54).

Expression of other regulatory kinases within the IRAK family is induced following TLR signaling. Increased expression of IRAK-M or alternatively spliced variants of IRAK1 (e.g., IRAK1c) suppresses TLR signaling (55, 56). IRAK-M was originally reported to prevent dissociation of IRAK and IRAK4 from MyD88, and to block engagement of TRAF6, inhibiting signaling (56); however, later studies demonstrated the ability of IRAK-M to engage a separate MEK kinase 3-dependent signaling pathway for NF-κB activation. This pathway leads to IκB-a phosphorylation, but not degradation, and controls a limited set of inflammatory cytokines and negative regulators [suppressor of cytokine signaling (SOCS) 1, SHIP-1, A20] whose expression is not controlled by mRNA stability (57). IRAK1c is an alternatively spliced form of IRAK that lacks a region encoded by exon 11 of the IRAK1 gene, resulting in a kinase-inactive form of IRAK (55, 56). IRAK1c can heterodimerize with IRAK, thereby fine-tuning the level of IRAK activity. Activator proteins engaged in MyD88- and TRIF-dependent signaling pathways also become phosphorylated, and these events play a critical role in TLR signaling.

Apoptotic cells were shown to downregulate TLR signaling events by us and by other groups, with effects on LPS-TLR4, zymosan-TLR2, and possibly other events (9, 37, 49, 50, 58, 59). Inhibition of TLR signaling after *in vivo* apoptotic cell administration has been clearly demonstrated in mouse models (36, 59, 60). Taken together, this strongly supports the impact of apoptotic cells on the inhibition of TLR signaling pathways in different innate immune cell subsets. On the other hand, in some abnormal conditions where pathogenic autoantibodies opsonize self-antigens and apoptotic debris, immune complexes are formed and bind to TLR 7 and TLR 9 and trigger the production of IFN-α, a hallmark of SLE (61, 62).

Nuclear Factor Kappa B

Nuclear factor kappa B is a major transcription factor that has been implicated as a critical regulator of gene expression in the setting of inflammation. NF-κB is ubiquitously expressed and is activated by a wide variety of stimuli, including pro-inflammatory cytokines such as tumor necrosis factor-α (TNFα) and IL-1, bacterial- or viral-derived PAMPs, and various types of stress (63, 64).

The NF-κB family consists of five DNA-binding members. NF-κB1 is synthesized as p105 and is processed into a DNA-binding subunit, p50 (65). Likewise, NF-κB2 is produced as p100, which serves as a precursor for the active transcription factor p52. p50 and p52 form various combinations of heterodimers with RelA (p65), c-Rel, and RelB. These DNA-binding complexes target distinct sets of genes for transcriptional activation (66). In addition, RelA and c-Rel can also activate gene transcription as homodimers (67). By contrast, p50 homodimers, which lack transactivation function, can instead suppress NF-κB target gene expression, for example, in response to stimulation by LPS (68, 69).

Depending on the extracellular stimuli and the receptors engaged, NF-κB activation mechanisms can be broadly classified into canonical and non-canonical pathways (70). In the absence of extracellular stimuli, transcriptional activity of NF-κB transcription factors is normally kept in check by sequestration in the cytoplasm (71). Following activation, the transient transcriptional activity of NF-κB is maintained by several mechanisms to prevent inflammation-induced tissue damage or malignancy associated with chronic NF-κB activation (72). IκBa is induced in an NF-κB-dependent manner, which contributes to the termination of NF-κB signaling in a negative feedback loop (73). In addition, p100, which also serves as an IκB-like protein termed "IκBd," plays a critical role in terminating NF-κB activity (71). Activation of the canonical NF-κB pathway depends on the IKK complex, which contains two catalytic subunits (IKKα and IKKβ) and a regulatory subunit NEMO/IKKc (74). Catalytically active IKKβ phosphorylates IκBα, signaling its ubiquitination and proteasomal degradation (75) in response to various stimuli, including TNF receptor 1, IL-1R, and TLRs. By contrast, non-canonical NF-κB pathway activation is mediated by IKKα, which phosphorylates p100. This is followed by partial p100 degradation to generate p52 in response to stimulation *via* certain TNFR family members such as B cell-activating factor receptor (BAFF-R), CD40R, and lymphotoxin-β receptor (76).

The ubiquitin-editing enzyme A20 complex is a negative regulator of canonical NF-κB signaling. Mice lacking A20 develop severe inflammation and cachexia, and die prematurely (77). These mice exhibit persistent NF-κB and IKK activation and severe systemic inflammation in response to TNFα and sublethal doses of LPS (77). Thus, A20 is an important negative feedback regulator of NF-κB required for immune homeostasis. Liberated NF-κB dimers from IκBα and IκBδ translocate to the nucleus and activate transcription of various genes involved in innate and adaptive immunity (78). Deletion of A20 in DC leads to the development of pathologies in mice similar to those seen in humans with inflammatory bowel disease (IBD) and SLE, including autoantibodies, splenomegaly, nephritis, colitis, and even ankylosing spondylitis (79–81). Similarly, humans who have been identified to have mutations in the *Tnfaip3* gene that encodes A20 show auto-inflammation (82, 83). Most important, apoptotic cell uptake by A20-deficient DC fails to inhibit pro-inflammatory cytokine production in response to LPS (79) and A20 expression is upregulated in small intestinal lamina propria

CD103+ DC in response to apoptotic IEC (84). Furthermore, small intestinal lamina propria CD103+ DC induction of A20 in response to apoptotic cells (84) shuts down both apoptotic cell phagocytosis and inflammation, and thus may limit the supply of self-antigen and its presentation in an inflammatory context (85).

Lipopolysaccharide-induced cytokine production is mainly mediated by activation of NF-κB, MAPKs, and IRF-3, and by induction of a type I IFN-mediated, signal transducer and activator of transcription 1 (STAT1)-dependent autocrine loop. Our group has suggested that the mechanism for apoptotic cell inhibition of pro-inflammatory cytokines, as originally showed by Fadok et al. and Voll et al. (49, 58), is due to inhibition of TLR and NF-κB signaling (37, 59) and inflammasome for IL-1β (86) (see below). Inhibition of NF-κB by apoptotic cells has been shown by others (87, 88) and by our team (37, 59, 86), and it is suggested that nuclear migration of p65 is inhibited at the transcriptional or post-transcriptional level (37, 86). In addition, Mer receptor tyrosine kinase (RTK) (MerTK, see below) was also found to activate the phosphatidylinositol 3-kinase/AKT pathway, which negatively regulates NF-κB (89).

Inflammasome

Inhibition of NF-κB could not explain very rapid inhibition of IL-1β secretion by apoptotic cells (37); thus, additional mechanisms remained unexplained.

IL-1β is a pro-inflammatory cytokine produced primarily by activated monocytes and macrophages that is involved in the regulation of immune responses as well as the pathogenesis of several acute and chronic inflammatory diseases. Release of IL-1β is mediated by a two-step process: first, transcriptional induction of pro-IL-1β, and then caspase 1-mediated cleavage for the generation and secretion of IL-1β (86). Inflammasomes are high-molecular-weight cytosolic complexes that mediate the activation of caspase 1 and therefore enable rapid secretion of IL-1β and IL-18, which already exist as pro-cytokines. There are many inflammasomes, and each is influenced by a unique pattern recognition receptor response. Two signals are typically involved in inflammasome pathways (90). Signal one involves recognition of PAMPs or DAMPs that interact with TLRs, thus inducing downstream production of pro-IL-1β. This is followed by signal two, which involves recognition of PAMPs or DAMPs made by cells such as uric acid or ATP *via* nucleotide-binding domain, leucine-rich-containing family (NLR) pyrin domain-containing-3 (NLRP3), which leads to caspase-1-dependent cleavage of pro-IL-1β to active IL-1β. Both PAMPs and DAMPs can be liberated by early insults. The consequence of inflammasome activation and IL-1β expression is the upregulation of adhesion molecules and chemokines, leading to neutrophil sequestration, mononuclear phagocyte recruitment, and T cell activation.

Apoptotic cells were shown by us to inhibit TLRs and the NF-κB pathway (37, 59). TLR triggering is important for enhanced transcription of pro-IL-1β and pro-IL-18, and is in fact needed for the effect, but is not sufficient for rapid IL-1β secretion. We were able to show that apoptotic cells inhibit secretion of activated IL-1β at both pre- and post-transcription levels, and have distinct inhibition effects on NF-κB and NLRP3 (86). The dextran sulfate sodium (DSS) colitis model is generally viewed as an epithelial damage model suited to investigate innate immune responses. Macrophages primed with LPS and subsequently exposed to DSS secrete high levels of IL-1β in an NLRP3-, ASC-, and caspase-1-dependent manner. The DSS murine model and *Nlrp3*-deficient mice were used by us to assess the effect apoptotic cells on colitis. Immunohistochemistry, flow-cytometry, and Western blots helped to explore the effect and mechanisms. Using a variety of NLRP3 triggering mechanisms, we showed that apoptotic cells negatively regulate NF-κB and NLRP3 activation at the pre- and post-transcription levels *via* inhibition of reactive oxygen species (ROS), lysosomal stabilization, and blocking potassium efflux. This property of apoptotic cells was associated with a dramatic clinical, histological, and immunological amelioration of DSS colitis in Balb/c and B6 mice following a single administration of apoptotic cells (86).

Apart from apoptotic cell opsonization, MFGE8 was also found to be an endogenous inhibitor of inflammasome-induced IL-1β production (91). MFGE8 inhibited necrotic cell-induced and ATP-dependent IL-1β production by macrophages through mediation of integrin β3 and P2X7 receptor interactions in primed cells. *itgb3* deficiency in macrophages abrogated the inhibitory effect of MFGE8 on ATP-induced IL-1β production. Furthermore, in a setting of post-ischemic cerebral injury in mice, MFGE8 deficiency was associated with enhanced IL-1β production and larger infarct size. The latter was abolished after treatment with IL-1β receptor antagonist. MFGE8 supplementation significantly dampened caspase-1 activation and IL-1β production and reduced infarct size in wild-type (WT) mice, but did not limit cerebral necrosis in *IL-1β*-, *Itgβ3*-, or *P2rx7*-deficient animals.

What is the mechanism by which apoptotic cells inhibit inflammasome? We could show the involvement of three mechanisms in the resolution by apoptotic cells of inflammasome-induced inflammation (86). First, other groups (92, 93) and our lab (86) have shown that apoptotic cells are able to reduce and inhibit the formation of ROS at rates similar to those shown for the chemical inhibitor *N*-acetyl cysteine. It is well established that macrophages make use of toxic ROS to control microbial pathogens as part of the innate immune response, and ROS were identified as major mediators of inflammatory signals believed to play a role in the development of IBD. Furthermore, generation of ROS was found to induce IL-1β *via* ERK phosphorylation (94). In addition, IL-1β signals may induce ROS generation (95). While it has been shown that DSS induces formation of ROS (96, 97), we were able to see a marked reduction in ROS generation, and consequently less IL-1β secretion, when macrophages were pretreated with apoptotic cells (86).

The second mechanism involves the lysosome. It was shown that lysosomal damage or leakage may serve as an endogenous danger signal and is sensed by the NLRP3 inflammasome (96, 98). We have analyzed involvement of the lysosome vacuole and were first to discover that lysosomes from peritoneal macrophages that were introduced to apoptotic cells were more stable to DSS challenge, and were not affected or damaged (86).

Inflammasomes were also suggested to be activated in response to signaling pathways that deplete intracellular potassium, such as the potassium ionophore nigericin pathway (99). We noted that when macrophages were pretreated with apoptotic cells, nigericin-induced IL-1β secretion was significantly inhibited. The mechanisms for apoptotic cell inhibition of this secretion are not clear.

Taken together, these results demonstrate a mechanism of inflammasome inhibition and resolution of inflammation stemming from apoptotic cell clearance and illustrate a mechanism for regulation of inflammation that could take place in both infectious and noninfectious inflammatory conditions.

Additional related possible mechanisms include a functional subgroup of NLRs that negatively regulate inflammation (100), including the possible effect of apoptotic cells on NLRP12. This suppressor of pro-inflammatory cytokine and chemokine production downstream of TLRs targets multiple points in the NF-κB pathway. However, it is clear that failure to clear apoptotic cells will trigger persistent inflammasome-dependent inflammation, as perhaps is seen in failure to clear intracellular organelles (101).

LIPID-ACTIVATED NUCLEAR RECEPTORS

Nuclear receptors are transcription factors that regulate gene transcription in response to their ligand and include lipids, vitamins, and hormones. They suppress or activate transcription, allowing regulation diverse biological functions that include cytokine production, lipid metabolism, and more (102). Following internalization of apoptotic cells, lipids, carbohydrates, protein, and nucleotides are acquired from the apoptotic cell. This content was suggested to be a significant metabolic burden on the phagocyte (103) and may also influence its immune response.

Several studies have shed light on the metabolic changes macrophages undergo to restore normal cellular function and homeostasis following apoptotic cell ingestion. Macrophages and DCs express several nuclear factors, most with a role in clearance of apoptotic cells, including peroxisome proliferator-activated receptor (PPAR)-α, -β/δ and -γ isotypes, liver x receptor (LXR) α and β isotypes, retinoid x receptor (RXR) α and β isotypes, retinoic acid receptor, vitamin D receptor, and glucocorticoid receptor (104). The putative natural ligands of LXR are oxysterols, which have been suggested to arise from the lipids derived from apoptotic cell/body membranes. This exemplifies the anti-inflammatory effects of the PPAR and LXR receptors (104).

Mukundan et al. (105) and A-Gonzalez et al. (106) showed how lipids in apoptotic cells induce activation of the transcription factors PPAR-δ and LXR, respectively. Activation of these transcription factors results in macrophage upregulation of cell surface receptors and soluble ligands that suppresses inflammatory cytokine production and promotes removal of apoptotic cells. In their absence, apoptotic cell clearance is impaired and a lupus-like autoimmunity develops.

Concerning the mechanism of action, like other members of the same nuclear receptor family, PPAR-δ and LXR heterodimerize with the RXR to regulate transcription of target genes. Depending on the presence of corepressors or coactivators, these dimeric transcription factors can either suppress or initiate transcription (107).

Mukundan et al. (105) showed that the lipids contained in apoptotic cells activate PPARs. They observed that PPAR-δ, but not PPAR-ϒ, was induced in macrophages after exposure to apoptotic cells. Furthermore, macrophages obtained from mice deficient in PPAR-δ showed reduced engulfment of apoptotic cells and impaired clearance of apoptotic cells *in vitro* and *in vivo*. To prove that these changes were due to defects in macrophages, they created mice that were PPAR-δ-deficient exclusively in macrophages and observed similar defects in clearance of apoptotic cells, but not of necrotic cells.

A-Gonzalez et al. (106) observed that knockout of the genes encoding both the α and β chains of LXR led to defective phagocytosis of apoptotic cells. They then identified candidate genes responsible for this defect by microarray. In contrast to the reduced expression of opsonins observed with PPAR-δ deficiency, loss of LXR led to a marked reduction in the expression of the macrophage receptor Mer (see below), which binds either the GAS6 or Pro S opsonins. Both of these opsonins attach to PtdSer on apoptotic cells, resulting in apoptotic cell engulfment. Experimental support for regulation of Mer by LXR was obtained by detection of LXR binding to the promoter of *Mertk* (which encodes Mer) by gain of function experiments and by demonstration of increased phagocytosis of apoptotic cells after macrophage stimulation with the synthetic LXR agonist GW3965. The findings suggest that, after the ingestion of apoptotic cells, oxysterols (oxidized derivatives of cholesterol) activate transcription of Mer by LXR and like PPAR-δ. This transcription factor enhances the clearance of dying cells (104).

Macrophage activation *via* adenosine receptors is followed by the upregulation of TSP-1 and Nr4a gene expression, especially of *Nur77*, NOR-1, and *Nurr1*. TSP-1 is the major activator of TGFβ (108) and Nr4a family members inhibit the expression of pro-inflammatory cytokines such as TNFα, IL-8, and IL-6 in macrophages (109) by recruiting a repressor complex to their promoter (110). Yamaguchi et al. (27) found that AMP was present at 30- to-100-fold higher concentrations than ATP in the culture supernatants of apoptotic thymocytes and a T cell line. When cells undergo apoptosis, ATP is quickly hydrolyzed to AMP, while its generating system is inactivated by caspases (111). The caspase-cleaved Pannexin channel also contributes to cellular loss of ATP by allowing ATP to exit cells through the plasma membrane. Chekeni et al. (112) showed that caspases cleave Pannexin 1 in the early stages of apoptosis, resulting in the release of ATP, which can serve as a “find me” signal to attract macrophages. Depending on the types of cells and apoptotic stimuli, intracellular ATP levels remain high (113–115) or rapidly decrease (116, 117).

INHIBITION OF INFLAMMATION BY *Tyro3*, *Axl*, AND *Mertk* (TAM) RECEPTORS

Early indications that these RTKs may have an important immune function came from observations made in studies by

Camenisch et al. (118), who used *Mertk*$^{-/-}$ mice. These mice had an approximately threefold enhancement of serum TNFα when administered 100 mg/kg of LPS *in vivo*. In 2001, Lu and Lemke (119) further characterized RTK's role in the immune system by using mice that were triple knockout for TAM receptors *Tyro3*$^{-/-}$ *Axl*$^{-/-}$ *Mertk*$^{-/-}$ (TAM RTK). These TAM RTK animals did not present with serious developmental anomalies and appeared normal (119). They had apparently normal immune systems with no differences in the size of the secondary lymphoid organs, and their role in inflammation was only clarified later.

The TAM receptors were originally discovered by Lai and Lemke (120). TAM receptors are key inhibitors of the immune system (121). Diverse immune cells in humans and mice express TAM components and are severely perturbed if their TAM-dependent cellular pathways are ablated (122). TAM signaling provides an indispensable inhibitory feedback mechanism responsible for safeguarding the shutdown of inflammation and promotion of tissue-repair processes. Blocking TAM signaling causes severe defects in apoptotic cell clearance, widespread inflammation, overactivation of the immune system, and development of systemic autoimmunity (123, 124).

TAM receptors are activated *via* two known mediators, ProS and GAS6 (125). Both bind to apoptotic cells and thereafter mediate ligation to TAM receptors. Both are Gla domain-containing proteins, i.e., proteins containing gamma-carboxylated glutamic acid residues. The gamma-carboxylation of glutamate residues vastly increases their ability to bind Ca^{2+}. GAS6 and ProS contain Gla domains consisting of ~50 amino acids stretched near their N termini. Gamma-carboxylation and PtdSer binding are essential for the maximal bioactivity of both full-length TAM ligands (126–129). For example, ProS has a cofactor activity; its ability to activate the TAM RTKs is dependent on gamma-carboxylation of the GLA domain and binding to PtdSer (130).

The current thought is that TAM RTKs are significantly upregulated as part of a pro-inflammatory response, for example, TLR engagement. However, in the case of apoptotic cell clearance there may be direct signaling following receptor binding. For example, and as mentioned earlier, A-Gonzalez et al. (106) observed that knockout of the genes encoding both the alpha and beta chains of LXR led to a marked reduction in the expression of Mer. Experimental support for regulation of Mer by LXR was obtained by detection of LXR binding to the promoter of *Mertk*, which encodes Mer, by gain of function experiments and demonstration of increased phagocytosis of apoptotic cells after macrophage stimulation with the synthetic LXR agonist. The findings suggest that, after the ingestion of apoptotic cells, oxysterols (oxidized derivatives of cholesterol) activate transcription of Mer by LXR and like PPAR-δ.

INHIBITION OF IFNs BY SOCS 1/3 UPREGULATION

Type II IFN

The phenotype that defines activated macrophages and DCs is characterized by increased microbicidal or autoimmune activity, high antigen-presenting activity associated with increased MHC class II expression, and increased production of IL-12 (131). These characteristics are promoted by an IFNγ-mediated Janus kinase signal transducer and activator of transcription (JAK-STAT) signaling. Stimulation of the IFNγ receptor triggers JAK-mediated tyrosine phosphorylation and subsequent dimerization of STAT1, which binds as a homodimer to elements known as gamma-activated sequences in the promoters of the genes encoding NOS2, the MHC class II transactivator, and IL-12, among others (132).

We were the first to show that clearance of apoptotic cells inhibits type II IFN (human γ-IFN) signaling by upregulation of SOCS (59). We showed that interaction of macrophages with apoptotic cells had no activation effect for MAPKs p38, JNK, or ERK1/2 (59). By contrast, apoptotic cells suppressed the LPS-induced IFN-mediated autocrine loop by attenuating STAT1 activation and suppressing IFN activation of STAT1-dependent genes such as CXCL10 (133, 134). It has been suggested that apoptotic cell induction of SOCS1 and SOCS3 expression contributes to suppression of IFN-induced gene expression (135), and thus suppresses JAK-STAT signaling and IFN-mediated responses downstream of TLR4.

Interferon-γ is a key activator of macrophages and is mainly produced by NK cells, Th1 cells at later stages of the immune response, and chimeric antigen receptor T cells. STAT1 mediates most of the IFN-γ activating effects on macrophages. We analyzed the effect of apoptotic cells on the IFN-γ signaling pathway in macrophages both *in vitro* and *in vivo*. We found that IFN-γ induced STAT1 activation at both the tyrosine phosphorylation and DNA-binding levels, and was significantly inhibited in macrophages that had interacted with apoptotic cells *in vitro* and *in vivo* in a chemically induced peritonitis murine model of inflammation. Inhibition of STAT activation was somewhat selective for STAT1 relative to STAT3, which is activated by IL-10 and is strongly anti-inflammatory. This selective inhibition pattern of cytokines and STATs would have the net effect of suppressing inflammatory macrophage activation while leaving deactivation pathways intact.

Suppressor of cytokine signaling ubiquitin ligases are responsible for downregulation of the immune response through the turnover of molecules that function in critical, positive, regulatory signaling cascades such as the TLR, NF-κB, and JAK-STAT pathways (136). Substrates of SOCS1 and SOCS3 include MAL, TRAFs, and JAKs.

Type I IFN

Rothlin and Lemke showed later that the TAM RTK-dependent upregulation of SOCS required the type I IFN receptor and STAT1 (89). Consistent with a central role for TAM RTK in the negative regulation of inflammation, the upregulation of SOCS by type I IFNs was contingent on TAM RTK. SOCS1 induction by IFN-α was significantly reduced in TAM RTK TKO DCs (89). Axl mRNA was induced by type I IFNs produced downstream of TLR activation (89, 137, 138), indicating that the braking mechanism is not available at the onset of the immune response but only following its initiation.

The mechanism of TAM RTK action in type I IFNs involves upregulation of the SOCS proteins SOCS1 and SOCS3 (89).

Consistent with a central role for TAM RTK in the negative regulation of inflammation, the upregulation of SOCS by type I IFNs was contingent on TAM RTK. SOCS1 induction by IFN-α was significantly reduced in TAM RTK TKO DCs (89). It was suggested that TAM RTKs can complex with type I IFN receptors and modify STAT function (89), potentially by altered phosphorylation. An additional mechanism of TAM RTK-mediated inhibition of inflammation includes upregulation of the transcription factor twist, which in turn leads to downregulation of TNFα (138).

Lipopolysaccharide-induced cytokine production is mainly mediated by activation of NF-κB, MAPKs, and IRF-3, and by induction of a type I IFN-mediated STAT1-dependent autocrine loop (139).

Taken together, these observations illustrate tight negative regulation of type I (89) and II (59) IFN pro-inflammatory signaling.

APOPTOTIC CELLS AND MAINTENANCE OF PERIPHERAL TOLERANCE

Recognition of an autoantigen by the T cell receptor (TCR) is the capability that distinguishes autoreactive T cells from other T cell subsets. The TCR repertoire is generated in immature T cells in a relatively random manner, and some TCRs recognize self-antigens. The majority of T cell clones with high-affinity TCRs that recognize self are deleted as a consequence of self-antigen presentation by thymic epithelial cells (140). Thymic selection is imperfect; therefore, autoreactive T cells are present in the peripheral T cell repertoire of healthy individuals (141). T cells that escape negative selection in the thymus must be held in check by additional peripheral tolerance mechanisms, and the ability to tightly control and avoid the activation of peripheral self-reactive T cells is crucial for avoiding autoimmunity.

Dendritic cells are the most potent antigen-presenting cells, and as such they are key regulators of the immune system (142). They share with macrophages many of the roles described earlier in the engulfment and clearance of apoptotic cells (85). Two main issues differentiating DCs from macrophages, in the context of this review, are the existence of more than two subpopulations of DCs with different roles and anatomical locations (85, 143), and their main role initiating the adaptive immune response classically described after migrating to a lymphoid organ (144). For example, CD8α$^+$ murine DCs and their suggested human analog, the CD141$^+$ DCs (145) are specialized in the uptake of apoptotic cells and the cross-presentation of their antigens to T cells (146). In this review, we will concentrate on myeloid DCs, although there are reports on the effects of plasmacytoid DCs after engulfment of apoptotic cells (147), and they can have anti-inflammatory effects *via* mechanisms similar to those used by myeloid DCs (148).

T cell activation requires a first signal provided by TCR ligation and a second signal provided by engagement of costimulatory molecules with their respective ligands on antigen-presenting cells. A third signal related to IL2/IL2R interactions should also sometimes be considered. The coordinated triggering of these two independent signaling systems ensures full T cell activation, including proliferation and acquisition of effector function. TCR occupancy in the absence of costimulatory signals leads to a sustained loss of antigen responsiveness called clonal anergy or T cell apoptosis, and therefore DCs interacting with apoptotic cells contribute to the maintenance of peripheral tolerance (9, 12, 149). We were able to show downregulation of costimulatory molecules, including CD40 and CD86 as well as MHC-DR, following apoptotic cell ingestion by DCs. Interestingly, DCs did not lose the ability to migrate to the lymph nodes following this effect, and upregulated CCR7, which is normally upregulated upon activation and allows tolerogenic DCs to migrate to naïve areas of the lymphatic system to encounter potential autoimmune T cells that may react to apoptotic cell antigens (9, 12).

Lung DCs are an excellent example of the dual role of DCs that can induce tolerance or activate naïve T cells, making these DCs well-suited to their role as lung sentinels. Lung DCs are supposed to serve as a functional signaling/sensing unit to maintain lung homeostasis and orchestrate host responses to benign and harmful foreign substances [reviewed in Ref. (150)]. *In vitro* observations demonstrating this role were further investigated using *in vivo* murine studies. CD11c+ CD11b+ CD8α+ DCs were found in the pancreatic lymph nodes, carrying fluorescently labeled dead cells that had been injected directly into the pancreas (151). Furthermore, apoptotic cells were phagocytosed by CD8α+ DCs within the spleen and apoptotic cell antigen was cross-presented to antigen-specific CD8 T cells, leading to their deletion as a mechanism of immune tolerance (10, 11).

Later, a role for specialized CD169+ macrophages in handling dead tumor cells and cross-presentation was also suggested (152), and identification of special subsets in different localizations within organs like lung, brain, and gut led to the suggestion that the nature of a phagocyte that recognizes, samples, and/or internalizes apoptotic cells is likely dependent on the tissue and its physiological state at any given time (84, 85). Further details are found elsewhere (85, 153).

With respect to apoptosis of antigen-presenting cells, we were able to show that apoptotic monocytes secrete TSP-1, which, by itself, induces a tolerogenic phenotype in DCs (26). While it has long been known that apoptotic cells change their phenotype during the death process (154), this has not been given much attention in the literature. Our group has shown that there are two subpopulations of human monocyte-derived DCs with different immune phenotypes and functions. Apoptotic cells are not necessarily homogeneous; thus, upon entering the process of apoptotic cell death, these two cell types differentially regulate their expression of cell surface antigens in a way that will dramatically influence interaction with T cells (149). Thus, even while dying, these DCs are explicitly signaling the cells they interact with and conveying information, their "immune will." We identified three general patterns of expression: Pattern 1, surface marker expression increases for both subpopulations as cell death progresses; Pattern 2, surface marker expression increases in one subpopulation; and Pattern 3, surface marker expression shows a mixed pattern as cell death progresses with behavior dependent on the stimuli used. Importantly, one subpopulation dramatically increased CD86 expression in correlation to advanced apoptosis, suggesting that even during the death process DCs can signal to T cells for immune responses.

In another example, it has been shown that the susceptibility of different DC subpopulations to apoptosis has significance for the immune response to viral infections (155). These processes highlight the importance of apoptosis of the antigen-presenting cells themselves as an immune regulatory event that is a less recognized way for apoptosis to affect the immune system.

One of the main effects of DCs after the uptake of apoptotic cells is secretion of TGF-β, IL-10, and retinoic acid, which promote the development of T-regulatory (Treg) cells (85, 156–159).

These could be induced *via* the abovementioned mechanism, but also by the induction of indoleamine 2,3 dioxygenase-1 (IDO-1), whose activity can be induced by TGF-β (160) and whose expression is induced by apoptotic cells. Both *in vitro* (161, 162) and *in vivo* administration of apoptotic cells to mice induce the expression of IDO in splenic marginal zone macrophages (163). IDO-1 promotes immune tolerance by the induction of Treg cell differentiation (163–168).

Other studies have shown that DC phagocytosis of apoptotic cells initiates naïve CD4 T cell differentiation into Treg cells (169–174).

In conclusion, apoptotic cells induce a tolerogenic DC phenotype and may directly (anergy- or activation-induced cell death) or indirectly (Tregs) inactivate potential autoreactive T cells. In this way they represent a potent peripheral tolerance mechanism.

APOPTOTIC CELLS AS THERAPEUTIC AGENTS

The *in vitro* and *in vivo* properties of apoptotic cells suggest their potential use in a broad range of inflammatory and immune-mediated conditions such as autoimmunity, graft rejection, post-ischemic injury, cytokine storm, and more.

Autoimmune and autoinflammatory conditions, including type 1 diabetes in non-obese diabetic mice, experimental autoimmune encephalomyelitis, arthritis, colitis, pulmonary fibrosis, fulminant hepatitis, contact hypersensitivity, acute- and chronic-graft rejection, hematopoietic cell engraftment, acute graft-versus-host disease (GvHD), and reduction of

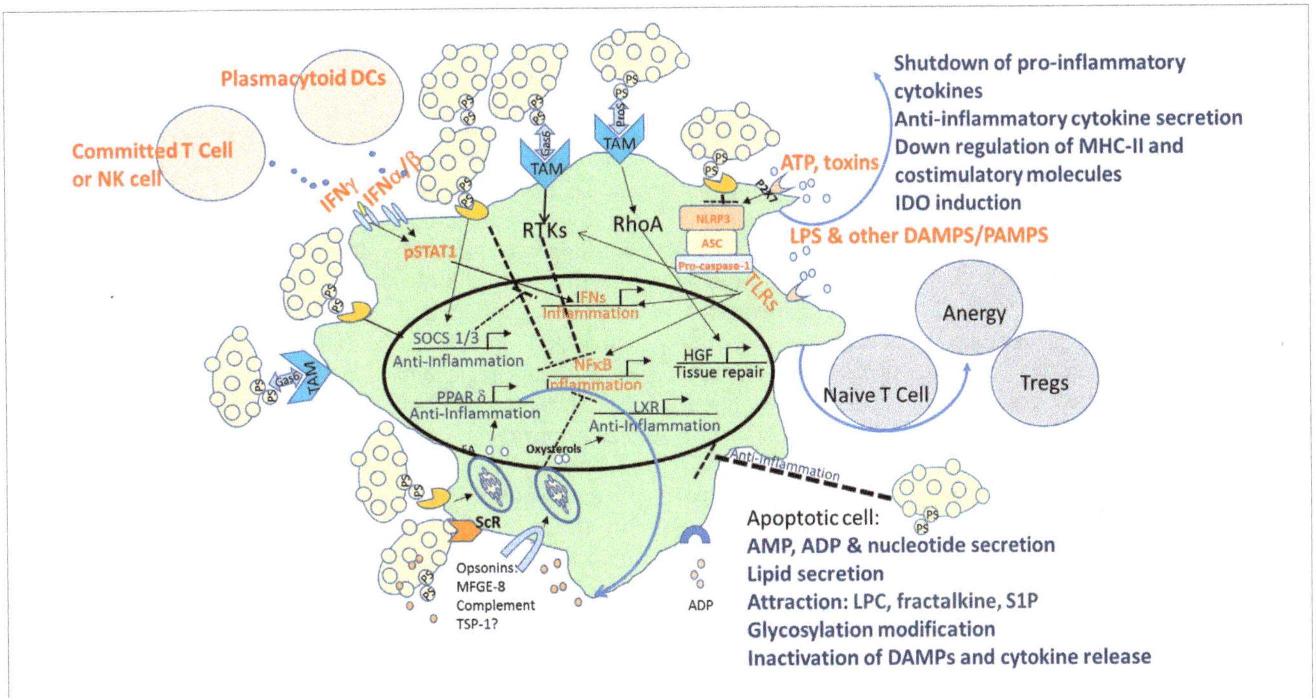

FIGURE 1 | Multiple mechanisms of immune modulation following interaction with apoptotic cells in macrophages and dendritic cells (DCs). Multiple mechanisms are used by apoptotic cells to create an immune homeostatic anti-inflammatory state in macrophages and DCs. In apoptotic cells themselves, in parallel with PtdSer exposure, caspase activation plays a critical role by deactivating potential danger-associated molecular patterns (DAMPs) and by releasing "find me" signals such as adenosine monophosphate (AMP), lysophosphatidylcholine (LPC), fractalkine, and sphingosine-1-phosphate (S1P). Apoptotic cells possess a direct immunosuppressive effect by the release of "calming" agents TGF-β, IL-10, adenosine diphosphate (ADP), thrombospondin-1 (TSP-1), and more. Direct binding to PtdSer receptors (PtsR) and indirect binding to TAM receptors, as well as signaling *via* opsonins/bridging molecules that use additional integrinsor scavenger receptors (ScRs) or complement receptors, act to reprogram the phagocyte, to inhibit toll-like receptors (TLRs), nuclear factor κB (NF-κB), signal transducer and activator of transcription 1 (STAT1), and interferon (IFN) signaling, and to activate liver X receptor (LXR), peroxisome proliferator-activated receptor delta (PPAR-δ), suppressors of cytokine signaling (SOCS) 1/3, and hepatic growth factor (HGF), and to downregulate costimulation and induce induction of indoleamine 2,3 dioxygenase-1 (IDO), that promote tolerogenic phenotype and the induction of T-regulatory (Treg) cell differentiation. The sum of these events leads to downregulation of the inflammatory characteristics of macrophages and DCs, repair, and peripheral tolerance. Pro-inflammatory pattern are marked in red and anti-inflammatory patterns in blue.

infarct size after acute myocardial infarction have all been treated quite successfully by apoptotic cell infusion (175). While these works have been performed in animal models, our group has shown a remarkable reduction in the occurrence of grade II–IV GvHD following heterologous hematopoietic stem cell transplantation in humans with apoptotic cell treatment (176).

In the IBD study mentioned earlier (86), a single infusion of apoptotic cells significantly ameliorated both the clinical score and histological appearance of DSS-induced colitis. We showed that apoptotic cell infusion is beneficial in murine models of IBD and inhibits both inflammasome- and NF-κB-dependent inflammation.

In another example of the use of apoptotic cells for the treatment of inflammatory conditions in post-ischemic cerebral injury in mice, MFGE8 deficiency was associated with enhanced IL-1β production and larger infarct size. The latter was abolished after treatment with IL-1β receptor antagonist. MFGE8 supplementation significantly dampened caspase-1 activation and IL-1β production and reduced infarct size in WT mice, but did not limit cerebral necrosis in *IL-1β*-, *Itgβ3*-, or *P2rx7*-deficient animals (91).

Lipopolysaccharide is a main causative agent of Gram-negative bacterial septic shock. Ren et al. (177) examined the possible protective effect of apoptotic cell infusion. They found that when apoptotic cells were administered 24 h after LPS challenge, B6 mice benefited, with a reduction in circulating pro-inflammatory cytokines, suppression of polymorphonuclear neutrophil infiltration in target organs, decreased serum LPS levels, and decreased mortality. Interestingly, LPS can quickly bind to apoptotic cells and these LPS-coated apoptotic cells can be recognized and cleared by macrophages accompanied with suppression of TNFα and enhancement of IL-10 expression. LPS-treated mice began to die at 8–12 h and all mice died within 3 days in the control group. By contrast, mice in the group injected intravenously (IV) with apoptotic cells (1×10^7/mouse) immediately after challenge with LPS exhibited fewer signs of sickness. Only 20% of treated mice died at day 7; that is, treatment with apoptotic cells resulted in 80% survival ($n = 12$, $p < 0.001$). Late deaths in the treatment group were not observed during the 3 weeks after LPS injection, indicating that apoptotic cell treatment conferred a complete and lasting protection against lethal endotoxemia.

To further examine whether the administration of apoptotic cells has a beneficial effect in another animal model, the authors induced sepsis in mice by cecal ligation and puncture (CLP) (177). Without any treatment, 52% of the mice (12 of 25) died within 5 days. Mice with apoptotic cell treatment 1 h after CLP exhibited fewer signs of sickness and less than 20% of treated mice (4 of 22) died in the first 3 days. The authors also investigated whether delayed administration of apoptotic cells would prevent mice from endotoxic lethality. Treatment with apoptotic cells was initiated 1, 3, 6, and 24 h, respectively, after the onset of endotoxemia. Delayed treatments at all time points significantly protected mice from lethal shock ($n = 6$/group, $p < 0.05$). No late deaths occurred during the subsequent 3-week period of observation. These results indicated that delayed administration of apoptotic cells in mice provided protection from LPS-induced lethal shock.

Other works showed that timing is important, and if apoptotic cells were given 5 days before sepsis induction, worsened survival was observed (178). Sepsis and septic conditions were also examined *in vivo* in *Mertk*$^{-/-}$ mice, by Camenisch et al. (118), who used *Mertk*$^{-/-}$ mice. *In vivo*, the LD_{50} of LPS for *Mertk*$^{-/-}$ mice was half of that for WT mice. *Mertk*$^{-/-}$ mice had an approximately threefold enhancement of serum TNFα when administered 100 mg/kg of LPS, and about 90% of mice died of endotoxic shock (118).

Taken together, these studies suggest that the best timing for apoptotic cell treatment during sepsis is after its onset. The treatment response mechanism is most probably a systemic increase in the ability to return to a homeostatic state and a reduction of the intensity of the initial unwanted immune response characterized by cytokine storm. Other possible mechanisms by which apoptotic cells could provide protection in sepsis include binding to toxins, promoting APC survival, and recovering APCs from their septic "reprogramming" (179, 180).

SUMMARY

Multiple mechanisms are used by apoptotic cells to create an immune homeostatic anti-inflammatory state in macrophages and DCs. As illustrated in **Figure 1**, these include direct binding to PtdSer and indirect binding to TAM receptors, as well as signaling *via* opsonins/bridging molecules that use additional integrins and ScRs to inhibit TLRs as well as NF-κb, STAT1, and IFN signaling, and to activate LXR, SOCS 1/3, PPAR-δ, and hepatic growth factor (HGF). The sum of these events leads to downregulation of the inflammatory characteristics of macrophages and DCs, repair, and peripheral tolerance. Despite establishment of a pattern recognition effect in the clearance of apoptotic cells, some controversies in the field exist, including the role of complement, the importance of different receptors for PtdSer, the "cross-talk" between different opsonins and receptors, and the specific conditions where "immunogenic" clearance is occurring.

AUTHOR CONTRIBUTIONS

DM: prepared the manuscript. UT: Helped prepare the manuscript.

FUNDING

Funding for the different scientific projects mentioned in this review was provided by the Israeli Science Foundation to DM. No specific source was provided for this specific manuscript.

REFERENCES

1. Botto M, Dell'Agnola C, Bygrave AE, Thompson EM, Cook HT, Petry F, et al. Homozygous C1q deficiency causes glomerulonephritis associated with multiple apoptotic bodies. *Nat Genet* (1998) 19(1):56–9. doi:10.1038/ng0598-56
2. Mevorach D, Zhou JL, Song X, Elkon KB. Systemic exposure to irradiated apoptotic cells induces autoantibody production. *J Exp Med* (1998) 188(2): 387–92. doi:10.1084/jem.188.2.387
3. Munoz LE, Lauber K, Schiller M, Manfredi AA, Herrmann M. The role of defective clearance of apoptotic cells in systemic autoimmunity. *Nat Rev Rheumatol* (2010) 6(5):280–9. doi:10.1038/nrrheum.2010.46
4. Asano K, Miwa M, Miwa K, Hanayama R, Nagase H, Nagata S, et al. Masking of phosphatidylserine inhibits apoptotic cell engulfment and induces autoantibody production in mice. *J Exp Med* (2004) 200(4):459–67. doi:10.1084/jem.20040342
5. Shoshan Y, Shapira I, Toubi E, Frolkis I, Yaron M, Mevorach D. Accelerated Fas-mediated apoptosis of monocytes and maturing macrophages from patients with systemic lupus erythematosus: relevance to in vitro impairment of interaction with iC3b-opsonized apoptotic cells. *J Immunol* (2001) 167(10):5963–9. doi:10.4049/jimmunol.167.10.5963
6. Hanayama R, Tanaka M, Miyasaka K, Aozasa K, Koike M, Uchiyama Y, et al. Autoimmune disease and impaired uptake of apoptotic cells in MFG-E8-deficient mice. *Science* (2004) 304(5674):1147–50. doi:10.1126/science.1094359
7. Kawane K, Ohtani M, Miwa K, Kizawa T, Kanbara Y, Yoshioka Y, et al. Chronic polyarthritis caused by mammalian DNA that escapes from degradation in macrophages. *Nature* (2006) 443(7114):998–1002. doi:10.1038/nature05245
8. Kawane K, Tanaka H, Kitahara Y, Shimaoka S, Nagata S. Cytokine-dependent but acquired immunity-independent arthritis caused by DNA escaped from degradation. *Proc Natl Acad Sci U S A* (2010) 107(45):19432–7. doi:10.1073/pnas.1010603107
9. Verbovetski I, Bychkov H, Trahtemberg U, Shapira I, Hareuveni M, Ben-Tal O, et al. Opsonization of apoptotic cells by autologous iC3b facilitates clearance by immature dendritic cells, down-regulates DR and CD86, and up-regulates CC chemokine receptor 7. *J Exp Med* (2002) 196(12):1553–61. doi:10.1084/jem.20020263
10. Iyoda T, Shimoyama S, Liu K, Omatsu Y, Akiyama Y, Maeda Y, et al. The CD8+ dendritic cell subset selectively endocytoses dying cells in culture and in vivo. *J Exp Med* (2002) 195(10):1289–302. doi:10.1084/jem.20020161
11. Liu K, Iyoda T, Saternus M, Kimura Y, Inaba K, Steinman RM. Immune tolerance after delivery of dying cells to dendritic cells in situ. *J Exp Med* (2002) 196(8):1091–7. doi:10.1084/jem.20021215
12. Mevorach D. The role of death-associated molecular patterns in the pathogenesis of systemic lupus erythematosus. *Rheum Dis Clin North Am* (2004) 30(3):487–504, viii. doi:10.1016/j.rdc.2004.04.011
13. Fuchs Y, Steller H. Programmed cell death in animal development and disease. *Cell* (2011) 147(4):742–58. doi:10.1016/j.cell.2011.10.033
14. Nagata S, Hanayama R, Kawane K. Autoimmunity and the clearance of dead cells. *Cell* (2010) 140(5):619–30. doi:10.1016/j.cell.2010.02.014
15. Ravichandran KS. Beginnings of a good apoptotic meal: the find-me and eat-me signaling pathways. *Immunity* (2011) 35(4):445–55. doi:10.1016/j.immuni.2011.09.004
16. Fadok VA, de Cathelineau A, Daleke DL, Henson PM, Bratton DL. Loss of phospholipid asymmetry and surface exposure of phosphatidylserine is required for phagocytosis of apoptotic cells by macrophages and fibroblasts. *J Biol Chem* (2001) 276(2):1071–7. doi:10.1074/jbc.M003649200
17. Leventis PA, Grinstein S. The distribution and function of phosphatidylserine in cellular membranes. *Annu Rev Biophys* (2010) 39:407–27. doi:10.1146/annurev.biophys.093008.131234
18. Suzuki J, Denning DP, Imanishi E, Horvitz HR, Nagata S. Xk-related protein 8 and CED-8 promote phosphatidylserine exposure in apoptotic cells. *Science* (2013) 341(6144):403–6. doi:10.1126/science.1236758
19. Janssen WJ, Bratton DL, Jakubzick CV, Henson PM. Myeloid cell turnover and clearance. *Microbiol Spectr* (2016) 4(6):1–24. doi:10.1128/microbiolspec.MCHD-0005-2015
20. Ravichandran KS. Find-me and eat-me signals in apoptotic cell clearance: progress and conundrums. *J Exp Med* (2010) 207(9):1807–17. doi:10.1084/jem.20101157
21. Panaretakis T, Kepp O, Brockmeier U, Tesniere A, Bjorklund AC, Chapman DC, et al. Mechanisms of pre-apoptotic calreticulin exposure in immunogenic cell death. *EMBO J* (2009) 28(5):578–90. doi:10.1038/emboj.2009.1
22. Lauber K, Bohn E, Krober SM, Xiao YJ, Blumenthal SG, Lindemann RK, et al. Apoptotic cells induce migration of phagocytes via caspase-3-mediated release of a lipid attraction signal. *Cell* (2003) 113(6):717–30. doi:10.1016/S0092-8674(03)00422-7
23. Truman LA, Ford CA, Pasikowska M, Pound JD, Wilkinson SJ, Dumitriu IE, et al. CX3CL1/fractalkine is released from apoptotic lymphocytes to stimulate macrophage chemotaxis. *Blood* (2008) 112(13):5026–36. doi:10.1182/blood-2008-06-162404
24. Gude DR, Alvarez SE, Paugh SW, Mitra P, Yu J, Griffiths R, et al. Apoptosis induces expression of sphingosine kinase 1 to release sphingosine-1-phosphate as a "come-and-get-me" signal. *FASEB J* (2008) 22(8): 2629–38. doi:10.1096/fj.08-107169
25. Elliott MR, Chekeni FB, Trampont PC, Lazarowski ER, Kadl A, Walk SF, et al. Nucleotides released by apoptotic cells act as a find-me signal to promote phagocytic clearance. *Nature* (2009) 461(7261):282–6. doi:10.1038/nature08296
26. Krispin A, Bledi Y, Atallah M, Trahtemberg U, Verbovetski I, Nahari E, et al. Apoptotic cell thrombospondin-1 and heparin-binding domain lead to dendritic-cell phagocytic and tolerizing states. *Blood* (2006) 108(10): 3580–9. doi:10.1182/blood-2006-03-013334
27. Yamaguchi H, Maruyama T, Urade Y, Nagata S. Immunosuppression via adenosine receptor activation by adenosine monophosphate released from apoptotic cells. *Elife* (2014) 3:e02172. doi:10.7554/eLife.02172
28. Kazama H, Ricci JE, Herndon JM, Hoppe G, Green DR, Ferguson TA. Induction of immunological tolerance by apoptotic cells requires caspase-dependent oxidation of high-mobility group box-1 protein. *Immunity* (2008) 29(1):21–32. doi:10.1016/j.immuni.2008.05.013
29. Rongvaux A, Jackson R, Harman CC, Li T, West AP, de Zoete MR, et al. Apoptotic caspases prevent the induction of type I interferons by mitochondrial DNA. *Cell* (2014) 159(7):1563–77. doi:10.1016/j.cell.2014.11.037
30. White MJ, McArthur K, Metcalf D, Lane RM, Cambier JC, Herold MJ, et al. Apoptotic caspases suppress mtDNA-induced STING-mediated type I IFN production. *Cell* (2014) 159(7):1549–62. doi:10.1016/j.cell.2014.11.036
31. Green DR, Ferguson T, Zitvogel L, Kroemer G. Immunogenic and tolerogenic cell death. *Nat Rev Immunol* (2009) 9(5):353–63. doi:10.1038/nri2545
32. Bondanza A, Zimmermann VS, Rovere-Querini P, Turnay J, Dumitriu IE, Stach CM, et al. Inhibition of phosphatidylserine recognition heightens the immunogenicity of irradiated lymphoma cells in vivo. *J Exp Med* (2004) 200(9):1157–65. doi:10.1084/jem.20040327
33. Gaipl US, Kuenkele S, Voll RE, Beyer TD, Kolowos W, Heyder P, et al. Complement binding is an early feature of necrotic and a rather late event during apoptotic cell death. *Cell Death Differ* (2001) 8(4):327–34. doi:10.1038/sj.cdd.4400826
34. Ramirez-Ortiz ZG, Pendergraft WF III, Prasad A, Byrne MH, Iram T, Blanchette CJ, et al. The scavenger receptor SCARF1 mediates the clearance of apoptotic cells and prevents autoimmunity. *Nat Immunol* (2013) 14(9):917–26. doi:10.1038/ni.2670
35. Son M, Santiago-Schwarz F, Al-Abed Y, Diamond B. C1q limits dendritic cell differentiation and activation by engaging LAIR-1. *Proc Natl Acad Sci U S A* (2012) 109(46):E3160–7. doi:10.1073/pnas.1212753109
36. Morelli AE, Larregina AT, Shufesky WJ, Zahorchak AF, Logar AJ, Papworth GD, et al. Internalization of circulating apoptotic cells by splenic marginal zone dendritic cells: dependence on complement receptors and effect on cytokine production. *Blood* (2003) 101(2):611–20. doi:10.1182/blood-2002-06-1769
37. Amarilyo G, Verbovetski I, Atallah M, Grau A, Wiser G, Gil O, et al. iC3b-opsonized apoptotic cells mediate a distinct anti-inflammatory response and transcriptional NF-kappaB-dependent blockade. *Eur J Immunol* (2010) 40(3):699–709. doi:10.1002/eji.200838951

38. Galvan MD, Foreman DB, Zeng E, Tan JC, Bohlson SS. Complement component C1q regulates macrophage expression of Mer tyrosine kinase to promote clearance of apoptotic cells. *J Immunol* (2012) 188(8):3716–23. doi:10.4049/jimmunol.1102920
39. Colonna L, Lood C, Elkon KB. Beyond apoptosis in lupus. *Curr Opin Rheumatol* (2014) 26(5):459–66. doi:10.1097/BOR.0000000000000083
40. Miksa M, Amin D, Wu R, Ravikumar TS, Wang P. Fractalkine-induced MFG-E8 leads to enhanced apoptotic cell clearance by macrophages. *Mol Med* (2007) 13(11–12):553–60. doi:10.2119/2007-00019.Miksa
41. Maceyka M, Harikumar KB, Milstien S, Spiegel S. Sphingosine-1-phosphate signaling and its role in disease. *Trends Cell Biol* (2012) 22(1):50–60. doi:10.1016/j.tcb.2011.09.003
42. Meyer zu Heringdorf D, Jakobs KH. Lysophospholipid receptors: signalling, pharmacology and regulation by lysophospholipid metabolism. *Biochim Biophys Acta* (2007) 1768(4):923–40. doi:10.1016/j.bbamem.2006.09.026
43. Trautmann A. Extracellular ATP in the immune system: more than just a "danger signal". *Sci Signal* (2009) 2(56):e6. doi:10.1126/scisignal.256pe6
44. Kepp O, Senovilla L, Vitale I, Vacchelli E, Adjemian S, Agostinis P, et al. Consensus guidelines for the detection of immunogenic cell death. *Oncoimmunology* (2014) 3(9):e955691. doi:10.4161/21624011.2014.955691
45. Nagata S, Tanaka M. Programmed cell death and the immune system. *Nat Rev Immunol* (2017) 17(5):333–40. doi:10.1038/nri.2016.153
46. Medzhitov R, Janeway C Jr. Innate immunity. *N Engl J Med* (2000) 343(5):338–44. doi:10.1056/NEJM200008033430506
47. Doyle SL, O'Neill LA. Toll-like receptors: from the discovery of NFkappaB to new insights into transcriptional regulations in innate immunity. *Biochem Pharmacol* (2006) 72(9):1102–13. doi:10.1016/j.bcp.2006.07.010
48. Pasare C, Medzhitov R. Toll-like receptors: linking innate and adaptive immunity. *Adv Exp Med Biol* (2005) 560:11–8. doi:10.1007/0-387-24180-9_2
49. Fadok VA, Bratton DL, Konowal A, Freed PW, Westcott JY, Henson PM. Macrophages that have ingested apoptotic cells in vitro inhibit proinflammatory cytokine production through autocrine/paracrine mechanisms involving TGF-beta, PGE2, and PAF. *J Clin Invest* (1998) 101(4):890–8. doi:10.1172/JCI1112
50. McDonald PP, Fadok VA, Bratton D, Henson PM. Transcriptional and translational regulation of inflammatory mediator production by endogenous TGF-beta in macrophages that have ingested apoptotic cells. *J Immunol* (1999) 163(11):6164–72.
51. Stuart LM, Lucas M, Simpson C, Lamb J, Savill J, Lacy-Hulbert A. Inhibitory effects of apoptotic cell ingestion upon endotoxin-driven myeloid dendritic cell maturation. *J Immunol* (2002) 168(4):1627–35. doi:10.4049/jimmunol.168.4.1627
52. Brubaker SW, Bonham KS, Zanoni I, Kagan JC. Innate immune pattern recognition: a cell biological perspective. *Annu Rev Immunol* (2015) 33:257–90. doi:10.1146/annurev-immunol-032414-112240
53. Lin SC, Lo YC, Wu H. Helical assembly in the MyD88-IRAK4-IRAK2 complex in TLR/IL-1R signalling. *Nature* (2010) 465(7300):885–90. doi:10.1038/nature09121
54. Burns K, Janssens S, Brissoni B, Olivos N, Beyaert R, Tschopp J. Inhibition of interleukin 1 receptor/Toll-like receptor signaling through the alternatively spliced, short form of MyD88 is due to its failure to recruit IRAK-4. *J Exp Med* (2003) 197(2):263–8. doi:10.1084/jem.20021790
55. Rao N, Nguyen S, Ngo K, Fung-Leung WP. A novel splice variant of interleukin-1 receptor (IL-1R)-associated kinase 1 plays a negative regulatory role in Toll/IL-1R-induced inflammatory signaling. *Mol Cell Biol* (2005) 25(15):6521–32. doi:10.1128/MCB.25.15.6521-6532.2005
56. Kobayashi K, Hernandez LD, Galan JE, Janeway CA Jr, Medzhitov R, Flavell RA. IRAK-M is a negative regulator of Toll-like receptor signaling. *Cell* (2002) 110(2):191–202. doi:10.1016/S0092-8674(02)00827-9
57. Zhou H, Yu M, Fukuda K, Im J, Yao P, Cui W, et al. IRAK-M mediates Toll-like receptor/IL-1R-induced NFkappaB activation and cytokine production. *EMBO J* (2013) 32(4):583–96. doi:10.1038/emboj.2013.2
58. Voll RE, Herrmann M, Roth EA, Stach C, Kalden JR, Girkontaite I. Immunosuppressive effects of apoptotic cells. *Nature* (1997) 390(6658):350–1. doi:10.1038/37022
59. Tassiulas I, Park-Min KH, Hu Y, Kellerman L, Mevorach D, Ivashkiv LB. Apoptotic cells inhibit LPS-induced cytokine and chemokine production and IFN responses in macrophages. *Hum Immunol* (2007) 68(3):156–64. doi:10.1016/j.humimm.2006.12.008
60. Perruche S, Saas P, Chen W. Apoptotic cell-mediated suppression of streptococcal cell wall-induced arthritis is associated with alteration of macrophage function and local regulatory T-cell increase: a potential cell-based therapy? *Arthritis Res Ther* (2009) 11(4):R104. doi:10.1186/ar2750
61. Mahajan A, Herrmann M, Munoz LE. Clearance deficiency and cell death pathways: a model for the pathogenesis of SLE. *Front Immunol* (2016) 7:35. doi:10.3389/fimmu.2016.00035
62. Marshak-Rothstein A. Toll-like receptors in systemic autoimmune disease. *Nat Rev Immunol* (2006) 6(11):823–35. doi:10.1038/nri1957
63. Takeuchi O, Akira S. Pattern recognition receptors and inflammation. *Cell* (2010) 140(6):805–20. doi:10.1016/j.cell.2010.01.022
64. Wellen KE, Hotamisligil GS. Inflammation, stress, and diabetes. *J Clin Invest* (2005) 115(5):1111–9. doi:10.1172/JCI25102
65. Zhang Q, Lenardo MJ, Baltimore D. 30 Years of NF-kappaB: a blossoming of relevance to human pathobiology. *Cell* (2017) 168(1–2):37–57. doi:10.1016/j.cell.2016.12.012
66. Hayden MS, Ghosh S. Shared principles in NF-kappaB signaling. *Cell* (2008) 132(3):344–62. doi:10.1016/j.cell.2008.01.020
67. Hansen SK, Baeuerle PA, Blasi F. Purification, reconstitution, and I kappa B association of the c-Rel-p65 (RelA) complex, a strong activator of transcription. *Mol Cell Biol* (1994) 14(4):2593–603. doi:10.1128/MCB.14.4.2593
68. Bohuslav J, Kravchenko VV, Parry GC, Erlich JH, Gerondakis S, Mackman N, et al. Regulation of an essential innate immune response by the p50 subunit of NF-kappaB. *J Clin Invest* (1998) 102(9):1645–52. doi:10.1172/JCI3877
69. Ishikawa H, Claudio E, Dambach D, Raventos-Suarez C, Ryan C, Bravo R. Chronic inflammation and susceptibility to bacterial infections in mice lacking the polypeptide (p)105 precursor (NF-kappaB1) but expressing p50. *J Exp Med* (1998) 187(7):985–96. doi:10.1084/jem.187.7.985
70. Gyrd-Hansen M, Meier P. IAPs: from caspase inhibitors to modulators of NF-kappaB, inflammation and cancer. *Nat Rev Cancer* (2010) 10(8):561–74. doi:10.1038/nrc2889
71. Gilmore TD. Introduction to NF-kappaB: players, pathways, perspectives. *Oncogene* (2006) 25(51):6680–4. doi:10.1038/sj.onc.1209954
72. Karin M, Greten FR. NF-kappaB: linking inflammation and immunity to cancer development and progression. *Nat Rev Immunol* (2005) 5(10):749–59. doi:10.1038/nri1703
73. Chen L, Fischle W, Verdin E, Greene WC. Duration of nuclear NF-kappaB action regulated by reversible acetylation. *Science* (2001) 293(5535):1653–7. doi:10.1126/science.1062374
74. Israel A. The IKK complex, a central regulator of NF-kappaB activation. *Cold Spring Harb Perspect Biol* (2010) 2(3):a000158. doi:10.1101/cshperspect.a000158
75. Harhaj EW, Dixit VM. Regulation of NF-kappaB by deubiquitinases. *Immunol Rev* (2012) 246(1):107–24. doi:10.1111/j.1600-065X.2012.01100.x
76. Sun SC. Non-canonical NF-kappaB signaling pathway. *Cell Res* (2011) 21(1):71–85. doi:10.1038/cr.2010.177
77. Lee EG, Boone DL, Chai S, Libby SL, Chien M, Lodolce JP, et al. Failure to regulate TNF-induced NF-kappaB and cell death responses in A20-deficient mice. *Science* (2000) 289(5488):2350–4. doi:10.1126/science.289.5488.2350
78. Solt LA, May MJ. The IkappaB kinase complex: master regulator of NF-kappaB signaling. *Immunol Res* (2008) 42(1–3):3–18. doi:10.1007/s12026-008-8025-1
79. Kool M, van Loo G, Waelput W, De Prijck S, Muskens F, Sze M, et al. The ubiquitin-editing protein A20 prevents dendritic cell activation, recognition of apoptotic cells, and systemic autoimmunity. *Immunity* (2011) 35(1):82–96. doi:10.1016/j.immuni.2011.05.013
80. Heger K, Fierens K, Vahl JC, Aszodi A, Peschke K, Schenten D, et al. A20-deficient mast cells exacerbate inflammatory responses in vivo. *PLoS Biol* (2014) 12(1):e1001762. doi:10.1371/journal.pbio.1001762
81. Hammer GE, Turer EE, Taylor KE, Fang CJ, Advincula R, Oshima S, et al. Expression of A20 by dendritic cells preserves immune homeostasis and prevents colitis and spondyloarthritis. *Nat Immunol* (2011) 12(12):1184–93. doi:10.1038/ni.2135

82. Zhou Q, Wang H, Schwartz DM, Stoffels M, Park YH, Zhang Y, et al. Loss-of-function mutations in TNFAIP3 leading to A20 haploinsufficiency cause an early-onset autoinflammatory disease. *Nat Genet* (2016) 48(1): 67–73. doi:10.1038/ng.3459
83. Mele A, Cervantes JR, Chien V, Friedman D, Ferran C. Single nucleotide polymorphisms at the TNFAIP3/A20 locus and susceptibility/resistance to inflammatory and autoimmune diseases. *Adv Exp Med Biol* (2014) 809: 163–83. doi:10.1007/978-1-4939-0398-6_10
84. Cummings RJ, Barbet G, Bongers G, Hartmann BM, Gettler K, Muniz L, et al. Different tissue phagocytes sample apoptotic cells to direct distinct homeostasis programs. *Nature* (2016) 539(7630):565–9. doi:10.1038/nature20138
85. Blander JM. The many ways tissue phagocytes respond to dying cells. *Immunol Rev* (2017) 277(1):158–73. doi:10.1111/imr.12537
86. Grau A, Tabib A, Grau I, Reiner I, Mevorach D. Apoptotic cells induce NF-kappaB and inflammasome negative signaling. *PLoS One* (2015) 10(3):e0122440. doi:10.1371/journal.pone.0122440
87. Cvetanovic M, Mitchell JE, Patel V, Avner BS, Su Y, van der Saag PT, et al. Specific recognition of apoptotic cells reveals a ubiquitous and unconventional innate immunity. *J Biol Chem* (2006) 281(29):20055–67. doi:10.1074/jbc.M603920200
88. Lucas M, Stuart LM, Zhang A, Hodivala-Dilke K, Febbraio M, Silverstein R, et al. Requirements for apoptotic cell contact in regulation of macrophage responses. *J Immunol* (2006) 177(6):4047–54. doi:10.4049/jimmunol.177.6.4047
89. Rothlin CV, Ghosh S, Zuniga EI, Oldstone MB, Lemke G. TAM receptors are pleiotropic inhibitors of the innate immune response. *Cell* (2007) 131(6):1124–36. doi:10.1016/j.cell.2007.10.034
90. Schroder K, Tschopp J. The inflammasomes. *Cell* (2010) 140(6):821–32. doi:10.1016/j.cell.2010.01.040
91. Deroide N, Li X, Lerouet D, Van Vre E, Baker L, Harrison J, et al. MFGE8 inhibits inflammasome-induced IL-1beta production and limits postischemic cerebral injury. *J Clin Invest* (2013) 123(3):1176–81. doi:10.1172/JCI65167
92. Johann AM, von Knethen A, Lindemann D, Brune B. Recognition of apoptotic cells by macrophages activates the peroxisome proliferator-activated receptor-gamma and attenuates the oxidative burst. *Cell Death Differ* (2006) 13(9):1533–40. doi:10.1038/sj.cdd.4401832
93. Serinkan BF, Gambelli F, Potapovich AI, Babu H, Di Giuseppe M, Ortiz LA, et al. Apoptotic cells quench reactive oxygen and nitrogen species and modulate TNF-alpha/TGF-beta1 balance in activated macrophages: involvement of phosphatidylserine-dependent and -independent pathways. *Cell Death Differ* (2005) 12(8):1141–4. doi:10.1038/sj.cdd.4401619
94. Fubini B, Hubbard A. Reactive oxygen species (ROS) and reactive nitrogen species (RNS) generation by silica in inflammation and fibrosis. *Free Radic Biol Med* (2003) 34(12):1507–16. doi:10.1016/S0891-5849(03)00149-7
95. Hwang YS, Jeong M, Park JS, Kim MH, Lee DB, Shin BA, et al. Interleukin-1beta stimulates IL-8 expression through MAP kinase and ROS signaling in human gastric carcinoma cells. *Oncogene* (2004) 23(39):6603–11. doi:10.1038/sj.onc.1207867
96. Bauer C, Duewell P, Mayer C, Lehr HA, Fitzgerald KA, Dauer M, et al. Colitis induced in mice with dextran sulfate sodium (DSS) is mediated by the NLRP3 inflammasome. *Gut* (2010) 59(9):1192–9. doi:10.1136/gut.2009.197822
97. Kwon KH, Ohigashi H, Murakami A. Dextran sulfate sodium enhances interleukin-1 beta release via activation of p38 MAPK and ERK1/2 pathways in murine peritoneal macrophages. *Life Sci* (2007) 81(5):362–71. doi:10.1016/j.lfs.2007.05.022
98. Hornung V, Bauernfeind F, Halle A, Samstad EO, Kono H, Rock KL, et al. Silica crystals and aluminum salts activate the NALP3 inflammasome through phagosomal destabilization. *Nat Immunol* (2008) 9(8):847–56. doi:10.1038/ni.1631
99. Mariathasan S, Weiss DS, Newton K, McBride J, O'Rourke K, Roose-Girma M, et al. Cryopyrin activates the inflammasome in response to toxins and ATP. *Nature* (2006) 440(7081):228–32. doi:10.1038/nature04515
100. Allen IC. Non-inflammasome forming NLRs in inflammation and tumorigenesis. *Front Immunol* (2014) 5:169. doi:10.3389/fimmu.2014.00169
101. Saitoh T, Fujita N, Jang MH, Uematsu S, Yang BG, Satoh T, et al. Loss of the autophagy protein Atg16L1 enhances endotoxin-induced IL-1beta production. *Nature* (2008) 456(7219):264–8. doi:10.1038/nature07383
102. Rigamonti E, Chinetti-Gbaguidi G, Staels B. Regulation of macrophage functions by PPAR-alpha, PPAR-gamma, and LXRs in mice and men. *Arterioscler Thromb Vasc Biol* (2008) 28(6):1050–9. doi:10.1161/ATVBAHA.107.158998
103. Han CZ, Ravichandran KS. Metabolic connections during apoptotic cell engulfment. *Cell* (2011) 147(7):1442–5. doi:10.1016/j.cell.2011.12.006
104. Kiss M, Czimmerer Z, Nagy L. The role of lipid-activated nuclear receptors in shaping macrophage and dendritic cell function: from physiology to pathology. *J Allergy Clin Immunol* (2013) 132(2):264–86. doi:10.1016/j.jaci.2013.05.044
105. Mukundan L, Odegaard JI, Morel CR, Heredia JE, Mwangi JW, Ricardo-Gonzalez RR, et al. PPAR-delta senses and orchestrates clearance of apoptotic cells to promote tolerance. *Nat Med* (2009) 15(11):1266–72. doi:10.1038/nm.2048
106. A-Gonzalez N, Bensinger SJ, Hong C, Beceiro S, Bradley MN, Zelcer N, et al. Apoptotic cells promote their own clearance and immune tolerance through activation of the nuclear receptor LXR. *Immunity* (2009) 31(2): 245–58. doi:10.1016/j.immuni.2009.06.018
107. Kidani Y, Bensinger SJ. Liver X receptor and peroxisome proliferator-activated receptor as integrators of lipid homeostasis and immunity. *Immunol Rev* (2012) 249(1):72–83. doi:10.1111/j.1600-065X.2012.01153.x
108. Crawford SE, Stellmach V, Murphy-Ullrich JE, Ribeiro SM, Lawler J, Hynes RO, et al. Thrombospondin-1 is a major activator of TGF-beta1 in vivo. *Cell* (1998) 93(7):1159–70. doi:10.1016/S0092-8674(00)81460-9
109. Bonta PI, van Tiel CM, Vos M, Pols TW, van Thienen JV, Ferreira V, et al. Nuclear receptors Nur77, Nurr1, and NOR-1 expressed in atherosclerotic lesion macrophages reduce lipid loading and inflammatory responses. *Arterioscler Thromb Vasc Biol* (2006) 26(10):2288–94. doi:10.1161/01.ATV.0000238346.84458.5d
110. Saijo K, Winner B, Carson CT, Collier JG, Boyer L, Rosenfeld MG, et al. A Nurr1/CoREST pathway in microglia and astrocytes protects dopaminergic neurons from inflammation-induced death. *Cell* (2009) 137(1): 47–59. doi:10.1016/j.cell.2009.01.038
111. Ricci JE, Munoz-Pinedo C, Fitzgerald P, Bailly-Maitre B, Perkins GA, Yadava N, et al. Disruption of mitochondrial function during apoptosis is mediated by caspase cleavage of the p75 subunit of complex I of the electron transport chain. *Cell* (2004) 117(6):773–86. doi:10.1016/j.cell.2004.05.008
112. Chekeni FB, Elliott MR, Sandilos JK, Walk SF, Kinchen JM, Lazarowski ER, et al. Pannexin 1 channels mediate "find-me" signal release and membrane permeability during apoptosis. *Nature* (2010) 467(7317):863–7. doi:10.1038/nature09413
113. Bossy-Wetzel E, Newmeyer DD, Green DR. Mitochondrial cytochrome c release in apoptosis occurs upstream of DEVD-specific caspase activation and independently of mitochondrial transmembrane depolarization. *EMBO J* (1998) 17(1):37–49. doi:10.1093/emboj/17.1.37
114. Eguchi Y, Shimizu S, Tsujimoto Y. Intracellular ATP levels determine cell death fate by apoptosis or necrosis. *Cancer Res* (1997) 57(10):1835–40.
115. Zamaraeva MV, Sabirov RZ, Maeno E, Ando-Akatsuka Y, Bessonova SV, Okada Y. Cells die with increased cytosolic ATP during apoptosis: a bioluminescence study with intracellular luciferase. *Cell Death Differ* (2005) 12(11):1390–7. doi:10.1038/sj.cdd.4401661
116. Gleiss B, Gogvadze V, Orrenius S, Fadeel B. Fas-triggered phosphatidylserine exposure is modulated by intracellular ATP. *FEBS Lett* (2002) 519(1–3): 153–8. doi:10.1016/S0014-5793(02)02743-6
117. Vander Heiden MG, Chandel NS, Schumacker PT, Thompson CB. Bcl-xL prevents cell death following growth factor withdrawal by facilitating mitochondrial ATP/ADP exchange. *Mol Cell* (1999) 3(2):159–67. doi:10.1016/S1097-2765(00)80307-X
118. Camenisch TD, Koller BH, Earp HS, Matsushima GK. A novel receptor tyrosine kinase, Mer, inhibits TNF-alpha production and lipopolysaccharide-induced endotoxic shock. *J Immunol* (1999) 162(6):3498–503.
119. Lu Q, Lemke G. Homeostatic regulation of the immune system by receptor tyrosine kinases of the Tyro 3 family. *Science* (2001) 293(5528):306–11. doi:10.1126/science.1061663

120. Lai C, Lemke G. An extended family of protein-tyrosine kinase genes differentially expressed in the vertebrate nervous system. *Neuron* (1991) 6(5):691–704. doi:10.1016/0896-6273(91)90167-X
121. Crosier PS, Lewis PM, Hall LR, Vitas MR, Morris CM, Beier DR, et al. Isolation of a receptor tyrosine kinase (DTK) from embryonic stem cells: structure, genetic mapping and analysis of expression. *Growth Factors* (1994) 11(2):125–36. doi:10.3109/08977199409001054
122. Fujimoto J, Yamamoto T. brt, a mouse gene encoding a novel receptor-type protein-tyrosine kinase, is preferentially expressed in the brain. *Oncogene* (1994) 9(3):693–8.
123. Bour-Jordan H, Esensten JH, Martinez-Llordella M, Penaranda C, Stumpf M, Bluestone JA. Intrinsic and extrinsic control of peripheral T-cell tolerance by costimulatory molecules of the CD28/B7 family. *Immunol Rev* (2011) 241(1):180–205. doi:10.1111/j.1600-065X.2011.01011.x
124. Dai W, Pan H, Hassanain H, Gupta SL, Murphy MJ Jr. Molecular cloning of a novel receptor tyrosine kinase, tif, highly expressed in human ovary and testis. *Oncogene* (1994) 9(3):975–9.
125. Manfioletti G, Brancolini C, Avanzi G, Schneider C. The protein encoded by a growth arrest-specific gene (gas6) is a new member of the vitamin K-dependent proteins related to protein S, a negative coregulator in the blood coagulation cascade. *Mol Cell Biol* (1993) 13(8):4976–85. doi:10.1128/MCB.13.8.4976
126. Anderson HA, Maylock CA, Williams JA, Paweletz CP, Shu H, Shacter E. Serum-derived protein S binds to phosphatidylserine and stimulates the phagocytosis of apoptotic cells. *Nat Immunol* (2003) 4(1):87–91. doi:10.1038/ni871
127. Benzakour O, Kanthou C. The anticoagulant factor, protein S, is produced by cultured human vascular smooth muscle cells and its expression is up-regulated by thrombin. *Blood* (2000) 95(6):2008–14.
128. Hasanbasic I, Rajotte I, Blostein M. The role of gamma-carboxylation in the anti-apoptotic function of gas6. *J Thromb Haemost* (2005) 3(12):2790–7. doi:10.1111/j.1538-7836.2005.01662.x
129. Nakano T, Kawamoto K, Kishino J, Nomura K, Higashino K, Arita H. Requirement of gamma-carboxyglutamic acid residues for the biological activity of Gas6: contribution of endogenous Gas6 to the proliferation of vascular smooth muscle cells. *Biochem J* (1997) 323(Pt 2):387–92. doi:10.1042/bj3230387
130. Lemke G, Rothlin CV. Immunobiology of the TAM receptors. *Nat Rev Immunol* (2008) 8(5):327–36. doi:10.1038/nri2303
131. Gordon S, Taylor PR. Monocyte and macrophage heterogeneity. *Nat Rev Immunol* (2005) 5(12):953–64. doi:10.1038/nri1733
132. Darnell JE Jr, Kerr IM, Stark GR. Jak-STAT pathways and transcriptional activation in response to IFNs and other extracellular signaling proteins. *Science* (1994) 264(5164):1415–21. doi:10.1126/science.8197455
133. Hoshino K, Kaisho T, Iwabe T, Takeuchi O, Akira S. Differential involvement of IFN-beta in Toll-like receptor-stimulated dendritic cell activation. *Int Immunol* (2002) 14(10):1225–31. doi:10.1093/intimm/dxf089
134. Kawai T, Takeuchi O, Fujita T, Inoue J, Muhlradt PF, Sato S, et al. Lipopolysaccharide stimulates the MyD88-independent pathway and results in activation of IFN-regulatory factor 3 and the expression of a subset of lipopolysaccharide-inducible genes. *J Immunol* (2001) 167(10):5887–94. doi:10.4049/jimmunol.167.10.5887
135. Yasukawa H, Sasaki A, Yoshimura A. Negative regulation of cytokine signaling pathways. *Annu Rev Immunol* (2000) 18:143–64. doi:10.1146/annurev.immunol.18.1.143
136. Yoshimura A, Naka T, Kubo M. SOCS proteins, cytokine signalling and immune regulation. *Nat Rev Immunol* (2007) 7(6):454–65. doi:10.1038/nri2093
137. Scutera S, Fraone T, Musso T, Cappello P, Rossi S, Pierobon D, et al. Survival and migration of human dendritic cells are regulated by an IFN-alpha-inducible Axl/Gas6 pathway. *J Immunol* (2009) 183(5):3004–13. doi:10.4049/jimmunol.0804384
138. Sharif MN, Sosic D, Rothlin CV, Kelly E, Lemke G, Olson EN, et al. Twist mediates suppression of inflammation by type I IFNs and Axl. *J Exp Med* (2006) 203(8):1891–901. doi:10.1084/jem.20051725
139. Akira S, Takeda K. Toll-like receptor signalling. *Nat Rev Immunol* (2004) 4(7):499–511. doi:10.1038/nri1391
140. Nikolich-Zugich J, Slifka MK, Messaoudi I. The many important facets of T-cell repertoire diversity. *Nat Rev Immunol* (2004) 4(2):123–32. doi:10.1038/nri1292
141. Hafler DA, Slavik JM, Anderson DE, O'Connor KC, De Jager P, Baecher-Allan C. Multiple sclerosis. *Immunol Rev* (2005) 204:208–31. doi:10.1111/j.0105-2896.2005.00240.x
142. Pulendran B. The varieties of immunological experience: of pathogens, stress, and dendritic cells. *Annu Rev Immunol* (2015) 33:563–606. doi:10.1146/annurev-immunol-020711-075049
143. Collin M, McGovern N, Haniffa M. Human dendritic cell subsets. *Immunology* (2013) 140(1):22–30. doi:10.1111/imm.12117
144. Worbs T, Hammerschmidt SI, Forster R. Dendritic cell migration in health and disease. *Nat Rev Immunol* (2017) 17(1):30–48. doi:10.1038/nri.2016.116
145. Jongbloed SL, Kassianos AJ, McDonald KJ, Clark GJ, Ju X, Angel CE, et al. Human CD141+ (BDCA-3)+ dendritic cells (DCs) represent a unique myeloid DC subset that cross-presents necrotic cell antigens. *J Exp Med* (2010) 207(6):1247–60. doi:10.1084/jem.20092140
146. Poulin LF, Salio M, Griessinger E, Anjos-Afonso F, Craciun L, Chen JL, et al. Characterization of human DNGR-1+ BDCA3+ leukocytes as putative equivalents of mouse CD8alpha+ dendritic cells. *J Exp Med* (2010) 207(6):1261–71. doi:10.1084/jem.20092618
147. Bonnefoy F, Perruche S, Couturier M, Sedrati A, Sun Y, Tiberghien P, et al. Plasmacytoid dendritic cells play a major role in apoptotic leukocyte-induced immune modulation. *J Immunol* (2011) 186(10):5696–705. doi:10.4049/jimmunol.1001523
148. Ceroi A, Delettre FA, Marotel C, Gauthier T, Asgarova A, Biichle S, et al. The anti-inflammatory effects of platelet-derived microparticles in human plasmacytoid dendritic cells involve liver X receptor activation. *Haematologica* (2016) 101(3):e72–6. doi:10.3324/haematol.2015.135459
149. Trahtemberg U, Grau A, Tabib A, Atallah M, Krispin A, Mevorach D. Identification and characterization of two human monocyte-derived dendritic cell subpopulations with different functions in dying cell clearance and different patterns of cell death. *PLoS One* (2016) 11(9):e0162984. doi:10.1371/journal.pone.0162984
150. Condon TV, Sawyer RT, Fenton MJ, Riches DW. Lung dendritic cells at the innate-adaptive immune interface. *J Leukoc Biol* (2011) 90(5):883–95. doi:10.1189/jlb.0311134
151. Turley S, Poirot L, Hattori M, Benoist C, Mathis D. Physiological beta cell death triggers priming of self-reactive T cells by dendritic cells in a type-1 diabetes model. *J Exp Med* (2003) 198(10):1527–37. doi:10.1084/jem.20030966
152. Asano K, Nabeyama A, Miyake Y, Qiu CH, Kurita A, Tomura M, et al. CD169-positive macrophages dominate antitumor immunity by crosspresenting dead cell-associated antigens. *Immunity* (2011) 34(1):85–95. doi:10.1016/j.immuni.2010.12.011
153. Swiecki M, Colonna M. The multifaceted biology of plasmacytoid dendritic cells. *Nat Rev Immunol* (2015) 15(8):471–85. doi:10.1038/nri3865
154. Frey T. Correlated flow cytometric analysis of terminal events in apoptosis reveals the absence of some changes in some model systems. *Cytometry* (1997) 28(3):253–63. doi:10.1002/(SICI)1097-0320(19970701)28:3<253::AID-CYTO10>3.0.CO;2-O
155. Silvin A, Yu CI, Lahaye X, Imperatore F, Brault JB, Cardinaud S, et al. Constitutive resistance to viral infection in human CD141+ dendritic cells. *Sci Immunol* (2017) 2(13). doi:10.1126/sciimmunol.aai8071
156. Curotto de Lafaille MA, Lafaille JJ. Natural and adaptive foxp3+ regulatory T cells: more of the same or a division of labor? *Immunity* (2009) 30(5):626–35. doi:10.1016/j.immuni.2009.05.002
157. Liu J, Cao X. Regulatory dendritic cells in autoimmunity: a comprehensive review. *J Autoimmun* (2015) 63:1–12. doi:10.1016/j.jaut.2015.07.011
158. Mahnke K, Ring S, Enk AH. Antibody targeting of "steady-state" dendritic cells induces tolerance mediated by regulatory T cells. *Front Immunol* (2016) 7:63. doi:10.3389/fimmu.2016.00063
159. Plitas G, Rudensky AY. Regulatory T cells: differentiation and function. *Cancer Immunol Res* (2016) 4(9):721–5. doi:10.1158/2326-6066.CIR-16-0193
160. Pallotta MT, Orabona C, Volpi C, Vacca C, Belladonna ML, Bianchi R, et al. Indoleamine 2,3-dioxygenase is a signaling protein in long-term tolerance by dendritic cells. *Nat Immunol* (2011) 12(9):870–8. doi:10.1038/ni.2077
161. Belladonna ML, Volpi C, Bianchi R, Vacca C, Orabona C, Pallotta MT, et al. Cutting edge: autocrine TGF-beta sustains default tolerogenesis

by IDO-competent dendritic cells. *J Immunol* (2008) 181(8):5194–8. doi:10.4049/jimmunol.181.8.5194
162. Williams CA, Harry RA, McLeod JD. Apoptotic cells induce dendritic cell-mediated suppression via interferon-gamma-induced IDO. *Immunology* (2008) 124(1):89–101. doi:10.1111/j.1365-2567.2007.02743.x
163. Ravishankar B, Liu H, Shinde R, Chandler P, Baban B, Tanaka M, et al. Tolerance to apoptotic cells is regulated by indoleamine 2,3-dioxygenase. *Proc Natl Acad Sci U S A* (2012) 109(10):3909–14. doi:10.1073/pnas.1117736109
164. Ravishankar B, Liu H, Shinde R, Chaudhary K, Xiao W, Bradley J, et al. The amino acid sensor GCN2 inhibits inflammatory responses to apoptotic cells promoting tolerance and suppressing systemic autoimmunity. *Proc Natl Acad Sci U S A* (2015) 112(34):10774–9. doi:10.1073/pnas.1504276112
165. Liu H, Huang L, Bradley J, Liu K, Bardhan K, Ron D, et al. GCN2-dependent metabolic stress is essential for endotoxemic cytokine induction and pathology. *Mol Cell Biol* (2014) 34(3):428–38. doi:10.1128/MCB.00946-13
166. Baban B, Chandler PR, Sharma MD, Pihkala J, Koni PA, Munn DH, et al. IDO activates regulatory T cells and blocks their conversion into Th17-like T cells. *J Immunol* (2009) 183(4):2475–83. doi:10.4049/jimmunol.0900986
167. Baban B, Hansen AM, Chandler PR, Manlapat A, Bingaman A, Kahler DJ, et al. A minor population of splenic dendritic cells expressing CD19 mediates IDO-dependent T cell suppression via type I IFN signaling following B7 ligation. *Int Immunol* (2005) 17(7):909–19. doi:10.1093/intimm/dxh271
168. Munn DH, Sharma MD, Baban B, Harding HP, Zhang Y, Ron D, et al. GCN2 kinase in T cells mediates proliferative arrest and anergy induction in response to indoleamine 2,3-dioxygenase. *Immunity* (2005) 22(5):633–42. doi:10.1016/j.immuni.2005.03.013
169. Paidassi H, Acharya M, Zhang A, Mukhopadhyay S, Kwon M, Chow C, et al. Preferential expression of integrin alphavbeta8 promotes generation of regulatory T cells by mouse CD103+ dendritic cells. *Gastroenterology* (2011) 141(5):1813–20. doi:10.1053/j.gastro.2011.06.076
170. Boucard-Jourdin M, Kugler D, Endale Ahanda ML, This S, De Calisto J, Zhang A, et al. beta8 integrin expression and activation of TGF-beta by intestinal dendritic cells are determined by both tissue microenvironment and cell lineage. *J Immunol* (2016) 197(5):1968–78. doi:10.4049/jimmunol.1600244
171. Worthington JJ, Czajkowska BI, Melton AC, Travis MA. Intestinal dendritic cells specialize to activate transforming growth factor-beta and induce Foxp3+ regulatory T cells via integrin alphavbeta8. *Gastroenterology* (2011) 141(5):1802–12. doi:10.1053/j.gastro.2011.06.057
172. Paidassi H, Acharya M, Lacy-Hulbert A. Alpha (v) integrins license regulatory T cells to apoptotic cells and self-associated antigens. *Ann N Y Acad Sci* (2010) 1209:68–76. doi:10.1111/j.1749-6632.2010.05783.x
173. Torchinsky MB, Garaude J, Martin AP, Blander JM. Innate immune recognition of infected apoptotic cells directs T(H)17 cell differentiation. *Nature* (2009) 458(7234):78–82. doi:10.1038/nature07781
174. Perruche S, Zhang P, Liu Y, Saas P, Bluestone JA, Chen W. CD3-specific antibody-induced immune tolerance involves transforming growth factor-beta from phagocytes digesting apoptotic T cells. *Nat Med* (2008) 14(5):528–35. doi:10.1038/nm1749
175. Saas P, Daguindau E, Perruche S. Concise review: apoptotic cell-based therapies-rationale, preclinical results and future clinical developments. *Stem Cells* (2016) 34(6):1464–73. doi:10.1002/stem.2361
176. Mevorach D, Zuckerman T, Reiner I, Shimoni A, Samuel S, Nagler A, et al. Single infusion of donor mononuclear early apoptotic cells as prophylaxis for graft-versus-host disease in myeloablative HLA-matched allogeneic bone marrow transplantation: a phase I/IIa clinical trial. *Biol Blood Marrow Transplant* (2014) 20(1):58–65. doi:10.1016/j.bbmt.2013.10.010
177. Ren Y, Xie Y, Jiang G, Fan J, Yeung J, Li W, et al. Apoptotic cells protect mice against lipopolysaccharide-induced shock. *J Immunol* (2008) 180(7):4978–85. doi:10.4049/jimmunol.180.7.4978
178. Hotchkiss RS, Chang KC, Grayson MH, Tinsley KW, Dunne BS, Davis CG, et al. Adoptive transfer of apoptotic splenocytes worsens survival, whereas adoptive transfer of necrotic splenocytes improves survival in sepsis. *Proc Natl Acad Sci U S A* (2003) 100(11):6724–9. doi:10.1073/pnas.1031788100
179. Delano MJ, Ward PA. The immune system's role in sepsis progression, resolution, and long-term outcome. *Immunol Rev* (2016) 274(1):330–53. doi:10.1111/imr.12499
180. van der Poll T, van de Veerdonk FL, Scicluna BP, Netea MG. The immunopathology of sepsis and potential therapeutic targets. *Nat Rev Immunol* (2017) 17(7):407–20. doi:10.1038/nri.2017.36

Conflict of Interest Statement: DM is the founder, chief scientist, and a option holder of Enlivex Ltd. The other co-author declares that the research was conducted in the absence of any commercial or financial relationships that could be construed as a potential conflict of interest.

6

Macrophage-Derived Protein S Facilitates Apoptotic Polymorphonuclear Cell Clearance by Resolution Phase Macrophages and Supports their Reprogramming

Delphine Lumbroso[1,2†], Soaad Soboh[1†], Avi Maimon[2], Sagie Schif-Zuck[1], Amiram Ariel[1] and Tal Burstyn-Cohen[2*]*

[1] Faculty of Dental Medicine, Institute for Dental Sciences, Hadassah Medical School, Hebrew University of Jerusalem, Jerusalem, Israel, [2] Department of Biology, The Faculty of Natural Sciences, University of Haifa, Haifa, Israel

Edited by:
Janos G. Filep,
Université de Montréal, Canada

Reviewed by:
Ian Dransfield,
University of Edinburgh,
United Kingdom
Joan Clària,
Hospital Clínic de Barcelona,
Spain

***Correspondence:**
Amiram Ariel
amiram@research.haifa.ac.il;
Tal Burstyn-Cohen
talbu@ekmd.huji.ac.il

†These authors have contributed equally to this work.

The complete resolution of inflammation requires the uptake of apoptotic polymorphonuclear cells (PMN) by local macrophages (efferocytosis) and the consequent reprogramming of the engulfing phagocytes to reparative and pro-resolving phenotypes. The tyrosine kinase receptors TYRO3, AXL, and MERTK (collectively named TAM) are fundamental mediators in regulating inflammatory responses and efferocytosis. Protein S (PROS1) is a ligand for all TAM receptors that mediates various aspects of their activity. However, the involvement of PROS1 in the resolution of inflammation is incompletely understood. Here, we report the upregulation of *Pros1* in macrophages during the resolution of inflammation. Selective knockout of *Pros1* in the myeloid lineage significantly downregulated macrophage pro-resolving properties. Hence, *Pros1*-deficient macrophages engulfed fewer apoptotic PMN remnants *in vivo*, and exogenous PROS1 rescued impaired efferocytosis *ex vivo*. Moreover, *Pros1*-deficient peritoneal macrophages secreted higher levels of the pro-inflammatory mediators TNFα and CCL3, while they secreted lower levels of the reparative/anti-inflammatory IL-10 following exposure to lipopolysaccharide in comparison to their WT counterparts. Moreover, *Pros1*-deficient macrophages expressed less of the anti-inflammatory/pro-resolving enzymes arginase-1 and 12/15-lipoxygenase and produced less of the specialized pro-resolving mediator resolvin D1. Altogether, our results suggest that macrophage-derived PROS1 is an important effector molecule in regulating the efferocytosis, maturation, and reprogramming of resolution phase macrophages, and imply that PROS1 could provide a new therapeutic target for inflammatory and fibrotic disorders.

Keywords: inflammation, macrophages, protein S, apoptosis, efferocytosis

INTRODUCTION

Inflammation normally resolves in an active process that eliminates the inflammatory effector components that harm the host (1–4). This is hallmarked by leukocyte apoptosis and clearance by macrophages (5–8). Apoptotic cell (AC) engulfment by phagocytes is mediated by signals that are expressed on the surface of ACs and their corresponding receptors [reviewed in Refs. (7, 9)]. Efferocytosis leads to macrophage reprogramming/immune-silencing (5, 10–13) in response to bacterial moieties through specific kinases, such as p38 MAPK, JNK, and cAMP production (14, 15). Macrophage reprogramming is defined by a reduction in the release of pro-inflammatory cytokines and chemokines, concomitant with the production of TGFβ and IL-10 (16–18), cytokines that can promote resolution and wound repair. In addition, the uptake of ACs promotes the expression of 15-lipoxygenase (LO)-1, which is involved in the generation of pro-resolving lipid mediators by macrophages (19–21). A new resolution phase macrophage phenotype—distinguishable from either M1 or M2—and characterized by low expression of CD11b, is generated from the CD11b^{high} phenotype upon the engulfment of threshold numbers of apoptotic polymorphonuclear cells (PMN) (21). This phenotypic conversion of macrophages results in significant immune-silencing in addition to the reduction in the expression of arginase-1, surface CD11b, and F4/80 (21). Specifically, CD11b^{low} macrophages stop producing TNFα and IL-1β, but increase the production of TGFβ and the expression of 12/15-LO, and emigrate to the lymphatics (21).

Protein S (PROS1) is a secreted multifunctional glycoprotein encoded by the *Pros1* gene and best known for its potent anticoagulant activity as a cofactor for activated protein C (22). Outside the coagulation system, PROS1 functions as an agonist for the TAM family of receptor tyrosine kinases, comprising TYRO3, AXL, and MERTK (23, 24), which are negative regulators of immunity (25–28). TAM receptors and their cognate ligands PROS1 and growth-arrest-specific-6 (GAS6) are expressed by immune cells, including macrophages and dendritic cells (DCs) (29, 30). This signaling axis dampens immune reactivity and contributes toward resolving inflammation through at least two distinct mechanisms: the molecular inhibition of the production and the secretion of pro-inflammatory cytokines and by the phagocytic clearance of ACs.

In macrophages, MERTK inhibits the production and secretion of the pro-inflammatory cytokine TNFα following lipopolysaccharide (LPS) exposure (31), while unchallenged macrophages isolated from TAM triple-mutant mice express aberrantly high levels of MHC class II molecules and IL-12 (32). The role of PROS1 as a TAM agonist *in vivo* was demonstrated in the phagocytic clearance of photoreceptor outer segments by retinal-pigment epithelial cells (33) and in T-cell–DC immune interaction (29). However, its role in macrophage-mediated resolution of inflammation has not been investigated.

To test the physiological role of PROS1 in resolution phase macrophages, we generated mice genetically ablated for *Pros1* expression in the myeloid lineage and assessed macrophage efferocytosis and reprogramming in a model of zymosan A-induced peritonitis. Here, we report that *Pros1* is produced by resolution phase macrophages and promotes key features of these macrophages. We show that *in vivo* efferocytosis is compromised in PROS1-deficient macrophages. Exogenous PROS1 was able to rescue the engulfment of ACs *ex vivo*. We also show that PROS1 ablation attenuates macrophage conversion toward anti-inflammatory/reparative phenotypes, as determined by their cytokine secretion balance. Moreover, PROS1 deficiency resulted in the hampered expression of the pro-resolving enzymes arginase-1 and 12/15-LO, as well as the latter's product, resolvin D1 (RvD1). Hence, PROS1 is a key mediator in successful resolution of inflammation.

MATERIALS AND METHODS

Reagents

ELISA kits for mouse TNFα, IL-10, and IL-6 were obtained from Biolegend; a mouse CCL3 detection kit was obtained from R&D Systems. LPS (from *Escherichia coli*, clone 055:B5) and zymosan A (from *Saccharomyces cerevisiae*) were purchased from Sigma-Aldrich (St. Louis, MI, USA). Docosahexaenoic acid (DHA) was purchased from Cayman Chemicals.

Murine Peritonitis

Male C57BL/6 LysM$^{Cre/+}$; *Pros1*$^{fl/fl}$ and *Pros1*$^{fl/fl}$ mice (8–10 week old), were injected intra-peritoneally (i.p.) with freshly prepared zymosan A (1 mg/25 g body weight; Sigma-Aldrich) from *S. cerevisiae* in sterile PBS or left unchallenged. After 24, 48, and 66 hrs, mice were euthanized and peritoneal exudates were collected by lavage with 5 ml of sterile PBS. All animal experiments were approved by the Hebrew University—Hadassah ethics committee.

Macrophages Isolation

Peritoneal exudates were recovered with 5 ml PBS and centrifuged (1300 rpm for 5 min). Then, cells and cell-free peritoneal fluids were separated. Peritoneal macrophages were isolated from exudate cells by EasySep™ mouse PE-positive selection magnetic beads kit (StemCell Technologies) using PE anti-mouse F4/80 antibody (#123110, Biolegend). Isolated macrophages were used for RT-qPCR, microscopic analysis, and *ex vivo* stimulation experiments.

Quantitative Real-time PCR (qPCR)

Isolated macrophages were harvested, washed once with PBS, and their RNA content was isolated with TRIZOL (Sigma-Aldrich). cDNA was synthesized with qScript cDNA synthesis kit (Quanta Biosciences). Real-time PCR reactions were performed in triplicates using KAPA SYBR FAST qPCR Kit (KAPA-Biosystems) following the manufacturer's instructions on a CFX96 Real-Time PCR Cycler (Bio-Rad). The reactions were normalized to *mGapdh* using the ΔΔ threshold cycle (Ct) method. Specificity of the primers was confirmed by dissociation curves. Mouse primer sequences were as follows: *mPros1* Forward 5'-TTC CGT GTT GGC TCA TTC C-3'; *mPros1* Reverse 5'-TTG GTC TGA GAT GGC TTT GAC A-3', *mGapdh*-Forward 5'-AGT TGG GAT AGG GCC TCT CTT-3', and *mGapdh*-Reverse 5'-TCC CAC TCT TCC ACC TTC GA-3'.

In Vivo Engulfment Assay

Peritoneal macrophages were isolated at the indicated time points and plated (2×10^5 cells/well) onto eight-well chamber glass slides for 2 h at 37°C in RPMI 1640 (Gibco) to allow adhesion. Then, the cells were rinsed briefly with PBS, fixed for 15 min with 4% paraformaldehyde (PFA) containing 5% sucrose in PBS, and washed twice in PBS. Fixed cells were incubated overnight at 4°C with phalloidin CF488A conjugate (5 units/ml, for F-actin, Biotium). Then, the cells were washed twice with PBS, stained with Hoechst (20 μg/ml for nuclear DNA, Invitrogen H3570) for 5 min, and washed thoroughly, but gently. Mounted slides were kept in the dark at 4°C until analyzed. Macrophages and engulfed apoptotic remnants were enumerated under a confocal fluorescent microscope (Zeiss) as previously described (21). Briefly, macrophages from five to eight fields per chamber (approx. 200 cells) were analyzed per mouse, and the average number of neutrophils engulfed per macrophage, as well as the number of macrophages that have actively engulfed apoptotic moieties, was calculated. Engulfed nuclear material was identified by a spherical bright nuclear Hoechst staining of 1–10 μm in diameter. Then, phagocytic efficiency was calculated as in Ref. (34). Briefly, the phagocytic efficiency index was calculated based on a weighted average of ingested apoptotic DNA-containing particles per macrophage and the number of macrophages containing a certain number of such particles. Phagocytic efficiency (%) = $[(1 \times X_{(1)} + 2 \times X_{(2)} + 3 \times X_{(3)} \ldots + n \times X_{(n)})/\text{total number of macrophages}] \times 100$. $X_{(n)}$ represents the number of macrophages containing n apoptotic particles ($n = 1, 2, 3, \ldots$, up to a maximum of six points for more than five apoptotic particles ingested per macrophage).

In Vitro Engulfment of ACs

Jurkat cells (10×10^6) were induced for apoptosis with staurosporine (Sigma-Aldrich, 1 μM/10^6 cells in 1 ml RPMI medium containing pen/strep/glutamine and 10% FBS) for 5 h. Cells were washed three times in PBS, labeled with CypHer5E (GE Healthcare; 1 μl CypHer5E/1 ml serum-free medium) for 30 min, and washed twice with PBS. Peritoneal macrophages were isolated as described, and 150×10^3 cells were plated on an eight-well glass chamber slide (Nunc) and fed with 750×10^3 pre-labeled apoptotic Jurkat cells for 4 h in a total of 150-μl medium with or without soluble PROS1 (25 nM; from Enzyme Research Laboratories). Next, the medium was aspirated, and bound cells were washed gently with PBS. Adherent cells were subsequently fixed in 200 μl of 4% PFA; 5% sucrose for 15 min. Cells were washed in PBS and incubated with 200 μl phalloidin (5 U/ml, CF488A conjugate, Biotium) at 4°C overnight. Then, cells were washed three times with PBS (10 min each) and stained in 200 μl of Hoechst 3570 (20 μg/ml, Invitrogen) for 5 min, and rinsed with PBS. The chambers were removed, mounted with Fluoromount G, and visualized under a Nikon A1 confocal microscope. The number of CypHer5E$^+$ engulfed ACs per macrophage was scored.

Cytokine and Chemokine Secretion *Ex Vivo*

Peritoneal macrophages were isolated using PE selection magnetic beads (StemCell Technologies) and incubated (5×10^6 cells in 5 ml of culture media) with LPS (1 μg/ml) or vehicle in RPMI 1640 under a humidified 5% CO_2 atmosphere at 37°C. After 24 h, the supernatants were collected, and their TNFα, IL-6, CCL3, and IL-10 contents were determined by standard ELISA (Biolegend kits for TNFα, IL-6, and IL-10 and R&D Systems kit for CCL3).

Western Blot Analysis

Protein extracts from equal volumes of peritoneal fluids or equal total protein content of isolated macrophages were subjected to SDS-PAGE using 10% polyacrylamide gels, transferred (1 h, 15 V) to PVDF membranes (Bio-Rad), and blocked for 1 h with 5% BSA in TBST (0.1% Tween 20 in Tris-buffered saline). Then, membranes were immuno-blotted overnight at 4°C with either goat anti-mouse CD11b (M-19, 1:200, SantaCruz Biotechnology), goat anti-mouse arginase-1 (ab60176, 1:20,000, Abcam), rabbit anti-mouse 12/15-LO (160704, 1:1,000, Cayman Chemical), rabbit anti-mouse PROS1 (AB15928, 1:1,000, Merck Millipore), and goat anti-mouse actin (I-19, 1:500, SantaCruz Biotechnology). Then, the membranes were washed three times with TBST and incubated with the appropriate HRP-conjugated secondary antibodies (1:10,000, 1 h, room temperature, Jackson ImmunoResearch). Blots were washed and developed using the EZ-ECL (Biological Industries) chemiluminescence kit and analyzed using the LAS-4000 luminescent image analyzer (Fujifilm) and the TotalLab TL-100 software (Nonlinear Dynamics). Band densitometric intensities among different samples were normalized to actin.

RvD1 Quantification

Peritoneal macrophages were isolated 66 h post peritonitis initiation from the indicated mice; 10^6 cells were resuspended in 1-ml medium and immediately supplemented with DHA (20 μM, Cayman Chemicals) for 4 h. Then, the incubation was stopped with cold MeOH, the supernatants were collected in a glass tube, and the MeOH was allowed to evaporate completely; 5 ml of ddH_2O and 200 μl of MeOH (pH 3.5) were added to the tubes and the samples were loaded through activated Sep-pak Vac 6cc (500 mg) C18 Cartridges (Waters, WAT043395) allowing for MeOH-activated lipid binding. Next, the bound lipids were released using methyl formate, which was then evaporated completely while adding small amounts of MeOH under nitrogen flow. Finally, RvD1 content in lipid-extracted samples was determined using a commercial ELISA kit (Cayman Chemicals), according to the manufacturer's instructions.

Statistical Analysis

Experiments were repeated at least three times with at least three replicates per experiment. Results were analyzed by two-way analysis of variance (for multiple groups) or Student's t-test (for comparison between two groups), unless otherwise mentioned. P-values (P), $^*P \leq 0.05$, $^{**}P \leq 0.01$, and $^{***}P \leq 0.001$, were considered statistically significant. Results are expressed as means ± SEM.

RESULTS

PROS1 Expression Is Upregulated in Resolution Phase Macrophages

The phagocytosis of apoptotic neutrophils is a key step during the resolution of inflammation. To investigate the role of PROS1 during the resolution of inflammation, we utilized the zymosan A-induced peritonitis as a prototypic model (35). Peritoneal macrophages were harvested from unchallenged mice or during the inflammatory (24 h) and resolving (48–66 h) phases of peritonitis. *Pros1* mRNA levels in isolated macrophages were quantified by RT-qPCR and normalized to *Gapdh*. Our results indicate that resident macrophages (present at 0 h), inflammatory, and early resolution phase macrophages (24 h) expressed very little *Pros1*, whereas a sharp upregulation of *Pros1* mRNA in resolving macrophages (66 h) was observed (**Figure 1A**). To verify *Pros1* upregulation in macrophages, we measured *Pros1* transcripts in peritoneal macrophages of LysM$^{Cre/+}$; *Pros1*$^{fl/fl}$ mice, in which *Pros1* expression is ablated following Cre-mediated excision in myeloid cells. We did not observe basal expression nor zymosan-induced upregulation of *Pros1* in peritoneal macrophages isolated from LysM$^{Cre/+}$; *Pros1*$^{fl/fl}$ mice (36, 37) (**Figure 1B**). During the resolution phase, PROS1 protein levels present in the peritoneal fluids of LysM$^{Cre/+}$; *Pros1*$^{fl/fl}$ mice, were also significantly reduced compared to *Pros1*$^{fl/fl}$ controls (**Figure 1C**). Thus, macrophages seem to express higher levels of PROS1 during the resolution of inflammation and contribute to its peritoneal levels.

PROS1 Deficiency in Resolution Phase Macrophages Impairs Their Ability to Engulf Apoptotic Remnants

The clearance of apoptotic neutrophils from resolving inflammation sites is essential for resolution and return to homeostasis (38). Considering the role of PROS1 in the phagocytosis of ACs (39), and its expression by resolution phase macrophages (**Figure 1**), we examined whether *Pros1*-deficient macrophages exert modified uptake of apoptotic neutrophils. To this end, we determined the phagocytic capacity *in vivo* of peritoneal resolution phase macrophages between LysM$^{Cre/+}$; *Pros1*$^{fl/fl}$ (cKO) and their control *Pros1*$^{fl/fl}$ littermates. Our results indicate that macrophages deficient in PROS1 display a reduced ability to phagocytose apoptotic particles *in vivo* (**Figures 2A–C**). Quantification of this phenomenon showed that PROS1-deficient macrophages had impaired phagocytic capacity, reflected by a lower phagocytic efficiency index compared to control cells (174 ± 16 and 271 ± 20%, respectively, **Figure 2B**). The majority of PROS1-deficient macrophages did not engulf any apoptotic particles (55 ± 4% compared to 37 ± 3% in controls) (**Figure 2C**). While a similar percentage of cKO and control cells had engulfed one or two apoptotic particles (25.9 ± 2 and 26.6 ± 2%, respectively), control cells were twice as active in engulfing three to seven apoptotic moieties compared to cKO cells (31.4 ± 2.5 and 16.2 ± 2.6%, respectively). Finally, the uptake efficiency declined for both cell types scored with 8 or more Hoechst-positive foci, with a nonsignificant trend pointing to decreased efferocytosis in cKO cells, with 5.4 ± 1% of control cells, but only 2.8 ± 1% of cKO cells (**Figure 2C**).

We next tested whether the addition of purified PROS1 would rescue the impaired phagocytosis exhibited by PROS1-deficient macrophages. For this, we performed an *ex vivo* phagocytosis assay, assessing the efferocytosis of resolution phase peritoneal macrophages that were fed with pre-labeled apoptotic Jurkat cells (**Figure 2D**). Akin to their *in vivo* performance, a decreased phagocytic index was recorded for PROS1-cKO macrophages (1 ± 0.008 compared to 1.7 ± 0.03 in controls) (**Figure 2E**). The supplementation of exogenous PROS1 rescued the phagocytic performance of cKO cells, bringing it to normal levels of untreated control macrophages, and augmented the phagocytic capacity of *Pros1*$^{fl/fl}$ control cells (phagocytic indices of 1.7 ± 0.14 and 2.2 ± 0.08, respectively) (**Figure 2E**).

Taken together, our results identify endogenously expressed PROS1 as an important mediator of efferocytosis in resolution phase macrophages. Given the importance of phagocytic macrophages in clearing apoptotic neutrophils during the resolution of inflammation, we conclude that *Pros1*-deficient macrophages exhibit a hampered pro-resolving phagocytic phenotype.

PROS1-Deficient Macrophages Display Reduced Reprogramming

Apoptotic neutrophil engulfment by resolution phase macrophages results in their conversion from pro-inflammatory cells

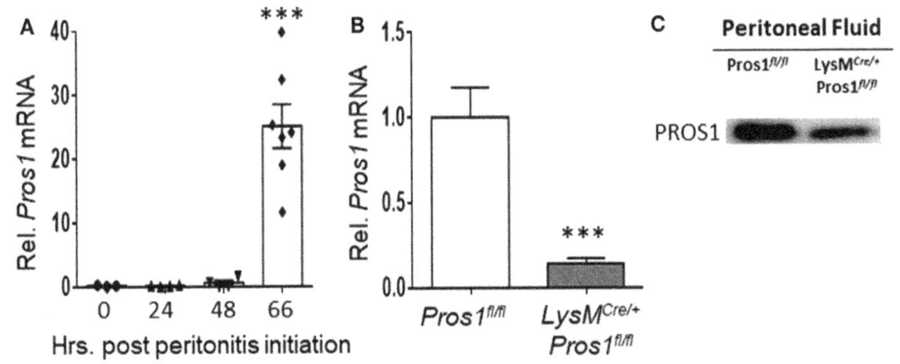

FIGURE 1 | Protein S (PROS1) mRNA expression in resolution phase macrophages. **(A)** Peritoneal cells were harvested from unchallenged *Pros1*$^{fl/fl}$ mice or mice undergoing zymosan A-initiated peritonitis for 24, 48, or 66 hrs. RNA was extracted from isolated macrophages, and *Pros1* transcript levels were quantified by RT-qPCR and normalized to GAPDH (n = 3–7 mice per time point). Results represent the distribution of individual mice and the mean ± SEM. Statistical significance by one-way ANOVA (***P-value < 0.0001) is indicated. **(B)** PROS1 expression in *Pros1*$^{fl/fl}$ and LysM$^{Cre/+}$; *Pros1*$^{fl/fl}$ resolution phase macrophages (66 h post peritonitis initiation) was quantified by RT-qPCR. Results are presented as the mean ± SEM from six mice/genotype. Student's t-test, ***$P \leq 0.0001$. **(C)** Peritoneal fluids from individual mice were collected 66 h after zymosan A injection and analyzed for PROS1 protein levels by Western blot. A representative blot of three experiments (three to five mice per experiment) is shown.

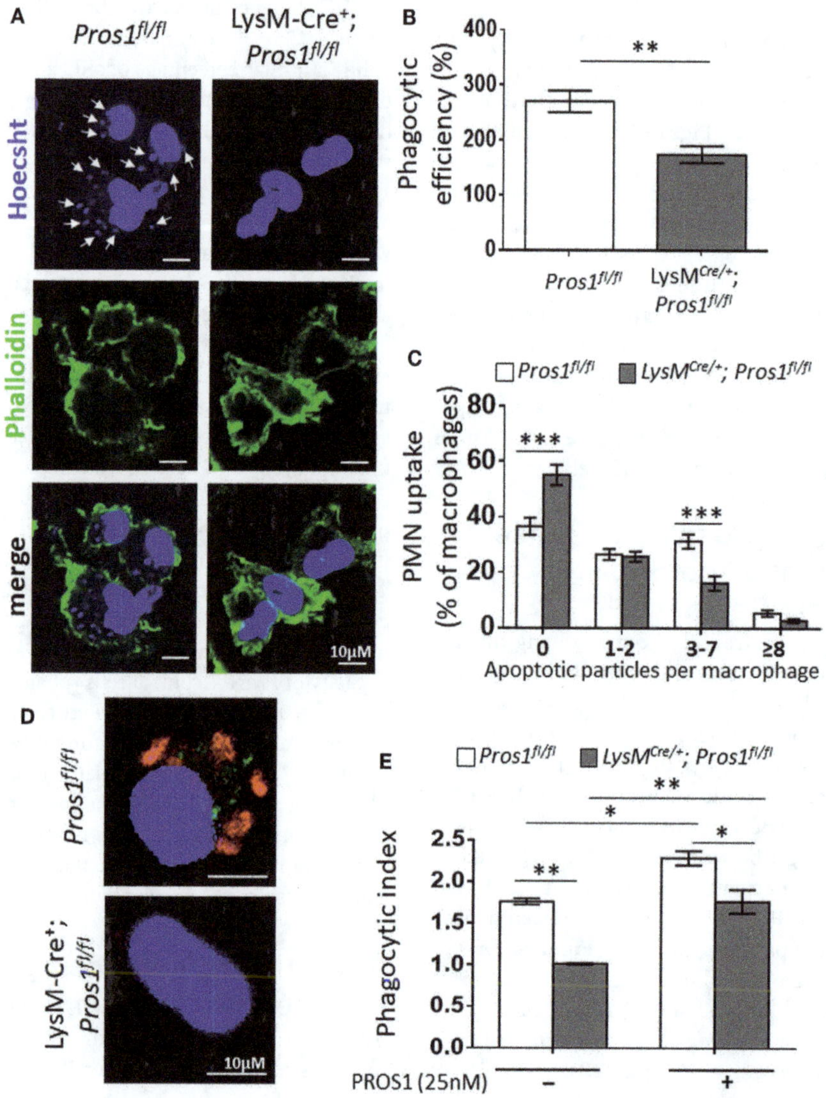

FIGURE 2 | Protein S (PROS1)-deficient macrophages demonstrate a decreased engulfment of apoptotic neutrophils *in vivo*. **(A)** Resolution phase peritoneal macrophages from *Pros1*$^{fl/fl}$ (left) and LysM$^{Cre/+}$; *Pros1*$^{fl/fl}$ (right) mice were isolated and stained with Hoechst (blue) and FITC-phaloidin (green). Z-stack 3D confocal images were taken with a Nikon A1-R microscope. **(B)** The number of PMN particles in each macrophage was enumerated using the Nikon NIS-Elements microscope imaging software, and the percentage of phagocytic macrophages and the number of apoptotic particles (arrows in **A**) per macrophage were calculated into the overall phagocytic efficiency index. $^{**}P = 0.004$ (*t*-test). **(C)** Engulfment according to thresholds of intracellular apoptotic particles was calculated. Two-way analysis of variance (ANOVA), $^{***}P \leq 0.0001$. **(D)** Confocal images of *in vitro* phagocytosis using pre-labeled apoptotic Jurkat cells (red). **(E)** The phagocytic index was calculated in the absence or presence of purified PROS1. Two-way ANOVA, $^{*}P \leq 0.02$, $^{**}P = 0.005$. Results are representative images **(A,D)** or means ± SEM **(B,C,E)** from at least three independent experiments (*n* = 8–11 mice; over 1,500 macrophages scored).

to anti-inflammatory/reparative ones (40). These M2-like macrophages prevent unwanted excessive inflammatory responses during the resolution phase of inflammation and promote tissue repair. To determine whether PROS1 expressed by resolution phase macrophages plays a role in this *in vivo* transition, termed reprogramming, we isolated resolution phase peritoneal macrophages 66 h post zymosan A treatment from *Pros1*-proficient or deficient mice. Isolated macrophages were then stimulated with LPS, and the secretion of pro-inflammatory cytokines and chemokines (TNFα, IL-6, and CCL3) or the anti-inflammatory cytokine IL-10 was determined (**Figures 3A–D**). Our results indicate that following LPS stimulation, PROS1-deficient macrophages secreted significantly elevated levels of TNFα in comparison to their *Pros1*$^{fl/fl}$ counterparts (820 ± 114.6 and 320.3 ± 61.3 pg/ml, respectively). The secretion of CCL3 and IL-6 was also elevated in cKO macrophages following stimulation with LPS in comparison to *Pros1*$^{fl/fl}$ ones, although the latter was not statistically significant (25.6 ± 3.3 versus 10.7 ± 1.8 pg/ml for CCL3 by cKO and controls, and 12,771 ± 4,432 versus 5,106 ± 2,684 pg/ml for IL-6 by cKO and control cells, respectively) (**Figures 3B,C**). Concomitantly, the secretion of IL-10 from LysM$^{Cre/+}$; *Pros1*$^{fl/fl}$ macrophages was significantly lower than its secretion by *Pros1*$^{fl/fl}$

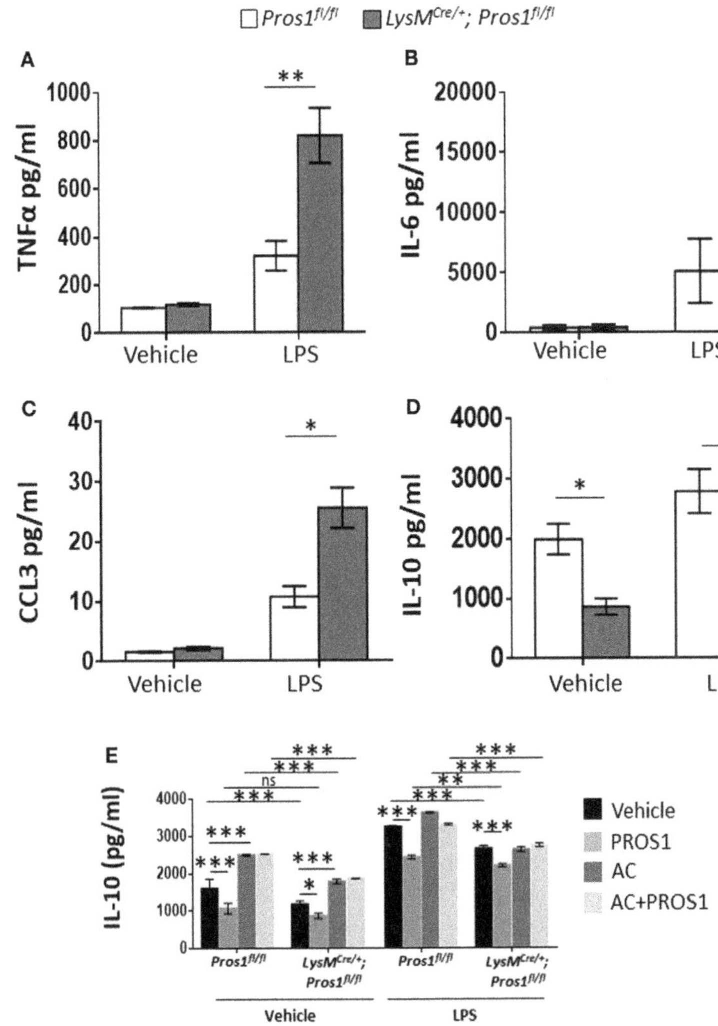

FIGURE 3 | Protein S (PROS1)-deficient macrophages display reduced reprogramming. Macrophages from $Pros1^{fl/fl}$ or $LysM^{Cre/+}$; $Pros1^{fl/fl}$ mice were isolated from peritoneal exudates 66 h after zymosan A injection and incubated for 24 h with vehicle or lipopolysaccharide (LPS, 1 µg/ml). Then, supernatants were collected and analyzed for their content of TNFα [**(A)**, $^{**}P = 0.003$], IL-6 **(B)**, CCL3 [**(C)**, $^{*}P = 0.02$], and IL-10 [**(D)**; $^{*}P \leq 0.03$] by standard ELISA. **(E)** IL-10 secretion by untreated or LPS-stimulated control and *Pros1*-cKO peritoneal macrophages, or by macrophages supplemented with either PROS1 (25 nM), apoptotic cells (AC) or both (AC + PROS1). Results represent the means ± SEM of at least three independent experiments. Two-way ANOVA, $^{*}P = 0.02$, $^{***}P \leq 0.0001$.

macrophages under baseline conditions and upon stimulation with LPS (861 ± 136 and 1,990 ± 258 pg/ml for unstimulated cKOs and controls, and 1,165.5 ± 431 and 2,783.2 ± 368 pg/ml for stimulated cKO and controls, respectively) (**Figure 3D**). Thus, *Pros1*-deficient macrophages present a shift toward exacerbated pro-inflammatory cytokine secretion, suggesting hampered reprogramming compared to their control counterparts. Since IL-10 is a key cytokine in macrophage reprogramming (5), we next determined the effect of exogenous PROS1 on AC-induced IL-10 secretion *ex vivo*. Our results indicate that ACs promoted the secretion of IL-10 in both control and *Pros1*-cKO mice when unstimulated, but upon LPS stimulation, this effect was absent in PROS1-deficient macrophages (**Figure 3E**). Unexpectedly, the addition of PROS1 was unable to enhance the effect of ACs on IL-10 secretion in PROS1-deficient mice and in fact reduced IL-10 secretion from untreated and LPS-stimulated macrophages (both WT and PROS1-deficient). This may be due to the biochemical nature of PROS1. PROS1 acts as a bridging molecule that binds phosphatidylserine (PstSer) exposed on the outer leaflets of ACs *via* its amino terminus and to the extracellular domain of TAM receptors on phagocytes and macrophages *via* its carboxy terminus (24). Thus, the addition of excess PROS1 to the combined culture of macrophages and ACs may saturate the binding sites on ACs and macrophages without physically bridging between them. It is conceivable that the controlled and sequential addition of PROS1 to macrophages and ACs would favor the bridging and subsequent reduction of IL-10 production. Nevertheless, our results indicate that compared to controls, PROS1-deficient macrophages present a hampered response to AC uptake that results in a pro-inflammatory imbalance with elevated TNFα and CCL3 as well as lower IL-10 production and could only partially be rescued by exogenous PROS1.

Pros1-Deficient Macrophages Express Reduced Levels of Pro-Resolving Enzymes

During the resolution of inflammation, the engulfment of apoptotic neutrophils leads to macrophage metamorphosis from an M1-like phenotype to an M2-like phenotype and then to a pro-resolving phenotype (Mres) highlighted by the expression of the functionally important enzyme 12/15-LO (40). These changes are associated with reprogramming of the engulfing macrophages (41). Since our results so far indicate that *Pros1* expression by resolution phase macrophages is required for their uptake of apoptotic neutrophils and reprogramming, we sought to analyze the expression of proteins that are instrumental to inflammation and its resolution (19, 42–44). 12/15-LO is a key enzyme involved in the synthesis of lipoxins, protectins, resolvins and other specialized lipid mediators that promote the resolution of inflammation by macrophages (45–47). Arginase-1 enzymatically inhibits nitric oxide (NO) production by inducible NO synthase (iNOS), thereby supporting an anti-inflammatory/reparative milieu (48). Since iNOS characterizes M1-like macrophages and is highly expressed by pro-inflammatory macrophages, the levels of arginase-1 expression reflect the maturation level of macrophages and their progression to the reparative phenotype during resolution (40). To evaluate macrophage maturation and differentiation, peritoneal macrophages from $LysM^{Cre/+}$; $Pros1^{fl/fl}$; and $Pros1^{fl/fl}$ mice were isolated 66 h post peritonitis initiation, and their protein content was immuno-blotted for the macrophage M2/maturation markers arginase-1 and CD11b as well as for the pro-resolving enzyme 12/15-LO (**Figure 4**). Our results indicate that $LysM^{Cre/+}$; $\mathit{Pros1}^{fl/fl}$ macrophages express significantly lower levels of all three proteins (86, 83, and 81% of controls, for 12/15-LO, arginase-1, and CD11b, respectively). Thus, $LysM^{Cre/+}$; $\mathit{Pros1}^{fl/fl}$ macrophages display a reduced capacity to mature and convert to the M2-like and Mres phenotypes during the resolution of inflammation. 12/15-LO catalyzes the derivation of DHA into resolvin (Rv) D1. Given the importance of RvD1 as a key mediator involved in the resolution of inflammation, we tested whether the decreased levels of 12/15-LO in PROS1-deficient macrophages (**Figure 4**) would also affect the potential production of RvD1 by these cells. We found a 25% decrease in the production of RvD1 by PROS1-defiecient cells compared to that of controls (1,899 ± 33.2 and 2,527 ± 57.8 pg/ml for cKO and controls, respectively) (Figure S1 in Supplementary Material). Thus, the reduced expression of 12/15-LO in PROS1-deficient resolution phase macrophages results in a reduced capacity to produce specialized pro-resolving lipid mediators. Taken together, our results indicate that macrophage-derived PROS1 is a molecular effector in the uptake of apoptotic neutrophils and participates in the consequent reprogramming and maturation of macrophages in resolving inflammation.

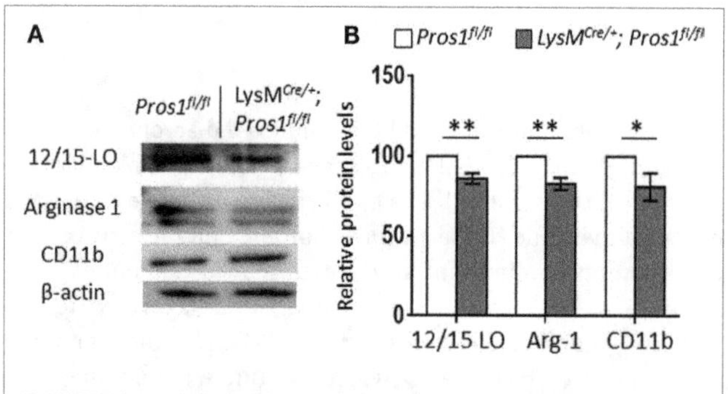

FIGURE 4 | Protein S (PROS1)-deficient macrophages express reduced levels of resolution phase enzymes. Peritoneal macrophages from $Pros1^{fl/fl}$ or $LysM^{Cre/+}$; $Pros1^{fl/fl}$ mice (n = 5/group) were isolated 66 h after zymosan A injection and lysed, and their protein content was analyzed by Western blot for the resolution phase markers 12/15-LO, arginase-1, and CD11b. A representative blot is shown **(A)**. Band intensities were quantified, and graphs represent the normalized means ± SEM from three independent experiments **(B)**. Student's *t*-test, $^{*}P = 0.03$, $^{**}P \leq 0.02$.

DISCUSSION

Protein S is a pleiotropic mediator involved in various processes, such as vasculogenesis, blood clotting, and immune regulation (29, 33, 37, 49). While hepatocytes are considered to be the major source of plasma PROS1, significant contributions to the volume and function of PROS1 were attributed to other sources, such as endothelial and T cells (29, 33, 37). Our findings now indicate that PROS1 is produced by macrophages during the resolution of murine peritonitis, and consequently its production and release by peritoneal macrophages as well as other cells contribute to peritoneal levels of PROS1 (**Figure 1**). While peritoneal PROS1 levels probably increase during the inflammatory phase of peritonitis, due to exudation of plasma proteins, it is unlikely that peritoneal PROS1 will remain at high levels in the absence of additional sources. Along these lines, PROS1 from macrophages seems to be an important component in its peritoneal levels during resolution. Notably, PROS1 produced by resolution phase macrophages is functionally important. This could be due to a critical concentration of the peritoneal protein or due to a local regulation of PROS1 function as seen in the lymph nodes (29) and the retina (33). Similarly, our results indicate that the discrete secretion of macrophage-generated PROS1 is essential for key-resolving features of efferocytosis and molecular reprogramming. Such localized secretion could be envisioned in the contact area between apoptotic PMN and the macrophages that engulf them (50).

Protein S was previously found to bind the tyrosine kinase receptors from the TAM family and mediate their interactions with ACs through the binding of PstSer (39, 51–53). As expected in this setting, the deficiency in PROS1 production by efferocytosis-competent macrophages resulted in a significant reduction in the uptake of apoptotic PMNs in $LysM^{Cre/+}$; $\mathit{Pros1}^{fl/fl}$ macrophages, which was rescued following the addition of purified PROS1 (**Figure 2**). The engulfment of apoptotic PMNs is essential for the resolution of inflammation [reviewed in Refs. (5, 40)]. This is in part due to the phenotypic changes that take place in the engulfing monocytes/macrophages (5, 10, 54). These changes, also termed macrophage reprogramming/immune-silencing, are characterized by a reduction in the production of pro-inflammatory cytokines and chemokines concomitant with an increase in the production of anti-inflammatory cytokines, such as IL-10, upon exposure to bacterial moieties (21, 41, 55). Notably, the TAM receptors were previously shown to block TNFα- and TLR-mediated inflammatory signals (25, 31, 56). In addition, the PROS1-mediated phagocytosis of ACs by peripheral blood monocytes contributes to the elimination of dying and dead cells in the circulation (39), thus avoiding the induction of harmful inflammatory responses. Our current results indicate that a myeloid-specific deficiency in *Pros1* culminates in increased amounts of the pro-inflammatory mediators TNFα and CCL3 and reduced amounts of the anti-inflammatory/pro-resolving cytokine IL-10, upon stimulation with LPS (**Figure 3**), a profile resembling pro-inflammatory rather than resolution phase macrophages.

During murine peritonitis, $CD11b^{med}$ monocytes differentiate to $CD11b^{high}$ macrophages that engulf apoptotic PMNs in a self-limiting fashion and convert to $CD11b^{low}$-satiated macrophages (21, 35). These phenotypic changes are associated with a temporal increase in the expression of arginase-1, a hallmark of M2 macrophages that is induced by AC uptake (19, 57). 12/15-LO, an enzyme that is involved in the production of pro-resolving lipid mediators including RvD1 (1, 58) and the uptake of ACs (59), is progressively upregulated by macrophages during the resolution of inflammation. Moreover, it is a hallmark of macrophage conversion from the $CD11b^{high}$ to the $CD11b^{low}$ phenotype (21). Although CD11b protein levels were reduced in *Pros1*-cKO-resolving macrophages, the expression of arginase-1 and 12/15-LO—two *bona fide* markers of resolution phase macrophages—was lower in efferocytosing macrophages lacking PROS1 (**Figure 4**). In line with the reduced levels of 12/15-LO, less RvD1 is produced by macrophages devoid of PROS1 (Figure S1 in Supplementary Material). Together, with their impaired efferocytosis and a pro-inflammatory cytokine profile, these results suggest that the conversion to resolution phase macrophages in the absence of PROS1 is either incomplete or hampered. Hence, PROS1 production by local macrophages seems to be important in their reprogramming during the resolution phase of inflammation.

In sum, our results indicate that *Pros1* production and action in pro-resolving macrophages are key events in the termination of inflammation. PROS1 acts in various modes as it regulates both the uptake of apoptotic PMNs and the consequent reprogramming of macrophages. These findings suggest that PROS1 might be harnessed as a new strategy for the treatment of inflammatory and autoimmune disorders.

ETHICS STATEMENT

All animal experiments were approved by the Hebrew University-Hadassah ethics committee.

AUTHOR CONTRIBUTIONS

DL, TB-C, and AA designed the research. DL, SS, AM, and SS-Z performed experiments, analyzed, and visualized data. TB-C and AA supervised the project. DL, TB-C, and AA wrote the manuscript.

FUNDING

This work was supported by grants from the Israel Science Foundation (grant number 678/13 to AA and grant number 984/2012 to TB-C), the Rosetrees Trust, and the Wolfson Family Charitable Trust to AA and the Israel Cancer Research Foundation to TB-C. DL was awarded with the Azrieli postdoctoral fellowship and supported by the Ministry of Aliya and Integration (MOIA) of the State of Israel.

REFERENCES

1. Levy BD, Clish CB, Schmidt B, Gronert K, Serhan CN. Lipid mediator class switching during acute inflammation: signals in resolution. *Nat Immunol* (2001) 2(7):612–9. doi:10.1038/89759
2. Majno G, Joris I. *Cells, Tissues, and Disease—Principles of General Pathology*. New York: Oxford University Press (2004).
3. Serhan CN, Brain SD, Buckley CD, Gilroy DW, Haslett C, O'Neill LAJ, et al. Resolution of inflammation: state of the art, definitions and terms. *FASEB J* (2007) 21(2):325–32. doi:10.1096/fj.06-7227rev
4. Ortega-Gómez A, Perretti M, Soehnlein O. Resolution of inflammation: an integrated view. *EMBO Mol Med* (2013) 5(5):661–74. doi:10.1002/emmm.201202382
5. Fadok VA, Bratton DL, Konowal A, Freed PW, Westcott JY, Henson PM. Macrophages that have ingested apoptotic cells *in vitro* inhibit proinflammatory cytokine production through autocrine/paracrine mechanisms involving TGF-beta, PGE2, and PAF. *J Clin Invest* (1998) 101(4):890–8. doi:10.1172/jci1112
6. Rossi AG, Haslett C. Inflammation, cell injury, and apoptosis. In: Said SI, editor. *Pro-Inflammatory and Anti-Inflammatory Peptides*. New York, NY: Marcel Dekker Inc. (1998). p. 9–24.
7. Savill J, Dransfield I, Gregory C, Haslett C. A blast from the past: clearance of apoptotic cells regulates immune responses. *Nat Rev Immunol* (2002) 2(12):965–75. doi:10.1038/nri957
8. Greenlee-Wacker MC. Clearance of apoptotic neutrophils and resolution of inflammation. *Immunol Rev* (2016) 273(1):357–70. doi:10.1111/imr.12453
9. Ravichandran KS, Lorenz U. Engulfment of apoptotic cells: signals for a good meal. *Nat Rev Immunol* (2007) 7(12):964–74. doi:10.1038/nri2214
10. Voll RE, Herrmann M, Roth EA, Stach C, Kalden JR, Girkontaite I. Immunosuppressive effects of apoptotic cells. *Nature* (1997) 390(6658):350–1. doi:10.1038/37022
11. Hanayama R, Tanaka M, Miwa K, Shinohara A, Iwamatsu A, Nagata S. Identification of a factor that links apoptotic cells to phagocytes. *Nature* (2002) 417(6885):182–7. doi:10.1038/417182a
12. Kim S, Elkon KB, Ma X. Transcriptional suppression of interleukin-12 gene expression following phagocytosis of apoptotic cells. *Immunity* (2004) 21(5):643–53. doi:10.1016/j.immuni.2004.09.009
13. Stables MJ, Shah S, Camon EB, Lovering RC, Newson J, Bystrom J, et al. Transcriptomic analyses of murine resolution-phase macrophages. *Blood* (2011) 118(26):e192–208. doi:10.1182/blood-2011-04-345330
14. Patel VA, Longacre A, Hsiao K, Fan H, Meng F, Mitchell JE, et al. Apoptotic cells, at all stages of the death process, trigger characteristic signaling events

that are divergent from and dominant over those triggered by necrotic cells: implications for the delayed clearance model of autoimmunity. *J Biol Chem* (2006) 281(8):4663–70. doi:10.1074/jbc.M508342200

15. Bystrom J, Evans I, Newson J, Stables M, Toor I, van Rooijen N, et al. Resolution-phase macrophages possess a unique inflammatory phenotype that is controlled by cAMP. *Blood* (2008) 112(10):4117–27. doi:10.1182/blood-2007-12-129767
16. Byrne A, Reen DJ. Lipopolysaccharide induces rapid production of IL-10 by monocytes in the presence of apoptotic neutrophils. *J Immunol* (2002) 168(4):1968–77. doi:10.4049/jimmunol.168.4.1968
17. Huynh M-LN, Fadok VA, Henson PM. Phosphatidylserine-dependent ingestion of apoptotic cells promotes TGF-β1 secretion and the resolution of inflammation. *J Clin Invest* (2002) 109(1):41–50. doi:10.1172/jci11638
18. Mitchell S, Thomas G, Harvey K, Cottell D, Reville K, Berlasconi G, et al. Lipoxins, aspirin-triggered Epi-lipoxins, lipoxin stable analogues, and the resolution of inflammation: stimulation of macrophage phagocytosis of apoptotic neutrophils *in vivo*. *J Am Soc Nephrol* (2002) 13(10):2497–507. doi:10.1097/01.asn.0000032417.73640.72
19. Freire-de-Lima CG, Xiao YQ, Gardai SJ, Bratton DL, Schiemann WP, Henson PM. Apoptotic cells, through transforming growth factor-β, coordinately induce anti-inflammatory and suppress pro-inflammatory eicosanoid and NO synthesis in murine macrophages. *J Biol Chem* (2006) 281(50):38376–84. doi:10.1074/jbc.M605146200
20. Schwab JM, Chiang N, Arita M, Serhan CN. Resolvin E1 and protectin D1 activate inflammation-resolution programmes. *Nature* (2007) 447(7146):869–74. doi:10.1038/nature05877
21. Schif-Zuck S, Gross N, Assi S, Rostoker R, Serhan CN, Ariel A. Saturated-efferocytosis generates pro-resolving CD11 blow macrophages: modulation by resolvins and glucocorticoids. *Eur J Immunol* (2011) 41(2):366–79. doi:10.1002/eji.201040801
22. Ten Kate MK, van der Meer J. Protein S deficiency: a clinical perspective. *Haemophilia* (2008) 14(6):1222–8. doi:10.1111/j.1365-2516.2008.01775.x
23. Stitt TN, Conn G, Gore M, Lai C, Bruno J, Radziejewski C, et al. The anticoagulation factor protein S and its relative, GAS6, are ligands for the Tyro 3/Axl family of receptor tyrosine kinases. *Cell* (1995) 80(4):661–70. doi:10.1016/0092-8674(95)90520-0
24. Lemke G. Biology of the TAM receptors. *Cold Spring Harb Perspect Biol* (2013) 5(11):a009076. doi:10.1101/cshperspect.a009076
25. Rothlin CV, Ghosh S, Zuniga EI, Oldstone MB, Lemke G. TAM receptors are pleiotropic inhibitors of the innate immune response. *Cell* (2007) 131(6):1124–36. doi:10.1016/j.cell.2007.10.034
26. Lemke G, Rothlin CV. Immunobiology of the TAM receptors. *Nat Rev Immunol* (2008) 8(5):327–36. doi:10.1038/nri2303
27. Lemke G, Burstyn-Cohen T. TAM receptors and the clearance of apoptotic cells. *Ann N Y Acad Sci* (2010) 1209(1):23–9. doi:10.1111/j.1749-6632.2010.05744.x
28. Rothlin CV, Carrera-Silva EA, Bosurgi L, Ghosh S. TAM receptor signaling in immune homeostasis. *Annu Rev Immunol* (2015) 33:355–91. doi:10.1146/annurev-immunol-032414-112103
29. Carrera Silva EA, Chan PY, Joannas L, Errasti AE, Gagliani N, Bosurgi L, et al. T cell-derived protein S engages TAM receptor signaling in dendritic cells to control the magnitude of the immune response. *Immunity* (2013) 39(1):160–70. doi:10.1016/j.immuni.2013.06.010
30. Zagorska A, Traves PG, Lew ED, Dransfield I, Lemke G. Diversification of TAM receptor tyrosine kinase function. *Nat Immunol* (2014) 15(10):920–8. doi:10.1038/ni.2986
31. Camenisch TD, Koller BH, Earp HS, Matsushima GK. A novel receptor tyrosine kinase, Mer, inhibits TNF-alpha production and lipopolysaccharide-induced endotoxic shock. *J Immunol* (1999) 162(6):3498–503.
32. Lu Q, Lemke G. Homeostatic regulation of the immune system by receptor tyrosine kinases of the Tyro 3 family. *Science* (2001) 293(5528):306–11. doi:10.1126/science.1061663
33. Burstyn-Cohen T, Lew ED, Traves PG, Burrola PG, Hash JC, Lemke G. Genetic dissection of TAM receptor-ligand interaction in retinal pigment epithelial cell phagocytosis. *Neuron* (2012) 76(6):1123–32. doi:10.1016/j.neuron.2012.10.015
34. Koenigsknecht J, Landreth G. Microglial phagocytosis of fibrillar beta-amyloid through a beta1 integrin-dependent mechanism. *J Neurosci* (2004) 24(44):9838–46. doi:10.1523/JNEUROSCI.2557-04.2004
35. Bannenberg GL, Chiang N, Ariel A, Arita M, Tjonahen E, Gotlinger KH, et al. Molecular circuits of resolution: formation and actions of resolvins and protectins. *J Immunol* (2005) 174(7):4345–55. doi:10.4049/jimmunol.174.7.4345
36. Clausen BE, Burkhardt C, Reith W, Renkawitz R, Forster I. Conditional gene targeting in macrophages and granulocytes using LysMcre mice. *Transgenic Res* (1999) 8(4):265–77. doi:10.1023/A:1008942828960
37. Burstyn-Cohen T, Heeb MJ, Lemke G. Lack of protein S in mice causes embryonic lethal coagulopathy and vascular dysgenesis. *J Clin Invest* (2009) 119(10):2942–53. doi:10.1172/JCI39325
38. Bratton DL, Henson PM. Neutrophil clearance: when the party is over, clean-up begins. *Trends Immunol* (2011) 32(8):350–7. doi:10.1016/j.it.2011.04.009
39. Anderson HA, Maylock CA, Williams JA, Paweletz CP, Shu H, Shacter E. Serum-derived protein S binds to phosphatidylserine and stimulates the phagocytosis of apoptotic cells. *Nat Immunol* (2003) 4(1):87–91. doi:10.1038/ni871
40. Ariel A, Serhan CN. New lives given by cell death: macrophage differentiation following their encounter with apoptotic leukocytes during the resolution of inflammation. *Front Immunol* (2012) 3:4. doi:10.3389/fimmu.2012.00004
41. Pashover-Schallinger E, Aswad M, Schif-Zuck S, Shapiro H, Singer P, Ariel A. The atypical chemokine receptor D6 controls macrophage efferocytosis and cytokine secretion during the resolution of inflammation. *FASEB J* (2012) 26(9):3891–900. doi:10.1096/fj.11-194894
42. Sica A, Mantovani A. Macrophage plasticity and polarization: *in vivo* veritas. *J Clin Invest* (2012) 122(3):787–95. doi:10.1172/jci59643
43. Chávez-Galán L, Olleros ML, Vesin D, Garcia I. Much more than M1 and M2 macrophages, there are also CD169$^+$ and TCR$^+$ macrophages. *Front Immunol* (2015) 6:263. doi:10.3389/fimmu.2015.00263
44. Roszer T. Understanding the mysterious M2 macrophage through activation markers and effector mechanisms. *Mediators Inflamm* (2015) 2015:16. doi:10.1155/2015/816460
45. Merched AJ, Ko K, Gotlinger KH, Serhan CN, Chan L. Atherosclerosis: evidence for impairment of resolution of vascular inflammation governed by specific lipid mediators. *FASEB J* (2008) 22(10):3595–606. doi:10.1096/fj.08-112201
46. Serhan CN, Chiang N, Van Dyke TE. Resolving inflammation: dual anti-inflammatory and pro-resolution lipid mediators. *Nat Rev Immunol* (2008) 8(5):349–61. doi:10.1038/nri2294
47. Krönke G, Katzenbeisser J, Uderhardt S, Zaiss MM, Scholtysek C, Schabbauer G, et al. 12/15-lipoxygenase counteracts inflammation and tissue damage in arthritis. *J Immunol* (2009) 183(5):3383–9. doi:10.4049/jimmunol.0900327
48. Rőszer T. Phagosomal and lysosomal NO synthesis. *The Biology of Subcellular Nitric Oxide*. Dordrecht: Springer (2012). p. 145–55.
49. Suleiman L, Negrier C, Boukerche H. Protein S: a multifunctional anticoagulant vitamin K-dependent protein at the crossroads of coagulation, inflammation, angiogenesis, and cancer. *Crit Rev Oncol Hematol* (2013) 88(3):637–54. doi:10.1016/j.critrevonc.2013.07.004
50. Poon IKH, Lucas CD, Rossi AG, Ravichandran KS. Apoptotic cell clearance: basic biology and therapeutic potential. *Nat Rev Immunol* (2014) 14(3):166–80. doi:10.1038/nri3607
51. Davra V, Kimani SG, Calianese D, Birge RB. Ligand activation of TAM family receptors-implications for tumor biology and therapeutic response. *Cancers (Basel)* (2016) 8(12):E107. doi:10.3390/cancers8120107
52. Kasikara C, Kumar S, Kimani S, Tsou W-I, Geng K, Davra V, et al. Phosphatidylserine sensing by TAM receptors regulates AKT-dependent chemoresistance and PD-L1 expression. *Mol Cancer Res* (2017) 15(6):753–64. doi:10.1158/1541-7786.mcr-16-0350
53. Lemke G. Phosphatidylserine is the signal for TAM receptors and their ligands. *Trends Biochem Sci* (2017) 42(9):738–48. doi:10.1016/j.tibs.2017.06.004
54. Golpon HA, Fadok VA, Taraseviciene-Stewart L, Scerbavicius R, Sauer C, Welte T, et al. Life after corpse engulfment: phagocytosis of apoptotic cells leads to VEGF secretion and cell growth. *FASEB J* (2004) 18(14):1716–8. doi:10.1096/fj.04-1853fje
55. Millet A, Martin KR, Bonnefoy F, Saas P, Mocek J, Alkan M, et al. Proteinase 3 on apoptotic cells disrupts immune silencing in autoimmune vasculitis. *J Clin Invest* (2015) 125(11):4107–21. doi:10.1172/jci78182

56. Sharif MN, Šošić D, Rothlin CV, Kelly E, Lemke G, Olson EN, et al. Twist mediates suppression of inflammation by type I IFNs and Axl. *J Exp Med* (2006) 203(8):1891–901. doi:10.1084/jem.20051725
57. Titos E, Rius B, González-Périz A, López-Vicario C, Morán-Salvador E, Martínez-Clemente M, et al. Resolvin D1 and its precursor docosahexaenoic acid promote resolution of adipose tissue inflammation by eliciting macrophage polarization toward an M2-like phenotype. *J Immunol* (2011) 187(10):5408–18. doi:10.4049/jimmunol.1100225
58. Ariel A, Li P-L, Wang W, Tang W-X, Fredman G, Hong S, et al. The docosatriene protectin D1 is produced by TH2 skewing and promotes human T cell apoptosis *via* lipid raft clustering. *J Biol Chem* (2005) 280(52):43079–86. doi:10.1074/jbc.M509796200
59. Uderhardt S, Herrmann M, Oskolkova OV, Aschermann S, Bicker W, Ipseiz N, et al. 12/15-lipoxygenase orchestrates the clearance of apoptotic cells and maintains immunologic tolerance. *Immunity* (2012) 36(5):834–46. doi:10.1016/j.immuni.2012.03.010

Conflict of Interest Statement: The authors declare that the research was conducted in the absence of any commercial or financial relationships that could be construed as a potential conflict of interest.

7

Iron Handling in Tumor-Associated Macrophages—Is there a New Role for Lipocalin-2?

*Michaela Jung[1], Andreas Weigert[1], Christina Mertens[1,2], Claudia Rehwald[1] and Bernhard Brüne[1,3]**

[1] Faculty of Medicine, Institute of Biochemistry I, Goethe University Frankfurt, Frankfurt, Germany, [2] Faculty 15, Biological Sciences, Goethe University Frankfurt, Frankfurt, Germany, [3] Project Group Translational Medicine and Pharmacology TMP, Fraunhofer Institute for Molecular Biology and Applied Ecology, IME, Frankfurt, Germany

Carcinogenesis is a multistep process. Besides somatic mutations in tumor cells, stroma-associated immunity is a major regulator of tumor growth. Tumor cells produce and secrete diverse mediators to create a local microenvironment that supports their own survival and growth. It is becoming apparent that iron acquisition, storage, and release in tumor cells is different from healthy counterparts. It is also appreciated that macrophages in the tumor microenvironment acquire a tumor-supportive, anti-inflammatory phenotype that promotes tumor cell proliferation, angiogenesis, and metastasis. Apparently, this behavior is attributed, at least in part, to the ability of macrophages to support tumor cells with iron. Polarization of macrophages by apoptotic tumor cells shifts the profile of genes involved in iron metabolism from an iron sequestering to an iron-release phenotype. Iron release from macrophages is supposed to be facilitated by ferroportin. However, lipid mediators such as sphingosine-1-phosphate, released form apoptotic tumor cells, upregulate lipocalin-2 (Lcn-2) in macrophages. This protein is known to bind siderophore-complexed iron and thus, may participate in iron transport in the tumor microenvironment. We describe how macrophages handle iron in the tumor microenvironment, discuss the relevance of an iron-release macrophage phenotype for tumor progression, and propose a new role for Lcn-2 in tumor-associated macrophages.

Edited by:
Amiram Ariel,
University of Haifa, Israel

Reviewed by:
Christopher Gregory,
University of Edinburgh, United Kingdom
Esther G. Meyron-Holtz,
Technion – Israel Institute of Technology, Israel

Correspondence:
Bernhard Brüne
b.bruene@biochem.uni-frankfurt.de

Keywords: apoptosis, phagocytosis, macrophage polarization, sphingosine-1-phosphate, lipocalin-2, tumor progression

MACROPHAGE UPTAKE OF DYING CELLS IN THE TUMOR MICROENVIRONMENT

Among factors of the tumor microenvironment that shapes the macrophage phenotype to promote cancer are dying cells (1). Tumor cells and other tumor-resident cells undergoing programmed, apoptotic, or necroptotic, as well as accidental necrotic cell death are sensed and removed by macrophages, which induces different functional macrophage programs (**Table 1**). While lytic forms

---

Abbreviations: ANXA1, annexin A1; BDH2, 3-hydroxybutyrate dehydrogenase, type 2; CX3CL1, fractalkine; DAMP, damage-associated molecular pattern; DHBA, dihydroxybenzoic acid; DMT-1, divalent metal transporter-1; FTH, ferritin heavy chain; FTL, ferritin light chain; FPN, ferroportin; HIF, hypoxia-inducible factor; HO-1, heme oxygenase-1; IL, interleukin; IRE, iron responsive element; IRP, iron-regulatory protein; Lcn-2, lipocalin-2; PS, phosphatidylserine; TAM, tumor-associated macrophage; Tf, transferring; TfR, transferrin receptor; S1P, sphingosine-1-phosphate.

TABLE 1 | Molecules involved in the attraction, recognition, and polarization of phagocytes by dying cells.

Mode of cell death	Mode of interaction with phagocytes	Dying cell-derived molecules	Outcome/goal
Apoptosis	Attraction	ATP/UTP (9), LPC (10), S1P (11), RPS19 (12), EMAPII (13), CX3CL1 (14)	Early phagocyte recruitment
	Recognition	PS (15), CRT (16), ANXA1 (17), PTX3 (18)	Corpse removal (phagocytosis)
	Polarization	Tolerogenic apoptosis: PS (19), S1P (20), IL-38 (21), ANXA1 (17)	Immuno-suppression
	Polarization	Immunogenic apoptosis: CRT (16), ATP (22)	Immune activation
Necrosis	Attraction	Primary necrosis: ATP? Secondary necrosis: ANXA1 (23)	Phagocyte recruitment
	Recognition	PS (24), complement (25), antibodies (25), pentraxins (25), F-actin (26)	Corpse removal (macropinocytosis)
	Polarization	HMGB1 (27), ATP (28), DNA (29), IL-1α (30), IL-33 (31)	Immune activation
Necroptosis	Attraction	ATP (32), others?	(Early) phagocyte recruitment
	Recognition	PS (33), others?	Corpse removal (mode unclear)
	Polarization	HMGB1, ATP, DNA, IL-1α, IL-33 + induced DAMPs? [reviewed in Ref. (34)]	Immune activation

LPC, lysophosphatidylcholine; S1P, sphingosine-1-phosphate; RPS19, ribosomal protein S19; EMAPII, endothelial monocyte-activating polypeptide 2; PTX3, pentraxin 3; IL, interleukin; ANXA-1, annexin A1; CRT, calreticulin; PS, phosphatidylserine; HMGB1, high mobility group box 1; DAMP, damage-associated molecular pattern.

of cell death such as necroptosis and necrosis predominantly induce inflammatory cascades that may promote tumor initiation through modifying DNA and triggering cytokine-induced survival pathways in tumor cells (2), apoptotic cells (AC) induce macrophage-dependent matrix remodeling, recruitment of vasculature, and inhibition of antitumor inflammation (1, 3). These properties of AC are seen in analogy to their function during wound healing and regeneration (4, 5), supporting the notion that tumors are "wounds that do not heal" (6). The interaction of macrophages and dying cells, however, does not only alter their functional response, it also comes with a high metabolic load after engulfment of cell debris that needs to be handled by macrophages (7, 8). Hereby, macrophages can be considered an extravascular relay station of tumor-associated metabolism, to acquire and redistribute metabolic intermediates and other (bio)chemical substances, including iron, as outlined in more detail below (see Macrophage Subsets and Iron Handling). To our knowledge, there are no detailed studies comparing the metabolic challenges macrophages face when taking up apoptotic versus necroptotic or necrotic cells, and whether redistribution of nutrients such as iron differs in these circumstances. Studies toward these directions will help to aid decisions, which mode of cell death should be initiated in pathologies such as cancer.

MACROPHAGE POLARIZATION BY DYING CELLS

Macrophage interactions with cells succumbed to different modes of cell death show overlapping and also discreet molecular features at the levels of attraction, recognition, and subsequent alteration of the macrophage phenotype (macrophage polarization) as summarized in **Table 1**.

In a first step, macrophages need to be alerted to their prey. In the case of AC, this is mediated by the active release of phagocyte-attracting molecules, so-called "find-me signals." Their functions are intrinsically coupled to the apoptotic machinery, i.e., demanding caspase activation (35, 36). The release of find-me signals serves to recruit macrophages with the goal to efficiently clear apoptotic corpses before they undergo secondary necrosis. To achieve this goal, AC produce a variety of different find-me signals, probably dependent on the respective apoptotic stimulus. These include the lipids lysophosphatidylcholine and sphingosine-1-phosphate (S1P), the nucleotides ATP and UTP as well as the proteins fractalkine (CX3CL1), ribosomal protein S19, and endothelial monocyte-activating polypeptide 2. Moreover, apoptosis in the context of an inflammatory environment generates a number of different chemokines (3, 37–39). Of these, only CX3CL1 (14), ATP/UTP (9), and S1P (40) have been connected to phagocyte recruitment to AC *in vivo*. The diverse biochemical nature of these find-me signals and their different production kinetics (35) suggests a remarkable degree of redundancy. This redundancy likely ensures efficient macrophage recruitment at different time-points during the apoptotic cascade and from different locations, i.e., local macrophages versus monocytes from the circulation, based on short half-life versus long half-life of the find-me signals and their concentration in local tissues versus the circulation (14, 41, 42). When looking at lytic forms of cell death, characterized by the loss of plasma membrane integrity, the picture appears less clear. Secondary necrotic cells that were formerly apoptotic generate distinct find-me signals such as annexin A1 (ANXA1) fragments to sustain their clearance (23). Whether such ANXA1 fragments or other specific find-me signals are actively produced during necrosis or cells undergoing accidental, primary necrosis is largely unknown. A recent report suggests the release of nucleotides from necroptotic cells that, *in vitro*, induced a rapid and immunological silent clearance (32). Importantly, lytic cell death will promote the passive release of a number of apoptotic cell-derived find-me signals such as lipids or nucleotides, likely in higher quantities. The contribution of these molecules to necrotic/necroptotic cell clearance remains to be determined.

Regardless the find-me signal, phagocytes need to discriminate dying cells from their living neighbors. This is accomplished by the recognition of "eat-me" signals that are exposed by dying cells, in concert with the absence of "don't eat-me" signals that restrict the uptake of living cells (43, 44). The most prominent eat-me signal that appears to be relevant to clear all dying cells, irrespective of the mode of cell death is the phospholipid phosphatidylserine

(PS) that is confined to the inner leaflet of the plasma membrane in living cells, but gets oxidized and redistributed to the outer leaflet during apoptosis (15, 45, 46). PS is also involved in clearance of necrotic or necroptotic cells, but whether oxidative modification is equally required remains unknown (24, 25, 33). Other eat-me signals that are specifically exposed on the plasma membrane of AC include calreticulin (CRT), ANXA1, and the long pentraxin PTX3 (37). These eat-me signals are recognized either directly by specific receptors including scavenger receptors, complement receptors, C-type lectin receptors, a number of PS-specific receptors, and the pattern recognition receptor (PRR) CD14, or indirectly *via* bridging molecules that mainly promote the recognition of PS by Tyro, Axl, and MerTK-family receptor tyrosine kinases or the vitronectin receptor (VnR, $\alpha_v\beta_3$ integrin) (3, 47). Some of the molecular interactions that mediate apoptotic cell recognition are employed for efficient uptake of necrotic debris, but specific pathways were also identified. Necrotic cell uptake is mediated by the classical complement pathway (C1q) and the mannose pathway (mannose-binding lectin and ficolins) recognized *via* complement receptors on phagocytes, antibodies, short pentraxins such as CRP, and serum amyloid protein *via* Fc-receptors, and also *via* PS recognition by VnR [reviewed in Ref. (25)]. Moreover, F-actin filaments in necrotic debris are recognized by the c-type lectin CLEC9A to facilitate clearance (26). The combination of different uptake receptors results in fundamental difference in the uptake mode. Whereas AC are engulfed *via* phagocytosis that is sensitive to PI3K inhibition, necrotic debris is taken up by PI3K-independent macropinocytosis (48). Specific uptake receptors and uptake mechanisms for necroptotic cells are so far not described.

Signals emanating from these numerous interactions between the phagocyte and its prey initiate not only corpse engulfment but also powerfully modulate inflammatory responses of the phagocytes [reviewed in Ref. (49)]. These responses are often fundamentally different when comparing apoptosis with cell death modes that are characterized by the loss of plasma membrane integrity and the subsequent spilling of intracellular components into the extracellular space. Many of such intracellularly confined molecules are considered as danger signals that are sensed by pathogen recognition receptors (PRRs) on phagocytes that also sense microbial components, and therefore trigger pro-inflammatory pathways. Such damage-associated molecular patterns (DAMPs) comprise high mobility group box 1, a protein usually found in complex with chromatin in the nucleus that activates toll-like receptors (TLR)-2, 4, 9 and the receptor for advanced glycation end products, the interleukins IL-1α and IL-33, DNA that is recognized by TLRs, and ATP that activates purine receptors including P2RX7. These pathways may be also exploited to trigger antitumor immunity (34, 50–53). Similar ligand receptor interactions may be assumed when considering the effect of necroptotic cells on the functional phenotype of their phagocytes based on the lytic nature of necroptotic cell death (53). An interesting difference is that necroptosis, as a form of regulated cell death, allows the transcriptional upregulation of additional DAMPs such as heat-shock proteins (34).

In contrast to necrosis or necroptosis, apoptotic cell death is usually considered as anti-inflammatory or immunologically silent. This is triggered at least partially through the recognition of PS, since mice lacking non-PS eat-me signal receptors such as CD14, CD36, and $\alpha_v\beta_3$ integrin do not show major signs of auto-inflammation (54, 55). PS recognition on AC by macrophages suppressed the production of inflammatory cytokines, dependent on the autocrine production of transforming growth factor-β, platelet-activating factor, and prostaglandin E_2 (PGE_2) (19). Inhibition of inflammatory cytokine release from macrophage interacting with AC was further linked to inhibition of the classical NF-κB pathway (p65/p50 heterodimers) through transcriptional repression *via* the nuclear hormone receptor peroxisome proliferator-activated receptor γ (PPARγ) (56–59), or through PPARγ-dependent upregulation of phagocytic receptors such as MerTK (60). MerTK activation in turn interferes with NF-κB signaling (61). Besides inhibiting NF-κB, PS recognition on AC reduced the formation of nitric oxide (NO) and NADPH oxidase-dependent reactive oxygen species in macrophages (62–64). Therapeutically, targeting PS in tumors induced inflammatory macrophage activation to suppress tumor growth and progression in prostate tumors (65). MerTK-deficient mice with autochthonous mammary carcinoma were protected from metastasis, which was initiated by the interaction of macrophages with AC during mammary gland involution after pregnancy (66). Thus, PS recognition creates a feed-forward loop to guarantee efficient corpse removal and blocks a number of inflammatory pathways. This likely promotes efficient and immunologically silent corpse removal during homeostasis, but is exploited by tumors to promote carcinogenesis.

Also soluble factors produced by AC contribute to limiting inflammation, including the find-me signals CX_3CL1 and S1P (3, 40, 67). Besides, AC release signals that exclusively limit destructive inflammation. The IL-1 family receptor antagonist IL-38 was proteolytically processed and released from AC and specifically inhibited the generation of Th17 cells, which are associated with chronic inflammation (21). In conclusion, apoptosis induces the production of membrane-bound and soluble cues that serve to ensure immunologically silent clearance by phagocytes.

The question remains why necrosis retains its inflammatory potential despite PS recognition being part of the program to remove necrotic debris. One explanation is that other signaling pathways are activated in necrotic versus AC due to alternative PS receptor usage and associated signaling pathways. For instance, apoptotic cell recognition activates the PI3K pathway, which is critical to limit inflammation (68), whereas removal of necrotic debris does not require PI3K (48). Alternatively, the context of PS recognition may matter, as phagocytes face necrotic cell-derived DAMPs before or simultaneously when engaging PS. Along this line, apoptotic cell recognition under conditions of danger, indicated by, e.g., ER stress or the presence of pathogens, promotes inflammation rather than restricting it. For instance, infected AC trigger autoimmune T cell (Th17) generation (69). Also, the potentially immunogenic molecule CRT on the surface of AC, which requires autophagy or ER stress linked to the execution of apoptosis, is recognized through the LDL-receptor-related protein on phagocytes (16, 70, 71). The find-me signal ATP, besides ensuring corpse clearance, can bind P2X purinoceptor 7 (P2RX7) to induce activation of the NOD-Like receptor family, pyrin domain containing 3 (NLRP3) inflammasome and subsequent IL-1β release. These examples illustrate that apoptosis does not always restrict inflammation, which may explain why necrosis can trigger inflammation despite PS recognition. In wounds, this ensures that inflammation proceeds as long as pathogens or other noxa are present. In tumors, the response to apoptosis likely depends on the microenvironment. Immunosuppressive apoptosis is exploited by the tumor to fuel its growth (1), whereas induction of immunogenic cell death may be exploited therapeutically to initiate protective immunity (72).

IRON HANDLING PROTEINS IN MACROPHAGES

In humans, approximately 60% of total body iron is present in erythrocytes, bound to heme in hemoglobin (73). About 2 million senescent red blood cells (RBCs) are cleared per second from the circulation by tissue-specific macrophages (74). Senescent erythrocytes get recognized by macrophages due to alterations in the membrane protein Band 3 that only is displayed by aged erythrocytes (75, 76). Additionally, PS is exposed on the outer leaflet of the cell membrane (77) and membrane rigidity is increased to foster recognition by macrophages (78). A daily turnover of about 20 mg iron makes macrophages essential players in iron metabolism, as we only take up 1–2 mg iron with our daily nutrition. In order to fulfill their essential roles in systemic iron homeostasis, macrophages evolved a variety of pathways to take up, recycle, store, or release iron. The majority of iron is delivered by transferrin (Tf), the main iron-transport protein in the blood, circulating between the reticuloendothelial system, and the bone marrow in order to guarantee hematopoiesis. Tf binds to the transferrin receptor (TfR) and following internalization, iron is released from Tf in the endosome. Subsequently, the divalent metal transporter-1 (DMT-1) shuttles iron from the endosome into the cytoplasm. In addition, macrophages recycle phagolysosomal iron through natural resistance-associated macrophage protein 1 (Nramp-1), a divalent metal transporter homologous to DMT-1 (79). If iron supply exceeds its demands, iron can be stored by the iron storage protein ferritin, consisting of ferritin heavy chain (FTH) and ferritin light chain (FTL) subunits. FTH and FTL differ in their function, as FTH has a ferroxidase activity and FTL is important for iron core nucleation (80). By transforming soluble and reactive ferrous iron (Fe^{2+}) of the cytoplasmic labile iron pool (LIP) into the insoluble ferric iron (Fe^{3+}) and store Fe^{3+} within the soluble ferritin shell, cells avoid the potential damage of redox active iron, i.e., Fenton chemistry. Consequently, the intracellular iron amount must be tightly regulated. Iron is sensed by the iron-regulatory proteins (IRPs) IRP1 and IRP2. When intracellular iron is low, IRPs bind to iron responsive elements (IREs) in the untranslated regions (UTRs) of certain mRNAs. Binding of IRPs to IREs in the 5′-UTR attenuates translation, whereas binding to IREs in the 3′-UTR stabilizes respective mRNAs and fosters translation. Mechanistically, when iron is high, mRNAs of TfR and DMT-1 are unstable, which decreases iron uptake and transport. Simultaneously, iron storage is supported by releasing a translational blockade of FTH and FTL. Iron export from cells is achieved through ferroportin (FPN), the only known ferrous iron exporter. Although details of the transport remain unclear, the transport of ferrous iron (Fe^{2+}) requires ferroxidase activity to convert it to ferric iron (Fe^{+3}) in order to load it onto Tf. The oxidation to ferric iron is catalyzed by either free or GPI-anchored ceruloplasmin (CP) or the transmembrane protein hephestin that are copper-containing ferroxidases in order to stabilize FPN to facilitate iron efflux (81) and assure efficient Tf-loading. This makes FPN an important checkpoint to adjust global and local iron homeostasis and to adjust iron storage versus its release (82). The protein amount of FPN is controlled by the peptide hormone hepcidin, which primarily is produced in the liver. Hepcidin regulates FPN stability by inducing its internalization and proteasomal degradation thus, affecting the systemic iron level (83) by reducing iron export. Consequently, hepcidin expression and release from the liver increases with the systemic iron amount and decreases under iron deprived conditions to guarantee macrophage iron supply for erythropoiesis. This is crucial as iron supply is the rate-limiting step during erythropoiesis (84). As shown in **Figure 1**, heme bound to the heme-sequestering protein hemopexin (Hpx) is taken up by the hemopexin receptor (CD91), while CD163 binds haptoglobin-transported hemoglobin (85).

Overall, iron metabolism is tightly regulated by a network of proteins to guarantee iron homeostasis with specific macrophage subsets in key positions to fulfill their role in iron recycling by erythrophagocytosis and iron release to sustain erythropoiesis.

MACROPHAGE SUBSETS AND IRON HANDLING

The major macrophages subsets involved in systemic iron homeostasis consist of red pulp macrophages in the spleen and Kupffer cells in the liver. Their main function is to phagocytose damaged or senescent erythrocytes to recover iron. Therefore, phagocytosed erythrocytes are exposed to reactive oxygen species and hydrolytic enzymes in the erythrolysosomal compartment, with the subsequent release of hemoglobin and heme. Heme is then degraded by heme oxygenase-1 (HO-1) to carbon monoxide, biliverdin, and free iron, which usually joins the chelatable LIP or is stored (86). Heme also induces the expression of the transcription factor Spi-C, which is essential for the differentiation of red pulp macrophages (87) and erythroid island macrophages in the bone marrow (88), with no other macrophage subset being affected. Consequently, Spi-C knockout mice accumulate iron in the red pulp of the spleen, as RBCs are trapped but inefficiently phagocytosed (87). Red pulp macrophages, compared to other macrophage subsets, show a specialized gene signature required for enhanced iron recycling in order to fulfill their crucial metabolic function in systemic iron homeostasis.

However, macrophages are highly plastic cells regarding their functional properties, responding to a great number of inflammatory stimuli (89, 90). As extremes within a continuum, two opposing states of macrophage activation were identified. Macrophages activated by T helper 1 (Th1) cell-derived interferon-γ (IFN-γ), in combination with TLR ligands such as lipopolysaccharide

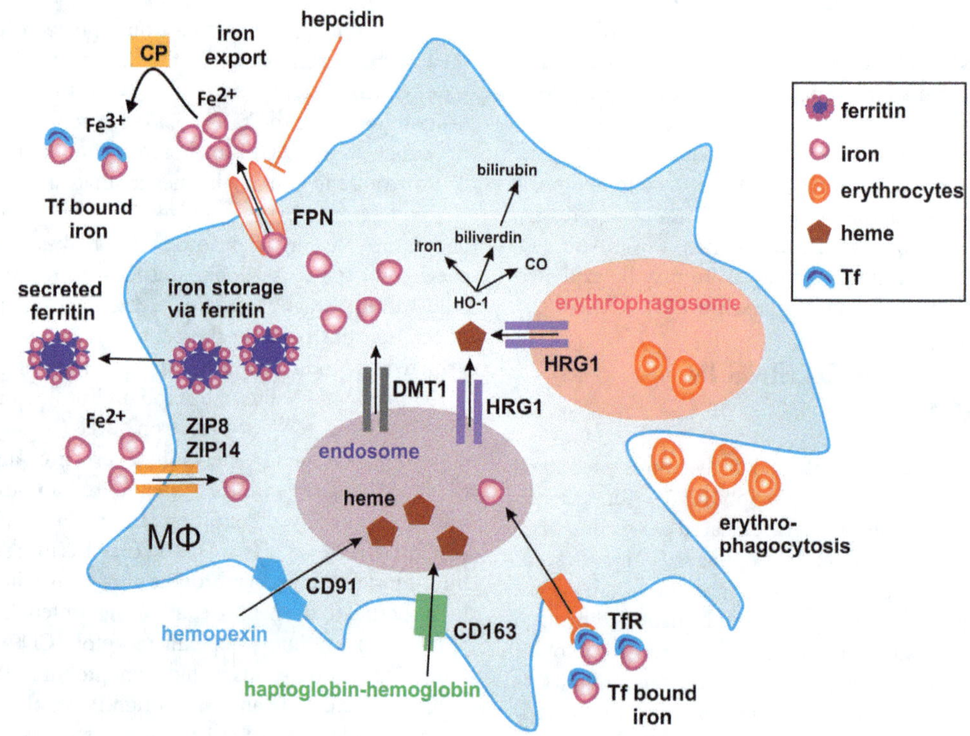

FIGURE 1 | Iron handling in macrophages. Macrophages take up, metabolize, store, and export iron. Classically activated macrophages sequester iron by taking up transferrin (Tf) bound iron *via* the Tf-receptor (TfR) or accumulating ferrous iron (Fe^{2+}) *via* zinc transporters ZIP8 and ZIP14. Iron export *via* ferroportin (FPN) is impaired by binding of hepcidin (HAMP), thus causing iron storage in ferritin. In alternatively activated macrophages, the uptake of hemopexin-heme (Hpx-heme) by CD91 or haptoglobin-hemoglobin by CD163 into endosomes as well as phagocytosis of senescent erythrocytes into erythrophagosomes results in the release of heme *via* the heme transporter HRG1 to the cytosol. The subsequent activation of heme oxygenase-1 (HO-1) further degrades heme into iron, CO, and biliverdin, which is further processed to bilirubin. Fe^{2+} is exported from alternatively activated macrophages through FPN and oxidized to ferric iron (Fe^{3+}) iron by ceruloplasmin (CP), which is essential for efficient binding to Tf. Secretion of ferritin-bound iron represents an alternative route of iron export.

(LPS) creates cells with a strong pro-inflammatory profile. These "classically activated" macrophages generate pro-inflammatory mediators such tumor necrosis factor-α, IL-1β, IL-6, IL-12, and IL-23, reactive oxygen and nitrogen species, and present antigens to T cells. Classically activated macrophages are efficient in microbial host defense and show antitumor activity. In contrast, macrophages stimulated by activated T helper 2 (Th2) cell-derived IL-4 or IL-13, or by IL-10, produce alternative sets of cytokines, functionally oppose the repertoire of classically activated macrophages, and help to resolve inflammation. Additionally, such "alternatively activated" macrophages express specific phagocytic receptors, combat extracellular parasites, and help to promote tissue remodeling by producing extracellular matrix and growth factors (89–91). Taking their functional diversity into account, it is not surprising that macrophages also show distinct properties in handling iron (92). Iron recycling by macrophages comprises the steps of uptake, storage, and release. These are critical features, as there is no way to get rid of body iron, except during bleeding or sloughing of mucosa and/or skin. As part of their functional repertoire upon activation of tissue-resident macrophages or differentiation of newly recruited, tissue-infiltrating monocytes, macrophages evolved multiple ways to handle iron (**Table 2**) according to diverse microenvironmental stimuli.

TABLE 2 | Iron regulated genes in classically and alternatively activated macrophages.

	Classically activated	Alternatively activated
Receptors		
Transferrin receptor (TfR)	↓	↑
CD91	↓	↑
CD163	↓	↑
Recycling		
Heme oxygenase-1 (HO-1)	↓	↑
Trafficking		
Ferroportin (FPN)	↓	↑
Divalent metal transporter 1 (DTM-1)	↑	↑
Transferrin	↑	↑
Storage		
Ferritin (FT)	↑	↓
Regulation		
Hepcidin (HAMP)	↑	↓
Iron-regulatory proteins (IRP)	↓	↑
Oxidreductase		
Ceruloplasmin (CP)	↓	↑

Regulation of genes related to iron metabolism in classically versus alternatively activated macrophages (93, 94). ↑ upregulation, ↓ downregulation.

The groups of Recalcati and Cairo (93) discovered that classically activated macrophages tend to accumulate iron, whereas alternatively activated macrophages provide recycled iron to their microenvironment. To do so, classically activated macrophages maximize iron uptake directly *via* the TfR, and indirectly *via* the Nramp-1 and DMT-1 as well as through storage by ferritin, whereas they downregulate FPN-mediated iron export. During inflammation, this serves to deplete invading pathogens from iron (95–99). In addition, bacteria-derived LPS and pro-inflammatory cytokines cause macrophages to express hepcidin (83), which degrades FPN and adds to restrain iron. Thus, under infectious/inflammatory conditions, macrophages are a major site for storing iron. Acute phase proteins as well as the formation of reactive oxygen species and NO join to induce a macrophage iron-sequestration phenotype, mainly achieved by downregulating FPN. Opposed to these functions, alternatively activated macrophages provide iron to their local microenvironment (93, 94). Taking into account that alternatively activated macrophages express scavenger receptors that not only serve as PRRs but also sense and clear AC these macrophages accumulate hemoglobin during hemodialysis or inflammation. The inflammatory-promoting actions of free heme are antagonized (100–102), thereby fostering the resolution of inflammation. Moreover, the redox-sensitive transcription factor nuclear factor erythoid 2-like 2 gets activated with concomitant transcription of the iron exporter FPN and the heme-degrading enzyme HO-1 (40). In alternatively activated macrophages, heme-recycled iron joins the LIP for a rapid release *via* FPN, while in inflammatory macrophages iron is stored in ferritin. Iron export may also add to stabilize hypoxia-inducible factor (HIF) in macrophages, as iron is a prerequisite for the activity prolyl hydroxylases (PHD) as part of the HIF-degrading machinery (103). Lowering iron deactivates PHD enzymes, which stabilizes HIF-1α (104). Active HIF-1 causes target gene activation of, e.g., arginase-1, which are part of the alternative macrophage signature (105). These considerations suggest that in alternatively activated macrophages iron from the LIP is preferentially provided to the local microenvironment. In turn, this source of extracellular iron, provided by macrophages, may add to promote tissue regeneration, but may also be part of the tumor-promoting capacity of tumor-associated macrophages (TAMs). Mechanistically, an increased pool of iron stimulates proliferation of fibroblasts or tumor cells in the neighborhood of macrophages. The question remains how cancer cells acquire iron from their local microenvironment that includes, among others, macrophages. Conclusively, macrophages residing in the tumor stroma may provide iron to their local microenvironment, which includes the expression of alternative iron-transport mechanisms.

TAMs AND IRON IN THE TUMOR MICROENVIRONMENT

Cell division, growth, and survival of malignant cells require iron. Therefore, tumor cells enhance iron import and storage mechanisms and decrease iron export. It is already known that tumor cells adopt an iron-utilization phenotype, tightly linked to intracellular iron sequestration (106). This is achieved by upregulating the TfR (107) and hepcidin (108) as well as downregulating FPN (109). In contrast to tumor cells, inflammatory cells of the stroma, e.g., infiltrating macrophages and lymphocytes, acquire an "iron-donor" phenotype (110), which is accomplished by upregulating the iron exporter FPN. Mechanistically, our and other labs noticed that IL-10- or IL-4-stimulated macrophages export iron to accelerate tumor growth by supplying iron to actively proliferating tumor cells (93, 94). This may point to a so far unappreciated facet of stromal cells in promoting tumor progression by supplying iron in order to facilitate the transition from pre-malignant lesions to invasive tumors. It was also speculated that tumor cells specifically hijack the process of erythrophagocytosis by macrophages to sustain their survival and proliferation. Knutson and coworkers illustrated the role of FPN in J774 macrophages in releasing ^{59}Fe after phagocytosis of ^{59}Fe-labeled RBCs (111). Erythrophagocytosis not necessarily is restricted to the systemic level, but may also occur locally under conditions of increased blood flow. In this respect, tumor cells may benefit from tumor angiogenesis and an increased migration of cells into tumor tissue (112), which fosters the convergence of TAMs with erythrocytes. Of note, tumor vessels are often leaky and it can be speculated that macrophages residing close to these vessels are particularly prone to recycle and donate iron to tumor cells (113). So far it remains unclear whether the physiological process of erythrophagocytosis alters genes that regulate iron metabolism in a way similar to those seen in alternatively activated macrophages or TAMs. TAMs alter their gene expression profile in favor of a tumor-supportive, iron-release phenotype (**Figure 2**). This is reflected by an increased expression of CD163, the high-affinity scavenger receptor for haptoglobin bound to hemoglobin (114). Subsequent to the uptake of hemoglobin or haptoglobin, released heme is then degraded *via* HO-1 and induces the downregulation of the transcription factor Bach-1, thereby allowing the transcription of FPN (115). Thus, at least in macrophages, heme may work as a modulator of FPN expression, independent of hepcidin. As a substantial part of plasma iron is provided by macrophages, recycled from senescent RBCs, it is surprising that mice carrying a knockout of FPN in macrophages only showed mild signs of anemia (116). One explanation could be that macrophages can export heme *via* the feline leukemia virus C receptor transporter (117), which might substitute for a loss of FPN. Moreover, in an inflammatory environment, ferritin can be secreted from lymphocytes and macrophages in the tumor stroma (118, 119). In line, extracellular ferritin causes proliferation of breast cancer cells, independent of its iron content. Another explanation, as discussed below, might be the development of additional or alternative iron-transport mechanisms.

Lcn-2 MAY FUNCTION AS AN IRON TRANSPORTER IN THE TUMOR MICROENVIRONMENT

Taking into consideration that cancer cells have a higher demand for iron, it is tempting to speculate that tumor cells hijack

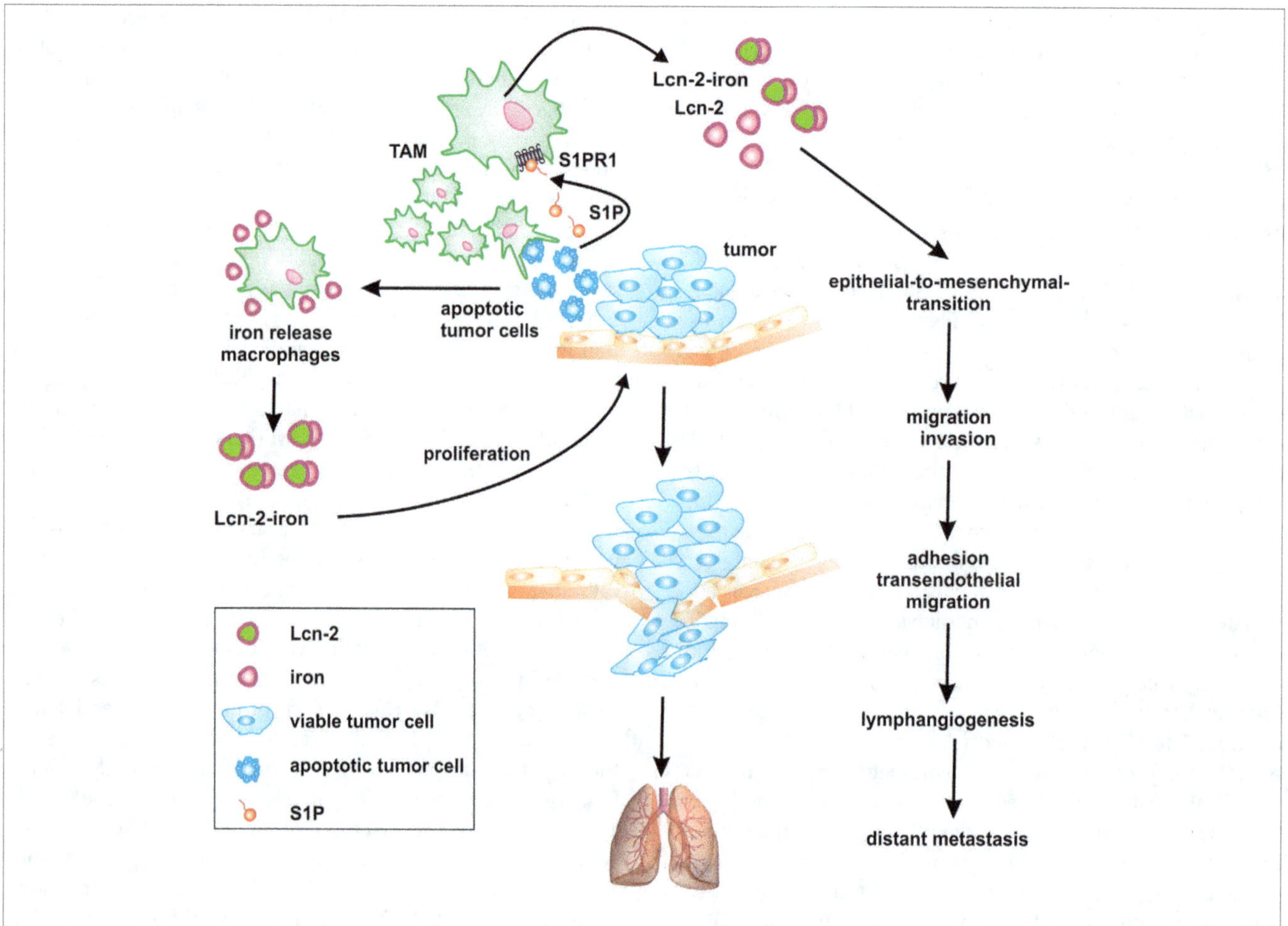

FIGURE 2 | Apoptotic cell-derived S1P induces lipocalin-2 (Lcn-2) to promote tumor progression. The cross talk between apoptotic tumor cells and macrophages creates a feed-forward mechanism, where macrophages become polarized toward a pro-tumorigenic, iron-release phenotype, including the formation of iron-loaded Lcn-2. Lcn-2 is activated in tumor-associated macrophages (TAMs) through apoptotic cell-derived sphingosine-1-phosphate (S1P) downstream of the S1P receptor 1. TAM-derived, iron-loaded Lcn-2 exerts pro-tumorigenic actions by promoting tumor cell proliferation and enhancing metastatic spread to the lung at various steps, including the induction of epithelial-to-mesenchymal transition, migration, and invasion as well as adhesion and transendothelial migration.

macrophages to turn them into an iron-delivery cell. In this regard, we noticed that the high-affinity iron-carrier Lcn-2 is part of the pro-tumor macrophage phenotype and upregulated following macrophage-apoptotic cell interactions.

Lipocalin-2 belongs to the lipocalin superfamily. These proteins are known for their bacteriostatic effects by capturing and depleting siderophores (120). Furthermore, recent evidence suggests that Lcn-2 stimulates growth and differentiation in various cells (121). Exogenous Lcn-2 causes marker gene expression profiles that reflect early epithelial progenitors and epithelial cell proliferation (122). Our work revealed that apoptotic tumor cells stimulated protein expression and secretion of Lcn-2 in macrophages along with their functional shift toward an alternative phenotype (123, 124). Macrophage-derived Lcn-2 stimulates cancer cell proliferation (124), tumor cell dissemination, metastasis (125), and tumor lymphangiogenesis (126). Mice lacking Lcn-2 developed significantly less tumors, while an impact on metastases was not consistently observed (127–129). The impact of Lcn-2 on metastasis might depend on the cellular source, with at least macrophage-derived Lcn2 promoting metastasis in mammary carcinoma (125, 126). Studies in humans pointed to Lcn-2 as a pro-tumorigenic factor in breast cancer, correlated with decreased survival and reduced responsiveness to neoadjuvant chemotherapy (130, 131). Mechanistically, overexpression of Lcn-2 in non-invasive human MCF-7 breast cancer cells elicits an aggressive phenotype that promotes growth and metastasis by inducing epithelial-to-mesenchymal transition (132). So far most studies focused on the role of Lcn-2 in stabilizing matrix metalloproteinase-9 to explain cancer metastasis, linked to extracellular matrix degradation, migration, and invasion. Furthermore, Lcn-2 coordinates the expression of vascular endothelial growth factor and causes the induction of angiogenesis in the tumor microenvironment (133). Interestingly, Lcn-2 was mainly examined in tumor cells. The possibility that Lcn-2 is also provided by tumor-infiltrating immune cells was not fully appreciated. However, own results provided evidence that TAM express increasing amounts of Lcn-2 (124, 134).

We previously showed that the interaction of macrophages with AC shapes the macrophage phenotype and function (11).

Importantly, macrophage activation upon their interaction with AC was independent of phagocytosis or cell-cell contact, but demanded the release of S1P from AC (135). Moreover, we demonstrated a critical involvement of sphingosine kinase 2 in the production of S1P during tumor cell apoptosis (20). We further obtained evidence that Lcn-2 was expressed in primary human macrophages in response to dying MCF-7 breast cancer cells (123). Mechanistically, Lcn-2 production was connected to S1P release from apoptotic cancer cells. S1P elicited signal transducer and activator of transcription 3-dependent induction of Lcn-2 in macrophages. siRNA studies in primary human macrophages and the use of bone marrow-derived macrophages from S1P receptor knockout mice suggested that the S1PR1 was required for Lcn-2 induction in macrophages (126). We substantiated Lcn-2 as a key macrophage phenotype determinant, with parallel actions during physiological tissue regeneration and repair mechanisms (123), but also under pathophysiological conditions such as tumor development. In human and experimental tumors, tumor-infiltrating macrophages are massively exposed to apoptotic tumor cells, since cells at core tumor regions undergo cell death as a consequence of oxygen and nutrient deprivation. Therefore, we speculate that dying tumor cells educate macrophages at core tumor regions in order to access additional iron *via* Lcn-2. However, it is presently unclear whether the pro-tumor actions of Lcn-2 depend on its iron loading or not. Previously, it was shown that iron-loaded holo-Lcn-2 favors cellular survival and proliferation by increasing the intracellular iron content and the induction of Bcl-2 (136). In contrast, the uptake to iron-free apo-Lcn-2 causes cell death, which was correlated to the expression of Bim.

Importantly, Lcn-2 does not directly bind iron. The iron-trafficking function of Lcn-2 largely depends on its association with bacterial or mammalian siderophores. Siderophores are iron-chelating molecules that were first described in bacteria (137–139). Devireddy et al. recently reported that mammals also produce iron-sequestring agents to enhance innate immune responses. Along this line, the mammalian siderophore 2,5-dihydroxybenzoic acid (2,5-DHBA) was characterized (140), which is structurally similar to the bacterial enterobactin. Lcn-2 interacts with siderophores in order to control bacterial growth as part of the innate immune response. Consequently, mice lacking Lcn-2 are more prone to a number of pathogens (141). Since siderophore-binding constitutes the limiting factor for Lcn-2-dependent iron handling, it is important to understand the function, regulation, and sources of mammalian siderophores. Several of the biological functions of Lcn-2 have already been linked to its association with the iron-loaded siderophore 2,5-DHBA (136). In the tumor, it might be speculated that tumor cells evolved a strategy to produce and secrete siderophores in order to sequester iron. Consequently, siderophore shuttling from tumor cells to TAM would allow Lcn-2 iron loading and the reverse transport of iron-loaded Lcn-2. Up to now, it is unclear how siderophores are taken up by mammalian cells. It also remains unknown whether 2,5-DHBA acts alone as the iron-chelating siderophore or whether it functions as the iron-binding moiety of a more complex siderophore structure as described for 2,3-DHBA in enterobactin. Regarding its production in mammalian cells, it was previously described that the mammalian enzyme 3-hydroxybutyrate dehydrogenase, type 2 (BDH2) synthesizes 2,5-DHBA. Unfortunately, the exact mechanism how BDH2 synthesizes 2,5-DHBA still remains elusive but the knockdown of BDH2 completely depleted cellular 2,5-DHBA (140). BDH2 knockout mice developed severe anemia and splenic iron overload (142) thus, confirming the requirement of 2,5-DHBA for iron transport (140). Additionally, the knockdown of BDH2 in mammalian cells points to an important role of 2,5-DHBA in balancing the LIP. A knockdown of BDH2 is linked to high cytoplasmic iron content and elevated levels of reactive oxygen species, whereas mitochondria became iron deficient (140). This increases an oxidative stress signature (143). Intriguingly, the expression of BDH2 negatively correlated with patient survival suffering from normal acute myeloid leukemia (144). However, regarding the clinical importance of siderophores, especially in tumors, more investigations are needed. Independent of their putative endogenous roles in pathology, sidrophores represent an attractive target for therapeutic approaches, e.g., as iron-chelating drugs in cancer therapy or iron-overload diseases, due to their high iron affinity (145, 146). Another possibility would be the use of siderophores as "trojan horse" in order to deliver antibiotics or other toxic compounds to resistant bacteria (147) and possibly tumor cells.

CONCLUSION

In the tumor microenvironment, macrophages are subjected to an intense cross talk with tumor cells. Signal exchange is facilitated by chemically diverse, soluble mediators as well as communication by cell-cell contacts. This comprises the release of S1P from apoptotic tumor cells. As a consequence of the liaison between innate immune and tumor cells, the phenotype of macrophages changes. They become less cytotoxic and their cytokine mediator profile supports rather than antagonizes tumor progression to basically support all hallmarks of cancer (**Figure 2**).

Tumor cells with their high capacity to proliferate show a strong demand for accumulating iron. Consequently, it seems rational that TAMs gain an iron-release phenotype, thereby allowing tumor cells to access additional sources of iron. In the tumor context, macrophages may be forced by S1P to upregulate a so far unappreciated iron export system, the key component being Lcn-2. Although highly speculative, one can envision that the siderophore 2,5-DHBA is produced and released from tumor cells, travels to macrophages to load iron and in turn shuttles back to tumor cells to unload its cargo. Surplus iron in tumor cells is now being used to foster growth and survival and to add to the distinct phases of tumor dissemination and metastasis. This unique iron distribution system may offer the advantage to interfere pharmacologically and thus, more selective than manipulating the overall iron homeostasis in our body. Selectivity may be obtained if we successfully chelate iron in TAMs, interfere with expression regulation of macrophage Lcn-2 or the proposed shuttling of the mammalian siderophore.

While the overarching role of TAMs during tumor progression is undisputed, underlying molecular mechanisms are less clear. We believe that altering mechanisms of iron handling in tumor and stroma cells, i.e., macrophages has to be added to the list of changes that occur in the tumor microenvironment and shape the unique macrophage phenotype found in tumors.

AUTHOR CONTRIBUTIONS

All authors added to designing the work, acquired, analyzed, and interpreted data, wrote parts of the manuscript, approved the final version, and discussed all questions concerning the accuracy and integrity of the work.

ACKNOWLEDGMENTS

We apologize to those researchers who added important original findings to the field that could only be cited indirectly due to space restrictions.

FUNDING

This work was supported by grants of the German Research Association (SFB815: TP08, SFB1039: TPB04 and TPB06), Deutsche Krebshilfe (109599, 70112451), Else Kröner-Fresenius-Stiftung (Else Kröner-Fresenius-Graduiertenkolleg), LOEWE (Research Center for Translational Medicine and Pharmacology TMP), and the Else Kröner-Fresenius Foundation (EKFS), Research Training Group Translational Research Innovation – Pharma (TRIP).

REFERENCES

1. Weigert A, Mora J, Sekar D, Syed S, Brune B. Killing is not enough: how apoptosis hijacks tumor-associated macrophages to promote cancer progression. *Adv Exp Med Biol* (2016) 930:205–39. doi:10.1007/978-3-319-39406-0_9
2. Kuraishy A, Karin M, Grivennikov SI. Tumor promotion via injury- and death-induced inflammation. *Immunity* (2011) 35(4):467–77. doi:10.1016/j.immuni.2011.09.006
3. Gregory CD, Pound JD. Cell death in the neighbourhood: direct microenvironmental effects of apoptosis in normal and neoplastic tissues. *J Pathol* (2011) 223(2):177–94. doi:10.1002/path.2792
4. Bosurgi L, Cao YG, Cabeza-Cabrerizo M, Tucci A, Hughes LD, Kong Y, et al. Macrophage function in tissue repair and remodeling requires IL-4 or IL-13 with apoptotic cells. *Science* (2017) 356(6342):1072–6. doi:10.1126/science.aai8132
5. Cummings RJ, Barbet G, Bongers G, Hartmann BM, Gettler K, Muniz L, et al. Different tissue phagocytes sample apoptotic cells to direct distinct homeostasis programs. *Nature* (2016) 539(7630):565–9. doi:10.1038/nature20138
6. Dvorak HF, Flier J, Frank H. Tumors – wounds that do not heal – similarities between tumor stroma generation and wound-healing. *N Engl J Med* (1986) 315(26):1650–9. doi:10.1056/NEJM198612253152606
7. Han CZ, Ravichandran KS. Metabolic connections during apoptotic cell engulfment. *Cell* (2011) 147(7):1442–5. doi:10.1016/j.cell.2011.12.006
8. Hochreiter-Hufford A, Ravichandran KS. Clearing the dead: apoptotic cell sensing, recognition, engulfment, and digestion. *Cold Spring Harb Perspect Biol* (2013) 5(1):a008748. doi:10.1101/cshperspect.a008748
9. Elliott MR, Chekeni FB, Trampont PC, Lazarowski ER, Kadl A, Walk SF, et al. Nucleotides released by apoptotic cells act as a find-me signal to promote phagocytic clearance. *Nature* (2009) 461(7261):282–6. doi:10.1038/nature08296
10. Lauber K, Bohn E, Krober SM, Xiao YJ, Blumenthal SG, Lindemann RK, et al. Apoptotic cells induce migration of phagocytes via caspase-3-mediated release of a lipid attraction signal. *Cell* (2003) 113(6):717–30. doi:10.1016/S0092-8674(03)00422-7
11. Weigert A, Johann AM, von Knethen A, Schmidt H, Geisslinger G, Brune B. Apoptotic cells promote macrophage survival by releasing the antiapoptotic mediator sphingosine-1-phosphate. *Blood* (2006) 108(5):1635–42. doi:10.1182/blood-2006-04-014852
12. Horino K, Nishiura H, Ohsako T, Shibuya Y, Hiraoka T, Kitamura N, et al. A monocyte chemotactic factor, S19 ribosomal protein dimer, in phagocytic clearance of apoptotic cells. *Lab Invest* (1998) 78(5):603–17.
13. Wakasugi K, Schimmel P. Two distinct cytokines released from a human aminoacyl-tRNA synthetase. *Science* (1999) 284(5411):147–51. doi:10.1126/science.284.5411.147
14. Truman LA, Ford CA, Pasikowska M, Pound JD, Wilkinson SJ, Dumitriu IE, et al. CX3CL1/fractalkine is released from apoptotic lymphocytes to stimulate macrophage chemotaxis. *Blood* (2008) 112(13):5026–36. doi:10.1182/blood-2008-06-162404
15. Fadok VA, Voelker DR, Campbell PA, Cohen JJ, Bratton DL, Henson PM. Exposure of phosphatidylserine on the surface of apoptotic lymphocytes triggers specific recognition and removal by macrophages. *J Immunol* (1992) 148(7):2207–16.
16. Obeid M, Tesniere A, Ghiringhelli F, Fimia GM, Apetoh L, Perfettini JL, et al. Calreticulin exposure dictates the immunogenicity of cancer cell death. *Nat Med* (2007) 13(1):54–61. doi:10.1038/nm1523
17. Scannell M, Flanagan MB, deStefani A, Wynne KJ, Cagney G, Godson C, et al. Annexin-1 and peptide derivatives are released by apoptotic cells and stimulate phagocytosis of apoptotic neutrophils by macrophages. *J Immunol* (2007) 178(7):4595–605. doi:10.4049/jimmunol.178.7.4595
18. Rovere P, Peri G, Fazzini F, Bottazzi B, Doni A, Bondanza A, et al. The long pentraxin PTX3 binds to apoptotic cells and regulates their clearance by antigen-presenting dendritic cells. *Blood* (2000) 96(13):4300–6.
19. Fadok VA, Bratton DL, Konowal A, Freed PW, Westcott JY, Henson PM. Macrophages that have ingested apoptotic cells in vitro inhibit proinflammatory cytokine production through autocrine/paracrine mechanisms involving TGF-beta, PGE2, and PAF. *J Clin Invest* (1998) 101(4):890–8. doi:10.1172/Jci1112
20. Weigert A, Tzieply N, von Knethen A, Johann AM, Schmidt H, Geisslinger G, et al. Tumor cell apoptosis polarizes macrophages role of sphingosine-1-phosphate. *Mol Biol Cell* (2007) 18(10):3810–9. doi:10.1091/mbc.E06-12-1096
21. Mora J, Schlemmer A, Wittig I, Richter F, Putyrski M, Frank AC, et al. Interleukin-38 is released from apoptotic cells to limit inflammatory macrophage responses. *J Mol Cell Biol* (2016) 8(5):426–38. doi:10.1093/jmcb/mjw006
22. Martins I, Tesniere A, Kepp O, Michaud M, Schlemmer F, Senovilla L, et al. Chemotherapy induces ATP release from tumor cells. *Cell Cycle* (2009) 8(22):3723–8. doi:10.4161/cc.8.22.10026
23. Blume KE, Soeroes S, Keppeler H, Stevanovic S, Kretschmer D, Rautenberg M, et al. Cleavage of annexin A1 by ADAM10 during secondary necrosis generates a monocytic "find-me" signal. *J Immunol* (2012) 188(1):135–45. doi:10.4049/jimmunol.1004073
24. Li Z, Venegas V, Nagaoka Y, Morino E, Raghavan P, Audhya A, et al. Necrotic cells actively attract phagocytes through the collaborative action of two distinct PS-exposure mechanisms. *PLoS Genet* (2015) 11(6):e1005285. doi:10.1371/journal.pgen.1005285
25. Poon IK, Hulett MD, Parish CR. Molecular mechanisms of late apoptotic/necrotic cell clearance. *Cell Death Differ* (2010) 17(3):381–97. doi:10.1038/cdd.2009.195
26. Ahrens S, Zelenay S, Sancho D, Hanc P, Kjaer S, Feest C, et al. F-actin is an evolutionarily conserved damage-associated molecular pattern recognized by DNGR-1, a receptor for dead cells. *Immunity* (2012) 36(4):635–45. doi:10.1016/j.immuni.2012.03.008
27. Scaffidi P, Misteli T, Bianchi ME. Release of chromatin protein HMGB1 by necrotic cells triggers inflammation. *Nature* (2002) 418(6894):191–5. doi:10.1038/nature00858

28. Bours MJ, Swennen EL, Di Virgilio F, Cronstein BN, Dagnelie PC. Adenosine 5'-triphosphate and adenosine as endogenous signaling molecules in immunity and inflammation. *Pharmacol Ther* (2006) 112(2):358–404. doi:10.1016/j.pharmthera.2005.04.013
29. Ishii KJ, Suzuki K, Coban C, Takeshita F, Itoh Y, Matoba H, et al. Genomic DNA released by dying cells induces the maturation of APCs. *J Immunol* (2001) 167(5):2602–7. doi:10.4049/jimmunol.167.5.2602
30. Chen CJ, Kono H, Golenbock D, Reed G, Akira S, Rock KL. Identification of a key pathway required for the sterile inflammatory response triggered by dying cells. *Nat Med* (2007) 13(7):851–6. doi:10.1038/nm1603
31. Cayrol C, Girard JP. The IL-1-like cytokine IL-33 is inactivated after maturation by caspase-1. *Proc Natl Acad Sci U S A* (2009) 106(22):9021–6. doi:10.1073/pnas.0812690106
32. Wang Q, Ju X, Zhou Y, Chen K. Necroptotic cells release find-me signal and are engulfed without proinflammatory cytokine production. *In Vitro Cell Dev Biol Anim* (2015) 51(10):1033–9. doi:10.1007/s11626-015-9926-7
33. Zargarian S, Shlomovitz I, Erlich Z, Hourizadeh A, Ofir-Birin Y, Croker BA, et al. Phosphatidylserine externalization, "necroptotic bodies" release, and phagocytosis during necroptosis. *PLoS Biol* (2017) 15(6): e2002711. doi:10.1371/journal.pbio.2002711
34. Yatim N, Cullen S, Albert ML. Dying cells actively regulate adaptive immune responses. *Nat Rev Immunol* (2017) 17(4):262–75. doi:10.1038/nri.2017.9
35. Elliott MR, Ravichandran KS. The dynamics of apoptotic cell clearance. *Dev Cell* (2016) 38(2):147–60. doi:10.1016/j.devcel.2016.06.029
36. Medina CB, Ravichandran KS. Do not let death do us part: 'find-me' signals in communication between dying cells and the phagocytes. *Cell Death Differ* (2016) 23(6):979–89. doi:10.1038/cdd.2016.13
37. Gregory CD, Pound JD. Microenvironmental influences of apoptosis in vivo and in vitro. *Apoptosis* (2010) 15(9):1029–49. doi:10.1007/s10495-010-0485-9
38. Lauber K, Blumenthal SG, Waibel M, Wesselborg S. Clearance of apoptotic cells: getting rid of the corpses. *Mol Cell* (2004) 14(3):277–87. doi:10.1016/S1097-2765(04)00237-0
39. Cullen SP, Henry CM, Kearney CJ, Logue SE, Feoktistova M, Tynan GA, et al. Fas/CD95-induced chemokines can serve as "find-me" signals for apoptotic cells. *Mol Cell* (2013) 49(6):1034–48. doi:10.1016/j.molcel.2013.01.025
40. Luo BW, Gan WT, Liu ZW, Shen ZG, Wang JS, Shi RC, et al. Erythropoeitin signaling in macrophages promotes dying cell clearance and immune tolerance. *Immunity* (2016) 44(2):287–302. doi:10.1016/j.immuni.2016.01.002
41. Lee HW, Choi HJ, Ha SJ, Lee KT, Kwon YG. Recruitment of monocytes/macrophages in different tumor microenvironments. *Biochim Biophys Acta* (2013) 1835(2):170–9. doi:10.1016/j.bbcan.2012.12.007
42. Green DR, Oguin TH, Martinez J. The clearance of dying cells: table for two. *Cell Death Differ* (2016) 23(6):915–26. doi:10.1038/cdd.2015.172
43. Poon IKH, Lucas CD, Rossi AG, Ravichandran KS. Apoptotic cell clearance: basic biology and therapeutic potential. *Nat Rev Immunol* (2014) 14(3):166–80. doi:10.1038/Nri3607
44. Ravichandran KS. Find-me and eat-me signals in apoptotic cell clearance: progress and conundrums. *J Exp Med* (2010) 207(9):1807–17. doi:10.1084/jem.20101157
45. Gardai SJ, Bratton DL, Ogden CA, Henson PM. Recognition ligands on apoptotic cells: a perspective. *J Leukoc Biol* (2006) 79(5):896–903. doi:10.1189/jlb.1005550
46. Matsura T, Serinkan BF, Jiang JF, Kagan VE. Phosphatidylserine peroxidation/externalization during staurosporine-induced apoptosis in HL-60 cells. *FEBS Lett* (2002) 524(1–3):25–30. doi:10.1016/S0014-5793(02)02990-3
47. Graham DK, DeRyckere D, Davies KD, Earp HS. The TAM family: phosphatidylserine-sensing receptor tyrosine kinases gone awry in cancer. *Nat Rev Cancer* (2014) 14(12):769–85. doi:10.1038/nrc3847
48. Krysko DV, Denecker G, Festjens N, Gabriels S, Parthoens E, D'Herde K, et al. Macrophages use different internalization mechanisms to clear apoptotic and necrotic cells. *Cell Death Differ* (2006) 13(12):2011–22. doi:10.1038/sj.cdd.4401900
49. Blander JM. The many ways tissue phagocytes respond to dying cells. *Immunol Rev* (2017) 277(1):158–73. doi:10.1111/imr.12537
50. Mora J, Weigert A. IL-1 family cytokines in cancer immunity – a matter of life and death. *Biol Chem* (2016) 397(11):1125–34. doi:10.1515/hsz-2016-0215
51. Biermann MH, Veissi S, Maueroder C, Chaurio R, Berens C, Herrmann M, et al. The role of dead cell clearance in the etiology and pathogenesis of systemic lupus erythematosus: dendritic cells as potential targets. *Expert Rev Clin Immunol* (2014) 10(9):1151–64. doi:10.1586/1744666X.2014.944162
52. Dehne N, Mora J, Namgaladze D, Weigert A, Brune B. Cancer cell and macrophage cross-talk in the tumor microenvironment. *Curr Opin Pharmacol* (2017) 35:12–9. doi:10.1016/j.coph.2017.04.007
53. Kaczmarek A, Vandenabeele P, Krysko DV. Necroptosis: the release of damage-associated molecular patterns and its physiological relevance. *Immunity* (2013) 38(2):209–23. doi:10.1016/j.immuni.2013.02.003
54. Devitt A, Parker KG, Ogden CA, Oldreive C, Clay MF, Melville LA, et al. Persistence of apoptotic cells without autoimmune disease or inflammation in CD14(-/-) mice. *J Cell Biol* (2004) 167(6):1161–70. doi:10.1083/jcb.200410057
55. Lucas M, Stuart LM, Zhang A, Hodivala-Dilke K, Febbraio M, Silverstein R, et al. Requirements for apoptotic cell contact in regulation of macrophage responses. *J Immunol* (2006) 177(6):4047–54. doi:10.4049/jimmunol.177.6.4047
56. Cvetanovic M, Ucker DS. Innate immune discrimination of apoptotic cells: repression of proinflammatory macrophage transcription is coupled directly to specific recognition. *J Immunol* (2004) 172(2):880–9. doi:10.4049/jimmunol.172.2.880
57. Jennewein C, Kuhn AM, Schmidt MV, Meilladec-Jullig V, von Knethen A, Gonzalez FJ, et al. Sumoylation of peroxisome proliferator-activated receptor gamma by apoptotic cells prevents lipopolysaccharide-induced NCoR removal from kappa B binding sites proinflammatory cytokines. *J Immunol* (2008) 181(8):5646–52. doi:10.4049/jimmunol.181.8.5646
58. von Knethen A, Sha LK, Kuchler L, Heeg AK, Fuhrmann D, Heide H, et al. 5-Lipoxygenase contributes to PPAR gamma activation in macrophages in response to apoptotic cells. *Cell Signal* (2013) 25(12):2762–8. doi:10.1016/j.cellsig.2013.08.045
59. Ramos GC, Fernandes D, Charao CT, Souza DG, Teixeira MM, Assreuy J. Apoptotic mimicry: phosphatidylserine liposomes reduce inflammation through activation of peroxisome proliferator-activated receptors (PPARs) in vivo. *Br J Pharmacol* (2007) 151(6):844–50. doi:10.1038/sj.bjp.0707302
60. Chen H, Shi R, Luo B, Yang X, Qiu L, Xiong J, et al. Macrophage peroxisome proliferator-activated receptor gamma deficiency delays skin wound healing through impairing apoptotic cell clearance in mice. *Cell Death Dis* (2015) 6:e1597. doi:10.1038/Cddis.2014.544
61. Sen P, Wallet MA, Yi ZA, Huang YS, Henderson M, Mathews CE, et al. Apoptotic cells induce Mer tyrosine kinase-dependent blockade of NTF-kappa B activation in dendritic cells. *Blood* (2007) 109(2):653–60. doi:10.1182/blood-2006-04-017368
62. Freire-de-Lima CG, Xiao YQ, Gardai SJ, Bratton DL, Schiemann WP, Henson PM. Apoptotic cells, through transforming growth factor-beta, coordinately induce anti-inflammatory and suppress pro-inflammatory eicosanoid and NO synthesis in murine macrophages. *J Biol Chem* (2006) 281(50):38376–84. doi:10.1074/jbc.M605146200
63. Johann AM, von Knethen A, Lindemann D, Brune B. Recognition of apoptotic cells by macrophages activates the peroxisome proliferator-activated receptor-gamma and attenuates the oxidative burst. *Cell Death Differ* (2006) 13(9):1533–40. doi:10.1038/sj.cdd.4401832
64. Serinkan BF, Gambelli F, Potapovich A, Babu H, Di Giuseppe M, Ortiz LA, et al. Apoptotic cells quench reactive oxygen and nitrogen species and modulate TNF-alpha/TGF-beta 1 balance in activated macrophages: involvement of phosphatidylserine-dependent and -independent pathways. *Cell Death Differ* (2005) 12(8):1141–4. doi:10.1038/sj.cdd.4401619
65. Yin Y, Huang XM, Lynn KD, Thorpe PE. Phosphatidylserine-targeting antibody induces M1 macrophage polarization and promotes myeloid-derived suppressor cell differentiation. *Cancer Immunol Res* (2013) 1(4): 256–68. doi:10.1158/2326-6066.CIR-13-0073
66. Stanford JC, Young C, Hicks D, Owens P, Williams A, Vaught DB, et al. Efferocytosis produces a prometastatic landscape during postpartum mammary gland involution. *J Clin Invest* (2014) 124(11):4737–52. doi:10.1172/JCI76375
67. Weigert A, Weichand B, Brune B. S1P regulation of macrophage functions in the context of cancer. *Anticancer Agents Med Chem* (2011) 11(9):818–29. doi:10.2174/187152011797655096

68. Kaneda MM, Messer KS, Ralainirina N, Li H, Leem CJ, Gorjestani S, et al. PI3Kgamma is a molecular switch that controls immune suppression. *Nature* (2016) 539(7629):437–42. doi:10.1038/nature19834
69. Torchinsky MB, Garaude J, Martin AP, Blander JM. Innate immune recognition of infected apoptotic cells directs T(H)17 cell differentiation. *Nature* (2009) 458(7234):78–82. doi:10.1038/nature07781
70. Gardai SJ, McPhillips KA, Frasch SC, Janssen WJ, Starefeldt A, Murphy-Ullrich JE, et al. Cell-surface calreticulin initiates clearance of viable or apoptotic cells through trans-activation of LRP on the phagocyte. *Cell* (2005) 123(2):321–34. doi:10.1016/j.cell.2005.08.032
71. Michaud M, Sukkurwala AQ, Di Sano F, Zitvogel L, Kepp O, Kroemer G. Synthetic induction of immunogenic cell death by genetic stimulation of endoplasmic reticulum stress. *Oncoimmunology* (2014) 3(3):e28276. doi:10.4161/onci.28276
72. Zitvogel L, Kepp O, Kroemer G. Immune parameters affecting the efficacy of chemotherapeutic regimens. *Nat Rev Clin Oncol* (2011) 8(3):151–60. doi:10.1038/nrclinonc.2010.223
73. Geissler C, Singh M. Iron, meat and health. *Nutrients* (2011) 3(3):283–316. doi:10.3390/nu3030283
74. Knutson M, Wessling-Resnick M. Iron metabolism in the reticuloendothelial system. *Crit Rev Biochem Mol Biol* (2003) 38(1):61–88. doi:10.1080/713609210
75. Low PS, Waugh SM, Zinke K, Drenckhahn D. The role of hemoglobin denaturation and band 3 clustering in red blood cell aging. *Science* (1985) 227(4686):531–3. doi:10.1126/science.2578228
76. Pantaleo A, Giribaldi G, Mannu F, Arese P, Turrini F. Naturally occurring anti-band 3 antibodies and red blood cell removal under physiological and pathological conditions. *Autoimmun Rev* (2008) 7(6):457–62. doi:10.1016/j.autrev.2008.03.017
77. Lee SJ, Park SY, Jung MY, Bae SM, Kim IS. Mechanism for phosphatidylserine-dependent erythrophagocytosis in mouse liver. *Blood* (2011) 117(19):5215–23. doi:10.1182/blood-2010-10-313239
78. Foller M, Huber SM, Lang F. Erythrocyte programmed cell death. *IUBMB Life* (2008) 60(10):661–8. doi:10.1002/iub.106
79. Soe-Lin S, Apte SS, Andriopoulos B Jr, Andrews MC, Schranzhofer M, Kahawita T, et al. Nramp1 promotes efficient macrophage recycling of iron following erythrophagocytosis in vivo. *Proc Natl Acad Sci U S A* (2009) 106(14):5960–5. doi:10.1073/pnas.0900808106
80. Theil EC. Ferritin: the protein nanocage and iron biomineral in health and in disease. *Inorg Chem* (2013) 52(21):12223–33. doi:10.1021/ic400484n
81. Harris ZL, Durley AP, Man TK, Gitlin JD. Targeted gene disruption reveals an essential role for ceruloplasmin in cellular iron efflux. *Proc Natl Acad Sci U S A* (1999) 96(19):10812–7. doi:10.1073/pnas.96.19.10812
82. Donovan A, Lima CA, Pinkus JL, Pinkus GS, Zon LI, Robine S, et al. The iron exporter ferroportin/Slc40a1 is essential for iron homeostasis. *Cell Metab* (2005) 1(3):191–200. doi:10.1016/j.cmet.2005.01.003
83. Nemeth E, Rivera S, Gabayan V, Keller C, Taudorf S, Pedersen BK, et al. IL-6 mediates hypoferremia of inflammation by inducing the synthesis of the iron regulatory hormone hepcidin. *J Clin Invest* (2004) 113(9):1271–6. doi:10.1172/JCI200420945
84. Ponka P. Tissue-specific regulation of iron metabolism and heme synthesis: distinct control mechanisms in erythroid cells. *Blood* (1997) 89(1):1–25.
85. Gordon S. Alternative activation of macrophages. *Nat Rev Immunol* (2003) 3(1):23–35. doi:10.1038/nri978
86. Poss KD, Tonegawa S. Heme oxygenase 1 is required for mammalian iron reutilization. *Proc Natl Acad Sci U S A* (1997) 94(20):10919–24. doi:10.1073/pnas.94.20.10919
87. Kohyama M, Ise W, Edelson BT, Wilker PR, Hildner K, Mejia C, et al. Role for Spi-C in the development of red pulp macrophages and splenic iron homeostasis. *Nature* (2009) 457(7227):318–21. doi:10.1038/nature07472
88. Haldar M, Kohyama M, So AYL, Wumesh KC, Wu XD, Briseno CG, et al. Heme-mediated SPI-C induction promotes monocyte differentiation into iron-recycling macrophages. *Cell* (2014) 156(6):1223–34. doi:10.1016/j.cell.2014.01.069
89. Brune B, Dehne N, Grossmann N, Jung M, Namgaladze D, Schmid T, et al. Redox control of inflammation in macrophages. *Antioxid Redox Sign* (2013) 19(6):595–637. doi:10.1089/ars.2012.4785
90. Murray PJ, Wynn TA. Protective and pathogenic functions of macrophage subsets. *Nat Rev Immunol* (2011) 11(11):723–37. doi:10.1038/nri3073
91. Ariel A, Ravichandran KS. 'This way please': apoptotic cells regulate phagocyte migration before and after engulfment. *Eur J Immunol* (2016) 46(7): 1583–6. doi:10.1002/eji.201646505
92. Zhu L, Zhao Q, Yang T, Ding W, Zhao Y. Cellular metabolism and macrophage functional polarization. *Int Rev Immunol* (2015) 34(1):82–100. doi:10.3109/08830185.2014.969421
93. Recalcati S, Locati M, Marini A, Santambrogio P, Zaninotto F, De Pizzol M, et al. Differential regulation of iron homeostasis during human macrophage polarized activation. *Eur J Immunol* (2010) 40(3):824–35. doi:10.1002/eji.200939889
94. Mertens C, Akam EA, Rehwald C, Brune B, Tomat E, Jung M. Intracellular iron chelation modulates the macrophage iron phenotype with consequences on tumor progression. *PLoS One* (2016) 11(11):e0166164. doi:10.1371/journal.pone.0166164
95. Paradkar PN, De Domenico I, Durchfort N, Zohn I, Kaplan J, Ward DM. Iron depletion limits intracellular bacterial growth in macrophages. *Blood* (2008) 112(3):866–74. doi:10.1182/blood-2007-12-126854
96. Chlosta S, Fishman DS, Harrington L, Johnson EE, Knutson MD, Wessling-Resnick M, et al. The iron efflux protein ferroportin regulates the intracellular growth of *Salmonella enterica*. *Infect Immun* (2006) 74(5): 3065–7. doi:10.1128/Iai.74.5.3065-3067.2006
97. Ganz T. Iron in innate immunity: starve the invaders. *Curr Opin Immunol* (2009) 21(1):63–7. doi:10.1016/j.coi.2009.01.011
98. Nairz M, Theurl I, Ludwiczek S, Theurl M, Mair SM, Fritsche G, et al. The co-ordinated regulation of iron homeostasis in murine macrophages limits the availability of iron for intracellular *Salmonella typhimurium*. *Cell Microbiol* (2007) 9(9):2126–40. doi:10.1111/j.1462-5822.2007.00942.x
99. Nairz M, Fritsche G, Brunner P, Talasz H, Hantke K, Weiss G. Interferon-gamma limits the availability of iron for intramacrophage *Salmonella typhimurium*. *Eur J Immunol* (2008) 38(7):1923–36. doi:10.1002/eji.200738056
100. Buehler PW, D'Agnillo F, Schaer DJ. Hemoglobin-based oxygen carriers: from mechanisms of toxicity and clearance to rational drug design. *Trends Mol Med* (2010) 16(10):447–57. doi:10.1016/j.molmed.2010.07.006
101. Kato GJ. Haptoglobin halts hemoglobin's havoc. *J Clin Invest* (2009) 119(8):2140–2. doi:10.1172/JCI40258
102. Larsen R, Gozzelino R, Jeney V, Tokaji L, Bozza FA, Japiassu AM, et al. A central role for free heme in the pathogenesis of severe sepsis. *Sci Transl Med* (2010) 2(51):51ra71. doi:10.1126/scitranslmed.3001118
103. Semenza GL. Hypoxia-inducible factor 1 (HIF-1) pathway. *Sci STKE* (2007) 2007(407):cm8. doi:10.1126/stke.4072007cm8
104. Baby SM, Roy A, Mokashi AM, Lahiri S. Effects of hypoxia and intracellular iron chelation on hypoxia-inducible factor-1alpha and -1beta in the rat carotid body and glomus cells. *Histochem Cell Biol* (2003) 120(5):343–52. doi:10.1007/s00418-003-0588-2
105. Takeda N, O'Dea EL, Doedens A, Kim JW, Weidemann A, Stockmann C, et al. Differential activation and antagonistic function of HIF-alpha isoforms in macrophages are essential for NO homeostasis. *Gene Dev* (2010) 24(5):491–501. doi:10.1101/gad.1881410
106. Cairo G, Recalcati S, Mantovani A, Locati M. Iron trafficking and metabolism in macrophages: contribution to the polarized phenotype. *Trends Immunol* (2011) 32(6):241–7. doi:10.1016/j.it.2011.03.007
107. Torti SV, Torti FM. Cellular iron metabolism in prognosis and therapy of breast cancer. *Crit Rev Oncog* (2013) 18(5):435–48. doi:10.1615/CritRevOncog.2013007784
108. Rivera S, Liu L, Nemeth E, Gabayan V, Sorensen OE, Ganz T. Hepcidin excess induces the sequestration of iron and exacerbates tumor-associated anemia. *Blood* (2005) 105(4):1797–802. doi:10.1182/blood-2004-08-3375
109. Pinnix ZK, Miller LD, Wang W, D'Agostino R Jr, Kute T, Willingham MC, et al. Ferroportin and iron regulation in breast cancer progression and prognosis. *Sci Transl Med* (2010) 2(43):43ra56. doi:10.1126/scisignal.3001127
110. Marques O, da Silva BM, Porto G, Lopes C. Iron homeostasis in breast cancer. *Cancer Lett* (2014) 347(1):1–14. doi:10.1016/j.canlet.2014.01.029
111. Knutson MD, Oukka M, Koss LM, Aydemir F, Wessling-Resnick M. Iron release from macrophages after erythrophagocytosis is up-regulated by ferroportin 1 overexpression and down-regulated by hepcidin. *Proc Natl Acad Sci U S A* (2005) 102(5):1324–8. doi:10.1073/pnas.0409409102

112. Folkman J. Tumor angiogenesis: therapeutic implications. *N Engl J Med* (1971) 285(21):1182–6. doi:10.1056/NEJM197111182852108
113. Ganz T, Nemeth E. Iron homeostasis in host defence and inflammation. *Nat Rev Immunol* (2015) 15(8):500–10. doi:10.1038/nri3863
114. Gnerlich JL, Deshpande AD, Jeffe DB, Seelam S, Kimbuende E, Margenthaler JA. Poorer survival outcomes for male breast cancer compared with female breast cancer may be attributable to in-stage migration. *Ann Surg Oncol* (2011) 18(7):1837–44. doi:10.1245/s10434-010-1468-3
115. Marro S, Chiabrando D, Messana E, Stolte J, Turco E, Tolosano E, et al. Heme controls ferroportin1 (FPN1) transcription involving Bach1, Nrf2 and a MARE/ARE sequence motif at position -7007 of the FPN1 promoter. *Haematologica* (2010) 95(8):1261–8. doi:10.3324/haematol.2009.020123
116. Zhang Z, Zhang F, An P, Guo X, Shen Y, Tao Y, et al. Ferroportin1 deficiency in mouse macrophages impairs iron homeostasis and inflammatory responses. *Blood* (2011) 118(7):1912–22. doi:10.1182/blood-2011-01-330324
117. Keel SB, Doty RT, Yang Z, Quigley JG, Chen J, Knoblaugh S, et al. A heme export protein is required for red blood cell differentiation and iron homeostasis. *Science* (2008) 319(5864):825–8. doi:10.1126/science.1151133
118. de Sousa M. An outsider's perspective – ecotaxis revisited: an integrative review of cancer environment, iron and immune system cells. *Integr Biol* (2011) 3(4):343–9. doi:10.1039/C0IB00116C
119. Alkhateeb AA, Han B, Connor JR. Ferritin stimulates breast cancer cells through an iron-independent mechanism and is localized within tumor-associated macrophages. *Breast Cancer Res Treat* (2013) 137(3):733–44. doi:10.1007/s10549-012-2405-x
120. Goetz DH, Holmes MA, Borregaard N, Bluhm ME, Raymond KN, Strong RK. The neutrophil lipocalin NGAL is a bacteriostatic agent that interferes with siderophore-mediated iron acquisition. *Mol Cell* (2002) 10(5):1033–43. doi:10.1016/S1097-2765(02)00708-6
121. Schmidt-Ott KM, Mori K, Li JY, Kalandadze A, Cohen DJ, Devarajan P, et al. Dual action of neutrophil gelatinase-associated lipocalin. *J Am Soc Nephrol* (2007) 18(2):407–13. doi:10.1681/ASN.2006080882
122. Mishra J, Mori K, Ma Q, Kelly C, Yang J, Mitsnefes M, et al. Amelioration of ischemic acute renal injury by neutrophil gelatinase-associated lipocalin. *J Am Soc Nephrol* (2004) 15(12):3073–82. doi:10.1097/01.ASN.0000145013.44578.45
123. Sola A, Weigert A, Jung M, Vinuesa E, Brecht K, Weis N, et al. Sphingosine-1-phosphate signalling induces the production of Lcn-2 by macrophages to promote kidney regeneration. *J Pathol* (2011) 225(4):597–608. doi:10.1002/path.2982
124. Jung M, Weigert A, Tausendschon M, Mora J, Oren B, Sola A, et al. Interleukin-10-induced neutrophil gelatinase-associated lipocalin production in macrophages with consequences for tumor growth. *Mol Cell Biol* (2012) 32(19):3938–48. doi:10.1128/MCB.00413-12
125. Oren B, Urosevic J, Mertens C, Mora J, Guiu M, Gomis RR, et al. Tumour stroma-derived lipocalin-2 promotes breast cancer metastasis. *J Pathol* (2016) 239(3):274–85. doi:10.1002/path.4724
126. Jung M, Oren B, Mora J, Mertens C, Dziumbla S, Popp R, et al. Lipocalin 2 from macrophages stimulated by tumor cell-derived sphingosine 1-phosphate promotes lymphangiogenesis and tumor metastasis. *Sci Signal* (2016) 9(434):ra64. doi:10.1126/scisignal.aaf3241
127. Berger T, Cheung CC, Elia AJ, Mak TW. Disruption of the Lcn2 gene in mice suppresses primary mammary tumor formation but does not decrease lung metastasis. *Proc Natl Acad Sci U S A* (2010) 107(7):2995–3000. doi:10.1073/pnas.1000101107
128. Shi H, Gu Y, Yang J, Xu L, Mi W, Yu W. Lipocalin 2 promotes lung metastasis of murine breast cancer cells. *J Exp Clin Cancer Res* (2008) 27:83. doi:10.1186/1756-9966-27-83
129. Leng X, Ding T, Lin H, Wang Y, Hu L, Hu J, et al. Inhibition of lipocalin 2 impairs breast tumorigenesis and metastasis. *Cancer Res* (2009) 69(22): 8579–84. doi:10.1158/0008-5472.CAN-09-1934
130. Bauer M, Eickhoff JC, Gould MN, Mundhenke C, Maass N, Friedl A. Neutrophil gelatinase-associated lipocalin (NGAL) is a predictor of poor prognosis in human primary breast cancer. *Breast Cancer Res Treat* (2008) 108(3):389–97. doi:10.1007/s10549-007-9619-3
131. Wenners AS, Mehta K, Loibl S, Park H, Mueller B, Arnold N, et al. Neutrophil gelatinase-associated lipocalin (NGAL) predicts response to neoadjuvant chemotherapy and clinical outcome in primary human breast cancer. *PLoS One* (2012) 7(10):e45826. doi:10.1371/journal.pone.0045826
132. Yang J, Bielenberg DR, Rodig SJ, Doiron R, Clifton MC, Kung AL, et al. Lipocalin 2 promotes breast cancer progression. *Proc Natl Acad Sci U S A* (2009) 106(10):3913–8. doi:10.1073/pnas.0810617106
133. Mira E, Lacalle RA, Buesa JM, de Buitrago GG, Jimenez-Baranda S, Gomez-Mouton C, et al. Secreted MMP9 promotes angiogenesis more efficiently than constitutive active MMP9 bound to the tumor cell surface. *J Cell Sci* (2004) 117(9):1847–56. doi:10.1242/jcs.01035
134. Jung M, Sola A, Hughes J, Kluth DC, Vinuesa E, Vinas JL, et al. Infusion of IL-10-expressing cells protects against renal ischemia through induction of lipocalin-2. *Kidney Int* (2012) 81(10):969–82. doi:10.1038/ki.2011.446
135. Weigert A, Cremer S, Schmidt MV, von Knethen A, Angioni C, Geisslinger G, et al. Cleavage of sphingosine kinase 2 by caspase-1 provokes its release from apoptotic cells. *Blood* (2010) 115(17):3531–40. doi:10.1182/blood-2009-10-243444
136. Devireddy LR, Gazin C, Zhu X, Green MR. A cell-surface receptor for lipocalin 24p3 selectively mediates apoptosis and iron uptake. *Cell* (2005) 123(7):1293–305. doi:10.1016/j.cell.2005.10.027
137. Corbin JL, Bulen WA. The isolation and identification of 2,3-dihydroxybenzoic acid and 2-N,6-N-di-92,3-dihydroxybenzoyl)-L-lysine formed by iron-deficient *Azotobacter vinelandii*. *Biochemistry* (1969) 8(3):757–62. doi:10.1021/bi00831a002
138. Gibson F, Magrath DI. The isolation and characterization of a hydroxamic acid (aerobactin) formed by *Aerobacter aerogenes* 62-I. *Biochim Biophys Acta* (1969) 192(2):175–84. doi:10.1016/0304-4165(69)90353-5
139. Pollack JR, Neilands JB. Enterobactin, an iron transport compound from *Salmonella typhimurium*. *Biochem Bioph Res Co* (1970) 38(5):989–92. doi:10.1016/0006-291X(70)90819-3
140. Devireddy LR, Hart DO, Goetz DH, Green MR. A mammalian siderophore synthesized by an enzyme with a bacterial homolog involved in enterobactin production. *Cell* (2010) 141(6):1006–17. doi:10.1016/j.cell.2010.04.040
141. Liu Z, Reba S, Chen WD, Porwal SK, Boom WH, Petersen RB, et al. Regulation of mammalian siderophore 2,5-DHBA in the innate immune response to infection. *J Exp Med* (2014) 211(6):1197–213. doi:10.1084/jem.20132629
142. Liu ZM, Ciocea A, Devireddy L. Endogenous siderophore 2,5-dihydroxybenzoic acid deficiency promotes anemia and splenic iron overload in mice. *Mol Cell Biol* (2014) 34(13):2533–46. doi:10.1128/Mcb.00231-14
143. Liu ZM, Velpula KK, Devireddy L. 3-Hydroxybutyrate dehydrogenase-2 and ferritin-H synergistically regulate intracellular iron. *FEBS J* (2014) 281(10):2410–21. doi:10.1111/febs.12794
144. Yang WC, Tsai WC, Lin PM, Yang MY, Liu YC, Chang CS, et al. Human BDH2, an anti-apoptosis factor, is a novel poor prognostic factor for de novo cytogenetically normal acute myeloid leukemia. *J Biomed Sci* (2013) 20:58. doi:10.1186/1423-0127-20-58
145. Miethke M, Marahiel MA. Siderophore-based iron acquisition and pathogen control. *Microbiol Mol Biol Rev* (2007) 71(3):413–51. doi:10.1128/Mmbr.00012-07
146. Pietrangelo A. Mechanism of iron toxicity. *Adv Exp Med Biol* (2002) 509:19–43. doi:10.1007/978-1-4615-0593-8_2
147. Saha M, Sarkar S, Sarkar B, Sharma BK, Bhattacharjee S, Tribedi P. Microbial siderophores and their potential applications: a review. *Environ Sci Pollut Res* (2016) 23(5):3984–99. doi:10.1007/s11356-015-4294-0

Conflict of Interest Statement: The authors declare that the research was conducted in the absence of any commercial or financial relationships that could be construed as a potential conflict of interest.

8

Harnessing Apoptotic Cell Clearance to Treat Autoimmune Arthritis

Philippe Saas[1,2], Francis Bonnefoy[1], Eric Toussirot[2,3,4] and Sylvain Perruche[1]*

[1] *INSERM, EFS BFC, UMR1098, Interactions Hôte-Greffon-Tumeur/Ingénierie Cellulaire et Génique, Fédération Hospitalo-Universitaire INCREASE, LabEx LipSTIC, Université Bourgogne Franche-Comté, Besançon, France,* [2] *INSERM CIC-1431, University Hospital of Besançon, Clinical Investigation Center in Biotherapy, Fédération Hospitalo-Universitaire INCREASE, LabEx LipSTIC, Besançon, France,* [3] *Department of Rheumatology, University Hospital of Besançon, Besançon, France,* [4] *Department of Therapeutics, Université Bourgogne Franche-Comté, UPRES EA 4266, Pathogenic Agents and Inflammation, Besancon, France*

Early-stage apoptotic cells possess immunomodulatory properties. Proper apoptotic cell clearance during homeostasis has been shown to limit subsequent immune responses. Based on these observations, early-stage apoptotic cell infusion has been used to prevent unwanted inflammatory responses in different experimental models of autoimmune diseases or transplantation. Moreover, this approach has been shown to be feasible without any toxicity in patients undergoing allogeneic hematopoietic cell transplantation to prevent graft-*versus*-host disease. However, whether early-stage apoptotic cell infusion can be used to treat ongoing inflammatory disorders has not been reported extensively. Recently, we have provided evidence that early-stage apoptotic cell infusion is able to control, at least transiently, ongoing collagen-induced arthritis. This beneficial therapeutic effect is associated with the modulation of antigen-presenting cell functions mainly of macrophages and plasmacytoid dendritic cells, as well as the induction of collagen-specific regulatory CD4$^+$ T cells (Treg). Furthermore, the efficacy of this approach is not altered by the association with two standard treatments of rheumatoid arthritis (RA), methotrexate and tumor necrosis factor (TNF) inhibition. Here, in the light of these observations and recent data of the literature, we discuss the mechanisms of early-stage apoptotic cell infusion and how this therapeutic approach can be transposed to patients with RA.

Edited by:
Amiram Ariel,
University of Haifa, Israel

Reviewed by:
Luis Enrique Munoz,
Friedrich-Alexander University, Germany
Markus H. Hoffmann,
University of Erlangen-Nuremberg, Germany

***Correspondence:**
Philippe Saas
philippe.saas@efs.sante.fr

Keywords: apoptotic cells, rheumatoid arthritis, collagen-induced arthritis, macrophages, regulatory T cells, efferocytosis, cell-based therapy, biologic DMARD

INTRODUCTION

Apoptotic cells, at least at their early stage, possess immunomodulatory properties [please refer to recent reviews (1, 2)]. These cells are generated by a process called apoptosis (primarily termed programmed cell death), initially defined on morphological features (3). Today, apoptotic cells can be characterized at different levels by biochemical and genetic methods (4). Different forms of cell death have been identified so far (4). Early-stage apoptotic cells, as defined in this review, correspond to cells characterized *in vitro* (*i.e.*, before *in vivo* administration) that express phosphatidylserine (PtdSer) at their cell surface and keep the ability to exclude vital dyes [propidium iodide (PI) or 7-aminoactinomycin D (7-AAD)]. This exclusion means that these early-stage apoptotic cells conserve their cell membrane integrity. Exposure of PtdSer on the surface of early-stage apoptotic cells allows their rapid removal by macrophages (5), thus preventing apoptotic cell "explosion" and the release of pro-inflammatory factors. At steady state, efficient apoptotic cell clearance by macrophages (a process called efferocytosis) has been shown to limit subsequent immune responses. Initially, this clearance of apoptotic cells by macrophages has been identified using apoptotic thymocytes (6). This observation has been then extended by Savill and colleagues to the removal of apoptotic neutrophils (7). This seminal work serves as a basis to explain later on, the resolution of inflammation (8). These interactions of apoptotic cells with monocytes or macrophages are associated with a decreased capacity to produce pro-inflammatory cytokines together with the ability to produce anti-inflammatory factors. This has been reported at the end of the nineties (9), and this process is now called macrophage reprogramming. For timelines of the history of apoptosis in inflammation, readers can refer to two recent reviews (10, 11). In contrast, altered efferocytosis has been associated with autoimmune diseases. For instance, a deficiency in the last step of efferocytosis, namely the digestion of apoptotic cell materials by macrophages (*i.e.*, a defect in intracellular DNase II), has been shown to be responsible for a polyarthritis syndrome similar to rheumatoid arthritis (RA) (12). Based on their immunomodulatory properties, early-stage apoptotic cells have been used to prevent unwanted inflammatory responses in different experimental models of autoimmune diseases or transplantation [for a recent review, please refer to Ref. (2)]. Hence, prevention of arthritis by early-stage apoptotic cell injection has been reported in different mouse and rat models (13–15). Moreover, this approach of apoptotic cell infusion has been shown to be feasible without any toxicity in patients undergoing allogeneic hematopoietic cell transplantation (16). However, whether early-stage apoptotic cell infusion can be used to treat ongoing inflammatory disorders has not been reported extensively. Xenogeneic human apoptotic cell administration 3 days after sepsis initiation in mouse models stimulates the resolution of acute inflammation (17). This therapeutic effect of apoptotic cell infusion in sepsis has been confirmed in lipopolysaccharide-induced endotoxic shock as well as in cecal ligation and puncture sepsis (18). Furthermore, donor apoptotic cell infusion can interfere with acute graft rejection in a mouse model of allogeneic cardiac transplantation (19). Recently, we have provided evidence that apoptotic cell infusion is able to control, at least transiently, ongoing collagen-induced arthritis (CIA) (20). Interestingly, the efficacy of this approach is not altered by the association with two standard treatments of RA, methotrexate (MTX) and tumor necrosis factor (TNF) inhibition (20). Here, in the light of these observations and recent data of the literature, we discuss the mechanisms of this therapeutic approach and how it can be transposed to patients with RA.

CURRENT KNOWLEDGE IN RA PATHOPHYSIOLOGY AND THERAPEUTIC APPROACHES

Rheumatoid arthritis is an autoimmune disorder characterized by a chronic inflammation of the synovial joints leading to the destruction of cartilage, bone, and ligaments (21). However, RA is a heterogeneous syndrome as attested by genetic studies (22, 23). The pathophysiology of RA implicates several immune cell subsets belonging to both innate (e.g., neutrophils, macrophages) and adaptive immunity (*i.e.*, T and B cells). At the inflammatory site, the synovial lining becomes thickened due to an infiltration of macrophages and the proliferation of resident synovial fibroblasts (also called fibroblast-like synoviocytes). At the end of the eighties, massive infiltration of neutrophils and macrophages was reported in the joints of patients suffering from acute sterile arthritis, among whom RA patients (7). These data serve as a basis for our current understanding of the resolution step of inflammation and identify neutrophils and macrophages as key players in RA pathogenesis. While the exact etiology of RA is still unknown, macrophage activation leading to local inflammatory cytokine secretion in the joints can be considered as one of these etiologies (12). Therapeutic approaches triggering these inflammatory cytokines (*i.e.*, TNF-α, IL-1β, or IL-6) have been used to treat RA patients (24, 25). Despite recent significant advances in the characterization of monocyte and macrophage subsets, the origin of macrophages infiltrating or present in the joint remains to be explored in RA (26). Indeed, the origin of joint macrophages (tissue-resident *versus* derived from blood Ly6C^{high} monocytes) depends on the considered arthritis models (26). Recently, it has been shown that neutrophils may participate in RA pathophysiology through the formation of neutrophil extracellular traps (NET), which consist of DNA fibers associated with a large amount of antimicrobial peptides (e.g., LL37) and nuclear proteins (e.g., high mobility group box-1). This has been reported in RA, as well as in experimental models such as CIA (27–29). Formation of NET by neutrophils during arthritis provides a pro-inflammatory loop *via* the secretion of pro-inflammatory cytokines (28). Dendritic cells (DC)—both conventional DC (cDC) and plasmacytoid DC (pDC)—may also play a role in RA pathophysiology. For instance, pDC are present in the synovial fluid of RA patients (30–32). Pro-inflammatory pDC aggravates ongoing CIA (33). Activation of cDC by NET may be also involved in arthritis pathogenesis (29). Pathogenic CD4$^+$ helper T (Th) and cytotoxic CD8$^+$ T cells have been also implicated in RA, while the exact target of these cells has not been fully characterized. However, autoreactive CD4$^+$ T cells specific to citrullinated epitopes with a memory and/or effector phenotype have been identified in some RA patients (34). Concerning CD8$^+$ T cells, Epstein–Barr virus (EBV)-derived antigens can be targeted antigens in RA since high expression of EBV markers is present in RA synovium (35). These cytotoxic T cells can mediate joint damage, but in all cases, inflammatory CD4$^+$ Th cells are required. Both interferon-γ (IFN-γ)-secreting Th1 and IL-17-producing Th17 cells (36) are involved in RA pathogenesis. They are driven mainly by macrophage cytokines consisting of TNF and IL-12 *versus* IL-23 for Th1 and Th17 cell polarization, respectively (26). These two Th cell polarization pathways occur in the absence of adequate immune regulation, since an altered regulatory CD4$^+$ T cell (Treg) response is another feature of RA (37). Finally, concerning B cell responses, a high frequency of circulating polyspecific B cell clones has been found in RA patients (23). However, it is unclear how such B cells contribute to RA disease. The reversion of anergic autoreactive B cells under inflammatory conditions has been suggested to participate in RA pathogenesis (23). Nevertheless, the implication of auto-antibodies in RA pathophysiology is highlighted by the two major biological tests performed for RA diagnosis: rheumatoid factor (RF) and anti-citrullinated protein antibody (ACPA) detection (35). RF is involved in the formation of immune complex (IC) that induces complement activation responsible for its consumption and generates non-resolving inflammation observed in RA (35, 38). Non-resolving inflammation significantly contributes to RA pathogenesis (38). Citrullinated proteins result from arginine-containing proteins modified by deimination mediated by intracellular enzymes, called peptidyl-arginine deiminases. NET produced by neutrophils can be an additional source of citrullinated autoantigens (28, 39). These resultant citrullinated proteins could be the antigenic component of IC driving RF production (35) and become the targets of autoantibody responses (35), as well as autoreactive CD4$^+$ T cells (34). Furthermore, ACPA are T cell-dependent immunoglobulin G auto-antibodies, and thus, follicular helper T cells may help B cell activation in ACPA-positive RA (34). Thus, several immune mechanisms and immune cell subsets participate in RA pathophysiology and represent targets for therapeutic strategies, such as apoptotic cell infusion.

Today, no causal treatment of RA is available, since RA is still a chronic inflammatory disorder of unknown cause. Hence, there is currently no curative treatment for RA and treatment has to be initiated for prolonged periods of time if not for life (40). The European League Against Rheumatism organization recommends that patient starts treatment with conventional synthetic disease-modifying anti-rheumatic drugs (csDMARD) in combination with corticosteroids, followed by biologic DMARD (bDMARD) in the case of a non-response to the initial regimen and the presence of poor prognosis markers (41). Treatment of RA aims to limit disease symptoms, delay or prevent future joint destruction, and target low disease activity (LDA) or remission. According to a recent review (40), LDA is a state in which the progression of joint damage is minimal and physical function, quality of life and work capacity are preserved. Low-dose MTX is the traditional csDMARD administered weekly either alone, or in combination with corticosteroid or bDMARD. While the precise molecular mechanism of MTX remains to be determined, MTX alone has been proven safe and efficient in RA (42). However, nearly a quarter of patients treated with MTX have to discontinue their treatment because of inadequate responses, adverse effects (e.g., hepatic, gastrointestinal, hematological, renal, or pulmonary toxicity), or both (43, 44). Biologic agents targeting inflammatory cytokines, such as anti-TNF therapy, combined with MTX have significantly improved the treatment of RA (24, 40). However, again, some RA patients are refractory or have contraindications to receive these agents (44, 45). The proportion of patients who do not respond favorably to TNF inhibitors is estimated between 30 and 40% (24). Only few RA patients achieve complete remission after such treatment (24). Moreover, adherence to treatment with biologic agents is moderate with only around 60% of RA patients respecting treatment regimens over a 1- or 2-year period (46). This requires frequently a switch to another form of treatment (40, 46). Biologic agents targeting different modes of action have been developed, and are now available in RA. This consists in TNF blocking agents, IL-6 or IL-1 inhibitors, T-cell costimulatory modulators, or B-cell depletion therapies (25). However, despite this multitude of treatments, treatment failure occurs and RA patients are still in need of new treatment modalities (25). Finally, it should be mentioned that combinations of bDMARD acting on different therapeutic targets (*i.e.,* TNF, IL-6, or B cells) usually do not increase efficacy, but are more toxic (47). Overall, new therapeutic strategies are needed in RA among which cell-based therapies could be proposed, such as apoptotic cell infusion.

THE DISEASE-MODIFYING ANTI-RHEUMATIC POTENTIAL OF APOPTOTIC CELL INFUSION

In this section, we will describe the mechanisms by which early-stage apoptotic cell infusion may treat ongoing arthritis. Based on our recent data (20), we will focus on the resolution of inflammation, antigen-presenting cells (APC), including DC subsets and macrophages, as well as CD4$^+$ T cell polarization. Data obtained using apoptotic cell infusion as prevention of arthritis (13–15, 48) will be also considered to shed light on these mechanisms.

Lessons from Preclinical Arthritis Models

Early-stage apoptotic cells have been injected in arthritis experimental models before the disease is fully established or at time of immunization with the autoantigen (13–15, 48). This is not relevant to the clinical situation, and contrasts with our recent study in which early-stage apoptotic cells are infused when CIA reaches a clinical score of 8 out of 16 (20). Therefore, one may distinguish the prophylactic *versus* the therapeutic effect of apoptotic cell infusion (**Table 1**). To date, one limitation is that only one experimental model has been tested for the therapeutic effect (20). For the prophylactic effect, several experimental models have been used (13–15). These models recapitulate differently RA pathophysiology. An absence of prevention has been reported in the serum transfer-induced arthritis (STIA) (13) in which arthritis is induced by the intraperitoneal (i.p.) injection of K/BxN serum in C57BL/6 mice (49) (**Table 1**). This STIA model recapitulates the effector phase of human RA, but is independent of the adaptive immune response (49, 50). Thus, this suggests that the prophylactic effect of apoptotic cell infusion modulates rather the adaptive immune response, such as CD4$^+$ T cell polarization. The model consisting in injecting streptococcal cell wall (SCW) in Lewis rats is induced by a single i.p. injection of SCW fragments (51). This results in a first T cell-independent phase followed by a chronic inflammatory phase that is T cell-dependent and associated with the production of high levels of inflammatory cytokines (51). This results in erosive cartilage damage in the joints (51). In the prophylactic approach using early-stage apoptotic cell infusion, both phases were significantly reduced but the effect was more impressive or pronounced on the chronic phase (14). This is consistent with an impact of apoptotic cell infusion on inflammatory cytokine secretion by macrophages affecting the first phase and on Treg increase modulating the second chronic phase (14) (**Table 1**). Methylated bovine serum albumin (mBSA)-induced arthritis in C57BL/6 mice belongs to antigen-induced arthritis.

TABLE 1 | Effects (therapeutic *versus* prophylactic) of early-stage apoptotic cell infusion in arthritis models.

Experimental models	Effects on disease	Administration route	Characteristics of infused apoptotic cells	Immune mechanisms	Reference
CIA (DBA/1)	Therapeutic	i.v.	Syngeneic thymocytes, 5 or 15 × 10^6, early-stage apoptotic cells (70–85% AxV^+/7-AAD^- and <10% 7-AAD^+)	Pro-Treg splenic macrophages; splenic cDC and pDC resistant to TLR ligand stimulation—pro-Treg splenic pDC; induction of auto-Ag-specific Treg in the DLN; reduction of pathogenic anti-collagen auto-Abs; depend on TGF-β	Bonnefoy *et al.* (20)
CIA (DBA/1)	Prophylactic	i.v. or i.p.	Syngeneic thymocytes, 2 × 10^7 (total 3 consecutive days), early-stage apoptotic cells (mean: 43% of AxV^+ and <5% PI^+)	IL-10-producing splenic and PLN $CD4^+$ T cells; reduction of IFN-γ secreting $CD4^+$ T cells; IL-10-producing MZB cells; reduction of pathogenic anti-collagen auto-Abs	Gray *et al.* (13)
STIA (C57BL/6)	No effect	i.v. or i.p.	Same as above	No prophylactic effect but T cell-independent model (50)	Gray *et al.* (13)
SCW (Lewis rats)	Prophylactic	i.p.	Mouse thymocytes, 2 × 10^8, early-stage apoptotic cells (90–95% AxV^+/7-AAD^-)	Decrease of peritoneal macrophage pro-inflammatory response (tumor necrosis factor); increase of blood and DLN Treg; depend on TGF-β	Perruche *et al.* (14)
mBSA (C57BL/6)	Prophylactic	i.v.	Syngeneic thymocytes, 3 × 10^7, 3 consecutive days, early-stage apoptotic cells (60–80% AxV^+/PI^-)	Decrease of DLN Th17, but not Th1 cells; increase of DLN IL-10-producing T cells; IL-10-producing MZB cells; depend on natural IgM	Notley *et al.*, 2011 (15)
mBSA (C57BL/6)	Prophylactic	i.v.	Syngeneic DC, 2 × 10^7, 3 consecutive days, early-stage apoptotic cells (60–75% AxV^+/PI^- and 8–11% PI^+)	Activated apoptotic cells induce IL-6 and prevent TGF-β-mediated prevention of arthritis	Notley *et al.*, 2015 (48)

7-AAD, 7-aminoactinomycin D; Ab, antibody; Auto-Ag, autoantigen; AxV, annexin-V; cDC, conventional DC; CIA, collagen-induced arthritis; DC, dendritic cell; DLN, inguinal draining lymph node; i.p., intraperitoneal; i.v., intravenous; mBSA, methylated bovine serum albumin; MZB, marginal B cells; pDC, plasmacytoid DC; PI, propidium iodide; PLN, peripheral lymph node; SCW, streptococcal cell wall; STIA, serum transfer-induced arthritis (i.e., intraperitoneal injection of K/BxN serum in C57BL/6 mice) (49); TLR, toll like receptor; Treg, regulatory CD4⁺ T cells.

In this model, arthritis results from IC-mediated inflammation followed by articular T cell-mediated responses. However, this model does not recapitulate the endogenous breach of tolerance that is typical of RA pathogenesis. This represents a limitation in applicability to RA (50). The prophylactic effect of apoptotic cell infusion has been observed in this model (15, 48) (**Table 1**). This effect is dependent on natural IgM and IL-10 secretion (15). Finally, the CIA model in DBA/1 mice has been used to evaluate the prophylactic and therapeutic effect of apoptotic cell infusion (13, 20) (**Table 1**). This mouse model shares with human RA several clinical (*i.e.*, erythema and edema), histopathological (*i.e.*, synovitis, pannus formation, cartilage, and bone erosion), as well as immunological features (51). These features consist in the breach of tolerance with the implication of pathogenic T cells associated with the production of inflammatory cytokines (e.g., TNF), as well as the production of auto-antibodies against self-antigens and collagen (50). Some drawbacks have been evoked for this CIA model. The main drawback is that CIA constitutes only an acute model in contrast to the SCW model (52). Nevertheless, all these models are relevant to some features of RA (49–51), and most of them have been used to test drugs now in clinical development for RA (51). Now, we will highlight some immune mechanisms (**Figures 1A–C**) and propose future investigations.

Effects on Macrophages and Resolution of Inflammation

One of our hypotheses concerning the use of early-stage apoptotic cell infusion to treat ongoing arthritis was that reintroducing apoptotic cells in a context of non-resolving inflammation—a key feature in RA (38)—may force macrophage reprogramming after apoptotic cell uptake and stimulate the resolution of inflammation with a decrease of pro-inflammatory cytokines (**Figure 1**). Macrophage reprogramming following efferocytosis stimulates anti-inflammatory factors (e.g., TGF-β or IL-10) and reduces the pro-inflammatory ones (e.g., TNF or IL-1β) (9). The importance of anti-inflammatory cytokines, such as TGF-β (14, 20, 48) and IL-10 (13, 15), was shown in the prevention or treatment of experimental arthritis (**Table 1**). In addition, reduction of TNF after apoptotic cell infusion in the SCW model has been also reported (14) (**Table 1**). In the therapeutic approach using intravenous (i.v.) apoptotic cell infusion, macrophages sorted from the spleen [*i.e.*, the main site where blood-borne apoptotic cells are eliminated (53, 54)] induce the polarization of naive $CD4^+$ T cells toward a Treg phenotype. Altogether, this sustains our initial hypothesis.

Now, we want to discuss the implications of macrophages in the beneficial effects of apoptotic cell infusion in the light of recent data from the literature. The critical macrophage subset for this beneficial effect can be: (*i*) splenic macrophages, as identified after i.v. apoptotic cell infusion (20), (*ii*) peritoneal macrophages, as shown after i.p. apoptotic cell infusion (14), or perhaps (*iii*) macrophages present in the joint (**Figure 1**). This may concern tissue-resident macrophages or monocyte-derived macrophages (26) (**Figure 1**).

Splenic Macrophages

Professional circulating phagocytes, and in particular monocyte-derived macrophages, are guided by "find-me" signals released by dying cells in order to remove apoptotic cells (55). But here, in the case of apoptotic cell infusion to prevent or treat arthritis, injections have been performed, not in the joint, but at distant sites, either i.v. (13, 15, 20) or i.p. (13, 14) (**Figure 1**). The spleen is the main blood filter (53, 54), and marginal zone macrophages of the spleen are specialized in the uptake of blood-borne apoptotic leukocytes (56). Thus, how can splenic macrophages act on the inflamed joint? First, it can occur by the release of anti-inflammatory cytokines that exert a systemic effect affecting inflamed joints (**Figure 1A**). Alternatively, immune cells generated in the spleen (e.g., Treg) may migrate to the inflamed joints and limit/control inflammation (**Figure 1A**). Second, the spleen is a site of immune tolerance induction and can be alerted—or

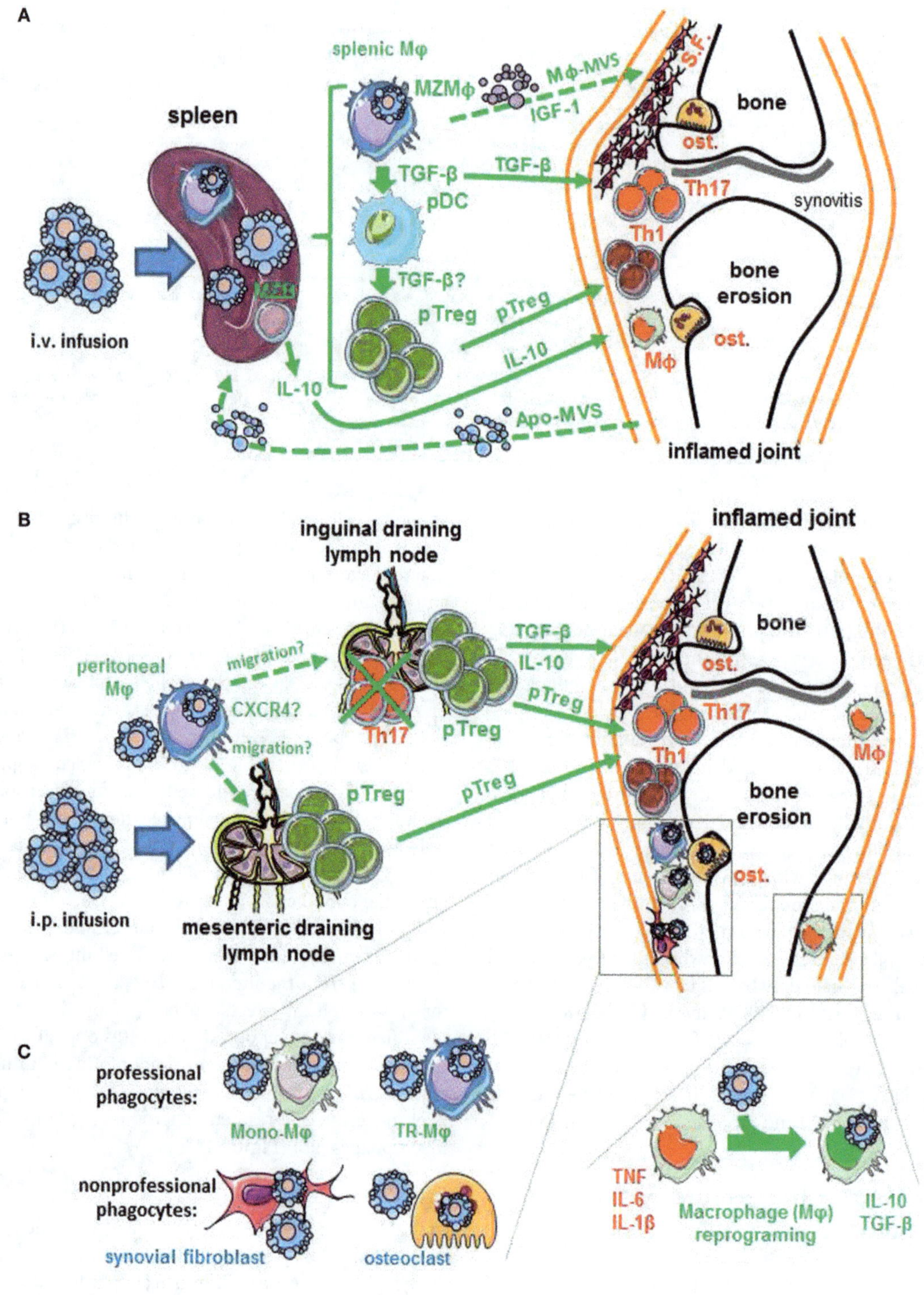

FIGURE 1 | Continued

FIGURE 1 | Continued
Potential immune mechanisms induced by early-stage apoptotic cell infusion in arthritis. Apoptotic cells are infused by two routes: the intravenous (i.v.) and the intraperitoneal (i.p.) routes. **(A)** Apoptotic cells infused intravenously are certainly eliminated by the spleen, and more specifically marginal zone (MZ) macrophages (MZMφ). Splenic Mφ plays a critical role in the effect of i.v. apoptotic cell infusion. These cells may act on inflamed joint by soluble factors (e.g., IL-10 or TGF-β), the generation of peripheral regulatory CD4+ T cells (pTreg) that migrate to the inflamed joints. Alternatively, the immunosuppressive mechanisms identified in the spleen (pro-Treg pDC, anti-inflammatory Mφ or pTreg) can reflect the transfer of tolerance generated in the joints by apoptotic cells to the spleen. Apoptotic materials, such as apoptotic-derived microvesicles (Apo-MVS) have been proposed to mediate this transfer of tolerance from peripheral tissues to the spleen. Finally, splenic Mφ may imprint local joint phagocytes *via* the release of insulin-like growth factor (IGF)-1 and macrophage-derived microvesicles (Mφ-MVS). **(B)** Apoptotic cells injected intraperitoneally can be eliminated by peritoneal Mφ. These cells may migrate to lymph nodes, including mesenteric lymph nodes, and maybe, inguinal draining lymph nodes to stimulate the generation of pTreg. This migration may be guided by the CXCR4/CXCL12 axis. Peripheral Treg generated in the draining lymph nodes are able to reach inflamed joints. **(C)** Infused apoptotic cells may reach the inflamed joints, and be eliminated by local joint Mφ. These Mφ can be either tissue-resident Mφ (TR-Mφ) that have colonized the joints during embryogenesis or blood monocyte-derived Mφ (Mono-Mφ) that have migrated in response to inflammatory signals. The uptake of apoptotic cells by these joint Mφ may be responsible for Mφ reprogramming, that corresponds to the capacity to produce anti-inflammatory factors (e.g., IL-10, TGF-β, or pro-resolving lipid mediators) and lose their ability to secrete pro-inflammatory cytokines [*i.e.*, tumor necrosis factor (TNF), IL-1β or IL-6]. Non-professional phagocytes, such as osteoclasts (ost.) or synovial fibroblasts (S.F.) may also remove apoptotic cells. Deleterious effectors (TNF, Mφ, osteoclasts, synovial fibroblasts, Th1, or Th17 cells) of arthritis present in the inflamed joints are written in red font, while factors or effectors triggered by apoptotic cell infusion are written in green font. Dotted arrows correspond to hypotheses, whereas solid arrows represent data obtained in experimental arthritis models. For references, see the text.

affected—*via* apoptotic "remnants" (including apoptotic cells, or apoptotic materials, such as apoptotic bodies or microvesicles) (54) released by distant tissues during normal cell turn-over (54). This transfer of tolerance from peripheral tissues to the spleen exists also under chronic inflammatory conditions (54). When the functions of splenic macrophages were assessed *ex vivo* after i.v. infusion of apoptotic cells in the setting of arthritis models (20), it is possible that we measured the consequence (*i.e.*, the transfer of tolerance from the joint to the spleen) and not the cause of clinical improvement. Third, based on data obtained in the lungs (57, 58), splenic macrophages phagocyting infused apoptotic cells may release insulin-like growth factor-1 (IGF-1) and macrophage-derived microvesicle (Mφ-MVS) (57) targeting joint-infiltrating immune cells. Extracellular vesicles emitted by macrophages have been shown to export an anti-inflammatory signal to distant cells (58). It remains to be determined which one of these three hypotheses (**Figure 1A**) is responsible for the beneficial effect in arthritis models.

Peritoneal Macrophages

Peritoneal macrophages are affected by i.p. infusion of early-stage apoptotic cells in the SCW (14) or the CIA (13) model (**Table 1**). Recently, it has been reported that macrophages phagocyting apoptotic cells acquire CXCR4 expression and the capacity to migrate in response to CXCL12 (59). This may explain the migration of the so-called "satiated" macrophages to draining lymph nodes after efferocytosis (60). This corresponds to an additional mechanism to export the anti-inflammatory response from tissues where cells die to draining lymph nodes and to participate to the maintenance of tolerance. It remains to be determined whether peritoneal macrophages migrate to draining lymph nodes in the setting of arthritis treatment by apoptotic cell infusion, and if they are then responsible for the modulation of T cell subsets in these lymph nodes. In support of this hypothesis, several modifications of T cell subsets in inguinal draining lymph nodes have been reported in arthritis models (13–15, 20) (**Table 1; Figure 1B**).

Joint Macrophages

In steady state, tissue-resident macrophages are the predominant phagocyting cells in the different tissues analyzed (*i.e.*, the bone marrow, spleen, intestine, liver, and the interstitial space of the lungs) (61). This may be related to the expression of an enzyme called, 12/15-Lipoxygenase (12/15-LOX), expressed by tissue-resident macrophages that confines apoptotic cell removal by these resident macrophages and blocks apoptotic cell uptake by inflammatory Ly6C^{high} monocyte-derived macrophages (62). This mechanism may be responsible for the non-immunogenic removal of apoptotic cell-derived antigens in steady state (62). The anti-inflammatory phenotype imprinted by apoptotic cell phagocytosis in resident macrophages is only partially preserved across the different tissues analyzed (61). Thus, based on this elegant study (61), it is not possible to predict the consequences for joint-infiltrating or joint-resident macrophages under inflammatory conditions. The origin of macrophages present in inflamed joint (tissue-resident *versus* derived from monocytes) has not been deciphered to date (26). Whatever the origin of joint macrophages contributing to apoptotic cell clearance in the therapeutic effect of early-stage apoptotic cells (**Figure 1C**), one may imagine that some mechanisms described for splenic or peritoneal macrophages (**Figures 1A,B**) may occur. Although the anti-inflammatory response imprinted by apoptotic cell phagocytosis in macrophages in steady state is partially preserved across the different tissues analyzed (61), one may postulate that certain mechanisms may be conserved, such as macrophage reprogramming associated with cytokine secretion. Indeed, the downregulation of *Il1b* transcripts in phagocyting macrophages has been found in all tissues analyzed so far (61). Nevertheless, this remains to be determined specifically in the inflamed joints.

The implication of 12/15-LOX in joint macrophages after apoptotic cell infusion is relevant in arthritis. Indeed, the expression of 12/15-LOX is not always confined to tissue-resident macrophages (63). While this observation is true during steady state, other macrophage subsets, in particular monocyte-derived macrophages, may acquire 12/15-LOX expression in response to cytokines (63) or after interactions with apoptotic cells (60, 63, 64). This is the case of the so-called "alternatively activated" M2 macrophages that express 12/15-LOX in response to the triggering of the IL-4 receptor-α signaling pathway, common to both IL-4 and IL-13 (63). An increase in *Alox15* (*i.e.*, the gene encoding 12/15-LOX) mRNA expression in macrophages during the resolution phase of inflammation has been reported in zymosan-induced peritonitis (65). Furthermore, CD11b^{low} "satiated" (*i.e.*, apoptotic cell ingesting) macrophages derived from monocytes have been also shown to express high levels of 12/15-LOX and to possibly promote efferocytosis by the production of pro-resolving lipid mediators, such as resolvin D1 (RvD1) (60). Interestingly, the induction of *Alox15* mRNA has been detected in the synovial tissue of inflamed joints of arthritic mice both in STIA (66) and CIA (67) models. The study of the kinetics of *Alox15* mRNA expression in the inflamed limbs is highly interesting, since *Alox15* mRNA transcripts increase during the CIA induction phase, returns to basal levels during the inflammatory phase, and then increase again during the resolution phase (67). This supports an acquisition of 12/15-LOX by macrophages during the resolution phase of inflammation (65), possibly after efferocytosis (60, 64). Moreover, increased *LOX-15* mRNA expression was found in synovial tissues of RA patients (68). The enzyme 12/15-LOX is the murine ortholog of human 15-LOX (63). These enzymes—human 15-LOX and mouse 12–15/LOX—mediate the oxidation of unsaturated fatty acids. Depending on its substrate (e.g., arachidonic, docosahexaenoic, or linoleic acid), 12/15-LOX generates different key lipid products with anti-inflammatory and pro-resolution properties, such as resolvins, protectins, or lipoxins (63, 69). Lipoxin A4 (LXA4) plays a major role in the resolution of inflammation mediated by 12/15-LOX in experimental arthritis (66, 67). Overall, this suggests that 12/15-LOX is expressed in inflamed joints during arthritis and that this enzyme present in joint macrophages may exert an anti-inflammatory role *via* the synthesis of pro-resolving lipid mediators (e.g., RvD1 or LXA4). One can hypothesize that this mechanism may participate in the local therapeutic effect after apoptotic cell infusion.

Non-Professional Phagocytes

Apoptotic cells can be eliminated by several subsets of phagocyting cells, including professional, but also non-professional phagocytes (70). The involvement of these phagocytes appears again to be tissue-dependent. For instance, five different professional phagocyte subsets (*i.e.*, macrophages and DC subsets) have been recently identified in the intestine (71). Each professional phagocyte subset is dedicated to a specific task (71). Macrophage subsets phagocyting apoptotic intestinal epithelial cells exert an anti-inflammatory response, while CD103^{+} cDC are rather dedicated to drive peripheral Treg (pTreg) in the draining mesenteric lymph nodes (71). However, non-professional phagocytes, mainly epithelial cells, are also important to control apoptotic cell-induced inflammatory responses in the intestine (72) or in the airway (57). In this latter site, non-professional phagocytes (*i.e.*, airway epithelial cells) are controlled by factors released by alveolar macrophages, including IGF-1 and Mφ-MVS (57). Thus, an interaction exists between different phagocytes, and thus, macrophages may affect non-professional phagocytes present in the joint when arthritic animals are treated by early-stage apoptotic cell infusion. Among the potential non-professional phagocytes present in the joint (**Figure 1C**), synovial fibroblasts can be considered as a candidate since fibroblasts are able to uptake apoptotic cells (73) and rabbit synovial fibroblasts have been reported to ingest latex beads in culture (74) or uptake soluble antigen when infused intravenously at high concentrations (75). Osteoclasts are another possibility of non-professional phagocytes for several reasons: (*i*) elevated osteaclast activities have been observed in RA patients (76); (*ii*) dead cells are found engulfed by osteoclasts *in vivo* (77); and (*iii*) osteoclasts are well capable of ingesting apoptotic thymocytes *in vitro* (78) (**Figure 1C**).

The immune consequences of apoptotic cell removal by non-professional phagocytes depend on the phagocytes considered. Apoptotic cell removal by non-professional phagocytes is usually slower than removal by professional phagocytes, and particularly macrophages. It requires apoptotic cells at a more advanced stage than early-stage apoptotic cells, and may be limited to subcellular fragments rather than the whole dying cell (73). In certain settings, pro-inflammatory chemokines (e.g., MCP-1) are released by these non-professional phagocytes, leading to inflammatory monocyte recruitment (73). In contrast, in other tissues, neighbor non-professional phagocytes participate efficiently in the control of inflammation after apoptotic removal (57, 72). Thus, the role of synovial fibroblasts and/or osteoclasts in the beneficial effect of apoptotic cell infusion has to be studied in the setting of experimental arthritis models.

Genetic manipulation of non-professional phagocytes (*i.e.*, epithelial cells) (72) attenuates inflammation *in vivo* at least in the intestine. This was performed in a non T cell-dependent disease, namely dextran sodium sulfate-induced colitis (72). Even if genetic manipulation is not easily transposable from experimental models to patients, this approach (72) does not appear to be appropriate since data obtained in STIA—a T cell-independent disease (50)—show that apoptotic cell infusion is inefficient to prevent this disease.

Effects on DC

Here, the interactions of apoptotic cells with cDC and pDC in the settings of arthritis will be discussed. The implication of pDC after i.v. apoptotic cell infusion has been initially shown in the bone marrow transplantation (BMT) model (79), and found again in the CIA model with the capacity of *ex vivo* sorted splenic pDC (20) to generate pTreg (**Table 1**; **Figure 1A**). Data on the interactions of early-stage apoptotic cells and pDC are certainly easier to interpret than data on cDC. cDC represent different heterogeneous cDC subsets (80), and tools used so far to analyze the impact of apoptotic cell infusion on cDC functions do not allow researchers to separate each subset. For instance, in CD11c/diphtheria toxin (DT) receptor (DTR) transgenic mice, all CD11c^{high} cells are depleted after DT infusion (81). These CD11c^{high} cells consist of cDC (81), but also of other APC subsets having the ability to eliminate apoptotic cells, such as marginal zone and metallophilic macrophages in the spleen (82), sinusoidal macrophages in the lymph node (82), or alveolar macrophages (83). In contrast, pDC have been shown to be spared by depletion after DT administration in CD11c/DTR mice (79). It is known that depending on the considered cDC subsets, the response against apoptotic cells can be the opposite, with splenic lymphoid-resident cDC implicated in tolerance induction (73) while a particular subset of cDC localized at barrier surfaces (e.g., the intestine, the lungs, and the skin) boosts inflammatory responses *via* a PtdSer receptor CD300a (84). Thus, the role of cDC in the therapeutic effect of apoptotic cell infusion in the setting of arthritis has to be further explored. Nevertheless, in the therapeutic approach using the CIA model, we observed that the addition of anti-TNF therapy to apoptotic cell infusion is able to generate *ex vivo* sorted splenic CD11c$^+$ cDC stimulating the polarization of Treg (20). The activation of draining lymph node cDC by NET exacerbated Th1-, but not Th17-, mediated autoimmune responses in CIA (29). This was confirmed *in vitro* by the maturation of human monocyte-derived DC or mouse bone marrow-derived DC in response to NET isolated from CIA mice or RA patients, respectively (29). Interestingly, it was reported that apoptotic cell clearance by neutrophils reduced NET formation (85). Apoptotic cell infusion may, therefore, limit cDC activation and subsequent Th1 responses by limiting NET formation.

Effects on CD4$^+$ T Cell Polarization

One of the salient consequences following apoptotic cell clearance is the induction of pTreg. This has been demonstrated after i.v. apoptotic cell infusion (86) or local apoptotic death of epithelial cells (87). The increase of Treg in the spleen following i.v. apoptotic cell infusion has been shown to require TGF-β (79, 86), splenic macrophages, and donor pDC in the setting of BMT (79). TGF-β is also required for Treg polarization after intestinal epithelial cell apoptosis (87). In arthritis models, the induction of pTreg after apoptotic cell infusion is also TGF-β-dependent (14, 20) (**Table 1**; **Figures 1A,B**). In the therapeutic CIA model, we took advantage of the presence of an infectious antigen, *Mycobacterium tuberculosis* (MBT), mixed with collagen in the complete Freund's adjuvant used for the induction of arthritis, to analyze T cell responses against another antigen than the autoantigen (*i.e.*, bovine type II collagen). In contrast to the response observed with collagen, a similar cell proliferation against MBT antigen is found between cells from apoptotic cell-treated and untreated CIA mice. Moreover, the suppressive activity of Treg sorted from apoptotic cell-treated arthritis mice is restricted to collagen and not extended to MBT (20). This strongly demonstrated that the infusion of apoptotic cells allows the induction of pTreg *in vivo* with an antigenic specificity restricted to the collagen autoantigen. The same effect (*i.e.*, induction of autoantigen-specific pTreg but not infectious antigen-specific Treg) has previously been reported in a similar therapeutic approach based on to the *in vivo* generation of apoptosis (88). Further works are necessary to explain why apoptotic cell infusion favors the induction of autoantigen-specific Treg. Nevertheless, other teams have reported the induction of IL-10-dependent Treg (13, 15) after the prophylactic infusion of apoptotic cells in arthritis models. Thus, this confirms the induction of pTreg after apoptotic cell infusion and may explain the anti-inflammatory effect in the joint whatever the administration route since the generated pTreg may migrate to inflamed joints.

Concerning safety reasons, one has to evoke the high plasticity of CD4$^+$ T cells, and in particular, pTreg that have in common with pro-inflammatory Th17 cells the requirement of TGF-β (36, 89). Apoptotic cell-induced pTreg polarization can be influenced by a simultaneous microbial infection providing IL-6 necessary for Th17 differentiation. The coincident production of IL-6 and TGF-β in response to bacteria and apoptotic epithelial cell death, during orogastric bacterial infection, leads to the generation of both bacteria-specific and autoreactive Th17 cells (87, 90). Similarly, the production of IL-6 together with TGF-β has been shown when "activated" apoptotic cells or apoptotic cells containing high amounts of demethylated DNA have been infused in mBSA arthritis model instead of "resting" apoptotic cells rather containing methylated DNA (48, 91). In the same model, the infusion of "resting" apoptotic thymocytes decreases Th17 cells in the inguinal draining lymph nodes (15). This dichotomy between anti-inflammatory pTreg and pro-inflammatory Th17 cells is not so simple, since different Th17 cell subsets have been now described including pro-inflammatory and anti-inflammatory Th17 cells (89). To date, these subsets have not been studied in the settings of apoptotic cell infusion.

Perspectives and Considerations for Therapeutic Apoptotic Cell Infusion

Here, we will evoke the clinical perspectives of apoptotic cell infusion. This is based on the preclinical data (**Table 1**), but also on data obtained in the field of cancer research. There is an extensive literature on the immunomodulation by dead and dying cells in the setting of cancer (92). We propose to discuss the critical points to achieve a beneficial therapeutic effect (73). These are the following: (*i*) peripheral blood leukocytes are the easiest and major source of apoptotic cells to consider in human settings, while apoptotic cells used in the preclinical studies were other apoptotic leukocytes [*i.e.*, thymocytes (13–15, 20) or DC (48), **Table 1**] from a practical point of view. In RA patients, cytapheresis has to be considered to achieve a sufficient number of apoptotic cells as it has been done in the clinical trial in the setting of hematopoietic cell transplantation (16). The highest number of apoptotic leukocytes planned to be infused is 210 million of cells per kilogram. This will require to pool two sequential cytaphereses. Donor-derived apoptotic cells (*i.e.*, allogeneic cells) will not be considered in the first instance for ethical/regulatory purposes; patient (*i.e.*, syngeneic) apoptotic leukocytes are considered as a cell-based product by the French regulatory agency, while apoptotic cells from healthy volunteers correspond to advanced therapy medicinal products. Nevertheless in experimental models, prevention of arthritis is observed independently of the apoptotic cell origin (*i.e.*, syngeneic, allogeneic, or even xenogeneic) (2); (*ii*) a tolerogenic signal inducing early-stage apoptotic cells, that is, leukocytes stained by annexin-V but little or no staining with PI or 7-AAD dyes (**Table 1**). These stimuli correspond to γ- or ultraviolet B (UVB)-irradiation (73, 93). Stimuli inducing apoptotic cell death have been particularly studied in the field of cancer research in order to generate immunogenic dying tumor cells to favor tumor rejection. A recent study performed *in vivo* using melanoma cells is particularly informative on these stimuli (93). The authors have compared three different apoptotic signals and have confirmed that UVB-irradiation generates non-immunogenic apoptotic cells. Furthermore, the authors have identified that the production of IL-27 and IL-1α by bone marrow-derived macrophages after *in vitro* incubation with apoptotic tumor cells predicts immunogenicity. In addition, this work shows that primary necrotic cells induced by tuberculosis-necrotizing toxin *in vivo* are also non-immunogenic

(93). This confirms that necrotic cells induced by a repeated freeze/thaw procedure or obtained by incubating apoptotic cells for 24 h before infusion are very poor inducers of CD8$^+$ T cell responses *in vivo* (94). The immunogenicity of necrotic cells remains, however, a matter of debate (92) that we do not want to comment further here; (*iii*) one infusion appears sufficient whereas multiple infusions may expose to a risk of immunization against apoptotic cell-derived antigens (95), as discussed in Ref. (92). The question arises as to how long the therapeutic effect will last. In the CIA model, the therapeutic effect of early-stage apoptotic cell infusion is transient but prolonged when associated with anti-TNF therapy (20). Only clinical studies will allow to answer to this question; and (*iv*) a systemic administration route can be considered while local administration is also possible. In experimental arthritis models, two distinct systemic routes of administration [*i.e.,* i.p. (13, 14) *versus* i.v. (13, 15, 20, 48)] have been tested with a similar efficacy (**Table 1**). However, no local administration has been evaluated so far. Another lesson coming from cancer research on dying/dead cells is the ability of apoptotic tumor cells to stimulate the proliferation of nearby viable tumor cells (96, 97). This apoptotic cell-mediated proliferation is not restricted to tumor cells (98). Relevant to the present review, primary human synovial fibroblasts isolated from knee joints of RA patients are also able to proliferate *in vitro* when these fibroblasts are in close contact with apoptotic tumor cells (97). Considering the therapeutic use of apoptotic cell infusion, it is, however, reassuring to see that when the number of apoptotic tumor cells is increased, the proliferative effect is limited (97). The percentage of infused early-stage apoptotic cells planned to be infused to RA patients is higher than 50%. However, one has to be cautious on this apoptosis-induced proliferative effect, since it is mainly mediated by a soluble factor, the nucleoside inosine (97).

Finally, one has to remain cautious, since a different effect can be obtained depending on the infusion of "resting" *versus* "activated" apoptotic CD11c$^+$ cDC (48). This may be related to the methylation status of DNA from apoptotic cells (91). This work found that, as apoptotic cDC, apoptotic CD4$^+$ T cells from RA patients exhibit a DNA demethylated status, suggesting a pro-inflammatory effect after infusion associated with IL-6 secretion (91). Whether this may impact on the therapeutic efficacy of apoptotic cell infusion remains to be determined. We used apoptotic splenic cells from arthritic mice (*i.e.,* containing multiple activated leukocytes) in the therapeutic CIA model, and we observed the same effects as apoptotic thymocytes (Bonnefoy F., Perruche S., unpublished results). An additional security can be also provided by the addition of csDMARD, such as low-dose MTX or bDMARD, such as TNF inhibitors. Indeed, these treatments do not inhibit the beneficial therapeutic effects of apoptotic cell infusion (20). MTX (at high-dose) has been also used as prophylaxis of graft-*versus*-host disease in the clinical trial testing the effects of donor early-stage apoptotic cell infusion (16). This confirms our experimental data in CIA: MTX does not affect the therapeutic effect of apoptotic cell infusion, and allows to preserve its beneficial effect on collagen (autoantigen)-specific Treg (20). In addition, the capacity of splenic pDC and macrophages to induce *ex vivo* pTreg polarization is not inhibited by MTX (20). Thus, MTX can be continued if an apoptotic cell-based therapy has to be proposed to patients. Compared with MTX, anti-TNF therapy has the advantage to synergize with apoptotic cell infusion to control ongoing arthritis (20). However, the exact mechanism responsible for this synergy has not been identified (20). In the future, IL-6 inhibitors, such as tocilizumab, can be also envisaged to be associated with apoptotic cell infusion in order to prevent the antagonistic effect of IL-6, previously reported in mBSA-induced arthritis (48, 91). This can be a way to neutralize the effects of the methylation status of DNA from apoptotic T cells obtained from RA patients (91).

Apoptotic cell infusion can potentially be associated with corticosteroids without any risk. Indeed, corticosteroids enhance apoptotic cell removal by inducing the expression of the PtdSer-binding protein, milk fat globule-EGF factor 8 (MFG-E8) selectively in human and mouse monocytes and macrophages (whatever their differentiation profile, M1 or M2) (99).

CONCLUSION/CONCLUDING REMARKS

Apoptotic cell infusion represents an additional potential bDMARD in RA, and more particularly a cell-based bDMARD. We plan to initiate a phase I/II clinical trial (ClinicalTrials.gov Identifier: NCT02903212) to achieve LDA in patients with RA who did not respond adequately to one previous bDMARD. Concerning the potential toxicity of this approach, one may build on the experience gained by the clinical trial performed in the setting of hematopoietic cell transplantation (16), but also those using extracorporeal photopheresis (ECP) in RA patients (100, 101). Even if ECP does not necessarily generate "proper/pure" early-stage apoptotic cells (102), this treatment introduces high amounts of dead cells in patients and no specific toxicity has been reported (100, 101). The efferocytosis capacity of monocyte-derived macrophages from 14 RA patients has been shown to be similar to those of healthy volunteers (103). A careful selection of patients should be done in order to avoid genetic alterations of molecules involved in efferocytosis (e.g., MFG-E8), as well as apoptotic cells carrying demethylated DNA (91). One advantage of cell-based therapies could be the adherence to treatment since we propose only one infusion in our clinical trial. This study is an opportunity to analyze in human the immune mechanisms triggered by infused apoptotic cells. Furthermore, association with biologic agents acting on different therapeutic targets (e.g., TNF) appears feasible to increase efficacy without the toxicity.

AUTHOR CONTRIBUTIONS

PS, FB, ET, and SP analyzed and discussed the literature and conceived the outline of the manuscript; PS wrote the manuscript. All authors edited the manuscript and provided valuable discussions and criticisms.

ACKNOWLEDGMENTS

This work is supported by the Agence Nationale de la Recherche (ANR) under the program "*Investissements d'Avenir*" with reference ANR-11-LABX-0021-LipSTIC, by the Ministère de la Santé (PHRC Interrégional-Est #PHRCI-15-037), by the Region

Bourgogne Franche-Comté (support to LipSTIC LabEX 2017 and MiMedI), the Arthritis Fondation Courtin (to SP), and the Fondation pour la Recherche Médicale (DBS20131128447 to SP). The authors would like to thank Sarah Odrion for her help in editing our manuscript, the members of our laboratory for their work, as well as both reviewers for their constructive comments.

REFERENCES

1. Poon IK, Lucas CD, Rossi AG, Ravichandran KS. Apoptotic cell clearance: basic biology and therapeutic potential. *Nat Rev Immunol* (2014) 14:166–80. doi:10.1038/nri3607
2. Saas P, Daguindau E, Perruche S. Concise review: apoptotic cell-based therapies-rationale, preclinical results and future clinical developments. *Stem Cells* (2016) 34:1464–73. doi:10.1002/stem.2361
3. Kerr JF, Wyllie AH, Currie AR. Apoptosis: a basic biological phenomenon with wide-ranging implications in tissue kinetics. *Br J Cancer* (1972) 26:239–57. doi:10.1038/bjc.1972.33
4. Galluzzi L, Vitale I, Abrams JM, Alnemri ES, Baehrecke EH, Blagosklonny MV, et al. Molecular definitions of cell death subroutines: recommendations of the Nomenclature Committee on Cell Death 2012. *Cell Death Differ* (2012) 19:107–20. doi:10.1038/cdd.2011.96
5. Fadok VA, Voelker DR, Campbell PA, Cohen JJ, Bratton DL, Henson PM. Exposure of phosphatidylserine on the surface of apoptotic lymphocytes triggers specific recognition and removal by macrophages. *J Immunol* (1992) 148:2207–16.
6. Duvall E, Wyllie AH, Morris RG. Macrophage recognition of cells undergoing programmed cell death (apoptosis). *Immunology* (1985) 56:351–8.
7. Savill JS, Wyllie AH, Henson JE, Walport MJ, Henson PM, Haslett C. Macrophage phagocytosis of aging neutrophils in inflammation. Programmed cell death in the neutrophil leads to its recognition by macrophages. *J Clin Invest* (1989) 83:865–75. doi:10.1172/JCI113970
8. Serhan CN, Savill J. Resolution of inflammation: the beginning programs the end. *Nat Immunol* (2005) 6:1191–7. doi:10.1038/ni1276
9. Voll RE, Herrmann M, Roth EA, Stach C, Kalden JR, Girkontaite I. Immunosuppressive effects of apoptotic cells. *Nature* (1997) 390:350–1. doi:10.1038/37022
10. Wallach D, Kang TB, Kovalenko A. Concepts of tissue injury and cell death in inflammation: a historical perspective. *Nat Rev Immunol* (2014) 14:51–9. doi:10.1038/nri3561
11. Nagata S, Tanaka M. Programmed cell death and the immune system. *Nat Rev Immunol* (2017) 17:333–40. doi:10.1038/nri.2016.153
12. Kawane K, Tanaka H, Kitahara Y, Shimaoka S, Nagata S. Cytokine-dependent but acquired immunity-independent arthritis caused by DNA escaped from degradation. *Proc Natl Acad Sci U S A* (2010) 107:19432–7. doi:10.1073/pnas.1010603107
13. Gray M, Miles K, Salter D, Gray D, Savill J. Apoptotic cells protect mice from autoimmune inflammation by the induction of regulatory B cells. *Proc Natl Acad Sci U S A* (2007) 104:14080–5. doi:10.1073/pnas.0700326104
14. Perruche S, Saas P, Chen W. Apoptotic cell-mediated suppression of streptococcal cell wall-induced arthritis is associated with alteration of macrophage function and local regulatory T-cell increase: a potential cell-based therapy? *Arthritis Res Ther* (2009) 11:R104. doi:10.1186/ar2750
15. Notley CA, Brown MA, Wright GP, Ehrenstein MR. Natural IgM is required for suppression of inflammatory arthritis by apoptotic cells. *J Immunol* (2011) 186:4967–72. doi:10.4049/jimmunol.1003021
16. Mevorach D, Zuckerman T, Reiner I, Shimoni A, Samuel S, Nagler A, et al. Single infusion of donor mononuclear early apoptotic cells as prophylaxis for graft-versus-host disease in myeloablative HLA-matched allogeneic bone marrow transplantation: a phase I/IIa clinical trial. *Biol Blood Marrow Transplant* (2014) 20:58–65. doi:10.1016/j.bbmt.2013.10.010
17. Huynh ML, Fadok VA, Henson PM. Phosphatidylserine-dependent ingestion of apoptotic cells promotes TGF-beta1 secretion and the resolution of inflammation. *J Clin Invest* (2002) 109:41–50. doi:10.1172/JCI11638
18. Ren Y, Xie Y, Jiang G, Fan J, Yeung J, Li W, et al. Apoptotic cells protect mice against lipopolysaccharide-induced shock. *J Immunol* (2008) 180:4978–85. doi:10.4049/jimmunol.180.7.4978
19. Wang Z, Larregina AT, Shufesky WJ, Perone MJ, Montecalvo A, Zahorchak AF, et al. Use of the inhibitory effect of apoptotic cells on dendritic cells for graft survival via T-cell deletion and regulatory T cells. *Am J Transplant* (2006) 6:1297–311. doi:10.1111/j.1600-6143.2006.01308.x
20. Bonnefoy F, Daoui A, Valmary-Degano S, Toussirot E, Saas P, Perruche S. Apoptotic cell infusion treats ongoing collagen-induced arthritis, even in the presence of methotrexate, and is synergic with anti-TNF therapy. *Arthritis Res Ther* (2016) 18:184. doi:10.1186/s13075-016-1084-0
21. McInnes IB, Schett G. The pathogenesis of rheumatoid arthritis. *N Engl J Med* (2011) 365:2205–19. doi:10.1056/NEJMra1004965
22. Kochi Y, Suzuki A, Yamada R, Yamamoto K. Genetics of rheumatoid arthritis: underlying evidence of ethnic differences. *J Autoimmun* (2009) 32:158–62. doi:10.1016/j.jaut.2009.02.020
23. Theofilopoulos AN, Kono DH, Baccala R. The multiple pathways to autoimmunity. *Nat Immunol* (2017) 18:716–24. doi:10.1038/ni.3731
24. Feldmann M, Maini RN. TNF defined as a therapeutic target for rheumatoid arthritis and other autoimmune diseases. *Nat Med* (2003) 9:1245–50. doi:10.1038/nm939
25. Burmester GR, Feist E, Dorner T. Emerging cell and cytokine targets in rheumatoid arthritis. *Nat Rev Rheumatol* (2014) 10:77–88. doi:10.1038/nrrheum.2013.168
26. Udalova IA, Mantovani A, Feldmann M. Macrophage heterogeneity in the context of rheumatoid arthritis. *Nat Rev Rheumatol* (2016) 12:472–85. doi:10.1038/nrrheum.2016.91
27. Angelotti F, Parma A, Cafaro G, Capecchi R, Alunno A, Puxeddu I. One year in review 2017: pathogenesis of rheumatoid arthritis. *Clin Exp Rheumatol* (2017) 35:368–78.
28. Khandpur R, Carmona-Rivera C, Vivekanandan-Giri A, Gizinski A, Yalavarthi S, Knight JS, et al. NETs are a source of citrullinated autoantigens and stimulate inflammatory responses in rheumatoid arthritis. *Sci Transl Med* (2013) 5:178ra40. doi:10.1126/scitranslmed.3005580
29. Papadaki G, Kambas K, Choulaki C, Vlachou K, Drakos E, Bertsias G, et al. Neutrophil extracellular traps exacerbate Th1-mediated autoimmune responses in rheumatoid arthritis by promoting DC maturation. *Eur J Immunol* (2016) 46:2542–54. doi:10.1002/eji.201646542
30. Van Krinks CH, Matyszak MK, Gaston JS. Characterization of plasmacytoid dendritic cells in inflammatory arthritis synovial fluid. *Rheumatology (Oxford)* (2004) 43:453–60. doi:10.1093/rheumatology/keh115
31. Lande R, Giacomini E, Serafini B, Rosicarelli B, Sebastiani GD, Minisola G, et al. Characterization and recruitment of plasmacytoid dendritic cells in synovial fluid and tissue of patients with chronic inflammatory arthritis. *J Immunol* (2004) 173:2815–24. doi:10.4049/jimmunol.173.4.2815
32. Cavanagh LL, Boyce A, Smith L, Padmanabha J, Filgueira L, Pietschmann P, et al. Rheumatoid arthritis synovium contains plasmacytoid dendritic cells. *Arthritis Res Ther* (2005) 7:R230–40. doi:10.1186/ar1467
33. Bonnefoy F, Couturier M, Clauzon A, Remy-Martin JP, Gaugler B, Tiberghien P, et al. TGF-beta-exposed plasmacytoid dendritic cells participate in Th17 commitment. *J Immunol* (2011) 186:6157–64. doi:10.4049/jimmunol.1002497
34. Malmstrom V, Catrina AI, Klareskog L. The immunopathogenesis of seropositive rheumatoid arthritis: from triggering to targeting. *Nat Rev Immunol* (2017) 17:60–75. doi:10.1038/nri.2016.124
35. Tan EM, Smolen JS. Historical observations contributing insights on etiopathogenesis of rheumatoid arthritis and role of rheumatoid factor. *J Exp Med* (2016) 213:1937–50. doi:10.1084/jem.20160792
36. Lubberts E. The IL-23-IL-17 axis in inflammatory arthritis. *Nat Rev Rheumatol* (2015) 11:415–29. doi:10.1038/nrrheum.2015.53
37. Ehrenstein MR, Evans JG, Singh A, Moore S, Warnes G, Isenberg DA, et al. Compromised function of regulatory T cells in rheumatoid arthritis and reversal by anti-TNFalpha therapy. *J Exp Med* (2004) 200:277–85. doi:10.1084/jem.20040165
38. Nathan C, Ding A. Nonresolving inflammation. *Cell* (2010) 140:871–82. doi:10.1016/j.cell.2010.02.029
39. Corsiero E, Pratesi F, Prediletto E, Bombardieri M, Migliorini P. NETosis as source of autoantigens in rheumatoid arthritis. *Front Immunol* (2016) 7:485. doi:10.3389/fimmu.2016.00485

40. Smolen JS, Aletaha D. Rheumatoid arthritis therapy reappraisal: strategies, opportunities and challenges. *Nat Rev Rheumatol* (2015) 11:276–89. doi:10.1038/nrrheum.2015.8
41. Smolen JS, Landewe R, Breedveld FC, Buch M, Burmester G, Dougados M, et al. EULAR recommendations for the management of rheumatoid arthritis with synthetic and biological disease-modifying antirheumatic drugs: 2013 update. *Ann Rheum Dis* (2014) 73:492–509. doi:10.1136/annrheumdis-2013-204573
42. Cronstein BN. Low-dose methotrexate: a mainstay in the treatment of rheumatoid arthritis. *Pharmacol Rev* (2005) 57:163–72. doi:10.1124/pr.57.2.3
43. Varatharajan N, Lim IG, Anandacoomarasamy A, Russo R, Byth K, Spencer DG, et al. Methotrexate: long-term safety and efficacy in an Australian consultant rheumatology practice. *Intern Med J* (2009) 39:228–36. doi:10.1111/j.1445-5994.2009.01800.x
44. van der Heijden JW, Dijkmans BA, Scheper RJ, Jansen G. Drug Insight: resistance to methotrexate and other disease-modifying antirheumatic drugs – from bench to bedside. *Nat Clin Pract Rheumatol* (2007) 3:26–34. doi:10.1038/ncprheum0380
45. Mok CC, van der Kleij D, Wolbink GJ. Drug levels, anti-drug antibodies, and clinical efficacy of the anti-TNFalpha biologics in rheumatic diseases. *Clin Rheumatol* (2013) 32:1429–35. doi:10.1007/s10067-013-2336-x
46. Simard JF, Arkema EV, Sundstrom A, Geborek P, Saxne T, Baecklund E, et al. Ten years with biologics: to whom do data on effectiveness and safety apply? *Rheumatology (Oxford)* (2011) 50:204–13. doi:10.1093/rheumatology/keq326
47. Smolen JS, Aletaha D. Forget personalised medicine and focus on abating disease activity. *Ann Rheum Dis* (2013) 72:3–6. doi:10.1136/annrheumdis-2012-202361
48. Notley CA, Brown MA, McGovern JL, Jordan CK, Ehrenstein MR. Engulfment of activated apoptotic cells abolishes TGF-beta-mediated immunoregulation via the induction of IL-6. *J Immunol* (2015) 194:1621–7. doi:10.4049/jimmunol.1401256
49. Christensen AD, Haase C, Cook AD, Hamilton JA. K/BxN serum-transfer arthritis as a model for human inflammatory arthritis. *Front Immunol* (2016) 7:213. doi:10.3389/fimmu.2016.00213
50. Asquith DL, Miller AM, McInnes IB, Liew FY. Animal models of rheumatoid arthritis. *Eur J Immunol* (2009) 39:2040–4. doi:10.1002/eji.200939578
51. Hegen M, Keith JC Jr, Collins M, Nickerson-Nutter CL. Utility of animal models for identification of potential therapeutics for rheumatoid arthritis. *Ann Rheum Dis* (2008) 67:1505–15. doi:10.1136/ard.2007.076430
52. Kollias G, Papadaki P, Apparailly F, Vervoordeldonk MJ, Holmdahl R, Baumans V, et al. Animal models for arthritis: innovative tools for prevention and treatment. *Ann Rheum Dis* (2011) 70:1357–62. doi:10.1136/ard.2010.148551
53. Mebius RE, Kraal G. Structure and function of the spleen. *Nat Rev Immunol* (2005) 5:606–16. doi:10.1038/nri1669
54. Bronte V, Pittet MJ. The spleen in local and systemic regulation of immunity. *Immunity* (2013) 39:806–18. doi:10.1016/j.immuni.2013.10.010
55. Ariel A, Ravichandran KS. 'This way please': apoptotic cells regulate phagocyte migration before and after engulfment. *Eur J Immunol* (2016) 46:1583–6. doi:10.1002/eji.201646505
56. Miyake Y, Asano K, Kaise H, Uemura M, Nakayama M, Tanaka M. Critical role of macrophages in the marginal zone in the suppression of immune responses to apoptotic cell-associated antigens. *J Clin Invest* (2007) 117:2268–78. doi:10.1172/JCI31990
57. Han CZ, Juncadella IJ, Kinchen JM, Buckley MW, Klibanov AL, Dryden K, et al. Macrophages redirect phagocytosis by non-professional phagocytes and influence inflammation. *Nature* (2016) 539:570–4. doi:10.1038/nature20141
58. Bourdonnay E, Zaslona Z, Penke LR, Speth JM, Schneider DJ, Przybranowski S, et al. Transcellular delivery of vesicular SOCS proteins from macrophages to epithelial cells blunts inflammatory signaling. *J Exp Med* (2015) 212:729–42. doi:10.1084/jem.20141675
59. Angsana J, Chen J, Liu L, Haller CA, Chaikof EL. Efferocytosis as a regulator of macrophage chemokine receptor expression and polarization. *Eur J Immunol* (2016) 46:1592–9. doi:10.1002/eji.201546262
60. Schif-Zuck S, Gross N, Assi S, Rostoker R, Serhan CN, Ariel A. Saturated-efferocytosis generates pro-resolving CD11b low macrophages: modulation by resolvins and glucocorticoids. *Eur J Immunol* (2011) 41:366–79. doi:10.1002/eji.201040801
61. A-Gonzalez N, Quintana JA, Garcia-Silva S, Mazariegos M, Gonzalez de la Aleja A, Nicolas-Avila JA, et al. Phagocytosis imprints heterogeneity in tissue-resident macrophages. *J Exp Med* (2017) 214:1281–96. doi:10.1084/jem.20161375
62. Uderhardt S, Herrmann M, Oskolkova OV, Aschermann S, Bicker W, Ipseiz N, et al. 12/15-lipoxygenase orchestrates the clearance of apoptotic cells and maintains immunologic tolerance. *Immunity* (2012) 36:834–46. doi:10.1016/j.immuni.2012.03.010
63. Ackermann JA, Hofheinz K, Zaiss MM, Kronke G. The double-edged role of 12/15-lipoxygenase during inflammation and immunity. *Biochim Biophys Acta* (2017) 1862:371–81. doi:10.1016/j.bbalip.2016.07.014
64. Godson C, Mitchell S, Harvey K, Petasis NA, Hogg N, Brady HR. Cutting edge: lipoxins rapidly stimulate nonphlogistic phagocytosis of apoptotic neutrophils by monocyte-derived macrophages. *J Immunol* (2000) 164:1663–7. doi:10.4049/jimmunol.164.4.1663
65. Stables MJ, Shah S, Camon EB, Lovering RC, Newson J, Bystrom J, et al. Transcriptomic analyses of murine resolution-phase macrophages. *Blood* (2011) 118:e192–208. doi:10.1182/blood-2011-04-345330
66. Kronke G, Katzenbeisser J, Uderhardt S, Zaiss MM, Scholtysek C, Schabbauer G, et al. 12/15-lipoxygenase counteracts inflammation and tissue damage in arthritis. *J Immunol* (2009) 183:3383–9. doi:10.4049/jimmunol.0900327
67. Chan MM, Moore AR. Resolution of inflammation in murine autoimmune arthritis is disrupted by cyclooxygenase-2 inhibition and restored by prostaglandin E2-mediated lipoxin A4 production. *J Immunol* (2010) 184:6418–26. doi:10.4049/jimmunol.0903816
68. Hashimoto A, Hayashi I, Murakami Y, Sato Y, Kitasato H, Matsushita R, et al. Antiinflammatory mediator lipoxin A4 and its receptor in synovitis of patients with rheumatoid arthritis. *J Rheumatol* (2007) 34:2144–53.
69. Serhan CN, Chiang N, Dalli J. The resolution code of acute inflammation: novel pro-resolving lipid mediators in resolution. *Semin Immunol* (2015) 27:200–15. doi:10.1016/j.smim.2015.03.004
70. Arandjelovic S, Ravichandran KS. Phagocytosis of apoptotic cells in homeostasis. *Nat Immunol* (2015) 16:907–17. doi:10.1038/ni.3253
71. Cummings RJ, Barbet G, Bongers G, Hartmann BM, Gettler K, Muniz L, et al. Different tissue phagocytes sample apoptotic cells to direct distinct homeostasis programs. *Nature* (2016) 539:565–9. doi:10.1038/nature20138
72. Lee CS, Penberthy KK, Wheeler KM, Juncadella IJ, Vandenabeele P, Lysiak JJ, et al. Boosting apoptotic cell clearance by colonic epithelial cells attenuates inflammation in vivo. *Immunity* (2016) 44:807–20. doi:10.1016/j.immuni.2016.02.005
73. Saas P, Kaminski S, Perruche S. Prospects of apoptotic cell-based therapies for transplantation and inflammatory diseases. *Immunotherapy* (2013) 5:1055–73. doi:10.2217/imt.13.103
74. Werb Z, Reynolds JJ. Stimulation by endocytosis of the secretion of collagenase and neutral proteinase from rabbit synovial fibroblasts. *J Exp Med* (1974) 140:1482–97. doi:10.1084/jem.140.6.1482
75. Mitnick H, Hoffstein S, Weissmann G. Fate of antigen after intravenous and intraarticular injection into rabbits. *Arthritis Rheum* (1978) 21:918–29. doi:10.1002/art.1780210808
76. Braun T, Zwerina J. Positive regulators of osteoclastogenesis and bone resorption in rheumatoid arthritis. *Arthritis Res Ther* (2011) 13:235. doi:10.1186/ar3380
77. Cerri PS, Boabaid F, Katchburian E. Combined TUNEL and TRAP methods suggest that apoptotic bone cells are inside vacuoles of alveolar bone osteoclasts in young rats. *J Periodontal Res* (2003) 38:223–6. doi:10.1034/j.1600-0765.2003.02006.x
78. Harre U, Keppeler H, Ipseiz N, Derer A, Poller K, Aigner M, et al. Moonlighting osteoclasts as undertakers of apoptotic cells. *Autoimmunity* (2012) 45:612–9. doi:10.3109/08916934.2012.719950
79. Bonnefoy F, Perruche S, Couturier M, Sedrati A, Sun Y, Tiberghien P, et al. Plasmacytoid dendritic cells play a major role in apoptotic leukocyte-induced immune modulation. *J Immunol* (2011) 186:5696–705. doi:10.4049/jimmunol.1001523
80. Sichien D, Lambrecht BN, Guilliams M, Scott CL. Development of conventional dendritic cells: from common bone marrow progenitors to multiple subsets in peripheral tissues. *Mucosal Immunol* (2017) 10:831–44. doi:10.1038/mi.2017.8
81. Jung S, Unutmaz D, Wong P, Sano G, De los Santos K, Sparwasser T, et al. In vivo depletion of CD11c+ dendritic cells abrogates priming of CD8+ T cells by exogenous cell-associated antigens. *Immunity* (2002) 17:211–20. doi:10.1016/S1074-7613(02)00365-5

82. Probst HC, Tschannen K, Odermatt B, Schwendener R, Zinkernagel RM, Van Den Broek M. Histological analysis of CD11c-DTR/GFP mice after in vivo depletion of dendritic cells. *Clin Exp Immunol* (2005) 141:398–404. doi:10.1111/j.1365-2249.2005.02868.x
83. van Rijt LS, Jung S, Kleinjan A, Vos N, Willart M, Duez C, et al. In vivo depletion of lung CD11c+ dendritic cells during allergen challenge abrogates the characteristic features of asthma. *J Exp Med* (2005) 201:981–91. doi:10.1084/jem.20042311
84. Nakahashi-Oda C, Udayanga KG, Nakamura Y, Nakazawa Y, Totsuka N, Miki H, et al. Apoptotic epithelial cells control the abundance of Treg cells at barrier surfaces. *Nat Immunol* (2016) 17:441–50. doi:10.1038/ni.3345
85. Manfredi AA, Covino C, Rovere-Querini P, Maugeri N. Instructive influences of phagocytic clearance of dying cells on neutrophil extracellular trap generation. *Clin Exp Immunol* (2015) 179:24–9. doi:10.1111/cei.12320
86. Kleinclauss F, Perruche S, Masson E, de Carvalho Bittencourt M, Biichle S, Remy-Martin JP, et al. Intravenous apoptotic spleen cell infusion induces a TGF-beta-dependent regulatory T-cell expansion. *Cell Death Differ* (2006) 13:41–52. doi:10.1038/sj.cdd.4401699
87. Torchinsky MB, Garaude J, Martin AP, Blander JM. Innate immune recognition of infected apoptotic cells directs T(H)17 cell differentiation. *Nature* (2009) 458:78–82. doi:10.1038/nature07781
88. Kasagi S, Zhang P, Che L, Abbatiello B, Maruyama T, Nakatsukasa H, et al. In vivo-generated antigen-specific regulatory T cells treat autoimmunity without compromising antibacterial immune response. *Sci Transl Med* (2014) 6:241ra78. doi:10.1126/scitranslmed.3008895
89. Muranski P, Restifo NP. Essentials of Th17 cell commitment and plasticity. *Blood* (2013) 121:2402–14. doi:10.1182/blood-2012-09-378653
90. Campisi L, Barbet G, Ding Y, Esplugues E, Flavell RA, Blander JM. Apoptosis in response to microbial infection induces autoreactive TH17 cells. *Nat Immunol* (2016) 17:1084–92. doi:10.1038/ni.3512
91. Notley CA, Jordan CK, McGovern JL, Brown MA, Ehrenstein MR. DNA methylation governs the dynamic regulation of inflammation by apoptotic cells during efferocytosis. *Sci Rep* (2017) 7:42204. doi:10.1038/srep42204
92. Maueroder C, Munoz LE, Chaurio RA, Herrmann M, Schett G, Berens C. Tumor immunotherapy: lessons from autoimmunity. *Front Immunol* (2014) 5:212. doi:10.3389/fimmu.2014.00212
93. Maueroder C, Chaurio RA, Dumych T, Podolska M, Lootsik MD, Culemann S, et al. A blast without power – cell death induced by the tuberculosis-necrotizing toxin fails to elicit adequate immune responses. *Cell Death Differ* (2016) 23:1016–25. doi:10.1038/cdd.2016.4
94. Yatim N, Jusforgues-Saklani H, Orozco S, Schulz O, Barreira da Silva R, Reis e Sousa C, et al. RIPK1 and NF-kappaB signaling in dying cells determines cross-priming of CD8(+) T cells. *Science* (2015) 350:328–34. doi:10.1126/science.aad0395
95. Mevorach D, Zhou JL, Song X, Elkon KB. Systemic exposure to irradiated apoptotic cells induces autoantibody production. *J Exp Med* (1998) 188: 387–92. doi:10.1084/jem.188.2.387
96. Chaurio R, Janko C, Schorn C, Maueroder C, Bilyy R, Gaipl U, et al. UVB-irradiated apoptotic cells induce accelerated growth of co-implanted viable tumor cells in immune competent mice. *Autoimmunity* (2013) 46:317–22. doi:10.3109/08916934.2012.754433
97. Chen J, Chaurio RA, Maueroder C, Derer A, Rauh M, Kost A, et al. Inosine released from dying or dead cells stimulates cell proliferation via adenosine receptors. *Front Immunol* (2017) 8:504. doi:10.3389/fimmu.2017.00504
98. Fogarty CE, Bergmann A. Killers creating new life: caspases drive apoptosis-induced proliferation in tissue repair and disease. *Cell Death Differ* (2017) 24:1390–400. doi:10.1038/cdd.2017.47
99. Lauber K, Keppeler H, Munoz LE, Koppe U, Schroder K, Yamaguchi H, et al. Milk fat globule-EGF factor 8 mediates the enhancement of apoptotic cell clearance by glucocorticoids. *Cell Death Differ* (2013) 20:1230–40. doi:10.1038/cdd.2013.82
100. Malawista SE, Trock D, Edelson RL. Photopheresis for rheumatoid arthritis. *Ann N Y Acad Sci* (1991) 636:217–26. doi:10.1111/j.1749-6632.1991.tb33453.x
101. Adamski J, Kinard T, Ipe T, Cooling L. Extracorporeal photopheresis for the treatment of autoimmune diseases. *Transfus Apher Sci* (2015) 52:171–82. doi:10.1016/j.transci.2015.02.005
102. Hannani D. Extracorporeal photopheresis: tolerogenic or immunogenic cell death? Beyond current dogma. *Front Immunol* (2015) 6:349. doi:10.3389/fimmu.2015.00349
103. Manoussakis MN, Fragoulis GE, Vakrakou AG, Moutsopoulos HM. Impaired clearance of early apoptotic cells mediated by inhibitory IgG antibodies in patients with primary Sjogren's syndrome. *PLoS One* (2014) 9:e112100. doi:10.1371/journal.pone.0112100

Conflict of Interest Statement: The authors declare that the research was conducted in the absence of any commercial or financial relationships that could be construed as a potential conflict of interest.

9

Pro-Resolving Mediators in Regulating and Conferring Macrophage Function

Jesmond Dalli[1]* *and Charles N. Serhan*[2]*

[1] Lipid Mediator Unit, William Harvey Research Institute, Bart's and the London School of Medicine, Queen Mary University of London, London, United Kingdom, [2] Center for Experimental Therapeutics and Reperfusion Injury, Harvard Institutes of Medicine, Brigham and Women's Hospital and Harvard Medical School, Boston, MA, United States

Macrophages are central in coordinating the host response to both sterile and infective insults. Clearance of apoptotic cells and cellular debris is a key biological action preformed by macrophages that paves the way to the resolution of local inflammation, repair and regeneration of damaged tissues, and re-establishment of function. The essential fatty acid-derived autacoids termed specialized pro-resolving mediators (SPM) play central roles in promoting these processes. In the present article, we will review the role of microvesicles in controlling macrophage efferocytosis and SPM production. We will also discuss the role of both apoptotic cells and microvesicles in providing substrate for transcellular biosynthesis of several SPM families during efferocyotsis. In addition, this article will discuss the biological actions of the recently uncovered macrophage-derived SPM termed maresins. These mediators are produced via 14-lipoxygenation of docosahexaenoic acid that is either enzymatically converted to mediators carrying two hydroxyl groups or to autacoids that are peptide-lipid conjugates, coined maresin conjugates in tissue regeneration. The formation of these mediators is temporally regulated during acute self-limited infectious-inflammation where they promote the uptake and clearance of apoptotic cells, regulate several aspects of the tissue repair and regeneration, and display potent anti-nociceptive actions.

Edited by:
Amiram Ariel, University of Haifa, Israel

Reviewed by:
David Greaves, University of Oxford, United Kingdom
Hatice Hasturk, The Forsyth Institute, United States

***Correspondence:**
Jesmond Dalli
j.dalli@qmul.ac.uk;
Charles N. Serhan
cnserhan@zeus.bwh.harvard.edu

Keywords: lipid mediators, omega 3, microvesicles, immunoresolvent, tissue regeneration

INTRODUCTION

Inflammation is mounted in response to injury and/or infection in vascularized tissues that results in edema formation and leukocyte trafficking to the injured site and/or point of bacterial invasion (1). This is a fundamental host defense process that ensures adequate and timely disposal of invading pathogens and the repair of damaged tissues, paving the way for organ/tissue regain of function. At a histological level, the resolution of inflammation is characterized by the clearance of infiltrated leukocytes from the site and regain of tissue architecture (1). For many years, it was thought that the inflammatory response is terminated when local inflammatory chemical messengers and cells were *passively* diluted at the site (dilution of chemotactic gradient), hence halting further leukocyte recruitment, resolving the exudate or battlefield of inflammation (1–3). Detailed studies of cellular trafficking at the site demonstrated that in self-resolving inflammatory exudates cellular trafficking was tightly coordinated, where tissue resident cells elaborated the inflammatory reaction when exposed to an inflammatory stimulus. This was rapidly followed by an influx of granuloctyes, primarily neutrophils, and subsequently monocytes (4). In self-contained exudates, these recruited monocytes change phenotype from an inflammatory to a tissue protective phenotype as they differentiate to macrophages. This specific macrophage subpopulation is referred to as a resolution phase macrophages (5) and is thought to play key roles in the clearance of cellular debris from the site of inflammation and may also be involved in promoting tissue repair and regeneration (6–8).

These trafficking studies also suggested that since resolution is a tightly coordinated process, it was unlikely that simple dissipation of inflammatory signals could be the underlying mechanism for such a fundamental process. Findings made using a systems approach, assessing cellular trafficking and function coupled with biochemical approaches for structure elucidation of previously unknown mediators, highlight that indeed resolution of inflammation is a biochemically active process. These studies demonstrate that within exudates the production of inflammatory mediators such as leukotriene (LT) B_4 and prostaglandin E_2 was temporally regulated and reached a maximum at peak leukocyte infiltration. These studies also demonstrate that the resolution phase is denoted by the formation of a novel genus of autacoids that actively counter-regulate the formation of pro-inflammatory mediators, cellular trafficking, and phenotype (2, 9, 10). Given their potent biological actions, this novel genus of mediators is termed specialized pro-resolving mediators (SPM). SPM encompass several families of structurally and chemically distinct mediators. These, include neuroprotectin D1/NPD1 (10*R*,17*S*-dihydroxy-4Z,7Z,11E,13E,15Z,19Z-docosahexaenoic acid), resolvin D2 (7*S*,16*R*,17*S*-trihydroxy-4*Z*,8*E*,10*Z*,12*E*,14*E*,19*Z*-docosahexaenoic acid), and resolvin E1 (5S,12R,18R-trihydroxy-6Z,8E,10E,14Z,16E-eicosapentaenoic acid) (11). In addition to confirming the original structural assignments and potent anti-inflammatory and pro-resolving actions *in vivo* of resolvins, lipoxins, and maresins (12), recent findings demonstrate their potent actions in experimental colitis (13), arthritis (14), arthritic pain (15), ocular diseases (16), resolving adipose tissue inflammation (17), and diabetes (18). SPM share defining actions in resolving local inflammation; they each enhance macrophage uptake of cellular debris and apoptotic cells and limit further neutrophil recruitment to the site of injury and/or microbial invasion to bring about resolution (19, 20). The placement of these mediators within the resolution of inflammation as well as the state-of-the art definitions are reviewed earlier in Ref. (20, 21).

Pioneering studies conducted by Elie Metchnikoff paved the way to understanding the important role that macrophages play in orchestrating the host response. In his initial observations, Metchnikoff observed phagocytes surrounding and attempting to devour a splinter he had introduced into the transparent body of a starfish larva. Since then, the role of this process in mammalian systems, has been extensively studied where it is appreciated to be critical in both the maintenance of homeostasis and clearance of cellular debris, bacteria (6, 22), and apoptotic cells, a process termed efferocytosis (23). One of the defining actions displayed by pro-resolving mediators is the regulation of this fundamental process. Indeed, these mediators upregulate the ability of macrophages to phagocytose and kill bacteria, as well as to clear apoptotic cells and cellular debris (24). While pharmacologically these actions appear to be overlapping, recent studies demonstrate that at a biological level they are not. This is because the production of the different mediator families is regulated both temporally and in a tissue-specific manner (25–27). Furthermore, expression of the specific receptors, which form part of the G-protein-coupled receptor family, for each of the pro-resolving mediators is regulated in a cell type specific manner. For a detailed recent review of the biological actions and expression profiles of SPM receptors, the reader is directed to the following review (28). Recent evidence suggests that the clearance of apoptotic cells by macrophages leads to changes in the phenotype and functions of these cells (6, 7, 20, 29, 30). Thus, the aim of the present article is to discuss the role of pro-resolving mediators in regulating efferocytosis and the impact that this process plays in regulating macrophage lipid mediator (LM) profiles in order to shed light on the change in biological function (see **Table 1** for a summary of the biological actions of SPM on macrophages).

Microvesicles as Regulators of Efferocytosis and Macrophage LM Profiles

First thought to be byproducts of platelet activation carrying no significant biological activity, microvesicles are now increasingly appreciated to regulate critical aspects of the host immune response. Since their first description in platelets by Wolf in 1967 (46), the production of these microstructures has been described in many cell types including leukocytes, muscle cells and endothelial cells (47). They also carry distinct functions in both the initiation and resolution of acute inflammation [reviewed in Ref. (48)]. Microvesicles carry a wide range of molecular cargos including miRNA that are implicated in the regulation of hematopoiesis (49) as well as in protection during ischemia-reperfusion-mediated kidney injury (50). The pro-inflammatory cytokine IL1-β is also carried by microvesicles and its loading into these structures is thought to be one of the mechanisms by which this cytokine, which does not possess a secretion motif, is released from cells (51). Morphogens, such as Sonic Hedgehog, are also part of the cargo carried by subsets of these microvesicles, where microvesicles enriched in Sonic Hedgehog were found to promote angiogenesis and thus may play a role in tumor growth (52).

Recently, we found that neutrophil-derived microvesicles also display anti-inflammatory actions (53), carry precursors for the biosynthesis of pro-resolving mediators (29, 54), and display potent pro-resolving and host protective actions (54, 55). Since evolving self-limited inflammatory exudates produce functional microvesicles that also signal to stimulate resolution of inflammation in mice (54), we investigated whether these microvesicles also regulated macrophage efferocytosis. Indeed these microvesicles dose dependently regulated macrophage efferocytosis. Assessment LM-SPM concentrations in these microstructures demonstrated that neutrophil microvesicles, in addition to carrying precursors for the production of the pro-resolving mediators, also carried bioactive mediators including Protectin (PD) 1 (29).

Using a LM profiling approach, we found that microvesicles regulate macrophage LM-SPM profiles. Incubation of macrophages

TABLE 1 | The role of SPM receptors in mediating the biological actions of these autacoids on macrophages.

Receptor	SPM	Biological action	Biological system	Reference
ALX/FPR2	RvD1	Suppression of AA-stimulated LTB_4	Mouse bone marrow-derived Mϕ and zymosan-elicited peritoneal Mϕ	Fredman et al. (31)
	RvD1	Enhanced zymosan phagocytosis	Mouse bio-gel elicited Mϕ	Norling et al. (32)
	RvD1	Phagocytosis of zymosan and apoptotic PMNs	Human monocyte-derived Mϕ	Krishnamoorthy et al. (33)
	RvD1	Increased M2 polarization during I/R	Murine Kupffer cells	Kang and Lee (34)
	RvD1	Increases IL-10 levels	ALX/FPR2-overexpressing transgenic mice Human monocyte-derived Mϕ	Krishnamoorthy et al. (35)
	RvD1	Reduces cigarette smoke extract promoted IL-6 and TNF-a	Human monocyte-derived Mϕ	Croasdell et al. (36)
	LXA_4 AT-LXA_4	Increased transforming growth factor–β1 Increased efferocytosis	Murine Mϕ	Mitchell et al. (37)
	LXA_4	Increase apoptotic PMN efferocytosis Did not increase IL-8 and MCP-1 Attenuated PGE_2-stimulated protein kinase A activation	Human monocyte-derived Mϕ	Godson et al. (38)
	LXA4	Reduced TNF-a production Zymosan phagocytosis	Human monocyte-derived Mϕ	Pierdomenico et al. (39)
GPR32/DRV1	RvD1	Phagocytosis of zymosan Efferocytosis of apoptotic PMNs	Human monocyte-derived Mϕ	Krishnamoorthy et al. (33)
	RvD1	Reduced IL-1β and IL-8 expression Reduced chemotaxis to chemerin, fMLF, and MCP-1	Human monocyte-derived Mϕ	Schmid et al. (40)
	RvD1 RvD5	Increase phagocytosis of bacteria	Human monocyte-derived Mϕ	Chiang et al. (41)
	RvD1	Reduces IL-6 and TNF-a expression in elicited by cigarette smoke extract	Monocyte-derived Mϕ	Croasdell et al. (36)
	RvD3 AT-RvD3	Upregulate macrophage efferocytosis	Monocyte-derived Mϕ	Dalli et al. (25)
ChemR23/ ERV1	RvE1	Increases IL-10 transcription and phagocytosis of microbial particles	Monocyte-derived Mϕ	Herova et al. (42)
	RvE1	Phagocytosis of zymosan A via AKT and ribosomal protein S6 phosphorylation	Monocyte-derived Mϕ	Ohira et al. (43)
	RvE1	Reduction of IL12p40 and TNF-a expression in cells incubated with LPS	Mouse peritoneal Mϕ	Ishida et al. (44)
GPR18/DRV2	RvD2	Enhanced phosphorylation of CREB, ERK1/2, and STAT3	Mouse exudate macrophages	Chiang et al. (45)
	RvD2	Enhanced phagocytosis of live *Escherichia coli* and apoptotic PMN	Human monocyte-derived macrophages	Chiang et al. (45)

with neutrophil-derived microvesicles increased the biosynthesis of lipoxygenase- and cyclooxygenase-derived LM. These increased the biosynthesis of DHA-derived D-series resolvins, EPA-derived E-series resolvins, and the AA-derived lipoxins and prostanoids (29). Among the SPM that are upregulated, we observed significant increases in RvD5, MaR1, PD1, RvE2, and RvE1. Of interest, incubation of macrophages with G-protein inhibitors (pertussis toxin and cholera toxin) reduced SPM biosynthesis without altering prostanoid levels. Together, these results demonstrated that microparticles selectively stimulate macrophage SPM production in a G-protein-coupled receptor-dependent manner (29).

Microvesicles Are a Nidus for Macrophage SPM Production during Efferocytosis

Having observed increases in SPM levels and efferocyotosis when macrophages were incubated with neutrophil-derived microvesicles and that these microstructures carried elevated concentrations of several SPM precursors, the question arose whether microvesicles contributed to macrophage SPM biosynthesis by donating specific precursors. The potential contribution of microvesicles to transcellular biosynthesis was assessed during macrophage efferocytosis (29). For this purpose, we employed precursurs labeled with deuterium, which can be distinguished from endogenous precursors, to investigate the contribution of essential fatty acids derived from neutrophil-derived microvesicles. During efferocytosis, microvesicles were found to contribute to the production of d_5-RvD2 and d_5-RvD5 from the D-series resolvins and d_5-PD1 from the protectin family (**Figure 1**). These results demonstrated that during macrophage efferocytosis transcellular biosynthesis contributes to LM production where both microvesicles contribute substrate utilized in the SPM production.

Apoptotic PMN and Microvesicles Stimulate Macrophage SPM Production

Recent studies suggest that the process of efferocytosis reprograms macrophage responses altering cytokine production (7). Our studies demonstrate that the regulation of functional responses also extends to the production of both pro-resolving and pro-inflammatory LM profiles. Efferocytosis of apoptotic PMN by macrophages increases SPM biosynthesis, primarily, RvD1, RvD2, and LXB_4. SPM biosynthesis during efferocytosis was further upregulated by microvesicles, increases that correlate with an enhancement in the uptake of apoptotic neutrophils by macrophages (29). Of note, addition of microvesicles to macrophages further upregulated macrophage biosynthesis of RvD2, LXB_4, and RvE2, while reducing $PGF_{2\alpha}$ and TXB_2 (29).

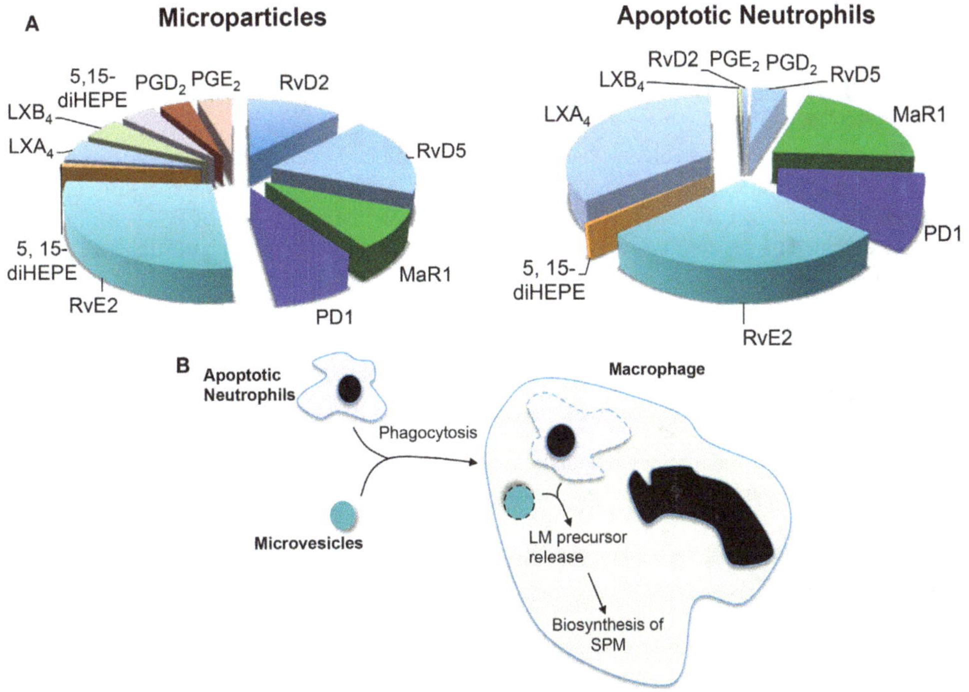

FIGURE 1 | Microvesicles and apoptotic neutrophils are a nidus for specialized pro-resolving mediator (SPM) biosynthesis during efferocytosis. Microvesicles or neutrophils were enriched in deuterium-labeled essential fatty acids and the conversion of these essential fatty acids to lipid mediators and their pathway markers/precursors was assessed during efferocytosis using lipid mediator profiling. **(A)** Relative contribution to lipid mediator biosynthesis by microvesicles and apoptotic neutrophils. **(B)** Cartoon depicting the process of trancellular biosynthesis during efferocytosis.

Apoptotic PMN, a Nidus for Macrophage SPM Biosynthesis

Apoptotic cells were also found to supply precursors to macrophages for the biosynthesis of both eicosanoids and SPM. This contribution was determined using a similar approach to that used for microvesicles where apoptotic neutrophils were enriched in deuterium-labeled (*d*) precursors. These studies demonstrate that substrates obtained from apoptotic neutrophils were also utilized for the production of pro-resolving mediators, whereas d_5-PD1 and d_5-RvE2 were produced when the substrate was obtained from either microvesicles or apoptotic cells d_5-MaR1 and d_8-LXB_4 were only identified in incubations with labeled apoptotic cells (29) (**Figure 1**). These results suggest that the origin of the substrate, and potentially its subcellular localization, may influence its contribution to specific mediator families.

Apoptotic Cells Differentially Regulate LMs in M1 and M2 Macrophages

Because apoptotic neutrophils stimulate LM biosynthesis in macrophages, we investigated whether this finding held for different macrophage subtypes. Assessment of LM biosynthesis after apoptotic cell efferocytosis demonstrated that the uptake of apoptotic neutrophils by classically activated macrophages upregulated the SPM production and reduced prostanoid biosynthesis. Of note, incubation of apoptotic neutrophils with M2 macrophages reduced overall LM production, including SPMs (29). These results indicated that the regulation of SPM profiles by apoptotic cells is also cell type dependent and may reflect the distinct biological actions of diverse cell types within the initiation-resolution spectrum.

Role the Role of G-Protein-Coupled Receptors in Mediating SPM Actions with Human Macrophages

SPM exert their biological actions via activating receptors of the G-protein-coupled receptor superfamily. RvD1, AT-RvD1, RvD3, AT-RvD3, LXA_4, and AT-LXA_4 activate the lipoxin receptor ALX/FPR2; in humans, these mediators also activate the orphan receptor DRV1/GPR32. RvD5 was also recently shown to activate DRV1/GPR32, while RvE1 binds to and activates ERV1/ChemR23; and DRV2/GPR18 mediates the biological actions of RvD2 (9, 23, 31, 39). Interested readers are referred to Ref. (9) for a detailed review on the biology of SPM receptors. In addition to displaying agonist actions to specific GPCR, RvE1, MaR1, and the MaR1 further metabolite 22-OH-MAR1 are also partial agonists/antagonists to LTB_4 receptor BLT1.

Activation of the pro-resolving receptors by their cognate mediators occurs in a stereospecific manner with even minor changes to their structure resulting in a significant loss in their ability to bind and activate these receptors. Recent studies suggest that the expression of SPM receptors differs between macrophage subsets thus suggesting that pro-resolving mediators differentially regulate the biological actions of distinct macrophage subsets. Murine peritoneal macrophages express the murine homologs of the ALX/FPR2, DRV2/GPR18, and ERV1/ChemR23 that mediate the biologic actions of their cognate SPM in regulating pro-inflammatory cytokine production, upregulating bacterial clearance and efferocytosis of apoptotic cells (9). In humans, expression of ERV1/Cherm23 was recently suggested to be restricted to macrophages obtained under conditions leading to a classically activated phenotype (42).

The Maresin Bioactive Metabolome in Human Macrophages

Studies using LM metabolomics and self-resolving exudates uncovered a new family pro-resolving mediators produced by macrophages, these mediators were coined macrophage mediators in resolving inflammation (maresins) (56). In the biosynthesis of maresins, DHA is lipoxygenated at carbon 14 yielding 14S-hydro(peroxy)-docosa-4Z,7Z,10Z,12E,16Z,19Z-hexaenoic acid (14-HpDHA) that is further converted to an allylic epoxide, reactions carried out by macrophage 12-lipoxgenase (30, 57). Using stereocontrolled total organic synthesis, enantiomerically pure 13*S*,14*S*-epoxy-docosa-4*Z*,7*Z*,9*E*,11*E*,16*Z*,19*Z*-hexaenoic acid (13*S*,14*S*-eMaR) was prepared and its stereochemistry confirmed by nuclear magnetic resonance spectroscopy. In macrophages, this intermediate is enzymatically converted to the pro-resolving mediator MaR1 (**Figure 2**). The stereochemistry of MaR1 was recently established as 7*R*,14*S*-dihydroxydocosa-4Z,8*E*,10*E*,12*Z*,16*Z*,19*Z*-hexaenoic acid, using a matching approach with material obtained from biological systems and total organic synthesis (11). The biological actions of MaR1 were also confirmed using synthetic material that included regulating leukocyte responses including limiting neutrophil infiltration in murine peritonitis (ng/mouse range) as well as enhancing human macrophage uptake of apoptotic neutrophils (11).

The biological actions of MaR1 extend beyond the regulation of cellular trafficking, where studies using planaria demonstrate that this mediator accelerates tissue regeneration following surgical injury (11). Upon injury, planaria also produce MaR1 from deuterium-labeled (d_5)-DHA in a lipoxygenase-dependent manner, since an inhibitor to this class of enzymes reduced MaR1 formation and also the ability of planaria to regenerate damaged tissues (11). Planaria are simple organisms capable of rapid regeneration, the process where in mammalian tissue macrophages play a central role. Factors involved in planaria tissue regeneration remained to be identified (58). Hence, we investigated whether MaR1 displays properties in controlling tissue regeneration. To this end, the anterior portions of planaria were surgically removed and the animals given MaR1. This accelerated tissue regeneration with the appearance of head regeneration was evident as early as 3 days post surgery. MaR1 addition at concentrations as low as 1 nM enhanced head regeneration as early as day 3 post injury (11).

MaR1 also possess potent anti-nociceptive actions, dose dependently inhibiting transient receptor potential cation channel subfamily V member 1 (TRPV1) currents in neurons, blocking capsaicin (CAP, 100 nM)-induced inward currents (IC_{50} = 0.49 ± 0.02) and reducing both inflammatory and chemotherapy-induced neuropathic pain in mice (11, 59, 60).

Of interest, 13*S*,14*S*-eMaR, the biosynthetic intermediate in the MaR1 metabolome, is also bioactive. Indeed this intermediate inhibits LTB_4 formation by human leukotriene A_4 hydrolase (LTA_4H) ~40% ($p < 0.05$) to a similar extent as LTA_4 (~50%, $p < 0.05$). Furthermore, LTA_4H was not involved in converting 13*S*,14*S*-eMaR to MaR1 pointing to the involvement of a yet unidentified epoxide hydrolase in catalyzing this biosynthetic step. 13*S*,14*S*-eMaR also reduced (~60%; $p < 0.05$) arachidonic acid conversion by human 12-LOX and promotes macrophage phenotype switch toward an M2 profile with similar potency as MaR1 (30). Incubation of M1 macrophages with either 13S,14S-eMaR (10 nM) or MaR1 (10 nM) led to significant reductions in the M1 lineage markers CD54 and CD80 expression and a concomitant upregulation of the M2 lineage markers CD163 and CD206 (30). We also investigated the conversion of 13S,14S-eMaR to MaR1 by different human macrophage subtypes. 13S,14S-eMaR (2 μM) with M2 macrophages gave higher MaR1 levels then when the 13S,14S-eMaR was incubated with M1 macrophages. These results suggest that the MaR1 metabolome is central in regulating macrophage function and mediating the biological actions of these phagocytes (30).

Identification of as Maresin Conjugates in Tissue Regeneration (MCTR) as Novel Regulators of Tissue Regeneration

Given the roles that macrophages play in the orchestrating wound healing, we questioned whether during the later stages of resolution these cells produce a distinct group of chemical signals that initiate tissue repair and regeneration. Using a systematic approach, we uncovered a group of peptide-lipid conjugated molecules that in addition to carrying the defining SPM pro-resolving actions, including the ability to regulate leukocyte trafficking and counter-regulate pro-inflammatory mediator production, also promote tissue repair and regeneration (8). The human macrophage 12-lipoxygenase is the initiating enzyme in the formation of these new signaling molecules, converting docosahexaenoic acid to 14S-HpDHA and then to 13S,14S-eMaR. Given that the initial biosynthetic steps are shared with MaR1, and carried a carbon 14 position alcohol, these bioactive molecules were coined as MCTR. The intermediate epoxide is converted to 13R-glutathionyl,14S-hydroxy-4Z,7Z,9E,11E,13R,14S,16Z,19Z-docosahexaenoic acid (MCTR1), a step that in human macrophages is catalyzed by Glutathione S-transferase MU 4 (GSTM4) and leukotriene C_4 synthase (LTC_4S) (61) (**Figure 2**). Of note, the two enzymes are shared with the cysteinyl LT pathway where they also catalyze the conversion of LTA_4 to leukotriene C_4 (LTC_4). Each of these enzymes displayed different affinities to the two substrates, where LTC_4S displayed a higher affinity to LTA_4, whereas GSTM4 displayed a higher affinity toward 13S,14S-eMAR. These findings suggest that in addition to substrate availability, the relative expression of the two enzymes in one cell type may determine the balance between the inflammation-, contraction-, and stress-initiating LTC_4 (7) in contrast with the tissue-regenerative pathway of

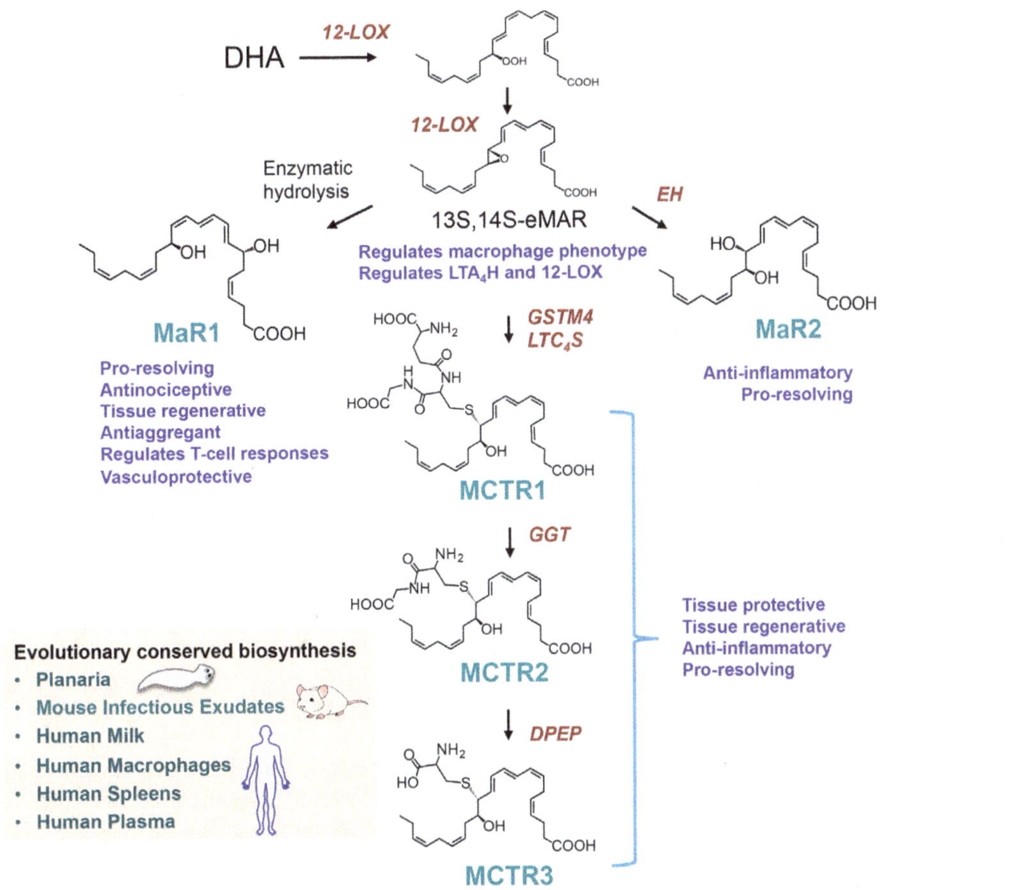

FIGURE 2 | Biosynthesis and actions of the macrophage-derived Maresins. The pathway is initiated by 14-lipoxygenation of DHA to yield 14S-hydro(peroxy)-4Z,7Z,10Z,12E,14S,16Z,19Z-docosahexaenoic acid and then to 13S,14S-epoxy-4Z,7Z,9E,11E,13S,14S,16Z,19Z-docosahexaenoic acid (13S,14S-eMaR) reactions that are catalyzed by 12-LOX. This intermediate is then enzymatically hydrolyzed to 7R,14S-dihydroxy-4Z,7R,8E,10E,12Z,14S,16Z,19Z-docosahexaenoic acid (MaR1) or *via* an epoxide hydrolase (EH) to 13,14S-epoxy-4Z,7Z,9,11,13,14S,16Z,19Z-docosahexaenoic acid (MaR2). 13S,14S-eMAR is also substrate for Glutathione S-transferase MU 4 (GSTM4) and leukotriene C_4 Synthase (LTC_4S) yielding MCTR1 (13R-glutathionyl,14S-hydroxy-4Z,7Z,9E,11E,13R,14S,16Z,19Z-docosahexaenoic acid), which is then converted to MCTR2 (13R-cysteinylglycinyl,14S-hydroxy-4Z,7Z,9E,11E,13R,14S,16Z,19Z-docosahexaenoic acid) by gamma-glutamyl transferase (GGT) and to MCTR3 (13R-cysteinyl,14S-hydroxy-4Z,7Z,9E,11E,13R,14S,16Z,19Z-docosahexaenoic acid) by dipeptidase (DPEP).

MCTRs. MCTR1 is precursor to 13R-cysteinylglycinyl,14S-hydroxy-4Z,7Z,9E,11E,13R,14S,16Z,19Z-docosahexaenoic acid (MCTR2), a conversion catalyzed by gamma-glutamyl transferase (**Figure 2**). This mediator in addition to carrying pro-resolving and tissue regenerative actions is also precursor to the third member of the MCTR family of mediators, 13R-cysteinyl,14S-hydroxy-4Z,7Z,9E,11E,13R,14S,16Z,19Z-docosahexaenoic acid (MCTR3), a reaction that is catalyzed by dipeptidases (61) (**Figure 2**). Of note, these enzymes are also shared with the cysteinyl LT biosynthetic pathway suggesting that processes regulating substrate availability may be critical in determining the macrophage LM phenotype and therefore function.

The production and biological actions of members of the maresin family are evolutionary conserved with identification in planaria mouse infectious exudates, human breast milk, spleen, and plasma from sepsis patients (**Figure 2**). Using deuterium-labeled docosahexaenoic acid, we found that MaR1 was produced by planaria following surgical injury (11). MCTR1 and MCTR2 were also identified in regenerating planaria and incubation of these mediators with planaria accelerated tissue regeneration (8). This increase in the rate of regeneration was associated with an early upregulation of a number of genes that in these animals are associated with head-to-tail differentiation suggesting that these molecules form part of the tissue regeneration process engaged following injury in planaria. This is further supported by the finding that planaria express genes that are homologous to MCTR biosynthetic enzymes including GSTM4, which are upregulated in regenerating tissues (8). Inhibition of these enzymes using both genetic approaches and small molecule inhibitors reduced MCTR levels as well as the ability of planaria to regenerate. In mice, MCTRs were also found to regulate tissue repair and regeneration in lung tissue where administration of these mediators during ischemia-reperfusion-mediated injury protected the lung from leukocyte-mediated damage and upregulated the expression of molecules that are associated with cell proliferation and tissue repair in the lung (8).

Therefore, SPM emerge both as potent regulators of macrophage responses of interest during the resolution phase of acute inflammatory responses and effectors in macrophage-mediated responses. Since the means and methodologies to identify these mediators have only recently become widely available, a number of aspects of their biology in humans have only just started to be explored such as the production of these pro-resolving mediators in human tissues (62–64). In addition, a number of questions still remain to be addressed such as the when and where these mediators are produced in human tissues in health and disease and the relevance of resolution processes that may fail giving rise to human disease. In this context, we recently found that following either a systemic insult or local inflammatory stimulus LM biosynthesis is differentially regulated between males and females. In these studies, vaccination with typhoid vaccine lead to peripheral blood leukocyte activation and endothelia dysfunction in males but not in females. This was associated with an increase peripheral blood E-series resolvin and a downregulation in the levels of LTB_4 in females (65). Local challenge using chantaridin led to a rapid resolution of the inflammatory response in females that was associated with an increased exudate RvD and decreased LTB_4 concentrations. Differences in tissue SPM concentrations were also recently reported in patients with inflammatory arthritis where a correlation between synovial RvE2 concentrations and pain was observed, with higher concentrations of RvE2 in these inflammatory exudates correlating with lower pain in arthritic patients (66). Thus, these results underscore the role of SPM in controlling tissue inflammation in humans and the utility of measuring these mediators as a potential diagnostic tool in patient stratification.

Since resolution of inflammation is a fundamental process in all human tissues and that phagocytes and specifically macrophages play a central role in orchestrating this response as well as tissue repair and regeneration, immunoresolvents (such as resolvins, protectins, and maresins) may provide a novel therapeutic approach for diseases characterized by uncontrolled inflammation and failed resolution. In this context, recent findings demonstrate that enriching microvesicles with SPM may represent a novel therapeutic approach to control chronic inflammation and promote tissue regeneration. In minipigs, administration of a single dose of a novel pro-resolving nanomedicine, produced by enriching microvesicles in a lipoxin analog, markedly reduced periodontal disease, and promoted bone regeneration (67). These results, together with many other recent studies, emphasize that using agonists to reprogram the immune cells rather then inhibiting the inflammatory response may represent a novel paradigm to controlling inflammation without compromising the host immune response.

AUTHOR CONTRIBUTIONS

JD and CS contributed to manuscript preparation.

FUNDING

Studies in the authors' laboratories were supported by a Sir Henry Dale Fellowship jointly funded by the Wellcome Trust and the Royal Society (Grant number: 107613/Z/15/Z), funding from the European Research Council (ERC) under the European Union's Horizon 2020 research and innovation program (Grant number: 677542), the Barts Charity (Grant number: MGU0343) to JD, and National Institutes of Health (grant numbers R01GM38765 and P01GM095467) to CNS.

REFERENCES

1. Majno G, Joris I. *Cells, Tissues, and Disease: Principles of General Pathology*. New York: Oxford University Press (2004).
2. Serhan CN, Clish CB, Brannon J, Colgan SP, Chiang N, Gronert K. Novel functional sets of lipid-derived mediators with antiinflammatory actions generated from omega-3 fatty acids via cyclooxygenase 2-nonsteroidal antiinflammatory drugs and transcellular processing. *J Exp Med* (2000) 192:1197–204. doi:10.1084/jem.192.8.1197
3. Willoughby DA, Moore AR, Colville-Nash PR, Gilroy D. Resolution of inflammation. *Int J Immunopharmacol* (2000) 22:1131–5. doi:10.1016/S0192-0561(00)00064-3
4. Bannenberg GL, Chiang N, Ariel A, Arita M, Tjonahen E, Gotlinger KH, et al. Molecular circuits of resolution: formation and actions of resolvins and protectins. *J Immunol* (2005) 174:4345–55. doi:10.4049/jimmunol.174.9.5884c
5. Stables MJ, Shah S, Camon EB, Lovering RC, Newson J, Bystrom J, et al. Transcriptomic analyses of murine resolution-phase macrophages. *Blood* (2011) 118:e192–208. doi:10.1182/blood-2011-04-345330
6. Gordon S. The macrophage: past, present and future. *Eur J Immunol* (2007) 37(Suppl 1):S9–17. doi:10.1002/eji.200737638
7. Ariel A, Serhan CN. New lives given by cell death: macrophage differentiation following their encounter with apoptotic leukocytes during the resolution of inflammation. *Front Immunol* (2012) 3:4. doi:10.3389/fimmu.2012.00004
8. Dalli J, Chiang N, Serhan CN. Identification of 14-series sulfido-conjugated mediators that promote resolution of infection and organ protection. *Proc Natl Acad Sci U S A* (2014) 111:E4753–61. doi:10.1073/pnas.1415006111
9. Chiang N, Serhan CN. Structural elucidation and physiologic functions of specialized pro-resolving mediators and their receptors. *Mol Aspects Med* (2017). doi:10.1016/j.mam.2017.03.005
10. Serhan CN. Treating inflammation and infection in the 21st century: new hints from decoding resolution mediators and mechanisms. *FASEB J* (2017) 31:1273–88. doi:10.1096/fj.201601222R
11. Serhan CN, Dalli J, Karamnov S, Choi A, Park CK, Xu ZZ, et al. Macrophage pro-resolving mediator maresin 1 stimulates tissue regeneration and controls pain. *FASEB J* (2012) 26:1755–65. doi:10.1096/fj.11-201442
12. Serhan CN, Petasis NA. Resolvins and protectins in inflammation-resolution. *Chem Rev* (2011) 111:5922–43. doi:10.1021/cr100396c
13. Bento AF, Claudino RF, Dutra RC, Marcon R, Calixto JB. Omega-3 fatty acid-derived mediators 17(R)-hydroxy docosahexaenoic acid, aspirin-triggered resolvin D1 and resolvin D2 prevent experimental colitis in mice. *J Immunol* (2011) 187:1957–69. doi:10.4049/jimmunol.1101305
14. Chan MM, Moore AR. Resolution of inflammation in murine autoimmune arthritis is disrupted by cyclooxygenase-2 inhibition and restored by prostaglandin E2-mediated lipoxin A4 production. *J Immunol* (2010) 184:6418–26. doi:10.4049/jimmunol.0903816
15. Bang S, Yoo S, Yang TJ, Cho H, Hwang SW. 17(R)-resolvin D1 specifically inhibits transient receptor potential ion channel vanilloid 3 leading to peripheral antinociception. *Br J Pharmacol* (2012) 165:683–92. doi:10.1111/j.1476-5381.2011.01568.x
16. Rajasagi NK, Reddy PB, Suryawanshi A, Mulik S, Gjorstrup P, Rouse BT. Controlling herpes simplex virus-induced ocular inflammatory lesions

with the lipid-derived mediator resolvin E1. *J Immunol* (2011) 186:1735–46. doi:10.4049/jimmunol.1003456
17. Titos E, Rius B, Gonzalez-Periz A, Lopez-Vicario C, Moran-Salvador E, Martinez-Clemente M, et al. Resolvin D1 and its precursor docosahexaenoic acid promote resolution of adipose tissue inflammation by eliciting macrophage polarization toward an M2-like phenotype. *J Immunol* (2011) 187:5408–18. doi:10.4049/jimmunol.1100225
18. White PJ, Arita M, Taguchi R, Kang JX, Marette A. Transgenic restoration of long-chain n-3 fatty acids in insulin target tissues improves resolution capacity and alleviates obesity-linked inflammation and insulin resistance in high-fat-fed mice. *Diabetes* (2010) 59:3066–73. doi:10.2337/db10-0054
19. Serhan CN. Novel resolution mechanisms in acute inflammation: to resolve or not? *Am J Pathol* (2010) 177:1576–91. doi:10.2353/ajpath.2010.100322
20. Serhan CN. Pro-resolving lipid mediators are leads for resolution physiology. *Nature* (2014) 510:92–101. doi:10.1038/nature13479
21. Serhan CN, Brain SD, Buckley CD, Gilroy DW, Haslett C, O'neill LAJ, et al. Resolution of inflammation: state of the art, definitions and terms. *FASEB J* (2007) 21:325–32. doi:10.1096/fj.06-7227rev
22. Tauber AI, Chernyak L. *Metchnikoff and the Origins of Immunology: From Metaphor to Theory*. New York: Oxford University Press (1991).
23. Savill JS, Wyllie AH, Henson JE, Walport MJ, Henson PM, Haslett C. Macrophage phagocytosis of aging neutrophils in inflammation. Programmed cell death in the neutrophil leads to its recognition by macrophages. *J Clin Invest* (1989) 83:865–75. doi:10.1172/JCI113970
24. Serhan CN, Chiang N, Dalli J. The resolution code of acute inflammation: novel pro-resolving lipid mediators in resolution. *Semin Immunol* (2015) 27:200–15. doi:10.1016/j.smim.2015.03.004
25. Dalli J, Winkler JW, Colas RA, Arnardottir H, Cheng CY, Chiang N, et al. Resolvin D3 and aspirin-triggered resolvin D3 are potent immunoresolvents. *Chem Biol* (2013) 20:188–201. doi:10.1016/j.chembiol.2012.11.010
26. Winkler JW, Orr SK, Dalli J, Cheng CY, Sanger JM, Chiang N, et al. Resolvin D4 stereoassignment and its novel actions in host protection and bacterial clearance. *Sci Rep* (2016) 6:18972. doi:10.1038/srep18972
27. Dalli J. Does promoting resolution instead of inhibiting inflammation represent the new paradigm in treating infections? *Mol Aspects Med* (2017). doi:10.1016/j.mam.2017.03.007
28. Serhan CN, Chiang N. Resolution phase lipid mediators of inflammation: agonists of resolution. *Curr Opin Pharmacol* (2013) 13:632–40. doi:10.1016/j.coph.2013.05.012
29. Dalli J, Serhan CN. Specific lipid mediator signatures of human phagocytes: microparticles stimulate macrophage efferocytosis and pro-resolving mediators. *Blood* (2012) 120:e60–72. doi:10.1182/blood-2012-04-423525
30. Dalli J, Zhu M, Vlasenko NA, Deng B, Haeggstrom JZ, Petasis NA, et al. The novel 13S,14S-epoxy-maresin is converted by human macrophages to maresin1 (MaR1), inhibits leukotriene A4 hydrolase (LTA4H), and shifts macrophage phenotype. *FASEB J* (2013) 27:2573–83. doi:10.1096/fj.13-227728
31. Fredman G, Ozcan L, Spolitu S, Hellmann J, Spite M, Backs J, et al. Resolvin D1 limits 5-lipoxygenase nuclear localization and leukotriene B4 synthesis by inhibiting a calcium-activated kinase pathway. *Proc Natl Acad Sci U S A* (2014) 111:14530–5. doi:10.1073/pnas.1410851111
32. Norling LV, Dalli J, Flower RJ, Serhan CN, Perretti M. Resolvin D1 limits polymorphonuclear leukocyte recruitment to inflammatory loci: receptor-dependent actions. *Arterioscler Thromb Vasc Biol* (2012) 32:1970–8. doi:10.1161/ATVBAHA.112.249508
33. Krishnamoorthy S, Recchiuti A, Chiang N, Yacoubian S, Lee CH, Yang R, et al. Resolvin D1 binds human phagocytes with evidence for proresolving receptors. *Proc Natl Acad Sci U S A* (2010) 107:1660–5. doi:10.1073/pnas.0907342107
34. Kang JW, Lee SM. Resolvin D1 protects the liver from ischemia/reperfusion injury by enhancing M2 macrophage polarization and efferocytosis. *Biochim Biophys Acta* (2016) 1861:1025–35. doi:10.1016/j.bbalip.2016.06.002
35. Krishnamoorthy S, Recchiuti A, Chiang N, Fredman G, Serhan CN. Resolvin D1 receptor stereoselectivity and regulation of inflammation and proresolving microRNAs. *Am J Pathol* (2012) 180:2018–27. doi:10.1016/j.ajpath.2012.01.028
36. Croasdell A, Thatcher TH, Kottmann RM, Colas RA, Dalli J, Serhan CN, et al. Resolvins attenuate inflammation and promote resolution in cigarette smoke-exposed human macrophages. *Am J Physiol Lung Cell Mol Physiol* (2015) 309:L888–901. doi:10.1152/ajplung.00125.2015
37. Mitchell S, Thomas G, Harvey K, Cottell D, Reville K, Berlasconi G, et al. Lipoxins, aspirin-triggered epi-lipoxins, lipoxin stable analogues, and the resolution of inflammation: stimulation of macrophage phagocytosis of apoptotic neutrophils in vivo. *J Am Soc Nephrol* (2002) 13:2497–507. doi:10.1097/01.ASN.0000032417.73640.72
38. Godson C, Mitchell S, Harvey K, Petasis NA, Hogg N, Brady HR. Cutting edge: lipoxins rapidly stimulate nonphlogistic phagocytosis of apoptotic neutrophils by monocyte-derived macrophages. *J Immunol* (2000) 164:1663–7. doi:10.4049/jimmunol.164.4.1663
39. Pierdomenico AM, Recchiuti A, Simiele F, Codagnone M, Mari VC, Davi G, et al. MicroRNA-181b regulates ALX/FPR2 receptor expression and proresolution signaling in human macrophages. *J Biol Chem* (2015) 290:3592–600. doi:10.1074/jbc.M114.592352
40. Schmid M, Gemperle C, Rimann N, Hersberger M. Resolvin D1 polarizes primary human macrophages toward a proresolution phenotype through GPR32. *J Immunol* (2016) 196:3429–37. doi:10.4049/jimmunol.1501701
41. Chiang N, Fredman G, Backhed F, Oh SF, Vickery T, Schmidt BA, et al. Infection regulates pro-resolving mediators that lower antibiotic requirements. *Nature* (2012) 484:524–8. doi:10.1038/nature11042
42. Herova M, Schmid M, Gemperle C, Hersberger M. ChemR23, the receptor for chemerin and resolvin E1, is expressed and functional on M1 but not on M2 macrophages. *J Immunol* (2015) 194:2330–7. doi:10.4049/jimmunol.1402166
43. Ohira T, Arita M, Omori K, Recchiuti A, Van Dyke TE, Serhan CN. Resolvin E1 receptor activation signals phosphorylation and phagocytosis. *J Biol Chem* (2010) 285:3451–61. doi:10.1074/jbc.M109.044131
44. Ishida T, Yoshida M, Arita M, Nishitani Y, Nishiumi S, Masuda A, et al. Resolvin E1, an endogenous lipid mediator derived from eicosapentaenoic acid, prevents dextran sulfate sodium-induced colitis. *Inflamm Bowel Dis* (2010) 16:87–95. doi:10.1002/ibd.21029
45. Chiang N, De La Rosa X, Libreros S, Serhan CN. Novel resolvin D2 receptor axis in infectious inflammation. *J Immunol* (2017) 198:842–51. doi:10.4049/jimmunol.1601650
46. Wolf P. The nature and significance of platelet products in human plasma. *Br J Haematol* (1967) 13:269–88. doi:10.1111/j.1365-2141.1967.tb08741.x
47. Boilard E, Nigrovic PA, Larabee K, Watts GF, Coblyn JS, Weinblatt ME, et al. Platelets amplify inflammation in arthritis via collagen-dependent microparticle production. *Science* (2010) 327:580–3. doi:10.1126/science.1181928
48. Norling LV, Dalli J. Microparticles are novel effectors of immunity. *Curr Opin Pharmacol* (2013) 13:570–5. doi:10.1016/j.coph.2013.05.008
49. Hunter MP, Ismail N, Zhang X, Aguda BD, Lee EJ, Yu L, et al. Detection of microRNA expression in human peripheral blood microvesicles. *PLoS One* (2008) 3:e3694. doi:10.1371/journal.pone.0003694
50. Cantaluppi V, Gatti S, Medica D, Figliolini F, Bruno S, Deregibus MC, et al. Microvesicles derived from endothelial progenitor cells protect the kidney from ischemia-reperfusion injury by microRNA-dependent reprogramming of resident renal cells. *Kidney Int* (2012) 82:412–27. doi:10.1038/ki.2012.105
51. Brown GT, McIntyre TM. Lipopolysaccharide signaling without a nucleus: kinase cascades stimulate platelet shedding of proinflammatory IL-1beta-rich microparticles. *J Immunol* (2011) 186:5489–96. doi:10.4049/jimmunol.1001623
52. Agouni A, Mostefai HA, Porro C, Carusio N, Favre J, Richard V, et al. Sonic hedgehog carried by microparticles corrects endothelial injury through nitric oxide release. *FASEB J* (2007) 21:2735–41. doi:10.1096/fj.07-8079com
53. Dalli J, Norling LV, Renshaw D, Cooper D, Leung K-Y, Perretti M. Annexin 1 mediates the rapid anti-inflammatory effects of neutrophil-derived microparticles. *Blood* (2008) 112:2512–9. doi:10.1182/blood-2008-02-140533
54. Norling LV, Spite M, Yang R, Flower RJ, Perretti M, Serhan CN. Cutting edge: humanized nano-proresolving medicines mimic inflammation-resolution and enhance wound healing. *J Immunol* (2011) 186:5543–7. doi:10.4049/jimmunol.1003865
55. Dalli J, Norling LV, Montero-Melendez T, Federici Canova D, Lashin H, Pavlov AM, et al. Microparticle alpha-2-macroglobulin enhances pro-resolving responses and promotes survival in sepsis. *EMBO Mol Med* (2014) 6:27–42. doi:10.1002/emmm.201303503
56. Serhan CN, Yang R, Martinod K, Kasuga K, Pillai PS, Porter TF, et al. Maresins: novel macrophage mediators with potent anti-inflammatory and pro-resolving actions. *J Exp Med* (2009) 206:15–23. doi:10.1084/jem.20081880

57. Deng B, Wang CW, Arnardottir HH, Li Y, Cheng CY, Dalli J, et al. Maresin biosynthesis and identification of maresin 2, a new anti-inflammatory and pro-resolving mediator from human macrophages. *PLoS One* (2014) 9:e102362. doi:10.1371/journal.pone.0102362
58. Pellettieri J, Fitzgerald P, Watanabe S, Mancuso J, Green DR, Alvarado AS. Cell death and tissue remodeling in planarian regeneration. *Dev Biol* (2010) 338:76–85. doi:10.1016/j.ydbio.2009.09.015
59. Serhan CN. The resolution of inflammation: the devil in the flask and in the details. *FASEB J* (2011) 25:1441–8. doi:10.1096/fj.11-0502ufm
60. Ji RR, Xu ZZ, Gao YJ. Emerging targets in neuroinflammation-driven chronic pain. *Nat Rev Drug Discov* (2014) 13:533–48. doi:10.1038/nrd4334
61. Dalli J, Vlasakov I, Riley IR, Rodriguez AR, Spur BW, Petasis NA, et al. Maresin conjugates in tissue regeneration biosynthesis enzymes in human macrophages. *Proc Natl Acad Sci U S A* (2016) 113:12232–7. doi:10.1073/pnas.1607003113
62. Markworth JF, Vella LD, Lingard BS, Tull DL, Rupasinghe TW, Sinclair AJ, et al. Human inflammatory and resolving lipid mediator responses to resistance exercise and ibuprofen treatment. *Am J Physiol Regul Integr Comp Physiol* (2013) 305:R1281–96. doi:10.1152/ajpregu.00128.2013
63. Colas RA, Shinohara M, Dalli J, Chiang N, Serhan CN. Identification and signature profiles for pro-resolving and inflammatory lipid mediators in human tissue. *Am J Physiol Cell Physiol* (2014) 307:C39–54. doi:10.1152/ajpcell.00024.2014
64. Sasaki A, Fukuda H, Shiida N, Tanaka N, Furugen A, Ogura J, et al. Determination of omega-6 and omega-3 PUFA metabolites in human urine samples using UPLC/MS/MS. *Anal Bioanal Chem* (2015) 407:1625–39. doi:10.1007/s00216-014-8412-5
65. Rathod KS, Kapil V, Velmurugan S, Khambata RS, Siddique U, Khan S, et al. Accelerated resolution of inflammation underlies sex differences in inflammatory responses in humans. *J Clin Invest* (2017) 127:169–82. doi:10.1172/JCI89429
66. Barden AE, Moghaddami M, Mas E, Phillips M, Cleland LG, Mori TA. Specialised pro-resolving mediators of inflammation in inflammatory arthritis. *Prostaglandins Leukot Essent Fatty Acids* (2016) 107:24–9. doi:10.1016/j.plefa.2016.03.004
67. Van Dyke TE, Hasturk H, Kantarci A, Freire MO, Nguyen D, Dalli J, et al. Proresolving nanomedicines activate bone regeneration in periodontitis. *J Dent Res* (2015) 94:148–56. doi:10.1177/0022034514557331

Conflict of Interest Statement: The authors declare that the research was conducted in the absence of any commercial or financial relationships that could be construed as a potential conflict of interest.

10

The "Phagocytic Synapse" and Clearance of Apoptotic Cells

*Nicole D. Barth, John A. Marwick, Marc Vendrell, Adriano G. Rossi and Ian Dransfield**

MRC Centre for Inflammation Research, Queen's Medical Research Institute, University of Edinburgh, Edinburgh, United Kingdom

Apoptosis and subsequent phagocytic clearance of apoptotic cells is important for embryonic development, maintenance of tissues that require regular cellular renewal and innate immunity. The timely removal of apoptotic cells prevents progression to secondary necrosis and release of cellular contents, preventing cellular stress and inflammation. In addition, altered phagocyte behavior following apoptotic cell contact and phagocytosis engages an anti-inflammatory phenotype, which impacts upon development and progression of inflammatory and immune responses. Defective apoptotic cell clearance underlies the development of various inflammatory and autoimmune diseases. There is considerable functional redundancy in the receptors that mediate apoptotic cell clearance, highlighting the importance of this process in diverse physiological processes. A single phagocyte may utilize multiple receptor pathways for the efficient capture of apoptotic cells by phagocytes (tethering) and the subsequent initiation of signaling events necessary for internalization. In this review, we will consider the surface alterations and molecular opsonization events associated with apoptosis that may represent a tunable signal that confers distinct intracellular signaling events and hence specific phagocyte responses in a context-dependent manner. Efficient molecular communication between phagocytes and apoptotic targets may require cooperative receptor utilization and the establishment of efferocytic synapse, which acts to stabilize adhesive interactions and facilitate the organization of signaling platforms that are necessary for controlling phagocyte responses.

Edited by:
Kirsten Lauber, Ludwig-Maximilians-Universität München, Germany

Reviewed by:
Raymond B. Birge, Rutgers University, The State University of New Jersey, United States
Michael R. Elliott, University of Rochester, United States

***Correspondence:**
Ian Dransfield
i.dransfield@ed.ac.uk

Keywords: macrophage, phagocytosis, apoptotic cells, cell–cell interactions, phagocytic receptor, phosphatidylserine, opsonin

INTRODUCTION

Elimination of injured or metabolically stressed cells in multicellular organisms is controlled *via* engagement of apoptotic programs together with efficient tissue clearance mechanisms (1–3). Phagocyte/apoptotic cell interactions also initiate anti-inflammatory reprogramming that regulates inflammation and immunity (4). Deficient clearance of apoptotic cells contributes to the development and/or exacerbation of many autoimmune and inflammatory diseases [reviewed in Ref. (5)].

The diversity of molecular pathways mediating recognition and phagocytosis of apoptotic cells (efferocytosis) reflects the fundamental importance of this process (4). There are several mechanisms by which viable cells avoid phagocytosis (6). However, altered plasma membrane lipid composition (7, 8) and/or oxidation status (9), together with changes in cell surface molecule repertoire and patterns of glycosylation (10) termed "apoptotic cell associated molecular patterns" (11) (**Figure 1**), allow phagocytes to distinguish viable and apoptotic cells. Here, we consider the formation of an "efferocytic synapse" and assembly of molecular platforms that facilitate phagocytosis and subsequent signaling events.

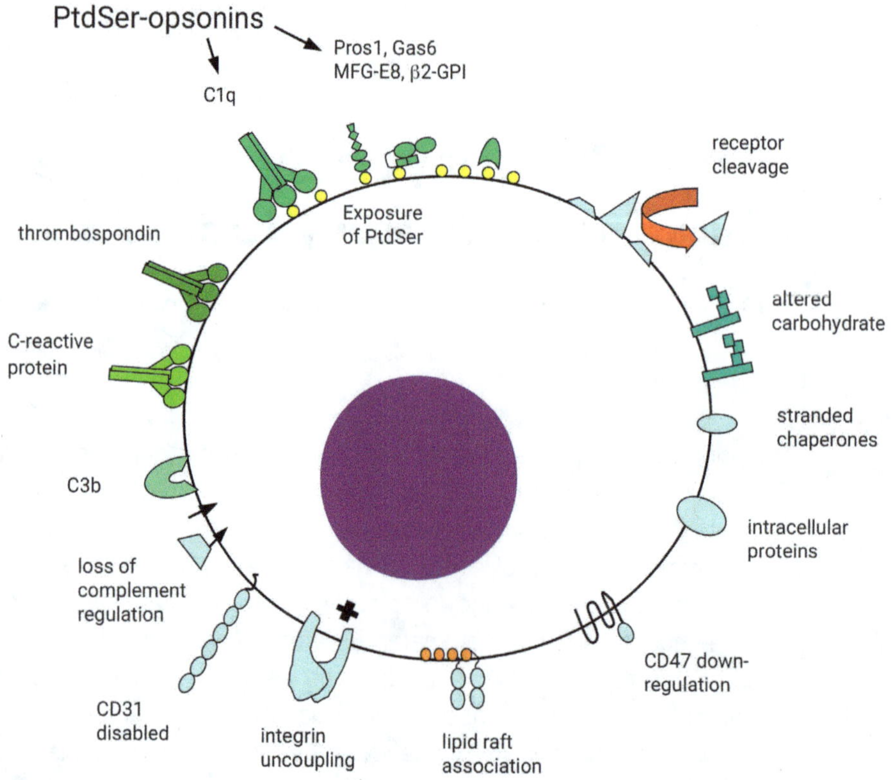

FIGURE 1 | Schematic representation of surface molecular changes associated with apoptosis. Reduced cell surface molecule expression may occur through metalloprotease-mediated proteolytic shedding. Reduced receptor expression may act to limit apoptotic cell function and generate a distinct cell surface profile from viable cells. For example, reduced expression of CD47 or disabled CD31 may lead to loss of signals preventing phagocytosis. In addition, loss of integrin regulation may result in functional uncoupling in apoptotic cells, leading to cell detachment. Altered association of cell surface molecules with lipid rafts may alter functional activity, including gain-of-function of some receptors, e.g., FcγRIIa on myeloid cells. Altered carbohydrate processing may result in reduced sialic acid exposure and appearance of accessible mannose residues. Downregulation of complement regulatory molecules (e.g., CD55 and CD46) may lead to opsonization with complement components including C3b. Exposure of anionic phospholipids, including phosphatidylserine (PtdSer), allows binding of a broad range of opsonins to apoptotic cells. Protein S and Gas6 bind to PtdSer in a Ca^{2+}-dependent manner, whereas milk fat globule EGF-factor 8 (MFG-E8) binds independent of Ca^{2+}. Other less well defined apoptotic cell surface changes may allow binding of other opsonins including thrombospondin, C-reactive protein, and surfactant protein A. Finally, proteins with intracellular localizations may appear on the surface of apoptotic cells, including heat-shock proteins and calreticulin. Apoptotic cell surface molecules are shaded blue whereas apoptotic cell opsonins are shaded green.

PHOSPHATIDYLSERINE (PtdSer) AS A LIGAND FOR APOPTOTIC CELL RECOGNITION

A near universal membrane alteration associated with apoptosis is the caspase-dependent exposure of PtdSer on the outer leaflet of the plasma membrane (12–14) *via* the XK-related protein 8 (15). Exposure of PtdSer affects the biophysical characteristics and organization of the plasma membrane through recruitment of proteins to PtdSer-enriched regions *via* electrostatic interactions (16). Phagocytes express transmembrane receptors that bind PtdSer directly, e.g., brain-specific angiogenesis inhibitor-1 (BAI-1) (17) and stabilin-2 (18). In addition, soluble molecules such as transthyretin-like protein TTR-52 (19), milk fat globule EGF-factor 8 (MFG-E8) (20, 21), protein S (Pros1), Gas6 (22) and C1q (23) also bind to (and opsonize) PtdSer, providing a scaffold for phagocyte recognition *via* a diverse array of counter-receptors.

Phagocytes fail to engulf viable cells that expose low levels of PtdSer during activation (24–26) or when PtdSer exposure is induced by overexpression of a phospholipid scramblase, transmembrane protein 16F (TMEM16F) (27), suggesting that additional signals are necessary to initiate efferocytosis. A critical threshold of PtdSer exposure on the cell surface may be required to trigger efferocytosis (28). For example, recognition of PtdSer *via* T-cell immunoglobulin and mucin-domain-containing molecule (TIM)-4 was dependent on ligand density, allowing phagocytes to distinguish between high and low level PtdSer exposure (28). Further modifications of PtdSer during apoptosis, e.g., oxidation or formation of lyso-PtdSer (29) may also be important.

CELL SURFACE RECEPTOR ALTERATIONS ASSOCIATED WITH APOPTOSIS

Apoptosis-dependent loss of cell surface receptors or appearance of "new" molecules may confer recognition by phagocytes. For example, signaling *via* Signal regulatory protein-α (SIRPα) inhibits myosin-II-mediated phagocytosis (30). Downregulation of ligands for SIRPα, e.g., CD47 (31), from the surface of apoptotic cells would be predicted to promote efferocytosis (32, 33). SIRPα-mediated signaling has also been reported to be triggered by binding of surfactant protein (SP)-A and SP-D to phagocytes. However, SP-A may have a dual role in regulation of phagocytosis as binding to apoptotic cells/debris results in promotion of phagocytosis *via* a calreticulin/CD91-mediated pathway (34). Early experiments identified a unique charge-sensitive mechanism for apoptotic cell recognition (35). The cell glycocalyx provides a negative surface charge "repulsive" force that counters cell–cell interactions (36). Loss of N-terminal sialic acid and exposure of mannose and fucose moieties during apoptosis reduces electrostatic forces that counter phagocyte recognition (37–39). In addition, the surface charge of apoptotic cells is further altered by reduced expression of heavily sialylated proteins (e.g., CD43, CD45, and CD162) (37). Consistent with this suggestion, removal of sialic acid from the cell surface by neuraminidase treatment enhances phagocytosis (37, 39).

Apoptosis is associated with loss of expression of complement regulatory proteins such as CD46/CD55 (40, 41). As a result, deposition of complement may occur, providing a cue for recognition by phagocytes. Additional signals for phagocytosis may occur as a result of exposure of intracellular proteins such as calreticulin (42) and annexin I (43). Following binding to PtdSer on apoptotic cells, oxidation of Pros1 induces oligomerization, which promotes Mer-dependent phagocytosis (44). Similarly, altered glycosylation of membrane proteins or oxidization of low-density lipoprotein-like moieties on apoptotic cells (8) may also contribute to the specific recognition by phagocytes. In addition to increased expression of ligands for phagocytic receptors on apoptotic cells, patching and/or clustering of surface molecules may also have important consequences for triggering phagocyte responses. Clustering might occur through specific association with membrane microdomains. For example, FcγRIIa redistributes to membrane microdomains during neutrophil apoptosis (45). In addition, specific proteolysis of adhesion molecules (e.g., CD62L) (46, 47) and uncoupled β_2 integrin-mediated adhesion (47) during apoptosis is likely to provide additional molecular cues for phagocytosis.

PHAGOCYTE MOLECULES THAT MEDIATE APOPTOTIC CELL RECOGNITION

Phagocytes are capable of direct recognition of PtdSer exposed on the apoptotic cell surface. BAI-1 binds PtdSer *via* thrombospondin (TSP) type 1 repeats present in the extracellular domain (48). Binding induces the formation of a trimeric complex of BAI-1 with the Rac-GEF ELMO and DOCK180 that promotes subsequent engulfment of apoptotic cells (17, 49). This pathway is homologous to the genetically defined pathway for removal of apoptotic cells in *Caenorhabditis elegans* (Ced2-CrkII, Ced5-DOCK180, Ced10-Rac, and Ced12-ELMO) (50).

Phosphatidylserine is also recognized by the CD300 family of molecules with an extracellular IgV-like domain and intracellular adaptor molecule binding sites (51). CD300b localizes to phagocytic cups and binds DAP12, activating Syk and PI3K/Akt to promote phagocytosis (52). Stabilin-2 binds PtdSer and also lacks direct signaling activity (18). However, the cytoplasmic domain of stabilin-2 can interact with GULP to facilitate phagocytosis (53). GULP also binds to NPxY motifs present in the cytoplasmic domains of CD91/LRP (low-density lipoprotein receptor-related protein) and the *C. elegans* scavenger receptor Ced-1 (54). In contrast, TIM-4 confers Ca^{2+}-dependent PtdSer-dependent apoptotic cell recognition, but lacks intracellular signaling potential (55). Thus, TIM-4 functions cooperatively with other receptors that trigger apoptotic cell internalization.

Indirect recognition of apoptotic cells by phagocytes is also achieved by phagocyte receptors that bind to soluble apoptotic cell opsonins. In *C. elegans*, TTR-52 bridges apoptotic cell-exposed PtdSer to phagocyte Ced-1 (19), which together with Ced-6 initiates rapid and efficient engulfment of apoptotic cell corpses by neighboring cells. This module of proteins (Ced1-MEGF-10; Ced6-GULP; and Ced7-ABCA1) has been defined genetically in *C. elegans* (50). Pros1 and Gas6 contain a Gla-domain that binds PtdSer in a Ca^{2+}-dependent manner (22), bridging to the Tyro3/Axl/Mer receptor tyrosine kinases that signal particle internalization *via* intrinsic kinase activity (56). By contrast, MFG-E8 binds PtdSer in a Ca^{2+}-independent manner and bridges to phagocyte integrins $\alpha_v\beta_{3/5}$ *via* arginine–glycine–aspartic acid (RGD) peptide motifs in the C1 and C2 domains (21). TSP-1 also bridges apoptotic cells *via* the phagocyte integrin $\alpha_v\beta_3$ and CD36 (57, 58).

COOPERATIVE RECEPTOR UTILIZATION IN PHAGOCYTOSIS OF APOPTOTIC CELLS

Phagocytes in different tissue settings or microenvironments express distinct repertoires of efferocytic receptors. Whether a single phagocytic cell utilizes multiple receptor pathways to recognize and internalize a single apoptotic target is not clear. However, the distinct molecular requirements for the capture and subsequent internalization may require that multiple receptors are involved (42). In addition, the complex topology of apoptotic cell surface molecules and co-opsonization of PtdSer with different proteins may determine the spectrum of signal transduction pathways engaged, controlling internalization and subsequent phagocyte responses in a context-dependent manner.

Tethering of IgG-opsonized particles to FcγR occurs at 4°C (59) whereas internalization requires cytoskeletal reorganization and metabolic activity (60). Similarly, apoptotic cells can also be tethered by phagocytes *via* Mer at low temperature (61). However, the avidity of low affinity receptors is influenced by receptor density and rapid lateral movement of receptors to facilitate target capture (62, 63). Receptor mobility is controlled by cytoskeletal constraint, association with membrane lipid microdomains and/or other membrane proteins (64). For example, cytoskeletal-associated CD44 restricts membrane lipid and receptor motility *via* interactions with hyaluronan, forming a glycosaminoglycan barrier that reduces binding of phagocytic targets (36). Interestingly, CD44 cross-linking with antibody augments macrophage phagocytosis of apoptotic cells, possibly as a result of changes in cytoskeletal regulation (65, 66). Phagocytic targets are bound at dynamic extensions of phagocytic cells, including filopodia and membrane ruffles (67), and receptor aggregation is required for orchestration of cytoskeletal alterations necessary for internalization [see Ref. (68) for a comprehensive review of cytoskeletal regulation in phagocytic synapses].

In contrast with IgG or complement attached to components of the microbial cell wall, molecules on the surface of apoptotic cells may exhibit unrestricted lateral mobility as a consequence of proteolytic cleavage of actin during apoptosis (69). Thus, there may be key mechanistic differences between efferocytosis and FcγR-dependent phagocytosis. Engagement of freely mobile molecules by phagocyte receptors may lead to assembly of receptor microclusters and significantly impact upon phagocytosis (33). Alternatively, opsonization of apoptotic cells may result in formation of immobile molecular complexes [like annexin V (70, 71)] that promote redistribution of phagocytic receptors necessary for signaling of internalization.

MOLECULAR SEGREGATION AND THE FORMATION OF A PHAGOCYTIC "SYNAPSE"

The cellular contact between phagocyte and apoptotic cells has parallels with those of antigen-specific T and B cells with antigen presenting cells that leads to establishment of an immunological synapse (72). This specialized intercellular contact zone stabilizes adhesion and facilitates efficient molecular communication following antigen-specific interactions. A number of different biophysical factors (including receptor density, ligand-binding affinity, molecular dimensions, and interactions with cytoskeletal elements) all contribute to the dynamic redistribution of adhesion and signaling receptors into distinct regions in the plasma membrane (73, 74).

Following phagocyte contact with IgG-coated surfaces or on supported lipid bilayers, formation of FcγRII nanoclusters suggests activation-driven organization of receptor redistribution (75). FcγR clusters are localized in front of ruffles on extending pseudopods, with rapid recruitment of Syk to advancing pseudopods and subsequent retrograde movement toward the cell center (76). PI3K co-localized with actin around FcγR clusters, suggestive of signal propagation from FcγR and consistent with PI3K-dependent control of actin cytoskeletal rearrangements (76).

Although phagocyte contact with apoptotic cells has not been examined with the high-resolution imaging techniques used in T cell–APC interactions, molecular segregation may also be a key feature of the formation of efferocytic synapses (see **Table 1**). For example, exclusion of phosphatases (e.g., CD45) from the immune synapse is a critical early event in the initiation of T cell receptor-mediated phosphorylation of Zap70 and Lck (77). Changes in the distribution of CD45 may also represent a general feature of membrane alterations that control signaling events associated with phagocytosis. The C-type lectin containing receptor for β-glucan, Dectin-1, mediates the recognition and phagocytosis of yeast particles. During Dectin-1-mediated phagocytosis, exclusion of CD45 and the receptor type protein tyrosine phosphatase CD148 from the nascent phagocytic cup (78) is important for signaling associated with particle internalization (79). Redistribution of phosphatases in the phagocyte membrane would likely be necessary for internalization of apoptotic cells.

TABLE 1 | Comparison of phagocytic synapse formation in FcγR-mediated and apoptotic cell phagocytosis.

	FcγR phagocytosis	**Apoptotic cell phagocytosis**
Receptor signaling[a]	ITAM or ITIM adaptors Src/Syk kinases PI3K	BAI-1:G protein-coupled (DOCK/ELMO Rac-GEF) Tyro3/Axl/Mer: receptor tyrosine kinase, PI3K, Rac Stabilins:GULP adaptor? CD300:ITIM or DAP12 adaptors α_v integrins (Rac?) TIMs:suppression of Src, none?
Ligand restraint	Restricted mobility	Not known Some ligands restrained by oligomerization?
Phosphatases	CD45/CD148 excluded from contact	Not known
Role of integrins	Formation of exclusion zone	Not known/co-receptors: $\alpha_v\beta_3$ and CD36 $\alpha_v\beta_5$ and Mer β_1 integrins and TIM-4
Receptor nanoclusters	Yes	Not known
Molecular dimensions	7–10 nm (FcγR) 15 nm (IgG) +Antigenic target	Very small (3–4 nm), e.g., CD300 to very large (40–50 nm), e.g., SCARF1/C1q

[a]See Elliott et al. (80) for a comprehensive overview of receptor signaling.
ITIM, immunoreceptor tyrosine-based inhibition motif; ITAM, immunoreceptor tyrosine-based activation motif; BAI-1, brain-specific angiogenesis inhibitor-1.

Analysis of the molecular basis of redistribution of CD45 following ligation of FcγR suggests non-linear pattern with formation of an exclusion barrier which restricts access of CD45 to the contact site (79). CD45 is a relatively rigid molecule that extends axially from the plasma membrane approximately 20 nm (81). Similar to the molecular redistribution that occurs during immune synapse formation, exclusion from phagocytic synapses was dependent on the axial molecular dimensions of CD45. In experiments using chimeric constructs in which the extracellular portion of CD45 was replaced by either CD43 (similar length) or CD2 (shorter length), the CD43/CD45 molecule was excluded from the phagocytic synapse, whereas CD2/CD45 was not (79). A requirement for integrins and cytoskeletal regulation was shown to be necessary to establish a CD45 exclusion zone that extended beyond the IgG layer (79). These results suggest that integrin-mediated contact between phagocyte and phagocytic targets facilitates engagement of phagocytic receptors at low ligand densities or when binding to larger particles.

For apoptotic cell recognition, it is intriguing that integrins have been proposed to act cooperatively with other receptors to mediate phagocytosis. For example, α_v integrins and CD36 are both required for the recognition of TSP-1 bound to apoptotic cells (58). Similarly, phagocytosis of apoptotic cells *via* TIM-4 requires β_1 integrins and activation of integrin-dependent signaling involving Src family kinases and FAK (82). Furthermore, interactions between the β_5 integrin and stabilin-2 were found to promote phagocytosis of apoptotic cells (83). Similarly, co-expression of $\alpha_v\beta_5$ with Mer increased activation of Rac-1, cytoskeletal regulation and the phagocytosis of apoptotic cells (84). Integrins can directly mediate the recognition of apoptotic cell opsonins, for example, RGD-dependent binding of MFG-E8 (21) or fibronectin (85). However, phagocytosis of apoptotic targets is increased following macrophage adhesion to extracellular matrix *via* β_1 integrins (86) and is compromised following exposure to oxidized extracellular matrix molecules (87). We would speculate that, as for FcγR-mediated phagocytosis, integrin signaling can regulate cytoskeletal organization and facilitate tethering/phagocytosis of apoptotic cells with low-level opsonization (36).

For human macrophage-like cells, CD47 expression acts to limit phagocytosis of IgG-opsonized erythrocytes (30). CD47 binds to phagocyte SIRPα, resulting in recruitment to the phagocytic synapse, decreasing the accumulation of non-muscle myosin IIa and levels of tyrosine phosphorylation (30). Localization of SIRPα to the site of cell contact would recruit inhibitory tyrosine phosphatases such as SHP-1 *via* ITIM motifs present in the SIRPα cytoplasmic domain (88). In the absence of CD47 or when CD47 was blocked with antibody, phagocytosis was increased. Specific membrane receptors are organized into protein islands in unactivated T cells that subsequently coalesce as a consequence of T cell receptor-mediated signaling (89). The membrane distribution of SIRPα and FcγR during phagocytosis has been further analyzed using super-resolution microscopy. When macrophages were plated onto poly-L-lysine-coated slides, molecular clusters containing both FcγRI and SIRPα were observed (75). These molecules were found to segregate into discrete nanoclusters when macrophages were plated onto IgG. Interestingly, IgG promoted the formation of concentric rings of FcγRI and FcγRII, with FcγRI redistributing more rapidly (<10 min). Similar results were obtained when macrophages interacted with IgG in a supported lipid bilayer. When recombinant CD47 was included into the supported lipid bilayer, segregation of FcγR and SIRPα and the formation of concentric rings of FcγR were blocked. Thus co-localization of SIRPα and FcγR inhibits cellular activation following FcγR ligation, whereas segregation of these two molecules leads to activation (75). We would speculate that efferocytic receptors would also be present in dynamically regulated nanoclusters in the phagocyte membrane.

MOLECULAR DIMENSIONS AND APOPTOTIC CELL RECOGNITION

Size-dependent redistribution of molecules within the phagocyte membrane may represent an important organizing principle for the assembly of molecular platforms that are essential for signaling the cytoskeletal alterations required for the internalization of apoptotic cells. Estimation of the molecular dimensions of receptors involved in efferocytosis using published structural data (51, 90–106) reveals considerable differences in axial dimensions (**Figure 2A**).

First, it is notable that CD300 (single Ig-like domain) and TIM-4 (single Ig-like domain with a potentially rigid mucin-like stalk), which mediate direct recognition of PtdSer, are predicted to span a relatively short intermembrane distance between phagocyte and target (4 and 7 nm, respectively, **Figure 2B**). By contrast, stabilin-2, which also binds to PtdSer, is a much longer molecule extending some 23 nm. Assuming a degree of structural flexibility, BAI-1 may be capable of functioning in a broad range of intermembrane distances. The extracellular region of BAI-1 contains 5 TSP type 1 repeats (around 5 nm in size) with LPS and PtdSer binding motifs, which together with the GAIN/HBD regions could extend to ~33 nm from the plasma membrane. Following initial tethering of PtdSer by the N-terminal TSP repeat, BAI-1 could align parallel to the apoptotic cell surface as additional TSP repeats become ligated (**Figure 2B**). Nevertheless, it seems likely that a degree of molecular segregation would be required for these different receptors to be involved in apoptotic cell uptake on the same phagocyte.

Second, receptors such as SR-A or LRP are predicted to be highly extended molecules (~40–50 nm). In addition, some of the well characterized opsonins for apoptotic cells are extremely large. For example, C1q is approximately 30 nm and TSP-1 is 40–50 nm. Together with the relatively large counter receptors for these opsonins [e.g., SCARF1 for C1q (107) and $\alpha_v\beta_3$ for TSP], the predicted intermembrane contact distance would likely be incompatible with those of BAI-1, $\alpha_v\beta_5$/MFG-E8, or Mer/Protein S. One possibility is that these extended structures are efficient in the initial capture of apoptotic targets, facilitating subsequent engagement of receptor/counter-receptor pairs that span a smaller intermembrane distance and are influenced by electrostatic repulsion between cells.

Third, receptors of the Tyro3/Axl/Mer family, BAI-1, and the integrin $\alpha_v\beta_5$ are able to initiate intracellular signaling that controls particle internalization. If there are parallels with immunological synapse formation, these receptors might be expected to become localized at the center of a contact zone. However, the intermembrane distance for TAMs and integrins to engage their counter-receptors (~30 nm) is considerably larger than that of the immunoreceptors that are responsible for signaling during the establishment of the immunological synapse (~14 nm for the TcR/MHC class II and co-stimulatory receptors such as CD80/CD28). We would speculate that the organizing principles for immune and efferocytic synapses would be different due to the distinct requirements for signal propagation following cognate interaction of receptors. In an immune synapse, delivery of a signal that controls T cell proliferation or target cell killing requires maintenance of intercellular adhesion and redistribution of antigen-specific recognition molecules to the contact zone. By contrast, an efferocytic synapse would likely be a more dynamic structure that facilitates particle tethering and allows dynamic regulation of cytoskeletal organization as particle internalization proceeds. As discussed earlier, the localization

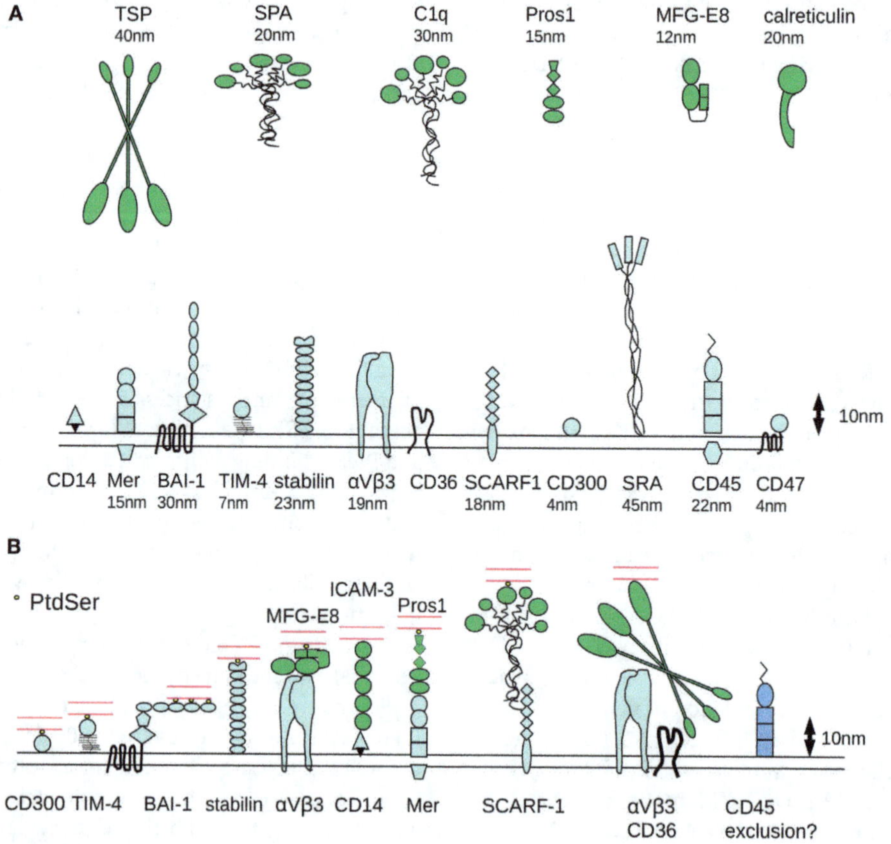

FIGURE 2 | Molecular dimensions of proteins involved in phagocytosis of apoptotic cells. **(A)** Some of the major families of receptors that have been identified as having a role in the phagocytosis of apoptotic cells are depicted, drawn approximately to scale based on available published crystallographic/NMR or cryo-electron microscopy data for various protein domains. One assumption made is that the molecules have a relatively rigid structure, although it is possible that flexibility would considerably alter axial length. There are considerable differences in the dimensions of apoptotic cell opsonins and in the receptors that mediate binding of phosphatidylserine (PtdSer) binding (either directly or indirectly). **(B)** Example receptor–counter-receptor pairings are shown to illustrate the likely differences in intermembrane "working" distance during phagocytosis, particularly for the molecules involved in the recognition of PtdSer (shown in yellow). Phagocyte receptors are shown in blue, opsonins are shown in green, and the apoptotic cell membrane is shown in red.

of receptors involved in phagocytosis may involve initial tethering mediated by larger molecules as a prerequisite for engagement of smaller PtdSer binding receptors. The mechanisms for exclusion of phosphatases from an efferocytic synapse might also be distinct.

Finally, although phagocytosis of intact apoptotic targets is readily observed *in vitro*, the tight apposition of the plasma membranes of phagocytes and apoptotic targets in certain tissues *in vivo* may require a mechanistically distinct clearance process. For example, in the retina clearance of the outer segments of photoreceptors by retinal pigment epithelial cells has been likened to a phagocytic "pruning" of the photoreceptor outer segments (108). The molecular basis of removal of photoreceptor outer segments by retinal pigment epithelial cells involved Mer, Pros1 and Gas6, and the integrin $\alpha_v\beta_5$ (108–110). Electron microscopy analysis of retinal cell architecture reveals the exquisitely close contact between RPE cells and photoreceptors (111). Elegant *in vivo* imaging studies have revealed the diurnal exposure of PtdSer in localized a manner, which then triggers the "pinching off" of the distal tips of the photoreceptors by the adjacent RPE cells (112). The exposure of PtdSer on viable photoreceptors may have some parallels with a process termed phagoptosis (113) in which viable cells are recognized by phagocytes. However, during phagoptosis, recognition triggers apoptosis in the target cell (114). Although similar molecular pathways are involved in the recognition of PtdSer exposure on viable cells (e.g., MFG-E8, stabilins, α_v integrins, and Mer) (115), the intercellular communication events that are involved are likely to be distinct from those required for efferocytosis.

In summary, the establishment of an efferocytic synapse may be required for efficient recognition of apoptotic cells by phagocytes. Cooperativity of receptor engagement may act to facilitate and stabilize adhesive interactions and lead to the assembly of signaling platforms that ultimately determine phagocyte responses to apoptotic cell binding and internalization.

AUTHOR CONTRIBUTIONS

ID wrote the article and generated the figures. NB, JM, MV, and AR wrote the article and critically appraised the figures.

FUNDING

MV acknowledges funding from the Biotechnology and Biological Sciences Research Council (BB/M025160/1). AR is part funded by a UK Medical Research Council (MR/K013386/1). NB acknowledges funding from the EPSRC and MRC Centre for Doctoral Training in Optical Medical Imaging, OPTIMA (EP/L016559/1).

REFERENCES

1. Persson C, Uller L. Transepithelial exit of leucocytes: inflicting, reflecting or resolving airway inflammation? *Thorax* (2010) 65:1111–5. doi:10.1136/thx.2009.133363
2. Poon IKH, Lucas CD, Rossi AG, Ravichandran KS. Apoptotic cell clearance: basic biology and therapeutic potential. *Nat Rev Immunol* (2014) 14:166–80. doi:10.1038/nri3607
3. Lauber K, Blumenthal SG, Waibel M, Wesselborg S. Clearance of apoptotic cells: getting rid of the corpses. *Mol Cell* (2004) 14:277–87. doi:10.1016/S1097-2765(04)00237-0
4. Savill J, Dransfield I, Gregory C, Haslett C. A blast from the past: clearance of apoptotic cells regulates immune responses. *Nat Rev Immunol* (2002) 2:965–75. doi:10.1038/nri957
5. Elliott MR, Ravichandran KS. Clearance of apoptotic cells: implications in health and disease. *J Cell Biol* (2010) 189:1059–70. doi:10.1083/jcb.201004096
6. Oldenborg PA, Gresham HD, Lindberg FP. CD47-signal regulatory protein alpha (SIRPalpha) regulates Fcgamma and complement receptor-mediated phagocytosis. *J Exp Med* (2001) 193:855–62. doi:10.1084/jem.193.7.855
7. Kagan VE, Gleiss B, Tyurina YY, Tyurin VA, Elenström-Magnusson C, Liu SX, et al. A role for oxidative stress in apoptosis: oxidation and externalization of phosphatidylserine is required for macrophage clearance of cells undergoing Fas-mediated apoptosis. *J Immunol* (2002) 169:487–99. doi:10.4049/jimmunol.169.1.487
8. Chang MK, Bergmark C, Laurila A, Hörkkö S, Han KH, Friedman P, et al. Monoclonal antibodies against oxidized low-density lipoprotein bind to apoptotic cells and inhibit their phagocytosis by elicited macrophages: evidence that oxidation-specific epitopes mediate macrophage recognition. *Proc Natl Acad Sci U S A* (1999) 96:6353–8. doi:10.1073/pnas.96.11.6353
9. Slater AF, Stefan C, Nobel I, van den Dobbelsteen DJ, Orrenius S. Signalling mechanisms and oxidative stress in apoptosis. *Toxicol Lett* (1995) 8(2–83): 149–53. doi:10.1016/0378-4274(95)03474-9
10. Morris RG, Hargreaves AD, Duvall E, Wyllie AH. Hormone-induced cell death. 2. Surface changes in thymocytes undergoing apoptosis. *Am J Pathol* (1984) 115:426–36.
11. Franc NC, White K, Ezekowitz RA. Phagocytosis and development: back to the future. *Curr Opin Immunol* (1999) 11:47–52. doi:10.1016/S0952-7915(99)80009-0
12. Martin SJ, Reutelingsperger CP, McGahon AJ, Rader JA, van Schie RC, LaFace DM, et al. Early redistribution of plasma membrane phosphatidylserine is a general feature of apoptosis regardless of the initiating stimulus: inhibition by overexpression of Bcl-2 and Abl. *J Exp Med* (1995) 182:1545–56. doi:10.1084/jem.182.5.1545
13. Fadok VA, Savill JS, Haslett C, Bratton DL, Doherty DE, Campbell PA, et al. Different populations of macrophages use either the vitronectin receptor or the phosphatidylserine receptor to recognize and remove apoptotic cells. *J Immunol* (1992) 149:4029–35.
14. McEvoy L, Williamson P, Schlegel RA. Membrane phospholipid asymmetry as a determinant of erythrocyte recognition by macrophages. *Proc Natl Acad Sci U S A* (1986) 83:3311–5. doi:10.1073/pnas.83.10.3311
15. Suzuki J, Denning DP, Imanishi E, Horvitz HR, Nagata S. Xk-related protein 8 and CED-8 promote phosphatidylserine exposure in apoptotic cells. *Science* (2013) 341:403–6. doi:10.1126/science.1236758
16. Yeung T, Gilbert GE, Shi J, Silvius J, Kapus A, Grinstein S. Membrane phosphatidylserine regulates surface charge and protein localization. *Science* (2008) 319:210–3. doi:10.1126/science.1152066
17. Park D, Tosello-Trampont AC, Elliott MR, Lu M, Haney LB, Ma Z, et al. BAI1 is an engulfment receptor for apoptotic cells upstream of the ELMO/Dock180/Rac module. *Nature* (2007) 450:430–4. doi:10.1038/nature06329
18. Park SY, Jung MY, Kim HJ, Lee SJ, Kim SY, Lee BH, et al. Rapid cell corpse clearance by stabilin-2, a membrane phosphatidylserine receptor. *Cell Death Differ* (2008) 15:192–201. doi:10.1038/sj.cdd.4402242
19. Wang X, Li W, Zhao D, Liu B, Shi Y, Chen B, et al. *Caenorhabditis elegans* transthyretin-like protein TTR-52 mediates recognition of apoptotic cells by the CED-1 phagocyte receptor. *Nat Cell Biol* (2010) 12:655–64. doi:10.1038/ncb2068
20. Andersen MH, Graversen H, Fedosov SN, Petersen TE, Rasmussen JT. Functional analyses of two cellular binding domains of bovine lactadherin. *Biochemistry (Mosc)* (2000) 39:6200–6. doi:10.1021/bi992221r
21. Hanayama R, Tanaka M, Miwa K, Shinohara A, Iwamatsu A, Nagata S. Identification of a factor that links apoptotic cells to phagocytes. *Nature* (2002) 417:182–7. doi:10.1038/417182a
22. Stitt TN, Conn G, Gore M, Lai C, Bruno J, Radziejewski C, et al. The anticoagulation factor protein S and its relative, Gas6, are ligands for the Tyro 3/Axl family of receptor tyrosine kinases. *Cell* (1995) 80:661–70.
23. Païdassi H, Tacnet-Delorme P, Garlatti V, Darnault C, Ghebrehiwet B, Gaboriaud C, et al. C1q binds phosphatidylserine and likely acts as a multiligand-bridging molecule in apoptotic cell recognition. *J Immunol* (2008) 180:2329–38. doi:10.4049/jimmunol.180.4.2329
24. Bevers EM, Comfurius P, Zwaal RF. Changes in membrane phospholipid distribution during platelet activation. *Biochim Biophys Acta* (1983) 736:57–66. doi:10.1016/0005-2736(83)90169-4
25. Dillon SR, Mancini M, Rosen A, Schlissel MS. Annexin V binds to viable B cells and colocalizes with a marker of lipid rafts upon B cell receptor activation. *J Immunol* (2000) 164:1322–32. doi:10.4049/jimmunol.164.3.1322
26. Fischer K, Voelkl S, Berger J, Andreesen R, Pomorski T, Mackensen A. Antigen recognition induces phosphatidylserine exposure on the cell surface of human CD8+ T cells. *Blood* (2006) 108:4094–101. doi:10.1182/blood-2006-03-011742
27. Segawa K, Suzuki J, Nagata S. Constitutive exposure of phosphatidylserine on viable cells. *Proc Natl Acad Sci U S A* (2011) 108:19246–51. doi:10.1073/pnas.1114799108
28. Tietjen GT, Gong Z, Chen CH, Vargas E, Crooks JE, Cao KD, et al. Molecular mechanism for differential recognition of membrane phosphatidylserine by the immune regulatory receptor Tim4. *Proc Natl Acad Sci U S A* (2014) 111:E1463–72. doi:10.1073/pnas.1320174111
29. Frasch SC, Berry KZ, Fernandez-Boyanapalli R, Jin HS, Leslie C, Henson PM, et al. NADPH oxidase-dependent generation of lysophosphatidylserine enhances clearance of activated and dying neutrophils via G2A. *J Biol Chem* (2008) 283:33736–49. doi:10.1074/jbc.M807047200
30. Tsai RK, Discher DE. Inhibition of 'self' engulfment through deactivation of myosin-II at the phagocytic synapse between human cells. *J Cell Biol* (2008) 180:989–1003. doi:10.1083/jcb.200708043
31. Gardai SJ, Bratton DL, Ogden CA, Henson PM. Recognition ligands on apoptotic cells: a perspective. *J Leukoc Biol* (2006) 79:896–903. doi:10.1189/jlb.1005550
32. Lawrence DW, King SB, Frazier WA, Koenig JM. Decreased CD47 expression during spontaneous apoptosis targets neutrophils for phagocytosis by monocyte-derived macrophages. *Early Hum Dev* (2009) 85:659–63. doi:10.1016/j.earlhumdev.2009.09.005
33. Lv Z, Bian Z, Shi L, Niu S, Ha B, Tremblay A, et al. Loss of cell surface CD47 clustering formation and binding avidity to SIRPα facilitate apoptotic cell clearance by macrophages. *J Immunol* (2015) 195:661–71. doi:10.4049/jimmunol.1401719
34. Gardai SJ, Xiao YQ, Dickinson M, Nick JA, Voelker DR, Greene KE, et al. By binding SIRPalpha or calreticulin/CD91, lung collectins act as dual function surveillance molecules to suppress or enhance inflammation. *Cell* (2003) 115:13–23. doi:10.1016/S0092-8674(03)00758-X
35. Savill JS, Henson PM, Haslett C. Phagocytosis of aged human neutrophils by macrophages is mediated by a novel 'charge-sensitive' recognition mechanism. *J Clin Invest* (1989) 84:1518–27. doi:10.1172/JCI114328
36. Ostrowski PP, Grinstein S, Freeman SA. Diffusion barriers, mechanical forces, and the biophysics of phagocytosis. *Dev Cell* (2016) 38:135–46. doi:10.1016/j.devcel.2016.06.023

37. Hart SP, Ross JA, Ross K, Haslett C, Dransfield I. Molecular characterization of the surface of apoptotic neutrophils: implications for functional downregulation and recognition by phagocytes. *Cell Death Differ* (2000) 7:493–503. doi:10.1038/sj.cdd.4400680
38. Franz S, Frey B, Sheriff A, Gaipl US, Beer A, Voll RE, et al. Lectins detect changes of the glycosylation status of plasma membrane constituents during late apoptosis. *Cytometry A* (2006) 69:230–9. doi:10.1002/cyto.a.20206
39. Meesmann HM, Fehr EM, Kierschke S, Herrmann M, Bilyy R, Heyder P, et al. Decrease of sialic acid residues as an eat-me signal on the surface of apoptotic lymphocytes. *J Cell Sci* (2010) 123:3347–56. doi:10.1242/jcs.066696
40. Jones J, Morgan BP. Apoptosis is associated with reduced expression of complement regulatory molecules, adhesion molecules and other receptors on polymorphonuclear leucocytes: functional relevance and role in inflammation. *Immunology* (1995) 86:651–60.
41. Takizawa F, Tsuji S, Nagasawa S. Enhancement of macrophage phagocytosis upon iC3b deposition on apoptotic cells. *FEBS Lett* (1996) 397:269–72. doi:10.1016/S0014-5793(96)01197-0
42. Ogden CA, deCathelineau A, Hoffmann PR, Bratton D, Ghebrehiwet B, Fadok VA, et al. C1q and mannose binding lectin engagement of cell surface calreticulin and CD91 initiates macropinocytosis and uptake of apoptotic cells. *J Exp Med* (2001) 194:781–95. doi:10.1084/jem.194.6.781
43. Arur S, Uche UE, Rezaul K, Fong M, Scranton V, Cowan AE, et al. Annexin I is an endogenous ligand that mediates apoptotic cell engulfment. *Dev Cell* (2003) 4:587–98. doi:10.1016/S1534-5807(03)00090-X
44. Uehara H, Shacter E. Auto-oxidation and oligomerization of protein S on the apoptotic cell surface is required for Mer tyrosine kinase-mediated phagocytosis of apoptotic cells. *J Immunol* (2008) 180:2522–30. doi:10.4049/jimmunol.180.4.2522
45. Hart SP, Alexander KM, Dransfield I. Immune complexes bind preferentially to Fc gamma RIIA (CD32) on apoptotic neutrophils, leading to augmented phagocytosis by macrophages and release of proinflammatory cytokines. *J Immunol* (2004) 172:1882–7. doi:10.4049/jimmunol.172.3.1882
46. Dransfield I, Buckle AM, Savill JS, McDowall A, Haslett C, Hogg N. Neutrophil apoptosis is associated with a reduction in CD16 (Fc gamma RIII) expression. *J Immunol* (1994) 153:1254–63.
47. Dransfield I, Stocks SC, Haslett C. Regulation of cell adhesion molecule expression and function associated with neutrophil apoptosis. *Blood* (1995) 85:3264–73.
48. Nishimori H, Shiratsuchi T, Urano T, Kimura Y, Kiyono K, Tatsumi K, et al. A novel brain-specific p53-target gene, BAI1, containing thrombospondin type 1 repeats inhibits experimental angiogenesis. *Oncogene* (1997) 15:2145–50. doi:10.1038/sj.onc.1201542
49. Tosello-Trampont AC, Kinchen JM, Brugnera E, Haney LB, Hengartner MO, Ravichandran KS. Identification of two signaling submodules within the CrkII/ELMO/Dock180 pathway regulating engulfment of apoptotic cells. *Cell Death Differ* (2007) 14:963–72. doi:10.1038/sj.cdd.4402094
50. Penberthy KK, Ravichandran KS. Apoptotic cell recognition receptors and scavenger receptors. *Immunol Rev* (2016) 269:44–59. doi:10.1111/imr.12376
51. Borrego F. The CD300 molecules: an emerging family of regulators of the immune system. *Blood* (2013) 121:1951–60. doi:10.1182/blood-2012-09-435057
52. Murakami Y, Tian L, Voss OH, Margulies DH, Krzewski K, Coligan JE. CD300b regulates the phagocytosis of apoptotic cells via phosphatidylserine recognition. *Cell Death Differ* (2014) 21:1746–57. doi:10.1038/cdd.2014.86
53. Park SY, Kang KB, Thapa N, Kim SY, Lee SJ, Kim IS. Requirement of adaptor protein GULP during stabilin-2-mediated cell corpse engulfment. *J Biol Chem* (2008) 283:10593–600. doi:10.1074/jbc.M709105200
54. Su HP, Nakada-Tsukui K, Tosello-Trampont AC, Li Y, Bu G, Henson PM, et al. Interaction of CED-6/GULP, an adapter protein involved in engulfment of apoptotic cells with CED-1 and CD91/low density lipoprotein receptor-related protein (LRP). *J Biol Chem* (2002) 277:11772–9. doi:10.1074/jbc.M109336200
55. Park D, Hochreiter-Hufford A, Ravichandran KS. The phosphatidylserine receptor TIM-4 does not mediate direct signaling. *Curr Biol* (2009) 19(4):346–51. doi:10.1016/j.cub.2009.01.042
56. Lemke G. Biology of the TAM receptors. *Cold Spring Harb Perspect Biol* (2013) 5:a009076. doi:10.1101/cshperspect.a009076
57. Savill J, Dransfield I, Hogg N, Haslett C. Vitronectin receptor-mediated phagocytosis of cells undergoing apoptosis. *Nature* (1990) 343:170–3. doi:10.1038/343170a0
58. Savill J, Hogg N, Ren Y, Haslett C. Thrombospondin cooperates with CD36 and the vitronectin receptor in macrophage recognition of neutrophils undergoing apoptosis. *J Clin Invest* (1992) 90:1513–22. doi:10.1172/JCI116019
59. Rabinovitch M. The dissociation of the attachment and ingestion phases of phagocytosis by macrophages. *Exp Cell Res* (1967) 46:19–28. doi:10.1016/0014-4827(67)90405-3
60. Griffin FM, Griffin JA, Leider JE, Silverstein SC. Studies on the mechanism of phagocytosis. I. Requirements for circumferential attachment of particle-bound ligands to specific receptors on the macrophage plasma membrane. *J Exp Med* (1975) 142:1263–82. doi:10.1084/jem.142.5.1263
61. Dransfield I, Zagórska A, Lew ED, Michail K, Lemke G. Mer receptor tyrosine kinase mediates both tethering and phagocytosis of apoptotic cells. *Cell Death Dis* (2015) 6:e1646. doi:10.1038/cddis.2015.18
62. Griffin FM, Mullinax PJ. Effects of differentiation in vivo and of lymphokine treatment in vitro on the mobility of C3 receptors of human and mouse mononuclear phagocytes. *J Immunol* (1985) 135:3394–7.
63. Segal G, Lee W, Arora PD, McKee M, Downey G, McCulloch CA. Involvement of actin filaments and integrins in the binding step in collagen phagocytosis by human fibroblasts. *J Cell Sci* (2001) 114:119–29.
64. Freeman SA, Grinstein S. Phagocytosis: receptors, signal integration, and the cytoskeleton. *Immunol Rev* (2014) 262:193–215. doi:10.1111/imr.12212
65. Hart SP, Rossi AG, Haslett C, Dransfield I. Characterization of the effects of cross-linking of macrophage CD44 associated with increased phagocytosis of apoptotic PMN. *PLoS One* (2012) 7:e33142. doi:10.1371/journal.pone.0033142
66. Hart SP, Dougherty GJ, Haslett C, Dransfield I. CD44 regulates phagocytosis of apoptotic neutrophil granulocytes, but not apoptotic lymphocytes, by human macrophages. *J Immunol* (1997) 159:919–25.
67. Flannagan RS, Harrison RE, Yip CM, Jaqaman K, Grinstein S. Dynamic macrophage 'probing' is required for the efficient capture of phagocytic targets. *J Cell Biol* (2010) 191:1205–18. doi:10.1083/jcb.201007056
68. Niedergang F, Di Bartolo V, Alcover A. Comparative anatomy of phagocytic and immunological synapses. *Front Immunol* (2016) 7:18. doi:10.3389/fimmu.2016.00018
69. Kayalar C, Ord T, Testa MP, Zhong LT, Bredesen DE. Cleavage of actin by interleukin 1 beta-converting enzyme to reverse DNase I inhibition. *Proc Natl Acad Sci U S A* (1996) 93:2234–8. doi:10.1073/pnas.93.5.2234
70. Concha NO, Head JF, Kaetzel MA, Dedman JR, Seaton BA. Annexin V forms calcium-dependent trimeric units on phospholipid vesicles. *FEBS Lett* (1992) 314:159–62. doi:10.1016/0014-5793(92)80964-I
71. van Genderen HO, Kenis H, Hofstra L, Narula J, Reutelingsperger CPM. Extracellular annexin A5: functions of phosphatidylserine-binding and two-dimensional crystallization. *Biochim Biophys Acta* (2008) 1783:953–63. doi:10.1016/j.bbamcr.2008.01.030
72. Grakoui A, Bromley SK, Sumen C, Davis MM, Shaw AS, Allen PM, et al. The immunological synapse: a molecular machine controlling T cell activation. *Science* (1999) 285:221–7. doi:10.1126/science.285.5425.221
73. Dustin ML, Chakraborty AK, Shaw AS. Understanding the structure and function of the immunological synapse. *Cold Spring Harb Perspect Biol* (2010) 2:a002311. doi:10.1101/cshperspect.a002311
74. Dustin ML, Groves JT. Receptor signaling clusters in the immune synapse. *Annu Rev Biophys* (2012) 41:543–56. doi:10.1146/annurev-biophys-042910-155238
75. Lopes FB, Bálint Š, Valvo S, Felce JH, Hessel EM, Dustin ML, et al. Membrane nanoclusters of FcγRI segregate from inhibitory SIRPα upon activation of human macrophages. *J Cell Biol* (2017) 216:1123–41. doi:10.1083/jcb.201608094
76. Lin J, Kurilova S, Scott BL, Bosworth E, Iverson BE, Bailey EM, et al. TIRF imaging of Fc gamma receptor microclusters dynamics and signaling on macrophages during frustrated phagocytosis. *BMC Immunol* (2016) 17:5. doi:10.1186/s12865-016-0143-2
77. Leupin O, Zaru R, Laroche T, Müller S, Valitutti S. Exclusion of CD45 from the T-cell receptor signaling area in antigen-stimulated T lymphocytes. *Curr Biol* (2000) 10:277–80. doi:10.1016/S0960-9822(00)00362-6
78. Goodridge HS, Reyes CN, Becker CA, Katsumoto TR, Ma J, Wolf AJ, et al. Activation of the innate immune receptor dectin-1 upon formation of a 'phagocytic synapse'. *Nature* (2011) 472:471–5. doi:10.1038/nature10071
79. Freeman SA, Goyette J, Furuya W, Woods EC, Bertozzi CR, Bergmeier W, et al. Integrins form an expanding diffusional barrier that coordinates phagocytosis. *Cell* (2016) 164:128–40. doi:10.1016/j.cell.2015.11.048

80. Elliott MR, Koster KM, Murphy PS. Efferocytosis signaling in the regulation of macrophage inflammatory responses. *J Immunol* (2017) 198:1387–94. doi:10.4049/jimmunol.1601520
81. Chang VT, Fernandes RA, Ganzinger KA, Lee SF, Siebold C, McColl J, et al. Initiation of T cell signaling by CD45 segregation at 'close contacts'. *Nat Immunol* (2016) 17:574–82. doi:10.1038/ni.3392
82. Flannagan RS, Canton J, Furuya W, Glogauer M, Grinstein S. The phosphatidylserine receptor TIM4 utilizes integrins as coreceptors to effect phagocytosis. *Mol Biol Cell* (2014) 25:1511–22. doi:10.1091/mbc.E13-04-0212
83. Kim S, Park SY, Kim SY, Bae DJ, Pyo JH, Hong M, et al. Cross talk between engulfment receptors stabilin-2 and integrin αvβ5 orchestrates engulfment of phosphatidylserine-exposed erythrocytes. *Mol Cell Biol* (2012) 32:2698–708. doi:10.1128/MCB.06743-11
84. Wu Y, Singh S, Georgescu M-M, Birge RB. A role for Mer tyrosine kinase in alphavbeta5 integrin-mediated phagocytosis of apoptotic cells. *J Cell Sci* (2005) 118:539–53. doi:10.1242/jcs.01632
85. Vernon-Wilson EF, Auradé F, Brown SB. CD31 promotes beta1 integrin-dependent engulfment of apoptotic Jurkat T lymphocytes opsonized for phagocytosis by fibronectin. *J Leukoc Biol* (2006) 79:1260–7. doi:10.1189/jlb.1005571
86. McCutcheon JC, Hart SP, Canning M, Ross K, Humphries MJ, Dransfield I. Regulation of macrophage phagocytosis of apoptotic neutrophils by adhesion to fibronectin. *J Leukoc Biol* (1998) 64:600–7.
87. Kirkham PA, Spooner G, Rahman I, Rossi AG. Macrophage phagocytosis of apoptotic neutrophils is compromised by matrix proteins modified by cigarette smoke and lipid peroxidation products. *Biochem Biophys Res Commun* (2004) 318:32–7. doi:10.1016/j.bbrc.2004.04.003
88. Barclay AN, Van den Berg TK. The interaction between signal regulatory protein alpha (SIRPα) and CD47: structure, function, and therapeutic target. *Annu Rev Immunol* (2014) 32:25–50. doi:10.1146/annurev-immunol-032713-120142
89. Lillemeier BF, Mörtelmaier MA, Forstner MB, Huppa JB, Groves JT, Davis MM. TCR and Lat are expressed on separate protein islands on T cell membranes and concatenate during activation. *Nat Immunol* (2009) 11:90–6. doi:10.1038/ni.1832
90. Araç D, Boucard AA, Bolliger MF, Nguyen J, Soltis SM, Südhof TC, et al. A novel evolutionarily conserved domain of cell-adhesion GPCRs mediates autoproteolysis. *EMBO J* (2012) 31:1364–78. doi:10.1038/emboj.2012.26
91. Campbell ID, Humphries MJ. Integrin structure, activation, and interactions. *Cold Spring Harb Perspect Biol* (2011) 3:A004994. doi:10.1101/cshperspect.a004994
92. Xiong JP, Stehle T, Diefenbach B, Zhang R, Dunker R, Scott DL, et al. Crystal structure of the extracellular segment of integrin alpha Vbeta3. *Science* (2001) 294:339–45. doi:10.1126/science.1064535
93. Chattopadhyay K, Lazar-Molnar E, Yan Q, Rubinstein R, Zhan C, Vigdorovich V, et al. Sequence, structure, function, immunity: structural genomics of costimulation. *Immunol Rev* (2009) 229:356–86. doi:10.1111/j.1600-065X.2009.00778.x
94. Dahlbäck B, Smith CA, Müller-Eberhard HJ. Visualization of human C4b-binding protein and its complexes with vitamin K-dependent protein S and complement protein C4b. *Proc Natl Acad Sci U S A* (1983) 80:3461–5. doi:10.1073/pnas.80.11.3461
95. Heiring C, Dahlbäck B, Muller YA. Ligand recognition and homophilic interactions in Tyro3: structural insights into the Axl/Tyro3 receptor tyrosine kinase family. *J Biol Chem* (2004) 279:6952–8. doi:10.1074/jbc.M311750200
96. Pratt KP, Shen BW, Takeshima K, Davie EW, Fujikawa K, Stoddard BL. Structure of the C2 domain of human factor VIII at 1.5 A resolution. *Nature* (1999) 402:439–42. doi:10.1038/46601
97. Rudenko G, Henry L, Henderson K, Ichtchenko K, Brown MS, Goldstein JL, et al. Structure of the LDL receptor extracellular domain at endosomal pH. *Science* (2002) 298:2353–8. doi:10.1126/science.1078124
98. Cao E, Zang X, Ramagopal UA, Mukhopadhaya A, Fedorov A, Fedorov E, et al. T cell immunoglobulin mucin-3 crystal structure reveals a galectin-9-independent ligand-binding surface. *Immunity* (2007) 26:311–21. doi:10.1016/j.immuni.2007.01.016
99. Cooke RM, Wilkinson AJ, Baron M, Pastore A, Tappin MJ, Campbell ID, et al. The solution structure of human epidermal growth factor. *Nature* (1987) 327:339–41. doi:10.1038/327339a0
100. Ellgaard L, Frickel E-M. Calnexin, calreticulin, and ERp57: teammates in glycoprotein folding. *Cell Biochem Biophys* (2003) 39:223–47. doi:10.1385/CBB:39:3:223
101. Leahy DJ, Aukhil I, Erickson HP. 2.0 A crystal structure of a four-domain segment of human fibronectin encompassing the RGD loop and synergy region. *Cell* (1996) 84:155–64. doi:10.1016/S0092-8674(00)81002-8
102. Lillis AP, Van Duyn LB, Murphy-Ullrich JE, Strickland DK. LDL receptor-related protein 1: unique tissue-specific functions revealed by selective gene knockout studies. *Physiol Rev* (2008) 88:887–918. doi:10.1152/physrev.00033.2007
103. Pike AC, Brzozowski AM, Roberts SM, Olsen OH, Persson E. Structure of human factor VIIa and its implications for the triggering of blood coagulation. *Proc Natl Acad Sci U S A* (1999) 96:8925–30. doi:10.1073/pnas.96.16.8925
104. Tan K, Duquette M, Liu JH, Dong Y, Zhang R, Joachimiak A, et al. Crystal structure of the TSP-1 type 1 repeats: a novel layered fold and its biological implication. *J Cell Biol* (2002) 159:373–82. doi:10.1083/jcb.200206062
105. Sasaki T, Knyazev PG, Clout NJ, Cheburkin Y, Göhring W, Ullrich A, et al. Structural basis for Gas6-Axl signalling. *EMBO J* (2006) 25:80–7. doi:10.1038/sj.emboj.7600912
106. Langenhan T, Piao X, Monk KR. Adhesion G protein-coupled receptors in nervous system development and disease. *Nat Rev Neurosci* (2016) 17:550–61. doi:10.1038/nrn.2016.86
107. Ramirez-Ortiz ZG, Pendergraft WF III, Prasad A, Byrne MH, Iram T, Blanchette CJ, et al. The scavenger receptor SCARF1 mediates the clearance of apoptotic cells and prevents autoimmunity. *Nat Immunol* (2013) 14:917–26. doi:10.1038/ni.2670
108. Burstyn-Cohen T, Lew ED, Través PG, Burrola PG, Hash JC, Lemke G. Genetic dissection of TAM receptor-ligand interaction in retinal pigment epithelial cell phagocytosis. *Neuron* (2012) 76:1123–32. doi:10.1016/j.neuron.2012.10.015
109. Finnemann SC. Role of alphavbeta5 integrin in regulating phagocytosis by the retinal pigment epithelium. *Adv Exp Med Biol* (2003) 533:337–42. doi:10.1007/978-1-4615-0067-4_42
110. Finnemann SC, Bonilha VL, Marmorstein AD, Rodriguez-Boulan E. Phagocytosis of rod outer segments by retinal pigment epithelial cells requires alpha(v)beta5 integrin for binding but not for internalization. *Proc Natl Acad Sci U S A* (1997) 94:12932–7. doi:10.1073/pnas.94.24.12932
111. Kevany BM, Palczewski K. Phagocytosis of retinal rod and cone photoreceptors. *Physiology (Bethesda)* (2010) 25:8–15. doi:10.1152/physiol.00038.2009
112. Ruggiero L, Connor MP, Chen J, Langen R, Finnemann SC. Diurnal, localized exposure of phosphatidylserine by rod outer segment tips in wild-type but not Itgb5-/- or Mfge8-/- mouse retina. *Proc Natl Acad Sci U S A* (2012) 109:8145–8. doi:10.1073/pnas.1121101109
113. Fricker M, Oliva-Martín MJ, Brown GC. Primary phagocytosis of viable neurons by microglia activated with LPS or Aβ is dependent on calreticulin/LRP phagocytic signalling. *J Neuroinflammation* (2012) 9:196. doi:10.1186/1742-2094-9-196
114. Hornik TC, Vilalta A, Brown GC. Activated microglia cause reversible apoptosis of pheochromocytoma cells, inducing their cell death by phagocytosis. *J Cell Sci* (2016) 129:65–79. doi:10.1242/jcs.174631
115. Nomura K, Vilalta A, Allendorf DH, Hornik TC, Brown GC. Activated microglia desialylate and phagocytose cells via neuraminidase, galectin-3, and Mer tyrosine kinase. *J Immunol* (2017) 198:4792–801. doi:10.4049/jimmunol.1502532

Conflict of Interest Statement: The authors declare that the research was conducted in the absence of any commercial or financial relationships that could be construed as a potential conflict of interest.

11

The Procoagulant Activity of Apoptotic Cells is Mediated by Interaction with Factor XII

Aizhen Yang[1†], Fengwu Chen[1†], Chao He[1], Junsong Zhou[1,2], Yi Lu[3], Jihong Dai[2,4], Raymond B. Birge[5] and Yi Wu[1,2]*

[1] *Cyrus Tang Hematology Center, Collaborative Innovation Center of Hematology, Soochow University, Suzhou, China,* [2] *The Sol Sherry Thrombosis Research Center, Temple University School of Medicine, Philadelphia, PA, United States,* [3] *Wuhan Thalys Medical Technology Inc., Wuhan, China,* [4] *Department of Pathology and Laboratory Medicine, Rutgers University-New Jersey Medical School, Newark, NJ, United States,* [5] *Department of Microbiology, Biochemistry and Molecular Genetics, Rutgers University-New Jersey Medical School, Newark, NJ, United States*

Edited by:
Amiram Ariel,
University of Haifa, Israel

Reviewed by:
Michal Amit Rahat,
Technion – Israel Institute of Technology, Israel
Tal Burstyn-Cohen,
Hebrew University of Jerusalem, Israel

***Correspondence:**
Yi Wu
yiwu@temple.edu

†*These authors have contributed equally to this work.*

Apoptotic cells, by externalizing phosphatidylserine (PS) as a hallmark feature, are procoagulant. However, the mechanism by which apoptotic cells activate coagulation system remains unknown. Intrinsic coagulation pathway is initiated by coagulation factor XII (FXII) of contact activation system. The purpose of this study was to determine whether FXII is involved in procoagulant activity of apoptotic cells. Using western blotting and chromogenic substrate assay, we found that incubation with apoptotic cells, but not with viable cells, resulted in rapid cleavage and activation of FXII in the presence of prekallikrein and high molecular weight kininogen (HK), other two components of contact activation system. As detected by flow cytometry, FXII bound to apoptotic cells in a concentration-dependent manner, which was inhibited by annexin V and PS liposome. Direct association of FXII with PS was confirmed in a surface plasmon resonance assay. Clotting time of FXII-deficient plasma induced by apoptotic cells was significantly prolonged, which was fully reversed by replenishment with FXII. Corn trypsin inhibitor, a FXII inhibitor, completely prevented apoptotic cells-induced intrinsic tenase complex formation. Consistently, apoptotic cells significantly increased thrombin production in normal plasma, which was not affected by an inhibitory anti-tissue factor antibody. However, blocking of PS by annexin V, inhibition of FXII, or the deficiency of FXII suppressed apoptotic cells-induced thrombin generation. Addition of purified FXII to FXII-deficient plasma recovered thrombin generation to the normal plasma level. In conclusion, FXII binds to apoptotic cells *via* PS and becomes activated, thereby constituting a novel mechanism mediating the procoagulant activity of apoptotic cells.

Keywords: apoptotic cells, factor XII, phosphatidylserine, coagulation, contact activation system

INTRODUCTION

Apoptosis, or programmed cell death, under physiologic conditions is an active process that is morphologically and biochemically different from necrosis. Apoptosis can be induced in a variety of pathological disorders, including inflammation, autoimmune diseases, atherosclerosis, tissue injury, and degeneration, as well as during radiation treatment and chemotherapy (1). When a cell undergoes apoptosis, phosphatidylserine (PS) typically becomes exposed on the cell membrane (2). If apoptotic cells are not rapidly cleared, they become procoagulant and are often associated with thrombotic disorders, such as atherothrombosis and Trousseau syndrome (3–5). Up to date, the mechanisms by which apoptotic cells activate the coagulation system and enhance blood clotting are largely unknown. Tissue factor (TF) is complexed with circulating coagulation factor VII and triggers the cascade that generates thrombin. Although TF is involved in the procoagulant activity of apoptotic cells induced by inflammatory mediators, such as LPS, only a portion of the thrombin generated during apoptosis is attributable to TF (3, 6, 7), suggesting the existence of additional mechanisms.

The plasma contact activation system, also called the intrinsic coagulation pathway, consists of factor XII (FXII), prekallikrein (PK), and high-molecular-weight kininogen (HK) (8, 9). FXII, or Hageman factor, is a zymogen of the serine protease factor XIIa (FXIIa). FXII zymogen is activated by limited proteolysis, involving cleavage of the Arg353-Val354 peptide bond, generating the two-chain molecule FXIIa (10). Two principal modes of FXII activation exist. In the first, FXII is activated by binding to negatively charged surfaces, which induces a conformational change (auto-activation). In the second, other proteases, such as kallikrein, cleave and activate FXII (hetero-activation) (10). FXIIa consists of a heavy chain and a light chain connected by a single disulfide bond between two cysteine residues (Cys340 and Cys367) (10, 11). For the last four decades, it has been known that FXII is activated by a variety of artificial and biological anionic surfaces (11, 12). These observations suggest that the contact activation system plays an important role in the enhancement of coagulation. However, whether FXII is involved in apoptotic cell-mediated procoagulant activity has never been studied.

In this study, we investigated whether FXII participates in the procoagulant activity of apoptotic cells. Our results indicate that FXII binds to apoptotic cells and rapidly becomes cleaved and activated. The binding of FXII to apoptotic cells is mediated by PS, and activation of FXII is required for apoptotic cell-mediated blood clotting and thrombin formation. Our study reveals a novel function for FXII and a new mechanism underlying apoptotic cell-mediated procoagulant activity.

MATERIALS AND METHODS

Materials

Human FXII, PK, HK, FIX, FX, and FXI were purchased from Enzyme Research Laboratories (South Bend, IN, USA). Recombinant human FVIII was obtained from American Diagnostica, Inc. The EZ-Link® Sulfo-NHS-LC-Biotinylation kit was purchased from Thermo Scientific, Inc. Chromogenic substrate for FXIIa and tissue factor blocking antibody (4501) were purchased from American Diagnostica. The chromogenic substrate for FXa was obtained from Chromogenix. Monoclonal antibody against FXII heavy chain (B7C9) was purchased from Santa Cruz Biotechnology. Monoclonal antibody against FXII light chain (C6B7) and PE-labeled anti-TF antibody were obtained from eBioscience. Polyclonal anti-HK heavy chain antibody was purchased from Abgent. Polyclonal anti-pKal antibody was obtained from Abcam (catalog number 43084). Corn trypsin inhibitor (CTI) was purchased from Merck Chemicals, Ltd. FXII-deficient plasma and pooled human normal plasma were obtained from George King Bio-Medical, Inc. Phosphatidylcholine (PC), PS, 1,2-dioleoyl-sn-glycero-3-phosphocholine(DOPC),and1-oleoyl-2-{6-[(7-nitro-2-1,3-benzoxadiazol-4-yl)amino]hexanoyl}-sn-glycero-3-phosphocholine (NBD-PC) were obtained from Avanti-Polar Lipids. The FITC-labeled Annexin V Apoptosis Detection kit was purchased from BD Biosciences. Magnetic annexin V microbeads were obtained from Miltenyi Biotech, Inc. Annexin V was obtained from BD PharMingen. Calf intestinal alkaline phosphatase (CIAP) and DNase were purchased from New England Biolabs. RPMI 1640 medium and fetal bovine serum (FBS) were purchased from Invitrogen. All other reagents were obtained from Sigma-Aldrich unless otherwise specified.

Cell Culture and Induction of Apoptosis

The human T lymphoblastoid cell line CCRF-CEM was purchased from American Type Culture Collection (ATCC® Number: CCL-119™). CEM cells were propagated in complete RPMI-1640 medium supplemented with 10% FBS, 1 mM L-glutamine, and 1% penicillin-streptomycin at 37°C in a humidified atmosphere with 5% CO_2 (13). To induce apoptosis, cells (2×10^5/mL) were cultured with 10 μM dexamethasone at 37°C in a humidified atmosphere containing 5% CO_2 for 24 h. Apoptosis was verified by the FITC Annexin V Apoptosis Detection kit according to the manufacturer's protocol (14). Apoptotic cells were purified with annexin V magnetic microbeads.

Preparation of Liposomes and Phospholipid-Coated Beads

To prepare liposomes containing 100% PC (PC liposomes) or PC and PS at 80:20 mol% (PS liposomes), the appropriate amounts of each phospholipid were dissolved with chloroform in a glass tube. Before use, the phospholipids were dried under nitrogen, suspended in PBS, and sonicated for 3 min (15, 16). The phospholipid concentration was determined by a phosphorus assay. To prepare phospholipid-coated beads, 6.5 mg/mL Nucleosil 120-3 C18 beads (Macherey-Nagel) in 1 mL of chloroform were incubated with 2.5 mg/mL DOPC (PC beads) or mixed DOPC/PS (80:20 mol%; PS beads). The phospholipid-coated beads were dried under nitrogen, resuspended in PBS, sonicated, and then labeled with 2.5 μM NBD-PC for 20 min at 37°C. After centrifugation at $13,000 \times g$ for 10 min, the beads were washed with ice-cold PBS twice to remove remaining NBD-PC, and then resuspended in PBS.

Flow Cytometric Analysis of FXII Binding to Apoptotic Cells

Human FXII was biotinylated using the EZ-Link® Sulfo-NHS-LC-Biotinylation Kit according to the manufacturer's protocol. Biotin-labeled FXII (B-FXII) at various concentrations was incubated with apoptotic cells at 4°C for 15 min. After washing, the cells were labeled with PE-avidin and analyzed by flow cytometry.

Measurement of FXII Activity and Cleavage

Cells were incubated with FXII in the presence or absence of PK and HK in HEPES buffer (137 mM NaCl, 5 mM HEPES, 2.7 mM

KCl, 2 mM $MgCl_2$, 0.42 mM NaH_2PO_4, and 1% BSA, pH 7.5) supplemented with 50 μM $ZnCl_2$ at 37°C for 30 min. After centrifugation at 2,500 rpm for 5 min, the supernatant was collected and a chromogenic substrate, Pefachrome FXIIa (0.5 mM), was added. The optical density of substrate hydrolysis was measured at 405 nm using a spectrophotometer (SpectraMax M5) (17). Cleavage of FXII was also detected by western blotting with a monoclonal antibody against FXII heavy chain. The density of the bands was measured by NIH Image J, and cleavage was defined as ratio of the percentage of cleaved FXIIa (48 kDa)/[uncleaved FXII (80 kDa) + cleaved FXIIa (48 kDa)].

Surface Plasmon Resonance Assay

The experiments were performed at 25°C using HBS-N buffer (20 nM HEPES and 0.15 M NaCl, pH 7.4) containing 50 μM Zn^{2+} as a running buffer. PS liposomes (PS:PC = 1:9) were diluted in running buffer and immobilized onto flow cell 2 (FL2) of the L1-sensor chip, and PC liposomes were immobilized onto flow cell 1 (FL1) of the L1-sensor chip. Subsequent measurements were obtained at a flow rate of 30 μL/min. A two-fold dilution series of FXII diluted in running buffer was generated (0, 2.5, 5, 10, 20, 40, 80, 100, 200, 400 nM) and was injected over the flow cells at a flow rate of 10 μL/min for 120 s and dissociated for 300 s in order of increasing concentration. Response to PS liposome binding curves was obtained by subtracting the FL2 curve from the FL1 curve and analysis using BIAevaluation software.

Intrinsic Tenase Complex Activity

Apoptotic cells were incubated with 200 μL of HEPES buffer containing 2.5 mM $CaCl_2$, 95 nM purified human FXII, 30 nM PK, 30 nM HK, 5.8 nM purified human FXI and FIX, 0.25 nM purified human FVIII, and (where indicated) 2 μM CTI at 37°C. Then, the reaction was started by the addition of purified human FX (170 nM). At various time points, a 25-μL aliquot of the mixture was removed, and 5 μL of 60 mM EDTA in PBS was added to stop FXa formation. FXa formation was monitored as the hydrolysis of the chromogenic substrate S-2222 (0.2 mM) over 30 min. Optical density at 405 nM was converted to FXa nM using a dilution curve of human FXa.

Clotting Time Assay

Blood drawn from drug-free healthy volunteers was anticoagulated by adding 1 part sodium citrate (110 mM) to 9 parts whole blood. Our study using blood from healthy volunteers was performed after approval by the IRBs of Temple University (IRB no. 20857) and Soochow University (IRB no. 2012037), obtaining informed consent, and in accordance with the Declaration of Helsinki. Platelet-poor plasma (PPP) was prepared by centrifugation at 2,000 × *g* for 30 min. In some experiments, commercial FXII-depleted plasma and normal plasma were used to examine the role of FXII in apoptotic cell-mediated clotting and thrombin generation. After viable or apoptotic cells were mixed with plasma and incubated at 37°C for 180 s, 50 μL of warmed 20 mM $CaCl_2$ was added to start the reaction, and clotting time was immediately recorded with an Amelung KC4A coagulometer (18).

Thrombin Generation Assay

Thrombin generation in plasma was measured over time with a coagulation analyzer Ceveron® alpha (Technoclone, Vienna, Austria), which employs a fluorogenic thrombin substrate (Z-Gly-Gly-Arg-AMC) to continuously monitor thrombin activity in plasma. Measurements were conducted in a total volume of 150 μL, including 40 μL of normal plasma or FXII-deficient plasma. Apoptotic cells were pretreated with 5 μg/mL BSA, annexin V, mouse IgG, and anti-TF antibody (4501) as indicated. The plasma was treated with mouse IgG and anti-FXII antibody (C6B7) when there was a need to block FXII. Then, an 80-μL aliquot of cells (1×10^5) was added to a 40-μL plasma sample. After incubation at 37°C for 15 min, 30 μL of fluorogenic buffer (2.5 mM fluorogenic substrate and 87 mM $CaCl_2$) was added to start the thrombin generation assay.

SDS-PAGE and Western Blotting

Cleavage of FXII into FXIIa was detected by SDS-PAGE (12%) under reducing conditions and immunoblotting. After incubation with apoptotic or viable cells as described above, the cells were pelleted by centrifugation, and samples containing FXII were collected (14). These samples and samples from a cell-free system were mixed with SDS-PAGE loading buffer and heated at 95°C. After the samples were separated by electrophoresis and transferred to a polyvinylidene difluoride membrane (Millipore), the membrane was blocked with 5% non-fat dry milk in blocking buffer. After extensive washing, the immunoblots were incubated for 2 h with the primary antibodies, including monoclonal anti-FXII (B7C9), polyclonal anti-HK heavy chain, and polyclonal anti-pKal antibodies. Antibody binding was detected by IRDye 800-conjugated goat anti-mouse IgG (LI-COR Bioscience) or IRDye 680-conjugated goat anti-rabbit IgG (LI-COR Bioscience) and visualized with the ODYSSEY infrared imaging system (LI-COR).

Identification of TF Antigen by Flow Cytometry

Cells (1×10^6/mL) were incubated with PE-conjugated monoclonal TF (CD142) antibody or isotype control IgG1 for 30 min at 4°C in the dark. Cells were resuspended in 400 μL of PBS before analysis. The mean fluorescence intensity of 10,000 events was determined for each sample.

Statistical Analysis

The results are expressed as the mean ± SEM of at least three experiments. One-way analysis of variance followed by Tukey's test (for multiple groups) or Student's *t*-test (for comparisons between two groups) was used, and a *p* value less than 0.05 was considered statistically significant. Unless stated otherwise, the data shown are from a single experiment that is representative of at least three separate experiments.

RESULTS

FXII Binds to Apoptotic Cells *via* Phosphatidylserine (PS)

To determine whether FXII binds to apoptotic cells, apoptotic cells were incubated with B-FXII. As detected by flow cytometry, the binding of B-FXII to apoptotic cells increased as the concentration increased (from 50 to 800 nM), and became saturated at 400 nM (**Figures 1A,B**). However, B-FXII did not bind to viable cells when added at the same concentrations (**Figure 1B**), suggesting that FXII specifically binds to apoptotic cells. PS is a phospholipid component that is usually maintained on the inner-leaflet (the cytosolic side) of the cell membrane by flippase (2). However, when a cell undergoes apoptosis, PS becomes exposed on the surface of the cell (2). Annexin V (or Annexin A5) is a member of the annexin family of intracellular proteins that binds to PS in a calcium-dependent manner. To examine whether FXII binds to apoptotic cells through PS, apoptotic cells were preincubated with annexin V or BSA in the presence or absence of 2.5 mM $CaCl_2$. As shown in **Figure 2A**, preincubation with annexin V markedly inhibited the binding of FXII to apoptotic cells in the presence of 2.5 mM $CaCl_2$, and its inhibitory effect was concentration dependent. In contrast, no inhibition by annexin V was observed in the

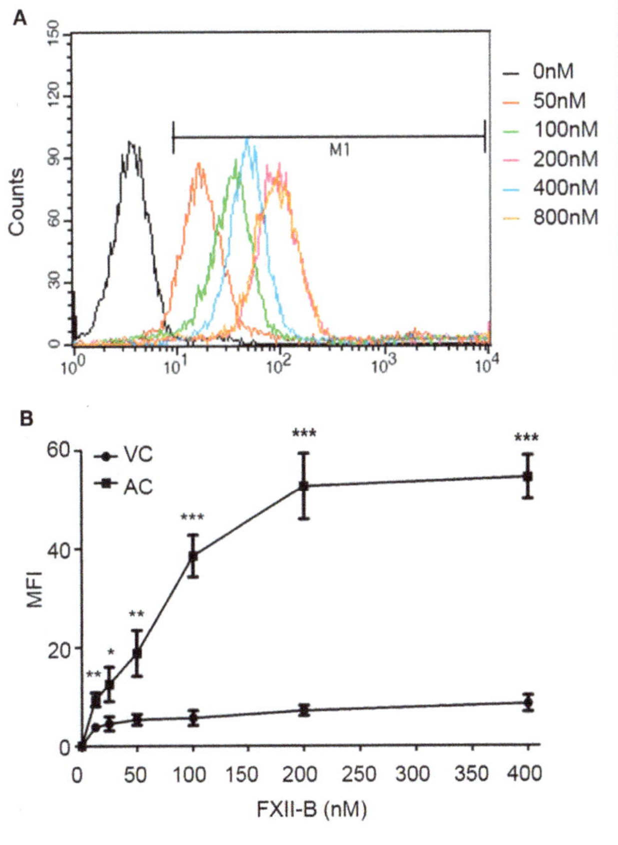

FIGURE 1 | FXII preferentially binds to apoptotic cells. **(A)** Biotin-labeled FXII (B-FXII) at the indicated concentrations was incubated with apoptotic CEM cells (2×10^5) at 4°C for 15 min. After washing, the cells were labeled with PE-avidin and analyzed by flow cytometry. The fluorescence tracings shown are representative data from three independent experiments. **(B)** Viable CEM cells (VC, closed circles) or apoptotic CEM cells (AC, squares) were incubated with B-FXII at the indicated concentrations at 4°C for 15 min. After washing, the cells were labeled with PE-avidin and analyzed by flow cytometry ($n = 5$). The data were analyzed by Student's t-test, $^*p < 0.05$; $^{**}p < 0.01$; $^{***}p < 0.001$.

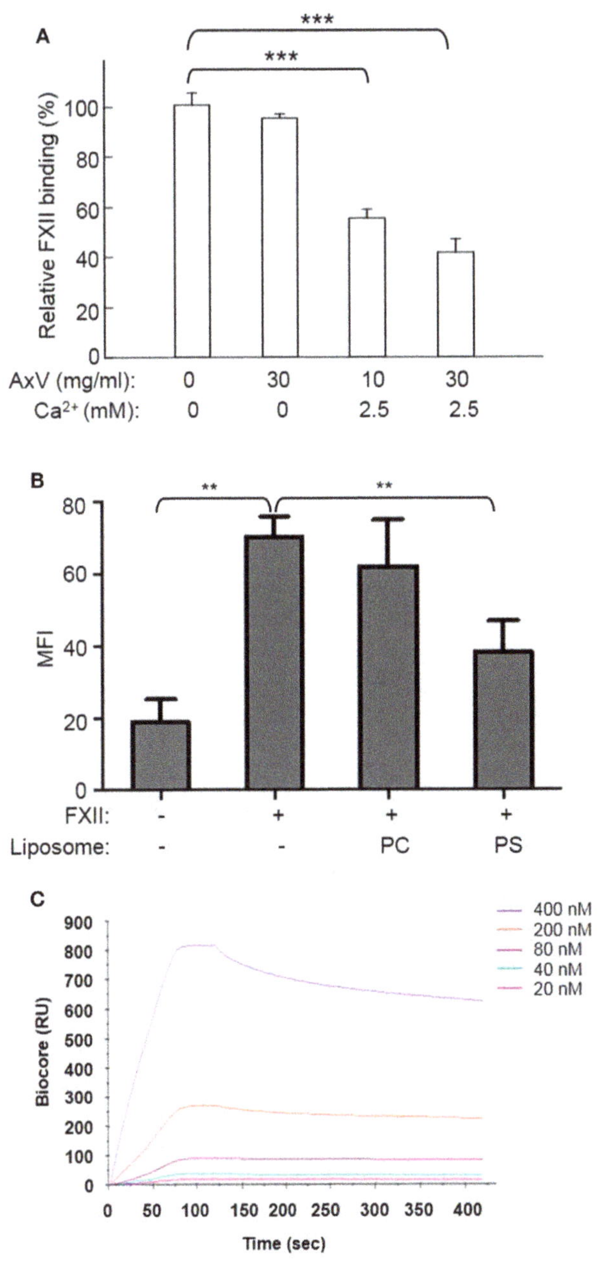

FIGURE 2 | Binding of FXII to apoptotic cells is dependent on phosphatidylserine (PS). **(A)** After preincubation with or without annexin V at the indicated concentrations in the presence of 2.5 mM $CaCl_2$ for 20 min, apoptotic CEM cells were labeled with 100 nM B-FXII and PE-avidin as described above. The binding of FXII was analyzed by flow cytometry and indicated as a percentage compared to the binding in the absence of annexin V and $CaCl_2$, which was set to 100% ($n = 4$). $^{***}p < 0.001$. **(B)** B-FXII (100 nM) was preincubated with or without 1 mM phosphatidylcholine (PC) or PS liposomes at 4°C for 15 min, and then apoptotic cells were added. After incubation for 60 min, the binding of B-FXII to apoptotic cells was analyzed by flow cytometry as described above, and shown as the mean fluorescence intensity (MFI). $^{**}p < 0.01$. **(C)** FXII binds to phosphatidylserine in a surface plasmon resonance assay. Serial concentrations of FXII were flowed over PS or PC liposomes immobilized on the Biacore sensor chip. The response curve of PS/FXII binding was obtained by subtracting the curve of PC from that of PS. Curves were analyzed with BIAevaluation software. The K_D was 3.857E-9 M, and the R_{max} was 787.0 Ru.

absence of $CaCl_2$ (**Figure 2A**), suggesting that the inhibitory effect of annexin V is specific. In addition, when FXII was preincubated with PS liposomes, its binding to apoptotic cells was significantly reduced by more than 55% (**Figure 2B**). However, preincubation of FXII with PC liposomes did not affect binding to apoptotic cells (**Figure 2B**). To evaluate the binding capacity of FXII to PS liposomes, a Biacore assay was used. The sensorgrams showed an increase in response units that was reflective of PS binding, and the binding response was concentration dependent, with a K_D of 3.857E-9 M (**Figure 2C**). These results suggest that FXII binds to apoptotic cells through a high-affinity association with PS.

The Binding of FXII to Apoptotic Cells Leads to Its Cleavage and Activation

We next examined whether FXII binding to apoptotic cells induces its cleavage and activation. FXII is converted to its active form, FXIIa, through auto-activation induced by contact with charged surfaces. Thus, the activation of FXII is dependent on its cleavage (8). As shown by western blotting using an antibody recognizing the heavy chain of FXII (**Figure 3A**, i and ii), incubation with apoptotic cells markedly induced the cleavage of full-length FXII zymogen (80 kDa) to produce a heavy chain fragment (48 kDa). However, FXII was not cleaved when incubated with viable cells (**Figure 3A**). More strikingly, the FXII cleavage induced by contact with apoptotic cells was significantly enhanced in the presence of HK and PK (**Figure 3A**). In the absence of cells, incubation of FXII with HK and PK induced cleavage of a small percentage of FXII (**Figure 3A**), suggesting that FXII may form a complex with HK and PK on the surface of test tubes, thus leading to its cleavage. As shown by densitometric measurement of band intensity, about 85% of the FXII was cleaved in the presence of HK, PK, and apoptotic cells, which was significantly higher than that induced in the presence of viable cells (**Figure 3A**, iii). Consistent with the observations in the presence of apoptotic cells, incubation with PS liposomes induced cleavage of >80% of the added FXII (**Figure 3B**, i and ii). In contrast, <15% of the FXII was cleaved when incubated with PC liposomes (PS liposome vs. PC liposomes, $p < 0.001$; **Figure 3B**). Taken together, PS-mediated FXII binding to apoptotic cells results in its cleavage. To further examine whether the cleaved FXII induced by apoptotic cells is proteolytically active, we performed an assay with a chromogenic substrate. As shown in **Figure 3C**, in the presence of HK and PK, incubation of FXII with apoptotic cells significantly increased FXIIa activity ($p < 0.001$). This increase in activity was completely prevented by the addition of a FXII inhibitor, corn trypsin inhibitor (CTI). This result suggests that binding of FXII to apoptotic cells promotes its activation.

FXII Is Required for Apoptotic Cell-Mediated Intrinsic Tenase Formation and Procoagulant Activity

The observation that apoptotic cells can activate FXII led us to hypothesize that this FXII activation is involved in apoptotic cell-mediated procoagulant activity. First, we determined whether the extrinsic coagulation pathway is involved in apoptotic cell-mediated procoagulant activity in our assay system. As evaluated by FACS analysis (**Figure 4A**), TF antigen remained undetectable on both viable and apoptotic CEMs. THP-1 served as a TF-positive control. Apoptotic CEM-mediated pro-coagulation is not dependent on the extrinsic coagulation pathway, which allowed us to evaluate the role of FXII in this process. In the clotting assay, the addition of apoptotic cells significantly shortened the clotting time of PPP triggered by the addition of $CaCl_2$, compared with the clotting time for viable cells (151.8 ± 2.1 vs. 281.9 ± 2.5 s, $p < 0.001$; **Figure 4B**). To evaluate the contribution of FXII, we measured the clotting time of PPP with apoptotic cells in the presence and absence of FXII. As shown in **Figure 4C**, compared to the clotting time of normal plasma with apoptotic cells, the clotting time of FXII-deficient plasma was significantly longer (150.8 ± 3.7 vs. 281.4 ± 7.9 s; $p < 0.001$), which was reversed by replenishment with a physiological concentration of purified FXII protein (165.7 ± 6.6 s), suggesting the requirement of FXII for apoptotic cell-mediated pro-coagulation. We next tested the capacity of FXII to initiate tenase formation, which contributes to the procoagulant activity of apoptotic cells. As shown in **Figure 4D**, in the presence of FXII, PK/HK, FIX, and FVIII, apoptotic cells increased the activity of FXa as a function of time, which was 7.3-fold higher than that with viable cells at 30 s. To examine whether apoptotic cell-mediated FX activation was FXII-dependent, we tested the inhibitory effect of CTI and found that 2 μM CTI reduced apoptotic cell-mediated FXa activation to basal levels (**Figure 4D**). The above results demonstrate that FXII is critical for intrinsic tenase complex formation, contributing to the procoagulant activity of apoptotic cells.

FXII Is Important for Apoptotic Cell-Induced Thrombin Generation

To further determine the contribution of PS to the FXII-dependent procoagulant activity of apoptotic cells, we employed a thrombin generation assay. As shown in **Figure 5A**, apoptotic cells dynamically induced thrombin generation, which was almost entirely inhibited by annexin V (AxV). Importantly, an inhibitory anti-FXII antibody (C6B7) greatly decreased apoptotic cell-mediated thrombin generation, whereas an anti-TF blocking antibody (TF Ab) did not have such an effect (**Figure 5B**). Consistently, apoptotic cells failed to induce thrombin generation in FXII-deficient plasma. However, the addition of 375 nM purified FXII to FXII-deficient plasma increased thrombin generation (**Figure 5C**). To determine the effect of PS on thrombin generation, PS or PC liposome-coated beads were used in a test of thrombin generation activity. As shown in **Figure 5D**, PS liposomes strongly induced thrombin generation, whereas PC liposomes had only a minor effect. Interestingly, the inhibitory anti-FXII antibody C6B7 greatly diminished PS-induced thrombin generation (**Figure 5D**), suggesting that FXII contributes to apoptotic cell-mediated thrombin generation by interacting with PS. It is important to note that there is no clear distinction between apoptosis and other cell types of cell death, such as necrosis, and all dying cells exhibit increased PS exposure and enhanced coagulation. Nucleic acids and polyphosphate that are released from necrotic cells can be also associated with FXII activation

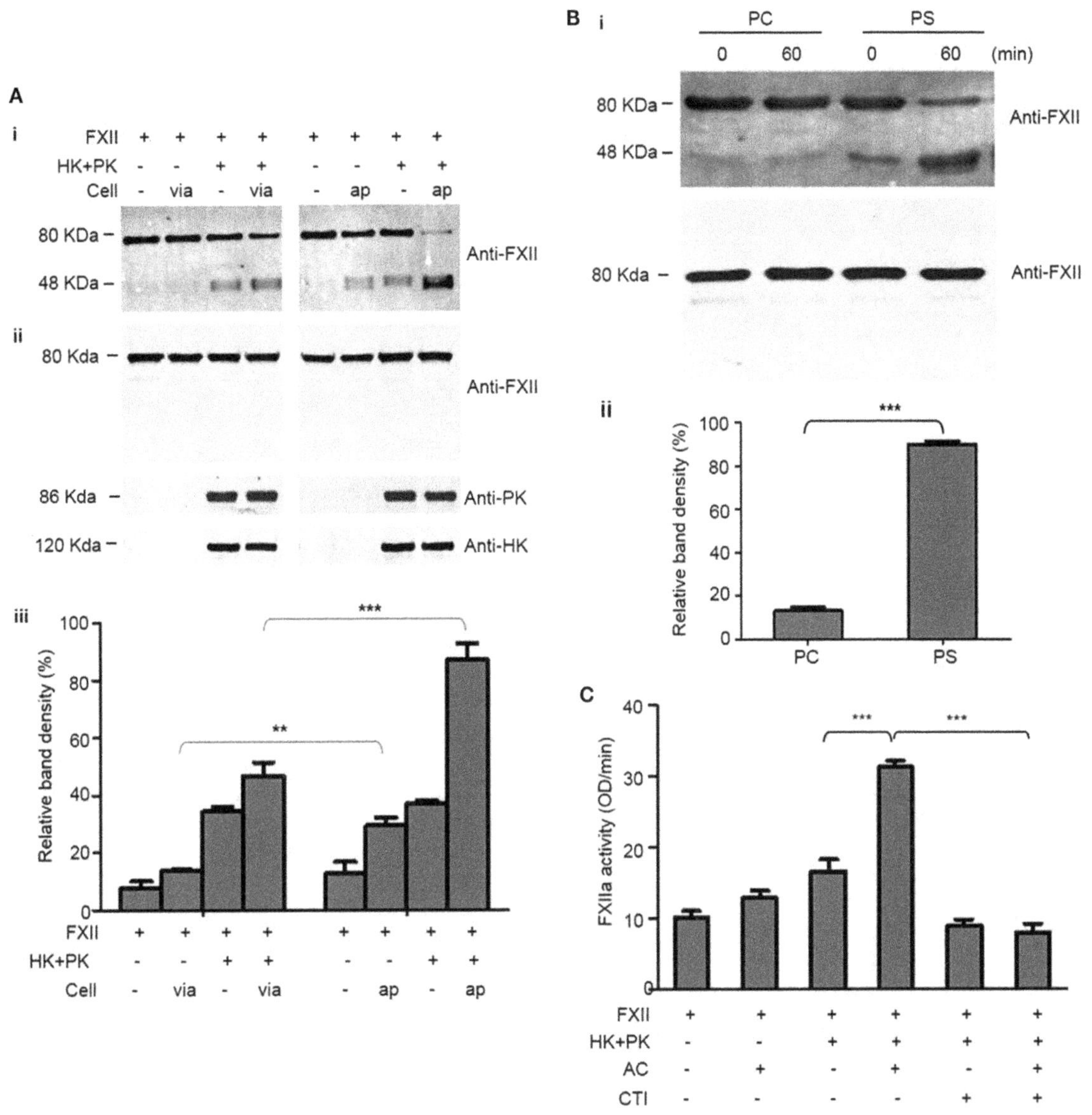

FIGURE 3 | Binding of FXII to apoptotic cells mediates its cleavage and activation. **(A)** As indicated, 95 nM FXII was incubated with viable cells (*via*) or apoptotic cells (ap) at density of 2×10^5 in the presence or absence of 30 nM prekallikrein (PK) and 30 nM high molecular weight kininogen (HK) at 37°C for 30 min. After centrifugation at 2,500 rpm for 5 min, the supernatant was collected and analyzed by western blot with an anti-FXII Ab (i). The levels of FXII, PK, and HK before incubation with cells are shown by western blotting (ii). In three independent experiments, the density of bands was measured by NIH Image J software, and the cleavage of FXII was defined as the ratio of [cleaved FXII chain (48 kDa)]/[uncleaved FXII (80 kDa) plus cleaved FXII chain (48 kDa)] shown as relative band density (iii). $^{**}p < 0.01$; $^{***}p < 0.001$. **(B)** In a cell-free system, 95 nM FXII was mixed with 30 nM PK and 30 nM HK followed by addition of 50 nM phosphatidylserine (PS) or phosphatidylcholine (PC) liposomes, one half of the mixture was immediately centrifuged (0 min), and the other half was incubated at 37°C for 60 min (60 min). After centrifugation at 55,000 rpm for 5 h, the supernatant was collected for analysis by western blotting using anti-FXII Ab (i, upper panel). The levels of FXII before incubation with liposomes are shown by western blotting (i, lower panel). In three independent experiments, the density of bands was measured, and the cleavage of FXII was calculated as described above and shown as relative band density (ii). $^{***}p < 0.001$. **(C)** Binding of FXII to apoptotic cells mediates its activation. As indicated, FXII (95 nM) was incubated with or without apoptotic cells (AC; 2×10^5), PK (30 nM) plus HK (30 nM), and corn trypsin inhibitor (CTI) (2 μM) at 37°C for 30 min. After centrifugation at 2,500 rpm for 5 min, the supernatant was collected and factor XIIa (FXIIa) activity was analyzed as the hydrolysis of a chromogenic substrate as described in the Section "Materials and Methods" ($n = 3$). $^{***}p < 0.001$.

(19, 20). However, after pretreatment with CIAP and DNase, which degrade polyphosphate and nucleic acids, respectively, there was no change in the level of thrombin generation induced by apoptotic cells (**Figure 5E**), indicating that polyphosphates and nucleic acids are not involved in the observed apoptotic cell-mediated thrombin generation.

DISCUSSION

Changes in phospholipid asymmetry, with outer surface exposure of PS, is a fundamental feature of apoptosis. On activated platelets, exposure of PS on the outer leaflet is essential for membrane assembly of the coagulation factor complexes, including the tenase complex, which are necessary for thrombin generation (21). Similarly, upon apoptotic cell death, PS is exposed on the membrane surface; concomitantly, they become procoagulant and activate the coagulation pathway. These enhanced procoagulant activities associated with membrane PS exposure have been widely observed in apoptotic endothelial cells, vascular smooth muscle cells, lymphocytes, monocytes, and cancer cells. It has been shown that tissue factor does not play a major role in this process, as there is no increase in antigen levels or the functional activity of tissue factor (3). Thus, the connection between apoptotic cells and coagulation was largely unknown. In this study, we showed that the FXII zymogen preferentially binds to apoptotic cells, leading to its rapid cleavage and activation, thereby contributing to apoptotic cell procoagulant activity.

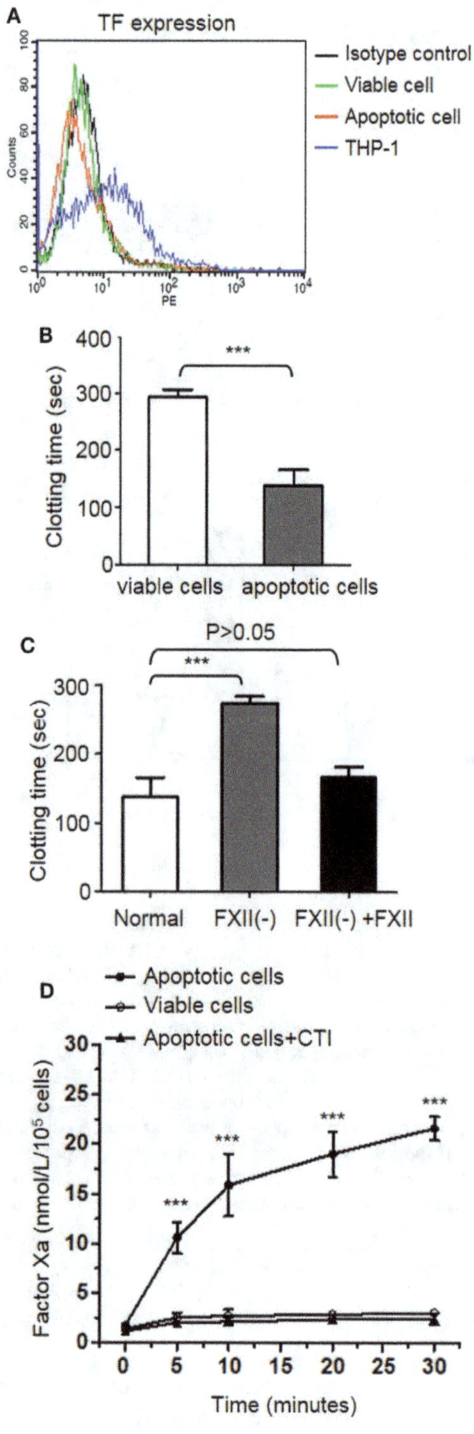

FIGURE 4 | Continued

FIGURE 4 | Continued
Apoptotic cell-mediated procoagulant activity and intrinsic tenase formation is dependent on FXII activation. **(A)** Tissue factor (TF) expression as determined by flow cytometry. CEM cells treated with (apoptotic cells) or without (viable cells) 10 μM dexamethasone for 24 h were stained with CD142-PE to analyze TF expression on the cell surface. THP-1 cells were used as a positive control. The background staining with an isotype control is shown in black. **(B)** After viable cells and apoptotic cells (2 × 10^5) were suspended in 150 μL of platelet-poor plasma, clotting was triggered by the addition of 20 mM $CaCl_2$ and measured as described in the Section "Materials and Methods" (n = 3). ***p < 0.001. **(C)** After apoptotic cells (2 × 10^5) were suspended in normal plasma (normal), FXII-deficient plasma [FXII(-)] and FXII-deficient plasma supplemented with 375 nM FXII [FXII(-) + FXII], respectively, clotting was triggered by the addition of 20 mM $CaCl_2$ and measured as described in the Section "Materials and Methods" (n = 3). **(D)** Effect of apoptotic cells on intrinsic tenase complex formation. As indicated, apoptotic or viable cells were incubated with FXII, prekallikrein, high molecular weight kininogen, FXI, FIX, and FVIII. Then FX was added, and tenase complex formation was analyzed using a chromogenic substrate (n = 3). Some samples were also treated with 2 μM corn trypsin inhibitor (CTI). ***p < 0.001.

Thrombin generation is initiated by two distinct pathways, and it can be triggered by exposure of blood to either a damaged vessel wall (extrinsic) or blood-borne factors (intrinsic). The intrinsic pathway of coagulation is initiated by FXII in a reaction involving HK and PK. These factors are collectively referred to as the contact activation system. The present findings of a correlation between thrombin generation and apoptosis showed increased intrinsic tenase activity; however, there is no available model that outlines how thrombin formation could be initiated by highly negatively charged cellular surfaces when functional TF is absent. It is known that FXII is activated by a variety of artificial or biological anionic surfaces, such as kaolin (22), ellagic acid (23), polymers (24), nucleotides (25), sulfatides (26), glycosaminoglycans (27), misfolded proteins (28), polyphosphates (29), and collagen (30). In this study, we showed, for the first time, that FXII zymogen directly binds to PS liposomes and PS on apoptotic cells (**Figures 1** and **2**). The binding of FXII to PS mediates its rapid cleavage and activation in the presence of HK and PK (**Figure 3**). These results suggest that PS on apoptotic cells serves as a docking site for FXII binding, which may induce its auto-activation. The PS on apoptotic cells may also recruit HK and PK to form a complex with FXII, thereby activating PK, and activated PK may increase FXII activation. In three assays, including APTT, intrinsic tenase formation, and thrombin generation assays, we employed FXII-deficient plasma and FXII inhibitor and provided evidence that FXII plays an important role in apoptotic cell-mediated coagulation. We further showed that TF is not involved in the FXII-mediated procoagulant activity of apoptotic cells, as a TF Ab did not affect thrombin generation by

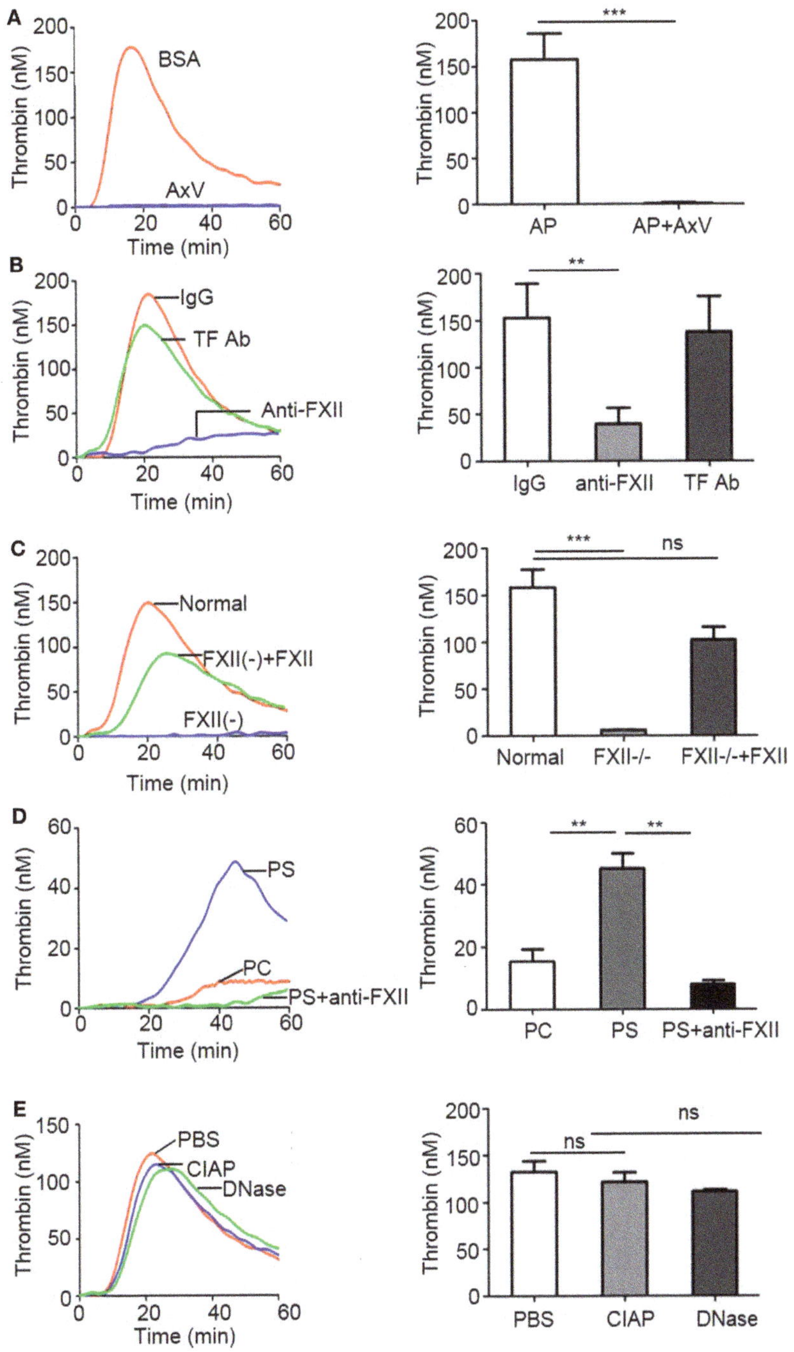

FIGURE 5 | FXII is important in thrombin generation induced by apoptotic cells. **(A–E)** Thrombin generation triggered by apoptotic cells in normal plasma was analyzed as described in the Section "Materials and Methods." Representative data are shown on the left, and accumulated data with statistical analysis are shown on the right. **(A)** Apoptotic cells were pretreated with 5 μg/mL BSA or annexin V, and then used in a thrombin generation assay ($n = 3$). **(B)** Apoptotic cells were pretreated with an anti-FXII antibody C6B7 or anti-tissue factor (TF) blocking antibody (4501; 5 μg/mL each), and then used in a thrombin generation assay ($n = 3$). **(C)** The thrombin generation assay was performed in normal plasma, FXII-deficient plasma, and FXII-deficient plasma reconstituted with human FXII ($n = 3$). **(D)** Phosphatidylcholine (PC) or phosphatidylserine (PS) liposome-coated beads were used to stimulate thrombin generation in normal plasma. Some samples using PS liposome-coated beads were also incubated with 5 μg/mL C6B7 ($n = 3$). **(E)** Apoptotic cells pretreated with PBS, 100 U/mL Calf intestinal alkaline phosphatase (CIAP), or 0.01% DNase, before thrombin generation was measured. **$p < 0.01$; ***$p < 0.001$.

apoptotic cells, which is consistent with the absence of TF expression on the surface of apoptotic cells (**Figure 4A**). In another study, we found that HK binds to apoptotic cells *via* PS, leading to cleavage and the production of bradykinin in the presence of FXII and PK (31). Therefore, PS on apoptotic cells may serve as a novel activator of FXII and a docking site for the assembly of the contact activation system, which may account for the role of FXII in apoptotic cell-mediated procoagulant activity.

This new function of FXII in apoptotic cell-mediated coagulation provides novel insight into the pathology of apoptotic cells. In pathological settings, such as autoimmune disease, chemotherapy, and inflammation, numerous cells frequently undergo apoptosis. Rapid clearance of these apoptotic cells is crucial for maintaining an anti-inflammatory and antithrombotic state (2). However, if these apoptotic cells are not removed efficiently, they may become procoagulant and proinflammatory. For example, in patients with acute leukemia, phagocytes are overwhelmed by the large numbers of apoptotic cells because of uncontrolled leukemic cell proliferation and cytotoxic chemotherapy. In patients with leukemia, coagulation parameters could be upregulated by chemotherapy; however, this cannot account for the hypercoagulable state, as the basal levels of contact activation system components, including FXII, are fairly high (>50 μg/mL). Patients with systemic lupus erythematosus (SLE) often develop microvessel thrombi, concomitant with the accumulation of apoptotic cells (5). The enhanced functional interaction of apoptotic cells with activation of coagulation system likely plays a major role. Kunzelmann et al. (32) demonstrated that malignant hematopoietic cells (HEL cells) trigger blood coagulation through phosphatidylserine exposure (32). When apoptosis occurs in a microenvironment in direct contact with circulating coagulation factors such as FXII, they may contribute to the initiation and enhancement of unique pro-coagulants on the apoptotic cell surface (33). In view of the increasing evidence for the occurrence of vascular cell apoptosis in the above pathological settings, it is important to characterize the mechanism underlying FXII-driven contact system activation when vascular cells become apoptotic. Because FXII can also activate plasminogen in the fibrinolytic pathway (34), how activated FXII integrates the intrinsic coagulation and fibrinolysis systems on the surface of apoptotic cells is an interesting topic for future investigation.

In conclusion, the current study demonstrates that FXII binds to apoptotic cells *via* PS, leading to activation of FXII. Activated FXII contributes to intrinsic tenase formation and blood clotting. This study not only revealed a novel mechanism underlying apoptotic cell-mediated procoagulant activity but also identified the apoptotic cell membrane as a new activation surface for the assembly and activation of the contact activation system. These findings suggest that the role of the contact activation system in apoptotic cell-related procoagulant events requires further investigation. Moreover, it will be worthwhile to explore whether inhibition of FXII could be a therapeutic strategy for the prevention and treatment of apoptotic cell-associated thrombotic diseases.

AUTHOR CONTRIBUTIONS

AY, FC, CH, JZ, and JD performed research, analyzed data. YL contributed critical reagent. RB and YW designed research and interpreted the data. YW wrote the paper.

ACKNOWLEDGMENTS

This work was supported by grants from the National Natural Scientific Foundation (91539122, 81770138, 30971491, 81270592, 81301534, and 81670133), the National Institute of Health (AR063290), and the Priority Academic Program Development of Jiangsu Higher Education Institutions (PAPD).

REFERENCES

1. Fuchs Y, Steller H. Programmed cell death in animal development and disease. *Cell* (2011) 147:742–58. doi:10.1016/j.cell.2011.11.045
2. Wu Y, Tibrewal N, Birge RB. Phosphatidylserine recognition by phagocytes: a view to a kill. *Trends Cell Biol* (2006) 16:189–97. doi:10.1016/j.tcb.2006.02.003
3. Bombeli T, Karsan A, Tait JF, Harlan JM. Apoptotic vascular endothelial cells become procoagulant. *Blood* (1997) 89:2429–42.
4. Flynn PD, Byrne CD, Baglin TP, Weissberg PL, Bennett MR. Thrombin generation by apoptotic vascular smooth muscle cells. *Blood* (1997) 89:4378–84.
5. Casciola-Rosen L, Rosen A, Petri M, Schlissel M. Surface blebs on apoptotic cells are sites of enhanced procoagulant activity: implications for coagulation events and antigenic spread in systemic lupus erythematosus. *Proc Natl Acad Sci U S A* (1996) 93:1624–9. doi:10.1073/pnas.93.4.1624
6. Wang J, Weiss I, Svoboda K, Kwaan HC. Thrombogenic role of cells undergoing apoptosis. *Br J Haematol* (2001) 115:382–91. doi:10.1046/j.1365-2141.2001.03095.x
7. Fu Y, Zhou J, Li H, Cao F, Su Y, Fan S, et al. Daunorubicin induces procoagulant activity of cultured endothelial cells through phosphatidylserine exposure and microparticles release. *Thromb Haemost* (2010) 104:1235–41. doi:10.1160/TH10-02-0102
8. Colman RW, Schmaier AH. Contact system: a vascular biology modulator with anticoagulant, profibrinolytic, antiadhesive, and proinflammatory attributes. *Blood* (1997) 90:3819–43.
9. Maas C, Renne T. Regulatory mechanisms of the plasma contact system. *Thromb Res* (2012) 129(Suppl 2):S73–6. doi:10.1016/j.thromres.2012.02.039
10. Björkqvist J, Nickel KF, Stavrou E, Renné T. In vivo activation and functions of the protease factor XII. *Thromb Haemost* (2014) 112:868–75. doi:10.1160/TH14-04-0311
11. Muller F, Renne T. Novel roles for factor XII-driven plasma contact activation system. *Curr Opin Hematol* (2008) 15:516–21. doi:10.1097/MOH.0b013e328309ec85
12. Renne T, Schmaier AH, Nickel KF, Blomback M, Maas C. In vivo roles of factor XII. *Blood* (2012) 120:4296–303. doi:10.1182/blood-2012-07-292094
13. Tibrewal N, Wu Y, D'Mello V, Akakura R, George TC, Varnum B, et al. Autophosphorylation docking site Tyr-867 in Mer receptor tyrosine kinase allows for dissociation of multiple signaling pathways for phagocytosis of apoptotic cells and down-modulation of lipopolysaccharide-inducible NF-κB transcriptional activation. *J Biol Chem* (2008) 283:3618–27. doi:10.1074/jbc.M706906200
14. Wu Y, Singh S, Georgescu MM, Birge RB. A role for Mer tyrosine kinase in alphavbeta5 integrin-mediated phagocytosis of apoptotic cells. *J Cell Sci* (2005) 118:539–53. doi:10.1242/jcs.01632
15. Yeung T, Gilbert GE, Shi J, Silvius J, Kapus A, Grinstein S. Membrane phosphatidylserine regulates surface charge and protein localization. *Science* (2008) 319:210–3. doi:10.1126/science.1152066
16. Martinez J, Almendinger J, Oberst A, Ness R, Dillon CP, Fitzgerald P, et al. Microtubule-associated protein 1 light chain 3 alpha (LC3)-associated phagocytosis is required for the efficient clearance of dead cells. *Proc Natl Acad Sci U S A* (2011) 108:17396–401. doi:10.1073/pnas.1113421108
17. van der Meijden PE, Munnix IC, Auger JM, Govers-Riemslag JW, Cosemans JM, Kuijpers MJ, et al. Dual role of collagen in factor XII-dependent thrombus formation. *Blood* (2009) 114:881–90. doi:10.1182/blood-2008-07-171066
18. Xie R, Gao C, Li W, Zhu J, Novakovic V, Wang J, et al. Phagocytosis by macrophages and endothelial cells inhibits procoagulant and fibrinolytic activity of acute promyelocytic leukemia cells. *Blood* (2012) 119:2325–34. doi:10.1182/blood-2011-06-362186

19. Geddings JE, Mackman N. Recently identified factors that regulate hemostasis and thrombosis. *Thromb Haemost* (2014) 111:570–4. doi:10.1160/TH13-10-0812
20. Smith SA, Morrissey JH. Polyphosphate: a novel modulator of hemostasis and thrombosis. *Arterioscler Thromb Vasc Biol* (2015) 35:1298–305. doi:10.1161/ATVBAHA.115.301927
21. Schoenwaelder SM, Yuan Y, Josefsson EC, White MJ, Yao Y, Mason KD, et al. Two distinct pathways regulate platelet phosphatidylserine exposure and procoagulant function. *Blood* (2009) 114:663–6. doi:10.1182/blood-2009-01-200345
22. Colman RW, Mattler L, Sherry S. Studies on the prekallikrein (kallikreinogen) – kallikrein enzyme system of human plasma. II. Evidence relating the kaolin-activated arginine esterase to plasma kallikrein. *J Clin Invest* (1969) 48:23–32. doi:10.1172/JCI105971
23. Ratnoff OD, Crum JD. Activation of Hageman factor by solutions of ellagic acid. *J Lab Clin Med* (1964) 63:359–77.
24. Nossel HL, Wilner GD, LeRoy EC. Importances of polar groups for initiating blood coagulation and aggregating platelets. *Nature* (1969) 221:75–6. doi:10.1038/221075a0
25. Paul A, Avci-Adali M, Neumann B, Guo K, Straub A, Dietz K, et al. Aptamers influence the hemostatic system by activating the intrinsic coagulation pathway in an in vitro Chandler-Loop model. *Clin Appl Thromb Hemost* (2010) 16:161–9. doi:10.1177/1076029608329580
26. Tans G, Griffin JH. Properties of sulfatides in factor-XII-dependent contact activation. *Blood* (1982) 59:69–75.
27. Renne T, Schuh K, Muller-Esterl W. Local bradykinin formation is controlled by glycosaminoglycans. *J Immunol* (2005) 175:3377–85. doi:10.4049/jimmunol.175.5.3377
28. Maas C, Govers-Riemslag JW, Bouma B, Schiks B, Hazenberg BP, Lokhorst HM, et al. Misfolded proteins activate factor XII in humans, leading to kallikrein formation without initiating coagulation. *J Clin Invest* (2008) 118:3208–18. doi:10.1172/JCI35424
29. Muller F, Mutch NJ, Schenk WA, Smith SA, Esterl L, Spronk HM, et al. Platelet polyphosphates are proinflammatory and procoagulant mediators in vivo. *Cell* (2009) 139:1143–56. doi:10.1016/j.cell.2009.11.001
30. Cheng Q, Tucker EI, Pine MS, Sisler I, Matafonov A, Sun MF, et al. A role for factor XIIa-mediated factor XI activation in thrombus formation in vivo. *Blood* (2010) 116:3981–9. doi:10.1182/blood-2010-02-270918
31. Yang A, Dai J, Xie Z, Colman RW, Wu Q, Birge RB, et al. High molecular weight kininogen binds phosphatidylserine and opsonizes urokinase plasminogen activator receptor-mediated efferocytosis. *J Immunol* (2014) 192:4398–408. doi:10.4049/jimmunol.1302590
32. Kunzelmann C, Freyssinet JM, Martínez MC. Rho A participates in the regulation of phosphatidylserine-dependent procoagulant activity at the surface of megakaryocytic cells. *J Thromb Haemost* (2004) 2:644–50. doi:10.1111/j.1538-7836.2004.00688.x
33. Romero-Díaz J, García-Sosa I, Sánchez-Guerrero J. Thrombosis in systemic lupus erythematosus and other autoimmune diseases of recent onset. *J Rheumatol* (2009) 36:68–75. doi:10.3899/jrheum.071244
34. Mitchell JL, Lionikiene AS, Georgiev G, Klemmer A, Brain C, Kim PY, et al. Polyphosphate colocalizes with factor XII on platelet-bound fibrin and augments its plasminogen activator activity. *Blood* (2016) 128:2834–45. doi:10.1182/blood-2015-10-673285

Conflict of Interest Statement: The authors declare that the research was conducted in the absence of any commercial or financial relationships that could be construed as a potential conflict of interest.

12

Elevated Serum Lysophosphatidylcholine in Patients with Systemic Lupus Erythematosus Impairs Phagocytosis of Necrotic Cells *In Vitro*

Gerhard E. Grossmayer[1], Hildegard Keppeler[2], Sebastian Boeltz[1], Christina Janko[1,3], Jürgen Rech[1], Martin Herrmann[1], Kirsten Lauber[4†] and Luis E. Muñoz[1†]*

[1] Department of Internal Medicine 3 – Rheumatology and Immunology, Universitätsklinikum Erlangen, Friedrich-Alexander-Universität Erlangen-Nürnberg, Erlangen, Germany, [2] Department of Internal Medicine II, University of Tübingen, Tübingen, Germany, [3] Department of Otorhinolaryngology, Head and Neck Surgery, Universitätsklinikum Erlangen, Else Kröner-Fresenius-Stiftung Professorship, Erlangen, Germany, [4] Department of Radiation Oncology and Radiotherapy, University Hospital, Ludwig-Maximilians-University, Munich, Germany

Edited by:
Uday Kishore,
Brunel University London,
United Kingdom

Reviewed by:
Angelo A. Manfredi,
Vita-Salute San Raffaele University,
Italy
Luuk Hilbrands,
Radboud University Nijmegen,
Netherlands

***Correspondence:**
Kirsten Lauber
kirsten.lauber@med.uni-muenchen.de

†These authors have contributed equally as senior authors.

Objectives: Impaired clearance of dying and dead cells by professional and amateur phagocytes plays a crucial role in the etiology of systemic lupus erythematosus (SLE). While dying, cells expose and release a plethora of eat-me and find-me signals to ensure their timely removal before entering the dangerous stage of secondary necrosis. A well-described chemoattractant for macrophages is dying cell-derived lysophosphatidylcholine (LPC). However, its implications for and/or its association with SLE disease, so far, have not been examined. In the present study, we analyzed the LPC serum concentrations of patients with SLE and rheumatoid arthritis (RA). Subsequently, we examined if and to which extent the measured serum concentrations of LPC and an LPC-rich environment can impact the phagocytosis of necrotic cells.

Methods: Sera from patients with SLE, RA, and normal healthy donors (NHD) were characterized for several parameters, including LPC concentrations. Phagocytosis of dead cells by human macrophages in the presence of SLE and NHD sera was quantified. Additionally, the impact of exogenously added, purified LPC on phagocytosis was analyzed.

Results: Patients with SLE had significantly increased LPC serum levels, and high serum LPC of SLE patients correlated significantly with impaired phagocytosis of dead cells in the presence of heat-inactivated serum. Phagocytosis in the presence of sera from NHD showed no correlation to LPC levels, but exogenous addition of purified LPC in the range as measured in SLE patients' sera led to a concentration-dependent decrease.

Conclusion: Our data show that high levels of LPC as observed in the sera of SLE patients have a negative impact on the clearance of dead cells by macrophages. Chemoattraction requires a concentration gradient. The higher the LPC concentration surrounding a dying or dead cell, the smaller the achievable gradient upon LPC release will be. Thus, it is feasible to assume that elevated LPC levels can interfere with the build-up of a local LPC gradient during cell death, and hence might play a role in the establishment and/or perpetuation of SLE disease.

Keywords: systemic lupus erythematosus, rheumatoid arthritis, efferocytosis, lysophosphatidylcholine, phagocytosis, necrotic cells, serum, clearance deficiency

INTRODUCTION

The process of apoptotic cell clearance—commonly referred to as efferocytosis—works with an impressively high degree of effectiveness and eliminates billions of dying cells in the context of tissue homeostasis and cell turnover every day (1). It is so efficient that, under physiological conditions, apoptotic corpses are rarely to be found. Defects and perturbations in this highly coordinate process result in the persistence of cellular debris and have been linked to the etiology and pathogenesis of systemic autoimmune diseases such as systemic lupus erythematosus (SLE) (2–5). Chronic, generalized inflammation, and multi-organ damage are the most typical manifestations of SLE (6).

It is widely accepted that uncleared secondary NEcrotic cell-derived material (SNEC) in germinal centers of secondary lymphatic organs serves as survival and proliferation signal for autoreactive B cells, which will then produce antinuclear antibodies (ANAs) in the periphery (7, 8). The co-existence of ANAs and accumulated SNEC leads to the formation of immune complexes in several tissues (9). This favors a pathological, pro-inflammatory mode of elimination leading to increased organ damage and accumulation of more dying cell debris, thus fueling a self-amplifying vicious circle (10). The result is a systemic type I interferon signature in response to the ANA–SNEC complexes analogous to a viral infection (3, 11).

Clinically, SLE presents as a very heterogeneous autoimmune disease with a broad range of manifestations. The prevalence of 20–150 cases per 100,000 has remained constant in the US in the last 20 years (12–14). Adult women are affected seven times more often than men and their life expectancy is considerable lower than that of the healthy population (15). Typically, the course of disease is characterized by remissions and relapses ranging from mild to severe with the latter ones also being known as flares (16). The classification as SLE of a patient with autoimmune manifestations is usually done by rheumatologists after meeting at least 4 out of 13 clinical and serological criteria of the American College of Rheumatology (17). Although efferocytosis is closely related with the etiopathogenesis of SLE, none of these criteria so far are associated with the clearance of dead and dying cells.

In order to attract phagocytes for their timely removal, dying cells release soluble find-me signals. Apart from nucleotides, the chemokine CX_3CL1, sphingosine-1-phosphate, and others, lysophosphatidylcholine (LPC) was one of the first dying cell-derived find-me signals that has been identified (3, 18, 19). We and others have shown that the release of LPC during apoptosis is orchestrated by caspase-3-mediated cleavage and activation of calcium-independent phospholipase A_2 ($iPLA_2$), and subsequent involvement of the ATP-binding-cassette transporter ABCA1 (18, 20). Once released, LPC stimulates chemotaxis of monocytes and macrophages with involvement of the G protein-coupled receptor G2A (21–24). Intriguingly, mice with genetic deletion of G2A develop a late-onset multi-organ autoimmune condition, which closely resembles human SLE (25), suggesting that interference with the LPC-G2A axis—presumably *via* disturbed phagocyte recruitment during efferocytosis—favors the development of autoimmunity. Therefore, the question arises if patients with autoimmune diseases display alterations in the LPC–G2A axis. To date, this has not been studied systematically. Previously, we performed transcriptomic profiling analyses in peripheral blood monocytes of SLE patients. G2A was among the examined candidate genes, but we did not detect any significant alterations in the SLE group in comparison to normal healthy donors (NHD) (26).

In the present study, we focused on LPC and analyzed the corresponding serum levels in SLE patients, patients with rheumatoid arthritis (RA), and NHD. We observed significantly increased LPC serum levels in the SLE group and particularly high levels in patients with involvement of vessels or kidneys, respectively. Since this can be a cause as well as a consequence of impaired efferocytosis, we next examined dead cell phagocytosis in the presence of SLE sera and in the presence of NHD sera supplemented with LPC concentrations that have been observed in the SLE group. A significant correlation of elevated LPC levels with impaired phagocytosis of dead cells in the presence of heat-inactivated SLE sera was detected and further confirmed by exogenous addition of purified LPC to NHD sera. Accordingly, we conclude that increased LPC levels can interfere with local dead cell-derived LPC, which functions as a phagocyte attraction signal. This might contribute to the impairment of clearance and consequently to the establishment and/or maintenance of SLE disease.

PATIENTS AND METHODS

Patients

We obtained sera from patients with SLE visiting our inpatient and outpatient departments. Demographic and clinical data available in 50 patients with SLE were summarized together with routine serology in **Tables 1** and **2**. Serologic parameters were measured in our routine laboratory immediately after blood drawing. Disease activity was recorded according to the SLE

TABLE 1 | Demographics of the cohorts studied.

Demographics	Normal healthy donors	Rheumatoid arthritis	Systemic lupus erythematosus
n	25	20	50
Sex M/F	13/12	13/7	5/45
Mean age (years)	31.5 ± 7.3	49.8 ± 17.8	43.2 ± 14.0

Disease Activity Index 2000 (SLEDAI-2K) or to the European Consensus Lupus Activity Measurement (ECLAM). All patients showing disease activity defined as a SLEDAI score of at least 5 or as an ECLAM score of at least 4 were considered as in flare. Organ involvement was determined by the affection of a specific system due to lupus pathology as ascertained by the treating physicians. Sera from further 41 patients with SLE, from 20 patients with RA, and from 25 NHD were used for the measurement of LPC levels (no clinical data available). All patients fulfilled the American College of Rheumatology criteria for the classification of SLE and RA, respectively (17, 27). A written informed consent was obtained from all blood donors and the study received the final approval from the ethics committee of the Friedrich-Alexander University Erlangen (Permit number 52_14 B).

TABLE 2 | Serological and clinical characteristics of the systemic lupus erythematosus cohort.

	In flare	No flare	*p*
Serology			
n	12	33	
CRP, median in mg/dl (IQR)	1.2 (1.7)	0.4 (1.5)	0.189[a]
C3, median in mg/dl (IQR)	72.5 (37.5)	93.0 (63.0)	0.098[a]
C4, median in mg/dl (IQR)	10.7 (10.2)	14.2 (10.2)	**0.038**[a]
Anti-dsDNA, median in U/ml (IQR)	33.8 (140.2)	11.8 (44.5)	0.097[a]
Antinuclear antibody titer, median^{-1} (IQR)	210.0 (280.0)	320.0 (980.0)	0.518[a]
APS, *n* (%)	4 (33.3)	6 (17.1)	0.237[b]
Treatment			
n	12	35	
Prednisone median in mg/day (IQR)	7.5 (39.4)	5.0 (10.0)	0.258[a]
Azathioprine, *n* (%)	3 (25.0)	15 (42.9)	0.272[b]
Chloroquine, *n* (%)	4 (33.3)	13 (37.1)	0.813[b]
Cyclophosphamide, *n* (%)	2 (16.7)	2 (5.7)	0.241[b]

IQR, interquartile range; APS, seropositive antiphospholipid syndrome.
[a]*Wilcoxon–Mann–Whitney test.*
[b]*Chi square test.*
Bold font indicates $p < 0.05$.

Handling of Serum Samples

Thirty nine serum samples from patients with SLE and 14 from NHD were used for phagocytosis assays in the presence of native serum. One aliquot of each serum was treated with heat (56°C) for 30 min for complement inactivation (28). Native and heat-inactivated samples were stored together at −20°C from the day of blood withdrawal and thawed once immediately before each phagocytosis assay. Additional 15 samples from patients with SLE and 4 from NHD were available for phagocytosis assays only in form of heat-inactivated serum.

Measurement of Serum LPC

Serum LPC concentrations were measured by a multi-step coupled enzymatic assay with slight modifications according to the protocol described (29). LPC was converted stepwise to glycerophosphocholine, choline, and betaine, which eventually was detected in a peroxidase reaction with Amplex Red (**Figure 1**).

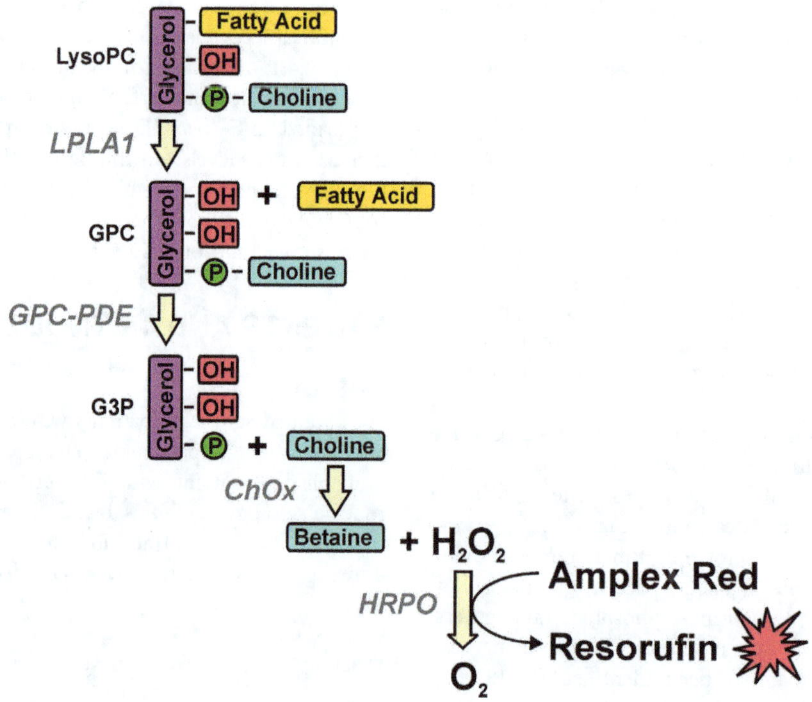

FIGURE 1 | Multi-step coupled enzymatic test for the detection of lysophosphatidylcholine (LPC). LPC is stepwise converted to glycerophosphocholine, choline, and betaine, which eventually is detected in a peroxidase reaction with Amplex Red. Methodologically, baseline fluorescence is recorded by incubating the samples with all ingredients except for the first enzyme (lysophospholipase A1) whose addition defines the start of LPC conversion and the value of fluorescence, which is subtracted from the final value of fluorescence for calculating the amount of LPC on the basis of a standard curve of purified LPC.

Briefly, 20 µl of diluted sera (1:100) were added to 96-well plates, 250 µl of prewarmed mix A consisting of 100 mM Tris–HCl pH 8.0, 0.01% Triton X-100, 1 mM calcium chloride, 20 µM Amplex Red, 1 U/ml horseradish peroxidase, 0.1 U/ml glycerophosphocholine phosphodiesterase, 1 U/ml choline oxidase (all from Sigma-Aldrich, Taufkirchen, Germany) were added, and baseline fluorescence was recorded for 15 min at 37°C in a Synergy Mx plate reader (BioTek Instruments, Bad Friedrichshall, Germany) at 563 nm excitation and 587 nm emission. Then, the reaction was started by adding 75 µl of mix B consisting of 100 mM Tris–HCl pH 8.0, 0.01% Triton X-100, and 20 µg/ml lysophospholipase A1 (cloned and purified from human THP-1 cells), and Amplex Red conversion was recorded over 40 min every minute. LPC concentrations were determined from the delta in fluorescence with a standard curve of purified LPC (Sigma-Aldrich).

Necrotic Cells

Necrotic Raji cells served as target cells in our phagocytosis experiments. Viable Raji cells were labeled with carboxyfluorescein succinimidyl ester-diacetate (CFSE-DA, Molecular Probes, Eugene, OR, USA) 1 day prior to the phagocytosis experiments according to manufacturer instructions and kept at 37°C in 5.5% CO_2 atmosphere in complete medium [RPMI 1640 medium containing 1% L-glutamine (Thermo Fisher Scientific, Munich, Germany)], 1% *N*-2-hydroxyethylpiperazine-*N'*-2-ethanesulfonic acid (HEPES; Merck, Darmstadt, Germany), and 1% penicillin–streptomycin (Thermo Fisher Scientific) supplemented with 10% fetal calf serum (FCS, Thermo Fisher Scientific). Necrosis was induced immediately before phagocytosis by incubation at 56°C for 30 min. Verification of necrosis was done by propidium iodide staining in phosphate-buffered saline (PBS) (Thermo Fisher Scientific). More than 99% of the cells were positive for propidium iodide.

In Vitro Differentiation of Human Monocyte-Derived Macrophages (HMDM)

Peripheral blood mononuclear cells (PBMC) of three healthy volunteers (blood group 0) were isolated from venous heparinized blood by Ficoll density-gradient centrifugation (Lymphoflot; Biotest, Dreieich, Germany). Platelets were depleted by centrifugation through a cushion of FCS. Monocytes were separated from PBMC by magnetic cell sorting employing human CD14 MicroBeads following the manufacturer's instructions (Miltenyi Biotec, Bergisch-Gladbach, Germany) and stained with the carbocyanine membrane dyes DiI or DiD (Thermo Fisher Scientific). Monocytes were differentiated to HMDM *in vitro* either at 50,000 cells/well (assays with patients' sera) or 100,000 cells/well (LPC interference assay) for 7 days at 37°C and 5.5% CO_2 humidified atmosphere in complete medium supplemented with 10% autologous serum and 10 ng/ml granulocyte-macrophage colony-stimulating factor (GM-CSF; Pharmaserv, Marburg, Germany). Fresh medium was added at days 3 and 5 after isolation.

Phagocytosis of Necrotic Cells

Medium from wells containing stained HMDM was removed and replaced by 100,000 necrotic Raji cells suspended in 200 µl complete medium containing 10% FCS, 10% serum samples from SLE patients or healthy controls, 10% autologous serum, and varying concentrations of LPC (Sigma-Aldrich), respectively. Medium containing up to 0.7% ethanol as LPC-vehicle was used as control in phagocytosis assays with added LPC. Phagocytosis was allowed for 1 h at 37°C in 5.5% CO_2 atmosphere. Adherent HMDM were detached by rinsing with cold PBS containing 2 mM ethylenediamine tetraacetic acid (EDTA; Merck).

Flow Cytometry and Data Management

The uptake of necrotic cells by HMDM was monitored by two-color flow cytometry. HMDM were identified by size and by positive staining for DiI or DiD. HMDM being double-positive for CFSE and DiI or DiD, respectively, were considered to have ingested necrotic cells, which was further confirmed by confocal laser microscopy as reported previously (30). The phagocytosis index (PhIx) was calculated as the standardized product of the percentage of double-positive HMDM and the mean CFSE fluorescence intensity (relative amount of engulfed material) of double-positive HMDM. Experiments with patients' sera were carried out in duplicates and assays with exogenously added LPC in triplicates. Standardization enabled fusion of results from different days and dyes and was achieved by normalizing the measured PhIx to phagocytosis in the absence of human serum of each experiment. In the LPC titration assays, normalization was done by defining the PhIx of the LPC-free samples as 100%.

Statistical Analyses

Statistical analyses were performed using Prism 5 for Windows (GraphPad Software, Inc., La Jolla, CA, USA) or SPSS Statistics 21 (IBM Corp., Armonk, NY, USA), respectively. Correlations were calculated according to the non-parametric Spearman algorithm. For comparisons between groups, normal distribution was confirmed by Kolmogorov–Smirnov tests followed by two-sided Student's *t*-tests. If normality was rejected, group comparisons were performed by two-sided Wilcoxon–Mann–Whitney tests. Independence of variables was examined by Chi square tests. Levels of $p < 0.05$ were considered as statistically significant.

RESULTS

In order to examine if SLE patients display alterations in LPC serum levels and if this might be of functional relevance for prey cell engulfment, we established a multi-step enzymatic test for the detection of LPC and applied the sera to *in vitro* phagocytosis assays with HMDM and dead Raji cells.

SLE Patients Exhibit Increased Levels of Serum LPC

The mean LPC serum concentration in the group of NHD as measured by the multi-step coupled enzymatic test was 886 ± 493.3 µM. In comparison, patients with SLE had significantly increased LPC serum levels (1,298 ± 687.9 µM; $p = 0.0132$; after Dunn's correction $p < 0.05$). No significant differences were observed when comparing RA patients (953.1 ± 233.4 µM) with NHD or with SLE patients, respectively (**Figure 2**). To our knowledge, this is the first detailed report about concentrations of LPC in sera of patients with SLE.

Demographical, Serological, and Clinical Characteristics of the SLE Cohort

Clinical and serological information about SLE patients are shown in **Tables 1** and **2**. Considering the actual clinical activity status of the patients, we observed that patients in flare (SLEDAI score ≥5 or ECLAM score ≥4, **Table 2**) showed significantly lower levels of serum complement C4 (in flare median: 10.7 mg/dl; no flare median 14.2 mg/dl; $p = 0.038$). The concentration of serum LPC was slightly elevated in patients with flares, but this difference did not reach statistical significance (not shown). We observed particularly high LPC serum levels in patients with vessel ($p = 0.036$) and kidney ($p = 0.030$) involvement (**Table 3**). Unfortunately, no complete data on renal function were available. Five out of 16 patients with renal involvement showed severely compromised renal function as manifested by high protein levels in urine and creatinine clearance lower than 50 ml/min (not shown), but no significant association with LPC serum levels was detected. Interestingly, serum LPC levels correlated significantly and inversely ($p = 0.014$) with antinuclear antibody (ANA) titers in patients with SLE (**Figure 3**). No other statistical association with laboratory parameters (**Tables 1** and **2**) or prednisone treatment was found (not shown).

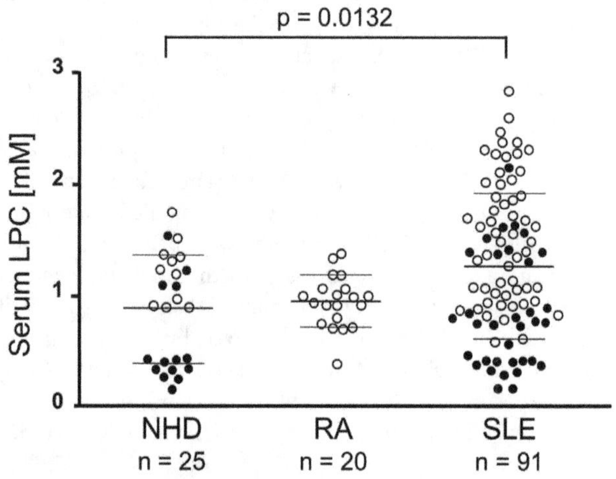

FIGURE 2 | Lysophosphatidylcholine (LPC) serum levels of patients with systemic lupus erythematosus (SLE) are elevated. In serum samples of normal healthy donors (NHD), patients with rheumatoid arthritis (RA) and SLE LPC concentrations were analyzed by a multi-step coupled enzymatic test as depicted in **Figure 1**. LPC levels from individual serum samples, their means, and SDs are shown. Open symbols (only LPC values available), filled symbols (LPC values and phagocytosis index available), bold bars (cohort means), thin bars (cohort SDs).

Destruction of Heat-Labile Serum Components Reveals the Distinct Effect of Thermostable Factors on Phagocytosis of Necrotic Cells *In Vitro*

In a first step to address the potential association of serum LPC levels with the process of dying cell clearance and SLE disease, we analyzed the effects of NHD and SLE sera in their native and heat-inactivated form on the phagocytosis of necrotic cells by HMDM (**Figure 4**). The uptake of necrotic cells was quantified by the PhIx. Phagocytosis in the presence of native sera from NHD (13.9 ± 4.3) was similar to that in the presence of native sera from patients with SLE (13.2 ± 4.7). In order to eliminate the influence of heat-labile serum factors on phagocytosis and to focus on the effects of thermally stable components (like LPC), sera were heat inactivated at 56°C for 30 min. This strongly reduced the uptake of necrotic cells in comparison to native serum in the NHD ($p < 0.0001$) as well as the SLE ($p < 0.0001$) cohort,

TABLE 3 | Serum concentrations of LPC and phagocytosis of necrotic cells in inactive serum stratified by organ involvement in patients with systemic lupus erythematosus.

Organ	Involvement	*n* (%)	LPC (μM) Mean ± SD	*p*	PhIx Mean ± SD	*p*
Vessels	Yes	6 (12.5)	**1,561.7 ± 519.0**	**0.036**	295.1 ± 104.4	0.051
	No	42 (87.5)	**952.2 ± 660.4**		427.8 ± 190.1	
Kidney	Yes	16 (33.3)	**1,322.4 ± 715.5**	**0.030**	371.3 ± 166.9	0.137
	No	32 (66.7)	**881.3 ± 607.0**		433.0 ± 194.8	
APS	Yes	10 (20.0)	1,091.1 ± 653.4	0.760	427.1 ± 119.8	0.386
	No	40 (80.0)	1,013.9 ± 672.8		407.6 ± 200.8	
Skin	Yes	25 (52.1)	1,107.9 ± 648.2	0.398	414.2 ± 178.2	0.459
	No	23 (47.9)	941.9 ± 699.3		408.6 ± 198.9	
Joints	Yes	18 (37.5)	998.1 ± 644.5	0.691	406.1 ± 192.1	0.351
	No	30 (62.5)	1,078.9 ± 729.7		414.5 ± 186.1	
Leukopenia	Yes	14 (29.2)	770.0 ± 490.4	0.088	477.2 ± 197.8	0.060
	No	34 (70.8)	1,134.6 ± 712.1		385.4 ± 177.5	
Central nervous system	Yes	6 (12.5)	945.7 ± 812.6	0.751	422.2 ± 130.2	0.448
	No	42 (87.5)	1,040.2 ± 659.4		410.4 ± 192.7	
Serosal membranes	Yes	11 (22.9)	926.2 ± 529.0	0.571	455.8 ± 247.3	0.189
	No	37 (77.1)	1,058.7 ± 711.5		398.8 ± 166.5	

LPC, lysophosphatidylcholine; PhIx, phagocytic index; APS, seropositive anti-phospholipid syndrome.
Bold font indicates p < 0.05

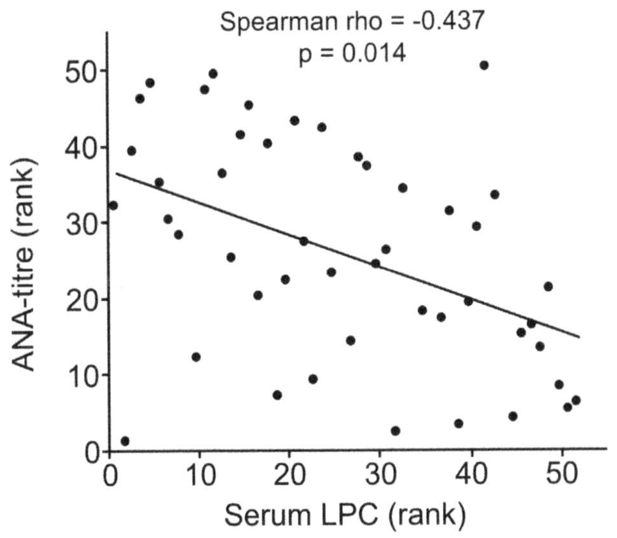

FIGURE 3 | Lysophosphatidylcholine (LPC) serum levels of patients with systemic lupus erythematosus inversely correlated with antinuclear antibody (ANA) titers. Correlation analysis of serum LPC values and ANA titers. The values of ANA titers and serum LPC concentrations were transformed to rank values, and the Spearman rho correlation coefficient was calculated ($p = 0.014$ after Bonferroni correction).

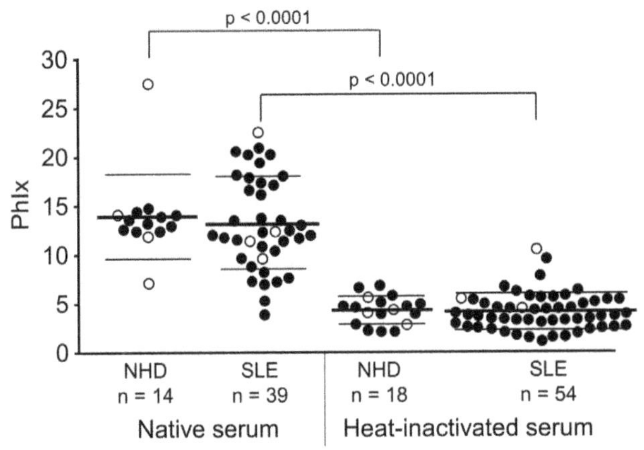

FIGURE 4 | Influence of heat-labile serum components on the phagocytosis of necrotic cells by human monocyte-derived macrophages. Uptake of necrotic cells was analyzed by flow cytometry and quantified by the phagocytosis index (PhIx). Phagocytosis was allowed in the presence of native or heat-inactivated (56°C for 30 min) serum from normal healthy donors (NHD) or patients with systemic lupus erythematosus (SLE). Phagocytosis in heat-inactivated serum was lower in both cohorts than in native serum (NHD and SLE each $p < 0.0001$). There was no significant difference in phagocytosis between the NHD and SLE cohort neither in native serum nor in heat-inactivated serum. Open symbols (only PhIx available), filled symbols (PhIx and lysophosphatidylcholine concentration available), bold bars (cohort means), thin bars (cohort SDs).

respectively—most likely due to inactivation of complement and other heat-sensitive opsonins. Again, addition of inactivated sera from NHD (4.3 ± 1.5) and SLE patients (4.1 ± 1.9) resulted in similar levels of phagocytosis, and no significant difference in PhIx was observed.

Serum LPC Levels of SLE Patients Negatively Correlate with Phagocytosis of Necrotic Cells *In Vitro*

In the next step, we analyzed the relationship between serum LPC concentrations and phagocytosis as quantified by the PhIx. No significant correlation was observed in case of native sera, neither in the NHD (not shown) nor in the SLE group (**Figure 5A**). In line with our previous findings and reports by others, complement C4 and ANA titers revealed significant correlations with phagocytosis in native serum (not shown) (30). Upon heat inactivation, the impact of thermosensitive components, including complement, is eliminated, and the influence of thermostable serum factors on the engulfment process can be examined. LPC belongs to the group of thermostable serum components due to its lipid nature. Whereas phagocytosis in the presence of heat-inactivated serum from NHD showed no significant association with LPC concentrations (not shown), a highly significant negative correlation ($r = -0.556$, $p < 0.0001$) between LPC levels and phagocytosis in case of heat-inactivated serum from patients with SLE was observed (**Figure 5B**).

We also performed group comparisons for all organ involvement groups. Sera from patients with kidney or vessel involvement with particularly high LPC levels displayed a tendency toward reduced phagocytosis activity. However, this did not reach statistical significance. In turn, sera from patients with leukopenia showed particularly low LPC levels and a trend toward increased phagocytosis activity (**Table 3**). This underlines the negative relationship between LPC serum concentration and phagocytosis activity and raises the question of causality.

Exogenously Added LPC Reduces Phagocytosis of Necrotic Cells *In Vitro*

As a first approach to analyze the causality between high serum LPC levels and decreased uptake of necrotic cells, NHD serum was supplemented with increasing concentrations of purified LPC (in the range that was measured in the SLE sera). **Figure 6** compares the phagocytosis results of LPC supplementation versus the corresponding vehicle control. All PhIxs were normalized to the PhIx value of the original serum (added LPC concentration = 0 mM). Addition of LPC resulted in a significant concentration-dependent decline in phagocytosis in native serum ($p < 0.0001$) as well as in heat-inactivated serum ($p = 0.0027$) in comparison to the LPC-free control (**Figures 6A,B**). The phagocytosis of necrotic cells was similarly inhibited by LPC to 35% in native and to 34% in heat-inactivated serum. The vehicle control showed no influence on phagocytosis, neither in native nor in heat-inactivated serum.

DISCUSSION

Billions of cells die in the context of normal tissue regeneration every day. If they are not cleared in time, the process of death enters the stage of secondary necrosis and SNEC accumulate in

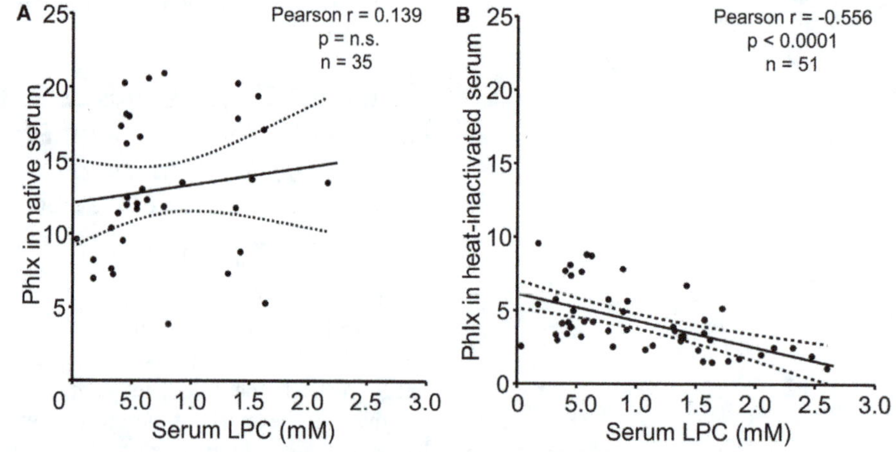

FIGURE 5 | Serum lysophosphatidylcholine (LPC) levels of systemic lupus erythematosus (SLE) patients correlate negatively with phagocytosis of necrotic cells *in vitro*. Serum LPC levels of SLE patients were correlated to phagocytosis of necrotic cells in the presence of the respective serum as determined by the phagocytosis index (PhIx). Each data point represents an individual serum value. **(A)** Phagocytosis assays in the presence of native serum. n.s., not significant. **(B)** Phagocytosis assays in the presence of heat-inactivated serum. Dashed lines indicate the 95% confidence interval.

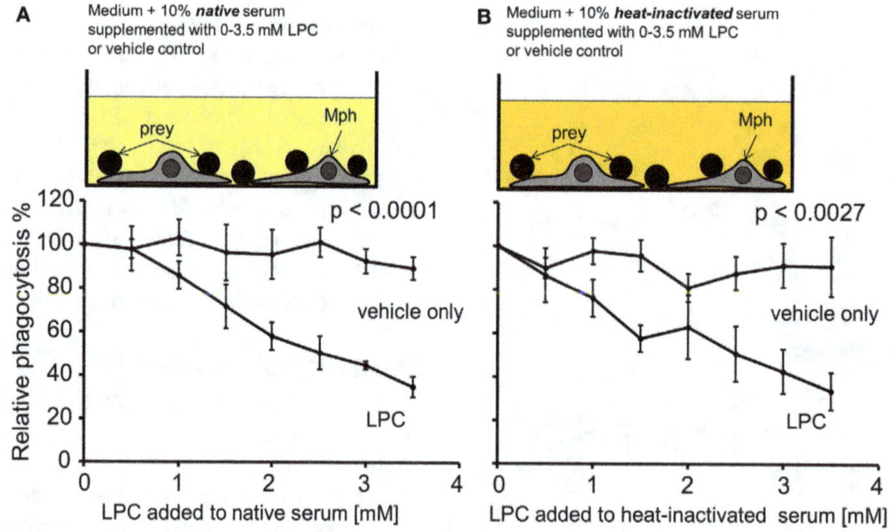

FIGURE 6 | Exogenously added lysophosphatidylcholine (LPC) reduces phagocytosis of necrotic cells *in vitro*. Normal healthy donor serum was supplemented with the indicated concentrations of purified LPC, added to phagocytosis assays with necrotic cells in its native or heat-inactivated form, and phagocytosis was measured as % of unsupplemented serum (100% control). **(A)** Data for native serum. **(B)** Data for heat-inactivated serum. The addition of LPC resulted in a significant concentration-dependent decline in phagocytosis in comparison to the LPC-free control (two-way ANOVA test). Means ± SEs of triplicates are shown.

peripheral tissues (3). Pioneering experimental evidence for the existence of this kind of non-engulfed cellular debris in tissue samples from patients with SLE was already provided in the late 1950s (31). Several studies reported so-called hematoxylin bodies defined as exclusively extracellular, single, clustered, or coalescent corpses found specifically in tissues of 70–90% of virtually untreated patients with SLE (32–35). Regrettably, at that time, these alterations could not be associated with cell death phenomena, because apoptosis was firstly described much later (36). More recently, increased amounts of apoptotic PBMC and neutrophils have been reported in patients with SLE (37, 38). The process of apoptosis itself turned out not to be affected (39). Instead, an engulfment dysfunction was identified as the underlying reason for the accumulation of uncleared, dead cell material in patients with SLE (40).

In the course of apoptosis, the plasma membrane of cells undergoes characteristic changes. Whereas viable cells exhibit a clear membrane phospholipid asymmetry, during apoptosis, redistribution of phospholipids between the membrane leaflets occurs (41, 42). Moreover, phospholipases (PLA_2) generate LPC by hydrolyzing the fatty acid at the sn-2 position of glycerophospholipids (43–45). Calcium-independent phospholipase A2 ($iPLA_2$) is processed during apoptosis by caspase-3-mediated cleavage leading to a truncated form with increased activity (46), and LPC appears to be the major product of lipid catabolism in apoptotic cells (45, 47). Its release under participation of the membrane transporter ABCA1 and its role as chemoattractant in the clearance of dead cells has been confirmed by us and others (18, 21, 24).

In the present study, we observed increased LPC levels in sera of patients with SLE, and the highest levels were detected in patients with vessel or kidney involvement (**Figure 2**). In principle, elevated serum LPC levels could be a cause as well as a consequence of SLE disease. It is feasible to assume that accumulating dying cell material due to known clearance deficiencies in SLE could give rise to accumulation of LPC in organs and tissues, which might finally "spill over" into the serum. This might be particularly strong in case of vasculitis since, here, cell death occurs in very close proximity to the blood. In case of kidney involvement, renal dysfunction and impaired lipid clearance might additionally contribute to the increase in LPC serum levels as has been reported in the context of type II diabetes and renal transplantation (48, 49). Vice versa, elevated LPC serum levels could affect dying cell clearance by interfering with proper phagocyte recruitment leading to delays in dying cell clearance and establishment as well as maintenance of SLE disease.

Lysophosphatidylcholine is a potent find-me signal of dying cells (50–53), provided that a concentration gradient between the dying/dead cell and the phagocyte can be established. Experimental neutralization of this gradient by adding exogenous LPC can efficiently interfere with chemoattraction of THP1-monocytes by supernatants of apoptotic cells (21), and sera from patients with SLE reportedly reduce chemoattraction of macrophages by supernatants of apoptotic cells (54). In contrast, NHD sera or sera from RA patients show no inhibition of phagocyte migration (54). Hence, the elevated LPC levels in SLE sera reported here might be the reason for the inhibition of chemoattraction in our previous study (54). Along the same lines, we now observed a significant correlation between high LPC levels in SLE sera and interference with phagocytosis of dead cells, but only in case of heat-inactivated sera, where the influence of ANAs, antiphospholipid antibodies, complement factors, and other heat-labile opsonins has been minimized (**Figure 5B**). ANAs, antiphospholipid antibodies, and complement factors are well-known to affect the internalization of dying cells, and their titers can vary strongly in sera of SLE patients (55–58). In turn, heat-inactivation of serum allows studying the phagocytic impact of heat-stable compounds, including LPC. This might be specifically relevant in tissues where the concentration of serum proteins is low. It should be noted that we observed this correlation between high LPC serum levels and interference with phagocytosis of dead cells only in SLE but not in NHD sera. Further subgroup analyses revealed that the correlation was particularly strong for LPC serum levels >1 mM (not shown). This might explain why no significant correlation for NHD sera (mean LPC concentration 886 ± 493.3 μM) was observed and suggests that a certain type of threshold does exist in this regard. Detailed analyses of the groups with different organ involvements revealed trends of high serum LPC levels and low phagocytosis activity in case of kidney and vessel involvement as well as low serum LPC level and high phagocytosis activity in case of leukopenia. Although these trends did not reach statistical significance, they argue in favor of a negative relationship between increased serum LPC concentrations and phagocytic activity.

Evidence for the causal involvement of LPC in phagocytosis inhibition derives from our add-in experiments: NHD serum supplemented with increasing concentrations of purified LPC hampered dead cell phagocytosis in an LPC concentration-dependent manner irrespective if it was heat-inactivated or not (**Figure 6**). Of note, the add-in phagocytosis experiments were performed with NHD serum which—in contrast to SLE sera—is of rather constant composition and lacks relevant titers of autoantibodies. This represents a plausible explanation why addition of LPC to native and inactivated NHD serum had similar effects on phagocytosis, while the correlation between LPC levels in SLE sera and phagocytic activity was only measurable upon heat inactivation.

Interestingly, we observed higher LPC serum levels in patients with flare than patients with non-flare. Although this trend was statistically not significant, it might be speculated that elevated levels of LPC function as a clearance inhibitor at a cellular level and lead to SLE disease exacerbation ensuing from reduced efferocytosis, augmented release of autoantigens, and immune complex formation, and resulting in enforced inflammatory conditions and clinical symptoms.

Eat-me signals presented on the membrane surface require a direct cell–cell contact in order to be functional. In contrast, find-me signals are released from dying and dead cells and operate over a certain distance [reviewed in Ref. (59)]. The existence of a chemotactic concentration gradient is a prerequisite for the attraction and migration of phagocytes. Under non-pathological conditions (as is the case in the presence of NHD sera), local gradients of find-me signals can be successfully built up around a dying cell (**Figure 7A**). Macrophages follow the gradient until they encounter dying cells and then can recognize and engulf them *via* exposed eat-me signals. In conditions of abnormally high LPC concentrations in the extracellular milieu (as we observed in SLE sera and upon supplementation of NHD sera in our phagocytosis assays), the LPC amount that is locally released by dying and dead cells appears not to be sufficient to establish an effective chemotactic gradient (**Figure 7B**). Consequently, macrophages cannot follow the gradient and do not find their way to the dying prey. Our data suggest that impaired chemoattraction decreases the likelihood of macrophages meeting dead cells, thus resulting in the observed decrease in phagocytosis.

Overall, our study is the first to provide experimental evidence for increased serum LPC levels in SLE patients. *In vitro*, this correlated with impaired phagocytosis of dead cells when the respective SLE sera were added to phagocytosis assays and when NHD sera were supplemented with LPC concentrations in the range as measured in SLE sera. Accordingly, we suggest that high extracellular LPC levels can neutralize local LPC gradients, which are required for phagocyte attraction in the context of

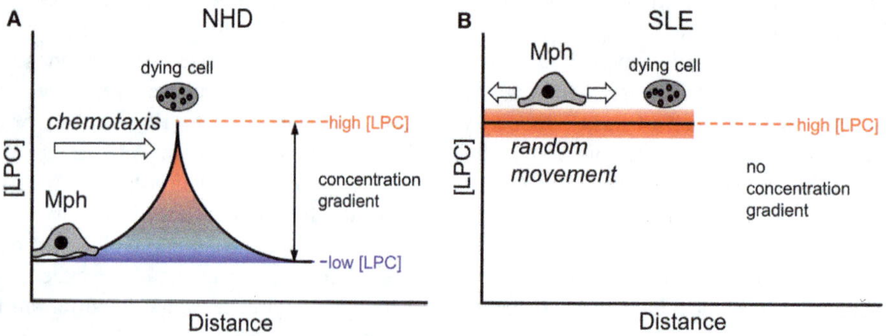

FIGURE 7 | Effect of abnormal high lysophosphatidylcholine (LPC) levels on chemotaxis of macrophages. A local chemotactic concentration gradient is required for the attraction and migration of phagocytes toward dying and dead cells. Under non-pathological conditions, an LPC concentration gradient is formed around a dying cell **(A)**. Macrophages follow the gradient until they encounter the dying cell and then start the phagocytic process. In conditions with increased extracellular LPC concentrations, like we observed in systemic lupus erythematosus (SLE) sera, the amount of LPC released by the dying cell might not be high enough to establish a functional chemotactic gradient **(B)**. Macrophages cannot be attracted, they move randomly. The likelihood of contacts between macrophages and dying cells is reduced resulting in decreased phagocytosis.

dying cell clearance, thus contributing to the establishment and/or perpetuation of SLE disease and rendering serum LPC a potential, mechanisms-based marker of SLE disease, which deserves further validation.

ETHICS STATEMENT

This study was carried out in accordance with the recommendations of the ethics committee of the Friedrich-Alexander University Erlangen with written informed consent from all subjects. All subjects gave written informed consent in accordance with the Declaration of Helsinki. The protocol was approved by the ethics committee of the Friedrich-Alexander University Erlangen (Permit number 52_14 B).

AUTHOR CONTRIBUTIONS

GG, LM, KL, and MH conceived and designed the study. GG, HK, CJ, LM, and KL performed experiments and analyzed the data. JR collected patients samples and clinical data. SB re-evaluated cases and analyzed the data LM, GG, and KL wrote the manuscript. All authors reviewed the manuscript.

FUNDING

This work was supported by the German Research Council project SFB685 to KL and projects SFB1181-C3 and KFO257 to MH, by the EU H2020-MSCE-RISE-2015 project Nr. 690836 PANG to LM.

REFERENCES

1. Elliott MR, Ravichandran KS. The dynamics of apoptotic cell clearance. *Dev Cell* (2016) 38(2):147–60. doi:10.1016/j.devcel.2016.06.029
2. Rekvig OP, Van der Vlag J. The pathogenesis and diagnosis of systemic lupus erythematosus: still not resolved. *Semin Immunopathol* (2014) 36(3):301–11. doi:10.1007/s00281-014-0428-6
3. Munoz LE, Lauber K, Schiller M, Manfredi AA, Herrmann M. The role of defective clearance of apoptotic cells in systemic autoimmunity. *Nat Rev Rheumatol* (2010) 6(5):280–9. doi:10.1038/nrrheum.2010.46
4. Munoz LE, Berens C, Lauber K, Gaipl US, Herrmann M. Apoptotic cell clearance and its role in the origin and resolution of chronic inflammation. *Front Immunol* (2015) 6:139. doi:10.3389/fimmu.2015.00139
5. Mahajan A, Herrmann M, Munoz LE. Clearance deficiency and cell death pathways: a model for the pathogenesis of SLE. *Front Immunol* (2016) 7:35. doi:10.3389/fimmu.2016.00035
6. Cervera R, Khamashta MA, Font J, Sebastiani GD, Gil A, Lavilla P, et al. Morbidity and mortality in systemic lupus erythematosus during a 10-year period: a comparison of early and late manifestations in a cohort of 1,000 patients. *Medicine* (2003) 82(5):299–308. doi:10.1097/01.md.0000091181.93122.55
7. Baumann I, Kolowos W, Voll RE, Manger B, Gaipl U, Neuhuber WL, et al. Impaired uptake of apoptotic cells into tingible body macrophages in germinal centers of patients with systemic lupus erythematosus. *Arthritis Rheum* (2002) 46(1):191–201. doi:10.1002/1529-0131(200201)46:1<191::AID-ART10027>3.0.CO;2-K
8. Sherer Y, Gorstein A, Fritzler MJ, Shoenfeld Y. Autoantibody explosion in systemic lupus erythematosus: more than 100 different antibodies found in SLE patients. *Semin Arthritis Rheum* (2004) 34(2):501–37. doi:10.1016/j.semarthrit.2004.07.002
9. Casciola-Rosen LA, Anhalt G, Rosen A. Autoantigens targeted in systemic lupus erythematosus are clustered in two populations of surface structures on apoptotic keratinocytes. *J Exp Med* (1994) 179(4):1317–30. doi:10.1084/jem.179.4.1317
10. Munoz LE, Janko C, Grossmayer GE, Frey B, Voll RE, Kern P, et al. Remnants of secondarily necrotic cells fuel inflammation in systemic lupus erythematosus. *Arthritis Rheum* (2009) 60(6):1733–42. doi:10.1002/art.24535
11. Biermann MH, Veissi S, Maueroder C, Chaurio R, Berens C, Herrmann M, et al. The role of dead cell clearance in the etiology and pathogenesis of systemic lupus erythematosus: dendritic cells as potential targets. *Expert Rev Clin Immunol* (2014) 10(9):1151–64. doi:10.1586/1744666x.2014.944162
12. Pons-Estel GJ, Alarcon GS, Scofield L, Reinlib L, Cooper GS. Understanding the epidemiology and progression of systemic lupus erythematosus. *Semin Arthritis Rheum* (2010) 39(4):257–68. doi:10.1016/j.semarthrit.2008.10.007
13. Chakravarty EF, Bush TM, Manzi S, Clarke AE, Ward MM. Prevalence of adult systemic lupus erythematosus in California and Pennsylvania in 2000: estimates obtained using hospitalization data. *Arthritis Rheum* (2007) 56(6):2092–4. doi:10.1002/art.22641
14. Lawrence RC, Helmick CG, Arnett FC, Deyo RA, Felson DT, Giannini EH, et al. Estimates of the prevalence of arthritis and selected musculoskeletal disorders in the United States. *Arthritis Rheum* (1998) 41(5):778–99. doi:10.1002/1529-0131(199805)41:5<778::AID-ART4>3.0.CO;2-V

15. Lahita RG. Emerging concepts for sexual predilection in the disease systemic lupus erythematosus. *Ann N Y Acad Sci* (1999) 876:64–9; discussion 9–70. doi:10.1111/j.1749-6632.1999.tb07623.x
16. Bengtsson AA, Ronnblom L. Systemic lupus erythematosus: still a challenge for physicians. *J Intern Med* (2017) 281(1):52–64. doi:10.1111/joim.12529
17. Hochberg MC. Updating the American College of Rheumatology revised criteria for the classification of systemic lupus erythematosus. *Arthritis Rheum* (1997) 40(9):1725. doi:10.1002/art.1780400928
18. Lauber K, Bohn E, Krober SM, Xiao YJ, Blumenthal SG, Lindemann RK, et al. Apoptotic cells induce migration of phagocytes via caspase-3-mediated release of a lipid attraction signal. *Cell* (2003) 113(6):717–30. doi:10.1016/S0092-8674(03)00422-7
19. Mueller RB, Sheriff A, Gaipl US, Wesselborg S, Lauber K. Attraction of phagocytes by apoptotic cells is mediated by lysophosphatidylcholine. *Autoimmunity* (2007) 40(4):342–4. doi:10.1080/08916930701356911
20. Peter C, Waibel M, Keppeler H, Lehmann R, Xu G, Halama A, et al. Release of lysophospholipid 'find-me' signals during apoptosis requires the ATP-binding cassette transporter A1. *Autoimmunity* (2012) 45(8):568–73. doi:10.3109/08916934.2012.719947
21. Peter C, Waibel M, Radu CG, Yang LV, Witte ON, Schulze-Osthoff K, et al. Migration to apoptotic "find-me" signals is mediated via the phagocyte receptor G2A. *J Biol Chem* (2008) 283(9):5296–305. doi:10.1074/jbc.M706586200
22. Radu CG, Yang LV, Riedinger M, Au M, Witte ON. T cell chemotaxis to lysophosphatidylcholine through the G2A receptor. *Proc Natl Acad Sci U S A* (2004) 101(1):245–50. doi:10.1073/pnas.2536801100
23. Wang L, Radu CG, Yang LV, Bentolila LA, Riedinger M, Witte ON. Lysophosphatidylcholine-induced surface redistribution regulates signaling of the murine G protein-coupled receptor G2A. *Mol Biol Cell* (2005) 16(5):2234–47. doi:10.1091/mbc.E04-12-1044
24. Yang LV, Radu CG, Wang L, Riedinger M, Witte ON. Gi-independent macrophage chemotaxis to lysophosphatidylcholine via the immunoregulatory GPCR G2A. *Blood* (2005) 105(3):1127–34. doi:10.1182/blood-2004-05-1916
25. Le LQ, Kabarowski JH, Weng Z, Satterthwaite AB, Harvill ET, Jensen ER, et al. Mice lacking the orphan G protein-coupled receptor G2A develop a late-onset autoimmune syndrome. *Immunity* (2001) 14(5):561–71. doi:10.1016/S1074-7613(01)00145-5
26. Lauber K, Keppeler H, Munoz LE, Koppe U, Schroder K, Yamaguchi H, et al. Milk fat globule-EGF factor 8 mediates the enhancement of apoptotic cell clearance by glucocorticoids. *Cell Death Differ* (2013) 20(9):1230–40. doi:10.1038/cdd.2013.82
27. Aletaha D, Neogi T, Silman AJ, Funovits J, Felson DT, Bingham CO III, et al. 2010 rheumatoid arthritis classification criteria: an American College of Rheumatology/European League Against Rheumatism collaborative initiative. *Arthritis Rheum* (2010) 62(9):2569–81. doi:10.1002/art.27584
28. Linscott WD, Triglia RP. The bovine complement system. *Adv Exp Med Biol* (1981) 137:413–30.
29. Mancini A, Del Rosso F, Roberti R, Caligiana P, Vecchini A, Binaglia L. Quantitation of glycerophosphorylcholine by flow injection analysis using immobilized enzymes. *Mol Cell Biochem* (1996) 162(2):83–7. doi:10.1007/BF00227533
30. Grossmayer GE, Munoz LE, Weber CK, Franz S, Voll RE, Kern PM, et al. IgG autoantibodies bound to surfaces of necrotic cells and complement C4 comprise the phagocytosis promoting activity for necrotic cells of systemic lupus erythaematosus sera. *Ann Rheum Dis* (2008) 67(11):1626–32. doi:10.1136/ard.2007.081828
31. Munoz LE, Leppkes M, Fuchs TA, Hoffmann M, Herrmann M. Missing in action – the meaning of cell death in tissue damage and inflammation. *Immunol Rev* (2017) 280(1):26–40. doi:10.1111/imr.12569
32. Gross L. The cardiac lesions in Libman-Sacks disease: with a consideration of its relationship to acute diffuse lupus erythematosus. *Am J Pathol* (1940) 16(4):375–408.11.
33. Gueft B, Laufer A. Further cytochemical studies in systemic lupus erythematosus. *AMA Arch Pathol* (1954) 57(3):201–26.
34. Klemperer P, Gueft B, Lee S, Leuchtenberger C, Pollister A. Cytochemical changes of acute lupus erythematosus. *Arch Pathol* (1950) 49(5):503–16.
35. Worthington JW Jr, Baggenstoss AH, Hargraves MM. Significance of hematoxylin bodies in the necropsy diagnosis of systemic lupus erythematosus. *Am J Pathol* (1959) 35:955–69.
36. Kerr JF, Wyllie AH, Currie AR. Apoptosis: a basic biological phenomenon with wide-ranging implications in tissue kinetics. *Br J Cancer* (1972) 26(4):239–57. doi:10.1038/bjc.1972.33
37. Courtney PA, Crockard AD, Williamson K, Irvine AE, Kennedy RJ, Bell AL. Increased apoptotic peripheral blood neutrophils in systemic lupus erythematosus: relations with disease activity, antibodies to double stranded DNA, and neutropenia. *Ann Rheum Dis* (1999) 58(5):309–14. doi:10.1136/ard.58.5.309
38. Perniok A, Wedekind F, Herrmann M, Specker C, Schneider M. High levels of circulating early apoptic peripheral blood mononuclear cells in systemic lupus erythematosus. *Lupus* (1998) 7(2):113–8. doi:10.1191/096120398678919804
39. Muñoz LE, Frey B, Appelt U, Janko C, Sarter K, Voll RE, et al. Peripheral blood stem cells of patients with systemic lupus erythematosus show altered differentiation into macrophages. *Open Autoimmun J* (2010) 2:11–6. doi:10.2174/1876894601002010011
40. Herrmann M, Voll RE, Zoller OM, Hagenhofer M, Ponner BB, Kalden JR. Impaired phagocytosis of apoptotic cell material by monocyte-derived macrophages from patients with systemic lupus erythematosus. *Arthritis Rheum* (1998) 41(7):1241–50. doi:10.1002/1529-0131(199807)41:7<1241:aid-art15>3.0.co;2-h
41. Martin SJ, Reutelingsperger CP, McGahon AJ, Rader JA, van Schie RC, LaFace DM, et al. Early redistribution of plasma membrane phosphatidylserine is a general feature of apoptosis regardless of the initiating stimulus: inhibition by overexpression of Bcl-2 and Abl. *J Exp Med* (1995) 182(5):1545–56. doi:10.1084/jem.182.5.1545
42. Verhoven B, Schlegel RA, Williamson P. Mechanisms of phosphatidylserine exposure, a phagocyte recognition signal, on apoptotic T lymphocytes. *J Exp Med* (1995) 182(5):1597–601. doi:10.1084/jem.182.5.1597
43. Atsumi G, Murakami M, Tajima M, Shimbara S, Hara N, Kudo I. The perturbed membrane of cells undergoing apoptosis is susceptible to type II secretory phospholipase A2 to liberate arachidonic acid. *Biochim Biophys Acta* (1997) 1349(1):43–54. doi:10.1016/S0005-2760(97)00082-9
44. Hack CE, Wolbink GJ, Schalkwijk C, Speijer H, Hermens WT, van den Bosch H. A role for secretory phospholipase A2 and C-reactive protein in the removal of injured cells. *Immunol Today* (1997) 18(3):111–5. doi:10.1016/S0167-5699(97)01002-5
45. Balsinde J, Perez R, Balboa MA. Calcium-independent phospholipase A2 and apoptosis. *Biochim Biophys Acta* (2006) 1761(11):1344–50. doi:10.1016/j.bbalip.2006.07.013
46. Atsumi G, Murakami M, Kojima K, Hadano A, Tajima M, Kudo I. Distinct roles of two intracellular phospholipase A2s in fatty acid release in the cell death pathway. Proteolytic fragment of type IVA cytosolic phospholipase A2alpha inhibits stimulus-induced arachidonate release, whereas that of type VI Ca2+-independent phospholipase A2 augments spontaneous fatty acid release. *J Biol Chem* (2000) 275(24):18248–58. doi:10.1074/jbc.M000271200
47. Perez R, Melero R, Balboa MA, Balsinde J. Role of group VIA calcium-independent phospholipase A2 in arachidonic acid release, phospholipid fatty acid incorporation, and apoptosis in U937 cells responding to hydrogen peroxide. *J Biol Chem* (2004) 279(39):40385–91. doi:10.1074/jbc.M402562200
48. Lin J, Hu FB, Rimm EB, Rifai N, Curhan GC. The association of serum lipids and inflammatory biomarkers with renal function in men with type II diabetes mellitus. *Kidney Int* (2006) 69(2):336–42. doi:10.1038/sj.ki.5000021
49. Stojanovic D, Cvetkovic T, Stojanovic M, Bojanic V, Stefanovic N, Radenkovic S, et al. Crosstalk of inflammatory mediators and lipid parameters as early markers of renal dysfunction in stable renal transplant recipients with regard to immunosuppression. *Ann Transplant* (2013) 18:414–23. doi:10.12659/AOT.889239
50. Quinn MT, Parthasarathy S, Steinberg D. Lysophosphatidylcholine: a chemotactic factor for human monocytes and its potential role in atherogenesis. *Proc Natl Acad Sci U S A* (1988) 85(8):2805–9. doi:10.1073/pnas.85.8.2805
51. Quinn MT, Kondratenko N, Parthasarathy S. Analysis of the monocyte chemotactic response to lysophosphatidylcholine: role of lysophospholipase C. *Biochim Biophys Acta* (1991) 1082(3):293–302. doi:10.1016/0005-2760(91)90205-V
52. Ryborg AK, Deleuran B, Thestrup-Pedersen K, Kragballe K. Lysophosphatidylcholine: a chemoattractant to human T lymphocytes. *Arch Dermatol Res* (1994) 286(8):462–5. doi:10.1007/BF00371572
53. Hoffman RD, Kligerman M, Sundt TM, Anderson ND, Shin HS. Stereospecific chemoattraction of lymphoblastic cells by gradients of lysophosphatidyl-

choline. *Proc Natl Acad Sci U S A* (1982) 79(10):3285–9. doi:10.1073/pnas.79.10.3285
54. Gaipl US, Munoz LE, Grossmayer G, Lauber K, Franz S, Sarter K, et al. Clearance deficiency and systemic lupus erythematosus (SLE). *J Autoimmun* (2007) 28(2–3):114–21. doi:10.1016/j.jaut.2007.02.005
55. Andreoli L, Fredi M, Nalli C, Franceschini F, Meroni PL, Tincani A. Antiphospholipid antibodies mediate autoimmunity against dying cells. *Autoimmunity* (2013) 46(5):302–6. doi:10.3109/08916934.2013.783025
56. Rovere P, Manfredi AA, Vallinoto C, Zimmermann VS, Fascio U, Balestrieri G, et al. Dendritic cells preferentially internalize apoptotic cells opsonized by anti-beta2-glycoprotein I antibodies. *J Autoimmun* (1998) 11(5):403–11. doi:10.1006/jaut.1998.0224
57. Kruse K, Janko C, Urbonaviciute V, Mierke CT, Winkler TH, Voll RE, et al. Inefficient clearance of dying cells in patients with SLE: anti-dsDNA autoantibodies, MFG-E8, HMGB-1 and other players. *Apoptosis* (2010) 15(9):1098–113. doi:10.1007/s10495-010-0478-8
58. Verbovetski I, Bychkov H, Trahtemberg U, Shapira I, Hareuveni M, Ben-Tal O, et al. Opsonization of apoptotic cells by autologous iC3b facilitates clearance by immature dendritic cells, down-regulates DR and CD86, and up-regulates CC chemokine receptor 7. *J Exp Med* (2002) 196(12):1553–61. doi:10.1084/jem.20020263
59. Ravichandran KS. Find-me and eat-me signals in apoptotic cell clearance: progress and conundrums. *J Exp Med* (2010) 207(9):1807–17. doi:10.1084/jem.20101157

Conflict of Interest Statement: The authors declare that the research was conducted in the absence of any commercial or financial relationships that could be construed as a potential conflict of interest.

13

Chronicles of Cell Death Foretold: Specificities in the Mechanism of Disposal

Lindsey D. Hughes[1†], Lidia Bosurgi[2,3†], Sourav Ghosh[4,5] and Carla V. Rothlin[1,4*]*

[1] *Department of Immunobiology, School of Medicine, Yale University, New Haven, CT, United States,* [2] *I. Medizinische Klinik und Poliklinik, Universitätsklinikum Hamburg-Eppendorf, Hamburg, Germany,* [3] *Bernard-Nocht-Institut für Tropenmedizin, Hamburg, Germany,* [4] *Department of Pharmacology, School of Medicine, Yale University, New Haven, CT, United States,* [5] *Department of Neurology, School of Medicine, Yale University, New Haven, CT, United States*

Massive turnover of cells occurs through apoptosis during the constant remodeling of our tissues at homeostasis, from the shedding of cells at exposed barrier surfaces to the elimination of autoreactive lymphocytes. However, a surge of apoptotic cells also accompanies tissue damage, infection, and inflammation. A salient feature of apoptosis in either scenario is the exposure of phosphatidylserine (PtdSer) on the outer leaflet of the plasma membrane. In response to this cue, a range of phagocytes are charged with the sizeable task of engulfing apoptotic bodies and disposing of the billions of cells that perish each day. The presence of apoptotic cells in the remarkably distinct immunological settings described above, therefore, raises the question of how phagocytes are able to coordinate appropriate responses to apoptotic cells—from their silent removal to the production of growth factors or tissue repair molecules—following such a ubiquitous signal as PtdSer exposure. Here, we consider several emergent properties of phagocytes and apoptotic cell clearance that may facilitate specification among this suite of potential responses.

Keywords: apoptosis, homeostasis, tissue repair, phagocytic receptors, signal integration

Edited by:
Amiram Ariel,
University of Haifa, Israel

Reviewed by:
Ariel Munitz,
Tel Aviv University, Israel
Yi Wu,
Temple University,
United States

Correspondence:
Sourav Ghosh
sourav.ghosh@yale.edu;
Carla V. Rothlin
carla.rothlin@yale.edu

†Co-first authors.

The title is inspired by the Gabriel Garcia Marquez novel "Chronicle of a death foretold."

INTRODUCTION

> "…but in this world nothing can be said to be certain, except death and taxes." (Benjamin Franklin)

Apoptotic cell death is a perennial component of tissues from embryonic development through to adulthood. Is death the great equalizer in which all apoptotic cells are handled identically by phagocytes? Or is every instance of apoptosis unique—the trigger and context of a particular cell death event granting specificity to the phagocyte response? During many instances of developmental apoptosis, such as digit formation or postnatal pruning of neuronal circuits, there is simply a need to cull. Conversely, in the case of apoptosis following tissue damage or injury, there is a fundamental need to replace the dying cells; disposal must be linked to regenerative signals. Viral infection of cells can also trigger death by apoptosis, such that the indiscriminate uptake of this cell could threaten the phagocyte with infection. Does knowledge of why cells are undergoing apoptosis enable us to foretell the nature of their removal? Are there distinct mechanisms of identification in effect to enable this functional diversity? In this review, we discuss the view that the array of phagocytes, their repertoire of receptors, the temporal expression pattern of these receptors, and coincidence detection mechanisms may confer specificity to the disposal of dead cells.

PHAGOCYTE AND RECEPTOR DIVERSITY: REDUNDANT OR REQUISITE?

Molecular Basis of Apoptotic Cell Recognition

Phagocytes rely on a set of specific, conserved morphological changes in apoptotic cells in order to recognize that these cells are dying and to initiate their selective and swift clearance. These permissive "eat-me" signals include, but are not limited to, oxidized low-density lipoproteins, surface-bound thrombospondin, ICAM-3, calreticulin, C1q, and phosphatidylserine (PtdSer) exposed on the outer leaflet of the plasma membrane (1). Most "eat-me" signals are byproducts of the activation of intracellular cysteine-dependent aspartate-directed proteases (caspases), found downstream of both the intrinsic and extrinsic initiators of apoptosis [molecular pathways reviewed in Ref. (2, 3)]. Additional modifications to these surface alterations, such as the oxidation of PtdSer lipids containing linoleic or arachidonic acids, may serve to minimize the aberrant removal of cells that transiently expose PtdSer while activated (4, 5).

Phagocytes express numerous receptors that identify each apoptotic "eat-me" signal and instigate the rearrangement of the cytoskeleton to engulf the apoptotic corpse. In mammals, a large cohort of receptors carry out apoptotic cell sensing, with upwards of 10 cell surface receptors capable of recognizing exposed PtdSer either directly or indirectly through bridging molecules (1). The importance of apoptotic cell clearance in preventing systemic autoimmunity may provide sufficient biological impetus for maintaining redundant receptors. Nevertheless, as discussed below, phagocytic receptors that appear interchangeable based on their activating signal may in fact harbor differences in their downstream signaling components, expression profiles, regulation, and interactions with other receptors that provide the basis for diverse phagocyte responses following exposure to apoptotic cells.

It should be noted that recent advances have also distinguished additional forms of programmed cell death, including necroptosis and pyroptosis, which engage distinct signaling pathways, but yet also parallel the lytic dissolution seen in necrosis (6, 7). Loss of cell membrane integrity in each of these cases results in the release of molecules typically restricted in the nucleus or cytosol, which can then be sensed as damage-associated molecular patterns by phagocytes. For example, SAP-130, a component of the U2 small ribonucleoprotein complex, and F-actin unleashed from necrotic cells have each been established as ligands for innate immune receptors (8–11). The identification of signals that are uniquely released or expressed by cells that die through these distinct modalities of necrosis remains an area of active research.

Dynamics of Phagocytic Receptor Expression

The expression profile of phagocytic receptors varies notably among phagocyte populations, and may be further modulated by the surrounding microenvironment. The developing embryo represents one setting of persistent apoptotic cell death and efferocytosis without concomitant inflammatory signals. Extensive cellular proliferation and apoptosis is required for proper limb remodeling, organogenesis, and formation of synapses throughout the nervous system. Mice lacking even single components of the apoptotic signaling pathways exhibit severe developmental defects, often leading to perinatal lethality. For instance, Hao *et al.* utilized a knock-in approach to express a specific point mutant of Cytochrome C, thereby selectively eliminating its activation of Apaf-1 and other apoptotic components, while preserving its activity in the electron transport chain (12). By embryonic day (E) 14.5, these mice had severe overgrowth in multiple regions of the brain and insufficient skull development in comparison to WT counterparts, with these malformations resulting in embryonic or perinatal death for a majority of the mutant pups. Satellite glial cell (SGC) precursors have subsequently been characterized as a primary phagocyte population that clears excess apoptotic neurons, specifically in the embryonic dorsal root ganglion (DRG) (13). Isolated SGC precursors from E12.5 were found to express *Megf10* and *Pear1*, two homologs of the CED-1 phagocytic receptor in *Caenorhabditis elegans*. shRNA-mediated knockdown of *Megf10* and *Pear1* expression in cultured SGC precursors significantly impeded their capacity to engulf dying DRG neurons *in vitro* (13). These findings indicate that MEGF10 and PEAR-1 may be required for proper clearance in the setting of the developing nervous system, though further *in vivo* studies would be necessary to confirm this model. In contrast, a distinct phagocytic receptor, MERTK, is not required in the embryo, although it is expressed in the brain at E14.5 (14). Genetic ablation of *Tyro3*, *Axl*, and *Mertk* (TAM TKO) in mice had no discernible effect on embryonic viability or development, with TAM TKO mice maintaining normal leukocyte development and numbers up until postnatal day 28 (15). Yet as mentioned above, these receptors are critical for efferocytosis in adult mice in a number of tissues, including the brain and thymus (16–18). A comprehensive analysis of the expression of and requirement for efferocytosis receptors in embryonic vs. adult phagocytes could further enhance our understanding of whether specific apoptotic cell sensors are dedicated to homeostatic clearance.

The TAM family of receptor tyrosine kinases, consisting of TYRO3, AXL, and MERTK, represent one group of apoptotic cell sensors that recognize the common "eat-me" signal of PtdSer, but are differentially employed for engulfment by diverse phagocyte populations (16). Systematic evaluation of the *in vitro* phagocytic capacity of resident and thioglycollate-induced peritoneal macrophages deficient in either *Mertk*, *Axl*, or *Tyro3* underscored that all three TAM receptors can contribute to engulfment of apoptotic cells by peritoneal macrophages, although to differing extents; macrophages lacking *Mertk* had the most significant impairment in phagocytosis (19). Interestingly, bone marrow-derived dendritic cells (DCs) demonstrated a distinct reliance on *Axl* and *Tyro3* for engulfment, rather than *Mertk* (19). Such divisions are also reflected *in vivo*—*Axl* and *Tyro3* were not required for homeostatic removal of apoptotic cells in the thymus or the photoreceptor outer segments in the retina, while *Mertk* was essential for proper clearance (19). Recent studies have demonstrated that the receptor expression profile of phagocyte subsets is also dynamic in response to specific stimuli. Zagorska *et al.* characterized the differential regulation and utilization of AXL and MERTK phagocytic receptors by macrophages in either inflammatory or tolerogenic contexts (20). Treatment of bone marrow-derived macrophages (BMDMs) with Dexamethasone, an anti-inflammatory corticosteroid, elevated expression of MERTK and another PtdSer sensor *Bai1*, but not AXL. In contrast, a range of inflammatory signaling molecules, including lipopolysaccharide (LPS), polyinosinic:polycytidylic acid, IFNα, and IFNγ, were sufficient to enhance the expression of AXL in BMDMs, which expressed AXL only at low levels under normal culture conditions. In line with this, BMDMs lacking *Mertk* displayed severe defects in their baseline capacity to engulf apoptotic thymocytes, whereas $Axl^{-/-}$ BMDMs only exhibited diminished uptake compared to WT BMDMs in inflammatory settings (20). This paradigm is also conserved in human phagocytes, as monocytes isolated from peripheral blood mononuclear cells upregulated MERTK expression in response to Dexamethasone or the combination of IL-10 and macrophage colony-stimulating factor (21). Such a "division of labor" is also not restricted to phagocytic receptor tyrosine kinases, as IL-4 was also shown to boost the expression of CD300f in a dose-dependent manner in macrophage subsets that typically do not express this receptor, including BMDMs cultured with IL-4 and peritoneal macrophages following *in vivo* administration of complexed IL-4 (22). CD300f levels in BMDMs were not sensitive to other cytokines tested, including the type II cytokine IL-13, which shares a receptor and has overlapping functions with IL-4 (23). Consequently, the sensitivity of apoptotic cell sensors to regulation by specific cytokines suggests that individual receptors may be equipped to help phagocytes respond to the presence of apoptotic cells in a given setting.

Intriguingly, additional signals in the tissue during infection may feedback to promote phagocytosis *via* specific phagocytic receptors, but without directly modulating receptor expression. Erdman *et al.* observed that human and murine macrophages displayed an increased capacity to clear either *Plasmodium falciparum*-infected or α-CD36-coated erythrocytes when pretreated with agonists to various Toll-like receptors (TLRs), including TLR2, TLR3, TLR4, and TLR9 (24). Uptake was dependent on CD36, a member of the class B scavenger receptor family; however, TLR stimulation did not alter surface expression of CD36 in the short timeframe of the assay (24). Thus, these results provide evidence for a level of cooperation between CD36 and various TLRs in order to potentiate engulfment. A similar synergy has also been described for uptake through integrins in the β_2 subfamily, which are important for the phagocytosis of complement-opsonized particles. Again, stimulation with TNFα, LPS, or platelet activating factor did not have an effect on $\alpha_M\beta_2$ expression by J774 murine macrophages (25). Instead, exposure to each of these inflammatory stimuli mediated activation of RAP1, a Ras-like GTPase that enhanced $\alpha_M\beta_2$ binding to C3bi-coated erythrocytes; expression of a dominant-negative form of RAP1 in macrophage cell lines abolished their capacity to phagocytose these target cells (25).

Signaling an Appropriate Response

While the downstream signaling pathways for phagocytic receptors are not yet fully defined, differences in the molecular players may enable distinct functional outputs by the phagocyte. BAI-1, TIM-4, and the TAM family of phagocytic receptors each recognize the same signal on apoptotic cargo (PtdSer), but partly diverge in the ensuing intracellular cascades that are activated. Immunoprecipitation assays have demonstrated that BAI-1, a G protein-coupled receptor, physically forms a complex with ELMO-1 and Dock-180. Altogether, these molecules function as a guanine nucleotide exchange factor for Rac, thereby driving phagosome formation. All components of this signaling complex were required for maximal engulfment, as silencing of endogenous *Elmo1* impaired the uptake of apoptotic thymocytes in *Bai1*-GFP transfected J774 murine macrophages (26). Other phagocytic receptors, including $\alpha_v\beta_5$ and MERTK, have also been shown to engage the ELMO-1/Dock-180 machinery (27, 28), with constitutively active MERTK only phosphorylating p130^{CAS}, an adaptor protein in this pathway, if $\alpha_v\beta_5$ is coexpressed.

As discussed in further detail in Section "Specification Through Integration," activated TAM receptors also have the capacity to trigger additional signaling cascades, like the phosphorylation of PLCγ2 (29), or the induction of *Socs1/3* to foster an active state of immunosuppression (30). In stark contrast, TIM-4 is thought to function solely as a tethering receptor, binding PtdSer *via* a metal-ion-dependent ligand binding site within its IgV domain (31) and securing apoptotic cargo on the surface of the phagocyte. In support of this, transient expression of either full-length *Tim4* or various versions of *Tim4* lacking the cytoplasmic tail all rendered LR73 fibroblasts more capable of engulfing apoptotic thymocytes than control fibroblasts (32). Moreover, ablation of *Tim4* prevented cultured peritoneal macrophages from binding FAS ligand-treated thymocytes, and transformation of the mouse B-cell line Ba/F3 with *Tim4* alone rescued the binding step of phagocytosis, but not internalization (33). Thus, TIM-4 requires pairing with other phagocytic receptors, such as MERTK, in order to mediate apoptotic cell engulfment in different phagocyte populations (33, 34). This cooperation adds a further layer of complexity to the apoptotic cell:phagocyte interface, as the activation of additional permutations of tethering and engulfing receptors could potentially generate distinctive combinations of intracellular signaling modules.

Phagocyte Identity and Localization

Differences in the identity of the engulfing phagocyte may also account for variation in the resulting response to apoptotic cell recognition. The clearance of dying cells is primarily carried out by professional phagocytes—tissue-resident macrophages, monocyte-derived macrophages, and DCs—that are responsible for taking up cellular debris, sensing for any molecular patterns associated with pathogens, and processing and presenting antigen to activate the adaptive immune response. Despite the maintenance of resident professional phagocyte populations within the tissues, dedicated phagocytes and non-professional phagocytes, including neighboring epithelial cells, endothelial cells, and fibroblasts, are also indispensible for clearing apoptotic cells (35, 36). Cummings *et al.* ascertained that within the small intestinal lamina propria alone, three subsets of resident professional phagocytes contribute to the basal clearance of apoptotic intestinal epithelial cells (IECs) that are not shed off into the intestinal lumen (37, 38). Using a mouse model in which the Villin promoter drives expression of the diphtheria toxin (DT) receptor and GFP, the authors administered low levels of DT to induce cell death in IECs without significant inflammation, and subsequently tracked dying IEC uptake. While $CD11b^{+}$ macrophages, $CD11b^{+}CD103^{+}$ macrophages, and $CD103^{+}$ DCs each adopted a broadly immunosuppressive transcriptional signature in response to IEC engulfment, microarray data revealed distinctions in the precise genetic program; only two genes were consistently up- or downregulated amongst all three groups of phagocytes (37). Comparable resident mononuclear phagocyte subpopulations have been classified in other tissues, such as the kidney, which at steady-state already differ in their phagocytic capacity and expression patterns of cytokines, growth factors, and chemokine receptors (39). Collectively, these findings indicate that subsets of phagocytes, despite being exposed to the same microenvironment, maintain an intrinsic capacity to respond differently to apoptotic corpse engulfment.

SPECIFICATION THROUGH INTEGRATION

Phagocytes rarely are exposed to a single stimulus at once nor, as considered, do they react equivalently to a given signal under all circumstances. The integration of multiple, contemporaneous signals therefore represents a general mechanism through which cells could generate distinct, tailored outputs in response to their microenvironment. This integration could occur at the molecular and/or cellular level, and then serve to further regulate the immune response, such as through amplification or inhibition of specific effectors.

One example of discrete signals combining to generate a specific output by phagocytic macrophages is the synergy between the sensing of apoptotic cells and of type 2 cytokines. Upon infection or injury, there is a significant increase in the magnitude of apoptotic cell death, with resident cells damaged by the insult dying off and infiltrating immune cells turning over. Multiple soluble ligands, like cytokines, are also produced at different phases of the inflammatory response, and phagocytes are also capable of sensing these factors. Work from our laboratory established that BMDMs, as well as macrophages in the lung, the intestine, the visceral adipose tissue, and the peritoneum, launch an efficient tissue repair response only when the sensing of IL-4/IL-13 occurs in the presence of apoptotic cells (**Figure 1**) (40). Importantly, this was not merely an additive effect of two parallel pathways—the presence of apoptotic cells alone did not induce the expression of some of the remodeling-associated genes analyzed. Integration

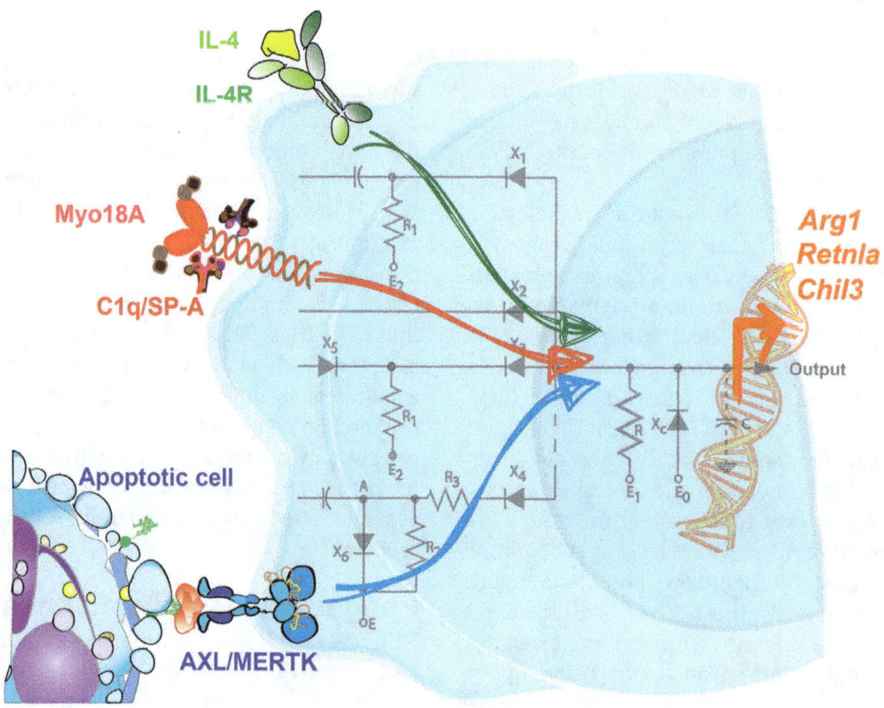

FIGURE 1 | Integration of signals is required to drive a specific tissue repair program in macrophages. Sensing of IL-4 in the presence of apoptotic cells promotes the expression of key tissue repair factors in macrophages. Neither signal on its own is sufficient to induce this genetic program, which includes the upregulation of *Arg1*, *Retnla*, and *Chil3*. Additionally, C1q and surfactant protein-A (SP-A) are each sensed through Myo18A in distinct macrophage populations, and in conjunction with IL-4, prompts a similar set of tissue repair genes. Mirroring the configuration of a circuit, multiple inputs are therefore necessary to coordinate a tailored output or response by phagocytes.

of these two signals was shown to be dependent on the detectors AXL and MERTK, two phagocytic receptors expressed at both steady state and/or upon damage by resident and monocyte-derived macrophages. Genetic ablation of *Axl* and *Mertk*, or the administration of Annexin-V, which coats and obscures exposed PtdSer on apoptotic cells, were each sufficient to curtail this specialized tissue repair genetic program in macrophages in response to IL-4/IL-13. These findings demonstrate that apoptotic cell sensing can influence phagocyte activity in ways beyond its described immunosuppressive effect.

Interestingly, while the maximal induction of remodeling-associated genes like *Arg1*, *Retnla*, and *Chil3* were reliant on macrophages receiving these two signals, other IL-4-mediated responses were not. Gene set enrichment analysis of BMDMs treated with IL-4 alone identified categories such as pattern recognition signaling pathways, regulation of cytokine production, chemotaxis, and the defense response, all of which were unaffected by an additional treatment of the BMDMs with exogenous apoptotic cells. Like other cytokines, IL-4 and IL-13 are produced at multiple phases of the immune response and have pleiotropic effects, including driving cell proliferation, antigen presentation, antimicrobial activity, and this induction of tissue remodeling (41–43). Collectively, these results shed light on a role for apoptotic cells in specifying the action of IL-4 on one of its target cell populations. In this way, the requirement of both signals functions as a checkpoint, restricts the expression of repair factors to both an appropriate time and location, and helps to avoid the potentially detrimental effects of aberrant remodeling. While the effects of this signal integration have been characterized, further investigation is needed to determine the precise molecular players that interact downstream in the IL-4/IL-13 and TAM signaling pathways.

It is important to note that additional tissue-specific factors have also been described to promote the resolution program in phagocytes in response to IL-4. Minutti *et al.* recently identified surfactant protein-A (SP-A) and complement component C1q as secondary stimuli that enhance the proliferative and tissue-remodeling capacity of IL-4-activated macrophages from the lung and the peritoneal cavity/liver, respectively (**Figure 1**) (44). In support of this, SP-A or C1q were each necessary for extensive repair responses in settings of acute and chronic damage, like with *Listeria monocytogenes* infection or in the Dineal PD-4 model of peritoneal fibrosis. *Sp-a*$^{-/-}$ mice, for example, exhibited severe immunopathology in the lung and increased worm burden in the intestine following infection with the helminth *Nippostrongylus brasiliensis* in comparison to their WT counterparts. Integration of these pathways is at least partially mediated through myosin 18 A (Myo18A); *in vivo* delivery of IL-4 was shown to increase the expression of this shared receptor for SP-A and C1q in macrophages isolated from the lung, liver, and peritoneal cavity (44).

As referenced before, uptake of apoptotic cells by phagocytes has been shown to suppress the response to inflammatory stimuli, like TLR ligands (45–51). Engulfment of apoptotic cells

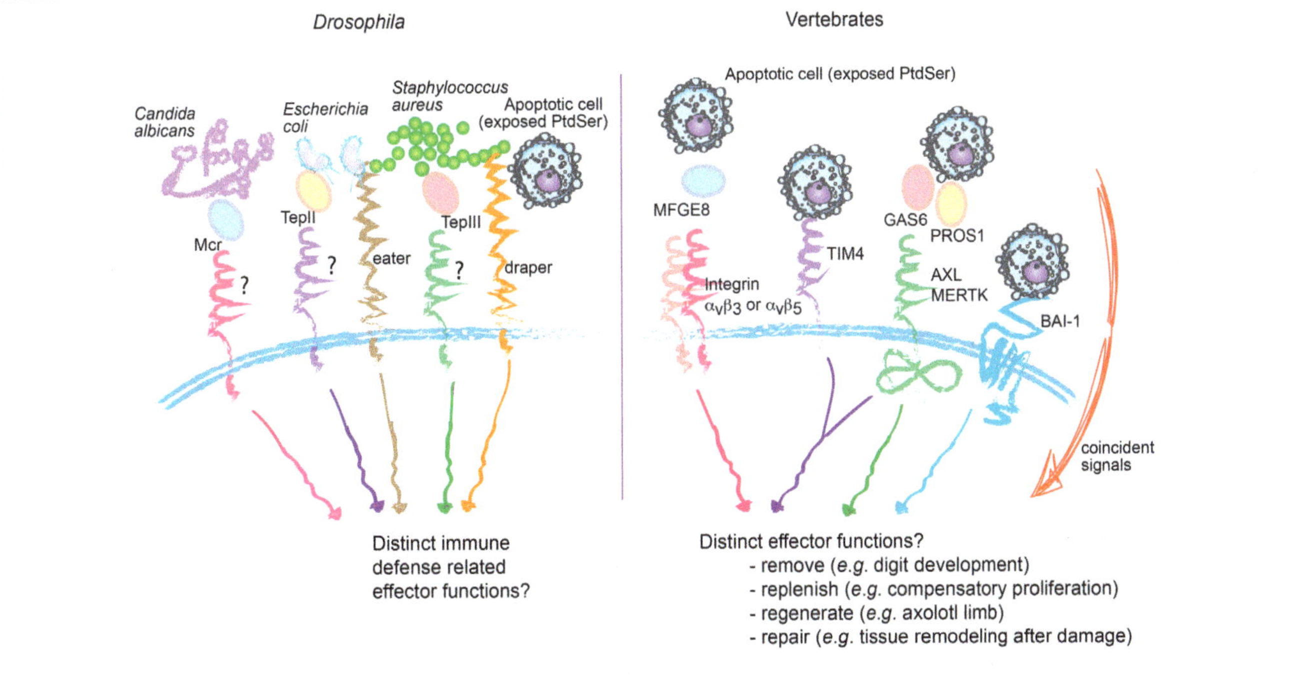

FIGURE 2 | Proposed specificity, rather than redundancy, in apoptotic cell recognition and engulfment. Comparative representation of molecular mediators involved in the recognition and engulfment of bacteria or apoptotic cells by phagocytes, in *Drosophila* (left) and vertebrates (right). In *Drosophila*, specific bacteria can be opsonized through secreted Mcr, Tep II, or Tep III before engulfment by hemocytes. Similarly, multiple phagocytic receptors in vertebrates, including integrins $\alpha_v\beta_3$, $\alpha_v\beta_5$ and the receptor tyrosine kinases AXL and MERTK (right) also require "opsonization" of the apoptotic cargo. This raises the possibility of selective disposal and tailored effector functions. Consistent with this proposition, in vertebrates, not only are the number of phagocytic receptors and ligands expanded but also their engagement is integrated with specific signals from the microenvironment.

by immature murine bone marrow-derived DCs was shown to dampen the response to LPS, with specific reduction of IL-12 secretion and CD86 expression, but not of other proinflammatory cytokines or costimulatory molecules, like TNFα and CD40 (50). In a similar manner, ingestion of apoptotic corpses by LPS-stimulated human or murine macrophages diminished secretion of proinflammatory molecules TNFα and IL-1β, but also enhanced release of anti-inflammatory signals, such as TGF-β1 and PGE-2 (45, 48, 49, 51). Thus, beyond suppressing proinflammatory events, integration of TLR signaling and apoptotic cell sensing actively promotes the generation of an anti-inflammatory environment.

The precise mechanism of integration of TLR signaling and uptake of apoptotic cells still needs to be fully elucidated. Intracellular lipid sensors have been described to contribute to this immunosuppressive effect of apoptotic cells (52, 53). Additionally, integration of phagocytic receptor and TLR signaling with cytokine signaling has been characterized as another strategy for dampening the inflammatory response by phagocytes. The receptor tyrosine kinase AXL, which is highly expressed by DCs (20, 30), was found to physically associate with the R1 chain of the type I IFN receptor (IFNAR1) upon administration of its ligand Gas6 (30). Engagement of and interaction between these two receptors led to the activation of STAT1, which in turn triggered SOCS1 and SOCS3 to limit both cytokine and TLR signaling (30). This cooperation highlights how the pairing of apoptotic cell sensors with other detectors not only can promote a specialized transcriptional program, but also can negatively regulate phagocyte activation in a specific manner. The extent to which phagocytic receptor signaling integrates with other cytokine pathways, and whether these are broader features of all phagocytic receptors, remain to be fully explored.

CONCLUDING THOUGHTS

The functional diversification of apoptotic cell removal in mammals is consistent with the evolutionary expansion of PtdSer receptors. While efferocytosis itself is conserved from *C. elegans* to mammals, at least some of the mammalian receptors such as BAI-1, TIM4, and the TAMs do not have orthologs in *C. elegans*. The argument that increased redundancy of these receptors simply ensures efficient removal of dead cells, as the risk for autoimmunity increases with evolution, cannot be formally ruled out. Intriguingly, TAM, integrin, and MEGF10-mediated efferocytosis require "opsonization" of the apoptotic cargo with their respective ligands, including PROS1/GAS6, MFGE8, or C1q (54) (**Figure 2**). This system seems analogous to that seen in *Drosophila* hemocytes, in which secreted Mcr, Tep II, or Tep III opsonize *Candida albicans*, *Escherichia coli*, or *Staphylococcus aureus*, respectively, during phagocytosis (55). The selective advantage of such a system is difficult to understand simply on the basis of redundancy; rather, it favors a paradigm wherein layers of specificities can be built in. Overall, evolutionary expansion of phagocytic receptors may have enabled cargo-selective disposal and the resulting specification in phagocyte response. Further inquiry of this topic may permit the establishment of a "code" (56), through which knowledge of the various input signals from the dying cell and the surrounding microenvironment can predict the functional output of the phagocyte.

AUTHOR CONTRIBUTIONS

All authors listed have made a substantial, direct, and intellectual contribution to the work and approved it for publication.

ACKNOWLEDGMENTS

The authors thank Macy Akalu and Silvina del Carmen for their helpful discussion and reading of this manuscript.

FUNDING

This work was supported by grants from the National Institutes of Health (NIH-NIAID R01 AI089824 to CR, NIH-NCI R01 CA212376 to CR and SG), the Deutsche Forschungsgemeinschaft (grant SFB841 to LB), Alliance for Lupus Research (grant 332789 to CR), the Yale Immunobiology Department (NIH-NIAID T32 AI007019 to LH), and the National Science Foundation (DGE-1122492 to LH). CR is a HHMI Faculty Scholar (grant 55108561).

REFERENCES

1. Arandjelovic S, Ravichandran KS. Phagocytosis of apoptotic cells in homeostasis. *Nat Immunol* (2015) 16(9):907–17. doi:10.1038/ni.3253
2. Taylor RC, Cullen SP, Martin SJ. Apoptosis: controlled demolition at the cellular level. *Nat Rev Mol Cell Biol* (2008) 9(3):231–41. doi:10.1038/nrm2312
3. Galluzzi L, Vitale I, Abrams JM, Alnemri ES, Baehrecke EH, Blagosklonny MV, et al. Molecular definitions of cell death subroutines: recommendations of the Nomenclature Committee on Cell Death 2012. *Cell Death Differ* (2012) 19(1):107–20. doi:10.1038/cdd.2011.96
4. Greenberg ME, Sun M, Zhang R, Febbraio M, Silverstein R, Hazen SL. Oxidized phosphatidylserine-CD36 interactions play an essential role in macrophage-dependent phagocytosis of apoptotic cells. *J Exp Med* (2006) 203(12):2613–25. doi:10.1084/jem.20060370
5. Tyurin VA, Balasubramanian K, Winnica D, Tyurina YY, Vikulina AS, He RR, et al. Oxidatively modified phosphatidylserines on the surface of apoptotic cells are essential phagocytic 'eat-me' signals: cleavage and inhibition of phagocytosis by Lp-PLA2. *Cell Death Differ* (2014) 21(5):825–35. doi:10.1038/cdd.2014.1
6. Weinlich R, Oberst A, Beere HM, Green DR. Necroptosis in development, inflammation and disease. *Nat Rev Mol Cell Biol* (2017) 18(2):127–36. doi:10.1038/nrm.2016.149
7. Shi J, Gao W, Shao F. Pyroptosis: gasdermin-mediated programmed necrotic cell death. *Trends Biochem Sci* (2017) 42(4):245–54. doi:10.1016/j.tibs.2016.10.004
8. Yamasaki S, Ishikawa E, Sakuma M, Hara H, Ogata K, Saito T. Mincle is an ITAM-coupled activating receptor that senses damaged cells. *Nat Immunol* (2008) 9(10):1179–88. doi:10.1038/ni.1651
9. Sancho D, Joffre OP, Keller AM, Rogers NC, Martinez D, Hernanz-Falcon P, et al. Identification of a dendritic cell receptor that couples sensing of necrosis to immunity. *Nature* (2009) 458(7240):899–903. doi:10.1038/nature07750
10. Ahrens S, Zelenay S, Sancho D, Hanc P, Kjaer S, Feest C, et al. F-actin is an evolutionarily conserved damage-associated molecular pattern recognized

by DNGR-1, a receptor for dead cells. *Immunity* (2012) 36(4):635–45. doi:10.1016/j.immuni.2012.03.008
11. Zhang JG, Czabotar PE, Policheni AN, Caminschi I, Wan SS, Kitsoulis S, et al. The dendritic cell receptor Clec9A binds damaged cells via exposed actin filaments. *Immunity* (2012) 36(4):646–57. doi:10.1016/j.immuni.2012.03.009
12. Hao Z, Duncan GS, Chang CC, Elia A, Fang M, Wakeham A, et al. Specific ablation of the apoptotic functions of cytochrome C reveals a differential requirement for cytochrome C and Apaf-1 in apoptosis. *Cell* (2005) 121(4):579–91. doi:10.1016/j.cell.2005.03.016
13. Wu HH, Bellmunt E, Scheib JL, Venegas V, Burkert C, Reichardt LF, et al. Glial precursors clear sensory neuron corpses during development via Jedi-1, an engulfment receptor. *Nat Neurosci* (2009) 12(12):1534–41. doi:10.1038/nn.2446
14. Mass E, Ballesteros I, Farlik M, Halbritter F, Gunther P, Crozet L, et al. Specification of tissue-resident macrophages during organogenesis. *Science* (2016) 353:6304. doi:10.1126/science.aaf4238
15. Lu Q, Lemke G. Homeostatic regulation of the immune system by receptor tyrosine kinases of the Tyro 3 family. *Science* (2001) 293(5528):306–11. doi:10.1126/science.1061663
16. Rothlin CV, Carrera-Silva EA, Bosurgi L, Ghosh S. TAM receptor signaling in immune homeostasis. *Annu Rev Immunol* (2015) 33:355–91. doi:10.1146/annurev-immunol-032414-112103
17. Fourgeaud L, Traves PG, Tufail Y, Leal-Bailey H, Lew ED, Burrola PG, et al. TAM receptors regulate multiple features of microglial physiology. *Nature* (2016) 532(7598):240–4. doi:10.1038/nature17630
18. Scott RS, McMahon EJ, Pop SM, Reap EA, Caricchio R, Cohen PL, et al. Phagocytosis and clearance of apoptotic cells is mediated by MER. *Nature* (2001) 411(6834):207–11. doi:10.1038/35075603
19. Seitz HM, Camenisch TD, Lemke G, Earp HS, Matsushima GK. Macrophages and dendritic cells use different Axl/Mertk/Tyro3 receptors in clearance of apoptotic cells. *J Immunol* (2007) 178(9):5635–42. doi:10.4049/jimmunol.178.9.5635
20. Zagorska A, Traves PG, Lew ED, Dransfield I, Lemke G. Diversification of TAM receptor tyrosine kinase function. *Nat Immunol* (2014) 15(10):920–8. doi:10.1038/ni.2986
21. Zizzo G, Hilliard BA, Monestier M, Cohen PL. Efficient clearance of early apoptotic cells by human macrophages requires M2c polarization and MerTK induction. *J Immunol* (2012) 189(7):3508–20. doi:10.4049/jimmunol.1200662
22. Moshkovits I, Karo-Atar D, Itan M, Reichman H, Rozenberg P, Morgenstern-Ben-Baruch N, et al. CD300f associates with IL-4 receptor alpha and amplifies IL-4-induced immune cell responses. *Proc Natl Acad Sci U S A* (2015) 112(28):8708–13. doi:10.1073/pnas.1507625112
23. Van Dyken SJ, Locksley RM. Interleukin-4- and interleukin-13-mediated alternatively activated macrophages: roles in homeostasis and disease. *Annu Rev Immunol* (2013) 31:317–43. doi:10.1146/annurev-immunol-032712-095906
24. Erdman LK, Cosio G, Helmers AJ, Gowda DC, Grinstein S, Kain KC. CD36 and TLR interactions in inflammation and phagocytosis: implications for malaria. *J Immunol* (2009) 183(10):6452–9. doi:10.4049/jimmunol.0901374
25. Caron E, Self AJ, Hall A. The GTPase Rap1 controls functional activation of macrophage integrin alphaMbeta2 by LPS and other inflammatory mediators. *Curr Biol* (2000) 10(16):974–8. doi:10.1016/S0960-9822(00)00641-2
26. Park D, Tosello-Trampont AC, Elliott MR, Lu M, Haney LB, Ma Z, et al. BAI1 is an engulfment receptor for apoptotic cells upstream of the ELMO/Dock180/Rac module. *Nature* (2007) 450(7168):430–4. doi:10.1038/nature06329
27. Albert ML, Kim JI, Birge RB. alphavbeta5 integrin recruits the CrkII-Dock180-rac1 complex for phagocytosis of apoptotic cells. *Nat Cell Biol* (2000) 2(12):899–905. doi:10.1038/35046549
28. Wu Y, Singh S, Georgescu MM, Birge RB. A role for Mer tyrosine kinase in alphavbeta5 integrin-mediated phagocytosis of apoptotic cells. *J Cell Sci* (2005) 118(Pt 3):539–53. doi:10.1242/jcs.01632
29. Todt JC, Hu B, Curtis JL. The receptor tyrosine kinase MerTK activates phospholipase C gamma2 during recognition of apoptotic thymocytes by murine macrophages. *J Leukoc Biol* (2004) 75(4):705–13. doi:10.1189/jlb.0903439
30. Rothlin CV, Ghosh S, Zuniga EI, Oldstone MB, Lemke G. TAM receptors are pleiotropic inhibitors of the innate immune response. *Cell* (2007) 131(6):1124–36. doi:10.1016/j.cell.2007.10.034
31. Santiago C, Ballesteros A, Tami C, Martinez-Munoz L, Kaplan GG, Casasnovas JM. Structures of T Cell immunoglobulin mucin receptors 1 and 2 reveal mechanisms for regulation of immune responses by the TIM receptor family. *Immunity* (2007) 26(3):299–310. doi:10.1016/j.immuni.2007.01.014
32. Park D, Hochreiter-Hufford A, Ravichandran KS. The phosphatidylserine receptor TIM-4 does not mediate direct signaling. *Curr Biol* (2009) 19(4):346–51. doi:10.1016/j.cub.2009.01.042
33. Nishi C, Toda S, Segawa K, Nagata S. Tim4- and MerTK-mediated engulfment of apoptotic cells by mouse resident peritoneal macrophages. *Mol Cell Biol* (2014) 34(8):1512–20. doi:10.1128/MCB.01394-13
34. Yanagihashi Y, Segawa K, Maeda R, Nabeshima YI, Nagata S. Mouse macrophages show different requirements for phosphatidylserine receptor Tim4 in efferocytosis. *Proc Natl Acad Sci U S A* (2017) 114(33):8800–5. doi:10.1073/pnas.1705365114
35. Monks J, Rosner D, Geske FJ, Lehman L, Hanson L, Neville MC, et al. Epithelial cells as phagocytes: apoptotic epithelial cells are engulfed by mammary alveolar epithelial cells and repress inflammatory mediator release. *Cell Death Differ* (2005) 12(2):107–14. doi:10.1038/sj.cdd.4401517
36. Parnaik R, Raff MC, Scholes J. Differences between the clearance of apoptotic cells by professional and non-professional phagocytes. *Curr Biol* (2000) 10(14):857–60. doi:10.1016/S0960-9822(00)00598-4
37. Cummings RJ, Barbet G, Bongers G, Hartmann BM, Gettler K, Muniz L, et al. Different tissue phagocytes sample apoptotic cells to direct distinct homeostasis programs. *Nature* (2016) 539(7630):565–9. doi:10.1038/nature20138
38. Eisenhoffer GT, Loftus PD, Yoshigi M, Otsuna H, Chien CB, Morcos PA, et al. Crowding induces live cell extrusion to maintain homeostatic cell numbers in epithelia. *Nature* (2012) 484(7395):546–9. doi:10.1038/nature10999
39. Kawakami T, Lichtnekert J, Thompson LJ, Karna P, Bouabe H, Hohl TM, et al. Resident renal mononuclear phagocytes comprise five discrete populations with distinct phenotypes and functions. *J Immunol* (2013) 191(6):3358–72. doi:10.4049/jimmunol.1300342
40. Bosurgi L, Cao YG, Cabeza-Cabrerizo M, Tucci A, Hughes LD, Kong Y, et al. Macrophage function in tissue repair and remodeling requires IL-4 or IL-13 with apoptotic cells. *Science* (2017) 356(6342):1072–6. doi:10.1126/science.aai8132
41. Ozaki K, Leonard WJ. Cytokine and cytokine receptor pleiotropy and redundancy. *J Biol Chem* (2002) 277(33):29355–8. doi:10.1074/jbc.R200003200
42. Varin A, Gordon S. Alternative activation of macrophages: immune function and cellular biology. *Immunobiology* (2009) 214(7):630–41. doi:10.1016/j.imbio.2008.11.009
43. Jenkins SJ, Ruckerl D, Cook PC, Jones LH, Finkelman FD, van Rooijen N, et al. Local macrophage proliferation, rather than recruitment from the blood, is a signature of TH2 inflammation. *Science* (2011) 332(6035):1284–8. doi:10.1126/science.1204351
44. Minutti CM, Jackson-Jones LH, Garcia-Fojeda B, Knipper JA, Sutherland TE, Logan N, et al. Local amplifiers of IL-4Ralpha-mediated macrophage activation promote repair in lung and liver. *Science* (2017) 356(6342):1076–80. doi:10.1126/science.aaj2067
45. Fadok VA, Bratton DL, Konowal A, Freed PW, Westcott JY, Henson PM. Macrophages that have ingested apoptotic cells in vitro inhibit proinflammatory cytokine production through autocrine/paracrine mechanisms involving TGF-beta, PGE2, and PAF. *J Clin Invest* (1998) 101(4):890–8. doi:10.1172/JCI1112
46. Meagher LC, Savill JS, Baker A, Fuller RW, Haslett C. Phagocytosis of apoptotic neutrophils does not induce macrophage release of thromboxane B2. *J Leukoc Biol* (1992) 52(3):269–73.
47. Stern M, Savill J, Haslett C. Human monocyte-derived macrophage phagocytosis of senescent eosinophils undergoing apoptosis. Mediation by alpha v beta 3/CD36/thrombospondin recognition mechanism and lack of phlogistic response. *Am J Pathol* (1996) 149(3):911–21.
48. Voll RE, Herrmann M, Roth EA, Stach C, Kalden JR, Girkontaite I. Immunosuppressive effects of apoptotic cells. *Nature* (1997) 390(6658):350–1. doi:10.1038/37022

49. Huynh ML, Fadok VA, Henson PM. Phosphatidylserine-dependent ingestion of apoptotic cells promotes TGF-beta1 secretion and the resolution of inflammation. *J Clin Invest* (2002) 109(1):41–50. doi:10.1172/JCI11638
50. Stuart LM, Lucas M, Simpson C, Lamb J, Savill J, Lacy-Hulbert A. Inhibitory effects of apoptotic cell ingestion upon endotoxin-driven myeloid dendritic cell maturation. *J Immunol* (2002) 168(4):1627–35. doi:10.4049/jimmunol.168.4.1627
51. Byrne A, Reen DJ. Lipopolysaccharide induces rapid production of IL-10 by monocytes in the presence of apoptotic neutrophils. *J Immunol* (2002) 168(4):1968–77. doi:10.4049/jimmunol.168.4.1968
52. A-Gonzalez N, Bensinger SJ, Hong C, Beceiro S, Bradley MN, Zelcer N, et al. Apoptotic cells promote their own clearance and immune tolerance through activation of the nuclear receptor LXR. *Immunity* (2009) 31(2): 245–58. doi:10.1016/j.immuni.2009.06.018
53. Mukundan L, Odegaard JI, Morel CR, Heredia JE, Mwangi JW, Ricardo-Gonzalez RR, et al. PPAR-delta senses and orchestrates clearance of apoptotic cells to promote tolerance. *Nat Med* (2009) 15(11):1266–72. doi:10.1038/nm.2048
54. Birge RB, Boeltz S, Kumar S, Carlson J, Wanderley J, Calianese D, et al. Phosphatidylserine is a global immunosuppressive signal in efferocytosis, infectious disease, and cancer. *Cell Death Differ* (2016) 23(6):962–78. doi:10.1038/cdd.2016.11
55. Stroschein-Stevenson SL, Foley E, O'Farrell PH, Johnson AD. Identification of *Drosophila* gene products required for phagocytosis of *Candida albicans*. *PLoS Biol* (2006) 4(1):e4. doi:10.1371/journal.pbio.0040004
56. Medzhitov R. Origin and physiological roles of inflammation. *Nature* (2008) 454(7203):428–35. doi:10.1038/nature07201

Conflict of Interest Statement: The authors declare that the research was conducted in the absence of any commercial or financial relationships that could be construed as a potential conflict of interest.

14

Clearance of Apoptotic Cells by Tissue Epithelia: A Putative Role for Hepatocytes in Liver Efferocytosis

*Scott P. Davies[1], Gary M. Reynolds[1,2,3] and Zania Stamataki[1]**

[1] Centre for Liver Research, College of Medical and Dental Sciences, Institute for Immunology and Immunotherapy, University of Birmingham, Birmingham, United Kingdom, [2] Centre for Liver Research and National Institute for Health Research (NIHR) Birmingham Liver Biomedical Research Unit, College of Medical and Dental Sciences, Institute of Immunology and Immunotherapy, University of Birmingham, Birmingham, United Kingdom, [3] University Hospitals Birmingham NHS Foundation Trust, Birmingham, United Kingdom

Toxic substances and microbial or food-derived antigens continuously challenge the liver, which is tasked with their safe neutralization. This vital organ is also important for the removal of apoptotic immune cells during inflammation and has been previously described as a "graveyard" for dying lymphocytes. The clearance of apoptotic and necrotic cells is known as efferocytosis and is a critical liver function to maintain tissue homeostasis. Much of the research into this form of immunological control has focused on Kupffer cells, the liver-resident macrophages. However, hepatocytes (and other liver resident cells) are competent efferocytes and comprise 80% of the liver mass. Little is known regarding the mechanisms of apoptotic and necrotic cell capture by epithelia, which lack key receptors that mediate phagocytosis in macrophages. Herein, we discuss recent developments that increased our understanding of efferocytosis in tissues, with a special focus on the liver parenchyma. We discuss the impact of efferocytosis in health and in inflammation, highlighting the role of phagocytic epithelia.

Keywords: efferocytosis, phagocytosis, liver, hepatocytes, regeneration, apoptosis, necrosis, cell death

Edited by:
Amiram Ariel,
University of Haifa, Israel

Reviewed by:
Christina Janko,
Universitätsklinikum Erlangen, Germany
Joan Clària,
Hospital Clinic-Barcelona University School of Medicine, Spain

****Correspondence:***
Zania Stamataki
z.stamataki@bham.ac.uk

KEY POINTS

- Efferocytosis is a vital process in tissues that can be carried out by multiple cell types, including blood derived and tissue resident phagocytes.
- Hepatocytes are competent efferocytes and play an important role in the clearance of dead cells in health and in inflammation.
- Epithelial cell efferocytosis is understudied and involves distinct mechanisms to professional phagocytes.
- Defects in efferocytosis have been linked to diseases such as autoimmunity, failure to prevent metastasis, failure to limit infection.
- Understanding molecular mechanisms of efferocytosis may reveal new pathways for therapeutic intervention to alleviate inflammation.

EFFEROCYTOSIS IN THE LIVER

Efferocytosis, the clearance of dead and dying cells, is important to prevent tissue damage and promote the resolution of inflammation (1). The liver has evolved into an expert in defusing biochemical threats emanating from food or microbial antigens, which reach the organ along with 75% of its blood supply that arrives through venous blood from the gut. Hepatocytes comprise

80% of liver cells and constitute the biochemical powerhouses of the liver parenchyma, and as a result they often perish in their duties to absorb toxic substances. To cope with loss of hepatic epithelia, the liver has evolved the remarkable ability to regenerate.

To perform their detoxification roles, hepatocytes are strategically organized roughly into two hepatocyte-thick cords, flanked by a thin layer of fenestrated endothelia (**Figure 1**). Nutrient-rich blood enters the liver *via* the portal vein and oxygen-rich blood *via* the hepatic artery, which, together with a bile duct, form the liver portal triad (**Figure 1A**). Blood from both sources mixes in the specialized hepatic capillaries termed sinusoids, and drains toward the central vein. Hepatocytes near the portal triads (designated zone 1) can be damaged by the inflammatory infiltrate during interface hepatitis, when immune cells cross the sinusoidal endothelia and reach the parenchyma. Zone 2 is found mid-distance from a portal triad and the draining central vein (zone 3). Periportal hepatocytes near zone 1 have access to oxygenated blood from the hepatic artery, and nutrients from the portal blood supply that arrives from the gut. Oxygen and nutrient levels reduce toward the central vein and hepatocytes in zone 3 are found in hypoxic conditions. Fenestrations in the sinusoids allow hepatocytes access to solutes and immune cells reaching through the fenestrations from the circulation (2, 3), but prevent unregulated migration of immune cells to the parenchyma (4).

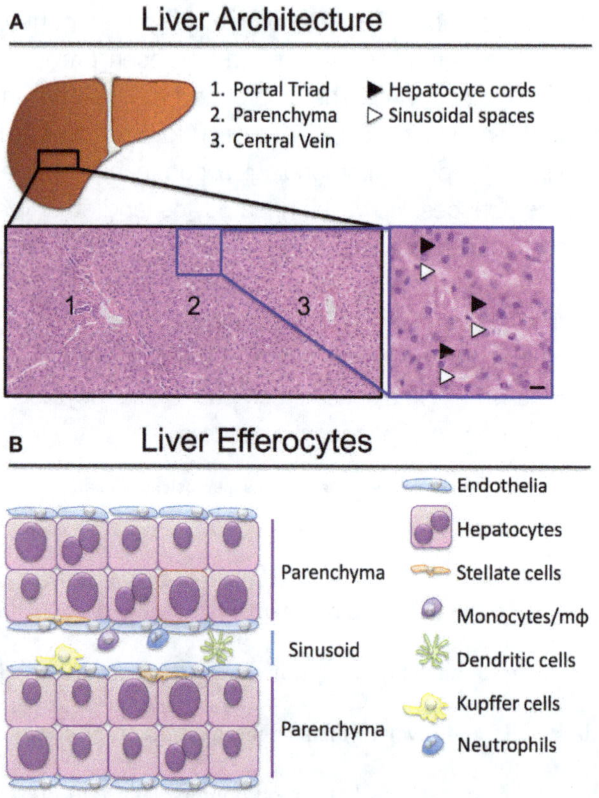

FIGURE 1 | Organization of liver-resident and recirculating efferocytes. **(A)** Hepatocytes are spread over three zones, exposed to different levels of oxygen and nutrients. Hepatocytes in zone 1 proximal to the portal triad (portal vein, hepatic artery, bile duct) have access to arterial and venous blood entering the liver through the circulation. Hepatocytes in zone 3 have less access to oxygen and nutrients and are exposed to blood draining into the central vein. Hematoxylin-eosin stain, scale bar represents 50 µm. **(B)** A plethora of liver resident and recirculating cells are able to engulf apoptotic and necrotic cells and clear them to maintain tissue homeostasis. Kupffer cells, monocytes and macrophages (mϕ) are the best-characterized efferocytes in the liver.

Cells that perish in the sinusoidal spaces are cleared by circulating phagocytes (monocytes, dendritic cells, neutrophils), liver-resident macrophages termed Kupffer cells, and by sinusoidal endothelia (**Figure 1B**). The best-characterized liver efferocytes are macrophages, both those derived from monocytes infiltrating from the circulation, and the self-renewing populations of Kupffer cells. It is understood that professional phagocytes are activated during injury and adapt their phenotype following the encounter of cellular debris, danger signals, and soluble mediators of the inflammatory milieu. The critical role of liver macrophages including Kupffer cells in the ebb and flow of inflammation was recently reviewed by Tacke's group (5, 6).

Activated hepatic stellate cells can also engulf apoptotic hepatocytes, which in turn lead to increases in tumor growth factor-β (TGF-β) secretion (7). Biliary epithelial cells (BECs) also take part in efferocytosis of neighboring apoptotic cells; an important adaptation for diseases associated with increased BEC apoptosis such as primary biliary cholangitis (8). The phagocytic activity of hepatocytes was noted in 1992 (9). Hepatocyte efferocytosis assists in parenchymal housekeeping to rapidly dispose of cell remnants and prevent excessive inflammation.

Hepatocyte death from biochemical toxicity (*necrosis*) occurs in health as part of normal homeostasis, however, liver damage is exacerbated in infection or in alcoholic, drug or ischemia-induced liver injury where large areas of necrotic lesions are evident (**Figure 2**). Acute-on-chronic liver failure is a syndrome associated with exacerbation of hepatitis B infection (HBV) and characterized by broad areas of hepatic necrosis in cirrhotic patients (**Figure 2A**). Lymphocyte infiltration is often seen in the parenchyma in chronic liver diseases. Crispe and others have elegantly put forward the "graveyard theory" where the liver is primary site for the disposal of spent immune cells (10). **Figure 2B** shows hepatic epithelia in the process of engulfing immune cells that have perished in the parenchyma, and this is seen predominantly near the portal regions. Conversely, in cases of acute liver injury such as paracetamol overdose (POD), hepatocyte necrosis due to loss of ATP is noted around the centrilobular regions (zone 3, **Figure 2C**). Histological features of necrotic hepatocytes include eosinophilic degradation and pyknotic nuclei, which are readily detectable by hematoxylin-eosin staining (inset, **Figure 2C**).

Hepatocytes also clear away cells that have triggered the molecular cascade of events of programmed cell death (*apoptosis*) (**Figures 2B,D**), but can actively destroy live autoreactive immune cells by direct engulfment as noted for CD8^{+} T cells undergoing *suicidal emperipolesis* (11). Immune cell death and liver damage are exacerbated in chronic liver inflammation of multiple

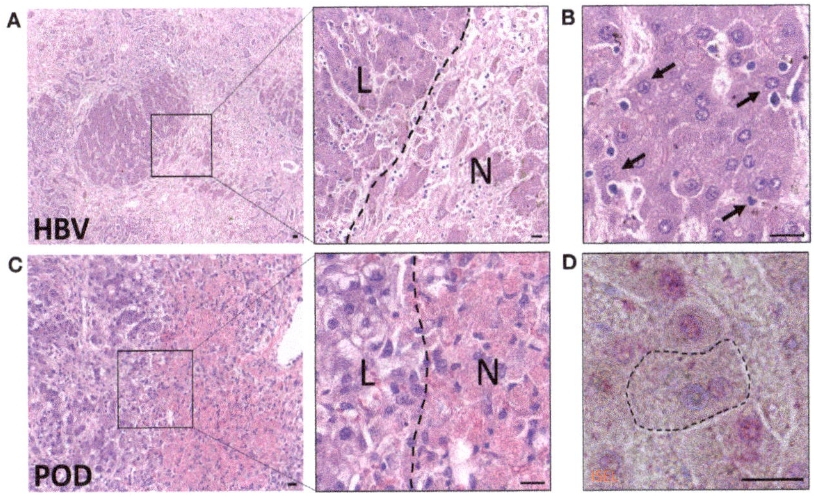

FIGURE 2 | Hepatocytes engulf necrotic and apoptotic cells in acute-on-chronic liver injury caused by hepatitis B infection (HBV) and in paracetamol injury (POD). **(A)** Hematoxylin–eosin staining of acute-on-chronic liver injury in a patient with HBV infection. Large areas of hepatocyte necrosis are evident. Inset image shows dark stained hepatocyte nuclei in live hepatocytes (L) and pyknotic or karyolytic nuclei in necrotic hepatocytes (N). **(B)** Healthy hepatocytes with clearly marked nuclei are seen phagocytosing small apoptotic cells (arrows). Note hepatocyte invaginations which have formed to enable capture of apoptotic cells. **(C)** Hematoxylin-eosin staining of liver with paracetamol-induced injury, which causes centrilobular necrosis. Inset shows pink cytoplasm in necrotic hepatocytes (N) compared to surviving non-discolored hepatocytes with clearly defined nuclei (L). **(D)** *In situ* end labeling (ISEL) of apoptotic cell nuclei is seen here in pink, in a liver with ischemia-reperfusion injury. The marked hepatocyte has a non-apoptotic nucleus seen in blue, and has engulfed an apoptotic cell with a pink nucleus. Neighboring apoptotic hepatocytes can be seen with pink nuclei, and non-apoptotic cells with blue nuclei. The bars show 20 μm.

etiologies, including autoimmune, metabolic, viral, and genetic diseases (12, 13). The rapid processing of dead and dying cells is vital to moderate inflammation (12, 14, 15).

It is remarkable how little we know about the molecular mechanisms that govern the ability of the largest internal organ in the body to mediate the clearance of damaged or dying cells, given that this is one of the liver's major functions. Herein, we bring together research on hepatocyte efferocytosis and place it into context with current molecular knowledge on the clearance of dead cells by immune phagocytes.

CLEARANCE OF APOPTOTIC AND NECROTIC CELLS

Cells die through a wide array of processes, each situational and requiring their own dedicated cascade of signaling events. The most frequent forms of cell death are attributed to apoptosis or necrosis. Apoptosis, an active form of programmed cell death, is characterized by the initiation of specific inducible pathways (16, 17). This includes the extrinsic pathway; the engagement of extracellular signals, including Fas ligand (FasL) (18) and tumor necrosis factor family cytokines (TNF) (19), amongst others, to their respective death receptors which initiates intracellular death signaling. Apoptosis can also be triggered intrinsically; certain signals, such as a lack of growth factors, endoplasmic reticulum stress or DNA damage, can induce a shift in expression of Bcl-2 family mitochondrial proteins (20). Increased activity of proapoptotic proteins lead to cytochrome C release and caspase 9 activation. Apoptosis pathways result in the activation of effector caspases (3, 6, and 7), which in turn begin to proteolytically degrade the cell's components. Apoptotic cells are generally smaller than live cells and can be identified by the formation of surface blebs (16).

Necrosis is considered a passive, unprogrammed type of cell death and is often incurred accidentally, although active mechanisms of necrosis have also been reported (21). While multiple mechanisms can induce necrosis, the major causes are attributed to compromising of the plasma membrane, or depletion of energy (22). Furthermore, apoptotic cells can be converted to necrotic cells (also known as *secondary necrosis*) if ATP levels fall below the quantity required to complete the active apoptotic process (23). The appearance of necrotic cells is often swollen with disrupted organelle and plasma membranes (24). The nucleus is often broken down and will be unstained by hematoxylin (**Figure 2**). As necrosis often occurs in areas of tissues, rather than the single cell death hallmark of apoptosis, often multiple necrotic cells can be identified in one area. Due to their lack of integrity, necrotic cells will often form cell debris, which can induce liver damage if not cleared swiftly, as we discuss in later sections.

Upon the death of a cell, its corpse must be cleared through efferocytosis. This is a specialist form of phagocytosis, whereby fragments of the dying cells are engulfed by other cells, which in turn degrade and recycle their components. Although both apoptotic and necrotic cells are often captured by the same efferocytes, each are recognized through different means and yield differing response in the predatory cell (25). Apoptotic cells are most commonly recognized through the display of the phospholipid phosphatidylserine (PtdSer) on the outer leaf of the plasma membrane that can be recognized by many receptors [phosphatidylserine receptors (PSRs)] directly (26) or *via* association with low-density lipoprotein (27, 28). Of note, in a rat liver model, it was shown that recognition of apoptotic cells from mice or humans was reduced compared to rat cells; it is therefore possible that species-specific recognition molecules can mediate efferocytosis (29).

A phenotypic aspect of apoptotic cells is that, although shriveled, the cell remains intact as a singular body. This allows for a clean removal of the dying cells by efferocytes, usually without provoking an inflammatory response. How intact a necrotic cell remains is reflected through the manner in which cell death was induced. As such, multiple modalities for necrotic cell recognition are necessary to guarantee their clearance. Some reports have suggested that necrotic cells can also be recognized by PSRs (30). However, due to the lack of integrity of most necrotic cells, they are often recognized through molecules exposed by necrotic death (25). The same mechanisms are also used to detect pathogens. For example, complement receptors and Fc receptors detect opsonized necrotic cells, and this recognition can trigger signaling events that activate the phagocyte (31–35). As such, necrotic cells are engulfed through the detection of autoantigens, which often increases the risk for autoimmune disease. Necrotic cells can also be indirectly recognized through opsonin engagement of other cellular components. For example, ficolin-2 and -3 have been shown to bind DNA, facilitating the clearance of late-apoptotic/necrotic cells through interactions with calreticulin (33, 36). A ubiquitous mechanism for clearance of necrotic cells remains uncertain.

The differences between recognition, and thus further downstream signaling of apoptotic and necrotic cells, result in conversing consequences for the efferocyte (37). Apoptotic cell clearance generally leads to the production of anti-inflammatory stimuli and pro-resolution signals for inflammation such as interleukin 10 (IL-10) and TGF-β (38). Conversely, necrotic clearance generally results in pro-inflammatory signaling, as many of the recognition receptors are also required for pathogen recognition. In the liver, the signals associated with hepatocyte death were recently reviewed by Brenner and colleagues (39). In this work the importance of the extent and duration of dead cell accumulation was highlighted, as mild and localized cell death can aid regeneration and exert hepatoprotective effects. Equally, prolonged and wide-spread cell death can exacerbate liver injury.

EXPERIMENTAL SYSTEMS TO STUDY EFFEROCYTOSIS

Multiple techniques have been described for both *in vivo* and *in vitro* studies of efferocytosis. Fluorescent dye-labeled efferocytes can be "fed" alternatively labeled dead cells under varying conditions and time courses. Early apoptosis can be confirmed by Annexin V labeling of the cell surface as it binds directly to PtdSer, although care must be taken when studying certain activated cell types or using calcium-sensitive protocols (40). Later stages of apoptosis or necrosis are often confirmed with cell impermeable DNA dyes such as 7AAD or TOPRO-3 iodide, which can enter cells once the membrane is compromised. Combined labeling with Annexin V and a membrane-impermeable DNA label was developed to identify the stages of apoptotic cells in more detail (41). Cells can then be assessed by flow cytometry, or imaged by fluorescent microscopy. Complete internalization of dead cells can be confirmed by lack of access to membrane dyes added to the culture media (such as CellMask Plasma Membrane Stains, Thermo Fisher Scientific) or demonstration of efferosome acidification using pH indicator dyes (**Figure 3**). Quantitative analyses by confocal and time-lapse microscopy can be useful to determine the frequency and kinetic of efferocytosis *in vitro*.

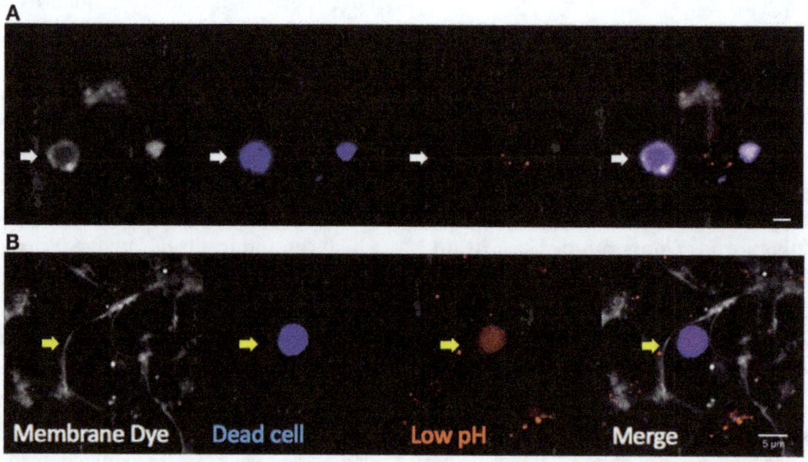

FIGURE 3 | Visualizing efferocytosis by confocal microscopy. Hepatic epithelia were cocultured with violet-labeled staurosporin-treated apoptotic Jurkat T cells in the presence of pHrodo red, which only fluoresces in conditions of low pH (Thermo Fisher Scientific). CellMask Plasma Membrane stain was added to the culture media to label all exposed cell membranes before imaging. **(A)** Non-internalized apoptotic cells (blue) attached to hepatocytes were labeled by CellMask Plasma Membrane in white, and they were not labeled by pHrodo red dye (white arrow). **(B)** Internalized dead cells were not accessible to the membrane dye, confirming internalization (yellow arrow). Complete internalization into an acidic compartment was confirmed by pHrodo red, which detected efferosome acidification as early as 3 hours following engulfment. The scale bar indicates 5 μm.

Fluorescent labeling of dead cells and efferocytes may also be adapted for flow cytometry-based studies, whereby double-positive cells represent efferocytes containing cargo. This form of analysis has been used to study the clearance of neuraminidase-treated red blood cells in mice (42). Other studies opt to analyze efferocytosis using downstream secreted molecules as proxy to utilize alternative techniques such as reporter assays. The capacity of Scavenger Receptor Class F Member 1 (SCARF1) to act as a dead-cell receptor on transfected HEK293T cells, for example, was confirmed using IL-8 mRNA production as a marker of NF-κB activation following apoptotic and necrotic cell efferocytosis (43).

Efferocytosis is not often as straightforward to detect *in vivo*. Fluorescent labeling can enable temporal measurements in mouse models by intravital imaging of the liver (44), but the technique remains to be adapted successfully for use in human tissues *ex vivo*. Molecular markers of cell death for use with fixed tissue are often important for the confirmation of efferocytosis. Caspase 3/7 activation or their effects can be measured to delineate apoptotic bodies by immunohistochemistry (IHC) or immunofluorescence (IF) (45). DNA end-labeling is frequently used to confirm the death of cells in tissues. End-labeling involves the addition of labeled nucleotides to DNA breaks induced throughout multiple modalities of death, using a DNA polymerase. This was historically used for *in situ* end labeling (ISEL) of fixed tissue sections as part of IHC chromagen staining (**Figure 2D**) (46). This was then adapted for the creation of terminal deoxynucleotidyl transferase (TdT) dUTP Nick-End Labeling (TUNEL) (47) which substitutes a polymerase for TdT. This adaptation allows for the use of many different modified forms of labeled nucleotides (often dUTP), such as non-reactive protein tags or fluorophores. TUNEL staining has been altered to specifically identify cells in late-stage apoptosis.

Further stains for cell membrane proteins or specific nucleic proteins can be used to determine complete engulfment of dead cells. Many of these techniques were exemplified in a recent study of macrophage/monocyte efferocytosis in models of acute liver injury (48). In this work, Antoniades and colleagues studied the mechanism of resolution of liver inflammation through apoptotic cell clearance by macrophages/monocytes *via* Mer tyrosine kinase receptor (MerTK). Staining for myeloperoxidase (activated neutrophils) combined with TUNEL allowed for the identification of apoptotic neutrophils in human liver, both through IHC and IF staining. Additionally, fluorescent monocytes were cocultured *in vitro* with alternatively fluorescent apoptotic hepatic cells or neutrophils. The ability of these monocytes to clear apoptotic cells was then assessed through fluorescent microscopy and flow cytometry. Both techniques were used to show the increased capacity of monocytes for dead cell clearance following stimulation with secretory leukocyte protease inhibitor.

THE MECHANISM OF DEAD CELL CAPTURE BY PROFESSIONAL AND NON-PROFESSIONAL EFFEROCYTES

Phagocytes express several receptors to recognize and subsequently clear dying cells from the tissues (49–51). In the case of professional phagocytes (e.g., macrophages) multiple apoptotic and necrotic cell receptors have been characterized and these remain relevant in the liver (37, 52) (**Table 1**). First described in 1992, it is now widely accepted that apoptotic cells are recognized through their expression of PtdSer on the outer leaf of the plasma membrane (53, 54). Several receptors directly recognize PtdSer, many of which are expressed by professional phagocytes (55). These include stabilin-1, stabilin-2, brain-specific angiogenesis inhibitor 1 (BAI1), and RAGE, as well as the TIM family of transmembrane glycoproteins, including TIM-1, -3, and -4 (56–61). Mammary, alveolar and mesangial epithelia recognize apoptotic cells *via* the PSR, CD36, the vitronectin receptor αvβ3, and CD91 (62–64). Of note, molecules that bind PtdSer such as high-mobility group box 1 (HMGB1) can also downregulate apoptotic cell clearance (65, 66).

It is common for PtdSer to be recognized in complex with certain bridging molecules. Some of the most well-studied PtdSer receptors, the TAM tyrosine kinases (Tyro3, Axl, and MerTK) work in this manner (102); notably, hepatocytes express Axl but not Tyro3 or MerTK (103). The earliest known examples of these are Gas6 and Protein S (104, 105). Gas6 is universally recognized by TAM receptors, whereas Protein S, which is expressed in hepatocytes, is not recognized by Axl. Similarly, integrins αvβ3 and αvβ5 have been shown to promote efferocytosis through the recognition of PtdSer in complex with lactadherin, also known as milk fat globule EGF factor 8 (MFG-E8) (67, 68, 106).

The entirety of apoptotic cell recognition does not lie with the detection of PtdSer expression. It was shown that Tubby protein and its relative Tubby-like protein 1 (TuLP1), which do not bind PtdSer, specifically localize at the surface of apoptotic cells and could act as TAM receptor bridging molecules in a similar manner to Gas6, which in turn promoted apoptotic cell clearance (89). All TAM tyrosine kinases recognized TuLP1, whereas Tubby was exclusively recognized by MerTK on macrophages and retinal pigment cells. Mechanisms of immune surveillance and signaling have also been shown to contribute to apoptotic cell clearance. Components of the complement pathway have been shown to induce phagocytosis in macrophages and DCs by opsonizing apoptotic cells, including C1q and C3 (34, 78, 107). Furthermore, SIGN-R1, a mouse analog of human mannose receptor DC-SIGN, was shown to bind apoptotic cells and induce their labeling with C3 and subsequent clearance by marginal zone macrophages (92).

Recognition of apoptotic cells, although important, is not sufficient for macrophages to engulf and clear them. Downstream intracellular signaling is necessary for load-processing following capture. An important, highly-conserved signaling pathway has been described downstream from most common PtdSer-receptors, involving GTPase Rac1 and ELMO1-DOCK180 interactions (50). TAM-family molecules, αvβ5 integrins and BAI1 act as docks for apoptotic cells, leading to intracellular signaling *via* this pathway (108, 109). Upon engagement of an apoptotic cell by these receptors, DOCK180 is recruited by ELMO1 (110, 111). In complex, these proteins act as guanine exchange factors, allowing for Rac1 activation, which induces necessary cytoskeletal arrangements required for complete engulfment of the prey cell. Stabilin 1 and 2 have also been shown to activate this

TABLE 1 | Efferocytosis receptors in professional phagocytes and tissue epithelia.

Name	Cell type	Target	Reference
Professional phagocyte receptors (macrophages/dendritic cells)			
αvβ3 integrins	Macrophages	Lactadherin (MFG-E8)—PtdSer, vitronectin	(67–69)
αvβ5 integrins	Macrophages Dendritic cells	Lactadherin (MFG-E8)—PtdSer, vitronectin	(67, 69–72)
Axl	Monocytes/macrophages	Gas6-PtdSer, Tubby-like protein 1 (GULP), Protein S	(73–76)
BAI1 (brain-specific angiogenesis inhibitor 1)	Macrophages	PtdSer	(77)
Calrecticulin/CD91	Monocytes, macrophages, neutrophils	Complement component C1q	(34)
CD11/b/c/CD18	Monocytes, macrophages, neutrophils, human DCs	Complement component C3bi	(78, 79)
CD14	Macrophages	Phospholipids (not PtdSer-dependent)	(80, 81)
CD36	Macrophages	Thrombospondin + PtdSer + oxLDLs	(82–84)
Clec9a	Dendritic cells	Necrotic cells, exposed actin filaments	(85, 86)
LOX1	Macrophages	oxLDLs—PtdSer	(28, 87)
MARCO	Macrophages	Uncertain	(88)
MerTK	Monocytes/macrophages	Gas6-PtdSer, Tubby, Protein S, Tubby-like protein 1, Protein S	(75, 89, 90)
Phosphatidylserine receptor (PSR)	Monocytes/macrophages	PtdSer	(91)
RAGE (receptor for advanced glycation end products)	Alveolar macrophages	PtdSer	(60)
SCARF1	Monocytes and dendritic cells	Complement component C1q—PtdSer	(43)
Scavenger receptor A (SR-A)	Macrophages	Uncertain	(88)
SIGN-R1 (specific intercellular adhesion molecule-3-grabbing nonintegrin-related 1) (murine)	Mouse marginal zone macrophages	Not confirmed for apoptotic cells	(92)
Stabilin-1 (CLEVER-1)	Tissue-specific, alternatively activated macrophages	PtdSer	(61)
Stabilin-2	Macrophages	PtdSer	(57)
TIM-3 (T cell/transmembrane, immunoglobulin, and mucin 3)	Dendritic cells	PtdSer	(93)
TIM-4	Monocytes/macrophages	PtdSer	(94)
Tyro3 (sky)	Monocytes/macrophages	Gas6-PtdSer, Protein S, Tubby-like protein 1, Protein S	(73–76)
Non-professional phagocytes			
αvβ5 integrins	Retinal epithelial cells	Lactadherin (MFG-E8)-PtdSer	(95)
ASGPR (asialoglycoprotein receptor)	Hepatocytes	asialoglycoprotein	(96, 97)
CD36	Retinal Pigment cells	PtdSer	(98)
KIM-1 (kidney injury molecule 1)/TIM1 T cell/ transmembrane, immunoglobulin, and mucin 1	Injured kidney endothelial cells	PtdSer	(99)
LOX1	Endothelial cells	oxLDLs—PtdSer. Ca^{2+}-dependent	(28, 100)
MerTK	Retinal pigment cells	Gas6-PtdSer, Tubby, Protein S, Tubby-like protein	(89, 90)
Phosphatidylserine receptor (PSR)	Fibroblasts Epithelial cells T and B lymphocytes (ectopic expression)	PtdSer	(91)
SCARF-1 (SREC-1)	Endothelial cells	Complement component C1q—PtdSer	(43)
Stabilin-1 (Clever-1)	Human sinusoidal endothelial cells	PtdSer	(101)
Stabilin-2	Human sinusoidal endothelial cells	PtdSer	(101)

pathway through the adaptor protein GULP (109, 112, 113). Completion of apoptotic cell engulfment also commonly involves the activation of nuclear receptors. Loose nucleotides released from dying cells commonly act as "*eat-me*" signals, and can engage purigenic P2 receptors (P2X and P2Y), leading to an increased capacity for efferocytosis in macrophages (114, 115). It was recently shown that liver X receptor (LXR) was necessary for the capture and processing of apoptotic cells by macrophages and dendritic cells (116, 117). LXR responds to oxysterols found in engulfed apoptotic cells. Stimulation of LXR upregulated MerTK and anti-inflammatory cytokines IL-10 and TGF-β, while also leading to the downregulation of proinflammatory cytokines such as IL-1β, CCL2, and MARCO. A-Gonzalez and Hidalgo reviewed nuclear receptors and their role in macrophage efferocytosis recently (118). LXRα mediates fatty acid regulation in hepatocytes (119), but its role in hepatocyte efferocytosis remains to be determined.

Non-professional phagocytes, such as epithelial cells express multifunctional scavenger receptors, or molecules that exert alternative functions in other cell types. For example, TIM-1, also known as kidney injury molecule 1 (KIM-1), is known to possess multiple immune functions, including $CD4^+$ T-cell and mast cell activation (59). However, TIM-1 was also upregulated in kidney epithelia following injury, allowing for a temporary efferocytic capability (99). Certain cell-exclusive receptors and modulators associated with apoptotic cell clearance have also been described. Apoptotic cell clearance in the liver has been shown to involve asialoglycoprotein receptor (ASGPR) on hepatocytes (96). This ASGPR1 and ASGPR2 complex is critical for receptor-mediated endocytosis of terminally desialylated glycoproteins and is restricted to the liver. Autoantibodies to ASGPR have been found in patients and models of autoimmune hepatitis (120–122). Resolvin D1 is also important in liver protection from ischemia/reperfusion injury, by enhancing efferocytosis by M2-polarizing macrophages (123). Furthermore, phagocytosis by retinal pigment cells, although mostly conducted through MerTK, was shown to be increased through recognition of ATP-binding cassette subfamily F member 1 (ABCF1) released from apoptotic photoreceptor outer segments (124). Overall, it appears that although the broad function of recognition and engulfment of apoptotic cells is conserved throughout many cell types, multiple mechanisms exist that conduct these processes across phagocytes, both homo- and heterotypically.

REGULATION OF EFFEROCYTOSIS

External stimuli are pertinent for regulation of dead cell clearance by efferocytosis. As such, "*find-me*" signals released by apoptotic cells are often necessary for the guidance of efferocytes to their prey (53). The best-characterized examples of these are extracellular nucleotides (115). It was shown that upon caspase 3/7 activation in apoptotic cells, ATP and UTP released from apoptotic cells could recruit monocytes/macrophages through recognition by $P2Y_2$. Conversely, molecules with the opposite effect known as "*don't eat-me*" signals have also been described. CD47 is the most notable, having been shown to provide resistance to clearance by macrophages on malignant cells and more recently on atherosclerotic plaques (125, 126). Similar "find-me" signals may be utilized by non-motile phagocytes, which extend protrusions to collect apoptotic cells for clearance but are restricted to targets within their tissue niche.

Due to the influence of dying cells on the immune response, cytokine and growth factor stimulation of both professional and non-professional phagocytes can regulate their capacity to clear dead cells. Apoptotic T-cell lymphomas release sphingosine-1-phosphate, a bioactive lipid often involved in immune cell recruitment, leading to the recruitment of macrophages and monocytes (127). Similarly, certain chemokines, tasked with immune cell recruitment have also been shown to increase phagocyte recruitment to areas of apoptotic cells. CX_3CL1 (fractalkine) was shown to recruit macrophages to its source, apoptotic Burkitt lymphoma cells (128).

Multiple cytokines have varying effects on efferocytosis (129). Most notably, secretion of IL-3 and IL-14 increased efferocytosis in macrophages through activation of PPAR and increase in CD36 expression (130, 131). IL-4 has been reported to upregulate expression of other PtdSer-receptors such as stabilin 1and 2 (61). IL-10 and TGF-β can also increase efferocytes by macrophages (132, 133). In contrast, pro-inflammatory cytokines reduce the capacity for dead cell engulfment: TNF-α has been shown to inhibit efferocytosis in macrophages (134) and both IFN-γ secretion and receptiveness were reversely correlated with anti-inflammatory cytokines and receptors including IL-4 and TIM receptors (59, 129, 135). However, this was not always the case for these cytokines. Both TNF-α and IFN-γ have been shown to increase LOX-1, which may recognize apoptotic cells by LDL-labeled PtdSer. Furthermore, IFN-γ activation of macrophages, in the absence of other pro-inflammatory stimuli, was shown to increase apoptotic uptake (135).

The ability of phagocytes to clear dead cells is also subject to regulation. This is the result of alterations in gene expression, which can function as negative feedback following initial engulfment of dying cells. For example, it has been shown that macrophages, upon engulfing apoptotic cells can undergo a form of activation and reprograming (136). As well as skewing the macrophage to a more anti-inflammatory phenotype, which in turn promotes inflammatory resolution, both mouse and human macrophages can upregulate CXCR4 during efferocytosis, which in turn encourages their recruitment to draining lymph nodes (137). These macrophages were also shown to subsequently reduce their efferocytosis capacity. Thus, apoptotic cells can reduce local levels of efferocytosis as well as promote them.

More recently macrophages were shown to regulate efferocytosis in surrounding non-circulating phagocytes, such as phagocytic airway epithelial cells (138). In response to IL-4 and IL-13, which are secreted by epithelia and stimulated Th2 cells, macrophages upregulated secretion of both insulin-like growth factor 1 (IGF-1) and microvesicles containing anti-inflammatory signals. Both microvesicles and IGF-1, in turn, fed back to epithelia, causing a reduction of apoptotic cell clearance in favor of microvesicle uptake.

In the context of the liver, some of the mechanisms described for the regulation of efferocytosis apply to circulating and resident macrophages/monocytes. Further, the neuronal guidance protein netrin-1 has been shown to promote resolution of ischemia/reperfusion injury, in part by increasing the capacity of Kupffer cells to engulf apoptotic cells (139). The same molecule was shown to promote liver regeneration (139). In a mouse model of colon carcinoma metastasis in the liver, intercellular cell adhesion molecule 1-deficient macrophages cocultured with tumor cells showed increased efferocytosis dependent on phosphatidylinositol 3 kinase (140).

Environmental factors can also affect phagocytosis, and this extends to the clearance of dead cells; studies in human skin have demonstrated that ethanol can reduce phagocytic function (141), and there have been reports on increased phagocytosis in ethanol-fed rats, which was modulated by diet (142). Hepatocyte phagocytosis of apoptotic cells was decreased in ethanol-fed rats compared to controls, therefore the effects of ethanol on efferocytosis may be cell type-dependent (97). It is unclear whether professional phagocytes play a role in the regulation of efferocytosis by hepatocytes and liver endothelial cells.

THE IMPACT OF EFFEROCYTOSIS BY TISSUE EPITHELIA

Non-professional efferocytes are important throughout all developmental stages of an organism and can take over the clearance of apoptotic cells in the absence of professional phagocytes (143). This was confirmed in PU.1 knockout mice that lack macrophages, and the removal of apoptotic cells required for foot-limb development was instead performed by mesenchymal cells (144). Non-professional phagocytes therefore contribute to efferocytosis, even at the earliest stages of development.

Some of the best-studied phagocytic epithelia are bronchial and alveolar epithelial cells (138, 145, 146). Epithelial cells lining the respiratory tract make first contact with airborne allergens such as house dust mite antigens. Subsequent inflammatory stimuli, including the recruitment of basophils, mast cells and lymphocytes, result in epithelial cell injury. Lung epithelia clear their dying neighbors through PtdSer and Rac1-dependent mechanisms, which can be modified experimentally (146). As with macrophages, apoptotic cell clearance by lung epithelia induced anti-inflammatory cytokines such as IL-10 and TGF-β. Conditional Rac1 deletion in mouse lung epithelia resulted in an exacerbated immune response and greater epithelial damage. These studies demonstrated the efficiency and importance for lung epithelial cell efferocytosis in the regulation of lung inflammation (146, 147).

Retinal epithelial cell efferocytosis has also been well characterized (148–151). Light-sensing cells of the retina are frequently turned over *via* programmed-cell death, often succumbing to autophagy-associated death, called *autolysis* (152, 153). Dysregulation of autophagy in these cells has been frequently reported to increase retinal pigment cell death (152, 154). Although a normal part of age-related macular degeneration, failure to clear these dying cells can accelerate retinal damage. Together with professional phagocytes, retinal pigment cells are also charged with the removal of dead cells, in a manner dependent on MerTK.

Throughout the lifecycle of an organism, the removal of immature cells or those with high turnover is necessary to maintain tissue homeostasis. Intravital microscopy has revealed how hair follicles in mice regress through programmed cell death of hair-producing basal epithelial cells (155). Neighboring cells of the same type then clear apoptotic cells through mechanisms requiring TGF-β signaling. In response to kidney damage, epithelial cells recognize and engulf PtdSer-positive apoptotic cells *via* KIM-1 or TIM-1 (99). Colonic epithelial cells have also been shown to engulf their apoptotic neighbors, which aids in maintaining low levels of inflammation (156).

Studies in multiple progenitor types have recently identified their importance in efferocytosis. Skeletal muscle progenitors recognizing PtdSer on neighboring apoptotic cells, receive the signal to differentiate and fuse into multinuclear myofibers (157). Mesenchymal stem cells take their cues from bone marrow apoptotic cells *via* efferocytosis and undergo osteogenic differentiation (158). Chondrogenic progenitor cells display macrophage-like abilities in that they react to "find-me" signals from apoptotic cells (159), and non-motile chondrocytes also have a role in efferocytosis (160). As previously discussed, neuronal progenitors which apoptose following failure to complete neural circuits throughout neurogenesis, were recognized and cleared by other progenitor cells *via* Rac1 activation following ELMO-1 signaling (161). Of note, neuronal and hepatic epithelia can be derived from common progenitor cells.

Through its cardinal role in the neutralization of toxic substances, to its frequent influx and arresting of leukocytes, the liver has evolved to cope well with cell death (10, 162). Although hepatocytes are somewhat resistant to intrinsic apoptotic pathways (163–165), many death receptors are ubiquitously expressed throughout the liver, increasing their susceptibility to extrinsic apoptosis by exposure to pro-inflammatory cytokines such as TNF family molecules including TNF-related apoptosis-inducing ligand (TRAIL) (166–170). Clearance of apoptotic cells by macrophages is a pro-resolution process, however, liver-infiltrating macrophages and Kupffer cells can upregulate death ligands in the liver, including FasL, TNF-α, and TRAIL, increasing the rate of local hepatocyte death and the risk of further inflammation (171, 172). Acute injury such as ischemia and the resulting trauma from hypoxia/reoxygenation can also result in similar sudden increases in necrotic cell death (164, 165, 169). Furthermore, steatosis—accumulation of lipids associated with a multitude of fatty liver diseases—can cause wide hepatocyte cell death *via* lipoapoptosis induced by ER stress-mediated intrinsic pathways (173, 174). Ethanol-induced injury can also have an impact on receptor-mediated endocytosis by the ASGPR and efferocytosis (175–178).

Failure to clear dead cells from the parenchyma is accumulatively detrimental to the liver; clearance of necrotic cells—both primary and secondary, resulting from uncleared apoptotic cells—results in increase in pro-inflammatory cell influx and cytokine secretion, leading to further damage to the liver (14). HMGB1 is important in liver protection from ischemia/reperfusion injury (179), yet in a sterile model it acted as a damage-associated molecular pattern that enhanced liver injury in both ischemia/reperfusion and POD models (180). Interactions between ASGPR on hepatocytes and B220 epitope of CD45 assist in the capture and trapping of apoptotic cells in the liver (96, 181, 182). The impact of hepatocyte efferocytosis on the inflammatory milieu remains to be established.

Beyond the capacity of hepatocytes for erythrocytosis (183), further evidence or insights into the mechanisms or anti-inflammatory impact of hepatocyte efferocytosis have not been elucidated. Hepatocytes express an array of immunomodulatory cytokines, including TNF-α and IL-10 (184–186); it is not known whether these are modulated during efferocytosis as in lung epithelia and in professional phagocytes. Understanding the molecular mechanisms, purpose and regulation of dead cell clearance by hepatocytes is vital to estimate its impact on the onset and resolution of inflammation, as elevation in hepatocyte apoptosis is key to the pathogenesis and progression of most forms of liver disease (14). Outstanding questions on hepatocyte efferocytosis include:

- *What are the molecules that mediate recognition and engulfment of apoptotic and/or necrotic cells by hepatocytes?* ASGPR is thus far the only receptor restricted to hepatocyte efferocytosis; despite its multiple roles in receptor-mediated efferocytosis, ASGPR-deficient mice develop normally yet have exacerbated pathology in liver injury models (177, 187).
- *Is efferocytosis by hepatocytes in portal and centrilobular regions mediated by the same molecular mechanisms?* These

regions have differential access to apoptotic and necrotic cells, respectively, as well as nutrient, inflammatory infiltrate and oxygenation levels that may all influence the capacity for efferocytosis.
- *How is hepatocyte efferocytosis regulated in health, infection, inflammation and cancer?*
- *Can hepatocyte efferocytosis be modulated by pharmacological interventions?*
- *Does efferocytosis affect the ability of hepatocytes to regenerate during injury?*

CLINICAL IMPLICATIONS OF DEFECTS IN EFFEROCYTOSIS

Failure to remove dying cells, both apoptotic and necrotic, have been connected to disease exacerbation (49). Accumulation of dying cells increases the availability of proimmunogenic factors and can increase the risk of autoimmunity, especially as death-recognition becomes skewed to proinflammatory recognition of secondary-necrotic cells. This topic was explored in a recent special issue in Frontiers in Immunology (188).

Defects in efferocytosis have also been shown to be beneficial for the longevity of tumors. Upregulation in the "don't-eat me" signal CD47 was reported in myeloid leukemia (189, 190) which was associated with increased tumor survival and poorer prognosis. Similar pathogenic consequences of aberrant efferocytosis have been exemplified through deficiencies in death receptors (1). Loss of axl, MerTK, and its associated ligand, Gas6, have all been shown to promote the growth of colon cancers (191, 192). Conversely, loss of stabilin-1 has shown to reduce growth of implanted tumors in knockout mice, due to reduced recruitment of tumor-associated lymphocytes and macrophages (193). As such, loss of death-receptor expression is not always beneficial for cancer vitality. However, loss of other receptors for dying cells has displayed varying phenotypes associated with the lack of apoptotic cell clearance. Loss of SCARF1 and axl has been reported to promote autoimmunity (43, 194).

Similar dangers to those mentioned above regarding deficiencies in dying cell clearance are apparent for many liver diseases. In the context of the liver, the effects of efferocytosis in autoimmune family disorders have not been established directly. Reports on efferocytosis in liver diseases are listed in **Table 2**. Clearance of dying cells in the liver is thought to reduce the risk of autoimmune hepatitis and promote reversal of fibrosis by macrophages (195, 196). In primary biliary cholangitis, efferocytosis by biliary epithelia may be important in defining the tissue specificity of the autoimmune response (8, 197). It is worth considering that standard of care treatments for autoimmune conditions include corticosteroid regimens, which have been shown to upregulate efferocytosis (49, 198). Prevention of efferocytosis may therefore exacerbate liver diseases.

As well as causing hepatocyte necrosis, chronic alcohol exposure was reported to reduce macrophage efferocytosis through diminishing MFG-E8 expression (209). Prevention of efferocytosis by macrophages in the liver could increase further inflammatory stimuli, although it is not clear how hepatocyte efferocytosis would be affected. Contrarily, reduced efferocytosis in certain disease models has been shown to be beneficial. Loss of the dead-cell receptor TIM4, for example, in a mouse model of ischemia/reperfusion injury reduced immune cell infiltration and hepatocyte damage (210). Understanding the protein-specific and situational benefits or detriments to reduced efferocytosis in diseases of the liver and other organs can give insights into possible therapeutics for tissue damage and autoimmunity.

TABLE 2 | The role of efferocytosis in liver diseases.

Liver disease	Efferocytosis relevance	Reference
Autoimmune hepatitis	Hepatocyte stress and correlations to disease Autoantibodies targeting ASGPR	(39, 121, 195)
Primary biliary cholangitis	Phagocytes were shown to contain PDC-E2 immunogen Biliary injury clearance is linked to autoimmunity Biliary injury clearance alleviates liver fibrosis	(8, 196, 199–201)
Primary sclerosing cholangitis	Collection of genome-wide studies that show a role of apoptosis	(202)
Alcohol injury	Ethanol exacerbates injury in ASGPR-deficient model	(97, 177, 178, 203, 204)
Fatty liver diseases	The role of specialized proresolving mediators in obese individuals (enhance efferocytosis)	(205–207)
Other liver injuries	Alpha 1 antitrypsin rescues macrophage efferocytosis Netrin 1 rescues efferocytosis in murine I/R injury model Efferocytosis and tissue remodeling in rat bile duct ligation model	(139, 196, 208)

Although efferocytosis is critical for liver homeostasis, there is limited information on specific efferocytosis pathways that contribute to liver disease pathogeneses. The importance of dead cell clearance is better established than the mechanisms that mediate efferocytosis in the inflamed or injured liver.

CONCLUSION

Recent advances in epithelial cell efferocytosis have highlighted the importance of tissue epithelia in the everyday clearance of billions of apoptotic cells. Compared to professional efferocytes, there is little known regarding the receptors and molecular processes involved in the recognition of apoptotic and necrotic cells by non-professional phagocytes, including molecules that may confer tissue-specific function. Given the impact of efferocytosis on the pathogenesis of autoimmunity, tissue injury and tumor biology (211), molecules driving efficient clearance of dead cells are valid therapeutic targets. Hepatocyte efferocytosis, accomplished at least in part by the liver-restricted ASGPR, is an attractive target for therapeutic intervention for a multitude of liver diseases.

ETHICS STATEMENT

This study was carried out in accordance with the recommendations of LREC 06/Q2708/11, South Birmingham, Birmingham,

UK. All subjects gave written informed consent in accordance with the Declaration of Helsinki.

AUTHOR CONTRIBUTIONS

SPD, GMR, and ZS performed stains for immunohistochemistry and immunofluorescence for illustrative purposes, and wrote the manuscript.

ACKNOWLEDGMENTS

We thank Janine Fear and Bridget Gunson for excellent laboratory and patient data management, and the staff and patients at the Queen Elizabeth Hospital Liver and Hepatobiliary Unit, Birmingham, UK.

FUNDING

SPD is supported by a PhD studentship from the Medical Research Council (MRC) Centre for Immune Regulation, University of Birmingham, UK. GMR is supported by the National Institute for Health Research (NIHR) Birmingham Liver Biomedical Research Unit, University of Birmingham, UK. ZS is supported by a Royal Society Dorothy Hodgkin Fellowship, the Wellcome Trust Institutional Strategic Support Fund, and MRC Confidence in Concept award.

REFERENCES

1. Arandjelovic S, Ravichandran KS. Phagocytosis of apoptotic cells in homeostasis. *Nat Immunol* (2015) 16:907–17. doi:10.1038/ni.3253
2. Guidotti LG, Inverso D, Sironi L, Di Lucia P, Fioravanti J, Ganzer L, et al. Immunosurveillance of the liver by intravascular effector CD8(+) T cells. *Cell* (2015) 161:486–500. doi:10.1016/j.cell.2015.03.005
3. Warren A, Le Couteur DG, Fraser R, Bowen DG, McCaughan GW, Bertolino P. T lymphocytes interact with hepatocytes through fenestrations in murine liver sinusoidal endothelial cells. *Hepatology* (2006) 44:1182–90. doi:10.1002/hep.21378
4. Edwards S, Lalor PF, Nash GB, Rainger GE, Adams DH. Lymphocyte traffic through sinusoidal endothelial cells is regulated by hepatocytes. *Hepatology* (2005) 41:451–9. doi:10.1002/hep.20585
5. Krenkel O, Tacke F. Liver macrophages in tissue homeostasis and disease. *Nat Rev Immunol* (2017) 17:306–21. doi:10.1038/nri.2017.11
6. Tacke F. Targeting hepatic macrophages to treat liver diseases. *J Hepatol* (2017) 66:1300–12. doi:10.1016/j.jhep.2017.02.026
7. Canbay A, Taimr P, Torok N, Higuchi H, Friedman S, Gores GJ. Apoptotic body engulfment by a human stellate cell line is profibrogenic. *Lab Invest* (2003) 83:655–63. doi:10.1097/01.LAB.0000069036.63405.5C
8. Rong GH, Yang GX, Ando Y, Zhang W, He XS, Leung PS, et al. Human intrahepatic biliary epithelial cells engulf blebs from their apoptotic peers. *Clin Exp Immunol* (2013) 172:95–103. doi:10.1111/cei.12046
9. Soji T, Murata Y, Ohira A, Nishizono H, Tanaka M, Herbert DC. Evidence that hepatocytes can phagocytize exogenous substances. *Anat Rec* (1992) 233:543–6. doi:10.1002/ar.1092330408
10. Crispe IN, Dao T, Klugewitz K, Mehal WZ, Metz DP. The liver as a site of T-cell apoptosis: graveyard, or killing field? *Immunol Rev* (2000) 174:47–62. doi:10.1034/j.1600-0528.2002.017412.x
11. Benseler V, Warren A, Vo M, Holz LE, Tay SS, Le Couteur DG, et al. Hepatocyte entry leads to degradation of autoreactive CD8 T cells. *Proc Natl Acad Sci U S A* (2011) 108:16735–40. doi:10.1073/pnas.1112251108
12. Guicciardi ME, Malhi H, Mott JL, Gores GJ. Apoptosis and necrosis in the liver. *Compr Physiol* (2013) 3:977–1010. doi:10.1002/cphy.c120020
13. Malhi H, Gores GJ, Lemasters JJ. Apoptosis and necrosis in the liver: a tale of two deaths? *Hepatology* (2006) 43:S31–44. doi:10.1002/hep.21062
14. Malhi H, Guicciardi ME, Gores GJ. Hepatocyte death: a clear and present danger. *Physiol Rev* (2010) 90:1165–94. doi:10.1152/physrev.00061.2009
15. Green DR. The end and after: how dying cells impact the living organism. *Immunity* (2011) 35:441–4. doi:10.1016/j.immuni.2011.10.003
16. Elmore S. Apoptosis: a review of programmed cell death. *Toxicol Pathol* (2007) 35:495–516. doi:10.1080/01926230701320337
17. Kerr JF, Wyllie AH, Currie AR. Apoptosis: a basic biological phenomenon with wide-ranging implications in tissue kinetics. *Br J Cancer* (1972) 26:239–57. doi:10.1038/bjc.1972.33
18. Strasser A, Jost PJ, Nagata S. The many roles of FAS receptor signaling in the immune system. *Immunity* (2009) 30:180–92. doi:10.1016/j.immuni.2009.01.001
19. Flusberg DA, Sorger PK. Surviving apoptosis: life-death signaling in single cells. *Trends Cell Biol* (2015) 25:446–58. doi:10.1016/j.tcb.2015.03.003
20. Wong WW, Puthalakath H. Bcl-2 family proteins: the sentinels of the mitochondrial apoptosis pathway. *IUBMB Life* (2008) 60:390–7. doi:10.1002/iub.51
21. Proskuryakov SY, Gabai VL, Konoplyannikov AG. Necrosis is an active and controlled form of programmed cell death. *Biochemistry (Mosc)* (2002) 67:387–408. doi:10.1023/A:1015289521275
22. Trump BF, Berezesky IK, Chang SH, Phelps PC. The pathways of cell death: oncosis, apoptosis, and necrosis. *Toxicol Pathol* (1997) 25:82–8. doi:10.1177/019262339702500116
23. Leist M, Single B, Castoldi AF, Kuhnle S, Nicotera P. Intracellular adenosine triphosphate (ATP) concentration: a switch in the decision between apoptosis and necrosis. *J Exp Med* (1997) 185:1481–6. doi:10.1084/jem.185.8.1481
24. Majno G, Joris I. Apoptosis, oncosis, and necrosis. An overview of cell death. *Am J Pathol* (1995) 146:3–15.
25. Poon IK, Hulett MD, Parish CR. Molecular mechanisms of late apoptotic/necrotic cell clearance. *Cell Death Differ* (2010) 17:381–97. doi:10.1038/cdd.2009.195
26. Li MO, Sarkisian MR, Mehal WZ, Rakic P, Flavell RA. Phosphatidylserine receptor is required for clearance of apoptotic cells. *Science* (2003) 302:1560–3. doi:10.1126/science.1087621
27. Fernandez-Castaneda A, Arandjelovic S, Stiles TL, Schlobach RK, Mowen KA, Gonias SL, et al. Identification of the low density lipoprotein (LDL) receptor-related protein-1 interactome in central nervous system myelin suggests a role in the clearance of necrotic cell debris. *J Biol Chem* (2013) 288:4538–48. doi:10.1074/jbc.M112.384693
28. Murphy JE, Tacon D, Tedbury PR, Hadden JM, Knowling S, Sawamura T, et al. LOX-1 scavenger receptor mediates calcium-dependent recognition of phosphatidylserine and apoptotic cells. *Biochem J* (2006) 393:107–15. doi:10.1042/BJ20051166
29. Chionna A, Panzarini E, Pagliara P, De Luca A, Caforio S, Abbro L, et al. Hepatic clearance of apoptotic lymphocytes: simply removal of waste cells? *Eur J Histochem* (2003) 47:97–104. doi:10.4081/813
30. Brouckaert G, Kalai M, Krysko DV, Saelens X, Vercammen D, Ndlovu MN, et al. Phagocytosis of necrotic cells by macrophages is phosphatidylserine dependent and does not induce inflammatory cytokine production. *Mol Biol Cell* (2004) 15:1089–100. doi:10.1091/mbc.E03-09-0668
31. Guilliams M, Bruhns P, Saeys Y, Hammad H, Lambrecht BN. The function of Fcgamma receptors in dendritic cells and macrophages. *Nat Rev Immunol* (2014) 14:94–108. doi:10.1038/nri3666
32. Fraser DA, Pisalyaput K, Tenner AJ. C1q enhances microglial clearance of apoptotic neurons and neuronal blebs, and modulates subsequent inflammatory cytokine production. *J Neurochem* (2010) 112:733–43. doi:10.1111/j.1471-4159.2009.06494.x
33. Honore C, Hummelshoj T, Hansen BE, Madsen HO, Eggleton P, Garred P. The innate immune component ficolin 3 (Hakata antigen) mediates the clearance of late apoptotic cells. *Arthritis Rheum* (2007) 56:1598–607. doi:10.1002/art.22564
34. Ogden CA, deCathelineau A, Hoffmann PR, Bratton D, Ghebrehiwet B, Fadok VA, et al. C1q and mannose binding lectin engagement of cell surface

calreticulin and CD91 initiates macropinocytosis and uptake of apoptotic cells. *J Exp Med* (2001) 194:781–95. doi:10.1084/jem.194.6.781

35. Paidassi H, Tacnet-Delorme P, Verneret M, Gaboriaud C, Houen G, Duus K, et al. Investigations on the C1q-calreticulin-phosphatidylserine interactions yield new insights into apoptotic cell recognition. *J Mol Biol* (2011) 408:277–90. doi:10.1016/j.jmb.2011.02.029
36. Jensen ML, Honore C, Hummelshoj T, Hansen BE, Madsen HO, Garred P. Ficolin-2 recognizes DNA and participates in the clearance of dying host cells. *Mol Immunol* (2007) 44:856–65. doi:10.1016/j.molimm.2006.04.002
37. Lawrence T, Willoughby DA, Gilroy DW. Anti-inflammatory lipid mediators and insights into the resolution of inflammation. *Nat Rev Immunol* (2002) 2:787–95. doi:10.1038/nri915
38. Szondy Z, Sarang Z, Kiss B, Garabuczi E, Koroskenyi K. Anti-inflammatory mechanisms triggered by apoptotic cells during their clearance. *Front Immunol* (2017) 8:909. doi:10.3389/fimmu.2017.00909
39. Brenner C, Galluzzi L, Kepp O, Kroemer G. Decoding cell death signals in liver inflammation. *J Hepatol* (2013) 59:583–94. doi:10.1016/j.jhep.2013.03.033
40. Birge RB, Boeltz S, Kumar S, Carlson J, Wanderley J, Calianese D, et al. Phosphatidylserine is a global immunosuppressive signal in efferocytosis, infectious disease, and cancer. *Cell Death Differ* (2016) 23:962–78. doi:10.1038/cdd.2016.11
41. Jiang L, Tixeira R, Caruso S, Atkin-Smith GK, Baxter AA, Paone S, et al. Monitoring the progression of cell death and the disassembly of dying cells by flow cytometry. *Nat Protoc* (2016) 11:655–63. doi:10.1038/nprot.2016.028
42. Bratosin D, Estaquier J, Ameisen JC, Aminoff D, Montreuil J. Flow cytometric approach to the study of erythrophagocytosis: evidence for an alternative immunoglobulin-independent pathway in agammaglobulinemic mice. *J Immunol Methods* (2002) 265:133–43. doi:10.1016/S0022-1759(02)00076-5
43. Ramirez-Ortiz ZG, Pendergraft WF III, Prasad A, Byrne MH, Iram T, Blanchette CJ, et al. The scavenger receptor SCARF1 mediates the clearance of apoptotic cells and prevents autoimmunity. *Nat Immunol* (2013) 14:917–26. doi:10.1038/ni.2670
44. Grandjean CL, Montalvao F, Celli S, Michonneau D, Breart B, Garcia Z, et al. Intravital imaging reveals improved Kupffer cell-mediated phagocytosis as a mode of action of glycoengineered anti-CD20 antibodies. *Sci Rep* (2016) 6:34382. doi:10.1038/srep34382
45. Leers MP, Bjorklund V, Bjorklund B, Jornvall H, Nap M. An immunohistochemical study of the clearance of apoptotic cellular fragments. *Cell Mol Life Sci* (2002) 59:1358–65. doi:10.1007/s00018-002-8513-8
46. Wheeldon EB, Williams SM, Soames AR, James NH, Roberts RA. Quantitation of apoptotic bodies in rat liver by in situ end labelling (ISEL): correlation with morphology. *Toxicol Pathol* (1995) 23:410–5. doi:10.1177/019262339502300317
47. Kyrylkova K, Kyryachenko S, Leid M, Kioussi C. Detection of apoptosis by TUNEL assay. *Methods Mol Biol* (2012) 887:41–7. doi:10.1007/978-1-61779-860-3_5
48. Triantafyllou E, Pop OT, Possamai LA, Wilhelm A, Liaskou E, Singanayagam A, et al. MerTK expressing hepatic macrophages promote the resolution of inflammation in acute liver failure. *Gut* (2018) 67(2):333–47. doi:10.1136/gutjnl-2016-313615
49. Poon IK, Lucas CD, Rossi AG, Ravichandran KS. Apoptotic cell clearance: basic biology and therapeutic potential. *Nat Rev Immunol* (2014) 14:166–80. doi:10.1038/nri3607
50. Penberthy KK, Ravichandran KS. Apoptotic cell recognition receptors and scavenger receptors. *Immunol Rev* (2016) 269:44–59. doi:10.1111/imr.12376
51. Armstrong A, Ravichandran KS. Phosphatidylserine receptors: what is the new RAGE? *EMBO Rep* (2011) 12:287–8. doi:10.1038/embor.2011.41
52. Elliott MR, Ravichandran KS. The dynamics of apoptotic cell clearance. *Dev Cell* (2016) 38:147–60. doi:10.1016/j.devcel.2016.06.029
53. Ravichandran KS. Find-me and eat-me signals in apoptotic cell clearance: progress and conundrums. *J Exp Med* (2010) 207:1807–17. doi:10.1084/jem.20101157
54. Fadok VA, Voelker DR, Campbell PA, Cohen JJ, Bratton DL, Henson PM. Exposure of phosphatidylserine on the surface of apoptotic lymphocytes triggers specific recognition and removal by macrophages. *J Immunol* (1992) 148:2207–16.
55. Medina CB, Ravichandran KS. Do not let death do us part: 'find-me' signals in communication between dying cells and the phagocytes. *Cell Death Differ* (2016) 23:979–89. doi:10.1038/cdd.2016.13
56. Park D, Tosello-Trampont AC, Elliott MR, Lu M, Haney LB, Ma Z, et al. BAI1 is an engulfment receptor for apoptotic cells upstream of the ELMO/Dock180/Rac module. *Nature* (2007) 450:430–4. doi:10.1038/nature06329
57. Park SY, Kim SY, Jung MY, Bae DJ, Kim IS. Epidermal growth factor-like domain repeat of stabilin-2 recognizes phosphatidylserine during cell corpse clearance. *Mol Cell Biol* (2008) 28:5288–98. doi:10.1128/MCB.01993-07
58. Park SY, Jung MY, Kim HJ, Lee SJ, Kim SY, Lee BH, et al. Rapid cell corpse clearance by stabilin-2, a membrane phosphatidylserine receptor. *Cell Death Differ* (2008) 15:192–201. doi:10.1038/sj.cdd.4402242
59. Freeman GJ, Casasnovas JM, Umetsu DT, DeKruyff RH. TIM genes: a family of cell surface phosphatidylserine receptors that regulate innate and adaptive immunity. *Immunol Rev* (2010) 235:172–89. doi:10.1111/j.0105-2896.2010.00903.x
60. He M, Kubo H, Morimoto K, Fujino N, Suzuki T, Takahasi T, et al. Receptor for advanced glycation end products binds to phosphatidylserine and assists in the clearance of apoptotic cells. *EMBO Rep* (2011) 12:358–64. doi:10.1038/embor.2011.28
61. Park SY, Jung MY, Lee SJ, Kang KB, Gratchev A, Riabov V, et al. Stabilin-1 mediates phosphatidylserine-dependent clearance of cell corpses in alternatively activated macrophages. *J Cell Sci* (2009) 122:3365–73. doi:10.1242/jcs.049569
62. Hughes J, Liu Y, Van Damme J, Savill J. Human glomerular mesangial cell phagocytosis of apoptotic neutrophils: mediation by a novel CD36-independent vitronectin receptor/thrombospondin recognition mechanism that is uncoupled from chemokine secretion. *J Immunol* (1997) 158:4389–97.
63. Sexton DW, Blaylock MG, Walsh GM. Human alveolar epithelial cells engulf apoptotic eosinophils by means of integrin- and phosphatidylserine receptor-dependent mechanisms: a process upregulated by dexamethasone. *J Allergy Clin Immunol* (2001) 108:962–9. doi:10.1067/mai.2001.119414
64. Monks J, Rosner D, Geske FJ, Lehman L, Hanson L, Neville MC, et al. Epithelial cells as phagocytes: apoptotic epithelial cells are engulfed by mammary alveolar epithelial cells and repress inflammatory mediator release. *Cell Death Differ* (2005) 12:107–14. doi:10.1038/sj.cdd.4401517
65. Liu G, Wang J, Park YJ, Tsuruta Y, Lorne EF, Zhao X, et al. High mobility group protein-1 inhibits phagocytosis of apoptotic neutrophils through binding to phosphatidylserine. *J Immunol* (2008) 181:4240–6. doi:10.4049/jimmunol.181.6.4240
66. Banerjee S, de Freitas A, Friggeri A, Zmijewski JW, Liu G, Abraham E. Intracellular HMGB1 negatively regulates efferocytosis. *J Immunol* (2011) 187:4686–94. doi:10.4049/jimmunol.1101500
67. Andersen MH, Berglund L, Rasmussen JT, Petersen TE. Bovine PAS-6/7 binds alpha v beta 5 integrins and anionic phospholipids through two domains. *Biochemistry* (1997) 36:5441–6. doi:10.1021/bi963119m
68. Hanayama R, Tanaka M, Miwa K, Shinohara A, Iwamatsu A, Nagata S. Identification of a factor that links apoptotic cells to phagocytes. *Nature* (2002) 417:182–7. doi:10.1038/417182a
69. Savill J, Dransfield I, Hogg N, Haslett C. Vitronectin receptor-mediated phagocytosis of cells undergoing apoptosis. *Nature* (1990) 343:170–3. doi:10.1038/343170a0
70. Akakura S, Singh S, Spataro M, Akakura R, Kim JI, Albert ML, et al. The opsonin MFG-E8 is a ligand for the alphavbeta5 integrin and triggers DOCK180-dependent Rac1 activation for the phagocytosis of apoptotic cells. *Exp Cell Res* (2004) 292:403–16. doi:10.1016/j.yexcr.2003.09.011
71. Albert ML, Kim JI, Birge RB. alphavbeta5 integrin recruits the CrkII-Dock180-rac1 complex for phagocytosis of apoptotic cells. *Nat Cell Biol* (2000) 2:899–905. doi:10.1038/35046549
72. Albert ML, Pearce SF, Francisco LM, Sauter B, Roy P, Silverstein RL, et al. Immature dendritic cells phagocytose apoptotic cells via alphavbeta5 and CD36, and cross-present antigens to cytotoxic T lymphocytes. *J Exp Med* (1998) 188:1359–68. doi:10.1084/jem.188.7.1359
73. Lemke G, Burstyn-Cohen T. TAM receptors and the clearance of apoptotic cells. *Ann N Y Acad Sci* (2010) 1209:23–9. doi:10.1111/j.1749-6632.2010.05744.x

74. Lu Q, Gore M, Zhang Q, Camenisch T, Boast S, Casagranda F, et al. Tyro-3 family receptors are essential regulators of mammalian spermatogenesis. *Nature* (1999) 398:723–8. doi:10.1038/19554
75. Nagata K, Ohashi K, Nakano T, Arita H, Zong C, Hanafusa H, et al. Identification of the product of growth arrest-specific gene 6 as a common ligand for Axl, Sky, and Mer receptor tyrosine kinases. *J Biol Chem* (1996) 271:30022–7.
76. Xiong W, Chen Y, Wang H, Wang H, Wu H, Lu Q, et al. Gas6 and the Tyro 3 receptor tyrosine kinase subfamily regulate the phagocytic function of Sertoli cells. *Reproduction* (2008) 135:77–87. doi:10.1530/REP-07-0287
77. Das S, Sarkar A, Ryan KA, Fox S, Berger AH, Juncadella IJ, et al. Brain angiogenesis inhibitor 1 is expressed by gastric phagocytes during infection with *Helicobacter pylori* and mediates the recognition and engulfment of human apoptotic gastric epithelial cells. *FASEB J* (2014) 28:2214–24. doi:10.1096/fj.13-243238
78. Mevorach D, Mascarenhas JO, Gershov D, Elkon KB. Complement-dependent clearance of apoptotic cells by human macrophages. *J Exp Med* (1998) 188:2313–20. doi:10.1084/jem.188.12.2313
79. Patel PC, Harrison RE. Membrane ruffles capture C3bi-opsonized particles in activated macrophages. *Mol Biol Cell* (2008) 19:4628–39. doi:10.1091/mbc.E08-02-0223
80. Devitt A, Moffatt OD, Raykundalia C, Capra JD, Simmons DL, Gregory CD. Human CD14 mediates recognition and phagocytosis of apoptotic cells. *Nature* (1998) 392:505–9. doi:10.1038/33169
81. Devitt A, Pierce S, Oldreive C, Shingler WH, Gregory CD. CD14-dependent clearance of apoptotic cells by human macrophages: the role of phosphatidylserine. *Cell Death Differ* (2003) 10:371–82. doi:10.1038/sj.cdd.4401168
82. Navazo MD, Daviet L, Savill J, Ren Y, Leung LL, McGregor JL. Identification of a domain (155–183) on CD36 implicated in the phagocytosis of apoptotic neutrophils. *J Biol Chem* (1996) 271:15381–5. doi:10.1074/jbc.271.26.15381
83. Savill J, Hogg N, Haslett C. Macrophage vitronectin receptor, CD36, and thrombospondin cooperate in recognition of neutrophils undergoing programmed cell death. *Chest* (1991) 99:6s–7s. doi:10.1378/chest.99.3_Supplement.6S-a
84. Savill J, Hogg N, Ren Y, Haslett C. Thrombospondin cooperates with CD36 and the vitronectin receptor in macrophage recognition of neutrophils undergoing apoptosis. *J Clin Invest* (1992) 90:1513–22. doi:10.1172/JCI116019
85. Zhang JG, Czabotar PE, Policheni AN, Caminschi I, Wan SS, Kitsoulis S, et al. The dendritic cell receptor Clec9A binds damaged cells via exposed actin filaments. *Immunity* (2012) 36:646–57. doi:10.1016/j.immuni.2012.03.009
86. Sancho D, Joffre OP, Keller AM, Rogers NC, Martinez D, Hernanz-Falcon P, et al. Identification of a dendritic cell receptor that couples sensing of necrosis to immunity. *Nature* (2009) 458:899–903. doi:10.1038/nature07750
87. Sambrano GR, Steinberg D. Recognition of oxidatively damaged and apoptotic cells by an oxidized low density lipoprotein receptor on mouse peritoneal macrophages: role of membrane phosphatidylserine. *Proc Natl Acad Sci U S A* (1995) 92:1396–400. doi:10.1073/pnas.92.5.1396
88. Wermeling F, Chen Y, Pikkarainen T, Scheynius A, Winqvist O, Izui S, et al. Class A scavenger receptors regulate tolerance against apoptotic cells, and autoantibodies against these receptors are predictive of systemic lupus. *J Exp Med* (2007) 204:2259–65. doi:10.1084/jem.20070600
89. Caberoy NB, Zhou Y, Li W. Tubby and tubby-like protein 1 are new MerTK ligands for phagocytosis. *EMBO J* (2010) 29:3898–910. doi:10.1038/emboj.2010.265
90. Caberoy NB, Alvarado G, Li W. Tubby regulates microglial phagocytosis through MerTK. *J Neuroimmunol* (2012) 252:40–8. doi:10.1016/j.jneuroim.2012.07.009
91. Fadok VA, Bratton DL, Rose DM, Pearson A, Ezekewitz RA, Henson PM. A receptor for phosphatidylserine-specific clearance of apoptotic cells. *Nature* (2000) 405:85–90. doi:10.1038/35011084
92. Prabagar MG, Do Y, Ryu S, Park JY, Choi HJ, Choi WS, et al. SIGN-R1, a C-type lectin, enhances apoptotic cell clearance through the complement deposition pathway by interacting with C1q in the spleen. *Cell Death Differ* (2013) 20:535–45. doi:10.1038/cdd.2012.160
93. Nakayama M, Akiba H, Takeda K, Kojima Y, Hashiguchi M, Azuma M, et al. Tim-3 mediates phagocytosis of apoptotic cells and cross-presentation. *Blood* (2009) 113:3821–30. doi:10.1182/blood-2008-10-185884
94. Miyanishi M, Tada K, Koike M, Uchiyama Y, Kitamura T, Nagata S. Identification of Tim4 as a phosphatidylserine receptor. *Nature* (2007) 450:435–9. doi:10.1038/nature06307
95. Nandrot EF, Anand M, Almeida D, Atabai K, Sheppard D, Finnemann SC. Essential role for MFG-E8 as ligand for alphavbeta5 integrin in diurnal retinal phagocytosis. *Proc Natl Acad Sci U S A* (2007) 104:12005–10. doi:10.1073/pnas.0704756104
96. Dini L, Autuori F, Lentini A, Oliverio S, Piacentini M. The clearance of apoptotic cells in the liver is mediated by the asialoglycoprotein receptor. *FEBS Lett* (1992) 296:174–8. doi:10.1016/0014-5793(92)80373-O
97. McVicker BL, Tuma DJ, Kubik JA, Hindemith AM, Baldwin CR, Casey CA. The effect of ethanol on asialoglycoprotein receptor-mediated phagocytosis of apoptotic cells by rat hepatocytes. *Hepatology* (2002) 36:1478–87. doi:10.1002/hep.1840360625
98. Ryeom SW, Sparrow JR, Silverstein RL. CD36 participates in the phagocytosis of rod outer segments by retinal pigment epithelium. *J Cell Sci* (1996) 109(Pt 2):387–95.
99. Ichimura T, Asseldonk EJ, Humphreys BD, Gunaratnam L, Duffield JS, Bonventre JV. Kidney injury molecule-1 is a phosphatidylserine receptor that confers a phagocytic phenotype on epithelial cells. *J Clin Invest* (2008) 118:1657–68. doi:10.1172/JCI34487
100. Oka K, Sawamura T, Kikuta K, Itokawa S, Kume N, Kita T, et al. Lectin-like oxidized low-density lipoprotein receptor 1 mediates phagocytosis of aged/apoptotic cells in endothelial cells. *Proc Natl Acad Sci U S A* (1998) 95:9535–40. doi:10.1073/pnas.95.16.9535
101. Lee SJ, Park SY, Jung MY, Bae SM, Kim IS. Mechanism for phosphatidylserine-dependent erythrophagocytosis in mouse liver. *Blood* (2011) 117:5215–23. doi:10.1182/blood-2010-10-313239
102. Lemke G. Biology of the TAM receptors. *Cold Spring Harb Perspect Biol* (2013) 5:a009076. doi:10.1101/cshperspect.a009076
103. Qi N, Liu P, Zhang Y, Wu H, Chen Y, Han D. Development of a spontaneous liver disease resembling autoimmune hepatitis in mice lacking tyro3, axl and mer receptor tyrosine kinases. *PLoS One* (2013) 8:e66604. doi:10.1371/journal.pone.0066604
104. Hafizi S, Dahlback B. Gas6 and protein S. Vitamin K-dependent ligands for the Axl receptor tyrosine kinase subfamily. *FEBS J* (2006) 273:5231–44. doi:10.1111/j.1742-4658.2006.05529.x
105. Stitt TN, Conn G, Gore M, Lai C, Bruno J, Radziejewski C, et al. The anticoagulation factor protein S and its relative, Gas6, are ligands for the Tyro 3/Axl family of receptor tyrosine kinases. *Cell* (1995) 80:661–70. doi:10.1016/0092-8674(95)90520-0
106. Andersen MH, Graversen H, Fedosov SN, Petersen TE, Rasmussen JT. Functional analyses of two cellular binding domains of bovine lactadherin. *Biochemistry* (2000) 39:6200–6. doi:10.1021/bi992221r
107. Nauta AJ, Trouw LA, Daha MR, Tijsma O, Nieuwland R, Schwaeble WJ, et al. Direct binding of C1q to apoptotic cells and cell blebs induces complement activation. *Eur J Immunol* (2002) 32:1726–36. doi:10.1002/1521-4141(200206)32:6<1726::AID-IMMU1726>3.0.CO;2-R
108. Elliott MR, Zheng S, Park D, Woodson RI, Reardon MA, Juncadella IJ, et al. Unexpected requirement for ELMO1 in clearance of apoptotic germ cells in vivo. *Nature* (2010) 467:333–7. doi:10.1038/nature09356
109. Kim S, Park SY, Kim SY, Bae DJ, Pyo JH, Hong M, et al. Cross talk between engulfment receptors stabilin-2 and integrin alphavbeta5 orchestrates engulfment of phosphatidylserine-exposed erythrocytes. *Mol Cell Biol* (2012) 32:2698–708. doi:10.1128/MCB.06743-11
110. Gumienny TL, Brugnera E, Tosello-Trampont AC, Kinchen JM, Haney LB, Nishiwaki K, et al. CED-12/ELMO, a novel member of the CrkII/Dock180/Rac pathway, is required for phagocytosis and cell migration. *Cell* (2001) 107:27–41. doi:10.1016/S0092-8674(01)00520-7
111. Brugnera E, Haney L, Grimsley C, Lu M, Walk SF, Tosello-Trampont AC, et al. Unconventional Rac-GEF activity is mediated through the Dock180-ELMO complex. *Nat Cell Biol* (2002) 4:574–82. doi:10.1038/ncb824
112. Park SY, Kang KB, Thapa N, Kim SY, Lee SJ, Kim IS. Requirement of adaptor protein GULP during stabilin-2-mediated cell corpse engulfment. *J Biol Chem* (2008) 283:10593–600. doi:10.1074/jbc.M709105200

113. Park SY, Kim SY, Kang KB, Kim IS. Adaptor protein GULP is involved in stabilin-1-mediated phagocytosis. *Biochem Biophys Res Commun* (2010) 398:467–72. doi:10.1016/j.bbrc.2010.06.101
114. Marques-da-Silva C, Burnstock G, Ojcius DM, Coutinho-Silva R. Purinergic receptor agonists modulate phagocytosis and clearance of apoptotic cells in macrophages. *Immunobiology* (2011) 216:1–11. doi:10.1016/j.imbio.2010.03.010
115. Elliott MR, Chekeni FB, Trampont PC, Lazarowski ER, Kadl A, Walk SF, et al. Nucleotides released by apoptotic cells act as a find-me signal to promote phagocytic clearance. *Nature* (2009) 461:282–6. doi:10.1038/nature08296
116. A-Gonzalez N, Bensinger SJ, Hong C, Beceiro S, Bradley MN, Zelcer N, et al. Apoptotic cells promote their own clearance and immune tolerance through activation of the nuclear receptor LXR. *Immunity* (2009) 31:245–58. doi:10.1016/j.immuni.2009.06.018
117. Chen Q, Chen J, Chen J, Lu XJ. Molecular and functional characterization of liver X receptor in ayu, *Plecoglossus altivelis*: regulator of inflammation and efferocytosis. *Dev Comp Immunol* (2016) 65:358–68. doi:10.1016/j.dci.2016.08.007
118. A-Gonzalez N, Hidalgo A. Nuclear receptors and clearance of apoptotic cells: stimulating the macrophage's appetite. *Front Immunol* (2014) 5:211. doi:10.3389/fimmu.2014.00211
119. Pawar A, Botolin D, Mangelsdorf DJ, Jump DB. The role of liver X receptor-alpha in the fatty acid regulation of hepatic gene expression. *J Biol Chem* (2003) 278:40736–43. doi:10.1074/jbc.M307973200
120. Diao J, Michalak TI. Composition, antigenic properties and hepatocyte surface expression of the woodchuck asialoglycoprotein receptor. *J Recept Signal Transduct Res* (1996) 16:243–71. doi:10.3109/10799899609039951
121. Poralla T, Treichel U, Lohr H, Fleischer B. The asialoglycoprotein receptor as target structure in autoimmune liver diseases. *Semin Liver Dis* (1991) 11:215–22. doi:10.1055/s-2008-1040439
122. Treichel U, Poralla T, Hess G, Manns M, Meyer zum Büschenfelde KH. Autoantibodies to human asialoglycoprotein receptor in autoimmune-type chronic hepatitis. *Hepatology* (1990) 11:606–12. doi:10.1002/hep.1840110413
123. Kang JW, Lee SM. Resolvin D1 protects the liver from ischemia/reperfusion injury by enhancing M2 macrophage polarization and efferocytosis. *Biochim Biophys Acta* (2016) 1861:1025–35. doi:10.1016/j.bbalip.2016.06.002
124. Guo F, Ding Y, Caberoy N, Alvarado G, Wang F, Chen R, et al. ABCF1 extrinsically regulates retinal pigment epithelial cell phagocytosis. *Mol Biol Cell* (2015) 26:2311–20. doi:10.1091/mbc.E14-09-1343
125. Kojima Y, Volkmer JP, McKenna K, Civelek M, Lusis AJ, Miller CL, et al. CD47-blocking antibodies restore phagocytosis and prevent atherosclerosis. *Nature* (2016) 536:86–90. doi:10.1038/nature18935
126. Majeti R, Chao MP, Alizadeh AA, Pang WW, Jaiswal S, Gibbs KD Jr, et al. CD47 is an adverse prognostic factor and therapeutic antibody target on human acute myeloid leukemia stem cells. *Cell* (2009) 138:286–99. doi:10.1016/j.cell.2009.05.045
127. Gude DR, Alvarez SE, Paugh SW, Mitra P, Yu J, Griffiths R, et al. Apoptosis induces expression of sphingosine kinase 1 to release sphingosine-1-phosphate as a "come-and-get-me" signal. *FASEB J* (2008) 22:2629–38. doi:10.1096/fj.08-107169
128. Truman LA, Ford CA, Pasikowska M, Pound JD, Wilkinson SJ, Dumitriu IE, et al. CX3CL1/fractalkine is released from apoptotic lymphocytes to stimulate macrophage chemotaxis. *Blood* (2008) 112:5026–36. doi:10.1182/blood-2008-06-162404
129. Korns D, Frasch SC, Fernandez-Boyanapalli R, Henson PM, Bratton DL. Modulation of macrophage efferocytosis in inflammation. *Front Immunol* (2011) 2:57. doi:10.3389/fimmu.2011.00057
130. Szanto A, Balint BL, Nagy ZS, Barta E, Dezso B, Pap A, et al. STAT6 transcription factor is a facilitator of the nuclear receptor PPARgamma-regulated gene expression in macrophages and dendritic cells. *Immunity* (2010) 33:699–712. doi:10.1016/j.immuni.2010.11.009
131. Berry A, Balard P, Coste A, Olagnier D, Lagane C, Authier H, et al. IL-13 induces expression of CD36 in human monocytes through PPAR-gamma activation. *Eur J Immunol* (2007) 37:1642–52. doi:10.1002/eji.200636625
132. Xu W, Roos A, Schlagwein N, Woltman AM, Daha MR, van Kooten C. IL-10-producing macrophages preferentially clear early apoptotic cells. *Blood* (2006) 107:4930–7. doi:10.1182/blood-2005-10-4144
133. Freire-de-Lima CG, Xiao YQ, Gardai SJ, Bratton DL, Schiemann WP, Henson PM. Apoptotic cells, through transforming growth factor-beta, coordinately induce anti-inflammatory and suppress pro-inflammatory eicosanoid and NO synthesis in murine macrophages. *J Biol Chem* (2006) 281:38376–84. doi:10.1074/jbc.M605146200
134. McPhillips K, Janssen WJ, Ghosh M, Byrne A, Gardai S, Remigio L, et al. TNF-alpha inhibits macrophage clearance of apoptotic cells via cytosolic phospholipase A2 and oxidant-dependent mechanisms. *J Immunol* (2007) 178:8117–26. doi:10.4049/jimmunol.178.12.8117
135. Fernandez-Boyanapalli RF, Frasch SC, McPhillips K, Vandivier RW, Harry BL, Riches DW, et al. Impaired apoptotic cell clearance in CGD due to altered macrophage programming is reversed by phosphatidylserine-dependent production of IL-4. *Blood* (2009) 113:2047–55. doi:10.1182/blood-2008-05-160564
136. Ariel A, Serhan CN. New lives given by cell death: macrophage differentiation following their encounter with apoptotic leukocytes during the resolution of inflammation. *Front Immunol* (2012) 3:4. doi:10.3389/fimmu.2012.00004
137. Angsana J, Chen J, Liu L, Haller CA, Chaikof EL. Efferocytosis as a regulator of macrophage chemokine receptor expression and polarization. *Eur J Immunol* (2016) 46:1592–9. doi:10.1002/eji.201546262
138. Han CZ, Juncadella IJ, Kinchen JM, Buckley MW, Klibanov AL, Dryden K, et al. Macrophages redirect phagocytosis by non-professional phagocytes and influence inflammation. *Nature* (2016) 539:570–4. doi:10.1038/nature20141
139. Schlegel M, Kohler D, Korner A, Granja T, Straub A, Giera M, et al. The neuroimmune guidance cue netrin-1 controls resolution programs and promotes liver regeneration. *Hepatology* (2016) 63:1689–705. doi:10.1002/hep.28347
140. Yang M, Liu J, Piao C, Shao J, Du J. ICAM-1 suppresses tumor metastasis by inhibiting macrophage M2 polarization through blockade of efferocytosis. *Cell Death Dis* (2015) 6:e1780. doi:10.1038/cddis.2015.144
141. Schopf RE, Trompeter M, Bork K, Morsches B. Effects of ethanol and acetaldehyde on phagocytic functions. *Arch Dermatol Res* (1985) 277:131–7. doi:10.1007/BF00414111
142. Earnest DL, Abril ER, Jolley CS, Martinez F. Ethanol and diet-induced alterations in Kupffer cell function. *Alcohol Alcohol* (1993) 28:73–83.
143. Schwegler M, Wirsing AM, Dollinger AJ, Abendroth B, Putz F, Fietkau R, et al. Clearance of primary necrotic cells by non-professional phagocytes. *Biol Cell* (2015) 107(10):372–87. doi:10.1111/boc.201400090
144. Wood W, Turmaine M, Weber R, Camp V, Maki RA, McKercher SR, et al. Mesenchymal cells engulf and clear apoptotic footplate cells in macrophageless PU.1 null mouse embryos. *Development* (2000) 127:5245–52.
145. Fond AM, Lee CS, Schulman IG, Kiss RS, Ravichandran KS. Apoptotic cells trigger a membrane-initiated pathway to increase ABCA1. *J Clin Invest* (2015) 125:2748–58. doi:10.1172/JCI80300
146. Juncadella IJ, Kadl A, Sharma AK, Shim YM, Hochreiter-Hufford A, Borish L, et al. Apoptotic cell clearance by bronchial epithelial cells critically influences airway inflammation. *Nature* (2013) 493:547–51. doi:10.1038/nature11714
147. Penberthy KK, Juncadella IJ, Ravichandran KS. Apoptosis and engulfment by bronchial epithelial cells. Implications for allergic airway inflammation. *Ann Am Thorac Soc* (2014) 11(Suppl 5):S259–62. doi:10.1513/AnnalsATS.201405-200AW
148. Burstyn-Cohen T, Lew ED, Traves PG, Burrola PG, Hash JC, Lemke G. Genetic dissection of TAM receptor-ligand interaction in retinal pigment epithelial cell phagocytosis. *Neuron* (2012) 76:1123–32. doi:10.1016/j.neuron.2012.10.015
149. Szatmari-Toth M, Kristof E, Vereb Z, Akhtar S, Facsko A, Fesus L, et al. Clearance of autophagy-associated dying retinal pigment epithelial cells – a possible source for inflammation in age-related macular degeneration. *Cell Death Dis* (2016) 7:e2367. doi:10.1038/cddis.2016.133
150. Petrovski G, Berenyi E, Moe MC, Vajas A, Fesus L, Berta A, et al. Clearance of dying ARPE-19 cells by professional and nonprofessional phagocytes in vitro—implications for age-related macular degeneration (AMD). *Acta Ophthalmol* (2011) 89:e30–4. doi:10.1111/j.1755-3768.2010.02047.x

151. Irschick EU, Sgonc R, Bock G, Wolf H, Fuchs D, Nussbaumer W, et al. Retinal pigment epithelial phagocytosis and metabolism differ from those of macrophages. *Ophthalmic Res* (2004) 36:200–10. doi:10.1159/000078778
152. Kaarniranta K, Sinha D, Blasiak J, Kauppinen A, Vereb Z, Salminen A, et al. Autophagy and heterophagy dysregulation leads to retinal pigment epithelium dysfunction and development of age-related macular degeneration. *Autophagy* (2013) 9:973–84. doi:10.4161/auto.24546
153. Kevany BM, Palczewski K. Phagocytosis of retinal rod and cone photoreceptors. *Physiology (Bethesda)* (2010) 25:8–15. doi:10.1152/physiol.00038.2009
154. Mitter SK, Song C, Qi X, Mao H, Rao H, Akin D, et al. Dysregulated autophagy in the RPE is associated with increased susceptibility to oxidative stress and AMD. *Autophagy* (2014) 10:1989–2005. doi:10.4161/auto.36184
155. Mesa KR, Rompolas P, Zito G, Myung P, Sun TY, Brown S, et al. Niche-induced cell death and epithelial phagocytosis regulate hair follicle stem cell pool. *Nature* (2015) 522:94–7. doi:10.1038/nature14306
156. Lee CS, Penberthy KK, Wheeler KM, Juncadella IJ, Vandenabeele P, Lysiak JJ, et al. Boosting apoptotic cell clearance by colonic epithelial cells attenuates inflammation in vivo. *Immunity* (2016) 44:807–20. doi:10.1016/j.immuni.2016.02.005
157. Hochreiter-Hufford AE, Arandjelovic S, Ravichandran KS. Using phosphatidylserine exposure on apoptotic cells to stimulate myoblast fusion. *Methods Mol Biol* (2015) 1313:141–8. doi:10.1007/978-1-4939-2703-6_10
158. Tso GH, Law HK, Tu W, Chan GC, Lau YL. Phagocytosis of apoptotic cells modulates mesenchymal stem cells osteogenic differentiation to enhance IL-17 and RANKL expression on CD4+ T cells. *Stem Cells* (2010) 28:939–54. doi:10.1002/stem.406
159. Seol D, McCabe DJ, Choe H, Zheng H, Yu Y, Jang K, et al. Chondrogenic progenitor cells respond to cartilage injury. *Arthritis Rheum* (2012) 64:3626–37. doi:10.1002/art.34613
160. Jiao K, Zhang J, Zhang M, Wei Y, Wu Y, Qiu ZY, et al. The identification of CD163 expressing phagocytic chondrocytes in joint cartilage and its novel scavenger role in cartilage degradation. *PLoS One* (2013) 8:e53312. doi:10.1371/journal.pone.0053312
161. Lu Z, Elliott MR, Chen Y, Walsh JT, Klibanov AL, Ravichandran KS, et al. Phagocytic activity of neuronal progenitors regulates adult neurogenesis. *Nat Cell Biol* (2011) 13:1076–83. doi:10.1038/ncb2299
162. Crispe IN. Hepatic T cells and liver tolerance. *Nat Rev Immunol* (2003) 3:51–62. doi:10.1038/nri981
163. Humphreys EH, Williams KT, Adams DH, Afford SC. Primary and malignant cholangiocytes undergo CD40 mediated Fas dependent apoptosis, but are insensitive to direct activation with exogenous Fas ligand. *PLoS One* (2010) 5:e14037. doi:10.1371/journal.pone.0014037
164. Bhogal RH, Weston CJ, Curbishley SM, Bhatt AN, Adams DH, Afford SC. Variable responses of small and large human hepatocytes to hypoxia and hypoxia/reoxygenation (H-R). *FEBS Lett* (2011) 585:935–41. doi:10.1016/j.febslet.2011.02.030
165. Bhogal RH, Weston CJ, Curbishley SM, Adams DH, Afford SC. Autophagy: a cyto-protective mechanism which prevents primary human hepatocyte apoptosis during oxidative stress. *Autophagy* (2012) 8:545–58. doi:10.4161/auto.19012
166. Faubion WA, Gores GJ. Death receptors in liver biology and pathobiology. *Hepatology* (1999) 29:1–4. doi:10.1002/hep.510290101
167. Afford SC, Hubscher S, Strain AJ, Adams DH, Neuberger JM. Apoptosis in the human liver during allograft rejection and end-stage liver disease. *J Pathol* (1995) 176:373–80. doi:10.1002/path.1711760408
168. Afford SC, Randhawa S, Eliopoulos AG, Hubscher SG, Young LS, Adams DH. CD40 activation induces apoptosis in cultured human hepatocytes via induction of cell surface fas ligand expression and amplifies fas-mediated hepatocyte death during allograft rejection. *J Exp Med* (1999) 189:441–6. doi:10.1084/jem.189.2.441
169. Bhogal RH, Curbishley SM, Weston CJ, Adams DH, Afford SC. Reactive oxygen species mediate human hepatocyte injury during hypoxia/reoxygenation. *Liver Transpl* (2010) 16:1303–13. doi:10.1002/lt.22157
170. Bhogal RH, Weston CJ, Curbishley SM, Adams DH, Afford SC. Activation of CD40 with platelet derived CD154 promotes reactive oxygen species dependent death of human hepatocytes during hypoxia and reoxygenation. *PLoS One* (2012) 7:e30867. doi:10.1371/journal.pone.0030867
171. Kiener PA, Davis PM, Starling GC, Mehlin C, Klebanoff SJ, Ledbetter JA, et al. Differential induction of apoptosis by Fas-Fas ligand interactions in human monocytes and macrophages. *J Exp Med* (1997) 185:1511–6. doi:10.1084/jem.185.8.1511
172. Canbay A, Feldstein AE, Higuchi H, Werneburg N, Grambihler A, Bronk SF, et al. Kupffer cell engulfment of apoptotic bodies stimulates death ligand and cytokine expression. *Hepatology* (2003) 38:1188–98. doi:10.1053/jhep.2003.50472
173. Wei Y, Wang D, Topczewski F, Pagliassotti MJ. Saturated fatty acids induce endoplasmic reticulum stress and apoptosis independently of ceramide in liver cells. *Am J Physiol Endocrinol Metab* (2006) 291:E275–81. doi:10.1152/ajpendo.00644.2005
174. Wang D, Wei Y, Pagliassotti MJ. Saturated fatty acids promote endoplasmic reticulum stress and liver injury in rats with hepatic steatosis. *Endocrinology* (2006) 147:943–51. doi:10.1210/en.2005-0570
175. McVicker BL, Casey CA. Ethanol-impaired hepatic protein trafficking: concepts from the asialoglycoprotein receptor system. *Clin Biochem* (1999) 32:557–61. doi:10.1016/S0009-9120(99)00055-7
176. McVicker BL, Casey CA. Effects of ethanol on receptor-mediated endocytosis in the liver. *Alcohol* (1999) 19:255–60. doi:10.1016/S0741-8329(99)00043-9
177. Lee SM, Casey CA, McVicker BL. Impact of asialoglycoprotein receptor deficiency on the development of liver injury. *World J Gastroenterol* (2009) 15:1194–200. doi:10.3748/wjg.15.1194
178. Casey CA, Lee SM, Aziz-Seible R, McVicker BL. Impaired receptor-mediated endocytosis: its role in alcohol-induced apoptosis. *J Gastroenterol Hepatol* (2008) 23(Suppl 1):S46–9. doi:10.1111/j.1440-1746.2007.05275.x
179. Huang H, Nace GW, McDonald KA, Tai S, Klune JR, Rosborough BR, et al. Hepatocyte-specific high-mobility group box 1 deletion worsens the injury in liver ischemia/reperfusion: a role for intracellular high-mobility group box 1 in cellular protection. *Hepatology* (2014) 59:1984–97. doi:10.1002/hep.26976
180. Huebener P, Pradere JP, Hernandez C, Gwak GY, Caviglia JM, Mu X, et al. The HMGB1/RAGE axis triggers neutrophil-mediated injury amplification following necrosis. *J Clin Invest* (2015) 125:539–50. doi:10.1172/JCI76887
181. Dalton SR, Wiegert RL, Baldwin CR, Kassel KM, Casey CA. Impaired receptor-mediated endocytosis by the asialoglycoprotein receptor in ethanol-fed mice: implications for studying the role of this receptor in alcoholic apoptosis. *Biochem Pharmacol* (2003) 65:535–43. doi:10.1016/S0006-2952(02)01555-1
182. Dini L, Pagliara P, Carla EC. Phagocytosis of apoptotic cells by liver: a morphological study. *Microsc Res Tech* (2002) 57:530–40. doi:10.1002/jemt.10107
183. Rosin A, Doljanski L. Erythrocytes in the cytoplsm and nuclei of liver cells. *Br J Exp Pathol* (1944) 25:111–5.
184. Alfrey EJ, Most D, Wang X, Lee LK, Holm B, Krieger NR, et al. Interferon-gamma and interleukin-10 messenger RNA are up-regulated after orthotopic liver transplantation in tolerant rats: evidence for cytokine-mediated immune dysregulation. *Surgery* (1995) 118:399–404; discussion 404–5. doi:10.1016/S0039-6060(05)80351-4
185. Rowell DL, Eckmann L, Dwinell MB, Carpenter SP, Raucy JL, Yang SK, et al. Human hepatocytes express an array of proinflammatory cytokines after agonist stimulation or bacterial invasion. *Am J Physiol* (1997) 273:G322–32.
186. Stonans I, Stonane E, Russwurm S, Deigner HP, Bohm KJ, Wiederhold M, et al. HepG2 human hepatoma cells express multiple cytokine genes. *Cytokine* (1999) 11:151–6. doi:10.1006/cyto.1998.0366
187. Dalton SR, Lee SM, King RN, Nanji AA, Kharbanda KK, Casey CA, et al. Carbon tetrachloride-induced liver damage in asialoglycoprotein receptor-deficient mice. *Biochem Pharmacol* (2009) 77:1283–90. doi:10.1016/j.bcp.2008.12.023
188. Munoz LE, Berens C, Lauber K, Gaipl US, Herrmann M. Apoptotic cell clearance and its role in the origin and resolution of chronic inflammation. *Front Immunol* (2015) 6:139. doi:10.3389/fimmu.2015.00139
189. Jaiswal S, Jamieson CH, Pang WW, Park CY, Chao MP, Majeti R, et al. CD47 is upregulated on circulating hematopoietic stem cells and leukemia cells to avoid phagocytosis. *Cell* (2009) 138:271–85. doi:10.1016/j.cell.2009.05.046
190. Chao MP, Alizadeh AA, Tang C, Myklebust JH, Varghese B, Gill S, et al. Anti-CD47 antibody synergizes with rituximab to promote phagocytosis and

eradicate non-Hodgkin lymphoma. *Cell* (2010) 142:699–713. doi:10.1016/j.cell.2010.07.044

191. Bosurgi L, Bernink JH, Delgado Cuevas V, Gagliani N, Joannas L, Schmid ET, et al. Paradoxical role of the proto-oncogene Axl and Mer receptor tyrosine kinases in colon cancer. *Proc Natl Acad Sci U S A* (2013) 110:13091–6. doi:10.1073/pnas.1302507110
192. Akitake-Kawano R, Seno H, Nakatsuji M, Kimura Y, Nakanishi Y, Yoshioka T, et al. Inhibitory role of Gas6 in intestinal tumorigenesis. *Carcinogenesis* (2013) 34:1567–74. doi:10.1093/carcin/bgt069
193. Karikoski M, Marttila-Ichihara F, Elima K, Rantakari P, Hollmen M, Kelkka T, et al. Clever-1/stabilin-1 controls cancer growth and metastasis. *Clin Cancer Res* (2014) 20:6452–64. doi:10.1158/1078-0432.CCR-14-1236
194. Weinger JG, Brosnan CF, Loudig O, Goldberg MF, Macian F, Arnett HA, et al. Loss of the receptor tyrosine kinase Axl leads to enhanced inflammation in the CNS and delayed removal of myelin debris during experimental autoimmune encephalomyelitis. *J Neuroinflammation* (2011) 8:49. doi:10.1186/1742-2094-8-49
195. Czaja AJ. Targeting apoptosis in autoimmune hepatitis. *Dig Dis Sci* (2014) 59:2890–904. doi:10.1007/s10620-014-3284-2
196. Popov Y, Sverdlov DY, Bhaskar KR, Sharma AK, Millonig G, Patsenker E, et al. Macrophage-mediated phagocytosis of apoptotic cholangiocytes contributes to reversal of experimental biliary fibrosis. *Am J Physiol Gastrointest Liver Physiol* (2010) 298:G323–34. doi:10.1152/ajpgi.00394.2009
197. Allina J, Hu B, Sullivan DM, Fiel MI, Thung SN, Bronk SF, et al. T cell targeting and phagocytosis of apoptotic biliary epithelial cells in primary biliary cirrhosis. *J Autoimmun* (2006) 27:232–41. doi:10.1016/j.jaut.2006.11.004
198. Garabuczi E, Sarang Z, Szondy Z. Glucocorticoids enhance prolonged clearance of apoptotic cells by upregulating liver X receptor, peroxisome proliferator-activated receptor-delta and UCP2. *Biochim Biophys Acta* (2015) 1853:573–82. doi:10.1016/j.bbamcr.2014.12.014
199. Sasaki M, Kakuda Y, Miyakoshi M, Sato Y, Nakanuma Y. Infiltration of inflammatory cells expressing mitochondrial proteins around bile ducts and in biliary epithelial layer may be involved in the pathogenesis in primary biliary cirrhosis. *J Clin Pathol* (2014) 67:470–6. doi:10.1136/jclinpath-2013-201917
200. Lleo A, Selmi C, Invernizzi P, Podda M, Coppel RL, Mackay IR, et al. Apotopes and the biliary specificity of primary biliary cirrhosis. *Hepatology* (2009) 49:871–9. doi:10.1002/hep.22736
201. Webb GJ, Hirschfield GM. Primary biliary cholangitis in 2016: high-definition PBC: biology, models and therapeutic advances. *Nat Rev Gastroenterol Hepatol* (2017) 14:76–8. doi:10.1038/nrgastro.2016.201
202. Karlsen TH, Folseraas T, Thorburn D, Vesterhus M. Primary sclerosing cholangitis – a comprehensive review. *J Hepatol* (2017) 67(6):1298–323. doi:10.1016/j.jhep.2017.07.022
203. McVicker BL, Thiele GM, Tuma DJ, Casey CA. Hepatocyte-mediated cytotoxicity and host defense mechanisms in the alcohol-injured liver. *Hepatol Int* (2014) 8(Suppl 2):432–8. doi:10.1007/s12072-013-9511-7
204. Casey CA, McVicker BL, Donohue TM Jr, McFarland MA, Wiegert RL, Nanji AA. Liver asialoglycoprotein receptor levels correlate with severity of alcoholic liver damage in rats. *J Appl Physiol* (2004) 96:76–80. doi:10.1152/japplphysiol.00375.2003
205. Lopategi A, Lopez-Vicario C, Alcaraz-Quiles J, Garcia-Alonso V, Rius B, Titos E, et al. Role of bioactive lipid mediators in obese adipose tissue inflammation and endocrine dysfunction. *Mol Cell Endocrinol* (2016) 419:44–59. doi:10.1016/j.mce.2015.09.033
206. Lopez-Vicario C, Rius B, Alcaraz-Quiles J, Garcia-Alonso V, Lopategi A, Titos E, et al. Pro-resolving mediators produced from EPA and DHA: overview of the pathways involved and their mechanisms in metabolic syndrome and related liver diseases. *Eur J Pharmacol* (2016) 785:133–43. doi:10.1016/j.ejphar.2015.03.092
207. Titos E, Rius B, Lopez-Vicario C, Alcaraz-Quiles J, Garcia-Alonso V, Lopategi A, et al. Signaling and immunoresolving actions of resolvin D1 in inflamed human visceral adipose tissue. *J Immunol* (2016) 197:3360–70. doi:10.4049/jimmunol.1502522
208. Serban KA, Petrusca DN, Mikosz A, Poirier C, Lockett AD, Saint L, et al. Alpha-1 antitrypsin supplementation improves alveolar macrophages efferocytosis and phagocytosis following cigarette smoke exposure. *PLoS One* (2017) 12:e0176073. doi:10.1371/journal.pone.0176073
209. Wang X, Bu HF, Zhong W, Asai A, Zhou Z, Tan XD. MFG-E8 and HMGB1 are involved in the mechanism underlying alcohol-induced impairment of macrophage efferocytosis. *Mol Med* (2013) 19:170–82. doi:10.2119/molmed.2012.00260
210. Ji H, Liu Y, Zhang Y, Shen XD, Gao F, Busuttil RW, et al. T-cell immunoglobulin and mucin domain 4 (TIM-4) signaling in innate immune-mediated liver ischemia-reperfusion injury. *Hepatology* (2014) 60:2052–64. doi:10.1002/hep.27334
211. Fond AM, Ravichandran KS. Clearance of dying cells by phagocytes: mechanisms and implications for disease pathogenesis. *Adv Exp Med Biol* (2016) 930:25–49. doi:10.1007/978-3-319-39406-0_2

Disclaimer: This article/paper/report presents independent research funded by the NIHR Birmingham Biomedical Research Centre at the University Hospitals Birmingham NHS Foundation Trust and the University of Birmingham. The views expressed are those of the author(s) and not necessarily those of the NHS, the NIHR or the Department of Health.

Conflict of Interest Statement: The authors declare that the research was conducted in the absence of any commercial or financial relationships that could be construed as a potential conflict of interest.

15

Exploitation of Apoptotic Regulation in Cancer

David S. Ucker[1] and Jerrold S. Levine[2]*

[1] *Department of Microbiology and Immunology, University of Illinois College of Medicine, Chicago, IL, United States,*
[2] *Department of Medicine, Division of Nephrology, University of Illinois College of Medicine, Jesse Brown Veterans Affairs Medical Center, Chicago, IL, United States*

Within an organism, environmental stresses can trigger cell death, particularly apoptotic cell death. Apoptotic cells, themselves, are potent regulators of their cellular environment, involved primarily in effecting homeostatic control. Tumors, especially, exist in a dynamic balance of cell proliferation and cell death. This special feature of the tumorous microenvironment—namely, the prominence and persistence of cell death—necessarily entails a magnification of the extrinsic, postmortem effects of dead cells. In both normal and malignant tissues, apoptotic regulation is exerted through immune as well as non-immune mechanisms. Apoptotic cells suppress the repertoire of immune reactivities, both by attenuating innate (especially inflammatory) responses and by abrogating adaptive responses. In addition, apoptotic cells modulate multiple vital cell activities, including survival, proliferation (cell number), and growth (cell size). While the microenvironment of the tumor may contribute to apoptosis, the postmortem effects of apoptotic cells feature prominently in the reciprocal acclimatization between the tumor and its environment. In much the same way that pathogens evade the host's defenses through exploitation of key aspects of innate and adaptive immunity, cancer cells subvert several normal homeostatic processes, in particular wound healing and organ regeneration, to transform and overtake their environment. In understanding this subversion, it is crucial to view a tumor not simply as a clone of malignant cells, but rather as a complex and highly organized structure in which there exists a multidirectional flow of information between the cancer cells themselves and the multiple other cell types and extracellular matrix components of which the tumor is comprised. Apoptotic cells, therefore, have the unfortunate consequence of facilitating tumorigenesis and tumor survival.

Edited by:
Amiram Ariel, University of Haifa, Israel

Reviewed by:
Tal Burstyn-Cohen, Hebrew University of Jerusalem, Israel
Raymond B. Birge, Rutgers, The State University of New Jersey, United States

***Correspondence:**
David S. Ucker
duck@uic.edu

Keywords: tumorigenesis, apoptosis, microenvironment, selective adaptation, inflammation, immunity

INTRODUCTION

A predisposition to apoptotic death among cells that have acquired a malignant mutation serves to facilitate one of the body's primary defenses against cancer. Induction of apoptosis is a failsafe and occurs as a result of the tight interweaving of the multiple genes and signaling pathways regulating survival, proliferation, and growth (1–4). Were the cell's demise the sole consequence of this defense against cancer, the benefits to the organism would be unambiguous and unopposed.

Apoptosis, while a cell-autonomous process, has postmortem consequences that are not simply cell-intrinsic. Through an array of mechanisms, both direct and indirect, dead or dying cells actively and potently influence other cells within their environment (5–8). Although earlier studies focused on the ability of apoptotic cells to suppress inflammation (9–12), apoptotic cells also affect a broad range of cellular functions, including such vital activities as survival (13–15), proliferation (13–15), differentiation (16), metabolism (17), and migration (5, 7, 8). Moreover, these effects are not only limited to the professional phagocytes charged with the clearance of apoptotic cells, but extend also to virtually every living cell in the vicinity of the apoptotic cell, regardless of its origin or lineage (14–20).

It is owing to these extrinsic, postmortem effects of apoptosis that dying cancer cells may act to promote, rather than retard, tumorigenesis. This facilitation comes about through a subversion of normal homeostatic mechanisms. In much the same way that pathogens evade the host's defenses through clever disguise and manipulation of key aspects of innate and adaptive immunity, cancer cells subvert several normal homeostatic processes, in particular wound healing and organ regeneration, to transform and overtake their environment (4, 16, 21–23). In understanding this subversion, it is crucial to view a tumor not simply as a clone of malignant cells, but rather as a complex and highly organized structure in which there exists a multidirectional flow of information between the cancer cells themselves and the multiple other cell types and extracellular matrix components of which the tumor is comprised (24). In a sense, a reciprocal process of acclimatization takes place, in which the environment becomes progressively more conducive to cancer cell growth, and the cancer cells themselves become progressively more adapted to their environment (**Figure 1**).

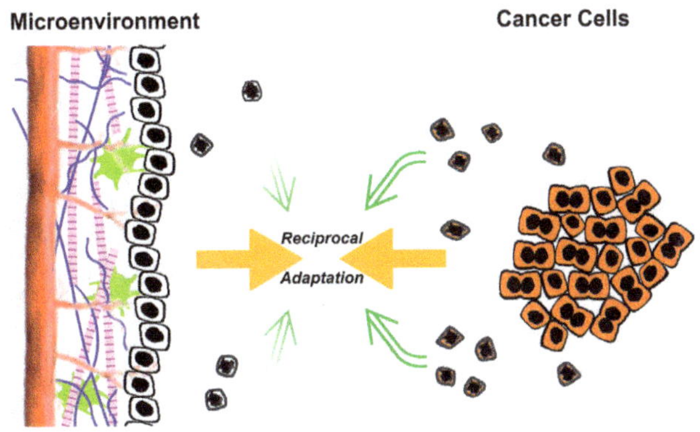

FIGURE 1 | The reciprocal acclimatization of cancer cells and their microenvironment is enhanced by apoptotic cells. A dynamic flow of information exists between transformed cancer cells and the tumorous microenvironment in which they reside. The microenvironment consists of non-transformed stromal cells, both resident and recruited, phagocytic cells including macrophages, as well as extracellular matrix components. As depicted in the figure, the ongoing interaction between transformed cancer cells (shaded in orange) and their microenvironment leads to a process of reciprocal adaptation, in which the environment becomes progressively more conducive to cancer cell growth, and the cancer cells themselves become progressively more adapted to their environment. Through their postmortem effects, apoptotic cells (schematized here as shrunken and misshapen, with extensive nuclear condensation) impact tumorigenesis and tumor growth. The relative effects of transformed cells and their microenvironment on tumorigenesis likely are in a continuously dynamic balance. Certainly, the balance of those inputs changes during the life of a tumor. One obvious shift occurs following antitumor therapy (e.g., chemotherapy and radiotherapy), in which the rates of cell death increase dramatically, including both transformed cancer cells and non-transformed stromal cells. These elevated levels of apoptosis can lead to further tumor-promoting enhancement of the microenvironment.

Several features unique to the milieu of the cancerous micro-environment facilitate this reciprocal adaptation, not least of which is the very prominence and persistence of cell death in nearly all cancers (4, 24, 25). As compared with normal tissues, tumors are characterized by increased rates of both proliferation and death, with tumor mass increasing if the rate of proliferation exceeds that of death (1–4). Increased cell death is not solely due to the failsafe induction of apoptosis during unregulated proliferation (1–4). Multiple other factors contribute to the death of cancer cells, including antitumor immune responses, competitive interactions among different clones, and metabolic stress as a result of limitations in growth factors, nutrients, and oxygen. While a variety of types of cell death arise, apoptosis is the predominant form of cell death in tumors, as it is in normal homeostatic physiology. Importantly, the rapidity and efficiency with which dead cells are cleared means that, in the absence of specific histologic labeling, cell death often goes undetected (4, 5, 7, 8). Even the heightened numbers of dead cells typically observed in cancerous tissues represent an underestimate of the actual number of dying cells. Indeed, with the possible exceptions of embryogenesis or the immature thymus, the degree of cell death in tumors far outstrips that found in any organ under physiological conditions. Moreover, the supply of dead cells undergoes more or less continuous renewal. This is unlike non-cancerous tissues, for which, even following severe injury, cell death rarely continues unabated for weeks to months, as occurs characteristically within growing tumors. This special feature of the tumorous microenvironment—namely, the prominence and persistence of cell death—necessarily entails a magnification of the extrinsic, postmortem effects of dead cells. In turn, this enables tumors to usurp and manipulate for their own advantage a number of the normal homeostatic processes initiated by the recognition and clearance of dead cells. As a result, cell death has the potential to be among the most crucial factors impacting the development and progression of cancers.

In this review, we will describe how, within the context of tumorigenesis, several of the normal homeostatic responses triggered by apoptotic cells are subverted, with the unfortunate consequence of facilitating tumorigenesis. At least in part, the reciprocal adaptation between cancer cells and their environment may be said to occur under the specific aegis of apoptotic cells. Because this seemingly paradoxical outcome of cell death derives mainly from the postmortem effects of apoptosis, we will focus on this aspect of cell death, rather than on the myriad genes involved in the regulation and execution of the apoptotic death program, many of whose mutations predispose to malignancy. Finally, we will highlight what, from our perspective, are some of the field's major unanswered questions and areas open for investigation.

PATHOGENIC EXPLOITATION OF APOPTOTIC IMMUNITY

Infections with microbial pathogens provide a useful context in which to appreciate the subversion of postmortem apoptotic effects. In the acute context of pathogenic infection, the repertoire of potent immunosuppressive responses elicited normally by apoptotic cells ("Innate Apoptotic Immunity"; refs. 9–12, 18, 19) appears to be specifically exploited as a means of enhancing pathogenicity (26). Multiple microbial pathogens subvert the processes of apoptosis and Innate Apoptotic Immunity in this way.

One of the hallmarks of this pathogenic sabotage is that pathogens trigger the apoptotic cell death of cells that are expendable for productive infection. In other words, while those viable cells are not essential to the pathogen, the apoptotic corpses of those cells serve to enhance pathogenicity. For example, in the case of the lethal food-borne bacterial pathogen *Listeria monocytogenes*, the extensive induction of apoptotic cell death, especially among lymphocytes (27), is of particular interest. Lymphocytes are not critical for the *in vivo* replication of the bacterium, do not serve as substantial bacterial reservoirs, and are not the primary cells of entry for productive infection (28, 29). Still immunocompromised mice genetically deficient in lymphocytes are less susceptible to *L. monocytogenes* infection than are lymphocyte-replete, wild-type mice (30). The reconstitution of normal lymphocyte populations in these mutants restores pathogen susceptibility to wild-type levels (30). Strikingly, exogenous apoptotic lymphocytes, including uninfected apoptotic lymphocytes, are as effective as viable lymphocytes (29). Thus, although viable lymphocytes are dispensable for *L. monocytogenes* replication, apoptotic lymphocytes are important for *L. monocytogenes* pathogenesis (29). Because apoptotic cells are not susceptible to *L. monocytogenes* infection (29), the uptake of those apoptotic cells cannot be responsible for pathogen spread. Similar results have been obtained with a sepsis model of bacterial pathogenicity (31, 32). The specific action of apoptotic lymphocytes in these cases appears to be the suppression of host inflammation *via* Innate Apoptotic Immunity.

Another hallmark of this process is that pathogen-induced host cell apoptosis is dissociable from the postmortem effects of the apoptotic cells. Again, in the case of *L. monocytogenes*, lymphocyte apoptosis depends upon pathogen-dependent stimulation of host innate immunity (and production of interferon-β; refs. 33–35), as well as a pathogen-encoded pore-forming protein (Listeriolysin O; ref. 36). The efficacy of apoptotic cells is fully independent of these mediators, however.

In the chronic setting of a tumor, the consequences of cell death (as judged by interference with that process; see below) allow the suggestion that apoptotic cells do similar things. Just as microbial pathogens exploit Innate Apoptotic Immunity, the extrinsic, postmortem effects of apoptotic cells also appear to be exploited in tumorigenesis. Indeed, it may even be that in the more chronic tumorigenic setting, a broader spectrum of postmortem apoptotic effects is involved.

CLINICAL EVIDENCE OF THE COUNTERVAILING ROLE OF APOPTOSIS IN CANCER

Initial evidence for a paradoxical, tumor-enhancing role of apoptosis in cancer arose from multiple studies of clinically related cohorts across a broad spectrum of cancers in which an association was observed between the extent of apoptosis and the aggressiveness of the underlying malignancy (22, 37–57). Cancers studied included non-Hodgkin's lymphoma (37, 38), synovial sarcoma (39), and carcinomas of the tongue (40), esophagus (41), bladder (42, 43), breast (22, 44–46), endometrium (47), prostate (48, 49), cervix (50–52), kidney (53), stomach (52, 54), liver (55), ovaries (52, 56), larynx (57), colorectum (52), and head and neck (22). In these studies, a statistical correlation was observed between the extent of apoptosis and the following parameters: histologic grade (37–39, 42–45, 53–55), cancer stage (39, 43, 55), mitotic/proliferative index (37, 38, 43–45, 50, 55, 56), metastasis (40, 45, 55), mortality (22, 37, 38, 43–46, 50, 52, 56, 57), recurrence following treatment (22, 40, 43–46, 49–51), local invasiveness (41, 55), tumor progression (43, 48), and tumor size (39, 45). Moreover, while a statistical association does not necessarily imply causality, it is noteworthy that in several of these studies, upon multivariate analysis, an index of the extent of apoptosis proved to be one of very few independent predictors (or even the only independent predictor) of overall or disease-free survival (37, 39, 43, 49–51, 55, 57).

THE LANGUAGE OF APOPTOTIC CELLS

The increased rate of cell death typical of nearly all cancers means that the specialized microenvironment of tumors, more so than other tissues under physiologic conditions, is characterized by a robust flow of information centered on dead or dying cells—*into* malignant cells from their environment that influences their decision whether to live or die, and *out of* them to live cells in their vicinity, both cancerous and non-cancerous (**Figure 1**). A sense of the vast extent of apoptosis observed in human malignancies can be informative. In most studies of human cancer, apoptosis has been quantified in the form of an apoptotic index, defined as the number of apoptotic nuclei per 100 intact neoplastic cells (37–42, 46, 47, 49–56). While rigor varied widely across these studies, the mean apoptotic indices in general fell in the range of 0.5–2.0% (37–39, 41, 42, 46, 49, 50, 56). With increasing markers of tumor aggressiveness, apoptotic indices reached as high as 5–10% (40, 49, 51, 54, 55), and at times even exceeded 10% (53). These numbers offer powerful evidence of the markedly increased rates of apoptosis characteristic of most tumors. While apoptotic cell death may be largely invisible under physiologic conditions (4, 5, 7, 8), it is not silent.

Transmission of information from apoptotic cells to the environment occurs in one of two fundamental ways, either directly, through physical interaction between dead and live cells, or indirectly, without physical interaction. Direct effects occur most commonly *via* receptor-mediated recognition by live cells of adjacent dead cells or their fragments (5–8, 11, 14, 18, 19). Indirect effects are most frequently the result of soluble mediators released from the dying cells, but can entail more subtle mechanisms (4–8). For example, apoptotic cells may adsorb soluble mediators and thereby lower effective concentrations, precluding viable cell responses (58). Dying cells also may shed various membrane-enclosed vesicles containing a combination of cytosolic proteins, RNA, and lipids (59–61) that can serve in information transmission. Depending upon the origin of these extracellular vesicles, whether from the plasma membrane or endosomes, they are referred to as microparticles or exosomes, respectively (62). Docking of these vesicles at the surface of live cells, followed by their fusion with the plasma membrane, or by their endocytosis and fusion within an endocytic compartment, leads to release of their contents and delivery of the message those contents represent (62).

A striking range and complexity characterize all steps of the information flow from apoptotic cells to live cells in their vicinity (4–8). Viewed as a language, dead cells carry a surprisingly large amount of information, with an extensive vocabulary and an intricate grammar. While this complexity has been shown predominantly for non-tumorous cells and tissues, some data suggest the same is true for tumors, especially since the ability to recognize and respond to dead cells seems to be ubiquitous across practically all organs and cell lineages (18, 63, 64). Death-related variables from which information can be extracted include, but are by no means limited to, the mode of cell death (11, 14, 15, 17, 65)—and, under certain circumstances, perhaps even the conditions and the particular inducer of that form of cell death (15, 66)—as well as the pattern, distribution, kinetics, rate, and extent of cell death (67–69). Moreover, the response by any given live cell, whether cancerous or not, depends as much on the identity of the responding cell itself—its lineage (13–15), organ of residence (13–15), and stage of differentiation (70, 71)—as on the specific nature of those death-related variables. It is easy to imagine that even live cells lacking direct physical interaction with apoptotic cells or their released mediators may be affected by the dynamic multidirectional flow of information. A ripple effect may ensue, in which apoptotic cells stimulate the synthesis and release of cytokines and mediators from live cells, and in turn these cytokines and mediators then modulate the activity of other live cells that reside at a distance from the dead cells, at the fringe or beyond the tumorous microenvironment, even in distal organs and tissues.

PARADIGM

As first formulated in an influential review on cancer (2, 3), the evolution of a cell from normal to neoplastic entails the acquisition of as many as eight discrete biological capabilities. These capabilities, or hallmarks, represent a series of steps, each of which confers a trait or selective advantage necessary to the emergence of a clone of highly malignant cells. Hallmarks include: (i) sustained proliferative signaling, (ii) evasion of growth suppressors, (iii) resistance to cell death, (iv) induction of angiogenesis, (v) tissue invasion and metastasis, (vi) reprogramming of energy metabolism, (vii) evasion of immune destruction, and (viii) replicative immortality. While the succession of hallmarks need not occur in any set order, the overall process itself proceeds in a step-by-step manner analogous to that of other cases of natural selection.

Integral to this evolution is the interaction between cancer cells themselves and the microenvironment in which they reside. The tumorous environment is a specialized structure, consisting not only of malignant clones, but also of multiple non-transformed cell types, both resident and recruited (2–4, 24, 25). These ostensibly normal cells, together with a surrounding extracellular matrix, comprise the tumor-associated stroma. A critical determinant underlying the acquisition of hallmark capabilities is the genomic instability of cancer cells (2, 3). The facilitated occurrence of genetic mutations and/or epigenetic changes affecting gene expression catalyzes the development of heritable phenotypes better adapted to the tumorous environment (2, 3, 72). In turn, cancer cells contribute to the favorable transformation of their environment (4, 24, 25). Through tissue remodeling and modulation of the function of stromal cells, such as macrophages (4, 73–75), cancer cells help make the environment more conducive to their outgrowth and supremacy.

Increasingly, apoptotic cells have been recognized as participants in the reciprocal adaptation between cancer cells and their microenvironment (4). While an established role for apoptotic cells has so far been limited to only a few of the hallmarks—most extensively, in sustained proliferative signaling and evasion of growth suppressors (22, 74–80)—it is striking how many of the biological capabilities of cancer cells are known to be impacted by apoptotic cells under non-cancerous conditions. Indeed, the only one of the eight hallmarks for which a convincing, or at least highly suggestive, example of the influence of apoptotic cells does not yet exist is the induction of replicative immortality.

Most of these activities of apoptotic cells fall under the rubric of two homeostatic processes in which the active role of apoptotic cells has been carefully explored, namely, wound healing and organ regeneration (4, 21, 67–69). Although these two processes probably differ more quantitatively than qualitatively, they do represent clearly separable stages in the life of a tumor, with treatment representing the boundary. As first pointed out over 30 years ago (21), and expanded upon by numerous reviewers since (4, 81, 82), growing tumors are like "wounds that do not heal," or even "wounds that do not stop repairing," whereas the repopulation of a tumor that occurs following therapy most resembles organ regeneration. These broad similarities dwarf the finer differences and provide a convenient lens through which to view the subversion of homeostatic processes by the apoptotic cells within tumors.

APOPTOTIC CELLS AND THE HALLMARKS OF CANCER

Postmortem apoptotic modulation targets almost every hallmark of cancer (**Figure 2**). The sole hallmark for which no evidence of apoptotic modulation yet exists is replicative immortality, in which cancer cells escape their inbuilt limitation on replicative doublings. This may not be so surprising since, given the increased numbers of apoptotic cells observed in nearly all cancers (4, 24, 25), a heightened effect of apoptotic signaling in cancerous as opposed to non-cancerous tissues would be expected. Moreover, as many of the hallmarks of cancer represent unregulated and essentially continuous manifestations of cellular functions normally modulated by apoptotic cells, it is tempting to hypothesize a prominent role for apoptotic cells in hallmark acquisition. As examples of the ways in which tumors can subvert the normal homeostatic functions of apoptotic cells, we examine the many parallels between tumor progression and the processes of wound healing and organ regeneration.

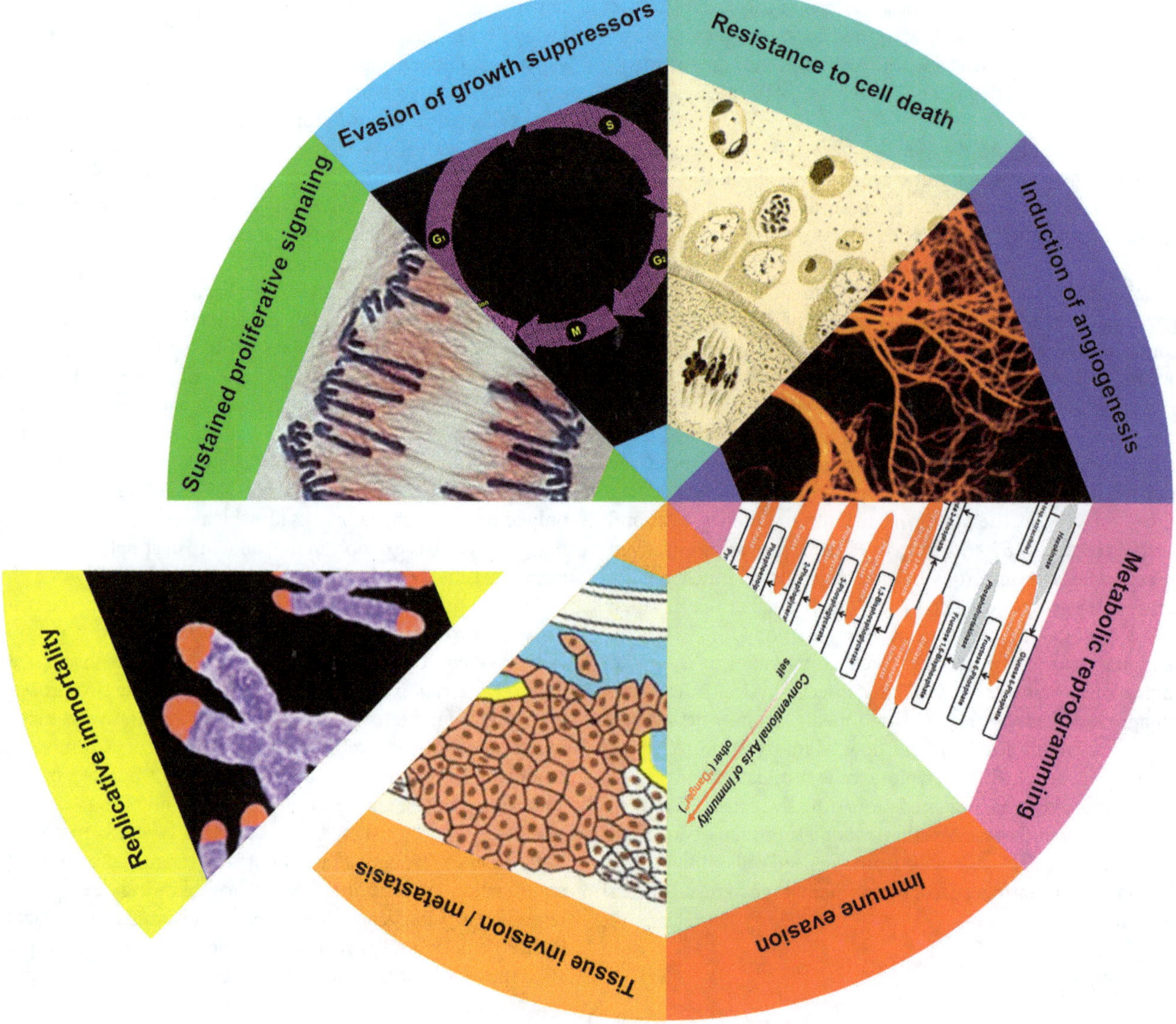

FIGURE 2 | Apoptotic cells enhance achievement of cancer hallmarks. Eight hallmarks of cancer (sustained proliferative signaling, evasion of growth suppressors, resistance to cell death, induction of angiogenesis, metabolic reprogramming, immune evasion, tissue invasion/metastasis, and replicative immortality) have been elaborated (3). With the exception of the last, apoptotic cells enhance each of these cancer attributes, as discussed in the text.

Tumor Growth and Progression: Wounds That Will Not Heal

The tissue repair that follows a wound can be divided into three broad and overlapping phases (**Figure 3**; see refs. 21, 81, 82). In the first inflammatory phase, a blood clot is formed and seals the wound; local inflammation, a direct consequence of injury, leads to the recruitment of inflammatory cells. In the second phase, new tissue, called granulation tissue, is formed and replaces the blood clot. Macrophages are key players in this phase, which entails the formation of new blood vessels (an especially critical event in the case of tumors), the laying down of new extracellular matrix, and an overall increased proliferation of multiple cell types, including fibroblasts and keratinocytes. During the final phase, granulation tissue is converted into a scar, with an overall decrease of cellularity and an extensive remodeling of the extracellular matrix. For each of these phases of wound healing, there are parallels with tumor growth, and for each of these parallels a plausible role for apoptotic cells exists.

The blood clot, formed during the initial inflammatory phase of wound healing, serves multiple purposes. It provides a protective barrier against infection and evaporative loss, it serves as a reservoir of growth factors, and it acts as a scaffold for the multiple cell types attracted to the wound. The two major matrix components of a clot are fibrin and fibronectin. Fibrin is generated as an end product of the coagulation cascade, whereas fibronectin leaks through capillary walls because of a local increase in vascular permeability. Although the origins of fibrin and fibronectin differ in the case of cancer, most tumors are also characterized by the

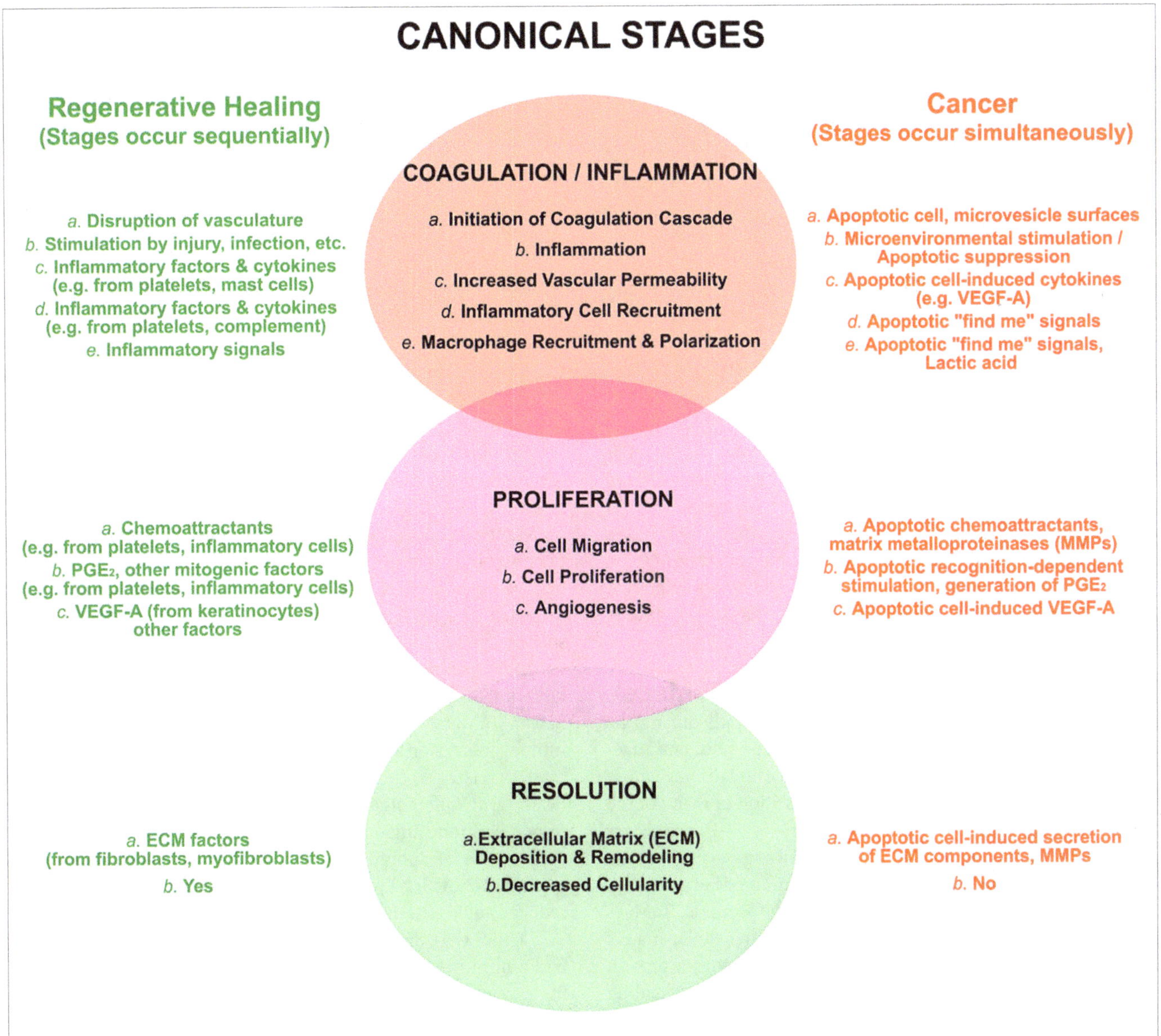

FIGURE 3 | The processes of regenerative wound healing are subverted by apoptotic cells in cancer. The canonical attributes of wound healing are categorized into overlapping and sequential phases, as discussed in the text. These are compared to similar, yet distinct, events occurring in an ongoing and simultaneous manner in cancer, resulting largely from the postmortem effects of apoptotic cells. The view that cancers are "wounds that do not resolve" emerges from this perspective.

presence of a fibrin and fibronectin matrix (21, 81). Apoptotic cells influence the generation and localization of both of these proteins. By virtue of reorganization of their membrane phospholipids (83, 84), apoptotic cells, and especially their shed microvesicles (85), are procoagulant, thereby providing a nidus for initiation and amplification of the coagulation cascade. In addition, exposure to apoptotic cells upregulates the expression by macrophages and endothelial cells of vascular endothelial growth factor (VEGF)-A, a factor that potently augments vascular permeability (73, 86, 87).

A key event in the initiation of coagulation is a loss of membrane asymmetry in platelets and other blood cells (88). Exoplasmic exposure of several phospholipids normally retained within the cytoplasmic leaflet of cell membranes, such as the anionic phospholipid phosphatidylserine (PS) and the amphipathic phospholipid phosphatidylethanolamine, supports the binding of clotting factors and initiates the coagulation cascade (88, 89). A similar loss of membrane asymmetry occurs in cells undergoing apoptosis (90, 91). Exposure of PS is mediated, generally, by activation of a phospholipid scramblase (an energy-independent bidirectional transporter that dissipates membrane asymmetry) and/or inhibition of a phospholipid flippase (a P4-ATPase that enhances asymmetry by transporting specific membrane phospholipids against their concentration gradient; refs. 89–91). The procoagulant activity of apoptotic cells, and especially their shed microvesicles (85), may, therefore, be attributed largely to their mimicking of the surface membranes of activated platelets and other blood cells. Although the normal physiologically handling of apoptotic cells does not appear to predispose to clotting disorders, the prominence and persistence of apoptotic death within tumors may upset the delicate balance between pro- and anti-coagulant activities.

Closely related to its procoagulant function, the exposure of PS on the outer leaflet of the apoptotic cell membrane has been identified as one element involved in the recognition of apoptotic cells. A related group of receptors with intrinsic tyrosine kinase activity, composed of the molecules *T*yro3, *A*xl, and *M*ertk (the "TAM receptor" family) are involved in PS-dependent apoptotic recognition (92, 93). TAM receptor-mediated recognition of apoptotic cells relies upon Vitamin K-dependent bridging molecules and is associated with the phagocytic clearance of apoptotic cells. TAM receptors are expressed primarily on myeloid cells of the immune system and exert a variety of immunoregulatory and other functions, including the suppression of inflammation and the enhancement of angiogenesis (92, 94, 95).

The recognition of apoptotic cells also occurs by PS-independent mechanisms (18, 96), relying on surface-exposed protein determinants (97). This recognition, which is not encumbered by serum-derived tethering molecules (18), triggers immediate-early responses independent of phagocytosis (11, 14, 15, 17). These include all of the homeostatic processes subverted in cancer (**Figure 2**). Most notably, PS-independent apoptotic recognition is ubiquitous among all cell types (18).

Tumor-Associated Macrophages (TaMacs)

Among the inflammatory cells recruited to wounds during the initial inflammatory phase are neutrophils, mast cells, and macrophages (21, 81, 82). Of these, the macrophages are the one whose pivotal role in wound healing and the acquisition of many of the hallmarks of cancer is best established (4, 73–75, 98–100). TaMacs acquire a phenotype that favors proliferation, angiogenesis, tissue invasion and metastasis, and evasion of immune destruction (4, 73–75, 98–100). Like macrophages in healing wounds, TaMacs are intimately involved in remodeling of the extracellular matrix and in creating an environment more favorable to tumor growth (4, 74, 75). Indeed, of the many parallels between wound healing and cancer, one of the strongest is the molecular concordance in gene expression patterns between TaMacs and repairing tissues (74, 75, 81).

Far more controversial is the origin and state of polarization of TaMacs, as compared with macrophages in other tissues and other pathological conditions. There is a general consensus that TaMacs are recruited from the blood as monocytic precursors (98–100). In the case of a murine mammary tumor model, TaMacs were shown to be derived predominantly from C–C chemokine receptor type 2 positive ($CCR2^+$) monocyte precursors, and were functionally and phenotypically distinct from the tissue macrophages of non-cancerous glands (101). As shown in two other models, murine breast cancer and xenografted Burkitt's lymphoma, TaMacs had a higher proliferative capacity than resident macrophages and, therefore, required less replenishing from the blood (74, 101). Although from a functional perspective, TaMacs found in most tumors resemble alternatively activated (M2-like) more than classically activated (M1 or $M_{(IFN-\gamma/LPS)}$) macrophages (98–100), their gene expression patterns, when carefully examined, fit the pattern of neither macrophage subtype (74, 101). For example, unlike alternatively activated macrophages, TaMacs from mouse mammary tumors lacked IL-4 dependence and failed to express several characteristic M2-like genes (101). Correspondingly, TaMacs from Burkitt's lymphoma expressed multiple classically activated genes (74). While macrophage activation is best viewed within a continuum of gene expression patterns rather than fixed in discrete polarized states, these data nonetheless suggest the presence within the tumor microenvironment of unique determinants influencing the phenotypic and genotypic state of TaMacs. One of these determinants, as demonstrated in murine models of melanoma and lung carcinoma, is lactic acid, which is produced by cancer cells as a result of their relatively hypoxic microenvironment (73). The effect of lactic acid is mediated by the transcription factor hypoxia-inducible factor 1α (HIF-1α), which induces strong expression of the genes for VEGF-A and arginase 1 (73). Again, non-tumor wounds also are relatively hypoxic environments, so that at least some of the similarities in gene expression patterns between TaMacs and wound tissue may be attributable to the effects of lactic acid and HIF-1α (21, 81, 82).

Significantly, exposure to apoptotic cells contributes to the phenotypic and genotypic expression of a number of the pro-oncogenic properties of TaMacs. *In vitro* exposure of classically activated $M_{(IFN-\gamma/LPS)}$ macrophages to apoptotic lymphoma cells shifted their gene expression pattern toward that of *in situ* TaMacs, as obtained by laser-capture microdissection of Burkitt's lymphoma xenografts (74, 75). Among the gene clusters identified in TaMacs *in situ* were several specifically associated with tumor progression and wound healing. These included functional clusters related to the key cancer hallmarks of proliferation, cell death, and differentiation; extracellular matrix deposition and remodeling; and angiogenesis (74, 75, 81, 82). In this same model, suppression of apoptosis, *via* expression of the anti-apoptotic genes Bcl-2 or $Bcl-X_L$, led to reduced TaMac accumulation and reduced angiogenesis when lymphoma cells were xenografted into SCID mice (74). Somewhat surprisingly, suppression of apoptosis had a minimal effect on tumor growth *in vivo*, despite promoting expansion *in vitro* (74). This may be related to the reduced angiogenesis observed in apoptosis-suppressed xenografts or to differences between *in vitro* and *in vivo* growth conditions. Similarly suggestive, though less complete, evidence for a role of apoptotic cell-dependent effects on TaMacs also was found in a model of melanoma (74).

Supportive data for the importance of apoptotic cells in the conditioning of TaMacs also comes from models of prostate and breast cancers (102, 103). Coculture of bone marrow derived macrophages with apoptotic cells from several prostate cancer cell lines induced an M2-like state of macrophage polarization characterized by increased expression of multiple M2-like-associated genes without alteration in the expression of several markers of classical activation (102). This occurred in a milk fat globule-EGF factor 8 (MFG-E8)-dependent manner, with significantly increased levels of MFG-E8 detected in exosomes from prostate cancer cells (102). Additional provocative findings were reported in a model of breast cancer, in which the widespread apoptosis of mammary epithelial cells that occurs in the postpartum period was shown to enhance tumor metastasis (103). Molecular or pharmacologic blockade of the clearance of apoptotic cells led to a reduction of M2-like TaMacs (without a change in the total number of macrophages) and a concomitant reduction in tumor metastasis as compared to nulliparous mice (103).

Tumors As Continually Regenerating Organs

The direct enhancement of cancer cell proliferation and tumor growth by apoptotic cells may be the most well-established postmortem apoptotic effect. More than 50 years ago, it was first reported that admixing lethally irradiated cancer cells with live cancer cells led to a higher incidence of rapidly growing tumors and shorter survival times when compared to injection of an equal number of live cancer cells (79). These results have been replicated both *in vitro* and *in vivo* in a variety of cancers and cancer cell lines (22, 74–80). In one study, the effect was shown to be specific to apoptotic cells, as necrotic cells had no effect (76). A strong parallel exists with wound healing in that the proliferative effect of apoptotic cells in both cancer and wound healing strongly depended on caspase-3-mediated activation of calcium-independent phospholipase A2 ($iPLA_2$) and the $iPLA_2$-mediated generation of prostaglandin E_2 (PGE_2; refs. 16, 22, 52, 77, 78). An important consideration here is that, in these experimental studies, the number of apoptotic cells typically was very large in comparison to that of viable cells, with the ratio of apoptotic to viable cells varying from 20–50:1 to as much as 1,000–10,000:1 (22, 76–80). In one study, the precise ratio proved important, as the effect was found to be highly dependent on apoptotic dose (80).

The later phases of wound healing involve extensive remodeling of the extracellular matrix (21, 81, 82). While the deposition and renovation of connective tissue is time-limited in wounds, eventually leading to the formation of a healed scar, the process is ongoing in tumors. Nevertheless, the genetic signatures of wounds and tumors—and, in particular, of TaMacs—are very similar, highlighting their many shared features (73–75, 81, 82). Among the categories of mutually expressed genes are proteinases and their regulators involved in the cleavage and restructuring of extracellular matrix components [metalloproteinase (MMP)-2, MMP-3, MMP-12, tissue inhibitor of MMP-2 (TIMP2), urokinase plasminogen activator], components of the extracellular matrix (fibronectin-1), and various cytokines regulating the activity and state of differentiation of fibroblasts, myofibroblasts, and other stromal cells (platelet derived growth factor, transforming growth factor-β, VEGF; refs. 73–75, 81, 82).

Antitumor therapies, of course, trigger massive cell death responses. While clearly an intended and vital outcome of therapy, the abrupt upsurge of cell death following radiation or chemotherapy is a double-edged sword, holding the potential to undermine the direct benefit of therapy by promoting the proliferation of surviving tumor cells (22, 23, 77, 78). The repopulation of tumors following therapy may be compared to the compensatory proliferation and organ regeneration observed in lower organisms following structural injury or amputation (67–69). In humans, the only organ with similar regenerative capacity is the liver (16). Many of the factors and signaling events involved in tumor repopulation mirror those driving organ regeneration in lower organisms as well as those promoting the pre-treatment growth and progression of tumors in mice and humans (22, 67–69, 74–80, 104–106). For example, caspase-3-mediated cleavage of $iPLA_2$, and the subsequent generation of PGE_2, which has been shown to be important for the proliferative effect of apoptotic cells in wound healing and tumor growth, also play a critical role in liver regeneration (16) and tumor repopulation (22, 23, 77, 78). The potential importance of this pathway in human cancer is highlighted by an association between elevated levels of caspase-3 and several markers of tumor aggressiveness, including shortened survival, in a variety of cancers (22, 52).

In many ways, the role of apoptotic cells in tumor regeneration may be viewed as an exaggeration of their role in pre-treatment tumor growth and progression. Still, there may be several notable differences. For example, the massive increase of cell death after therapy shifts the proportions of live and dead cells, so that apoptotic cells almost certainly pass from a minority to a majority of cells. The extremely elevated ratios of apoptotic to viable cells used in most studies describing a proliferative effect of apoptotic cancer cells actually may reflect the post-therapy situation (22, 76–80). Additionally, since therapy-induced death is not limited to cancerous cells, the bystander death of non-cancerous cells within the tumor's microenvironment may have further deleterious consequences. It may be that the activities of cancerous and non-cancerous apoptotic cells differ. For example, apoptotic human umbilical vein endothelial cells (HUVEC) stimulated the growth *in vitro* of glioma cell lines; as with apoptotic cancer cells, the effect was linked to PGE_2 released from apoptotic HUVEC (78). Similar results were obtained *in vivo* in murine models of fibrosarcoma and melanoma (107). Following therapy, tumors not only grew twice as fast but also became resistant to further radiation in mice whose endothelial cells were rendered more sensitive to radiation-induced apoptosis (107). Further, non-cancerous cells need not die to have a profound impact on tumor repopulation (108, 109). Following chemotherapy, secreted factors produced by stressed stromal cells within the tumor's microenvironment, which have sustained sublethal DNA damage, may enhance tumorigenesis and promote resistance to future therapy. For example, transcripts of the Wnt family member, wingless-type MMTV integration site family member 16 B (WNT16B), were increased approximately sixfold in prostate stroma after chemotherapy (109). Augmented expression of WNT16B within the prostate tumor microenvironment *in vivo* promoted cancer cell proliferation, migration, and tumor invasiveness, and attenuated the effects of chemotherapy (109). While this study did not specifically address the role of apoptotic cells, the induction of WNT16B by genotoxic stress and the subsequent role of Wnt-dependent signaling events in accelerated repopulation are reminiscent of the roles of apoptotic cells in organ regeneration and compensatory proliferation of lower multicellular organisms (67–69, 104–106).

Immunity, Apoptosis, and Tumorigenesis

The impact of apoptotic cells in suppressing inflammatory responses ("Innate Apoptotic Immunity") is well established (9–12, 18, 19). The contribution of inflammation appears to be critical in tumorigenesis, and the potential for apoptotic cell intervention on this level is obvious. A recent elegant series of studies sheds particular light on the role of inflammation in tumor initiation in the context of the liver (110–114). The combination of chronic inflammation and continuously increased rates of apoptotic hepatocellular death leads to the spontaneous development of hepatocellular carcinoma. While the liver may be unique in its susceptibility, perhaps because of its enormous regenerative capacity, abundant evidence supports a role for inflammation as an enabling feature of tumor development and as an essential characteristic of the tumor microenvironment (3, 4, 74, 75, 98, 100). On the surface, it may appear difficult to reconcile ongoing inflammation with the known potent anti-inflammatory properties of apoptotic cells (9–11), especially as they are present in tumors in increased numbers (4, 24, 25). In the case of wounds, a plausible resolution is that macrophages recruited during the initial inflammatory phase start out in a classically activated state [M1 or $M_{(IFN-\gamma/LPS)}$] state, but, under the influence of apoptotic cells and other factors, they eventually transition to an alternatively activated M2-like state. Given that tumors "never heal," it may be that pro- and anti-inflammatory forces are each continuously and potently present within tumors, and that their continuous opposition helps to explain the lack of a clear state of polarization among TaMacs (4, 73–75, 98–101).

While robust adaptive immunity certainly can function effectively against established tumors (as demonstrated dramatically with the recent success of T cell stimulatory treatments targeting so-called checkpoint inhibitors [such as PD-1 and its ligands]; ref. 115), the role of adaptive immunity normally in tumorigenic initiation and propagation is less well-defined. A plethora of studies spanning almost the entirety of the field of tumor immunology has led to the striking axiom that tumors, generally, are poorly immunogenic, eliciting feeble immune responses. The abundant presence of anti-inflammatory apoptotic cells within a tumor gives rise to the notion that apoptotic tumor cells may be responsible for triggering tumor-specific immunosuppression. However, in contrast to their well-described suppression of innate immune responses, the effects of apoptotic cells on adaptive immune responses are uncertain, even in the non-tumorigenic setting (116). Further, it is not clear whether a common basis exists for the poor immunogenicity of tumors. This might reflect a deficit in antigen presentation, a failure to activate antigen-specific T lymphocyte responders (for example due to a defect in co-stimulation), and/or the stimulation of antigen-specific inhibitory (e.g., T-regulatory) cells.

Under experimental conditions (especially *in vitro*), it is clear that apoptotic tumor cells can be a source of antigen and elicit antigen-specific T cell responses (116). Whether this is meaningful physiologically, and what the relative potency of apoptotic immunostimulatory activity might be, remains unresolved, although other work suggests that non-apoptotic corpses may be more immunogenic (117, 118). It is worth noting that, in cases in which apoptotic immunostimulatory activity has been observed, those apoptotic cells also have provided an innate immune stimulus (due to viral infection; refs. 119, 120). On the other hand, compelling data indicating that apoptotic cells interfere with the process of antigen presentation or skew T cell responses toward a regulatory phenotype are lacking, and the possibility that co-stimulatory molecule expression is modulated by apoptotic cells is contentious. Published studies have variously reported that the expression of one or more co-stimulatory molecules (CD40, C80, and CD86) is diminished, increased, or unaltered (118, 121–123).

Perhaps the most parsimonious view is that, within the tumor microenvironment, the preponderance of apoptotic cells, which shift TaMacs away from a classically activated profile and toward an alternatively activated one, has the potential to attenuate inflammatory responsiveness generally. In this context, requisite innate immune triggers for adaptive immune responsiveness may be insufficient. This perspective also suggests that the paradigmatic conviction that the tumor microenvironment is frankly pro-inflammatory may be in need of re-evaluation.

OPEN QUESTIONS FOR INVESTIGATION

Our discussion of the multifaceted effects of apoptotic cells on tumors rests on the recognition of the unique and ongoing place of apoptotic cells in the tumor microenvironment. Implicit in our discussion is the assumption that the potent postmortem regulatory activities of apoptotic cells are comparable between transformed and non-transformed apoptotic cells. This, however, remains to be tested experimentally, not only in terms of the apoptotic cells eliciting a response, but also in terms of the responding cells themselves. In particular, the possibility exists that subtle differences distinguish the postmortem activities of transformed and non-transformed apoptotic cells. Similarly, it remains to be explored whether subtle differences also distinguish the repertoire of responses of viable transformed and non-transformed cells to apoptotic ones. We have noted that apoptotic cells modulate multiple vital cell activities of untransformed cells, including their survival, proliferation, and growth, but that no evidence exists for the apoptotic modulation of the cancer hallmark of replicative immortality. This is an interesting issue for exploration.

Just as apoptosis is ongoing throughout the life of a tumor, so too is inflammation. We have suggested that the preponderance of apoptotic cells in the tumor microenvironment may shift subtly its balance away from a pro-inflammatory one. While our understanding of the anti-inflammatory, homeostatic effect of apoptotic cells is best contextualized in terms of wound healing, there are important differences between healing wounds and tumors. For example, what are the long-term consequences of an "unhealing" tumor for apoptotic modulation? Might the resolving activity of apoptotic cells eventually become "exhausted" in this setting?

The basis of the insubstantial immunogenicity of tumors remains puzzling and is a critical issue for resolution. Certainly, the state of inflammation within the tumor microenvironment, and the question of whether apoptotic suppression of innate immune responsiveness is the basis of adaptive immune unresponsiveness, must be major considerations. A direct evaluation of the efficacy with which antigen-specific T cell responses (including cross-primed responses) are elicited by apoptotic tumor cells, as compared with other dead tumor cell forms, is urgently needed. In addition, it is important to know whether apoptotic modulation of adaptive immune responsiveness, like that of innate immune responsiveness, is manifest in a dominant manner.

While apoptosis is the primary mechanism by which cells die physiologically and, as we have discussed, is ongoing throughout tumor life, other forms of cell death can occur as well. Notably, and especially post-chemotherapy, this may include "immunogenic cell death" (6, 66). Cells dying in that way, in contrast to typical immunosuppressive apoptosis on which we have focused, can elicit immune responses. It will be interesting to consider the postmortem effects of those non-immunosuppressive cell death forms and the roles that they play in tumorigenesis.

AUTHOR CONTRIBUTIONS

DU and JL collaborated on the thesis of this review article, and wrote it together.

FUNDING

This work was supported in part by an NIH/NIA grant (AG029633) to DU and by institutional funds from Dr. José A. Arruda and the Section of Nephrology, University of Illinois at Chicago, to JL.

REFERENCES

1. Evan GI, Vousden KH. Proliferation, cell cycle and apoptosis in cancer. *Nature* (2001) 411:342–8. doi:10.1038/35077213
2. Hanahan D, Weinberg RA. The hallmarks of cancer. *Cell* (2000) 100:57–70. doi:10.1016/S0092-8674(00)81683-9
3. Hanahan D, Weinberg RA. Hallmarks of cancer: the next generation. *Cell* (2011) 144:646–74. doi:10.1016/j.cell.2011.02.013
4. Gregory CD, Ford CA, Voss JJLP. Microenvironmental effects of cell death in malignant disease. *Adv Exp Med Biol* (2016) 930:51–88. doi:10.1007/978-3-319-39406-0_3
5. Peter C, Wesselborg S, Herrmann M, Lauber K. Dangerous attraction: phagocyte recruitment and danger signals of apoptotic and necrotic cells. *Apoptosis* (2010) 15:1007–28. doi:10.1007/s10495-010-0472-1
6. Zitvogel L, Kepp O, Kroemer G. Decoding cell death signals in inflammation and immunity. *Cell* (2010) 140:798–804. doi:10.1016/j.cell.2010.02.015
7. Hochreiter-Hufford A, Ravichandran KS. Clearing the dead: apoptotic cell sensing, recognition, engulfment, and digestion. *Cold Spring Harb Perspect Biol* (2013) 5:a008748. doi:10.1101/cshperspect.a008748
8. Medina CB, Ravichandran KS. Do not let death do us part: 'find-me' signals in communication between dying cells and the phagocytes. *Cell Death Differ* (2016) 23:979–89. doi:10.1038/cdd.2016.13
9. Voll RE, Herrmann M, Roth EA, Stach C, Kalden JR, Girkontaite I. Immunosuppressive effects of apoptotic cells. *Nature* (1997) 390:350–1. doi:10.1038/37022
10. Fadok VA, Bratton DL, Konowal A, Freed PW, Westcott JY, Henson PM. Macrophages that have ingested apoptotic cells in vitro inhibit proinflammatory cytokine production through autocrine/paracrine mechanisms involving TGF-β, PGE2, and PAF. *J Clin Invest* (1998) 101:890–8. doi:10.1172/JCI1112
11. Cvetanovic M, Ucker DS. Innate immune discrimination of apoptotic cells: repression of proinflammatory macrophage transcription is coupled directly to specific recognition. *J Immunol* (2004) 172:880–9. doi:10.4049/jimmunol.172.2.880
12. Ucker DS. Innate apoptotic immunity: a potent immunosuppressive response repertoire elicited by specific apoptotic cell recognition. In: Krysko DV, Vandenabeele P, editors. *Phagocytosis of Dying Cells: From Molecular Mechanisms to Human Diseases*. Berlin, NY: Springer (2009). p. 163–87.
13. Reddy SM, Hsiao KH, Abernethy VE, Fan H, Longacre A, Lieberthal W, et al. Phagocytosis of apoptotic cells by macrophages induces novel signaling events leading to cytokine-independent survival and inhibition of proliferation: activation of Akt and inhibition of extracellular signal-regulated kinases 1 and 2. *J Immunol* (2002) 169:702–13. doi:10.4049/jimmunol.169.2.702
14. Patel VA, Lee DJ, Feng L, Antoni A, Lieberthal W, Schwartz JH, et al. Recognition of apoptotic cells by epithelial cells: conserved versus tissue-specific signaling responses. *J Biol Chem* (2010) 285:1829–40. doi:10.1074/jbc.M109.018440
15. Patel VA, Feng L, Lee DJ, Massenburg D, Pattabiraman G, Antoni A, et al. Recognition-dependent signaling events in response to apoptotic targets inhibit epithelial cell viability by multiple mechanisms: implications for non-immune tissue homeostasis. *J Biol Chem* (2012) 287:13761–77. doi:10.1074/jbc.M112.350843
16. Li F, Huang Q, Chen J, Peng Y, Roop DR, Bedford JS, et al. Apoptotic cells activate the "phoenix rising" pathway to promote wound healing and tissue regeneration. *Sci Signal* (2010) 3:ra13. doi:10.1126/scisignal.2000634
17. Patel VA, Massenburg D, Vujicic S, Feng L, Tang M, Litbarg N, et al. Apoptotic cells activate AMP-activated protein kinase (AMPK) and inhibit epithelial cell growth without change in intracellular energy stores. *J Biol Chem* (2015) 290:22352–69. doi:10.1074/jbc.M115.667345
18. Cvetanovic M, Mitchell JE, Patel V, Avner BS, Su Y, van der Saag PT, et al. Specific recognition of apoptotic cells reveals a ubiquitous and unconventional innate immunity. *J Biol Chem* (2006) 281:20055–67. doi:10.1074/jbc.M603920200
19. Birge RB, Ucker DS. Innate apoptotic immunity: the calming touch of death. *Cell Death Differ* (2008) 15:1096–102. doi:10.1038/cdd.2008.58
20. Weihua Z, Tsan R, Schroit AJ, Fidler IJ. Apoptotic cells initiate endothelial cell sprouting via electrostatic signaling. *Cancer Res* (2005) 65:11529–35. doi:10.1158/0008-5472.CAN-05-2718
21. Dvorak HF. Tumors: wounds that do not heal. Similarities between tumor stroma generation and wound healing. *N Engl J Med* (1986) 315:1650–9. doi:10.1056/NEJM198612253152606
22. Huang Q, Li F, Liu X, Li W, Shi W, Liu F-F, et al. Caspase 3-mediated stimulation of tumor cell repopulation during cancer radiotherapy. *Nat Med* (2011) 17:860–6. doi:10.1038/nm.2385
23. Lauber K, Munoz LE, Berens C, Jendrossek V, Belka C, Herrmann M. Apoptosis induction and tumor cell repopulation: the yin and yang of radiotherapy. *Radiat Oncol* (2011) 6:176. doi:10.1186/1748-717X-6-176
24. Egeblad M, Nakasone ES, Werb Z. Tumors as organs: complex tissues that interface with the entire organism. *Dev Cell* (2010) 18:884–901. doi:10.1016/j.devcel.2010.05.012
25. Yaacoub K, Pedeux R, Tarte K, Guillaudeux T. Role of the tumor microenvironment in regulating apoptosis and cancer progression. *Cancer Lett* (2016) 378:150–9. doi:10.1016/j.canlet.2016.05.012
26. Ucker DS. Exploiting death: apoptotic immunity in microbial pathogenesis. *Cell Death Differ* (2016) 23:990–6. doi:10.1038/cdd.2016.17
27. Merrick JC, Edelson BT, Bhardwaj V, Swanson PE, Unanue ER. Lymphocyte apoptosis during early phase of *Listeria* infection in mice. *Am J Pathol* (1997) 151:785–92.
28. Edelson BT, Bradstreet TR, Hildner K, Carrero JA, Frederick KE, Kc W, et al. CD8α^+ dendritic cells are an obligate cellular entry point for productive infection by *Listeria monocytogenes*. *Immunity* (2011) 35:236–48. doi:10.1016/j.immuni.2011.06.012
29. Pattabiraman G, Palasiewicz K, Visvabharathy L, Freitag NE, Ucker DS. Apoptotic cells enhance pathogenesis of *Listeria monocytogenes*. *Microb Pathog* (2017) 105:218–25. doi:10.1016/j.micpath.2017.02.036
30. Carrero JA, Calderon B, Unanue ER. Lymphocytes are detrimental during the early innate immune response against *Listeria monocytogenes*. *J Exp Med* (2006) 203:933–40. doi:10.1084/jem.20060045
31. Hotchkiss RS, Chang KC, Grayson MH, Tinsley KW, Dunne BS, Davis CG, et al. Adoptive transfer of apoptotic splenocytes worsens survival, whereas adoptive transfer of necrotic splenocytes improves survival in sepsis. *Proc Natl Acad Sci U S A* (2003) 100:6724–9. doi:10.1073/pnas.1031788100
32. Hotchkiss RS, McConnell KW, Bullok K, Davis CG, Chang KC, Schwulst SJ, et al. TAT-BH4 and TAT-Bcl-x_L peptides protect against sepsis-induced lymphocyte apoptosis in vivo. *J Immunol* (2006) 176:5471–7. doi:10.4049/jimmunol.176.9.5471
33. Auerbuch V, Brockstedt DG, Meyer-Morse N, O'Riordan M, Portnoy DA. Mice lacking the type I interferon receptor are resistant to *Listeria monocytogenes*. *J Exp Med* (2004) 200:527–33. doi:10.1084/jem.20040976

34. Carrero JA, Calderon B, Unanue ER. Type I interferon sensitizes lymphocytes to apoptosis and reduces resistance to *Listeria* infection. *J Exp Med* (2004) 200:535–40. doi:10.1084/jem.20040769
35. O'Connell RM, Saha SK, Vaidya SA, Bruhn KW, Miranda GA, Zarnegar B, et al. Type I interferon production enhances susceptibility to *Listeria monocytogenes* infection. *J Exp Med* (2004) 200:437–45. doi:10.1084/jem.20040712
36. Carrero JA, Calderon B, Unanue ER. Listeriolysin O from *Listeria monocytogenes* is a lymphocyte apoptogenic molecule. *J Immunol* (2004) 172:4866–74. doi:10.4049/jimmunol.172.8.4866
37. Leoncini L, Del Vecchio MT, Megha T, Barbini P, Galieni P, Pileri S, et al. Correlations between apoptotic and proliferative indices in malignant non-Hodgkin's lymphomas. *Am J Pathol* (1993) 142:755–63.
38. Spina D, Leoncini L, Del Vecchio MT, Megha T, Minacci C, Poggi SA, et al. Low versus high cell turnover in diffusely growing non-Hodgkin's lymphomas. *J Pathol* (1995) 177:335–41. doi:10.1002/path.1711770403
39. Sun B, Sun Y, Wang J, Zhao X, Wang X, Hao X. Extent, relationship and prognostic significance of apoptosis and cell proliferation in synovial sarcoma. *Eur J Cancer Prev* (2006) 15:258–65. doi:10.1097/01.cej.0000198896.02185.68
40. Naresh KN, Lakshminarayanan K, Pai SA, Borges AM. Apoptosis index is a predictor of metastatic phenotype in patients with early stage squamous carcinoma of the tongue: a hypothesis to support this paradoxical association. *Cancer* (2001) 91:578–84. doi:10.1002/1097-0142(20010201)91:3<578::AID-CNCR1037>3.0.CO;2-W
41. Ohbu M, Saegusa M, Okayasu I. Apoptosis and cellular proliferation in oesophageal squamous cell carcinomas: differences between keratinizing and nonkeratinizing types. *Virchows Arch* (1995) 427:271–6. doi:10.1007/BF00203394
42. Jalali Nadoushan MR, Peivareh H, Azizzadeh Delshad A. Correlation between apoptosis and histological grade of transitional cell carcinoma of urinary bladder. *Urol J* (2004) 1:177–9. doi:10.22037/uj.v1i3.314
43. Lipponen PK, Aaltomaa S. Apoptosis in bladder cancer as related to standard prognostic factors and prognosis. *J Pathol* (1994) 173:333–9. doi:10.1002/path.1711730408
44. Lipponen P, Aaltomaa S, Kosma VM, Syrjänen K. Apoptosis in breast cancer as related to histopathological characteristics and prognosis. *Eur J Cancer* (1994) 30A:2068–73. doi:10.1016/0959-8049(94)00342-3
45. Zhang G-J, Kimijima I, Abe R, Watanabe T, Kanno M, Hara K, et al. Apoptotic index correlates to bcl-2 and p53 protein expression, histological grade and prognosis in invasive breast cancers. *Anticancer Res* (1998) 18:1989–98.
46. Vakkala M, Lähteenmäki K, Raunio H, Pääkkö P, Soini Y. Apoptosis during breast carcinoma progression. *Clin Cancer Res* (1999) 5:319–24.
47. Heatley MK. Association between the apoptotic index and established prognostic parameters in endometrial adenocarcinoma. *Histopathology* (1995) 27:469–72. doi:10.1111/j.1365-2559.1995.tb00312.x
48. Aihara M, Scardino PT, Truong LD, Wheeler TM, Goad JR, Yang G, et al. The frequency of apoptosis correlates with the prognosis of Gleason Grade 3 adenocarcinoma of the prostate. *Cancer* (1995) 75:522–9. doi:10.1002/1097-0142(19950115)75:2<522::AID-CNCR2820750215>3.0.CO;2-W
49. Stapleton AMF, Zbell P, Kattan MW, Yang G, Wheeler TM, Scardino PT, et al. Assessment of the biologic markers p53, Ki-67, and apoptotic index as predictive indicators of prostate carcinoma recurrence after surgery. *Cancer* (1998) 82:168–75. doi:10.1002/(SICI)1097-0142(19980101)82:1<168::AID-CNCR21>3.0.CO;2-#
50. Levine EL, Renehan A, Gossiel R, Davidson SE, Roberts SA, Chadwick C, et al. Apoptosis, intrinsic radiosensitivity and prediction of radiotherapy response in cervical carcinoma. *Radiother Oncol* (1995) 37:1–9. doi:10.1016/0167-8140(95)01622-N
51. Tsang RW, Wong CS, Fyles AW, Levin W, Manchul LA, Milosevic M, et al. Tumour proliferation and apoptosis in human uterine cervix carcinoma II: correlations with clinical outcome. *Radiother Oncol* (1999) 50:93–101. doi:10.1016/S0167-8140(98)00119-4
52. Hu Q, Peng J, Liu W, He X, Cui L, Chen X, et al. Elevated cleaved caspase-3 is associated with shortened overall survival in several cancer types. *Int J Clin Exp Pathol* (2014) 7:5057–70.
53. Hindermann W, Berndt A, Wunderlich H, Katenkamp D, Kosmehl H. Quantitative evaluation of apoptosis and proliferation in renal cell carcinoma. Correlation to tumor subtype, cytological grade according to thoenes-classification and the occurrence of metastasis. *Pathol Res Pract* (1997) 193:1–7. doi:10.1016/S0344-0338(97)80088-X
54. Kasagi N, Gomyo Y, Shirai H, Tsujitani S, Ito H. Apoptotic cell death in human gastric carcinoma: analysis by terminal deoxynucleotidyl transferase-mediated dUTP-biotin nick end labeling. *Jpn J Cancer Res* (1994) 85:939–45. doi:10.1111/j.1349-7006.1994.tb02972.x
55. Ito Y, Matsuura N, Sakon M, Takeda T, Umeshita K, Nagano H, et al. Both cell proliferation and apoptosis significantly predict shortened disease-free survival in hepatocellular carcinoma. *Br J Cancer* (1999) 81:747–51. doi:10.1038/sj.bjc.6690758
56. McMenamin ME, O'Neill AJ, Gaffney EF. Extent of apoptosis in ovarian serous carcinoma: relation to mitotic and proliferative indices, p53 expression, and survival. *Mol Pathol* (1997) 50:242–6. doi:10.1136/mp.50.5.242
57. Hirvikoski P, Kumpulainen E, Virtaniemi J, Pirinen R, Salmi L, Halonen P, et al. Enhanced apoptosis correlates with poor survival in patients with laryngeal cancer but not with cell proliferation, bcl-2 or p53 expression. *Eur J Cancer* (1999) 35:231–7. doi:10.1016/S0959-8049(98)00347-5
58. Ariel A, Fredman G, Sun Y-P, Kantarci A, Van Dyke TE, Luster AD, et al. Apoptotic neutrophils and T cells sequester chemokines during immune response resolution through modulation of CCR5 expression. *Nat Immunol* (2006) 7:1209–16. doi:10.1038/ni1392
59. Zernecke A, Bidzhekov K, Noels H, Shagdarsuren E, Gan L, Denecke B, et al. Delivery of microRNA-126 by apoptotic bodies induces CXCL12-dependent vascular protection. *Sci Signal* (2009) 2:ra81. doi:10.1126/scisignal.2000610
60. Castellana D, Zobairi F, Martinez MC, Panaro MA, Mitolo V, Freyssinet J-M, et al. Membrane microvesicles as actors in the establishment of a favorable prostatic tumoral niche: a role for activated fibroblasts and CX3CL1-CX3CR1 axis. *Cancer Res* (2009) 69:785–93. doi:10.1158/0008-5472.CAN-08-1946
61. Lopatina T, Gai C, Deregibus MC, Kholia S, Camussi G. Cross talk between cancer and mesenchymal stem cells through extracellular vesicles carrying nucleic acids. *Front Oncol* (2016) 6:125. doi:10.3389/fonc.2016.00125
62. Raposo G, Stoorvogel W. Extracellular vesicles: exosomes, microvesicles, and friends. *J Cell Biol* (2013) 200:373–83. doi:10.1083/jcb.201211138
63. Parnaik R, Raff MC, Scholes J. Differences between the clearance of apoptotic cells by professional and non- professional phagocytes. *Curr Biol* (2000) 10:857–60. doi:10.1016/S0960-9822(00)00598-4
64. Devitt A, Marshall LJ. The innate immune system and the clearance of apoptotic cells. *J Leukoc Biol* (2011) 90:447–57. doi:10.1189/jlb.0211095
65. Krysko DV, Denecker G, Festjens N, Gabriels S, Parthoens E, D'Herde K, et al. Macrophages use different internalization mechanisms to clear apoptotic and necrotic cells. *Cell Death Differ* (2006) 13:2011–22. doi:10.1038/sj.cdd.4401900
66. Obeid M, Tesniere A, Ghiringhelli F, Fimia GM, Apetoh L, Perfettini J-L, et al. Calreticulin exposure dictates the immunogenicity of cancer cell death. *Nat Med* (2007) 13:54–61. doi:10.1038/nm1523
67. Bergmann A, Steller H. Apoptosis, stem cells, and tissue regeneration. *Sci Signal* (2010) 3:re8. doi:10.1126/scisignal.3145re8
68. Suzanne M, Steller H. Shaping organisms with apoptosis. *Cell Death Differ* (2013) 20:669–75. doi:10.1038/cdd.2013.11
69. King RS, Newmark PA. The cell biology of regeneration. *J Cell Biol* (2012) 196:553–62. doi:10.1083/jcb.201105099
70. Esmann L, Idel C, Sarkar A, Hellberg L, Behnen M, Möller S, et al. Phagocytosis of apoptotic cells by neutrophil granulocytes: diminished proinflammatory neutrophil functions in the presence of apoptotic cells. *J Immunol* (2010) 184:391–400. doi:10.4049/jimmunol.0900564
71. Bournazou I, Pound JD, Duffin R, Bournazos S, Melville LA, Brown SB, et al. Apoptotic human cells inhibit migration of granulocytes via release of lactoferrin. *J Clin Invest* (2009) 119:20–32. doi:10.1172/JCI36226
72. Flavahan WA, Gaskell E, Bernstein BE. Epigenetic plasticity and the hallmarks of cancer. *Science* (2017) 357:eaal2380. doi:10.1126/science.aal2380
73. Colegio OR, Chu N-Q, Szabo AL, Chu T, Rhebergen AM, Jairam V, et al. Functional polarization of tumour-associated macrophages by tumour-derived lactic acid. *Nature* (2014) 513:559–63. doi:10.1038/nature13490
74. Ford CA, Petrova S, Pound JD, Voss JJLP, Melville L, Paterson M, et al. Oncogenic properties of apoptotic tumor cells in aggressive B cell lymphoma. *Curr Biol* (2015) 25:577–88. doi:10.1016/j.cub.2014.12.059
75. Voss JJLP, Ford CA, Petrova S, Melville L, Paterson M, Pound JD, et al. Modulation of macrophage antitumor potential by apoptotic lymphoma cells. *Cell Death Differ* (2017) 24:971–83. doi:10.1038/cdd.2016.132

76. Reiter I, Krammer B, Schwamberger G. Differential effect of apoptotic versus necrotic tumor cells on macrophage antitumor activities. *J Immunol* (1999) 163:1730–2.
77. Donato AL, Huang Q, Liu X, Li F, Zimmerman MA, Li C-Y. Caspase 3 promotes surviving melanoma tumor cell growth after cytotoxic therapy. *J Invest Dermatol* (2014) 134:1686–92. doi:10.1038/jid.2014.18
78. Mao P, Smith L, Xie W, Wang M. Dying endothelial cells stimulate proliferation of malignant glioma cells via a caspase 3-mediated pathway. *Oncol Lett* (2013) 5:1615–20. doi:10.3892/ol.2013.1223
79. Révész L. Effect of tumour cells killed by x-rays upon the growth of admixed viable cells. *Nature* (1956) 178:1391–2. doi:10.1038/1781391a0
80. Chaurio R, Janko C, Schorn C, Maueröder C, Bilyy R, Gaipl U, et al. UVB-irradiated apoptotic cells induce accelerated growth of co-implanted viable tumor cells in immune competent mice. *Autoimmunity* (2013) 46:317–22. doi:10.3109/08916934.2012.754433
81. Schäfer M, Werner S. Cancer as an overhealing wound: an old hypothesis revisited. *Nat Rev Mol Cell Biol* (2008) 9:628–38. doi:10.1038/nrm2455
82. Byun JS, Gardner K. Wounds that will not heal: pervasive cellular reprogramming in cancer. *Am J Pathol* (2013) 182:1055–64. doi:10.1016/j.ajpath.2013.01.009
83. Fadok VA, Voelker DR, Campbell PA, Cohen JJ, Bratton DL, Henson PM. Exposure of phosphatidylserine on the surface of apoptotic lymphocytes triggers specific recognition and removal by macrophages. *J Immunol* (1992) 148:2207–16.
84. Wang J, Kwaan HC. The pathogenetic role of apoptosis in hypercoagulable states. *Hematology* (2001) 6:143–52. doi:10.1080/10245332.2001.11746565
85. Muhsin-Sharafaldine M-R, Kennedy BR, Saunderson SC, Buchanan CR, Dunn AC, Faed JM, et al. Mechanistic insight into the procoagulant activity of tumor-derived apoptotic vesicles. *Biochim Biophys Acta* (2017) 1861:286–95. doi:10.1016/j.bbagen.2016.11.020
86. Taraseviciene-Stewart L, Kasahara Y, Alger L, Hirth P, Mc Mahon G, Waltenberger J, et al. Inhibition of the VEGF receptor 2 combined with chronic hypoxia causes cell death-dependent pulmonary endothelial cell proliferation and severe pulmonary hypertension. *FASEB J* (2001) 15:427–38. doi:10.1096/fj.00-0343com
87. Golpon HA, Fadok VA, Taraseviciene-Stewart L, Scerbavicius R, Sauer C, Welte T, et al. Life after corpse engulfment: phagocytosis of apoptotic cells leads to VEGF secretion and cell growth. *FASEB J* (2004) 18:1716–8. doi:10.1096/fj.04-1853fje
88. Lentz BR. Exposure of platelet membrane phosphatidylserine regulates blood coagulation. *Prog Lipid Res* (2003) 42:423–38. doi:10.1016/S0163-7827(03)00025-0
89. Andersen JP, Vestergaard AL, Mikkelsen SA, Mogensen LS, Chalat M, Molday RS. P4-ATPases as phospholipid flippases—structure, function, and enigmas. *Front Physiol* (2016) 7:275. doi:10.3389/fphys.2016.00275
90. Suzuki J, Denning DP, Imanishi E, Horvitz HR, Nagata S. Xk-related protein 8 and CED-8 promote phosphatidylserine exposure in apoptotic cells. *Science* (2013) 341:403–6. doi:10.1126/science.1236758
91. Segawa K, Kurata S, Nagata S. Human type IV P-type ATPases that work as plasma membrane phospholipid flippases and their regulation by caspase and calcium. *J Biol Chem* (2016) 291:762–72. doi:10.1074/jbc.M115.690727
92. Rothlin CV, Carrera-Silva EA, Bosurgi L, Ghosh S. TAM receptor signaling in immune homeostasis. *Annu Rev Immunol* (2015) 33:355–91. doi:10.1146/annurev-immunol-032414-112103
93. Graham DK, DeRyckere D, Davies KD, Earp HS. The TAM family: phosphatidylserine sensing receptor tyrosine kinases gone awry in cancer. *Nat Rev Cancer* (2014) 14:769–85. doi:10.1038/nrc3847
94. Rothlin CV, Ghosh S, Zuniga EI, Oldstone MB, Lemke G. TAM receptors are pleiotropic inhibitors of the innate immune response. *Cell* (2007) 131:1124–36. doi:10.1016/j.cell.2007.10.034
95. Linger RM, Keating AK, Earp HS, Graham DK. TAM receptor tyrosine kinases: biologic functions, signaling, and potential therapeutic targeting in human cancer. *Adv Cancer Res* (2008) 100:35–83. doi:10.1016/S0065-230X(08)00002-X
96. Cocco RE, Ucker DS. Distinct modes of macrophage recognition for apoptotic and necrotic cells are not specified exclusively by phosphatidylserine exposure. *Mol Biol Cell* (2001) 12:919–30. doi:10.1091/mbc.12.4.919
97. Ucker DS, Jain MR, Pattabiraman G, Palasiewicz K, Birge RB, Li H. Externalized glycolytic enzymes are novel, conserved, and early biomarkers of apoptosis. *J Biol Chem* (2012) 287:10325–43. doi:10.1074/jbc.M111.314971
98. Mantovani A, Sica A. Macrophages, innate immunity and cancer: balance, tolerance, and diversity. *Curr Opin Immunol* (2010) 22:231–7. doi:10.1016/j.coi.2010.01.009
99. Noy R, Pollard JW. Tumor-associated macrophages: from mechanisms to therapy. *Immunity* (2014) 41:49–61. doi:10.1016/j.immuni.2014.06.010
100. Mantovani A, Allavena P. The interaction of anticancer therapies with tumor-associated macrophages. *J Exp Med* (2015) 212:435–45. doi:10.1084/jem.20150295
101. Franklin RA, Liao W, Sarkar A, Kim MV, Bivona MR, Liu K, et al. The cellular and molecular origin of tumor-associated macrophages. *Science* (2014) 344:921–5. doi:10.1126/science.1252510
102. Soki FN, Koh AJ, Jones JD, Kim YW, Dai J, Keller ET, et al. Polarization of prostate cancer-associated macrophages is induced by milk fat globule-EGF factor 8 (MFG-E8)-mediated efferocytosis. *J Biol Chem* (2014) 289:24560–72. doi:10.1074/jbc.M114.571620
103. Stanford JC, Young C, Hicks D, Owens P, Williams A, Vaught DB, et al. Efferocytosis produces a prometastatic landscape during postpartum mammary gland involution. *J Clin Invest* (2014) 124:4737–52. doi:10.1172/JCI76375
104. Tseng A-S, Adams DS, Qiu D, Koustubhan P, Levin M. Apoptosis is required during early stages of tail regeneration in *Xenopus laevis*. *Dev Biol* (2007) 301:62–9. doi:10.1016/j.ydbio.2006.10.048
105. Chera S, Ghila L, Dobretz K, Wenger Y, Bauer C, Buzgariu W, et al. Apoptotic cells provide an unexpected source of Wnt3 signaling to drive *Hydra* head regeneration. *Dev Cell* (2009) 17:279–89. doi:10.1016/j.devcel.2009.07.014
106. Goessling W, North TE, Loewer S, Lord AM, Lee S, Stoick-Cooper CL, et al. Genetic interaction of PGE2 and Wnt signaling regulates developmental specification of stem cells and regeneration. *Cell* (2009) 136:1136–47. doi:10.1016/j.cell.2009.01.015
107. Garcia-Barros M, Paris F, Cordon-Cardo C, Lyden D, Rafii S, Haimovitz-Friedman A, et al. Tumor response to radiotherapy regulated by endothelial cell apoptosis. *Science* (2003) 300:1155–9. doi:10.1126/science.1082504
108. Östman A. The tumor microenvironment controls drug sensitivity. *Nat Med* (2012) 18:1332–4. doi:10.1038/nm.2938
109. Sun Y, Campisi J, Higano C, Beer TM, Porter P, Coleman I, et al. Treatment-induced damage to the tumor microenvironment promotes prostate cancer therapy resistance through WNT16B. *Nat Med* (2012) 18:1359–68. doi:10.1038/nm.2890
110. Kondylis V, Kumari S, Vlantis K, Pasparakis M. The interplay of IKK, NF-κB and RIPK1 signaling in the regulation of cell death, tissue homeostasis and inflammation. *Immunol Rev* (2017) 277:113–27. doi:10.1111/imr.12550
111. Luedde T, Beraza N, Kotsikoris V, van Loo G, Nenci A, De Vos R, et al. Deletion of NEMO/IKKγ in liver parenchymal cells causes steatohepatitis and hepatocellular carcinoma. *Cancer Cell* (2007) 11:119–32. doi:10.1016/j.ccr.2006.12.016
112. Liedtke C, Bangen J-M, Freimuth J, Beraza N, Lambertz D, Cubero FJ, et al. Loss of caspase-8 protects mice against inflammation-related hepatocarcinogenesis but induces non-apoptotic liver injury. *Gastroenterology* (2011) 141:2176–87. doi:10.1053/j.gastro.2011.08.037
113. Ehlken H, Krishna-Subramanian S, Ochoa-Callejero L, Kondylis V, Nadi NE, Straub BK, et al. Death receptor-independent FADD signalling triggers hepatitis and hepatocellular carcinoma in mice with liver parenchymal cell-specific NEMO knockout. *Cell Death Differ* (2014) 21:1721–32. doi:10.1038/cdd.2014.83
114. Kondylis V, Polykratis A, Ehlken H, Ochoa-Callejero L, Straub BK, Krishna-Subramanian S, et al. NEMO prevents steatohepatitis and hepatocellular carcinoma by inhibiting RIPK1 kinase activity-mediated hepatocyte apoptosis. *Cancer Cell* (2015) 28:582–98. doi:10.1016/j.ccell.2015.10.001
115. Vesely MD, Kershaw MH, Schreiber RD, Smyth MJ. Natural innate and adaptive immunity to cancer. *Annu Rev Immunol* (2011) 29:235–71. doi:10.1146/annurev-immunol-031210-101324
116. Yatim N, Cullen S, Albert ML. Dying cells actively regulate adaptive immune responses. *Nat Rev Immunol* (2017) 17:262–75. doi:10.1038/nri.2017.9
117. Sauter B, Albert ML, Francisco L, Larsson M, Somersan S, Bhardwaj N. Consequences of cell death: exposure to necrotic tumor cells, but not primary

tissue cells or apoptotic cells, induces the maturation of immunostimulatory dendritic cells. *J Exp Med* (2000) 191:423–34. doi:10.1084/jem.191.3.423
118. Barker RN, Erwig L-P, Hill KS, Devine A, Pearce WP, Rees AJ. Antigen presentation by macrophages is enhanced by the uptake of necrotic, but not apoptotic, cells. *Clin Exp Immunol* (2002) 127:220–5. doi:10.1046/j.1365-2249.2002.01774.x
119. Albert ML, Pearce SFA, Francisco LM, Sauter B, Roy P, Silverstein RL, et al. Immature dendritic cells phagocytose apoptotic cells via $\alpha_V\beta_5$ and CD36, and cross-present antigens to cytotoxic T lymphocytes. *J Exp Med* (1998) 188:1359–68. doi:10.1084/jem.188.7.1359
120. Torchinsky MB, Garaude J, Martin AP, Blander JM. Innate immune recognition of infected apoptotic cells directs T_H17 cell differentiation. *Nature* (2009) 458:78–82. doi:10.1038/nature07781
121. Clayton AR, Prue RL, Harper L, Drayson MT, Savage CO. Dendritic cell uptake of human apoptotic and necrotic neutrophils inhibits CD40, CD80, and CD86 expression and reduces allogeneic T cell responses: relevance to systemic vasculitis. *Arthritis Rheum* (2003) 48:2362–74. doi:10.1002/art.11130
122. Parameswaran S, Khalil M, Ahmed KA, Sharma RK, Xiang J. Enhanced protective immunity derived from dendritic cells with phagocytosis of CD40 ligand transgene-engineered apoptotic tumor cells via increased dendritic cell maturation. *Tumori* (2015) 101:637–43. doi:10.5301/tj.5000297
123. Yassin LM, Rojas M, Ramírez LA, García LF, Vásquez G. Monocyte activation by apoptotic cells removal in systemic lupus erythematosus patients. *Cell Immunol* (2010) 266:52–60. doi:10.1016/j.cellimm.2010.08.012

Conflict of Interest Statement: The authors declare that the research was conducted in the absence of any commercial or financial relationships that could be construed as a potential conflict of interest.

16

Scrambled Eggs: Apoptotic Cell Clearance by Non-Professional Phagocytes in the *Drosophila* Ovary

Sandy B. Serizier[1,2] *and Kimberly McCall*[1*]

[1] *Department of Biology, Boston University, Boston, MA, United States,* [2] *Graduate Program in Molecular Biology, Cell Biology and Biochemistry, Boston University, Boston, MA, United States*

For half of a century, it has been known that non-professional phagocytes, such as fibroblasts, endothelial, and epithelial cells, are capable of efferocytosis (engulfment of apoptotic cells). Non-professional phagocytes differ from professional phagocytes in the range and efficiency of engulfment. Much of the recognition and underlying signaling machinery between non-professional and professional phagocytes is the same, but it is not known how the engulfment capacity of non-professional phagocytes is controlled. Moreover, the signaling networks involved in cell corpse recognition, engulfment, and phagosome maturation are only partially understood. The *Drosophila* ovary provides an excellent system to investigate the regulation of phagocytic activity by epithelial cells, a major class of non-professional phagocytes. During *Drosophila* oogenesis, mid-stage egg chambers undergo apoptosis of the germline in response to nutrient deprivation. Epithelial follicle cells then undergo major cell shape changes and concomitantly engulf the germline material. Our previous work has established that Draper and the integrin α-PS3/β-PS heterodimer are required in follicle cells for germline cell clearance. In addition, we have characterized phagosome maturation pathways, and found that the JNK pathway amplifies the engulfment response. In this review, we discuss recent advances on the interplay between engulfment pathways in the follicular epithelium for cell clearance in the *Drosophila* ovary. We also provide a comparison to apoptotic cell clearance mechanisms in *C. elegans* and mammals, illustrating strong conservation of efferocytosis mechanisms by non-professional phagocytes.

Edited by:
Estee Kurant,
University of Haifa, Israel

Reviewed by:
Barbara Conradt,
Ludwig-Maximilians-Universität München, Germany
Denise Montell,
University of California, Santa Barbara, United States

***Correspondence:**
Kimberly McCall
kmccall@bu.edu

Keywords: cell death, apoptosis, engulfment, phagocytosis, efferocytosis, epithelial cells, *Drosophila*, oogenesis

INTRODUCTION

Apoptotic cell clearance by phagocytic cells is critical for organismal homeostasis. Professional phagocytes are cells whose main task in the milieu is to efficiently clear dead cells. Non-professional phagocytes, on the other hand, have other tissue-resident functions, but can engulf when needed. Differences between professional and non-professional phagocytes are not well understood. In this review, we present the *Drosophila* ovary as an outstanding model to investigate engulfment by non-professional phagocytes. We first discuss the diversity of apoptotic cell clearance pathways across *Drosophila*, *C. elegans*, and mammals. We next discuss professional and non-professional phagocytes in different organisms with an emphasis on the molecular biology of apoptotic cell clearance in the *Drosophila* ovary by epithelial follicle cells. We compare the follicle cell model to examples of phagocytosis by epithelial cells in mammals and their clinical relevance to health and disease.

APOPTOTIC CELL CLEARANCE MECHANISMS IN *C. elegans*, *D. melanogaster*, AND MAMMALS

The CED-2, -5, and -12 and CED-1, -6, and -7 pathways were first identified in *C. elegans* as the major pathways that control engulfment. Both pathways act in parallel and converge on CED-10 (Rac1) to promote the cytoskeletal rearrangements required for engulfment (1). Rho family GTPases, such as Rac1 and Cdc42, function downstream of apoptotic cell recognition to induce cytoskeletal shape changes to form a phagocytic cup. Rac1 functions across all model systems and is the best characterized cytoskeletal modulator of engulfment (2–4). In *C. elegans*, CDC42 acts in parallel to CED-2/5/12 and CED-10 and downstream of CED-1/6/7 (5). The first *cell death abnormal* (*ced*) genes in *C. elegans*, *ced-1* and *ced-2*, were identified in a screen for mutants with an abnormal persistence of embryonic cell corpses. While all corpses are cleared in late stages of embryogenesis in wild-type strains, *ced-1* and *-2* mutants have persisting corpses (6). Ellis et al. later conducted a mutagenesis screen to isolate maternal effect mutations that prevent corpse clearance. In this screen, analysis of CED mutant progeny of egg laying defective mothers found additional alleles of *ced-1* and *-2*, and identified *ced-5*, *-6*, *-7*, *-8*, and *-10* genes as regulators of corpse clearance (7). Electron microscopy revealed that these mutants specifically exhibit defects in engulfment at the uptake step. Double mutant analysis determined that CED-2, -5, and -12 and the CED-1, -6, and -7 signaling axes act in parallel.

One of the earliest experiments that supported the conservation of apoptotic cell clearance mechanisms across organisms was a study whereby the expression of human orthologs was shown to rescue the CED mutant clearance defects. Specifically, Dock180, the mammalian ortholog of CED-5, was shown to be capable of rescuing the *ced-5* mutant phenotype (8). These early studies in *C. elegans* complemented the discovery of signaling machinery that control apoptotic cell clearance in mammals (1, 8–14).

The engulfment machinery in *C. elegans* is conserved in mammals (**Table 1**). Mammalian Multiple EGF-Like Domains 10 (MEGF-10) is homologous to CED-1, a transmembrane receptor that binds to phosphatidylserine, an aminophospholipid that is exposed on the surface of apoptotic cells and functions as an "eat me" signal (15). The immunoreceptor tyrosine-based activation motifs (ITAMs) of MEGF-10 are phosphorylated by the Src family kinases, and this mediates interaction with Syk tyrosine kinase for the activation of downstream effectors. Engulfment Adaptor PTB Domain Containing 1 (GULP), the CED-6 ortholog, is an adaptor protein that binds to the NPXY motif of the intracellular domain of MEGF-10 *via* its PTB binding domain (12, 16). ABCA1/7, the CED-7 ortholog, is an ABC transporter that has been shown to function in both the apoptotic and engulfing cell. ABCA1 has recently been shown to function in homeostasis to increase cholesterol efflux during apoptotic cell clearance by macrophages (17). CrkII (CED-2 ortholog), Dock180 (CED-5 ortholog), and ELMO (CED-12 ortholog), all encode cytoplasmic signaling proteins that help propagate the engulfment process by activating Rac1 (CED-10 ortholog). The SH3 domain of Dock180 interacts with the PxxP motif and PH domain of ELMO. This ELMO contact with Rac1 and Dock180 stabilizes the Rac1/Dock180 interaction, allowing for Rac1 activation (17). The functional contribution of Cdc42 in mammals is more elusive and context dependent. Specifically, dominant negative Cdc42 blocks F-actin recruitment to phagocytic cups in BMM and NIH3T3 cells (18, 19), but has no effect on photoreceptor outer segment uptake in the retinal pigment epithelium (20). Surprisingly, overexpression of Cdc42 does not induce more phagocytosis in NIH3T3 cells (19). Several additional engulfment receptors have been identified in mammals including BAI1, Tim4, Stablin-2, and MERTK (21–25).

Like mammals, conservation of the underlying engulfment machinery has also been demonstrated in *Drosophila* (**Table 1**). For example, Draper, the *Drosophila* ortholog of MEGF-10/CED-1, requires Src42A (Src ortholog) and Shark (Syk ortholog) kinase activity for the clearance of severed axons in the brain, similar to mammals (26, 27). *Drosophila* glia also require Crk/Mbc (CED-2/CED-5) for the clearance of axonal debris (28). Crk and Mbc, but not ELMO/Ced-12, are required downstream of integrins αPS3/ßν, for apoptotic cell priming for efficient engulfment by hemocytes (29). Studies in *Drosophila* hemocytes have uncovered new players in engulfment such as Pallbearer, an E3-ubiquitin ligase, ribosomal protein S6, and Rac2 (30). In addition, Undertaker, a junctophilin, responds to calcium flux to mediate clearance by hemocytes (31). Many engulfment genes have been shown to have other functions in *Drosophila*. For example, Mbc and Ced-12/ELMO are required for ommatidial development, myoblast fusion, and cell migration (32–38).

TABLE 1 | Engulfment machinery in professional and non-professional phagocytes in *C. elegans*, *D. melanogaster*, and mammals.

	C. elegans	*D. melanogaster*	Mammals
Engulfment receptors	CED-7	ND	ABCA1/7
	αPAT-3/INA-1	αPS3βPS/βν	αvβ5/β3 Integrin
	CED-1	Draper	Megf10/SCARF1
	C03F11.3	Croquemort	CD36
	ND	NimC4/SIMU	ND
	ND	ND	BAI-1
	ND	ND	BAI-3
	ND	ND	TIM-1
	ND	ND	TIM-4
	ND	ND	MerTK
	ND	ND	Fc Receptor
	ND	ND	Stablin 2
	ND	ND	LRP/1
	ND	ND	KIM-1
Adaptor proteins	CED-2	ND	CrkII
	CED-5	Mbc/Sponge	Dock180
	CED-6	Ced-6	GULP
	CED-12	Ced-12	ELMO 1(2)
GTPases	Ced-10	Rac1	Rac1
	Cdc-42	Rac2	Cdc42
		Cdc42	

Orthologs based on literature review. ND indicates ortholog not determined.

PROFESSIONAL AND NON-PROFESSIONAL PHAGOCYTES IN *C. elegans, D. melanogaster*, AND MAMMALS

Professional phagocytes, such as macrophages and monocytes, function to maintain tissue homeostasis by removing dying and infected cells from the milieu (39). Professional phagocytes engulf with high efficiency and have been shown to use several engulfment receptors to complete the task (17, 40–43). Integrins, CD36, and MEGF-10 have been shown to function in professional phagocytes (**Table 1**) (44–47). In *Drosophila*, hemocytes are macrophage-like cells that circulate to clear apoptotic cells and pathogens (48–50). They have also been shown to be critical for proper patterning of the central nervous system, innate immunity, and wound healing (51). Similar to mammals, multiple receptors and bridging proteins, including Draper, Six-microns-under, Croquemort, and Integrins function in apoptotic cell clearance by hemocytes.

Non-professional phagocytes have other resident functions, but can engulf when needed. Examples of non-professional phagocytes include epithelial and endothelial cells, and astrocyte glia (**Table 2**). Interestingly, *C. elegans* do not have professional phagocytes and, therefore, rely solely on neighboring non-professional phagocytes to clear apoptotic debris. The diverse engulfment receptors used by professional phagocytes for clearance of apoptotic cells are conserved in non-professional phagocytes (16). In *Drosophila*, several cell types, including epithelial follicle cells and imaginal disk cells have been shown to function as non-professional phagocytes (52, 53). Thus, like mammals, *Drosophila* utilizes both professional and non-professional phagocytes.

The mechanisms by which professional and non-professional phagocytes communicate and coordinate their activities are currently under investigation. In mammals, alveolar macrophages release IGF-1 when mice are exposed to house dust mites and IGF-1 activates IGF-1R on the surface of airway epithelial cells. This interaction functions to redirect airway epithelial cells from engulfing to initiating an inflammatory response (54). It also stimulates airway epithelial cells to takeup macrophage-derived multivesicular bodies (MVBs) that contain anti-inflammatory cytokines. The anti-inflammatory cytokine containing MVBs suppress the expression of pro-inflammatory gene PTX3, suggesting a mechanism whereby macrophages signal to non-professional phagocytes to resolve inflammation (55).

TABLE 2 | Examples of non-professional phagocytes in *C. elegans, D. melanogaster*, and mammals.

C. elegans	*D. melanogaster*	Mammals
Hypodermal cells	Follicle cells	Mammary gland
Gonadal sheath cells	Ensheathing glia	Retinal pigment epithelium
Pharyngeal muscle cells	Cortex glia	Gut epithelium
Endothelial cells	Astrocyte glia	Liver endothelium
	Imaginal disks	Airway epithelium
	Epidermal cells	Kidney epithelium

THE *Drosophila* OVARY AS A MODEL FOR STUDYING NON-PROFESSIONAL PHAGOCYTES

The *Drosophila* ovary (**Figure 1A**, left) comprises 20 strands of progressively developing egg chambers called ovarioles. Egg chambers arise from the germarium, the anteriormost region of each ovariole that houses germline and follicle stem cells. The germline stem cells produce cystoblasts that undergo four rounds of mitosis to generate cysts containing 16 interconnected cells. Due to incomplete cytokinesis of the dividing cysts, each egg chamber is arranged in a syncytium, where each germline cell is connected to the next through ring canals. One of the 16 cyst cells is specified as the oocyte and the remaining 15 cells differentiate into polyploid nurse cells, whose main task is to provide organelles, RNA, proteins, and nutrients necessary for oocyte growth and embryogenesis. During the division, specification, and differentiation steps, the germarium produces somatically derived epithelial follicle cells that surround the 15 nurse cells and oocyte, which together constitute an egg chamber (**Figure 1A**, right, **Figures 1B,C**). The epithelial follicle cells serve as a protective barrier for the growing germline. Follicle cells also synthesize and secrete yolk and proteins that mediate the formation of vitelline membrane and chorion. As described below, the follicle cells also act as non-professional phagocytes.

Each egg chamber progresses through 14 well-characterized stages of oogenesis (56). Vitellogenesis begins at stage 8 of oogenesis, which has a characteristic loss of proportion between nurse cells and oocyte, where the oocyte becomes noticeably larger than the nurse cells. During vitellogenesis, the follicle cells begin to synthesize and transport yolk proteins into the oocyte. Oocyte growth culminates in a process called dumping, whereby the nurse cells rapidly dump all of their cytoplasmic contents into the oocyte, leaving little nurse cell cytoplasm behind (57). Nurse cells then undergo programmed cell death with nuclear condensation, fragmentation, acidification, and clearance by the follicle cells.

CHECKPOINTS OF CELL DEATH IN THE OVARY

Aside from the developmental cell death that naturally occurs in all egg chambers at the end of oogenesis, insults such as starvation have been shown to induce cell death earlier in oogenesis. These cell deaths occur in response to checkpoints during specific stages of oogenesis when the tissue senses and responds to environmental changes. The earliest checkpoint occurs in the germarium and the second occurs in mid-oogenesis at the onset of vitellogenesis between stages 7 and 9. It is thought that mid-stage egg chambers are primed to respond to environmental stimuli before investing in the energetically expensive vitellogenic process (58, 59).

Several factors have been shown to induce cell death at mid-oogenesis, including temperature, mating, daylength, developmental abnormalities, chemical treatment, cocaine exposure,

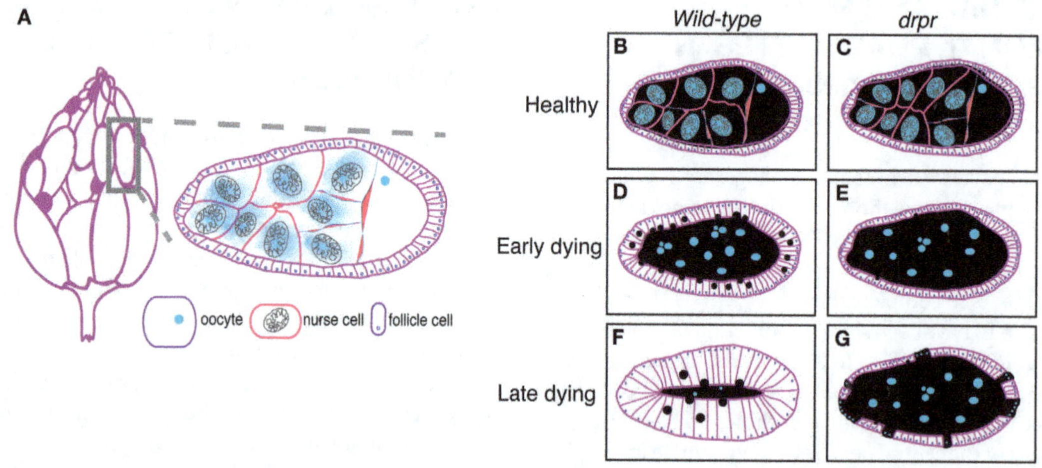

FIGURE 1 | The *Drosophila* ovary as a model for engulfment. **(A)** Each ovary (left) comprises strands of developing eggs called egg chambers. Stage 8 egg chamber (gray box) zoom (right) has three main cell types: germline-derived nurse cells and oocyte and somatically derived follicle cells. Dark purple egg chambers (left) correspond to the egg chambers that spontaneously die in the ovary. Ovary drawing adapted from Ref. (52). **(B–G)** Schematics of healthy and degenerating egg chambers in wild-type and *draper*$^{\Delta 5}$ mutant flies. Cell membranes are shown in purple, nuclei in blue, nurse cell cytoplasm in black, and follicular epithelium in white. Wild-type starvation-induced apoptotic nurse cells are cleared by enlarged follicle cells **(D)** until there is little to no germline material left **(F)**. *draper* mutant follicle cells do not enlarge and fail to clear germline material **(E)**. Follicle cells prematurely die (cyan dots) **(G)**.

cellular phone irradiation, and starvation, but only starvation-induced cell death mechanisms are well characterized (59–64). Apoptosis, the best characterized type of programmed cell death, is defined by caspase activation, chromatin condensation, DNA fragmentation, membrane blebbing, cell shrinkage, and the formation of apoptotic bodies (65). Nurse cell death in mid-oogenesis has been shown to be apoptotic using TUNEL assays to detect fragmented DNA, active effector caspase Dcp-1/Casp3 staining, and DNA morphology analysis that detects highly condensed chromatin (60, 61, 66, 67). The canonical apoptotic pathway in *Drosophila* consists of the activation of pro-apoptotic proteins Reaper, Hid, and Grim (RHG). RHG proteins facilitate activation of an apoptosome-like complex (involving Dark and Dronc) and inactivation of the anti-apoptotic protein Drosophila inhibitor of apoptosis 1 (Diap-1), which leads to activation of effector caspases (Drice and Dcp-1) to dismantle the cell (68). Unusually, nurse cells in mid-oogenesis use mechanisms independent of RHG proteins and the apoptosome to initiate death (69). What is known during the mid-oogenesis checkpoint is that the effector caspase Dcp-1 is essential. Rather than undergoing apoptosis, nurse cell nuclei of Dcp-1 null or Diap-1 overexpressing flies fail to condense, and surrounding follicle cells prematurely die (67, 70, 71).

Buszczak et al. proposed that ecdysone-responsive genes function as part of a surveillance mechanism to detect environmental conditions before progressing to later stages of oogenesis (72). In support of this hypothesis, there are several lines of evidence that ecdysteroid signaling determines whether an egg chamber will progress past mid-oogenesis. Ecdysteroid concentrations increase in ovary extracts upon starvation (73). Early ecdysone genes are expressed in follicle cells at the developmentally sensitive mid-oogenesis timepoint (72, 74, 75). *E75*^{e213} germline mutant clones arrest at stages 8 and 9 of oogenesis, and adrenodoxin reductase, the enzyme required for steroid hormone synthesis, is required in the germline for egg chambers to progress beyond vitellogenic stages (72).

CHARACTERIZATION OF ENGULFMENT BY EPITHELIAL FOLLICLE CELLS

Germline cell death in mid-oogenesis is coupled with engulfment by the surrounding follicle cells, providing a powerful model for engulfment by epithelial cells. Engulfment by follicle cells was first visualized by electron microscopy (52), and later by observations of uptake of fluorescent germline markers (71, 76). In an effort to closely investigate the morphological changes that take place during engulfment by the follicle cells, Etchegaray et al. characterized engulfment in response to starvation-induced apoptosis (77). They found that the underlying follicular epithelial cells synchronously enlarged to engulf the apoptotic germline and the growth in the follicle cells correlated with nurse cell nuclear condensation and fragmentation (**Figures 1B,D,F**). The epithelial follicle cells proceeded to engulf the apoptotic germline until there was no material left (**Figures 1D,F**).

Draper, the *Drosophila* CED-1 ortholog, had been shown to be required for engulfment in multiple tissues, and it was also found to be required in the follicle cells. Draper protein expression levels increases throughout the progression of engulfment in mid-oogenesis (**Figure 2**), indicating that follicle cells may modulate their phagocytic capacity by Draper upregulation (77). *draper* null mutants have severe defects in the uptake of germline material, showing a lack of follicle cell enlargement, premature follicle cell death, and apoptotic germline material that fails to be cleared [**Figures 1E,G**; (77)]. Draper RNAi knockdown in the epithelial follicle cells, but not the germline showed the same phenotype, and overexpression of

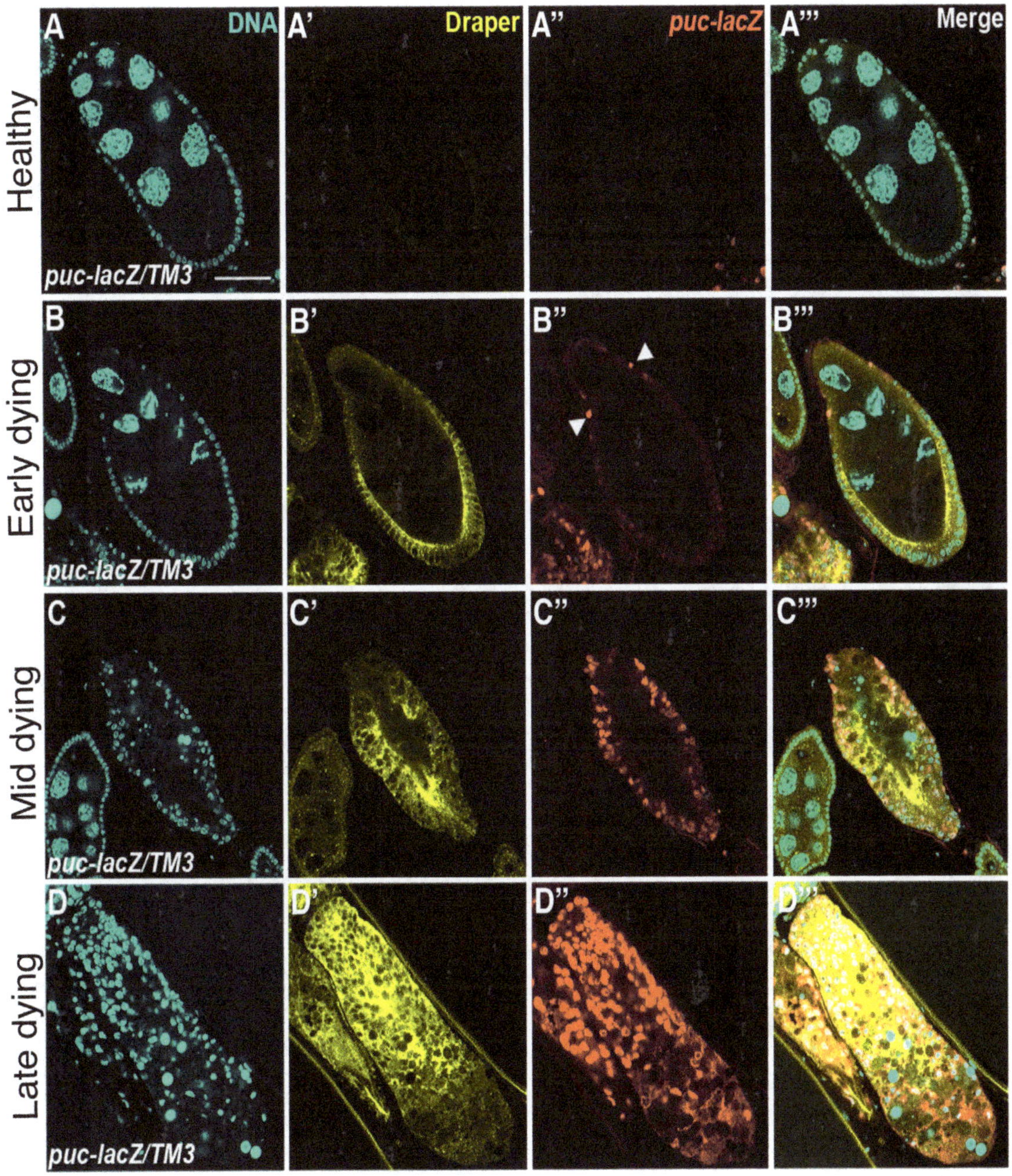

FIGURE 2 | Draper and the JNK signaling pathway are activated in follicle cells during engulfment. Healthy **(A)** and degenerating **(B–D)** egg chambers expressing the JNK reporter *puc-lacZ,* stained with anti-Draper (yellow), anti-βgal (red) and DAPI (cyan). Draper is expressed at low levels in healthy egg chambers **(A′)** and increases in expression throughout the progression of engulfment **(B′–D′)**. *puckered* is not expressed in healthy egg chambers **(A″)**, but is activated during engulfment **(B″–D″)**, merged images are shown in panels **(A′″–D′″)**. Adapted with permission from Ref. (77).

draper in the follicle cells in the *draper* null mutant background rescues these defects, demonstrating that Draper is required in the underlying follicular epithelium for clearance of the dying germline.

To determine the mechanism by which Draper engulfs the apoptotic germline, candidate screens were conducted to identify other genes that are required for Draper-mediated engulfment (77–79). Draper-associated Shark kinase propagates intracellular signaling by recognizing and binding to the phosphorylated YXXL motif of the intracellular domain of Draper (27). Engulfment by follicle cells fails to proceed in the absence of Shark or Src42A, another Draper receptor kinase (78). CED-6 function has been implicated in *C. elegans* and mammals as an adaptor molecule, but RNA interference that targets *Ced-6* in follicle cells has no phenotype, suggesting that another adaptor may function in this context or RNAi knockdown was incomplete (80). Sponge, one of the *Drosophila* CED-5 orthologs, and Ced-12 have both been confirmed to be required for engulfment by follicle cells (78, 80). Consistent with engulfment pathways in other tissues and organisms, expression of dominant negative Rac1, the mediator of cytoskeletal shape changes required for engulfment, blocks engulfment by epithelial follicle cells.

The integrin α-PS3/β-PS heterodimer is induced and required for engulfment by the follicular epithelium during starvation-induced germline death (78). One might speculate that integrins function in tandem with Draper in follicle cells for efficient clearance of apoptotic germline corpses. The following hypotheses could explain the need for multiple receptors: (1) one receptor may function to signal downstream to activate Rac1, while the other receptor functions as a tethering molecule by binding "eat me" signals on the apoptotic cell surface; (2) both receptors may function to activate downstream signals, which converge at Rac1 for efficient engulfment; and (3) both receptors may function to activate downstream signals, one of which will converge onto Rac1 and another that has other cellular functions. *α-PS3 draper* double mutants manifested more severe engulfment defects than single mutants, but were not completely defective in engulfment, which suggests that other receptors may contribute to this engulfment process. Mutant analyses of Croquemort, a scavenger receptor previously implicated in cell clearance and known to be expressed in the ovary, were also investigated in combination with Draper and integrins, but did not worsen the defects in engulfment (78), which suggests another engulfment receptor may function in the ovary. The engulfment machinery required for clearance by the folllicular epithelium in the Drosophila ovary and *C. elegans* and mammalian orthologs are listed in **Table 3**.

TABLE 3 | Required engulfment machinery in *Drosophila* follicular epithelium and orthologs.[a]

D. melanogaster	Mammals	*C. elegans*
Eiger	TNF-α	ND
Draper	Megf-10	CED-1
Integrin α-PS3	Integrin subunit α4[a]	ND
Integrin β-PS	Integrin subunit β1[a]	ND
Crumbs	Crumbs 1	CRUMBS-1
Mekk1	MAP3k4	MTK-1
Basket	MAPK8	JNK-1
Kayak	Fos	FOS-1
Cka	ND	ND
Shark	Syk	ND
Shibire	Dynamin 1	DYNAMIN-1
Deep orange	Vps18	VPS18
Sponge	Dock-3[a]	CED-5
Ced-12	ELMO	CED-12
Rac1	Rac1	CED-10
Dhc64C	Dync1h1	DHC-1
Bazooka	PARD3	PAR-3
aPKC	PRKCI	PKC-3
Par-6	Par-6	PAR-6
Cdc42	Cdc42	CDC-42

[a]*Orthologs were identified using the DRSC Integrative Ortholog Prediction Tool. ND indicates ortholog not determined.*

APOPTOTIC CELL CLEARANCE IS AFFECTED BY CELL POLARITY

Based on observations of enriched expression of Draper and integrins on the apical surface of the follicular epithelium, Meehan et al. (78) sought to determine whether proper cell polarity was required for engulfment. *aPKC, baz, par-6, crb,* and *Dhc64C* were all found to be required for the progression of engulfment, suggesting that an underlying directionality is required for proper localization of engulfment receptors. Indeed, *Dhc64C* (Dynein heavy chain) knockdown in follicle cells by RNAi blocks α-PS3/β-PS apical enrichment. Moreover, *aPKC* knockdown in follicle cells prevents Draper enrichment to the apical surface, suggesting a novel mechanism whereby polarity genes regulate localization of multiple engulfment receptors in epithelial cells. Cdc42 is also required for the progression of engulfment in the *Drosophila* ovary. Follicle cells of engulfing egg chambers form a double layer at the posterior end, similar to cell polarity mutant follicle cells (78, 79), suggesting that Cdc42 functions in cell polarity.

The follicle cell model system shows interesting similarities to the mammalian retinal pigment epithelium. Mouse integrin αv, β5, MFGE8, and MERTK mutants exhibit defects in the clearance of shed outer segments of the retina (81–85). Mouse myosin VIIa mutants have abnormal apical localization of engulfed phagosomes in retinal pigment epithelial cells, suggesting that cell polarity or phagosome trafficking contributes to outer segment disk clearance (86).

PHAGOSOME MATURATION IN THE *Drosophila* OVARY

Phagosome maturation is the last step of corpse clearance where the apoptotic corpse is internalized into a phagosome, or vesicle that matures to its final degradation. Membrane modifications change throughout the maturation process and can be used as markers to visualize how far along the corpse is in the steps of degradation. The earliest modification known to occur is an increase in PtdIns(4,5)P_2 in early phagocytic cups (87). As the phagocytic cup progresses, PtdIns(4,5)P is depleted and PtdIns(3,4)P_2 and PIP3 concentrations increase (87). Much of what is known about phosphoinositol changes during phagosome maturation has been characterized in Fc-receptor-mediated clearance by macrophages (88–90). In *C. elegans*, the sequence of protein recruitment is conserved whereby PtdIns3P and DYNAMIN-1 function at the phagocytic cup followed by RAB2, RAB5, and RAB7 recruitment and fusion with lysosomes (91–95). Upon internalization, the phagosome transitions into an increasingly acidic organelle and fuses with lysosomes to complete corpse degradation. In addition to lipid modifications, the phagosome first associates with RAB5 GTPase and as the phagosome matures, RAB5 is replaced by RAB7. RAB7 is finally replaced by LAMP-1, targeting the corpse for degradation *via* lysosomal fusion (96). PtdIns3P, Rab5, and Rab7 phagosome maturation markers have all been found to occur in the same sequence in the *Drosophila* ovary. *draper*$^{\Delta 5}$ mutants were found to not only have defects in uptake, as shown by a reduced number of engulfed Dcp-1 positive particles, but also could not process the little material engulfed, as shown by the lack of the Rab7 phagosome maturation marker and LysoTracker-positive vesicles compared to controls. *Ced-12* and *Src42A* mutants exhibited fewer vesicles taken up and matured, but there were only mild defects in acidification. α-PS3 mutants had defects in uptake, but Dcp-1 positive vesicles that were able to form, did get coated with Rab7, and were acidified. These findings suggest that the α-PS3/β-PS integrin heterodimer is required for the uptake of dying germline corpses, but Draper has a dual requirement for uptake and phagosome maturation (78).

DRAPER SIGNALS TO THE JNK SIGNALING PATHWAY IN MANY *Drosophila* TISSUES

The JNK signaling pathway is pleiotropic and can induce a variety of downstream effectors that include apoptotic machinery. To determine whether JNK signaling played a role in mid-oogenesis cell death, the expression of *puckered (puc)*, a JNK signaling pathway responsive gene, was investigated (77). Rather than being activated in the dying germline, Puc was specifically induced in follicle cells of apoptotic egg chambers and its expression increased with the progression of engulfment (**Figure 2**). Knockdown of JNK signaling pathway components *Mekk1* (JNKKK), *Bsk* (JNK), *Jra* (jun), or *kayak* (fos) by RNA interference in the follicle cells results in engulfment defects, demonstrating a requirement for JNK signaling in engulfment in the ovary. However, RNA interference of *hemipterous* (JNKK), *slipper* (JNKKK), or *misshapen* in the follicle cells results in normal engulfment, suggesting that only some of the canonical pathway members are required in this context. Similar to the ovary, *Mekk1* and *kayak* are also required in the adult brain for axonal debris clearance, but in contrast to the ovary, Tak1 and Slipper activate Mekk1 in the brain (97). Furthermore, Misshapen loss of function in the adult brain blocks engulfment progression, suggesting context dependent mechanisms for JNK axis activation (98). The JNKKKK and JNKKK acting in the ovary remain to be determined. One activator of JNK signaling is Eiger, the TNF-α ortholog. Eiger interacts with receptors Wengen and Grindelwald, a recently identified TNFR ortholog (99). Eiger/Wengen interaction results in the activation of Misshapen (Msn, JNKKKK), Tak1 (JNKKK), Hemipterous (Hep, JNKK), and Basket (Bsk, JNK). Eiger loss of function results in defects in engulfment in the ovary, suggesting that Eiger may activate JNK in the follicle cells (77). A summary of the signaling pathways in the follicle cells is shown in **Figure 3**.

In the ovary, JNK activity is markedly delayed in *draper* null mutants, suggesting that Draper and the JNK cascade act in the same signaling pathway (77). When the JNKK Hep is constitutively expressed in the *draper* null background, engulfment defects are suppressed, suggesting that JNK acts downstream of Draper, and can activate other targets that facilitate engulfment. Activation of *hep* promotes Draper expression, suggesting JNK is required for *draper* induction during engulfment. Consistent with findings in the ovary, studies in the *Drosophila* embryonic central nervous system, adult brain, and wing epithelium have shown a requirement for the JNK pathway downstream of Draper (27, 100, 101). Studies in the embryonic central nervous system show that while expression levels of Draper and the bridging molecule Six-microns-under remain the same in glia that have active JNK signaling, *hep*^{CA} gain of function in *draper* loss-of-function mutants restores defects in apoptotic clearance (101). In the adult brain, Traf4, a Misshapen binding partner in the JNK signaling pathway, is required for Draper-mediated JNK activation in glial cells for axonal clearance (102). More recent clonal analyses in the brain revealed that TRE, a JNK signaling reporter, fails to turn on in *draper*$^{\Delta 5}$ clones compared to wild type, suggesting a cell autonomous requirement for *draper* activation of JNK in glia for clearance of axons in response to axonal injury, like the ovary (102). In macrophages, JNK is also required for the induction of *draper* in response to corpses (103). Surprisingly, overexpression of Draper II, the inhibitory isoform of Draper, results in increased JNK pathway activity in the wing (100). Taken together, these studies in *Drosophila* indicate that Draper and JNK regulate each other in multiple cellular contexts. JNK is activated in professional and non-professional phagocytes in mammals (104, 105), but whether JNK is required remains to be determined. Perhaps JNK signaling machinery is a prerequisite for phagocytes to increase phagocytic capacity.

Apoptotic cells are thought to signal to phagocytes for their removal by exposing caspase-dependent "eat me" signals. Dcp-1, an effector caspase, is required for germline cell death in response to starvation in the ovary, and mutants block engulfment progression but surprisingly do not affect Draper localization to the follicle cell membrane or activation of the JNK pathway during engulfment. Moreover, overexpression of Dcp-1 in the germline induces cell death but there is delayed JNK activity and Draper expression in the engulfing epithelial follicle cells. These findings suggest that the caspase Dcp-1 is required for an "eat me" signal that acts independently of Draper and JNK. Eat me signals from the germline have not been identified.

MURDERERS BY NATURE: DEATH BY PHAGOPTOSIS

Interestingly, Draper overexpression in epithelial follicle cells was found to induce nurse cell death in the absence of starvation (77). Egg chambers induced to die by *draper* expression have an underlying epithelium with active JNK signaling, the key regulator of engulfment in starvation-induced cell death. This suggests that engulfment machinery has the ability to induce the death of an otherwise healthy cell, a form of cell death coined "phagoptosis" (106). The defining characteristic of phagoptosis is that the loss of function of engulfment machinery blocks cell death. Other characteristics of phagoptosis include the induction of "eat me" signals or the loss of "don't eat me" signals on the surface of healthy cells. Intriguingly, *draper* loss-of-function mutants show a delay in subtle aspects of

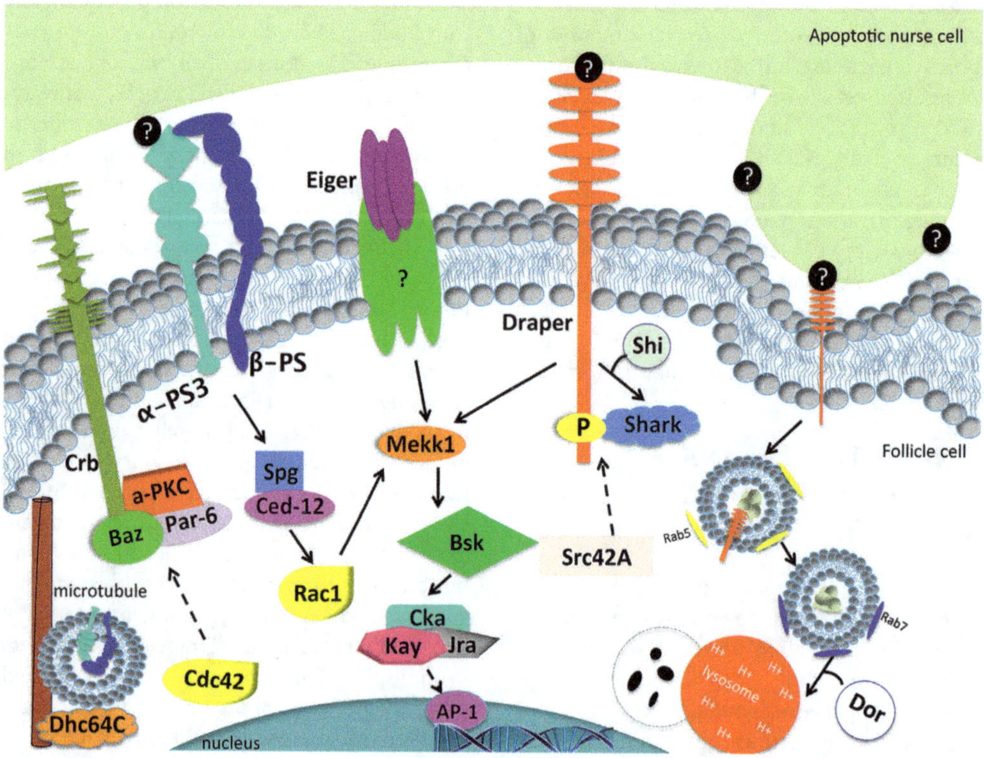

FIGURE 3 | Model of engulfment by follicle cells in the *Drosophila* ovary. Apoptotic cell recognition (left side): Draper binds to an unknown "eat me" signal (black circles) and activates the JNK signaling cascade (Mekk1, Bsk, Cka, Kay, Jra) through Shark to mediate engulfment. Eiger may or may not interact with a TNF receptor to activate the JNK signaling pathway. α-PS3/β-PS integrin heterodimer signals through Spg and Ced-12 to activate Rac1. Rac1 also activates the JNK signaling pathway. The Crb/Baz, Par-6, and aPKC polarity proteins are also required for engulfment. Dhc64C assists in the trafficking of α-PS3/β-PS. Phagosome maturation (right side): Draper is required for nurse cell phagosome maturation. Draper and corpse material first become enclosed in Rab-5 positive phagosomes and mature into Rab-7 positive phagosomes, until the phagosome fuses with lysosomes for degradation. Shi is required for early phagosome maturation. Dor is required for phagosome and lysosome fusion. Dashed arrows correspond to proteins that were tested without epistasis analysis and have engulfment defects. Question marks correspond to unknown proteins.

germline death, including chromatin fragmentation and loss of the oocyte nuclear lamina (80, 107), suggesting that Draper has a normal function in promoting death of the germline. Indeed, Draper directly contributes to cell death during nurse cell developmental cell death in late oogenesis (108).

In *C. elegans*, it has long been thought that engulfment machinery contributes to cell death (109). Engulfment mutants show enhanced survival of cells destined to die in a *ced* mutant background (109, 110). Interestingly, CED-1 has been shown to promote asymmetric localization of the caspase CED-3 in mothers of apoptotic cells (111). Another example where phagocytic machinery is required for death in *C. elegans* occurs in the developing male tail during larval stages (112). In mammals, phagoptosis has been reported in several processes, such as blood cell clearance (113–115), which allows for rapid turnover of blood cells during tissue homeostasis. During mammalian eye development, macrophages induce phagoptosis of vascular endothelial cells by locally releasing WNT7b ligands (116, 117). This newly studied cell death mechanism has implications in therapy, as a recent study found that malignant B cell cancer cell lines die by phagoptosis (118). Furthermore, amyloid β induces superoxide release from microglia to murder neurons (119). Characterization of the mechanisms of phagoptosis may help understand how cancer cells evade death and how amyloid β leads to neurodegeneration.

Fly examples of phagoptosis have also come to light. Kuraishi et al. modified the ER retention motif of Pretaporter, a Draper "eat me" signal, to contain the sequence of a glycosylphosphatidylinositol anchor, which artificially induced Pretaporter exposure on the surface of healthy cells (120). They found that this exposure increases the phagocytic capacity of engulfing S2 cells. This result suggests that the exposure of an "eat me" signal on an otherwise healthy cell can result in its degradation, supporting the conservation of phagoptosis mechanisms across organisms. Interestingly, cell competition studies in wing imaginal disks determined that Draper, Wasp, Mbc/Dock180, and Rac1 engulfment proteins are required for the elimination of adjacent Minute mutant cells (121). *scrib* and *dlg*-induced imaginal disk tumors are eliminated by the activation of a JNK/PVR/Mbc engulfment signaling axis in wild-type adjacent cells (53). These examples indicate that phagoptosis might be more widespread than previously thought.

EPITHELIAL CELL MECHANISMS OF ENGULFMENT ACROSS ORGANISMS

For many years, it has been thought that non-professional phagocytes could not clear apoptotic cells from tissue with the same efficiency as professional phagocytes (122). For example, macrophage-free mice have mesenchymal cells that engulf to a lesser extent (123). Microglia need to contact an apoptotic cell once to initiate clearance, but non-professional phagocytes can take hours to respond post-recognition (124). This suggests that non-professional phagocytes require more time to assemble engulfment machinery and may not be primed to the same extent as professional phagocytes. Studies in kidney 293 cells, however, have shown that the efficiency in the ability to clear apoptotic cells may not always be the result of the type of phagocyte that clears, but the mechanism by which apoptosis is induced (125).

Many non-professional phagocytes are epithelial cells that are found in a wide range of tissues of the human body, including the mammary gland, gastrointestinal tract, and eye (16). Mammary epithelial cells play a key role in phagocytosis during involution post-lactation where mammary alveolar epithelial cells undergo apoptosis and are cleared by adjacent alveolar epithelial cells and macrophages (126). Defects in apoptotic mammary alveolar epithelial cell clearance coincide with inflammation that results in inefficient redevelopment of mammary glands (127). Mfge-8, Dock180, and Rac1 are all required for alveolar epithelial cell death and clearance (127, 128). Although Monks et al. (129) could not detect macrophages or neutrophil presence during involution, other immunohistochemistry and microarray experimental findings demonstrate a change in macrophage gene expression profile changes in mice post weaning (129–132). When macrophages are depleted from mice prior to milk fat globule production for weaning, mammary epithelial cell death, adipocyte repopulation, and other postpartum involution events fail to occur, which suggests that macrophages contribute to the signaling that drives these events post weaning (133).

An example of phagocytic epithelial cells in the eye are the retinal pigment epithelial cells (RPE), a single layer of epithelial cells housed on the retina that line the outer segment of the eye between the photoreceptor cells and choroid. RPE have microvilli that project into the outer segment layer to pinch off and clear the shed distal outer segment ends of photoreceptor cells as part of the daily circadian schedule. Daily clearance of turned over outer segment prevents oxidative toxicity by photoreceptor cells and is critical for vision. Many receptors are required for clearance by RPE including integrins, CD36, and MERTK (81, 85).

A critical role for phagocytic epithelial cells in the colon was recently shown in a mouse dextran sodium sulfate (DSS)-induced colitis mouse model where BAI1 mRNA loss positively correlates with disease progression (134). Members of the TAM engulfment receptor family, interestingly, are either lowly expressed or when expressed are incapable of compensating for the increase in disease progression of BAI1 null mutants. Reintroduction of recombinant BAI1 to DSS-treated BAI1 null mice resulted in a decrease in colitis severity index. Strikingly, BAI1 overexpression in colon epithelium, but not in the myeloid cell lineage, alleviates colitis disease progression. This suggests a critical role for non-professional phagocytes in the colon. This study has shed light on some of the local tissue-specific contributions of epithelium independent of professional phagocytes and may help reveal other disease contexts where professional phagocytes are incapable of clearing tissue secondarily to engulfment defects in the milieu (134).

Asthma patients typically have excessive apoptotic airway epithelial cells in their mucus. Although professional macrophages, such as macrophages, neutrophils, and dendritic cells, circulate in the bronchus, because epithelial cells are in close proximity to apoptotic airway cells, Juncadella et al. asked whether the engulfment machinery of airway epithelial cells contribute to the clearance of apoptotic cells. Indeed, Rac1 loss of function in tracheal and lung epithelium in airway allergen- exposed mice led to increased inflammation and mucus buildup reminiscent of asthma (16, 135). Airway epithelial cells can clear apoptotic epithelial cells and function to dampen the inflammation caused by allergen exposure. These studies illustrate that non-professional phagocytes have a greater function in apoptotic cell clearance and relevance to human disease than previously appreciated.

OPEN QUESTIONS IN THE FIELD ABOUT APOPTOTIC CELL CLEARANCE

While much progress has been made in understanding the molecular biology of apoptotic cell clearance, much is still unknown. Many systems of engulfment across model systems require multiple engulfment receptors for cell clearance to occur properly, but little is known about why more than one is necessary for the execution of clearance. In cases where non-professional phagocytes rapidly remove apoptotic cells, how do they propagate the signal to macrophages once the task is under control? Furthermore, there are many examples where both professional and non-professional phagocytes contribute to apoptotic cell clearance, but how do these two cell populations communicate to ensure that the apoptotic cells are cleared and inflammation is dampened? Draper has been shown to function in multiple steps of engulfment in follicle cells. What signaling mechanisms allow for this molecular switch allowing for Draper to not only initiate cytoskeletal shape changes but also promote apoptotic corpse maturation to lysosomes?

Non-professional phagocytes need to undergo molecular changes that allow them to increase their phagocytic capacity for apoptotic cell clearance. It would be of interest to determine how epithelial cells enhance their phagocytic capacity and whether the JNK signaling pathway promotes such capacity. Moreover, how the Rho family GTPases affect polarity during engulfment is unknown. Whether all epithelial cells require underlying polarization for enhanced phagocytic capacity is also of interest. Interestingly, when follicle cells cannot engulf, they prematurely die. What signaling cues are responsible for premature follicle cell death and what are the environmental cues that control follicle cell death? Expression of Draper in follicle cells promotes germline cell death. What are the mechanisms of phagoptosis in the absence of nutrient deprivation? What factors does Draper rely on to induce apoptotic versus non-apoptotic cell death? The *Drosophila* ovary is an exceptional model for understanding engulfment by non-professional phagocytes and can be used to address pressing questions in the field.

AUTHOR CONTRIBUTIONS

SS prepared the figures. SS and KM wrote the manuscript.

ACKNOWLEDGMENTS

Special thanks to the McCall lab members for providing helpful comments.

FUNDING

This manuscript was supported by NIH grant R01 GM060574 to KM.

REFERENCES

1. Kinchen JM, Cabello J, Klingele D, Wong K, Feichtinger R, Schnabel H, et al. Two pathways converge at CED-10 to mediate actin rearrangement and corpse removal in *C. elegans*. *Nature* (2005) 434:93–9. doi:10.1038/nature03263
2. Erwig LP, Henson PM. Clearance of apoptotic cells by phagocytes. *Cell Death Differ* (2008) 15:243–50. doi:10.1038/sj.cdd.4402184
3. Poon IK, Lucas CD, Rossi AG, Ravichandran KS. Apoptotic cell clearance: basic biology and therapeutic potential. *Nat Rev Immunol* (2014) 14:166–80. doi:10.1038/nri3607
4. Reddien PW, Horvitz HR. The engulfment process of programmed cell death in *Caenorhabditis elegans*. *Annu Rev Cell Dev Biol* (2004) 20:193–221. doi:10.1146/annurev.cellbio.20.022003.114619
5. Neukomm LJ, Zeng S, Frei AP, Huegli PA, Hengartner MO. Small GTPase CDC-42 promotes apoptotic cell corpse clearance in response to PAT-2 and CED-1 in *C. elegans*. *Cell Death Differ* (2014) 21:845–53. doi:10.1038/cdd.2014.23
6. Hedgecock EM, Sulston JE, Thomson JN. Mutations affecting programmed cell deaths in the nematode *Caenorhabditis elegans*. *Science* (1983) 220: 1277–9. doi:10.1126/science.6857247
7. Ellis RE, Jacobson DM, Horvitz HR. Genes required for the engulfment of cell corpses during programmed cell death in *Caenorhabditis elegans*. *Genetics* (1991) 129:79–94.
8. Wu YC, Horvitz HR. *C. elegans* phagocytosis and cell-migration protein CED-5 is similar to human DOCK180. *Nature* (1998) 392:501–4. doi:10.1038/33163
9. Banerjee H, Hawkins Z, Johnson T, Eley S, Alikhan A, McDaniel M, et al. Identification of a mouse orthologue of the CED-6 gene of *Caenorhabditis elegans*. *Plasmid* (2003) 49:30–3. doi:10.1016/S0147-619X(02)00106-3
10. Luciani MF, Chimini G. The ATP binding cassette transporter ABC1, is required for the engulfment of corpses generated by apoptotic cell death. *EMBO J* (1996) 15:226–35.
11. Smits E, Van Criekinge W, Plaetinck G, Bogaert T. The human homologue of *Caenorhabditis elegans* CED-6 specifically promotes phagocytosis of apoptotic cells. *Curr Biol* (1999) 9:1351–4. doi:10.1016/S0960-9822(00)80062-7
12. Su HP, Nakada-Tsukui K, Tosello-Trampont AC, Li Y, Bu G, Henson PM, et al. Interaction of CED-6/GULP, an adapter protein involved in engulfment of apoptotic cells with CED-1 and CD91/low density lipoprotein receptor-related protein (LRP). *J Biol Chem* (2002) 277:11772–9. doi:10.1074/jbc.M109336200
13. Tosello-Trampont AC, Brugnera E, Ravichandran KS. Evidence for a conserved role for CRKII and Rac in engulfment of apoptotic cells. *J Biol Chem* (2001) 276:13797–802. doi:10.1074/jbc.M011238200
14. Wu YC, Horvitz HR. The *C. elegans* cell corpse engulfment gene ced-7 encodes a protein similar to ABC transporters. *Cell* (1998) 93:951–60. doi:10.1016/S0092-8674(00)81201-5
15. Zhou Z, Hartwieg E, Horvitz HR. CED-1 is a transmembrane receptor that mediates cell corpse engulfment in *C. elegans*. *Cell* (2001) 104:43–56. doi:10.1016/S0092-8674(01)00190-8
16. Arandjelovic S, Ravichandran KS. Phagocytosis of apoptotic cells in homeostasis. *Nat Immunol* (2015) 16:907–17. doi:10.1038/ni.3253
17. Penberthy KK, Ravichandran KS. Apoptotic cell recognition receptors and scavenger receptors. *Immunol Rev* (2016) 269:44–59. doi:10.1111/imr.12376
18. Leverrier Y, Ridley AJ. Requirement for Rho GTPases and PI 3-kinases during apoptotic cell phagocytosis by macrophages. *Curr Biol* (2001) 11:195–9. doi:10.1016/S0960-9822(01)00047-1
19. Nakaya M, Kitano M, Matsuda M, Nagata S. Spatiotemporal activation of Rac1 for engulfment of apoptotic cells. *Proc Natl Acad Sci U S A* (2008) 105:9198–203. doi:10.1073/pnas.0803677105
20. Mao Y, Finnemann SC. Essential diurnal Rac1 activation during retinal phagocytosis requires alphavbeta5 integrin but not tyrosine kinases focal adhesion kinase or Mer tyrosine kinase. *Mol Biol Cell* (2012) 23:1104–14. doi:10.1091/mbc.E11-10-0840
21. Miyanishi M, Tada K, Koike M, Uchiyama Y, Kitamura T, Nagata S. Identification of Tim4 as a phosphatidylserine receptor. *Nature* (2007) 450:435–9. doi:10.1038/nature06307
22. Park D, Hochreiter-Hufford A, Ravichandran KS. The phosphatidylserine receptor TIM-4 does not mediate direct signaling. *Curr Biol* (2009) 19:346–51. doi:10.1016/j.cub.2009.01.042
23. Park D, Tosello-Trampont AC, Elliott MR, Lu M, Haney LB, Ma Z, et al. BAI1 is an engulfment receptor for apoptotic cells upstream of the ELMO/Dock180/Rac module. *Nature* (2007) 450:430–4. doi:10.1038/nature06329
24. Park SY, Jung MY, Kim HJ, Lee SJ, Kim SY, Lee BH, et al. Rapid cell corpse clearance by stabilin-2, a membrane phosphatidylserine receptor. *Cell Death Differ* (2008) 15:192–201. doi:10.1038/sj.cdd.4402242
25. Scott RS, McMahon EJ, Pop SM, Reap EA, Caricchio R, Cohen PL, et al. Phagocytosis and clearance of apoptotic cells is mediated by MER. *Nature* (2001) 411:207–11. doi:10.1038/35079659
26. MacDonald JM, Beach MG, Porpiglia E, Sheehan AE, Watts RJ, Freeman MR. The *Drosophila* cell corpse engulfment receptor Draper mediates glial clearance of severed axons. *Neuron* (2006) 50:869–81. doi:10.1016/j.neuron.2006.04.028
27. Ziegenfuss JS, Biswas R, Avery MA, Hong K, Sheehan AE, Yeung YG, et al. Draper-dependent glial phagocytic activity is mediated by Src and Syk family kinase signalling. *Nature* (2008) 453:935–9. doi:10.1038/nature06901
28. Lu TY, Doherty J, Freeman MR. DRK/DOS/SOS converge with Crk/Mbc/dCed-12 to activate Rac1 during glial engulfment of axonal debris. *Proc Natl Acad Sci U S A* (2014) 111:12544–9. doi:10.1073/pnas.1403450111
29. Nonaka S, Ando Y, Kanetani T, Hoshi C, Nakai Y, Nainu F, et al. Signaling pathway for phagocyte priming upon encounter with apoptotic cells. *J Biol Chem* (2017) 292:8059–72. doi:10.1074/jbc.M116.769745
30. Xiao H, Wang H, Silva EA, Thompson J, Guillou A, Yates JR Jr, et al. The Pallbearer E3 ligase promotes actin remodeling via RAC in efferocytosis by degrading the ribosomal protein S6. *Dev Cell* (2015) 32:19–30. doi:10.1016/j.devcel.2014.11.015
31. Cuttell L, Vaughan A, Silva E, Escaron CJ, Lavine M, Van Goethem E, et al. Undertaker, a *Drosophila* Junctophilin, links Draper-mediated phagocytosis and calcium homeostasis. *Cell* (2008) 135:524–34. doi:10.1016/j.cell.2008.08.033
32. Bianco A, Poukkula M, Cliffe A, Mathieu J, Luque CM, Fulga TA, et al. Two distinct modes of guidance signalling during collective migration of border cells. *Nature* (2007) 448:362–5. doi:10.1038/nature05965

33. Cote JF, Vuori K. GEF what? Dock180 and related proteins help Rac to polarize cells in new ways. *Trends Cell Biol* (2007) 17:383–93. doi:10.1016/j.tcb.2007.05.001
34. Duchek P, Somogyi K, Jekely G, Beccari S, Rorth P. Guidance of cell migration by the *Drosophila* PDGF/VEGF receptor. *Cell* (2001) 107:17–26. doi:10.1016/S0092-8674(01)00502-5
35. Hakeda-Suzuki S, Ng J, Tzu J, Dietzl G, Sun Y, Harms M, et al. Rac function and regulation during *Drosophila* development. *Nature* (2002) 416:438–42. doi:10.1038/416438a
36. Katoh H, Negishi M. RhoG activates Rac1 by direct interaction with the Dock180-binding protein Elmo. *Nature* (2003) 424:461–4. doi:10.1038/nature01817
37. Luo L, Liao YJ, Jan LY, Jan YN. Distinct morphogenetic functions of similar small GTPases: *Drosophila* Drac1 is involved in axonal outgrowth and myoblast fusion. *Genes Dev* (1994) 8:1787–802. doi:10.1101/gad.8.15.1787
38. Nolan KM, Barrett K, Lu Y, Hu KQ, Vincent S, Settleman J. Myoblast city, the *Drosophila* homolog of DOCK180/CED-5, is required in a Rac signaling pathway utilized for multiple developmental processes. *Genes Dev* (1998) 12:3337–42. doi:10.1101/gad.12.21.3337
39. Rabinovitch M. Professional and non-professional phagocytes: an introduction. *Trends Cell Biol* (1995) 5:85–7. doi:10.1016/S0962-8924(00)88955-2
40. Gordon S. Phagocytosis: an immunobiologic process. *Immunity* (2016) 44:463–75. doi:10.1016/j.immuni.2016.02.026
41. Manaka J, Kuraishi T, Shiratsuchi A, Nakai Y, Higashida H, Henson P, et al. Draper-mediated and phosphatidylserine-independent phagocytosis of apoptotic cells by *Drosophila* hemocytes/macrophages. *J Biol Chem* (2004) 279:48466–76. doi:10.1074/jbc.M408597200
42. Nagaosa K, Okada R, Nonaka S, Takeuchi K, Fujita Y, Miyasaka T, et al. Integrin betanu-mediated phagocytosis of apoptotic cells in *Drosophila* embryos. *J Biol Chem* (2011) 286:25770–7. doi:10.1074/jbc.M110.204503
43. Nonaka S, Nagaosa K, Mori T, Shiratsuchi A, Nakanishi Y. Integrin alphaPS3/betanu-mediated phagocytosis of apoptotic cells and bacteria in *Drosophila*. *J Biol Chem* (2013) 288:10374–80. doi:10.1074/jbc.M113.451427
44. Ferracini M, Rios FJ, Pecenin M, Jancar S. Clearance of apoptotic cells by macrophages induces regulatory phenotype and involves stimulation of CD36 and platelet-activating factor receptor. *Mediators Inflamm* (2013) 2013:950273. doi:10.1155/2013/950273
45. Greenberg ME, Sun M, Zhang R, Febbraio M, Silverstein R, Hazen SL. Oxidized phosphatidylserine-CD36 interactions play an essential role in macrophage-dependent phagocytosis of apoptotic cells. *J Exp Med* (2006) 203:2613–25. doi:10.1084/jem.20060370
46. Savill J, Dransfield I, Hogg N, Haslett C. Vitronectin receptor-mediated phagocytosis of cells undergoing apoptosis. *Nature* (1990) 343:170–3. doi:10.1038/343170a0
47. Scheib JL, Sullivan CS, Carter BD. Jedi-1 and MEGF10 signal engulfment of apoptotic neurons through the tyrosine kinase Syk. *J Neurosci* (2012) 32:13022–31. doi:10.1523/JNEUROSCI.6350-11.2012
48. Kocks C, Cho JH, Nehme N, Ulvila J, Pearson AM, Meister M, et al. Eater, a transmembrane protein mediating phagocytosis of bacterial pathogens in *Drosophila*. *Cell* (2005) 123:335–46. doi:10.1016/j.cell.2005.08.034
49. Nainu F, Tanaka Y, Shiratsuchi A, Nakanishi Y. Protection of insects against viral infection by apoptosis-dependent phagocytosis. *J Immunol* (2015) 195:5696–706. doi:10.4049/jimmunol.1500613
50. Philips JA, Rubin EJ, Perrimon N. *Drosophila* RNAi screen reveals CD36 family member required for mycobacterial infection. *Science* (2005) 309:1251–3. doi:10.1126/science.1116006
51. Vlisidou I, Wood W. *Drosophila* blood cells and their role in immune responses. *FEBS J* (2015) 282:1368–82. doi:10.1111/febs.13235
52. Giorgi F, Deri P. Cell death in ovarian chambers of *Drosophila melanogaster*. *J Embryol Exp Morphol* (1976) 35:521–33.
53. Ohsawa S, Sugimura K, Takino K, Xu T, Miyawaki A, Igaki T. Elimination of oncogenic neighbors by JNK-mediated engulfment in *Drosophila*. *Dev Cell* (2011) 20:315–28. doi:10.1016/j.devcel.2011.02.007
54. Han CZ, Juncadella IJ, Kinchen JM, Buckley MW, Klibanov AL, Dryden K, et al. Macrophages redirect phagocytosis by non-professional phagocytes and influence inflammation. *Nature* (2016) 539:570–4. doi:10.1038/nature20141
55. Arango Duque G, Descoteaux A. Macrophages tell the non-professionals what to do. *Dev Cell* (2016) 39:633–5. doi:10.1016/j.devcel.2016.12.009
56. King RC. *Ovarian Development in Drosophila melanogaster*. New York: Academic Press (1970).
57. Cooley L, Verheyen E, Ayers K. Chickadee encodes a profilin required for intercellular cytoplasm transport during *Drosophila* oogenesis. *Cell* (1992) 69:173–84. doi:10.1016/0092-8674(92)90128-Y
58. Buszczak M, Cooley L. Eggs to die for: cell death during *Drosophila* oogenesis. *Cell Death Differ* (2000) 7:1071–4. doi:10.1038/sj.cdd.4400755
59. Drummond-Barbosa D, Spradling AC. Stem cells and their progeny respond to nutritional changes during *Drosophila* oogenesis. *Dev Biol* (2001) 231:265–78. doi:10.1006/dbio.2000.0135
60. Chao S, Nagoshi RN. Induction of apoptosis in the germline and follicle layer of *Drosophila* egg chambers. *Mech Dev* (1999) 88:159–72. doi:10.1016/S0925-4773(99)00183-5
61. De Lorenzo C, Strand D, Mechler BM. Requirement of *Drosophila* I(2)gl function for survival of the germline cells and organization of the follicle cells in a columnar epithelium during oogenesis. *Int J Dev Biol* (1999) 43:207–17.
62. Panagopoulos DJ, Chavdoula ED, Nezis IP, Margaritis LH. Cell death induced by GSM 900-MHz and DCS 1800-MHz mobile telephony radiation. *Mutat Res* (2007) 626:69–78. doi:10.1016/j.mrgentox.2006.08.008
63. Pritchett TL, Tanner EA, McCall K. Cracking open cell death in the *Drosophila* ovary. *Apoptosis* (2009) 14:969–79. doi:10.1007/s10495-009-0369-z
64. Steller H. Regulation of apoptosis in *Drosophila*. *Cell Death Differ* (2008) 15:1132–8. doi:10.1038/cdd.2008.50
65. Kerr JF, Wyllie AH, Currie AR. Apoptosis: a basic biological phenomenon with wide-ranging implications in tissue kinetics. *Br J Cancer* (1972) 26:239–57. doi:10.1038/bjc.1972.33
66. Nezis IP, Stravopodis DJ, Papassideri I, Robert-Nicoud M, Margaritis LH. Stage-specific apoptotic patterns during *Drosophila* oogenesis. *Eur J Cell Biol* (2000) 79:610–20. doi:10.1078/0171-9335-00088
67. Peterson JS, Barkett M, McCall K. Stage-specific regulation of caspase activity in *Drosophila* oogenesis. *Dev Biol* (2003) 260:113–23. doi:10.1016/S0012-1606(03)00240-9
68. McCall K. Eggs over easy: cell death in the *Drosophila* ovary. *Dev Biol* (2004) 274:3–14. doi:10.1016/j.ydbio.2004.07.017
69. Peterson JS, Bass BP, Jue D, Rodriguez A, Abrams JM, McCall K. Noncanonical cell death pathways act during *Drosophila* oogenesis. *Genesis* (2007) 45:396–404. doi:10.1002/dvg.20306
70. Laundrie B, Peterson JS, Baum JS, Chang JC, Fileppo D, Thompson SR, et al. Germline cell death is inhibited by P-element insertions disrupting the dcp-1/pita nested gene pair in *Drosophila*. *Genetics* (2003) 165:1881–8.
71. Mazzalupo S, Cooley L. Illuminating the role of caspases during *Drosophila* oogenesis. *Cell Death Differ* (2006) 13:1950–9. doi:10.1038/sj.cdd.4401892
72. Buszczak M, Freeman MR, Carlson JR, Bender M, Cooley L, Segraves WA. Ecdysone response genes govern egg chamber development during mid-oogenesis in *Drosophila*. *Development* (1999) 126:4581–9.
73. Terashima J, Takaki K, Sakurai S, Bownes M. Nutritional status affects 20-hydroxyecdysone concentration and progression of oogenesis in *Drosophila melanogaster*. *J Endocrinol* (2005) 187:69–79. doi:10.1677/joe.1.06220
74. Terashima J, Bownes M. Translating available food into the number of eggs laid by *Drosophila melanogaster*. *Genetics* (2004) 167:1711–9. doi:10.1534/genetics.103.024323
75. Terashima J, Bownes M. E75A and E75B have opposite effects on the apoptosis/development choice of the *Drosophila* egg chamber. *Cell Death Differ* (2006) 13:454–64. doi:10.1038/sj.cdd.4401745
76. Tanner EA, Blute TA, Brachmann CB, McCall K. Bcl-2 proteins and autophagy regulate mitochondrial dynamics during programmed cell death in the *Drosophila* ovary. *Development* (2011) 138:327–38. doi:10.1242/dev.057943
77. Etchegaray JI, Timmons AK, Klein AP, Pritchett TL, Welch E, Meehan TL, et al. Draper acts through the JNK pathway to control synchronous

engulfment of dying germline cells by follicular epithelial cells. *Development* (2012) 139:4029–39. doi:10.1242/dev.082776

78. Meehan TL, Joudi TF, Timmons AK, Taylor JD, Habib CS, Peterson JS, et al. Components of the engulfment machinery have distinct roles in corpse processing. *PLoS One* (2016) 11:e0158217. doi:10.1371/journal.pone.0158217
79. Meehan TL, Kleinsorge SE, Timmons AK, Taylor JD, McCall K. Polarization of the epithelial layer and apical localization of integrins are required for engulfment of apoptotic cells in the *Drosophila* ovary. *Dis Model Mech* (2015) 8:1603–14. doi:10.1242/dmm.021998
80. Timmons AK. *Phagocytosis Genes and the JNK Signaling Pathway Promote Developmental Programmed Cell Death Are Essential for Engulfment of the Dying Germline in the Drosophila ovary [Dissertation]*. Boston University. (2015).
81. Burstyn-Cohen T, Lew ED, Traves PG, Burrola PG, Hash JC, Lemke G. Genetic dissection of TAM receptor-ligand interaction in retinal pigment epithelial cell phagocytosis. *Neuron* (2012) 76:1123–32. doi:10.1016/j.neuron.2012.10.015
82. Feng W, Yasumura D, Matthes MT, LaVail MM, Vollrath D. Mertk triggers uptake of photoreceptor outer segments during phagocytosis by cultured retinal pigment epithelial cells. *J Biol Chem* (2002) 277:17016–22. doi:10.1074/jbc.M107876200
83. Mullen RJ, LaVail MM. Inherited retinal dystrophy: primary defect in pigment epithelium determined with experimental rat chimeras. *Science* (1976) 192:799–801. doi:10.1126/science.1265483
84. Nandrot EF, Anand M, Almeida D, Atabai K, Sheppard D, Finnemann SC. Essential role for MFG-E8 as ligand for alphavbeta5 integrin in diurnal retinal phagocytosis. *Proc Natl Acad Sci U S A* (2007) 104:12005–10. doi:10.1073/pnas.0704756104
85. Nandrot EF, Kim Y, Brodie SE, Huang X, Sheppard D, Finnemann SC. Loss of synchronized retinal phagocytosis and age-related blindness in mice lacking alphavbeta5 integrin. *J Exp Med* (2004) 200:1539–45. doi:10.1084/jem.20041447
86. Gibbs D, Kitamoto J, Williams DS. Abnormal phagocytosis by retinal pigmented epithelium that lacks myosin VIIa, the Usher syndrome 1B protein. *Proc Natl Acad Sci U S A* (2003) 100:6481–6. doi:10.1073/pnas.1130432100
87. Swanson JA. Phosphoinositides and engulfment. *Cell Microbiol* (2014) 16:1473–83. doi:10.1111/cmi.12334
88. Botelho RJ, Teruel M, Dierckman R, Anderson R, Wells A, York JD, et al. Localized biphasic changes in phosphatidylinositol-4,5-bisphosphate at sites of phagocytosis. *J Cell Biol* (2000) 151:1353–68. doi:10.1083/jcb.151.7.1353
89. Kamen LA, Levinsohn J, Swanson JA. Differential association of phosphatidylinositol 3-kinase, SHIP-1, and PTEN with forming phagosomes. *Mol Biol Cell* (2007) 18:2463–72. doi:10.1091/mbc.E07-01-0061
90. Vieira OV, Botelho RJ, Rameh L, Brachmann SM, Matsuo T, Davidson HW, et al. Distinct roles of class I and class III phosphatidylinositol 3-kinases in phagosome formation and maturation. *J Cell Biol* (2001) 155:19–25. doi:10.1083/jcb.200107069
91. Kitano M, Nakaya M, Nakamura T, Nagata S, Matsuda M. Imaging of Rab5 activity identifies essential regulators for phagosome maturation. *Nature* (2008) 453:241–5. doi:10.1038/nature06857
92. Lu N, Zhou Z. Membrane trafficking and phagosome maturation during the clearance of apoptotic cells. *Int Rev Cell Mol Biol* (2012) 293:269–309. doi:10.1016/B978-0-12-394304-0.00013-0
93. Yu X, Lu N, Zhou Z. Phagocytic receptor CED-1 initiates a signaling pathway for degrading engulfed apoptotic cells. *PLoS Biol* (2008) 6:e61. doi:10.1371/journal.pbio.0060061
94. Yu X, Odera S, Chuang CH, Lu N, Zhou Z. *C. elegans* Dynamin mediates the signaling of phagocytic receptor CED-1 for the engulfment and degradation of apoptotic cells. *Dev Cell* (2006) 10:743–57. doi:10.1016/j.devcel.2006.04.007
95. Zhou Z, Yu X. Phagosome maturation during the removal of apoptotic cells: receptors lead the way. *Trends Cell Biol* (2008) 18:474–85. doi:10.1016/j.tcb.2008.08.002
96. Kinchen JM, Ravichandran KS. Phagosome maturation: going through the acid test. *Nat Rev Mol Cell Biol* (2008) 9:781–95. doi:10.1038/nrm2515
97. Macdonald JM, Doherty J, Hackett R, Freeman MR. The c-Jun kinase signaling cascade promotes glial engulfment activity through activation of draper and phagocytic function. *Cell Death Differ* (2013) 20:1140–8. doi:10.1038/cdd.2013.30
98. Doherty J, Sheehan AE, Bradshaw R, Fox AN, Lu TY, Freeman MR. PI3K signaling and Stat92E converge to modulate glial responsiveness to axonal injury. *PLoS Biol* (2014) 12:e1001985. doi:10.1371/journal.pbio.1001985
99. Andersen DS, Colombani J, Palmerini V, Chakrabandhu K, Boone E, Rothlisberger M, et al. The *Drosophila* TNF receptor Grindelwald couples loss of cell polarity and neoplastic growth. *Nature* (2015) 522:482–6. doi:10.1038/nature14298
100. Fullard JF, Baker NE. Signaling by the engulfment receptor draper: a screen in *Drosophila melanogaster* implicates cytoskeletal regulators, Jun N-terminal Kinase, and Yorkie. *Genetics* (2015) 199:117–34. doi:10.1534/genetics.114.172544
101. Shklover J, Mishnaevski K, Levy-Adam F, Kurant E. JNK pathway activation is able to synchronize neuronal death and glial phagocytosis in *Drosophila*. *Cell Death Dis* (2015) 6:e1649. doi:10.1038/cddis.2015.27
102. Lu TY, MacDonald JM, Neukomm LJ, Sheehan AE, Bradshaw R, Logan MA, et al. Axon degeneration induces glial responses through Draper-TRAF4-JNK signalling. *Nat Commun* (2017) 8:14355. doi:10.1038/ncomms14355
103. Weavers H, Evans IR, Martin P, Wood W. Corpse engulfment generates a molecular memory that primes the macrophage inflammatory response. *Cell* (2016) 165:1658–71. doi:10.1016/j.cell.2016.04.049
104. Patel VA, Lee DJ, Feng L, Antoni A, Lieberthal W, Schwartz JH, et al. Recognition of apoptotic cells by epithelial cells: conserved versus tissue-specific signaling responses. *J Biol Chem* (2010) 285:1829–40. doi:10.1074/jbc.M109.018440
105. Patel VA, Longacre A, Hsiao K, Fan H, Meng F, Mitchell JE, et al. Apoptotic cells, at all stages of the death process, trigger characteristic signaling events that are divergent from and dominant over those triggered by necrotic cells: implications for the delayed clearance model of autoimmunity. *J Biol Chem* (2006) 281:4663–70. doi:10.1074/jbc.M508342200
106. Brown GC, Neher JJ. Eaten alive! Cell death by primary phagocytosis: 'phagoptosis'. *Trends Biochem Sci* (2012) 37:325–32. doi:10.1016/j.tibs.2012.05.002
107. Kleinsorge SE. *The Role of Cell Polarity during Cell Fate Specification and Programmed Cell Death in the Drosophila Ovary [Dissertation]*. Boston University (2015).
108. Timmons AK, Mondragon AA, Schenkel CE, Yalonetskaya A, Taylor JD, Moynihan KE, et al. Phagocytosis genes nonautonomously promote developmental cell death in the *Drosophila* ovary. *Proc Natl Acad Sci U S A* (2016) 113:E1246–55. doi:10.1073/pnas.1522830113
109. Reddien PW, Cameron S, Horvitz HR. Phagocytosis promotes programmed cell death in *C. elegans*. *Nature* (2001) 412:198–202. doi:10.1038/35084096
110. Hoeppner DJ, Hengartner MO, Schnabel R. Engulfment genes cooperate with ced-3 to promote cell death in *Caenorhabditis elegans*. *Nature* (2001) 412:202–6. doi:10.1038/35084103
111. Chakraborty S, Lambie EJ, Bindu S, Mikeladze-Dvali T, Conradt B. Engulfment pathways promote programmed cell death by enhancing the unequal segregation of apoptotic potential. *Nat Commun* (2015) 6:10126. doi:10.1038/ncomms10126
112. Johnsen HL, Horvitz HR. Both the apoptotic suicide pathway and phagocytosis are required for a programmed cell death in *Caenorhabditis elegans*. *BMC Biol* (2016) 14:39. doi:10.1186/s12915-016-0262-5
113. Albacker LA, Karisola P, Chang YJ, Umetsu SE, Zhou M, Akbari O, et al. TIM-4, a receptor for phosphatidylserine, controls adaptive immunity by regulating the removal of antigen-specific T cells. *J Immunol* (2010) 185:6839–49. doi:10.4049/jimmunol.1001360
114. Brown GC, Vilalta A, Fricker M. Phagoptosis – cell death by phagocytosis – plays central roles in physiology, host defense and pathology. *Curr Mol Med* (2015) 15:842–51. doi:10.2174/156652401509151105130628
115. Olsson M, Oldenborg PA. CD47 on experimentally senescent murine RBCs inhibits phagocytosis following Fcgamma receptor-mediated but not scavenger receptor-mediated recognition by macrophages. *Blood* (2008) 112:4259–67. doi:10.1182/blood-2008-03-143008

116. Diez-Roux G, Lang RA. Macrophages induce apoptosis in normal cells in vivo. *Development* (1997) 124:3633–8.
117. Lobov IB, Rao S, Carroll TJ, Vallance JE, Ito M, Ondr JK, et al. WNT7b mediates macrophage-induced programmed cell death in patterning of the vasculature. *Nature* (2005) 437:417–21. doi:10.1038/nature03928
118. Metayer LE, Vilalta A, Burke GAA, Brown GC. Anti-CD47 antibodies induce phagocytosis of live, malignant B cells by macrophages via the Fc domain, resulting in cell death by phagoptosis. *Oncotarget* (2017) 8:60892–903. doi:10.18632/oncotarget.18492
119. Neniskyte U, Neher JJ, Brown GC. Neuronal death induced by nanomolar amyloid beta is mediated by primary phagocytosis of neurons by microglia. *J Biol Chem* (2011) 286:39904–13. doi:10.1074/jbc.M111.267583
120. Kuraishi T, Nakagawa Y, Nagaosa K, Hashimoto Y, Ishimoto T, Moki T, et al. Pretaporter, a *Drosophila* protein serving as a ligand for Draper in the phagocytosis of apoptotic cells. *EMBO J* (2009) 28:3868–78. doi:10.1038/emboj.2009.343
121. Li W, Baker NE. Engulfment is required for cell competition. *Cell* (2007) 129:1215–25. doi:10.1016/j.cell.2007.03.054
122. van Furth R, Cohn ZA, Hirsch JG, Humphrey JH, Spector WG, Langevoort HL. The mononuclear phagocyte system: a new classification of macrophages, monocytes, and their precursor cells. *Bull World Health Organ* (1972) 46:845–52.
123. Wood W, Turmaine M, Weber R, Camp V, Maki RA, McKercher SR, et al. Mesenchymal cells engulf and clear apoptotic footplate cells in macrophageless PU.1 null mouse embryos. *Development* (2000) 127:5245–52.
124. Parnaik R, Raff MC, Scholes J. Differences between the clearance of apoptotic cells by professional and non-professional phagocytes. *Curr Biol* (2000) 10:857–60. doi:10.1016/S0960-9822(00)00598-4
125. Wiegand UK, Corbach S, Prescott AR, Savill J, Spruce BA. The trigger to cell death determines the efficiency with which dying cells are cleared by neighbours. *Cell Death Differ* (2001) 8:734–46. doi:10.1038/sj.cdd.4400867
126. Fadok VA. Clearance: the last and often forgotten stage of apoptosis. *J Mammary Gland Biol Neoplasia* (1999) 4:203–11. doi:10.1023/A:1011384009787
127. Hanayama R, Nagata S. Impaired involution of mammary glands in the absence of milk fat globule EGF factor 8. *Proc Natl Acad Sci U S A* (2005) 102:16886–91. doi:10.1073/pnas.0508599102
128. Bagci H, Laurin M, Huber J, Muller WJ, Cote JF. Impaired cell death and mammary gland involution in the absence of Dock1 and Rac1 signaling. *Cell Death Dis* (2014) 5:e1375. doi:10.1038/cddis.2014.338
129. Monks J, Smith-Steinhart C, Kruk ER, Fadok VA, Henson PM. Epithelial cells remove apoptotic epithelial cells during post-lactation involution of the mouse mammary gland. *Biol Reprod* (2008) 78:586–94. doi:10.1095/biolreprod.107.065045
130. Clarkson RW, Wayland MT, Lee J, Freeman T, Watson CJ. Gene expression profiling of mammary gland development reveals putative roles for death receptors and immune mediators in post-lactational regression. *Breast Cancer Res* (2004) 6:R92–109. doi:10.1186/bcr754
131. O'Brien J, Lyons T, Monks J, Lucia MS, Wilson RS, Hines L, et al. Alternatively activated macrophages and collagen remodeling characterize the postpartum involuting mammary gland across species. *Am J Pathol* (2010) 176:1241–55. doi:10.2353/ajpath.2010.090735
132. Stein T, Morris JS, Davies CR, Weber-Hall SJ, Duffy MA, Heath VJ, et al. Involution of the mouse mammary gland is associated with an immune cascade and an acute-phase response, involving LBP, CD14 and STAT3. *Breast Cancer Res* (2004) 6:R75–91. doi:10.1186/bcr753
133. O'Brien J, Martinson H, Durand-Rougely C, Schedin P. Macrophages are crucial for epithelial cell death and adipocyte repopulation during mammary gland involution. *Development* (2012) 139:269–75. doi:10.1242/dev.071696
134. Lee CS, Penberthy KK, Wheeler KM, Juncadella IJ, Vandenabeele P, Lysiak JJ, et al. Boosting apoptotic cell clearance by colonic epithelial cells attenuates inflammation in vivo. *Immunity* (2016) 44:807–20. doi:10.1016/j.immuni.2016.02.005
135. Juncadella IJ, Kadl A, Sharma AK, Shim YM, Hochreiter-Hufford A, Borish L, et al. Apoptotic cell clearance by bronchial epithelial cells critically influences airway inflammation. *Nature* (2013) 493:547–51. doi:10.1038/nature11714

Conflict of Interest Statement: The authors declare that the research was conducted in the absence of any commercial or financial relationships that could be construed as a potential conflict of interest.

17

Efferocytosis and Outside-In Signaling by Cardiac Phagocytes. Links to Repair, Cellular Programming, and Intercellular Crosstalk in Heart

*Matthew DeBerge, Shuang Zhang, Kristofor Glinton, Luba Grigoryeva, Islam Hussein, Esther Vorovich, Karen Ho, Xunrong Luo and Edward B. Thorp**

Department of Pathology, Feinberg Cardiovascular Research Institute, Feinberg School of Medicine, Northwestern University, Chicago, IL, United States

Edited by:
Kirsten Lauber,
Ludwig-Maximilians-Universität München, Germany

Reviewed by:
Luis Enrique Munoz,
University of Erlangen-Nuremberg, Germany
Ian Dransfield,
University of Edinburgh, United Kingdom

****Correspondence:***
Edward B. Thorp
ebthorp@northwestern.edu

Phagocytic sensing and engulfment of dying cells and extracellular bodies initiate an intracellular signaling cascade within the phagocyte that can polarize cellular function and promote communication with neighboring non-phagocytes. Accumulating evidence links phagocytic signaling in the heart to cardiac development, adult myocardial homeostasis, and the resolution of cardiac inflammation of infectious, ischemic, and aging-associated etiology. Phagocytic clearance in the heart may be carried out by professional phagocytes, such as macrophages, and non-professional cells, including myofibrolasts and potentially epithelial cells. During cardiac development, phagocytosis initiates growth cues for early cardiac morphogenesis. In diseases of aging, including myocardial infarction, heightened levels of cell death require efficient phagocytic debridement to salvage further loss of terminally differentiated adult cardiomyocytes. Additional risk factors, including insulin resistance and other systemic risk factors, contribute to inefficient phagocytosis, altered phagocytic signaling, and delayed cardiac inflammation resolution. Under such conditions, inflammatory presentation of myocardial antigen may lead to autoimmunity and even possible rejection of transplanted heart allografts. Increased understanding of these basic mechanisms offers therapeutic opportunities.

Keywords: efferocytosis, heart, phagocytosis, macrophage, cardiomyocyte

INTRODUCTION

Each day billions of cells per person must be cleared during homeostatic cellular turnover (1). Inefficiencies of phagocytic clearance lead to exposure of self-antigen, which is a precursor to autoimmune reactivity (2). In contrast to unicellular organisms that utilize phagocytosis primarily to ingest nutrients, herein we focus on metazoans, and specifically heart tissue, in which the process of phagocytosis by professional phagocytes, particularly macrophages, has evolved additional organ-specific significance, the latter a topic of significant interest in understanding how the local environment shapes cellular identity and tissue homeostasis (3). Prior to just 2014, traditional views held that cardiac phagocytes arose from blood monocytes. However, discoveries on the contributions of cardiac macrophages that have taken up residence prenatally has evolved this view (4) and elevated our appreciation of the diversity of cardiac phagocyte subsets. Herein, this review will discuss our own evolving understanding of phagocyte function in the heart, with particular attention paid to phagocytic signaling, a core phagocyte function with consequences during cardiac development, homeostasis, and disease.

PHAGOCYTOSIS IN THE DEVELOPING HEART

The inextricably linked pathways of programed cell death and cellular removal, in turn contribute to tissue remodeling that is integral to embryonic and postnatal organ development. During cardiac development, human fetal cardiocytes appear to ingest other apoptotic cardiocytes, and failure to do so is a hallmark of fetal congenital heart block (CHB) and associated maternal antibodies to ribonucleoproteins (5). For example, binding of antiribonucleoprotein antibodies to apoptotic cardiocytes modifies the distribution of urokinase plasminogen activator receptors, which serves as an antiphagocytic "*don't eat me*" signal and prevents phagocytosis of apoptotic cardiocytes by neighboring viable cardiocytes (6). Accumulation of these opsonized apoptotic cardiocytes triggers proinflammatory cytokine secretion by macrophages, leading to the fibrosis characteristic of CHB (7, 8). In the case of myeloid phagocytes, and in contrast to the roles of cardiac macrophages in the adult heart, far less is understood on the function of phagocytes during embryonic, fetal, and neonatal stages. Loss of macrophage differentiation or function in the developing heart of *Xenopus* embryos arrests heart formation with targeted depletion of *spib*, a transcription factor essential for primitive macrophage differentiation, or *lurp1*, a protein secreted by macrophages that is linked to embryogenesis, through preventing formation of the fused, wedge-shaped trough that is a precursor to heart tube formation (9). Thus, macrophages are positioned to shape the myocardial layer and remain in proximity during remodeling of the developing heart. In the case of rodents, apoptotic debris has been observed in macrophages of the developing rat heart, likely acquired through the physiological processes of vestigial structure deletion, cell number control, and structure remodeling, suggesting that phagocytic signaling could modulate growth cues for early cardiac morphogenesis (10). In mice, the developing heart contains multiple macrophage subsets, which can be classified into distinct populations based on the expression of C-C chemokine receptor 2 (CCR2) and are derived from yolk sac, recombination activating gene 1^+ lymphomyeloid, and fetal Fms-like tyrosine kinase 3^+ monocyte lineages (11). Functionally, $CCR2^-$ yolk sac-derived macrophages were found to be required for coronary development and maturation, whereas macrophages derived from lymphomyeloid and fetal monocyte lineages appeared dispensable for normal heart development. Mechanistically, embryonic $CCR2^-$ macrophages demonstrated increased expression of insulin-like growth factor (IGF) ligands, a proangiogenic signal, compared to $CCR2^+$ macrophages, and were selectively recruited to perfused vasculature where they functioned to remodel the developing coronary vascular plexus by promoting expansion of perfused blood vessels. Despite no overt role for lymphomyeloid and fetal monocyte lineages for heart development in this study, the differences in timing of recruitment, location within the developing heart, and transcriptional profiles indicate the need for additional studies to understand whether these distinct macrophage lineages contribute to embryonic and postnatal organ development, or in response to embryonic cardiac developmental insults.

NONPHLOGISTIC PHAGOCYTIC CLEARANCE DURING THE CARDIAC STEADY STATE

In adults, efficient cell removal is critical for ensuring that the daily turnover of senescent cells does not disturb the steady state by inciting inflammation. During steady state, both professional phagocytes and non-professional "bystander" cells may participate in removal and metabolism of dead cells through the process of efferocytosis (12). Apoptosis eliminates senescent cells in the absence of inflammation, as efferocytic mechanisms suppress proinflammatory cytokines (13). In the heart, a recent analysis of cell generation and turnover revealed that cardiomyocyte numbers are initially established perinatally and appear to be constant throughout human life; cardiomyocyte turnover was estimated at <1% per year in adulthood (14). This was in contrast to endothelial and mesenchymal cells, including fibroblasts and smooth muscle cells, which exchanged at a high rate. Thus, in comparison to other cell types, cardiomyocyte apoptosis is not a likely significant factor in daily macrophage phagocytic programming. This does not rule out however, that cardiomyocytes, through release of degradation products through lysosomal exocytosis (15), exosomes, or ectosomes, in turn may stimulate receptor mediated endocytosis or phagocyte signaling toward the maintenance of the steady state. For example, ectosomes released by some cell types can induce phagocytic receptor anti-inflammatory pathways in macrophages (16). In the adult murine heart at steady state, resident macrophages are maintained by local proliferation to populate and replicate within the myocardium (4). Resident $CCR2^-$ macrophages can be divided into $MHCII^{HI}$ and $MHCII^{LO}$ subsets, which differ significantly in gene ontology of antigen processing pathways (4). In a scenario where mice express the fluorescent TdTom reporter, strictly in cardiomyocytes (Rosa-TdTom x Mlc2V-cre), increased fluorescence can be found associated with resident cardiac macrophages. Although this could be associated with macrophage phagocytosis during preparation of cardiac extracts, it does support the prospect of cardiomyocyte sampling as a means for communication between myocytes and macrophages. *In vitro*, $MHCII^{LO}$ cardiac macrophages were most efficient at taking up dead cell cargo (4, 17). Many interesting questions remain in terms of homeostatic clearance in the heart. For example, it is unclear whether specific receptors are utilized in the steady state, as opposed to during cardiac inflammation or injury, as well as the associated relationships to anti-inflammatory pathways. Furthermore, the classification and characterization of cardiac resident macrophages in these studies and others discussed herein are derived from the murine heart. Whether these observations will translate to the human heart remains to be determined.

PHAGOCYTIC CLEARANCE DURING CARDIAC INFECTION

A potential yet understudied role for resident cardiac macrophages in the steady state is the defense against infection. As proof of principle, injection of fluorescently labeled bacteria leads to uptake by cardiac phagocytes (18). *Streptococcus pneumoniae* enters the myocardium and forms damaging microlesions (19); however, these lesions exhibit relatively low levels of inflammatory infiltrate that only increase after antimicrobial therapy. In the case of Chagas heart disease, the leading cause of infectious myocarditis and caused by the protozoan parasite *Trypanosoma cruzi*, infection can be characterized by cardiomyocyte necrosis throughout the course of disease (20), likely inducing activation of cardiac macrophages. Another important feature of experimental infection with *T. cruzi* is the massive increase in apoptotic, activation-induced cell death in $CD4^+$ T lymphocytes (21). Phagocytosis of these apoptotic lymphocytes by macrophages results in macrophage secretion of TGF-β leading to suppressive TGF-β signaling and increased growth of *T. cruzi* in the macrophage (22, 23). Interestingly, in patients with cardiac clinical forms of Chagas disease, there is an increase in the expression of CCR5 on $CD4^+$ T cells, which controls leukocyte migration into the inflamed heart (24), and while CCR5 expression is required during the acute phase for protection against experimental *T. cruzi* infection in mice, it is dispensable for the chronic phase of infection (25). Thus, during the chronic phase of infection, continuous recruitment of $CD4^+$ T cells to the infected heart followed by their apoptosis and engulfment by cardiac macrophages could contribute to an immunosuppressive environment to allow *T. cruzi* to escape host responses leading to chronic cardiomyopathies. Similar to Chagas disease, patients with infectious endocartitis due to *Coxiella burnetti*, can exhibit valvuopathy with increased levels of apoptotic leukocytes. This has also been linked to efferocytic anti-inflammatory macrophage polarization, thereby permitting increased bacterial replication (26). In contrast, anti-inflammatory macrophages play an important role in limiting excessive inflammation during viral myocarditis (27). Following coxsackievirus B3 infection, viral myocarditis was milder in female mice, which displayed enhanced expression of anti-inflammatory mediators by macrophages, compared to male mice, which displayed higher levels of proinflammatory macrophage markers. Adoptive transfer of *ex vivo* alternatively activated macrophages alleviated the excessive inflammation in male mice, consistent with macrophage polarization contributing to the extent of myocardial inflammation. These studies highlight that cardiac macrophages likely play an important role in shaping host defense against a variety of pathogens in the heart and this is further supported by the ability of pathogens, such a *T. cruzi*, to exploit essential phagocyte function to evade clearance.

PHAGOCYTIC CLEARANCE AS AN INDUCER OF PHAGOCYTE PROGRAMMING OF CARDIAC REPAIR AFTER ACUTE ISCHEMIC INJURY AND CLINICAL REPERFUSION

In Western Societies, including the United States, heart disease and stroke remain leading causes of death (28). Patients who survive their first heart attack have an increased risk of secondary MI, heart failure, and stroke, and secondary risk is linked to the local and systemic inflammation that occurs after first MI (29). A key function of recruited and mobilized leukocytes at site of infarction is the degradation and phagocytosis of dying and necrotic cells, and extracellular matrix. Inhibition of innate immune cells is associated with adverse outcomes post-MI (30). Similar to inflammatory atherosclerosis (31, 32), the infarction consists of a necrotic core (33) that can expand between the endocardium and epicardium. Bordering this necrotic core are endangered cardiomyocytes that may be either salvaged or not, dependent in part on the efficiency of the repair process. This is necessary for subsequent fibrogenic responses and remodeling to compensate for lost cardiomyocytes, as well as angiogenesis to reperfuse the tissue. Recent data directly link efferocytosis by inflammatory immune cells (12) to wound healing in the myocardium and implicate phagocytosis receptors on monocytes and macrophages as a key link between inflammation resolution and organ function (34, 35). In the elderly, suboptimal dying-cell clearance may lead to a non-resolving inflammation (36), and maladaptive cardiac repair, thereby accelerating heart failure (37). An additional clinical component is the contribution of reperfusion, which although restores oxygen supply, can also itself facilitate reperfusion-associated injury (38). Below we expand on key steps surrounding phagocytic clearance after cardiac wound injury.

CHEMOTAXIS SIGNALS FOR PHAGOCYTES TO SITES OF MYOCARDIAL INJURY

Directed chemotaxis to the infarction proceeds by trafficking through a gradient of reducing oxygen tension and a chemotactic gradient of local so-called apoptotic *find-me* signals (39), including lipids, such as lyso-phosphatidyl-choline (LPC) and sphingosine-1-phosphate (S1P). LPC is externalized and excreted during apoptosis (40, 41) and amasses during ischemia in the heart *via* thrombin activation of $Ca2^+$-independent phospholipases (42), consistent with its role as a *find-me* signal in the damaged heart. S1P, another lipid *find-me* signal is produced by sphingosine kinase 1 (SPHK1). Apoptotic stress induces SPHK1 activation, which can then promote S1P secretion (43). In addition to lipid *find-me* signals, proteinaceous tissue recruitment factors include fractalkine (CX3CL1), which is cleaved by caspase-3 during apoptosis. Released fractalkine interacts with CX3CR1 on macrophages for cell recruitment (44). Fas/CD95-induced chemokines, which includes monocyte chemoattractant protein 1/C-C chemokine ligand 2 (MCP-1/CCL2), can recruit monocytes and monocyte-derived macrophages for phagocytosis *via* CCR2 (45). Nucleotides ATP and UTP from apoptotic and necrotic cells also likely act as *find-me* signals in the myocardium. In apoptotic cells, the plasma membrane channel pannexin 1 (PANX1) may act as a portal for nucleotide release (46). During ischemia, cellular stress increases glycosylation of PANX1 and increased ATP release from myocytes to activate fibroblast transformation (47). ATP can also serve as a signal for neutrophil chemotaxis *via* purinergic P2Y2 and A3 adenosine receptors *in vitro* and *in vivo* (48). Knockdown of *P2y2* inhibits migration (49), all consistent with ATP released from PANX1 acting as a *find-me* signal in the heart. Taken together, a variety of "find me" signals may be released by apoptotic cells in the heart but whether these signals cooperate or are distinct and how they direct the phagocytic response during cardiac injury require further investigation.

UNIQUE FUNCTIONS OF PHAGOCYTE SUBSETS DURING CARDIAC REPAIR POST MI

Following MI, innate immune cells are recruited and mobilized to heart to clear damaged tissue and initiate cardiac repair. Neutrophils, the recruitment of which is linked to circadian oscillation (50), accumulate in the ischemic myocardium in large numbers within a few hours, acting as first responders after the onset of injury (51). Neutrophils act to clear necrotic debris, but are also capable of clearing apoptotic cells in other circumstances (52). While neutrophils are among the first to arrive in the injured heart, their function has often been associated with detrimental effects on heart healing. For example, blockade of neutrophil function has been shown to limit adverse ventricular remodeling and preserve systolic function (53, 54) and the magnitude of the neutrophil response was predictive of adverse outcomes in both mice (50) and humans (55, 56). Initial studies using antibody-mediated depletion of neutrophils revealed protective effects during myocardial ischemia–reperfusion injury (IRI) (57–59). However, in the context of the prolonged ischemia that occurs after experimental, permanent coronary ligation, neutrophil depletion led to increased cardiac fibrosis and progressively worsened cardiac function with increased markers of heart failure (57). The worsened outcome following neutrophil depletion was attributed to reduced phagocytic receptor *Mertk* gene expression on cardiac macrophages, preventing efficient clearance of dying cardiomyocytes and proper inflammation resolution. Mechanistically, neutrophil gelatinase-associated lipocalin was identified as a neutrophil secreted molecule that was capable of programming macrophages toward a highly phagocytic, MerTK-expressing, proreparative phenotype. In addition to secreted factors, neutrophils represent a large population of short-lived inflammatory cells that undergo apoptosis in the infarcted myocardium. Phagocytosis of apoptotic neutrophils by macrophages directs inflammation resolution by promoting an anti-inflammatory program leading to the release of proresolving mediators such as IL-10, TGF-β, lipoxins, and resolvins (13, 60) and also contributes to the maintenance of homeostasis by imprinting tissue resident macrophages with an anti-inflammatory phenotype in various tissues throughout the body (61). It has been proposed that phagocytosis of apoptotic neutrophils by cardiac macrophages promotes inflammation resolution in the infarcted myocardium (51). Therefore, depletion of neutrophils might be expected to worsen repair by limiting phagocytosis-dependent reprogramming of macrophages toward a reparative phenotype. However, additional studies are required to directly assess this in the heart. Thus, neutrophils likely contribute to repair after myocardial infarction through the secretion of soluble mediators, which promote the differentiation of reparative macrophages, but also by acting as a direct trigger for phagocytosis-dependent, anti-inflammatory pathways in macrophages.

After neutrophil numbers peak in the infarcted mouse myocardium, Ly-6C^{hi} monocytes (**Figure 1** Working Model) accumulate in response to CCL2 and exhibit proteolytic and phagocytic functions to degrade and clear the damaged myocardium (62). Ly-6C^{hi} monocytes engulf dying cardiomyocytes (63) and in other contexts, are able to efferocytose and cross-present cell-associated antigens (64). In the heart, Ly6C^{hi} monocytes give rise to proliferative Ly6C^{low} macrophages, and this requires the nuclear receptor protein NR4A1 (65). Interestingly, NR4A1 is also linked to *Mertk* gene expression (63) and therefore as an expected consequence, NR4A1 deficiency in macrophages has been shown to impair engulfment and clearance of apoptotic cells (66). Macrophage polarization may also be important in cardiac wound healing, as alternatively activated macrophages have been linked to enhanced efferocytosis (67) and repair of the infarcted adult murine heart (68). There are multiple inducers of anti-inflammatory macrophage polarization, including cytokines IL-4 and IL-13 (69), which transduce their effects through IL-4 and IL-13 receptors, including the common IL-4Rα subunit (70). Administration of IL-4 increases survival and improves cardiac function after MI, however, in *Trib1*-deficient mice, which exhibit impaired alternative macrophage polarization, these mice are not protected by IL-4 (68). Complete deficiency of IL-13 in male mice decreases survival and impairs cardiac remodeling after myocardial infarction (71). Additionally, the combination of IL-4 or IL-13, together with apoptotic cells, promotes macrophage tissue repair (72).

While these observations have advanced our understanding on how apoptotic cell engulfment reprograms macrophage function and how this in turn informs phagocyte function during cardiac injury, many of these studies require moving beyond the generalized M1/M2 macrophage polarization paradigm to comprehend deeper relationships between macrophage polarization and function (73). This is further emphasized by the identification of a variety of macrophage subsets residing in the myocardium of differing developmental origins, which changes over the course of development and aging or following cardiac injury (4, 74, 75). Of the cardiac resident macrophage populations examined to date, all appear capable of phagocytosing cardiomyocytes (4, 17), highlighting the potential of these different phagocytes to participate in the wound healing process. However, many questions remain on whether the different populations have distinct or overlapping function and whether their function differs under varied pathophysiological conditions, such as sterile would healing or host defense. Emerging evidence indicates

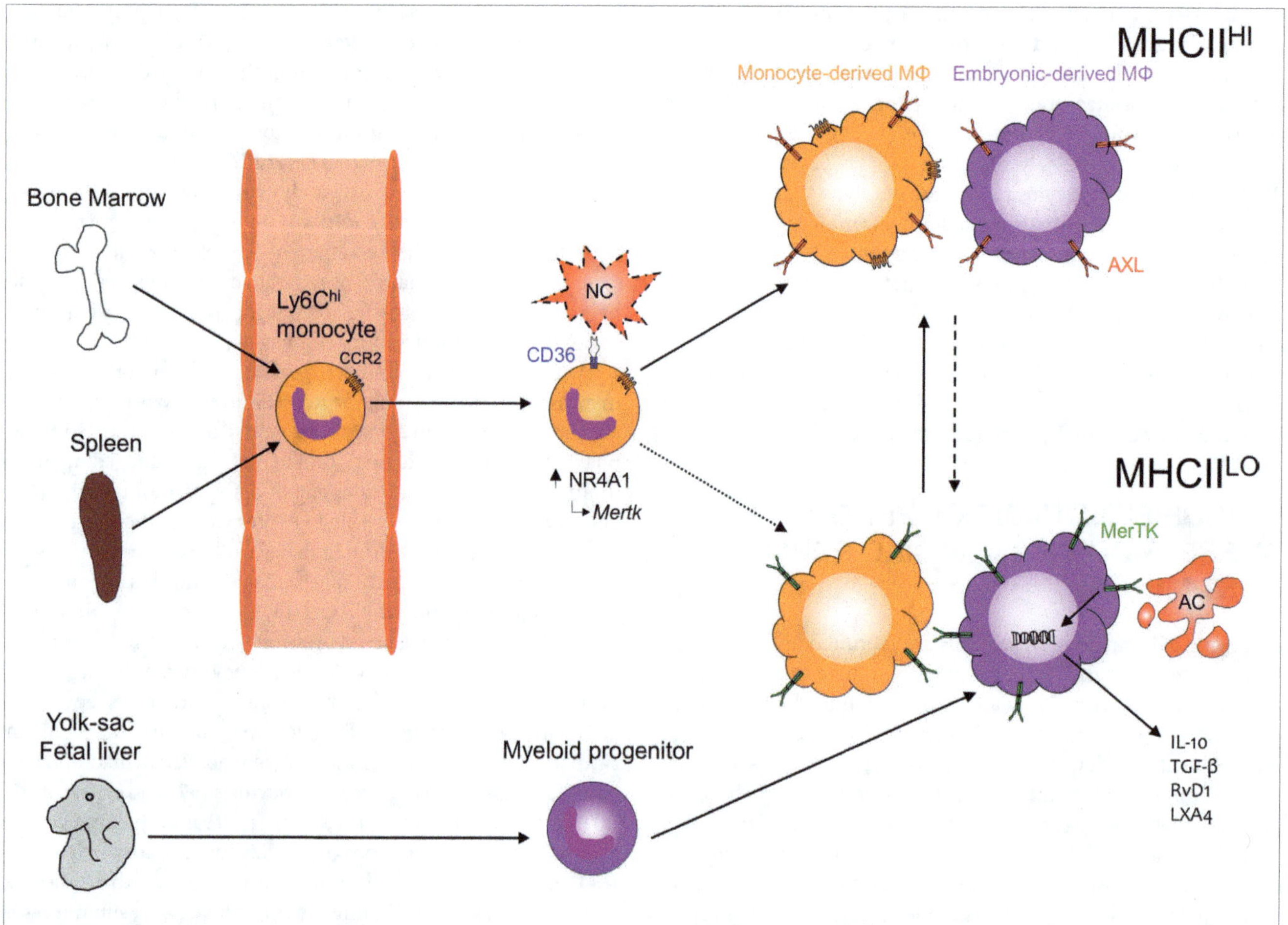

FIGURE 1 | Phagocytosis in the heart. During development, embryonic yolk-sac-derived and then fetal liver-derived macrophages seed the heart with MHCII^{LO} MΦ that can later differentiate into MHCII^{HI} MΦ. Following cardiac injury, both recruited phagocytes (Ly6C^{hi} monocytes) and resident (embryonic-derived MΦ) contribute to tissue repair and inflammation resolution. CCR2 mediates recruitment of Ly6C^{hi} monocytes to the injured myocardium where CD36 expression on Ly6C^{hi} monocytes leads to the recognition and clearance of necrotic cardiomyocyte (NC) debris and the subsequent induction of NR4A1-dependent transcription and reprogramming to MerTK-expressing, monocyte-derived MΦ. Peripheral blood monocytes preferentially differentiate into MHCII^{HI}CCR2^{+} MΦ in the heart but whether these cells also become MHCH^{LO}CCR2^{-} MΦ (dotted line) requires further investigation. Concurrently, MHCII^{LO} embryonic-derived MΦ recognize apoptotic cardiomyocytes (ACs) through MerTK leading to the production of anti-inflammatory cytokines (IL-10, TGF-β) and specialized, proresolving lipid mediators (RvD1, LXA4). MHCII^{HI} embryonic-derived MΦ may also recognize ACs through MerTK and other phagocytic receptors, such as AXL and differentiate into proresolving MHCH^{LO} MΦ (dashed line). How monocyte-derived MΦ and embryonic-derived MΦ interact and the mechanisms regulating differentiation between the different MHCII-expressing populations remain understudied and will likely have important consequences toward recovery after cardiac injury.

that embryonic-derived cardiac macrophages may be superior at mediating inflammation resolution and tissue repair following cardiac injury (74). The progressive loss of these cells with age may also explain in part the adverse outcomes that occur in adult humans during cardiovascular disease (75), and lead to the identification of novel therapeutic avenues to reverse the clock and recapture the protective responses of embryonic-derived macrophages. For example, embryonic-derived macrophages rely on macrophage colony-stimulating factor (M-CSF) signaling through CSFR1 as both a survival and self-renewal signal (76), and injection of M-CSF, but not granulocyte colony-stimulating factor (G-CSF), increases collagen content to accelerate infarct repair and attenuate left ventricular dysfunction (77), suggesting that M-CSF-mediated preservation of embryonic-derived macrophages may improve repair after cardiac injury. At the time of this publication, significantly more studies are warranted on examining the different macrophage subsets in the heart both at steady-state and during the many forms of cardiac disease.

PHAGOCYTOSIS BY NON-PROFESSIONAL PHAGOCYTES

While phagocytosis of microbes or apoptotic cells in the heart is predominantly promoted by macrophages, other non-professional phagocytes have been shown to participate in this process.

Interestingly, cardiomyocytes themselves can phagocytose latex particles *in vitro* (78) and potentially cardiomyocyte debris *in vivo* (79, 80), the latter of which may have an important role in the developing heart. Recently, myofibroblasts were identified as another non-professional phagocyte that were capable of engulfing apoptotic cardiomyocytes (81). Myofibroblast-mediated clearance of dying cells after myocardial infarction was dependent on milk fat globule epidermal growth factor (MFG-E8), which was produced in part by myofibroblasts, and mice lacking MFG-E8 displayed increased inflammation and adverse tissue remodeling. Furthermore, macrophages, through the release of microvesicles, can alter the type of particles engulfed by non-professional phagocytes, to in turn affect the inflammatory response. For example, efferocytes release IGF-1, which upon recognition by non-professional phagocytes such as epithelial cells, reduces the size of particulate uptake (82). While the presence of professional cardiac phagocytes, such as macrophages, minimalizes the necessity for non-professional phagocytes, such as myofibroblasts, to phagocytose a dying neighbor, the contribution of non-professional phagocytes to the clearance of apoptotic and necrotic debris is underexplored and whether these cells cooperate with macrophages in the heart to promote cardiac repair requires additional study.

PHAGOCYTIC RECEPTOR TARGETS AND STIMULI IN HEART

There are multiple targets in addition to dying cardiomyocytes that can activate phagocytic receptor signaling in the heart, including clearance of red blood cells, a process known as erythrophagocytosis. Intramyocardial hemorrhage is a frequent complication in ST-Elevation Myocardial Infarction (STEMI) patients reperfused by primary percutaneous coronary intervention. STEMI patients with intramyocardial hemorrhage also frequently present with residual myocardial iron that is associated with adverse left ventricular remodeling and suggestive of ongoing inflammation (83). In canines, myocardial iron deposits were directly related to proinflammatory burden with iron deposits found directly in cardiac macrophages (84). The accumulation of iron in macrophages is likely a direct consequence of excessive erythrophagocytosis in the hemorrhage. Iron overloading of macrophages has been shown to induce a proinflammatory activation state characterized by TNF-α and toxic hydroxyl radicals release, which can then lead to premature senescence of resident fibroblasts and impaired wound healing (85). Similar to intramyocardial hemorrhage, erythrocyte-rich thrombi also contain more inflammatory cells leading to impaired reperfusion in STEMI patients (86). In addition to erythrophagocytosis, the thrombus includes platelets and fibrin that are processed by phagocytes and for which the mechanisms of removal remain unclear. Macrophages are capable of phagocytosing platelets leading to the induction of iNOS (87), which may contribute to matrix degradation and adverse ventricular remodeling. A recent report also identified macrophages as important mediators of fibrin clearance with $CCR2^+$ macrophages constituting the majority of cells engulfing fibrin (88). This has important implications for the heart, where $CCR2^+$ macrophages are present early in cardiac development and expand in numbers after injury.

In the case of necroptosis, a regulated, nonapoptotic form of necrotic cell death, signaling through receptor-interacting protein kinase-3 and mixed lineage kinase-like proteins leads to externalization of phosphatidylserine, a prophagocytic "*eat me*" signal (89), and an opportunity for phagocytes to recognize and clear these "necrotic bodies" to limit inflammation in the injured heart (90). Given the relatively long-lived life cycle of adult differentiated cardiomyocytes, it is logical to speculate that antiphagocytic, or so-called "*don't-eat-me*" signals (39), may be important in warding off macrophage-mediated elimination. Indeed, *don't-eat-me* signals, which include CD31 and plasminogen activator inhibitor I, prevent viable cells from engulfment by phagocytes (91). The most widely studied *don't-eat-me* signal is CD47, which is a membrane protein expressed on the surface of most cells and has been shown to prevent tumor cells from immunologic removal (92). CD47 interacts with SIRPα on phagocytes, recruits phosphatases, and inhibits downstream activation of the phagocyte actin cytoskeleton, thereby preventing engulfment (93, 94), and has been associated with blocking recognition of prophagocytic molecules, such as calreticulin (93). It has been shown that CD47 is expressed in abundance on apoptotic neonatal cardiocytes (95), and mice lacking thrombospondin-2, a CD47 ligand, exhibit impaired cardiomyocyte survival and dilated cardiomyopathy leading to higher mortality (96). However, whether the aforementioned requires CD47, or whether CD47 is directly involved in removal of apoptotic cells in the heart, is unknown. Recently, CD47-blocking antibodies have been effective at restoring defective atherosclerotic phagocytosis (97, 98) and preventing atherosclerosis in experimental mouse models (99). More recent studies (100), suggest that early targeting of CD47 in the myocardium after infarction may be a new viable strategy, in combination with current standards of care, to enhance the efficacy of wound repair in the ischemic heart, and specifically through promotion of enhanced cardiomyocyte phagocytosis. However, the titration of anti-CD47 antibodies will likely need to be optimized to prevent phagoptosis of live cells (101).

The clearance of apoptotic cells and cellular debris is also mediated by soluble mediators of the acute-phase response including pentraxins and complement. The long pentraxin, PTX3, has been observed in the myocardium and increases in the blood of both humans (102) and mice (103) after MI. PTX3 has been shown to bind to apoptotic cells limiting activation of the first component of the classical complement pathway, C1q, and inhibiting their phagocytosis by dendritic cells (104). In contrast to dendritic cells, PTX3 increased macrophage phagocytosis of apoptotic cells (105), indicating that PTX3 may redirect apoptotic cell phagocytosis during injury to promote inflammation resolution and limit self-antigen presentation. The cumulative effect for the actions of PTX3 are cardioprotective as PTX3-deficient mice display exacerbated heart damage with increased cardiomyocyte apoptosis and complement activation after MI (103), and administration of exogenous PTX3 ameliorated cardiomyocyte apoptosis and inflammation in a heart transplantation model (106). Circulating levels of the classical short pentraxin, C-reactive protein (CRP), are also elevated in the blood of humans after MI (102). Like

PTX3, CRP is also able to promote apoptotic cell clearance by binding to oxidized phosphorylcholine on the apoptotic cell surface leading to recognition and phagocytosis by macrophages (107, 108). While both elevated and insufficient levels of CRP have been linked with disease progression in a variety of autoimmune disorders (109), the increased levels of CRP observed after MI in humans is believed to promote complement activation in the infarct leading to increased cardiomyocyte death (110). Consistent with a detrimental role for CRP after cardiac injury, selective apheresis of CRP reduced infarct size in pigs after MI (111), and administration of human CRP, which binds to damaged cells and activates complement, enhanced infarct size in rats after MI (112). Inhibition of complement activation in rabbits reduced infarct size after cardiac IRI (113), suggesting that regulation of complement activation by PTX3 and CRP may control the extent of damage after cardiac injury.

Finally, the biodegradation of collagen by phagocytes and the deposition of new extracellular matrix is formative during the final stages of tissue remodeling. Macrophages are capable of phagocytosing collagen with M2-like macrophages predominating collagen uptake *in vivo* in a mannose receptor (CD206)-dependent pathway (114). Whether collagen phagocytosis stimulates macrophages to promote extracellular matrix deposition remains unclear; however, loss of $CD206^{+}$ M2-like macrophages during MI and the resultant catastrophic decrease in collagen deposition (68), underscores the importance of macrophages shaping the extracellular matrix during the final stages of tissue remodeling. Importantly, phagocytes can fine-tune their response according to the size and source of the phagocytic target. A recent finding indicated that reactive oxygen species localization may be one signal that regulates this response with smaller microbes triggering ROS intracellularly in neutrophils and larger microbes triggering extracellular release of ROS, effectively adapting the immune response to the microbe size (115). This may be particularly relevant in the heart where macrophages encounter apoptotic targets of varying size during wound healing ranging from the diminutive red blood cell to the relatively larger cardiomyocyte, which is many fold larger in surface area relative to the macrophage.

RECOGNITION OF THE CARDIAC PARENCHYMA BY PHAGOCYTE RECEPTORS

The recognition of "eat-me" signals on apoptotic cells is performed by a variety of conserved recognition receptors, which either directly or indirectly recognize the apoptotic cell and often display redundancy in the "eat-me" signals recognized. In the heart, early reports have linked apoptotic cell recognition by scavenger receptors (SRs) in cardiac repair. For example, mice deficient in class A scavenger receptor (SR-A) exhibit increased myocardial rupture after infarction resulting in part from excessive inflammation (116). Whether SR-A deficiency impairs phagocytosis of dying cardiomyocytes by macrophages is unclear; however, the hearts of SR-A-deficient mice display evidence of increased cardiomyocyte necrosis (117), which could be the consequence of secondary necrosis following impaired apoptotic cell clearance. Interestingly, SR-A deficiency reduced myocardial IRI and this was associated with increased microRNA-125b expression and reduced apoptosis in macrophages (118). In contrast to permanent occlusion MI, reperfusion spares resident cardiac macrophages that would otherwise be subject to ischemic-induced cell death (119), so the attenuated injury in SR-A-deficient mice after IRI may be due to the actions of preserved resident macrophage function. CD36, another SR, also appears important for wound healing after myocardial injury (120), particularly early after the onset of injury (63). Within hours after MI, uptake of apoptotic and necrotic cardiomyocyte debris was mediated by CD36 on $Ly6C^{hi}$ monocytes and the importance of CD36-mediated clearance by $Ly6C^{hi}$ monocytes was revealed in CD36-deficient bone marrow recipients, which displayed increased infarct size early after MI compared to WT recipients (63). CD36-mediated engulfment was also found to induce the expression of NR4A1, which is required to mediate the differentiation of $Ly6C^{hi}$ monocytes into reparative $Ly6C^{lo}$ macrophages (65). The protective effects mediated by CD36 may be limited by its proteolytic degradation as CD36 levels decreased after MI in WT but not in matrix metalloproteinase (MMP)9-deficient mice (120). Preservation of CD36 in MMP-9-deficient mice increased macrophage phagocytosis of apoptotic neutrophils, improving inflammation resolution and LV function. Efferocytosis of apoptotic cardiomyocytes has been shown to require MerTK to resolve acute inflammation and permit cardiac repair after permanent occlusion (34, 35) and clinically relevant myocardial reperfusion (17). Additionally, combined deficiency of MerTK and MFG-E8 in macrophages impaired efferocytosis-linked vascular endothelial growth factor (VEGF)-A secretion, worsening angiogenesis and cardiac repair after MI (34). MerTK and additional receptor tyrosine kinase family members, Tyro3 and AXL, indirectly recognize apoptotic cells through bridging molecules growth-arrest-specific 6 and protein S, which bind phosphatidylserine. Galectin-3 has been suggested as a new, putative MerTK ligand (121), and consistent with this role, Galectin-3-deficient mice had increased infarct size and worsened ventricular function after MI (122). While the role of MerTK in cardiac repair is well characterized, roles for either Tyro3 or AXL in the heart are currently unknown. Overall, the phagocyte is equipped with a variety of receptors capable of recognizing apoptotic cells. How these receptors mediate engulfment, the signals that regulate their expression in the heart during both homeostasis and disease, and in many cases, the ligands these receptors recognize on the surface of apoptotic cells remain unknown and are the focus of current investigations.

CARDIAC CONSEQUENCES OF PHAGOCYTOSIS-DEPENDENT INTRACELLULAR SIGNALING AND REPROGRAMMING

The engulfment of foreign bodies by phagocytes triggers signal transduction cascades beyond the necessary cytoskeletal and phago-lyosomal processing pathways that are required to physically internalize and digest extracellular-derived material. In the case of microbial phagocytosis, phagosomes have been shown to recruit pH-lowering caspase-1, which was activated by the NLRP3 inflammasome and ROS signaling, leading to cross-presentation of phagocytosed bacterial antigens (123). In the case of macrophages that have ingested apoptotic cells, intracellular signaling culminates in inhibition of proinflammatory cytokine production and secretion of anti-inflammatory mediators (13). Such signaling pathways remain an active area of investigation. One key family of efferocytosis-signaling molecules are the nuclear receptors. For example, the nuclear receptor, liver X receptor (LXR), is activated upon apoptotic cell engulfment and can in turn promote further efferocytic events through the induction of efferocytic receptors (124). *In vitro*, loss of LXR reduced macrophage-mediated efferocytosis and subsequently impaired the tolerogenic effects that result from apoptotic cell engulfment, and *in vivo*, LXR-deficient mice exhibited a break in self-tolerance, developing autoantibodies and autoimmune glomerulonephritis. Some of this reprogramming may occur through so-called apoptotic cell response elements (ACREs) (125). With respect to IL-10, apoptotic cell engulfment induces binding of the transcription factor, pre-B cell leukemia transcription factor (Pbx)-1, to the IL-10 promoter and deletion of the Pbx-1 promoter binding site reduces promoter activity and IL-10 production. Interestingly, Pbx-1 deficiency did not completely ablate apoptotic cell-induced IL-10 production, indicating the likelihood of additional transcription factors or ACREs regulating apoptotic cell-induced IL-10 expression. Additional signals from the local milieu also translates into both phenotypic and functional properties of phagocytes. For example, tissue-resident macrophages display unique enhancer landscapes beyond what may be explained by developmental origin and this is determined in part by the tissue microenvironment (3). Transfer of mature, peritoneal macrophages into the lung resulted in upregulation of lung macrophage-specific genes and downregulation of peritoneal macrophage-specific genes in the transferred macrophages, indicating macrophages can be reprogramed by the tissue microenvironment. Phagocytosis itself imprints phagocyte heterogeneity in a tissue-specific context, and though tissue residence defines core macrophage signatures, the function of phagocytosis overlays an additional anti-inflammatory profile (61). Relative to other tissues, such as the lung, cardiac-specific imprinting after phagocytosis has not been fully explored.

METABOLIC PROCESSING OF CLEARED MYOCARDIAL TISSUE BY IMMUNE CELLS, AND LINKS TO CARDIAC REPAIR

Tissue injury generates heightened levels of apoptotic and necrotic debris and matrix remnants that once cleared by phagocytes, must be metabolized. Despite the current interest in immunometabolism, the relevance of this process in the heart by immune cells is largely unexplored. For example, emerging roles for metabolism have been linked to stem cell development (126), cell proliferation (127), and T-cell activation (128). In particular, mitochondrial metabolism has been linked to many key macrophage functions, including inflammasome activation (129), bacterial defense (130), and polarization (131). Given that macrophages can engulf cardiomyocytes and associated debris and cardiomyocytes may have both denser cellular and elevated mitochondria content (132), it is reasonable to suspect that following engulfment, macrophages need to increase cellular metabolism to process this large metabolic load and that this in turn influences phagocyte intracellular signaling and reprogramming.

In contrast to traditional viewpoints that metabolic reprogramming occurs solely in response to nutrient or oxygen availability, newer studies reveal that intracellular metabolism is further linked to receptors of damage-associated molecular patterns (DAMPs), which are present in abundance after myocardial infarction (133). In response to LPS, macrophages increase glycolysis and the pentose phosphate pathway, and reduce oxidative phosphorylation despite the presence of abundant molecular oxygen (131, 134, 135). Approaches utilizing glucose tracers demonstrate conservation of this glycolytic shift in response to other proinflammatory stimuli such as IFN-γ and DAMPs (136). Mechanistically, integrated transcriptional and metabolic network analyzes revealed that proinflammatory macrophages have a so-called "broken TCA cycle," where the truncation of isocitrate dehydrogenase and succinate dehydrogenase (SDH) leads to an accumulation of succinate (137). The increase in succinate stabilizes hypoxia inducible factor (HIF)-1α resulting in an increase in reverse electron transport and ROS production from complex I of the electron transport chain and favoring glycolysis by promoting phosphofructokinase isoform conversion (135, 138). Metabolomic studies also revealed that itaconate modulates proinflammatory macrophage metabolism and effector function by inhibiting the oxidation of succinate to fumarate by SDH (139). Furthermore, HIF-1α may also directly be stabilized by ROS generated during IRI driving a metabolic shift in macrophages toward glycolysis and the subsequent proinflammatory polarization (140).

While DAMPs and hypoxia may polarize cardiac macrophages toward a glycolysis-dominated, proinflammatory profile early during myocardial injury, increased oxygen tensions due to angiogenesis and increased levels of lipids from engulfed apoptotic debris may promote a metabolic shift toward fatty acid oxidation (FAO) through the mitochondria. In this context, alternatively activated macrophages induced by IL-4 consumed more oxygen (141) and this increase in oxidative metabolism was required for the anti-inflammatory phenotype, as inhibition of FAO with the carnitine palmitoyltransferase (CPT)-1 inhibitor, etomoxir, inhibited IL-4 induced alternative macrophage polarization (131). However, another group contrasted CTP-2 requirements by showing that CPT-2-deficient macrophages can still fully polarize toward an alternatively activated macrophage phenotype after IL-4 stimulation, despite inhibition of FAO. Thus, the effect of etomoxir on macrophage polarization might be partially due to off target effects (142). Additionally, few processes are all or none and another recent study reported glucose requirements during alternative macrophage polarization, which was dependent on a mTORC2/Stat6/IRF4 signaling axis (143). Still the evidence to date largely supports a role for mitochondrial oxidative phosphorylation in anti-inflammatory responses as

IL-10 can alter macrophage function by promoting mitophagy of damaged mitochondria to support oxidative phosphorylation and limiting glucose uptake and glyocylosis to oppose inflammatory metabolic reprogramming (144). As IL-10 is actively produced in macrophages after efferocytosis, it is worth exploring whether efferocytosis influences cellular metabolism to promote IL-10 production or whether macrophage secretion of IL-10 after efferocytosis functions in an autocrine manner to affect macrophage metabolism. Metabolism of small molecules such as amino acids and vitamins are also involved in macrophage activation. For example, L-arginine-derived metabolites are important mediators for inhibiting the production of TNF-α in mouse splenic macrophages after intestinal obstruction (145). Vitamin A has also been shown to be required for the phenotypic conversion of IL-4 activated macrophages within tissue resident macrophages of the peritoneal cavity (146). Besides its contribution to alternative macrophage activities, lipid metabolism also likely contributes to macrophage phagocytosis by fulfilling its energetic needs and regulating the membrane fluidity that is required for phagocytosis (147). Other links to mitochondrial pathways include mitochondrial UCP2, which is required for continuous uptake of apoptotic cells (148). Taken together, many of the metabolic links between phagocytosis and macrophage function remain unknown, especially in the heart, and discoveries made in the field of immunometabolism as it pertains to the macrophage will likely influence our understanding of inflammation resolution after cardiac injury and inform new therapeutic strategies.

CARDIAC LYMPHATICS IN IMMUNE SURVEILLANCE AND TISSUE HOMEOSTASIS

Recent evidence has demonstrated crucial roles for the lymphatic vasculature of the heart in both immune surveillance and tissue-fluid homeostasis. Under steady-state conditions, the lymphatic network provides a path for dendritic cells to constitutively phagocytose apoptotic cell remnants and transport this self-antigen to T-cell areas in draining lymph nodes, contributing to peripheral self-tolerance (149). In the heart, IRF8-dependent conventional dendritic cells phagocytose the cardiac self-antigen, α-myosin, and transport it to, and present it in, the heart-draining mediastinal lymph node (MLN) where it promotes the induction of α-myosin-specific $CD4^+$ T regulatory cells to maintain tolerance (150). Indeed, and in our own hands, trafficking of cardiac antigen to lymph nodes appears to be found in phagocytes (**Figure 2** Lymphatics). During inflammation, an elevated number of phagocytes can traffic from the site of injury and carry phagocytosed antigen to the draining lymph nodes (151). Although the cessation of the phagocyte response in inflamed infarct tissue may occur primarily through local cell death, some phagocytes traffic from the infarct tissue to lymphatic organs (119). With respect to DCs, MI results in massive maturation and expansion of all DC subsets in the heart, including monocyte-derived DCs, followed by trafficking and presentation of cardiac-derived antigens to $CD4^+$ T cells in the MLN (150). Macrophages also utilize lymphatic vessels to traffic antigen and modulate inflammatory responses in draining lymph nodes and distal sites (152). For example, macrophages take part in reverse cholesterol transport through lymphatics with ablation of these pathways leading to heightened atherosclerosis and inflammatory disease (153). In the heart, direct labeling of cardiac resident macrophages, but not $Ly6C^{hi}$ monocytes, by intramyocardial injection of a cell tracking dye demonstrated that cardiac macrophages constitutively traffic to the MLN, spleen, and bone marrow under steady-state conditions (18). Following MI, the percentage of labeled macrophages doubles in both the spleen and the bone marrow; however, the significance of this migration remains unknown.

With respect to tissue-fluid homeostasis, recent research has demonstrated that VEGF-C-dependent lymphatics expand in the border zone post-MI with further induction of lymphangiogenesis, through VEGF-C injection, leading to increased measures of cardiac function and repair (154, 155). VEGF-C-induced lymphangiogenesis led not only to improved myocardial fluid balance through resolution of tissue edema but also to attenuated cardiac inflammation, in part through egress of DCs and macrophages from the wounded heart. Interestingly, VEGF-C secreting macrophages are implicated in several pathologies and inflammatory processes. In anti-inflammatory tumor environments, macrophages are able to secrete VEGF-C and increase lymphatics, causing downstream tumor metastasis (156). As demonstrated in murine models of lung damage, intestinal bowel disease, and in corneal inflammation, macrophages are able to secrete VEGF-C to induce lymphangiogenesis, regulate lymphatics, and migrate through lymphatics into lymph nodes (157, 158). In the heart specifically, macrophages secrete VEGF-C in murine models of hypertension and when activated by tonicity enhanced binding protein, an osmotic stress responsive transcription factor, contribute to the adaptive response in maintaining interstitial fluid and blood pressure homeostasis (159). Furthermore, macrophages were shown to not only modulate lymphangiogenesis, but also directly interact and remodel lymphatic vessel structure and function. In such cases, macrophages closely interact with lymphatics, possibly incorporating into the vessels, and ultimately augmenting the branching of newly forming lymphatics (160). While efferocytosis has been linked to macrophage production of VEGF-A after MI (34), it is currently unknown whether efferocytosis plays a role in VEGF-C production by macrophages.

PHAGOCYTOSIS LINKS TO CARDIAC-SPECIFIC T-CELL RESPONSES

Professional phagocytes, such as macrophages and dendritic cells, play a critical role in bridging innate and adaptive immunity, which is important in the context of host defense (discussed above). However, after sterile inflammation, such as myocardial infarction, anticardiac T- and B-cell responses can develop suggesting that phagocytosis by cardiac macrophages and DCs can initiate autoimmune responses. Myocardial infarction induces activation and proliferation of $CD4^+$ T cells in a cardiac

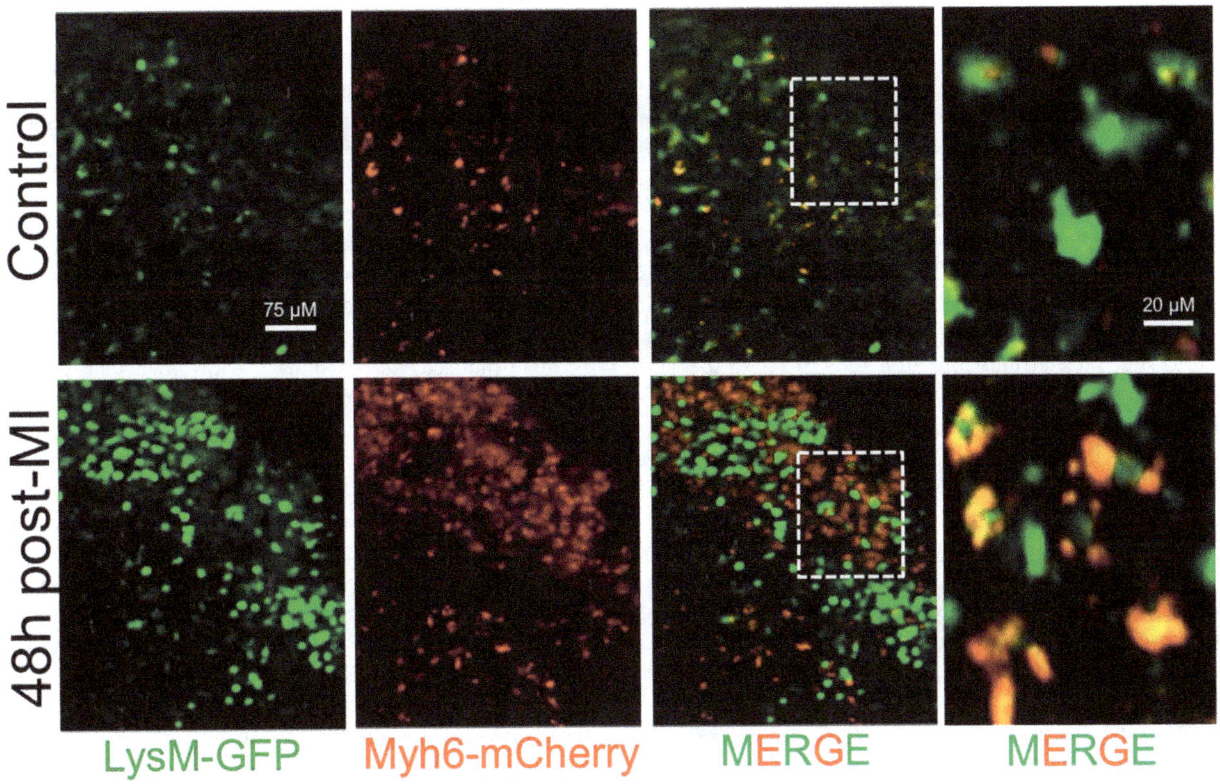

FIGURE 2 | Accumulation of phagocyte-associated cardiac antigen in the mediastinal lymph nodes (MLN) after MI. Myeloid cells (LysM-GFP+) traffic cardiac antigen derived from α-myosin heavy chain in cardiomyocytes (Myh6-mCherry+) to the MLN during steady-state in mice. Following MI, there is an increase in the number of phagocytes and the amount of phagocyte engulfed cardiac antigen in the MLN.

antigen-specific manner, as mice with CD4+ T cells specific to an irrelevant antigen fail to mount a response (161). This process is exacerbated in the presence of additional pathophysiology, such as type I diabetes. In both mice susceptible to type I diabetes and type I diabetic patients, myocardial infarction induces postinfarction autoimmunity specific for cardiac antigens such as α-myosin heavy chain, α-actinin-2, and troponin I (162). This is due in part to a lack of central tolerance to cardiac antigens (163), and impaired efferocytosis in individuals susceptible to type I diabetes. High glucose conditions impair macrophage-mediated efferocytosis (164), likely contributing to liberation of self-antigens in an inflammatory context. More recently, infarct lysate-primed, tolerogenic dendritic cells improved remodeling and cardiac function after MI by affecting regulatory T-cell and macrophage polarization (165). Interestingly, tolerogenic dendritic cells migrated only to the regional lymph node near the site of injection but were still able to induce a systemic activation of MI-specific regulatory T cells. In contrast to tolerogenic DCs, conventional and myeloid-derived DCs infiltrate and mature in the infarcted heart and migrate to the MLN, where it was demonstrated that IRF4-dependent conventional DCs were superior in presenting α-myosin to CD4+ T cells (150). Despite the importance of IRF4- and IRF8-dependent conventional DCs in presenting cardiac antigens in the MLN, loss of either subset did not impair α-myosin-specific CD4+ T-cell responses. Perhaps this is due to the massive presence of monocyte-derived DCs in the MLN after MI. While monocyte-derived DCs were shown to be inferior in generating α-myosin-specific CD4+ T-cell responses at either steady-state or after MI, it has been previously reported that this can be overcome by MHC class I/peptide transfer to bystander DCs (166). Importantly, transient autoimmune reactions to cardiac myosin after MI appear to be relatively common among the general population (167), necessitating a better understanding of how phagocytes drive postinfarction autoimmune responses.

PHAGOCYTE-LINKED MYOCYTE REGENERATION

Although not a feature of the adult mammalian heart, the ability to regenerate damaged tissue is common to many multicellular organisms and tissues. For example in the skin and liver, apoptotic cells, prior to engulfment, release growth signals to stimulate the proliferation of progenitor cells (168), and during skeletal muscle injury, cooperation between skeletal phagocytes and satellite cells leads to myocyte regeneration (169). The latter results in part through recognition of phosphatidylserine on apoptotic myoblasts by brain-specific angiogenesis inhibitor 1 on healthy myoblasts promoting fusion between healthy myoblasts to form myotubes. Similar to the heart, macrophages have also been demonstrated to participate in the tissue repair process of damaged skeletal muscle with macrophage depletion by clodronate-containing liposomes leading to prolonged clearance of necrotic myofibers and impaired skeletal muscle regeneration (170). Some of the mechanisms can be attributed to generalized tissue repair processes, where macrophages acquire an anti-inflammatory phenotype characterized by gene expression of IL-10, IL-13 receptor, arginase 1 (in mice), and other factors. In skeletal muscle, macrophage polarized gene expression requires CREB, as mice with conditionally mutated promoters exhibit severe defects in muscle fiber regeneration (171). Additionally, both heart and skeletal muscle injury leads to recruitment of Ly6C^{hi} monocytes that ultimately give rise to anti-inflammatory Ly6C^{lo} macrophages (65, 172). However, tissue-specific differences likely exist that shape the role for phagocytes in repair after injury as CCR2 deletion or antagonism reduces adverse ventricular remodeling and improves ventricular function after MI (173, 174), but impairs myogenesis following skeletal muscle injury (172, 175). For example, recruited phagocytes are a critical source of IGF-1 that is needed to promote muscle regeneration following skeletal muscle injury (175), but in the heart, embryonic-derived resident macrophages may be the critical source for this growth factor (11).

Despite some similarities in the tissue repair process, the adult heart possesses poor regenerative potential in contrast to the regenerative capacity of skeletal muscle and relative to reports in the neonatal mouse heart, where injury can stimulate cardiomyocyte proliferation (176). In the adult heart, immune-mechanisms of tissue replacement largely leads to fibrosis and therefore loss of full cardiac contractile potential. In a genetic model of cardiomyocyte cell death, neonatal mice expanded a population of embryonic-derived resident cardiac macrophages, which generated marginal inflammation and promoted cardiac recovery after cardiomyocyte proliferation and angiogenesis (74). Similarly, macrophages were required for neonatal heart regeneration and neoangiogenesis after MI with macrophages from P1 hearts promoting angiogenesis essential for cardiac regeneration compared to macrophages from P14 hearts, which produced factors repairing the damaged tissue but also stimulating fibrotic scar formation (177). Thus, understanding the context specific molecular cues that empower regenerative potential versus scarring is of critical clinical importance in the heart. Cardiospheres and cardiosphere-derived exosomes have shown promise for cardiac regeneration and some of these pathways may signal through apoptotic cell receptors (178, 179). Recent clinical trials suggest that yet more work is to be done in the field of cardiosphere-derived therapy to translate findings from mouse to man (180).

PHAGOCYTE FUNCTION DURING HEART FAILURE AND ASSOCIATED SYSTEMIC FACTORS

Nonresolving inflammation is a driver of disease and a hallmark of many cardiovascular syndromes including heart failure (36). In both animal models of heart failure and in humans with end-stage heart failure, there is evidence of ongoing cardiomyocyte apoptosis indicating that continued clearance of dying cardiomyocytes by phagocytes and the subsequent reprogramming of these efferocytes may influence the progression of this disease (181–183). During chronic heart failure, macrophages continue to increase in numbers due to increased local macrophage proliferation and differentiation of recruited monocytes into macrophages with each population displaying distinct gene expression patterns (73). Limiting the expansion of monocyte-derived macrophages through blockade of monocyte recruitment preserves ejection fraction after MI (73), indicating that altered phagocyte function contributes to heart failure. Similarly, in a model of pressure overload-induced heart failure, ICAM1-deficient mice have decreased monocyte recruitment and exhibit no overt signs of cardiac fibrosis and minimal ventricular dysfunction (184). ICAM1 has been linked to suppression of efferocytosis with ICAM1 deficiency in macrophages promoting efferocytosis of apoptotic cells (185). Increased efferocytosis by ICAM1-deficient macrophages led to increased expression of IL-10, which has been shown to attenuate pressure overload-induced hypertrophic remodeling (186), indicating that enhanced efferocytosis by cardiac resident macrophages may contribute to the protective response to heart failure in ICAM1-deficient mice. Galectin-3 represents another marker of altered phagocyte phenotype during heart failure as it is expressed only by myocardial macrophages in failure-prone hypertrophied hearts but not normal hearts, where it has been shown to contribute to cardiac dysfunction in rats (187) and be predictive of adverse events in human heart failure patients (188). In mice lacking Galectin-3, myocardial fibrosis and macrophage infiltration were reduced with preservation of left ventricular function during chronic angiotensin II-induced hypertension demonstrating a cardiac-deleterious role for Galectin-3 (189). Galectin-3 plays a critical role in phagocytosis by macrophages (190), but it can also be proteolytically cleaved by MMP to release a soluble protein (191), which is capable of inducing fibroblast proliferation and collagen production (187). Whether Galectin-3-dependent phagocytosis or production of soluble Galectin-3 by macrophages contributes to the progression of heart failure remains to be determined; however, Galectin-3 influences macrophage polarization *in vitro* (192), suggesting that Galectin-3 alters cardiac macrophage function in the failing heart. Additional phagocytic receptors, such as SR-A, have been implicated in regulating phagocyte function during heart failure with SR-A-deficient macrophages displaying increased expression of proinflammatory genes following LPS-stimulation *in vitro* and adverse vascular remodeling during angiotensin II-induced hypertension *in vivo* (193).

The link between phagocyte function and heart failure is likely a consequence of both local pathological changes within the myocardium itself and pathophysiologies in distant organ systems that feedback on the heart and also manifest as systemic changes. For example, cardiorenal syndrome manifests as impaired renal function following MI characterized by increased infiltration of macrophages into the kidney and elevated renal levels of TGF-β and T-cell immunoglobulin and mucin domain (TIM)-1 associated with the onset of renal fibrosis (194). TIM-1 has been linked to efferocytosis (195), and overexpression of TIM-1 in mice leads to the development of spontaneous and progressive interstitial kidney inflammation with fibrosis (196), demonstrating that enhanced expression of apoptotic receptors distal to the site of injury may have deleterious effects. In the context of hypertension, a recent report has demonstrated an interplay between neurohormonal modulation of phagocyte function before the onset of hypertension, leading to excessive inflammation by the phagocyte system and contributing to the development of hypertension (197). Heart failure also has systemic consequences and in a recent report, mice subjected to myocardial pressure overload in turn activated a heart–brain–kidney network that required phagocyte function in both the heart and kidney and culminated in activation of cardiac-resident Ly6C^{lo} macrophages to mediate the adaptive response (198). Following pressure overload in the heart, sympathetic nerve activation led to activation of renal collecting-duct cells, which through interactions with renal macrophages led to the release of CSF2 into the circulation by endothelial cells within the kidney. Within the overloaded heart, CSF2 expanded and activated Ly6C^{lo} macrophages to secrete amphiregulin inducing a cardiac hypertrophic response. Interestingly, a parabiosis model revealed that these cardiac-resident Ly6C^{lo} macrophages increased in number largely through *in situ* proliferation; however, further examination of the Ly6C^{lo} resident macrophage population using well-established markers MHCII and CCR2 or lineage tracing was not performed. This is of particular importance as embryonic-derived resident cardiac macrophages have been shown to decline with age (75) and embryonic-derived resident cardiac macrophages may be superior at mediating adaptive responses in the heart (74). How the different macrophage subsets change during the course of disease and whether phagocytosis of ongoing cardiomyocyte death alter the progression of heart disease and its related sequela remain to be determined.

PHAGOCYTOSIS POST HEART TRANSPLANTATION AND NEW THERAPEUTIC OPPORTUNITIES FOR TOLERANCE

Transplant rejection involves both innate and adaptive immune responses. Clinical progress has reduced acute cardiac transplant rejection, however, beyond ten years, complications of immunologic intervention often lead to significant comorbidities, particularly posttransplant vasculopathy. Continuous immunesuppression during transplant raises risks of opportunistic infections, and of hematologic (199), metabolic, and nephrotoxic side effects (200). Interestingly, acute phagocytosis and innate inflammation during allograft IRI has been linked to chronic pathophysiology. For example, perioperative and acute inflammation are prognostic for worse long-term transplant outcome (201, 202). Graft reperfusion may trigger reperfusion-associated cell death (38) and cell necrosis occurs during allograft cold storage and continues in allograft reperfusion. Both of these processes liberates allo-antigens in an inflammatory context. Efficient clearance of dead cells by macrophages prevents these self-antigens from becoming immunogenic debris and can actively initiate tissue-reparative and tolerogenic signaling (203) and as a consequence, natural defects in phagocytosis have been correlated with, but not yet causally linked, to poor outcomes posttransplant (204). Similar to post-MI, cardiac allograft rejection and tolerance are regulated by phagocyte subsets. After IRI, cardiac graft rejection is linked to elevated Ly6C^{hi} monocytes (205, 206) with both alloantigen-dependent and -independent factors contributing to immune cell activation (207). Ly6C^{hi} monocytes differentiate into Ly6C^{lo} macrophages and antigen presenting cells, which recognize allogenic non-self and contribute to graft injury (208, 209) through cytokines and T-cell activation (210, 211). However, not all macrophage function is detrimental, as some macrophage subsets belong to the heterogeneous classification of myeloid-derived suppressor cells (MDSCs), which can accumulate in allografts, suppress effector T cells, and induce tolerance (212–214). For example, anti-CD40L mAbs (clone MR1) promote experimental cardiac tolerance through suppressive DC-SIGN^{+} macrophages (215). Separately, a unique strategy harnesses *natural* immune-regulatory properties of efferocytosis: apoptotic donor *sp*lenocytes, fixed with the chemical cross-linker 1-*e*thyl-3-(3′-dimethylaminopropyl)-carbo*dii*mide (ECDI-SPs) (216), are engulfed by macrophages to induce transplant tolerance (217). In the heart, transfusion of ECDI-SPs from the donor strain prior to heart transplantation dramatically prolongs survival of the heart graft and tolerance induction is dependent on phagocytosis of the apoptotic cells (218) and signaling through apoptotic cell receptors. This process enhances the accumulation of MDSCs in both the spleen and the cardiac allograft, which limit the activation and recruitment of antiallograft CD8^{+} T cells (219), and also appears to involve alterations of antigen presenting cell costimulatory ligands (220). Alloantigen-presenting plasmacytoid dendritic cells have also been shown to mediate tolerance to vascularized grafts (221). The complete molecular mechanisms by which uptake of apoptotic splenocytes induce cardiac allograft survival remain unclear.

MOLECULAR MODULATORS AND INHIBITORS OF CARDIAC PHAGOCYTOSIS

Accumulating molecular evidence suggests that the clearance efficiency of the cardiac infarct is not optimal and therefore amenable to potential therapeutic intervention. For example, cardiomyocytes induce macrophage receptor shedding to suppress phagocytosis (222). The mechanism involves the activity of ADAM proteases, which recognize the efferocytosis receptor, MerTK and execute its proteolytic shedding from the surface of macrophages (223). ADAM17-mediated proteolytic degradation releases a soluble protein (solMER), which is believed to compete with membrane-bound MerTK in the binding of bridging molecules on the surface of apoptotic cells antagonizing MerTK-mediated efferocytosis. Additionally, loss of MerTK from the cell surface also eliminates MerTK-induced anti-inflammatory responses. For example, MerTK signaling increases the ratio of cytoplasmic-to-nuclear 5-lipoxygenase promoting the production of specialized proresolving mediators (SPMs), including

LXA_4 and RvD1 (60). In the heart, production of SPMs initiates a proresolving response actively promoting inflammation resolution with administration of RvD1 during MI reducing neutrophil recruitment to the spleen and infarct and promoting anti-inflammatory polarization of macrophages culminating in reduced fibrosis and preserved ventricular function (224). Efferocytosis and SPM production is impaired under conditions where MerTK is cleaved and introduction of a cleavage-resistant MerTK in mice improves inflammation resolution during both peritonitis (60) and myocardial reperfusion (17). Previous studies have also shown that the SR CD36 is susceptible to cleavage during atherosclerosis (225) and by MMP-9 after myocardial infarction (120) and our own work recapitulates these findings (63). Of course proteolysis is not specific to macrophages as TLR7/8 activation in neutrophils reduces immune complex phagocytosis through shedding of FcgRIIA (226). On the target cell side, molecules of the CD47/SIRP1α axis have also been associated with proteolysis susceptibility. In vascular smooth muscle cells, CD47 was shown to be a target of MMP-2-mediated proteolytic degradation (227) and in mice, MMP-2 deficiency led to enhanced survival after MI (228). While CD47 cleavage was not examined in the MMP-2-deficient mice, a greater number of cardiomyocytes persisted in the infarct and border zone.

Other common comorbidities such as diabetes mellitus and hyperlipidemia in patients with heart failure have also been linked to both apoptotic receptor shedding and impaired phagocytosis. Macrophage exposure to diabetic conditions *in vitro*, leads to reduced miR-126 expression and a concomitant increase in its direct target, ADAM9 (164). Similar to ADAM17, ADAM9 is also able to mediate the cleavage destruction of MerTK and impair efferocytosis of apoptotic cardiomyocytes. Human heart tissue from diabetic patients with heart failure recapitulated the *in vitro* findings with human diabetic failing heart tissue exhibiting reduced miR126 and increased ADAM9 expression with a reduction in phosphorylated MerTK, a surrogate marker for MerTK signaling, indicating that antagonism of MerTK-mediated processes under diabetic conditions could translate to an increase in adverse clinical outcomes in heart failure patients. With respect to hyperlipidemia, apolipoprotein E (apoE)-deficient mice, which develop hyperlipidemia and atheroscleorosis when maintained on a high-fat diet, exhibit impaired wound healing after MI (62). The dysregulated healing response may be due in part to effects of hyperlipidemia on apoptotic cell clearance (229). In vascular smooth muscle cells, oxidized low-density lipoprotein *in vitro* or hyperlipidemia *in vivo* impaired phagocytosis of apoptotic cells leading to the development of secondary necrotic cells capable of releasing both IL-1α and IL-1β further propagating the inflammatory response. Mice maintained on a high-fat diet also displayed increased activation of ADAM17 on the vascular endothelium, indicating that proteolytic degradation of apoptotic receptors may contribute to impaired phagocytosis during obesity and hyperlipidemia (230). Targeting hyperlipidemia using atorvastatin, restored phagocytic function to retinal pigment epithelial cells treated with cholesterol crystals or oxidized low-density lipoproteins (231), demonstrating a proof-of-principle approach to mitigating detrimental effects of hyperlipidemia on phagocytosis in the heart.

Hypoxia is another common environmental stress after coronary artery occlusion. The activation of proteases such as ADAM17 have been linked to hypoxia and HIFs (232), indicating that the hypoxic myocardium may also antagonize phagocytosis by promoting ADAM17-mediated proteolytic degradation of apoptotic receptors. However, in the absence of detectable receptor shedding, hypoxia has also been shown to suppress efferocytosis by macrophages and this was due in part to reduced gene expression of efferocytic receptors during hypoxia exposure (233). While phagocytes encounter varying oxygen tensions during development, migration, and infiltration of tissues, hypoxia was shown to be dispensable for macrophage differentiation, at least in the hypoxic tumor microenvironment (234), suggesting that oxygen tension may act to fine-tune macrophage function. HIFα subunits are the critical regulators of phagocyte function during hypoxia and inflammation, controlling metabolism, cytokine production, migration, and survival and thus, likely shape the phagocyte response during wound healing in the hypoxic heart. In the infarcted myocardium, macrophages have been shown to express both HIF-1α and HIF-2α (235); however, our understanding of the timing and functional significance of HIFα subunits in phagocytes during cardiac injury is incomplete. Knockdown of HIF-1α in the hematopoietic compartment improved LV function after MI and this was attributed to reduced recruitment of neutrophils and monocytes to the infarcted myocardium (236). In contrast to the more widely studied HIF-1α isoform, less is known about HIF-2α function in macrophages. A recent study implicated a role for HIF-2α in redox control and phagocytosis by macrophages during normoxic conditions (237). Elevations in mitochondrial ROS in HIF-2α-deficient macrophages led to nuclear translocation of NRF2 and NRF2-dependent transcriptional induction of the phagocytic receptor, MARCO, which translated into increased macrophage phagocytic function. Similar to HIF-1α, HIF-2α has also been implicated in regulating macrophage LPS-induced secretion of proinflammatory cytokines and chemokines (238), though this occurred independent of changes in cellular energy homeostasis, in contrast to the connection of HIF-1α to glycolysis. Given the importance of both HIF-1α and HIF-2α in secretion of proinflammatory mediators by macrophages, it is tempting to speculate that myeloid cell expression of HIFα subunits may play a detrimental role in the hypoxic heart. However, HIFα isoforms show differential activation in macrophages with HIF-1α induction following M1 polarization and HIF-2α induction following M2 polarization (239), the latter of which leads to macrophage expression of arginase 1 and attenuated inflammation during obesity-induced insulin resistance in adipose tissue (240). HIFα subunits also play an important role in the adaptive angiogenic and lymphangiogenic responses to hypoxia, both of which have been implicated in heart healing after MI. Here again, HIF-1α and HIF-2α may have opposing roles in phagocytes with HIF-1α promoting and HIF-2α antagonizing the angiogenic effects of VEGF (241). Additional studies are needed to fill the large gaps that remain in our understanding on how oxygen tension and HIFα subunits regulate phagocytosis in the heart. Taken together, a number of risk factors promote inefficient clearance of dying cells during cardiovascular disease, leading to secondary necrosis and prolonged inflammation. Future studies identifying the mechanisms of how these risk factors conspire to antagonize phagocytosis in the heart will lead to the identification of new targets for the development of novel therapeutics to promote wound healing in the heart.

FUTURE PROSPECTS FOR THE THERAPEUTIC TARGETING OF PHAGOCYTIC REPAIR AND REGENERATIVE PATHWAYS IN THE HEART

Improvements in clinical treatment have led to reduced mortality after first MI. Nevertheless, the incidence of heart failure, including after MI, is on the rise. Current clinical trials, including the CANTOS trial, have demonstrated the merits of anti-inflammatory therapy for the reduction of secondary events (242, 243). Given the critical role of efferocytosis and phagocytosis after MI, this pathway represents a tractable target amenable to modulation during the time patients are in hospital and during the formative stages of disease progression. Approaches that strategize to enhance phagocytic efficiency are appealing due to the multiple checks and balances that are naturally built in to mechanisms of phagocytosis and therefore minimize off target effects; that is engulfment requires both the downregulation of "don't eat me" signals, as well the converse presentation of prophagocytic "eat me" ligands. One approach is to use cardiosphere-derived cells or exosomes, which are heart cell products with antifibrotic, anti-inflammatory, and angiogenic properties. Uptake of CDCs by macrophages induces a proresolving phenotype leading to reductions in left ventricular fibrosis and inflammation with improved left ventricular function (178, 179, 244). Alternative strategies have employed liposomes carrying phosphatidylserine, an "eat me" signal that directs efferocytosis, to increase anti-inflammatory cytokine production by macrophages after myocardial infarction (245). Other macrophage-targeted lipid based drug carriers are able to reprogram macrophages to a proresolving phenotype and improve tissue repair and limit infarct expansion (246). With respect to the proteolytic degradation of efferocytic receptors, such as MerTK and CD36 (17, 225), strategies which block the cleavage degradation including cleavage-blocking peptides, may be a viable approach for enhancing protection after cardiac insult. Additionally, since the soluble forms of these efferocytic receptors are increased in the serum after MI, monitoring their levels in humans may serve as a useful biomarker for novel therapeutic interventions. Taken together, targeting efferocytosis and phagocytosis pathways in the heart represents a promising therapeutic strategy to limit inflammation and promote reparative functions in a variety of cardiovascular disease settings.

AUTHOR CONTRIBUTIONS

MD, SZ, and ET wrote the manuscript. MD, LG, and ET designed and prepared the figures. KG, IH, EV, KH, and XL provided their expertise in critically reviewing and revising the manuscript. All authors edited the manuscript and approved the final version for publication.

ACKNOWLEDGMENTS

This work was supported by National Institute of Health grants R01HL122309 (to ET) and F32HL127958 (to MD). Also, this work was funded by the Chicago Biomedical Consortium. Publication of this research was supported by the Sidney & Bess Eisenberg Memorial Fund.

REFERENCES

1. Arandjelovic S, Ravichandran KS. Phagocytosis of apoptotic cells in homeostasis. *Nat Immunol* (2015) 16:907–17. doi:10.1038/ni.3253
2. Wermeling F, Karlsson MC, McGaha TL. An anatomical view on macrophages in tolerance. *Autoimmun Rev* (2009) 9:49–52. doi:10.1016/j.autrev.2009.03.004
3. Lavin Y, Winter D, Blecher-Gonen R, David E, Keren-Shaul H, Merad M, et al. Tissue-resident macrophage enhancer landscapes are shaped by the local microenvironment. *Cell* (2014) 159:1312–26. doi:10.1016/j.cell.2014.11.018
4. Epelman S, Lavine KJ, Beaudin AE, Sojka DK, Carrero JA, Calderon B, et al. Embryonic and adult-derived resident cardiac macrophages are maintained through distinct mechanisms at steady state and during inflammation. *Immunity* (2014) 40:91–104. doi:10.1016/j.immuni.2013.11.019
5. Clancy RM, Neufing PJ, Zheng P, O'Mahony M, Nimmerjahn F, Gordon TP, et al. Impaired clearance of apoptotic cardiocytes is linked to anti-SSA/Ro and -SSB/La antibodies in the pathogenesis of congenital heart block. *J Clin Invest* (2006) 116:2413–22. doi:10.1172/JCI27803
6. Briassouli P, Komissarova EV, Clancy RM, Buyon JP. Role of the urokinase plasminogen activator receptor in mediating impaired efferocytosis of anti-SSA/Ro-bound apoptotic cardiocytes: implications in the pathogenesis of congenital heart block. *Circ Res* (2010) 107:374–87. doi:10.1161/CIRCRESAHA.109.213629
7. Clancy RM, Kapur RP, Molad Y, Askanase AD, Buyon JP. Immunohistologic evidence supports apoptosis, IgG deposition, and novel macrophage/fibroblast crosstalk in the pathologic cascade leading to congenital heart block. *Arthritis Rheum* (2004) 50:173–82. doi:10.1002/art.11430
8. Miranda-Carus ME, Askanase AD, Clancy RM, Di Donato F, Chou TM, Libera MR, et al. Anti-SSA/Ro and anti-SSB/La autoantibodies bind the surface of apoptotic fetal cardiocytes and promote secretion of TNF-alpha by macrophages. *J Immunol* (2000) 165:5345–51. doi:10.4049/jimmunol.165.9.5345
9. Smith SJ, Mohun TJ. Early cardiac morphogenesis defects caused by loss of embryonic macrophage function in *Xenopus*. *Mech Dev* (2011) 128:303–15. doi:10.1016/j.mod.2011.04.002
10. van den Hoff MJ, van den Eijnde SM, Viragh S, Moorman AF. Programmed cell death in the developing heart. *Cardiovasc Res* (2000) 45:603–20. doi:10.1016/S0008-6363(99)00401-0
11. Leid J, Carrelha J, Boukarabila H, Epelman S, Jacobsen SE, Lavine KJ. Primitive embryonic macrophages are required for coronary development and maturation. *Circ Res* (2016) 118:1498–511. doi:10.1161/CIRCRESAHA.115.308270
12. Vandivier RW, Henson PM, Douglas IS. Burying the dead: the impact of failed apoptotic cell removal (efferocytosis) on chronic inflammatory lung disease. *Chest* (2006) 129:1673–82. doi:10.1378/chest.129.6.1673
13. Fadok VA, Bratton DL, Konowal A, Freed PW, Westcott JY, Henson PM. Macrophages that have ingested apoptotic cells in vitro inhibit proinflammatory cytokine production through autocrine/paracrine mechanisms involving TGF-beta, PGE2, and PAF. *J Clin Investig* (1998) 101:890–8. doi:10.1172/JCI1112
14. Bergmann O, Zdunek S, Felker A, Salehpour M, Alkass K, Bernard S, et al. Dynamics of cell generation and turnover in the human heart. *Cell* (2015) 161:1566–75. doi:10.1016/j.cell.2015.05.026

15. Medina DL, Fraldi A, Bouche V, Annunziata F, Mansueto G, Spampanato C, et al. Transcriptional activation of lysosomal exocytosis promotes cellular clearance. *Dev Cell* (2011) 21:421–30. doi:10.1016/j.devcel.2011.07.016
16. Eken C, Martin PJ, Sadallah S, Treves S, Schaller M, Schifferli JA. Ectosomes released by polymorphonuclear neutrophils induce a MerTK-dependent anti-inflammatory pathway in macrophages. *J Biol Chem* (2010) 285:39914–21. doi:10.1074/jbc.M110.126748
17. DeBerge M, Yeap XY, Dehn S, Zhang S, Grigoryeva L, Misener S, et al. MerTK cleavage on resident cardiac macrophages compromises repair of post-myocardial infarction reperfusion injury. *Circ Res* (2017) 121(8):930–40. doi:10.1161/CIRCRESAHA.117.311327
18. Heidt T, Courties G, Dutta P, Sager HB, Sebas M, Iwamoto Y, et al. Differential contribution of monocytes to heart macrophages in steady-state and after myocardial infarction. *Circ Res* (2014) 115:284–95. doi:10.1161/CIRCRESAHA.115.303567
19. Brown AO, Mann B, Gao G, Hankins JS, Humann J, Giardina J, et al. *Streptococcus pneumoniae* translocates into the myocardium and forms unique microlesions that disrupt cardiac function. *PLoS Pathog* (2014) 10:e1004383. doi:10.1371/journal.ppat.1004383
20. Bonney KM, Engman DM. Chagas heart disease pathogenesis: one mechanism or many? *Curr Mol Med* (2008) 8:510–8. doi:10.2174/156652408785748004
21. Lopes MF, da Veiga VF, Santos AR, Fonseca ME, DosReis GA. Activation-induced CD4+ T cell death by apoptosis in experimental Chagas' disease. *J Immunol* (1995) 154:744–52.
22. Freire-de-Lima CG, Nascimento DO, Soares MB, Bozza PT, Castro-Faria-Neto HC, de Mello FG, et al. Uptake of apoptotic cells drives the growth of a pathogenic trypanosome in macrophages. *Nature* (2000) 403:199–203. doi:10.1038/35003208
23. Mendonca PHB, da Rocha R, Moraes JBB, LaRocque-de-Freitas IF, Logullo J, Morrot A, et al. Canine macrophage DH82 cell line as a model to study susceptibility to *Trypanosoma cruzi* infection. *Front Immunol* (2017) 8:604. doi:10.3389/fimmu.2017.00604
24. Gomes JA, Bahia-Oliveira LM, Rocha MO, Busek SC, Teixeira MM, Silva JS, et al. Type 1 chemokine receptor expression in Chagas' disease correlates with morbidity in cardiac patients. *Infect Immun* (2005) 73:7960–6. doi:10.1128/IAI.73.12.7960-7966.2005
25. Hardison JL, Wrightsman RA, Carpenter PM, Kuziel WA, Lane TE, Manning JE. The CC chemokine receptor 5 is important in control of parasite replication and acute cardiac inflammation following infection with *Trypanosoma cruzi*. *Infect Immun* (2006) 74:135–43. doi:10.1128/IAI.74.1.125-134.2006
26. Benoit M, Ghigo E, Capo C, Raoult D, Mege JL. The uptake of apoptotic cells drives *Coxiella burnetii* replication and macrophage polarization: a model for Q fever endocarditis. *PLoS Pathog* (2008) 4:e1000066. doi:10.1371/journal.ppat.1000066
27. Li K, Xu W, Guo Q, Jiang Z, Wang P, Yue Y, et al. Differential macrophage polarization in male and female BALB/c mice infected with coxsackievirus B3 defines susceptibility to viral myocarditis. *Circ Res* (2009) 105:353–64. doi:10.1161/CIRCRESAHA.109.195230
28. Benjamin EJ, Blaha MJ, Chiuve SE, Cushman M, Das SR, Deo R, et al. Heart disease and stroke statistics-2017 update: a report from the American Heart Association. *Circulation* (2017) 135:e146–603. doi:10.1161/CIR.0000000000000485
29. Dutta P, Courties G, Wei Y, Leuschner F, Gorbatov R, Robbins CS, et al. Myocardial infarction accelerates atherosclerosis. *Nature* (2012) 487:325–9. doi:10.1038/nature11260
30. van Amerongen MJ, Harmsen MC, van Rooijen N, Petersen AH, van Luyn MJ. Macrophage depletion impairs wound healing and increases left ventricular remodeling after myocardial injury in mice. *Am J Pathol* (2007) 170:818–29. doi:10.2353/ajpath.2007.060547
31. Libby P, Tabas I, Fredman G, Fisher EA. Inflammation and its resolution as determinants of acute coronary syndromes. *Circ Res* (2014) 114:1867–79. doi:10.1161/CIRCRESAHA.114.302699
32. Virmani R, Burke AP, Farb A, Kolodgie FD. Pathology of the vulnerable plaque. *J Am Coll Cardiol* (2006) 47:C13–8. doi:10.1016/j.jacc.2005.10.065
33. Whelan RS, Kaplinskiy V, Kitsis RN. Cell death in the pathogenesis of heart disease: mechanisms and significance. *Annu Rev Physiol* (2010) 72:19–44. doi:10.1146/annurev.physiol.010908.163111
34. Howangyin KY, Zlatanova I, Pinto C, Ngkelo A, Cochain C, Rouanet M, et al. Myeloid-epithelial-reproductive receptor tyrosine kinase and milk fat globule epidermal growth factor 8 coordinately improve remodeling after myocardial infarction via local delivery of vascular endothelial growth factor. *Circulation* (2016) 133:826–39. doi:10.1161/CIRCULATIONAHA.115.020857
35. Wan E, Yeap XY, Dehn S, Terry R, Novak M, Zhang S, et al. Enhanced efferocytosis of apoptotic cardiomyocytes through myeloid-epithelial-reproductive tyrosine kinase links acute inflammation resolution to cardiac repair after infarction. *Circ Res* (2013) 113:1004–12. doi:10.1161/CIRCRESAHA.113.301198
36. Nathan C, Ding A. Nonresolving inflammation. *Cell* (2010) 140:871–82. doi:10.1016/j.cell.2010.02.029
37. Chen W, Frangogiannis NG. The role of inflammatory and fibrogenic pathways in heart failure associated with aging. *Heart Fail Rev* (2010) 15:415–22. doi:10.1007/s10741-010-9161-y
38. Eltzschig HK, Eckle T. Ischemia and reperfusion – from mechanism to translation. *Nat Med* (2011) 17:1391–401. doi:10.1038/nm.2507
39. Ravichandran KS. Beginnings of a good apoptotic meal: the find-me and eat-me signaling pathways. *Immunity* (2011) 35:445–55. doi:10.1016/j.immuni.2011.09.004
40. Lauber K, Bohn E, Krober SM, Xiao YJ, Blumenthal SG, Lindemann RK, et al. Apoptotic cells induce migration of phagocytes via caspase-3-mediated release of a lipid attraction signal. *Cell* (2003) 113:717–30. doi:10.1016/S0092-8674(03)00422-7
41. Peter C, Waibel M, Radu CG, Yang LV, Witte ON, Schulze-Osthoff K, et al. Migration to apoptotic "find-me" signals is mediated via the phagocyte receptor G2A. *J Biol Chem* (2008) 283:5296–305. doi:10.1074/jbc.M706586200
42. Daleau P. Lysophosphatidylcholine, a metabolite which accumulates early in myocardium during ischemia, reduces gap junctional coupling in cardiac cells. *J Mol Cell Cardiol* (1999) 31:1391–401. doi:10.1006/jmcc.1999.0973
43. Matloubian M, Lo CG, Cinamon G, Lesneski MJ, Xu Y, Brinkmann V, et al. Lymphocyte egress from thymus and peripheral lymphoid organs is dependent on S1P receptor 1. *Nature* (2004) 427:355–60. doi:10.1038/nature02284
44. Truman LA, Ford CA, Pasikowska M, Pound JD, Wilkinson SJ, Dumitriu IE, et al. CX3CL1/fractalkine is released from apoptotic lymphocytes to stimulate macrophage chemotaxis. *Blood* (2008) 112:5026–36. doi:10.1182/blood-2008-06-162404
45. Cullen SP, Henry CM, Kearney CJ, Logue SE, Feoktistova M, Tynan GA, et al. Fas/CD95-induced chemokines can serve as "find-me" signals for apoptotic cells. *Mol Cell* (2013) 49:1034–48. doi:10.1016/j.molcel.2013.01.025
46. Chekeni FB, Elliott MR, Sandilos JK, Walk SF, Kinchen JM, Lazarowski ER, et al. Pannexin 1 channels mediate 'find-me' signal release and membrane permeability during apoptosis. *Nature* (2010) 467:863–7. doi:10.1038/nature09413
47. Dolmatova E, Spagnol G, Boassa D, Baum JR, Keith K, Ambrosi C, et al. Cardiomyocyte ATP release through pannexin 1 aids in early fibroblast activation. *Am J Physiol Heart Circ Physiol* (2012) 303:14. doi:10.1152/ajpheart.00251.2012
48. Ayata CK, Ganal SC, Hockenjos B, Willim K, Vieira RP, Grimm M, et al. Purinergic P2Y(2) receptors promote neutrophil infiltration and hepatocyte death in mice with acute liver injury. *Gastroenterology* (2012) 143:1620–9. doi:10.1053/j.gastro.2012.08.049
49. Chen Y, Corriden R, Inoue Y, Yip L, Hashiguchi N, Zinkernagel A, et al. ATP release guides neutrophil chemotaxis via P2Y2 and A3 receptors. *Science* (2006) 314:1792–5. doi:10.1126/science.1132559
50. Schloss MJ, Horckmans M, Nitz K, Duchene J, Drechsler M, Bidzhekov K, et al. The time-of-day of myocardial infarction onset affects healing through oscillations in cardiac neutrophil recruitment. *EMBO Mol Med* (2016) 8:937–48. doi:10.15252/emmm.201506083
51. Frangogiannis NG. Regulation of the inflammatory response in cardiac repair. *Circ Res* (2012) 110:159–73. doi:10.1161/CIRCRESAHA.111.243162
52. Esmann L, Idel C, Sarkar A, Hellberg L, Behnen M, Moller S, et al. Phagocytosis of apoptotic cells by neutrophil granulocytes: diminished proinflammatory neutrophil functions in the presence of apoptotic cells. *J Immunol* (2010) 184:391–400. doi:10.4049/jimmunol.0900564
53. Ali M, Pulli B, Courties G, Tricot B, Sebas M, Iwamoto Y, et al. Myeloperoxidase inhibition improves ventricular function and remodeling after experimental myocardial infarction. *JACC Basic Transl Sci* (2016) 1:633–43. doi:10.1016/j.jacbts.2016.09.004

54. Carbone F, Crowe LA, Roth A, Burger F, Lenglet S, Braunersreuther V, et al. Treatment with anti-RANKL antibody reduces infarct size and attenuates dysfunction impacting on neutrophil-mediated injury. *J Mol Cell Cardiol* (2016) 94:82–94. doi:10.1016/j.yjmcc.2016.03.013
55. Chia S, Nagurney JT, Brown DF, Raffel OC, Bamberg F, Senatore F, et al. Association of leukocyte and neutrophil counts with infarct size, left ventricular function and outcomes after percutaneous coronary intervention for ST-elevation myocardial infarction. *Am J Cardiol* (2009) 103:333–7. doi:10.1016/j.amjcard.2008.09.085
56. Kyne L, Hausdorff JM, Knight E, Dukas L, Azhar G, Wei JY. Neutrophilia and congestive heart failure after acute myocardial infarction. *Am Heart J* (2000) 139:94–100. doi:10.1016/S0002-8703(00)90314-4
57. Horckmans M, Ring L, Duchene J, Santovito D, Schloss MJ, Drechsler M, et al. Neutrophils orchestrate post-myocardial infarction healing by polarizing macrophages towards a reparative phenotype. *Eur Heart J* (2017) 38:187–97. doi:10.1093/eurheartj/ehw002
58. Kohtani T, Abe Y, Sato M, Miyauchi K, Kawachi K. Protective effects of anti-neutrophil antibody against myocardial ischemia/reperfusion injury in rats. *Eur Surg Res* (2002) 34:313–20. doi:10.1159/000063073
59. Romson JL, Hook BG, Kunkel SL, Abrams GD, Schork MA, Lucchesi BR. Reduction of the extent of ischemic myocardial injury by neutrophil depletion in the dog. *Circulation* (1983) 67:1016–23. doi:10.1161/01.CIR.67.5.1016
60. Cai B, Thorp EB, Doran AC, Subramanian M, Sansbury BE, Lin CS, et al. MerTK cleavage limits proresolving mediator biosynthesis and exacerbates tissue inflammation. *Proc Natl Acad Sci U S A* (2016) 113:6526–31. doi:10.1073/pnas.1524292113
61. A-Gonzalez N, Quintana JA, Garcia-Silva S, Mazariegos M, Gonzalez de la Aleja A, Nicolas-Avila JA, et al. Phagocytosis imprints heterogeneity in tissue-resident macrophages. *J Exp Med* (2017) 214:1281–96. doi:10.1084/jem.20161375
62. Nahrendorf M, Swirski FK, Aikawa E, Stangenberg L, Wurdinger T, Figueiredo JL, et al. The healing myocardium sequentially mobilizes two monocyte subsets with divergent and complementary functions. *J Exp Med* (2007) 204:3037–47. doi:10.1084/jem.20070885
63. Dehn S, Thorp EB. Myeloid receptor CD36 is required for early phagocytosis of myocardial infarcts and induction of Nr4a1-dependent mechanisms of cardiac repair. *FASEB J* (2017). doi:10.1096/fj.201700450R
64. Larson SR, Atif SM, Gibbings SL, Thomas SM, Prabagar MG, Danhorn T, et al. Ly6C(+) monocyte efferocytosis and cross-presentation of cell-associated antigens. *Cell Death Differ* (2016) 23:997–1003. doi:10.1038/cdd.2016.24
65. Hilgendorf I, Gerhardt LM, Tan TC, Winter C, Holderried TA, Chousterman BG, et al. Ly-6Chigh monocytes depend on Nr4a1 to balance both inflammatory and reparative phases in the infarcted myocardium. *Circ Res* (2014) 114:1611–22. doi:10.1161/CIRCRESAHA.114.303204
66. Tacke R, Hilgendorf I, Garner H, Waterborg C, Park K, Nowyhed H, et al. The transcription factor NR4A1 is essential for the development of a novel macrophage subset in the thymus. *Sci Rep* (2015) 5:10055. doi:10.1038/srep10055
67. Korns D, Frasch SC, Fernandez-Boyanapalli R, Henson PM, Bratton DL. Modulation of macrophage efferocytosis in inflammation. *Front Immunol* (2011) 2:57. doi:10.3389/fimmu.2011.00057
68. Shiraishi M, Shintani Y, Shintani Y, Ishida H, Saba R, Yamaguchi A, et al. Alternatively activated macrophages determine repair of the infarcted adult murine heart. *J Clin Invest* (2016) 126:2151–66. doi:10.1172/JCI85782
69. Martinez FO, Gordon S. The M1 and M2 paradigm of macrophage activation: time for reassessment. *F1000Prime Rep* (2014) 6:13. doi:10.12703/P6-13
70. Wynn TA. Type 2 cytokines: mechanisms and therapeutic strategies. *Nat Rev Immunol* (2015) 15:271–82. doi:10.1038/nri3831
71. Hofmann U, Knorr S, Vogel B, Weirather J, Frey A, Ertl G, et al. Interleukin-13 deficiency aggravates healing and remodeling in male mice after experimental myocardial infarction. *Circ Heart Fail* (2014) 7:822–30. doi:10.1161/CIRCHEARTFAILURE.113.001020
72. Bosurgi L, Cao YG, Cabeza-Cabrerizo M, Tucci A, Hughes LD, Kong Y, et al. Macrophage function in tissue repair and remodeling requires IL-4 or IL-13 with apoptotic cells. *Science* (2017) 356:1072–6. doi:10.1126/science.aai8132
73. Sager HB, Hulsmans M, Lavine KJ, Moreira MB, Heidt T, Courties G, et al. Proliferation and recruitment contribute to myocardial macrophage expansion in chronic heart failure. *Circ Res* (2016) 119:853–64. doi:10.1161/CIRCRESAHA.116.309001
74. Lavine KJ, Epelman S, Uchida K, Weber KJ, Nichols CG, Schilling JD, et al. Distinct macrophage lineages contribute to disparate patterns of cardiac recovery and remodeling in the neonatal and adult heart. *Proc Natl Acad Sci U S A* (2014) 111:16029–34. doi:10.1073/pnas.1406508111
75. Molawi K, Wolf Y, Kandalla PK, Favret J, Hagemeyer N, Frenzel K, et al. Progressive replacement of embryo-derived cardiac macrophages with age. *J Exp Med* (2014) 211:2151–8. doi:10.1084/jem.20140639
76. Davies LC, Jenkins SJ, Allen JE, Taylor PR. Tissue-resident macrophages. *Nat Immunol* (2013) 14:986–95. doi:10.1038/ni.2705
77. Yano T, Miura T, Whittaker P, Miki T, Sakamoto J, Nakamura Y, et al. Macrophage colony-stimulating factor treatment after myocardial infarction attenuates left ventricular dysfunction by accelerating infarct repair. *J Am Coll Cardiol* (2006) 47:626–34. doi:10.1016/j.jacc.2005.09.037
78. Garfield RE, Chacko S, Blose S. Phagocytosis by muscle cells. *Lab Invest* (1975) 33:418–27.
79. Hurle JM, Lafarga M, Ojeda JL. Cytological and cytochemical studies of the necrotic area of the bulbus of the chick embryo heart: phagocytosis by developing myocardial cells. *J Embryol Exp Morphol* (1977) 41:161–73.
80. Hurle JM, Lafarga M, Ojeda JL. In vivo phagocytosis by developing myocardial cells: an ultrastructural study. *J Cell Sci* (1978) 33:363–9.
81. Nakaya M, Watari K, Tajima M, Nakaya T, Matsuda S, Ohara H, et al. Cardiac myofibroblast engulfment of dead cells facilitates recovery after myocardial infarction. *J Clin Invest* (2017) 127:383–401. doi:10.1172/JCI83822
82. Han CZ, Juncadella IJ, Kinchen JM, Buckley MW, Klibanov AL, Dryden K, et al. Macrophages redirect phagocytosis by non-professional phagocytes and influence inflammation. *Nature* (2016) 539:570–4. doi:10.1038/nature20141
83. Bulluck H, Rosmini S, Abdel-Gadir A, White SK, Bhuva AN, Treibel TA, et al. Residual myocardial iron following intramyocardial hemorrhage during the convalescent phase of reperfused ST-segment-elevation myocardial infarction and adverse left ventricular remodeling. *Circ Cardiovasc Imaging* (2016) 9(10):ii:e004940. doi:10.1161/CIRCIMAGING.116.004940
84. Kali A, Cokic I, Tang R, Dohnalkova A, Kovarik L, Yang HJ, et al. Persistent microvascular obstruction after myocardial infarction culminates in the confluence of ferric iron oxide crystals, proinflammatory burden, and adverse remodeling. *Circ Cardiovasc Imaging* (2016) 9:ii:e004996. doi:10.1161/CIRCIMAGING.115.004996
85. Sindrilaru A, Peters T, Wieschalka S, Baican C, Baican A, Peter H, et al. An unrestrained proinflammatory M1 macrophage population induced by iron impairs wound healing in humans and mice. *J Clin Invest* (2011) 121:985–97. doi:10.1172/JCI44490
86. Yunoki K, Naruko T, Sugioka K, Inaba M, Iwasa Y, Komatsu R, et al. Erythrocyte-rich thrombus aspirated from patients with ST-elevation myocardial infarction: association with oxidative stress and its impact on myocardial reperfusion. *Eur Heart J* (2012) 33:1480–90. doi:10.1093/eurheartj/ehr486
87. De Meyer GR, De Cleen DM, Cooper S, Knaapen MW, Jans DM, Martinet W, et al. Platelet phagocytosis and processing of beta-amyloid precursor protein as a mechanism of macrophage activation in atherosclerosis. *Circ Res* (2002) 90:1197–204. doi:10.1161/01.RES.0000020017.84398.61
88. Motley MP, Madsen DH, Jurgensen HJ, Spencer DE, Szabo R, Holmbeck K, et al. A CCR2 macrophage endocytic pathway mediates extravascular fibrin clearance in vivo. *Blood* (2016) 127:1085–96. doi:10.1182/blood-2015-05-644260
89. Fadok VA, Voelker DR, Campbell PA, Cohen JJ, Bratton DL, Henson PM. Exposure of phosphatidylserine on the surface of apoptotic lymphocytes triggers specific recognition and removal by macrophages. *J Immunol* (1992) 148:2207–16.
90. Zargarian S, Shlomovitz I, Erlich Z, Hourizadeh A, Ofir-Birin Y, Croker BA, et al. Phosphatidylserine externalization, "necroptotic bodies" release, and phagocytosis during necroptosis. *PLoS Biol* (2017) 15:e2002711. doi:10.1371/journal.pbio.2002711
91. Park YJ, Liu G, Lorne EF, Zhao X, Wang J, Tsuruta Y, et al. PAI-1 inhibits neutrophil efferocytosis. *Proc Natl Acad Sci U S A* (2008) 105:11784–9. doi:10.1073/pnas.0801394105
92. Jaiswal S, Jamieson CH, Pang WW, Park CY, Chao MP, Majeti R, et al. CD47 is upregulated on circulating hematopoietic stem cells and leukemia cells to avoid phagocytosis. *Cell* (2009) 138:271–85. doi:10.1016/j.cell.2009.05.046
93. Gardai SJ, McPhillips KA, Frasch SC, Janssen WJ, Starefeldt A, Murphy-Ullrich JE, et al. Cell-surface calreticulin initiates clearance of viable or

apoptotic cells through trans-activation of LRP on the phagocyte. *Cell* (2005) 123:321–34. doi:10.1016/j.cell.2005.08.032

94. Tsai RK, Discher DE. Inhibition of "self" engulfment through deactivation of myosin-II at the phagocytic synapse between human cells. *J Cell Biol* (2008) 180:989–1003. doi:10.1083/jcb.200708043
95. Izmirly PM, Saxena A, Kim MY, Wang D, Sahl SK, Llanos C, et al. Maternal and fetal factors associated with mortality and morbidity in a multi-racial/ethnic registry of anti-SSA/Ro-associated cardiac neonatal lupus. *Circulation* (2011) 124:1927–35. doi:10.1161/CIRCULATIONAHA.111.033894
96. Swinnen M, Vanhoutte D, Van Almen GC, Hamdani N, Schellings MW, D'Hooge J, et al. Absence of thrombospondin-2 causes age-related dilated cardiomyopathy. *Circulation* (2009) 120:1585–97. doi:10.1161/CIRCULATIONAHA.109.863266
97. Schrijvers DM, De Meyer GR, Kockx MM, Herman AG, Martinet W. Phagocytosis of apoptotic cells by macrophages is impaired in atherosclerosis. *Arterioscler Thromb Vasc Biol* (2005) 25:1256–61. doi:10.1161/01.ATV.0000166517.18801.a7
98. Tabas I. Macrophage death and defective inflammation resolution in atherosclerosis. *Nat Rev Immunol* (2010) 10:36–46. doi:10.1038/nri2675
99. Kojima Y, Volkmer JP, McKenna K, Civelek M, Lusis AJ, Miller CL, et al. CD47-blocking antibodies restore phagocytosis and prevent atherosclerosis. *Nature* (2016) 536:86–90. doi:10.1038/nature18935
100. Zhang S, Yeap XY, DeBerge M, Naresh NK, Wang K, Jiang Z, et al. Acute CD47 blockade during ischemic myocardial reperfusion enhances phagocytosis-associated cardiac repair. *JACC Basic Transl Sci* (2017) 2(4):386–97. doi:10.1016/j.jacbts.2017.03.013
101. Metayer LE, Vilalta A, Burke GAA, Brown GC. Anti-CD47 antibodies induce phagocytosis of live, malignant B cells by macrophages via the Fc domain, resulting in cell death by phagoptosis. *Oncotarget* (2017) 8(37):60892–903. doi:10.18632/oncotarget.18492
102. Peri G, Introna M, Corradi D, Iacuitti G, Signorini S, Avanzini F, et al. PTX3, A prototypical long pentraxin, is an early indicator of acute myocardial infarction in humans. *Circulation* (2000) 102:636–41. doi:10.1161/01.CIR.102.6.636
103. Salio M, Chimenti S, De Angelis N, Molla F, Maina V, Nebuloni M, et al. Cardioprotective function of the long pentraxin PTX3 in acute myocardial infarction. *Circulation* (2008) 117:1055–64. doi:10.1161/CIRCULATIONAHA.107.749234
104. Baruah P I, Dumitriu E, Peri G, Russo V, Mantovani A, Manfredi AA, et al. The tissue pentraxin PTX3 limits C1q-mediated complement activation and phagocytosis of apoptotic cells by dendritic cells. *J Leukoc Biol* (2006) 80:87–95. doi:10.1189/jlb.0805445
105. Guo T, Ke L, Qi B, Wan J, Ge J, Bai L, et al. PTX3 is located at the membrane of late apoptotic macrophages and mediates the phagocytosis of macrophages. *J Clin Immunol* (2012) 32:330–9. doi:10.1007/s10875-011-9615-6
106. Zhu H, Cui D, Liu K, Wang L, Huang L, Li J. Long pentraxin PTX3 attenuates ischemia reperfusion injury in a cardiac transplantation model. *Transpl Int* (2014) 27:87–95. doi:10.1111/tri.12197
107. Chang MK, Binder CJ, Torzewski M, Witztum JL. C-reactive protein binds to both oxidized LDL and apoptotic cells through recognition of a common ligand: Phosphorylcholine of oxidized phospholipids. *Proc Natl Acad Sci U S A* (2002) 99:13043–8. doi:10.1073/pnas.192399699
108. Kim SJ, Gershov D, Ma X, Brot N, Elkon KB. Opsonization of apoptotic cells and its effect on macrophage and T cell immune responses. *Ann N Y Acad Sci* (2003) 987:68–78. doi:10.1111/j.1749-6632.2003.tb06034.x
109. Vogt B, Fuhrnrohr B, Muller R, Sheriff A. CRP and the disposal of dying cells: consequences for systemic lupus erythematosus and rheumatoid arthritis. *Autoimmunity* (2007) 40:295–8. doi:10.1080/08916930701358925
110. Lagrand WK, Niessen HW, Wolbink GJ, Jaspars LH, Visser CA, Verheugt FW, et al. C-reactive protein colocalizes with complement in human hearts during acute myocardial infarction. *Circulation* (1997) 95:97–103. doi:10.1161/01.CIR.95.1.97
111. Sheriff A, Schindler R, Vogt B, Abdel-Aty H, Unger JK, Bock C, et al. Selective apheresis of C-reactive protein: a new therapeutic option in myocardial infarction? *J Clin Apher* (2015) 30:15–21. doi:10.1002/jca.21344
112. Griselli M, Herbert J, Hutchinson WL, Taylor KM, Sohail M, Krausz T, et al. C-reactive protein and complement are important mediators of tissue damage in acute myocardial infarction. *J Exp Med* (1999) 190:1733–40. doi:10.1084/jem.190.12.1733
113. Ito W, Schafer HJ, Bhakdi S, Klask R, Hansen S, Schaarschmidt S, et al. Influence of the terminal complement-complex on reperfusion injury, no-reflow and arrhythmias: a comparison between C6-competent and C6-deficient rabbits. *Cardiovasc Res* (1996) 32:294–305. doi:10.1016/0008-6363(96)00082-X
114. Madsen DH, Leonard D, Masedunskas A, Moyer A, Jurgensen HJ, Peters DE, et al. M2-like macrophages are responsible for collagen degradation through a mannose receptor-mediated pathway. *J Cell Biol* (2013) 202:951–66. doi:10.1083/jcb.201301081
115. Warnatsch A, Tsourouktsoglou TD, Branzk N, Wang Q, Reincke S, Herbst S, et al. Reactive oxygen species localization programs inflammation to clear microbes of different size. *Immunity* (2017) 46:421–32. doi:10.1016/j.immuni.2017.02.013
116. Tsujita K, Kaikita K, Hayasaki T, Honda T, Kobayashi H, Sakashita N, et al. Targeted deletion of class A macrophage scavenger receptor increases the risk of cardiac rupture after experimental myocardial infarction. *Circulation* (2007) 115:1904–11. doi:10.1161/CIRCULATIONAHA.106.671198
117. Hu Y, Zhang H, Lu Y, Bai H, Xu Y, Zhu X, et al. Class A scavenger receptor attenuates myocardial infarction-induced cardiomyocyte necrosis through suppressing M1 macrophage subset polarization. *Basic Res Cardiol* (2011) 106:1311–28. doi:10.1007/s00395-011-0204-x
118. Ren D, Wang X, Ha T, Liu L, Kalbfleisch J, Gao X, et al. SR-A deficiency reduces myocardial ischemia/reperfusion injury; involvement of increased microRNA-125b expression in macrophages. *Biochim Biophys Acta* (2013) 1832:336–46. doi:10.1016/j.bbadis.2012.10.012
119. Leuschner F, Rauch PJ, Ueno T, Gorbatov R, Marinelli B, Lee WW, et al. Rapid monocyte kinetics in acute myocardial infarction are sustained by extramedullary monocytopoiesis. *J Exp Med* (2012) 209:123–37. doi:10.1084/jem.20111009
120. DeLeon-Pennell KY, Tian Y, Zhang B, Cates CA, Iyer RP, Cannon P, et al. CD36 is a matrix metalloproteinase-9 substrate that stimulates neutrophil apoptosis and removal during cardiac remodeling. *Circ Cardiovasc Genet* (2016) 9:14–25. doi:10.1161/CIRCGENETICS.115.001249
121. Caberoy NB, Alvarado G, Bigcas JL, Li W. Galectin-3 is a new MerTK-specific eat-me signal. *J Cell Physiol* (2012) 227:401–7. doi:10.1002/jcp.22955
122. Gonzalez GE, Cassaglia P, Noli Truant S, Fernandez MM, Wilensky L, Volberg V, et al. Galectin-3 is essential for early wound healing and ventricular remodeling after myocardial infarction in mice. *Int J Cardiol* (2014) 176:1423–5. doi:10.1016/j.ijcard.2014.08.011
123. Sokolovska A, Becker CE, Ip WK, Rathinam VA, Brudner M, Paquette N, et al. Activation of caspase-1 by the NLRP3 inflammasome regulates the NADPH oxidase NOX2 to control phagosome function. *Nat Immunol* (2013) 14:543–53. doi:10.1038/ni.2595
124. A-Gonzalez N, Bensinger SJ, Hong C, Beceiro S, Bradley MN, Zelcer N, et al. Apoptotic cells promote their own clearance and immune tolerance through activation of the nuclear receptor LXR. *Immunity* (2009) 31:245–58. doi:10.1016/j.immuni.2009.06.018
125. Chung EY, Liu J, Homma Y, Zhang Y, Brendolan A, Saggese M, et al. Interleukin-10 expression in macrophages during phagocytosis of apoptotic cells is mediated by homeodomain proteins Pbx1 and Prep-1. *Immunity* (2007) 27:952–64. doi:10.1016/j.immuni.2007.11.014
126. Anso E, Weinberg SE, Diebold LP, Thompson BJ, Malinge S, Schumacker PT, et al. The mitochondrial respiratory chain is essential for haematopoietic stem cell function. *Nat Cell Biol* (2017) 19:614–25. doi:10.1038/ncb3529
127. Weinberg F, Hamanaka R, Wheaton WW, Weinberg S, Joseph J, Lopez M, et al. Mitochondrial metabolism and ROS generation are essential for Kras-mediated tumorigenicity. *Proc Natl Acad Sci U S A* (2010) 107:8788–93. doi:10.1073/pnas.1003428107
128. Sena LA, Li S, Jairaman A, Prakriya M, Ezponda T, Hildeman DA, et al. Mitochondria are required for antigen-specific T cell activation through reactive oxygen species signaling. *Immunity* (2013) 38:225–36. doi:10.1016/j.immuni.2012.10.020
129. Usui F, Shirasuna K, Kimura H, Tatsumi K, Kawashima A, Karasawa T, et al. Inflammasome activation by mitochondrial oxidative stress in macrophages leads to the development of angiotensin II-induced aortic aneurysm. *Arterioscler Thromb Vasc Biol* (2015) 35:127–36. doi:10.1161/ATVBAHA.114.303763
130. Garaude J, Acin-Perez R, Martinez-Cano S, Enamorado M, Ugolini M, Nistal-Villan E, et al. Mitochondrial respiratory-chain adaptations in macrophages

contribute to antibacterial host defense. *Nat Immunol* (2016) 17:1037–45. doi:10.1038/ni.3509

131. Huang SC, Everts B, Ivanova Y, O'Sullivan D, Nascimento M, Smith AM, et al. Cell-intrinsic lysosomal lipolysis is essential for alternative activation of macrophages. *Nat Immunol* (2014) 15:846–55. doi:10.1038/ni.2956
132. Piquereau J, Caffin F, Novotova M, Lemaire C, Veksler V, Garnier A, et al. Mitochondrial dynamics in the adult cardiomyocytes: which roles for a highly specialized cell? *Front Physiol* (2013) 4:102. doi:10.3389/fphys.2013.00102
133. Zheng Y, Gardner SE, Clarke MC. Cell death, damage-associated molecular patterns, and sterile inflammation in cardiovascular disease. *Arterioscler Thromb Vasc Biol* (2011) 31:2781–6. doi:10.1161/ATVBAHA.111.224907
134. Haschemi A, Kosma P, Gille L, Evans CR, Burant CF, Starkl P, et al. The sedoheptulose kinase CARKL directs macrophage polarization through control of glucose metabolism. *Cell Metab* (2012) 15:813–26. doi:10.1016/j.cmet.2012.04.023
135. Tannahill GM, Curtis AM, Adamik J, Palsson-McDermott EM, McGettrick AF, Goel G, et al. Succinate is an inflammatory signal that induces IL-1beta through HIF-1alpha. *Nature* (2013) 496:238–42. doi:10.1038/nature11986
136. Rodriguez-Prados JC, Traves PG, Cuenca J, Rico D, Aragones J, Martin-Sanz P, et al. Substrate fate in activated macrophages: a comparison between innate, classic, and alternative activation. *J Immunol* (2010) 185:605–14. doi:10.4049/jimmunol.0901698
137. Jha AK, Huang SC, Sergushichev A, Lampropoulou V, Ivanova Y, Loginicheva E, et al. Network integration of parallel metabolic and transcriptional data reveals metabolic modules that regulate macrophage polarization. *Immunity* (2015) 42:419–30. doi:10.1016/j.immuni.2015.02.005
138. Chouchani ET, Pell VR, Gaude E, Aksentijevic D, Sundier SY, Robb EL, et al. Ischaemic accumulation of succinate controls reperfusion injury through mitochondrial ROS. *Nature* (2014) 515:431–5. doi:10.1038/nature13909
139. Lampropoulou V, Sergushichev A, Bambouskova M, Nair S, Vincent EE, Loginicheva E, et al. Itaconate links inhibition of succinate dehydrogenase with macrophage metabolic remodeling and regulation of inflammation. *Cell Metab* (2016) 24:158–66. doi:10.1016/j.cmet.2016.06.004
140. Mills EL, Kelly B, Logan A, Costa AS, Varma M, Bryant CE, et al. Succinate dehydrogenase supports metabolic repurposing of mitochondria to drive inflammatory macrophages. *Cell* (2016) 167(457–470):e413. doi:10.1016/j.cell.2016.08.064
141. Wu Z, Puigserver P, Andersson U, Zhang C, Adelmant G, Mootha V, et al. Mechanisms controlling mitochondrial biogenesis and respiration through the thermogenic coactivator PGC-1. *Cell* (1999) 98:115–24. doi:10.1016/S0092-8674(00)80611-X
142. Nomura M, Liu J, Rovira II, Gonzalez-Hurtado E, Lee J, Wolfgang MJ, et al. Fatty acid oxidation in macrophage polarization. *Nat Immunol* (2016) 17:216–7. doi:10.1038/ni.3366
143. Huang SC, Smith AM, Everts B, Colonna M, Pearce EL, Schilling JD, et al. Metabolic reprogramming mediated by the mTORC2-IRF4 signaling axis is essential for macrophage alternative activation. *Immunity* (2016) 45:817–30. doi:10.1016/j.immuni.2016.09.016
144. Ip WKE, Hoshi N, Shouval DS, Snapper S, Medzhitov R. Anti-inflammatory effect of IL-10 mediated by metabolic reprogramming of macrophages. *Science* (2017) 356:513–9. doi:10.1126/science.aal3535
145. Quirino IE, Carneiro MB, Cardoso VN, das Gracas Carvalho Dos Santos R, Vieira LQ, Fiuza JA, et al. Arginine supplementation induces arginase activity and inhibits TNF-alpha synthesis in mice spleen macrophages after intestinal obstruction. *JPEN J Parenter Enteral Nutr* (2016) 40:417–22. doi:10.1177/0148607114546374
146. Gundra UM, Girgis NM, Gonzalez MA, San Tang M, Van Der Zande HJP, Lin JD, et al. Vitamin A mediates conversion of monocyte-derived macrophages into tissue-resident macrophages during alternative activation. *Nat Immunol* (2017) 18:642–53. doi:10.1038/ni.3734
147. Biswas SK, Mantovani A. Orchestration of metabolism by macrophages. *Cell Metab* (2012) 15:432–7. doi:10.1016/j.cmet.2011.11.013
148. Park D, Han CZ, Elliott MR, Kinchen JM, Trampont PC, Das S, et al. Continued clearance of apoptotic cells critically depends on the phagocyte Ucp2 protein. *Nature* (2011) 477:220–4. doi:10.1038/nature10340
149. Huang FP, Platt N, Wykes M, Major JR, Powell TJ, Jenkins CD, et al. A discrete subpopulation of dendritic cells transports apoptotic intestinal epithelial cells to T cell areas of mesenteric lymph nodes. *J Exp Med* (2000) 191:435–44. doi:10.1084/jem.191.3.435
150. Van der Borght K, Scott CL, Nindl V, Bouche A, Martens L, Sichien D, et al. Myocardial infarction primes autoreactive T cells through activation of dendritic cells. *Cell Rep* (2017) 18:3005–17. doi:10.1016/j.celrep.2017.02.079
151. Randolph GJ, Inaba K, Robbiani DF, Steinman RM, Muller WA. Differentiation of phagocytic monocytes into lymph node dendritic cells in vivo. *Immunity* (1999) 11:753–61. doi:10.1016/S1074-7613(00)80149-1
152. Girard JP, Moussion C, Forster R. HEVs, lymphatics and homeostatic immune cell trafficking in lymph nodes. *Nat Rev Immunol* (2012) 12:762–73. doi:10.1038/nri3298
153. Kuan EL, Ivanov S, Bridenbaugh EA, Victora G, Wang W, Childs EW, et al. Collecting lymphatic vessel permeability facilitates adipose tissue inflammation and distribution of antigen to lymph node-homing adipose tissue dendritic cells. *J Immunol* (2015) 194:5200–10. doi:10.4049/jimmunol.1500221
154. Henri O, Pouehe C, Houssari M, Galas L, Nicol L, dwards-Levy FE, et al. Selective stimulation of cardiac lymphangiogenesis reduces myocardial edema and fibrosis leading to improved cardiac function following myocardial infarction. *Circulation* (2016) 133:1484–97; discussion 1497. doi:10.1161/CIRCULATIONAHA.115.020143
155. Klotz L, Norman S, Vieira JM, Masters M, Rohling M, Dube KN, et al. Cardiac lymphatics are heterogeneous in origin and respond to injury. *Nature* (2015) 522:62–7. doi:10.1038/nature14483
156. Zumsteg A, Christofori G. Myeloid cells and lymphangiogenesis. *Cold Spring Harb Perspect Med* (2012) 2:a006494. doi:10.1101/cshperspect.a006494
157. Cimpean AM, Raica M. Lymphangiogenesis and inflammation-looking for the "Missing Pieces" of the puzzle. *Arch Immunol Ther Exp (Warsz)* (2015) 63:415–26. doi:10.1007/s00005-015-0349-7
158. Granata F, Frattini A, Loffredo S, Staiano RI, Petraroli A, Ribatti D, et al. Production of vascular endothelial growth factors from human lung macrophages induced by group IIA and group X secreted phospholipases A2. *J Immunol* (2010) 184:5232–41. doi:10.4049/jimmunol.0902501
159. Jones D, Min W. An overview of lymphatic vessels and their emerging role in cardiovascular disease. *J Cardiovasc Dis Res* (2011) 2:141–52. doi:10.4103/0975-3583.85260
160. Harvey NL, Gordon EJ. Deciphering the roles of macrophages in developmental and inflammation stimulated lymphangiogenesis. *Vasc Cell* (2012) 4:15. doi:10.1186/2045-824X-4-15
161. Hofmann U, Beyersdorf N, Weirather J, Podolskaya A, Bauersachs J, Ertl G, et al. Activation of CD4+ T lymphocytes improves wound healing and survival after experimental myocardial infarction in mice. *Circulation* (2012) 125:1652–63. doi:10.1161/CIRCULATIONAHA.111.044164
162. Gottumukkala RV, Lv H, Cornivelli L, Wagers AJ, Kwong RY, Bronson R, et al. Myocardial infarction triggers chronic cardiac autoimmunity in type 1 diabetes. *Sci Transl Med* (2012) 4:138ra180. doi:10.1126/scitranslmed.3003551
163. Lv H, Havari E, Pinto S, Gottumukkala RV, Cornivelli L, Raddassi K, et al. Impaired thymic tolerance to alpha-myosin directs autoimmunity to the heart in mice and humans. *J Clin Invest* (2011) 121:1561–73. doi:10.1172/JCI44583
164. Suresh Babu S, Thandavarayan RA, Joladarashi D, Jeyabal P, Krishnamurthy S, Bhimaraj A, et al. MicroRNA-126 overexpression rescues diabetes-induced impairment in efferocytosis of apoptotic cardiomyocytes. *Sci Rep* (2016) 6:36207. doi:10.1038/srep36207
165. Choo EH, Lee JH, Park EH, Park HE, Jung NC, Kim TH, et al. Infarcted myocardium-primed dendritic cells improve remodeling and cardiac function after myocardial infarction by modulating the regulatory T cell and macrophage polarization. *Circulation* (2017) 135:1444–57. doi:10.1161/CIRCULATIONAHA.116.023106
166. Qu C, Nguyen VA, Merad M, Randolph GJ. MHC class I/peptide transfer between dendritic cells overcomes poor cross-presentation by monocyte-derived APCs that engulf dying cells. *J Immunol* (2009) 182:3650–9. doi:10.4049/jimmunol.0801532
167. Moraru M, Roth A, Keren G, George J. Cellular autoimmunity to cardiac myosin in patients with a recent myocardial infarction. *Int J Cardiol* (2006) 107:61–6. doi:10.1016/j.ijcard.2005.02.036
168. Li F, Huang Q, Chen J, Peng Y, Roop DR, Bedford JS, et al. Apoptotic cells activate the "phoenix rising" pathway to promote wound healing and tissue regeneration. *Sci Signal* (2010) 3:ra13. doi:10.1126/scisignal.2000634
169. Hochreiter-Hufford AE, Lee CS, Kinchen JM, Sokolowski JD, Arandjelovic S, Call JA, et al. Phosphatidylserine receptor BAI1 and apoptotic cells as

new promoters of myoblast fusion. *Nature* (2013) 497:263–7. doi:10.1038/nature12135
170. Summan M, Warren GL, Mercer RR, Chapman R, Hulderman T, Van Rooijen N, et al. Macrophages and skeletal muscle regeneration: a clodronate-containing liposome depletion study. *Am J Physiol Regul Integr Comp Physiol* (2006) 290:R1488–95. doi:10.1152/ajpregu.00465.2005
171. Ruffell D, Mourkioti F, Gambardella A, Kirstetter P, Lopez RG, Rosenthal N, et al. A CREB-C/EBPbeta cascade induces M2 macrophage-specific gene expression and promotes muscle injury repair. *Proc Natl Acad Sci U S A* (2009) 106:17475–80. doi:10.1073/pnas.0908641106
172. Arnold L, Henry A, Poron F, Baba-Amer Y, van Rooijen N, Plonquet A, et al. Inflammatory monocytes recruited after skeletal muscle injury switch into antiinflammatory macrophages to support myogenesis. *J Exp Med* (2007) 204:1057–69. doi:10.1084/jem.20070075
173. Kaikita K, Hayasaki T, Okuma T, Kuziel WA, Ogawa H, Takeya M. Targeted deletion of CC chemokine receptor 2 attenuates left ventricular remodeling after experimental myocardial infarction. *Am J Pathol* (2004) 165:439–47. doi:10.1016/S0002-9440(10)63309-3
174. Majmudar MD, Keliher EJ, Heidt T, Leuschner F, Truelove J, Sena BF, et al. Monocyte-directed RNAi targeting CCR2 improves infarct healing in atherosclerosis-prone mice. *Circulation* (2013) 127:2038–46. doi:10.1161/CIRCULATIONAHA.112.000116
175. Lu H, Huang D, Saederup N, Charo IF, Ransohoff RM, Zhou L. Macrophages recruited via CCR2 produce insulin-like growth factor-1 to repair acute skeletal muscle injury. *FASEB J* (2011) 25:358–69. doi:10.1096/fj.10-171579
176. Porrello ER, Mahmoud AI, Simpson E, Hill JA, Richardson JA, Olson EN, et al. Transient regenerative potential of the neonatal mouse heart. *Science* (2011) 331:1078–80. doi:10.1126/science.1200708
177. Aurora AB, Porrello ER, Tan W, Mahmoud AI, Hill JA, Bassel-Duby R, et al. Macrophages are required for neonatal heart regeneration. *J Clin Invest* (2014) 124:1382–92. doi:10.1172/JCI72181
178. de Couto G, Gallet R, Cambier L, Jaghatspanyan E, Makkar N, Dawkins JF, et al. Exosomal MicroRNA transfer into macrophages mediates cellular postconditioning. *Circulation* (2017) 136:200–14. doi:10.1161/CIRCULATIONAHA.116.024590
179. Gallet R, de Couto G, Simsolo E, Valle J, Sun B, Liu W, et al. Cardiosphere-derived cells reverse heart failure with preserved ejection fraction (HFpEF) in rats by decreasing fibrosis and inflammation. *JACC Basic Transl Sci* (2016) 1:14–28. doi:10.1016/j.jacbts.2016.01.003
180. Gyongyosi M, Wojakowski W, Lemarchand P, Lunde K, Tendera M, Bartunek J, et al. Meta-analysis of cell-based CaRdiac stUdiEs (ACCRUE) in patients with acute myocardial infarction based on individual patient data. *Circ Res* (2015) 116:1346–60. doi:10.1161/CIRCRESAHA.116.304346
181. Li Z, Bing OH, Long X, Robinson KG, Lakatta EG. Increased cardiomyocyte apoptosis during the transition to heart failure in the spontaneously hypertensive rat. *Am J Physiol* (1997) 272:H2313–9.
182. Narula J, Haider N, Virmani R, DiSalvo TG, Kolodgie FD, Hajjar RJ, et al. Apoptosis in myocytes in end-stage heart failure. *N Engl J Med* (1996) 335:1182–9. doi:10.1056/NEJM199610173351603
183. Sharov VG, Sabbah HN, Shimoyama H, Goussev AV, Lesch M, Goldstein S. Evidence of cardiocyte apoptosis in myocardium of dogs with chronic heart failure. *Am J Pathol* (1996) 148:141–9.
184. Salvador AM, Nevers T, Velazquez F, Aronovitz M, Wang B, Abadia Molina A, et al. Intercellular adhesion molecule 1 regulates left ventricular leukocyte infiltration, cardiac remodeling, and function in pressure overload-induced heart failure. *J Am Heart Assoc* (2016) 5:e003126. doi:10.1161/JAHA.115.003126
185. Yang M, Liu J, Piao C, Shao J, Du J. ICAM-1 suppresses tumor metastasis by inhibiting macrophage M2 polarization through blockade of efferocytosis. *Cell Death Dis* (2015) 6:e1780. doi:10.1038/cddis.2015.144
186. Verma SK, Krishnamurthy P, Barefield D, Singh N, Gupta R, Lambers E, et al. Interleukin-10 treatment attenuates pressure overload-induced hypertrophic remodeling and improves heart function via signal transducers and activators of transcription 3-dependent inhibition of nuclear factor-kappaB. *Circulation* (2012) 126:418–29. doi:10.1161/CIRCULATIONAHA.112.112185
187. Sharma UC, Pokharel S, van Brakel TJ, van Berlo JH, Cleutjens JP, Schroen B, et al. Galectin-3 marks activated macrophages in failure-prone hypertrophied hearts and contributes to cardiac dysfunction. *Circulation* (2004) 110:3121–8. doi:10.1161/01.CIR.0000147181.65298.4D
188. de Boer RA, Lok DJ, Jaarsma T, van der Meer P, Voors AA, Hillege HL, et al. Predictive value of plasma galectin-3 levels in heart failure with reduced and preserved ejection fraction. *Ann Med* (2011) 43:60–8. doi:10.3109/07853890.2010.538080
189. Gonzalez GE, Rhaleb NE, D'Ambrosio MA, Nakagawa P, Liao TD, Peterson EL, et al. Cardiac-deleterious role of galectin-3 in chronic angiotensin II-induced hypertension. *Am J Physiol Heart Circ Physiol* (2016) 311:H1287–96. doi:10.1152/ajpheart.00096.2016
190. Sano H, Hsu DK, Apgar JR, Yu L, Sharma BB, Kuwabara I, et al. Critical role of galectin-3 in phagocytosis by macrophages. *J Clin Invest* (2003) 112:389–97. doi:10.1172/JCI200317592
191. Ochieng J, Fridman R, Nangia-Makker P, Kleiner DE, Liotta LA, Stetler-Stevenson WG, et al. Galectin-3 is a novel substrate for human matrix metalloproteinases-2 and -9. *Biochemistry* (1994) 33:14109–14. doi:10.1021/bi00251a020
192. MacKinnon AC, Farnworth SL, Hodkinson PS, Henderson NC, Atkinson KM, Leffler H, et al. Regulation of alternative macrophage activation by galectin-3. *J Immunol* (2008) 180:2650–8. doi:10.4049/jimmunol.180.4.2650
193. Qian L, Li X, Fang R, Wang Z, Xu Y, Zhang H, et al. Class A scavenger receptor deficiency augments angiotensin II-induced vascular remodeling. *Biochem Pharmacol* (2014) 90:254–64. doi:10.1016/j.bcp.2014.05.015
194. Lekawanvijit S, Kompa AR, Zhang Y, Wang BH, Kelly DJ, Krum H. Myocardial infarction impairs renal function, induces renal interstitial fibrosis, and increases renal KIM-1 expression: implications for cardiorenal syndrome. *Am J Physiol Heart Circ Physiol* (2012) 302:H1884–93. doi:10.1152/ajpheart.00967.2011
195. Gandhi R, Yi J, Ha J, Shi H, Ismail O, Nathoo S, et al. Accelerated receptor shedding inhibits kidney injury molecule-1 (KIM-1)-mediated efferocytosis. *Am J Physiol Renal Physiol* (2014) 307:F205–21. doi:10.1152/ajprenal.00638.2013
196. Humphreys BD, Xu F, Sabbisetti V, Grgic I, Movahedi Naini S, Wang N, et al. Chronic epithelial kidney injury molecule-1 expression causes murine kidney fibrosis. *J Clin Invest* (2013) 123:4023–35. doi:10.1172/JCI45361
197. Harwani SC, Chapleau MW, Legge KL, Ballas ZK, Abboud FM. Neurohormonal modulation of the innate immune system is proinflammatory in the prehypertensive spontaneously hypertensive rat, a genetic model of essential hypertension. *Circ Res* (2012) 111:1190–7. doi:10.1161/CIRCRESAHA.112.277475
198. Fujiu K, Shibatà M, Nakayama Y, Ogata F, Matsumoto S, Noshita K, et al. A heart-brain-kidney network controls adaptation to cardiac stress through tissue macrophage activation. *Nat Med* (2017) 23:611–22. doi:10.1038/nm.4326
199. Sykes M, Levy G. Advances in transplantation. *Semin Immunol* (2011) 23:222–3. doi:10.1016/j.smim.2011.08.013
200. Lechler RI, Sykes M, Thomson AW, Turka LA. Organ transplantation – how much of the promise has been realized? *Nat Med* (2005) 11:605–13. doi:10.1038/nm1251
201. De Santo LS, Torella M, Romano G, Maiello C, Buonocore M, Bancone C, et al. Perioperative myocardial injury after adult heart transplant: determinants and prognostic value. *PLoS One* (2015) 10:e0120813. doi:10.1371/journal.pone.0120813
202. Shen H, Heuzey E, Mori DN, Wong CK, Colangelo CM, Chung LM, et al. Haptoglobin enhances cardiac transplant rejection. *Circ Res* (2015) 116:1670–9. doi:10.1161/CIRCRESAHA.116.305406
203. Serhan CN, Savill J. Resolution of inflammation: the beginning programs the end. *Nat Immunol* (2005) 6:1191–7. doi:10.1038/ni1276
204. Hodge S, Dean M, Hodge G, Holmes M, Reynolds PN. Decreased efferocytosis and mannose binding lectin in the airway in bronchiolitis obliterans syndrome. *J Heart Lung Transplant* (2011) 30:589–95. doi:10.1016/j.healun.2011.01.710
205. Christen T, Nahrendorf M, Wildgruber M, Swirski FK, Aikawa E, Waterman P, et al. Molecular imaging of innate immune cell function in transplant rejection. *Circulation* (2009) 119:1925–32. doi:10.1161/CIRCULATIONAHA.108.796888
206. Swirski FK, Wildgruber M, Ueno T, Figueiredo JL, Panizzi P, Iwamoto Y, et al. Myeloperoxidase-rich Ly-6C+ myeloid cells infiltrate allografts and contribute to an imaging signature of organ rejection in mice. *J Clin Invest* (2010) 120:2627–34. doi:10.1172/JCI42304
207. Vassalli G, Gallino A, Weis M, von Scheidt W, Kappenberger L, von Segesser LK, et al. Alloimmunity and nonimmunologic risk factors in

cardiac allograft vasculopathy. *Eur Heart J* (2003) 24:1180–8. doi:10.1016/S0195-668X(03)00237-9
208. Jiang X, Tian W, Sung YK, Qian J, Nicolls MR. Macrophages in solid organ transplantation. *Vasc Cell* (2014) 6:5. doi:10.1186/2045-824X-6-5
209. Salehi S, Reed EF. The divergent roles of macrophages in solid organ transplantation. *Curr Opin Organ Transplant* (2015) 20:446–53. doi:10.1097/MOT.0000000000000209
210. Girlanda R, Kleiner DE, Duan Z, Ford EA, Wright EC, Mannon RB, et al. Monocyte infiltration and kidney allograft dysfunction during acute rejection. *Am J Transplant* (2008) 8:600–7. doi:10.1111/j.1600-6143.2007.02109.x
211. Oberbarnscheidt MH, Zeng Q, Li Q, Dai H, Williams AL, Shlomchik WD, et al. Non-self recognition by monocytes initiates allograft rejection. *J Clin Invest* (2014) 124:3579–89. doi:10.1172/JCI74370
212. Boros P, Ochando JC, Chen SH, Bromberg JS. Myeloid-derived suppressor cells: natural regulators for transplant tolerance. *Hum Immunol* (2010) 71:1061–6. doi:10.1016/j.humimm.2010.08.001
213. Garcia MR, Ledgerwood L, Yang Y, Xu J, Lal G, Burrell B, et al. Monocytic suppressive cells mediate cardiovascular transplantation tolerance in mice. *J Clin Invest* (2010) 120:2486–96. doi:10.1172/JCI41628
214. Ochando JC, Chen SH. Myeloid-derived suppressor cells in transplantation and cancer. *Immunol Res* (2012) 54(1–3):275–85. doi:10.1007/s12026-012-8335-1
215. Conde P, Rodriguez M, van der Touw W, Jimenez A, Burns M, Miller J, et al. DC-SIGN(+) Macrophages control the induction of transplantation tolerance. *Immunity* (2015) 42:1143–58. doi:10.1016/j.immuni.2015.05.009
216. Miller SD, Turley DM, Podojil JR. Antigen-specific tolerance strategies for the prevention and treatment of autoimmune disease. *Nat Rev Immunol* (2007) 7:665–77. doi:10.1038/nri2153
217. Chen G, Kheradmand T, Bryant J, Wang S, Tasch J, Wang JJ, et al. Intragraft CD11b(+) IDO(+) cells mediate cardiac allograft tolerance by ECDI-fixed donor splenocyte infusions. *Am J Transplant* (2012) 12:2920–9. doi:10.1111/j.1600-6143.2012.04203.x
218. Sun E, Gao Y, Chen J, Roberts AI, Wang X, Chen Z, et al. Allograft tolerance induced by donor apoptotic lymphocytes requires phagocytosis in the recipient. *Cell Death Differ* (2004) 11:1258–64. doi:10.1038/sj.cdd.4401500
219. Bryant J, Lerret NM, Wang JJ, Kang HK, Tasch J, Zhang Z, et al. Preemptive donor apoptotic cell infusions induce IFN-gamma-producing myeloid-derived suppressor cells for cardiac allograft protection. *J Immunol* (2014) 192:6092–101. doi:10.4049/jimmunol.1302771
220. Luo X, Pothoven KL, McCarthy D, DeGutes M, Martin A, Getts DR, et al. ECDI-fixed allogeneic splenocytes induce donor-specific tolerance for long-term survival of islet transplants via two distinct mechanisms. *Proc Natl Acad Sci U S A* (2008) 105:14527–32. doi:10.1073/pnas.0805204105
221. Ochando JC, Homma C, Yang Y, Hidalgo A, Garin A, Tacke F, et al. Alloantigen-presenting plasmacytoid dendritic cells mediate tolerance to vascularized grafts. *Nat Immunol* (2006) 7:652–62. doi:10.1038/ni1333
222. Zhang S, Yeap XY, Grigoryeva L, Dehn S, DeBerge M, Tye M, et al. Cardiomyocytes induce macrophage receptor shedding to suppress phagocytosis. *J Mol Cell Cardiol* (2015) 87:171–9. doi:10.1016/j.yjmcc.2015.08.009
223. Thorp E, Vaisar T, Subramanian M, Mautner L, Blobel C, Tabas I. Shedding of the Mer tyrosine kinase receptor is mediated by ADAM17 protein through a pathway involving reactive oxygen species, protein kinase Cdelta, and p38 mitogen-activated protein kinase (MAPK). *J Biol Chem* (2011) 286:33335–44. doi:10.1074/jbc.M111.263020
224. Kain V, Ingle KA, Colas RA, Dalli J, Prabhu SD, Serhan CN, et al. Resolvin D1 activates the inflammation resolving response at splenic and ventricular site following myocardial infarction leading to improved ventricular function. *J Mol Cell Cardiol* (2015) 84:24–35. doi:10.1016/j.yjmcc.2015.04.003
225. Driscoll WS, Vaisar T, Tang J, Wilson CL, Raines EW. Macrophage ADAM17 deficiency augments CD36-dependent apoptotic cell uptake and the linked anti-inflammatory phenotype. *Circ Res* (2013) 113:52–61. doi:10.1161/CIRCRESAHA.112.300683
226. Lood C, Arve S, Ledbetter J, Elkon KB. TLR7/8 activation in neutrophils impairs immune complex phagocytosis through shedding of FcgRIIA. *J Exp Med* (2017) 214:2103–19. doi:10.1084/jem.20161512
227. Maile LA, Capps BE, Miller EC, Allen LB, Veluvolu U, Aday AW, et al. Glucose regulation of integrin-associated protein cleavage controls the response of vascular smooth muscle cells to insulin-like growth factor-I. *Mol Endocrinol* (2008) 22:1226–37. doi:10.1210/me.2007-0552
228. Matsumura S, Iwanaga S, Mochizuki S, Okamoto H, Ogawa S, Okada Y. Targeted deletion or pharmacological inhibition of MMP-2 prevents cardiac rupture after myocardial infarction in mice. *J Clin Invest* (2005) 115:599–609. doi:10.1172/JCI22304
229. Clarke MC, Talib S, Figg NL, Bennett MR. Vascular smooth muscle cell apoptosis induces interleukin-1-directed inflammation: effects of hyperlipidemia-mediated inhibition of phagocytosis. *Circ Res* (2010) 106:363–72. doi:10.1161/CIRCRESAHA.109.208389
230. Dou H, Feher A, Davila AC, Romero MJ, Patel VS, Kamath VM, et al. Role of adipose tissue endothelial ADAM17 in age-related coronary microvascular dysfunction. *Arterioscler Thromb Vasc Biol* (2017) 37:1180–93. doi:10.1161/ATVBAHA.117.309430
231. Tian B, Al-Moujahed A, Bouzika P, Hu Y, Notomi S, Tsoka P, et al. Atorvastatin promotes phagocytosis and attenuates pro-inflammatory response in human retinal pigment epithelial cells. *Sci Rep* (2017) 7:2329. doi:10.1038/s41598-017-02407-7
232. Charbonneau M, Harper K, Grondin F, Pelmus M, McDonald PP, Dubois CM. Hypoxia-inducible factor mediates hypoxic and tumor necrosis factor alpha-induced increases in tumor necrosis factor-alpha converting enzyme/ADAM17 expression by synovial cells. *J Biol Chem* (2007) 282:33714–24. doi:10.1074/jbc.M704041200
233. Marsch E, Theelen TL, Demandt JA, Jeurissen M, van Gink M, Verjans R, et al. Reversal of hypoxia in murine atherosclerosis prevents necrotic core expansion by enhancing efferocytosis. *Arterioscler Thromb Vasc Biol* (2014) 34:2545–53. doi:10.1161/ATVBAHA.114.304023
234. Laoui D, Van Overmeire E, Di Conza G, Aldeni C, Keirsse J, Morias Y, et al. Tumor hypoxia does not drive differentiation of tumor-associated macrophages but rather fine-tunes the M2-like macrophage population. *Cancer Res* (2014) 74:24–30. doi:10.1158/0008-5472.CAN-13-1196
235. Jurgensen JS, Rosenberger C, Wiesener MS, Warnecke C, Horstrup JH, Grafe M, et al. Persistent induction of HIF-1alpha and -2alpha in cardiomyocytes and stromal cells of ischemic myocardium. *FASEB J* (2004) 18:1415–7. doi:10.1096/fj.04-1605fje
236. Dong F, Khalil M, Kiedrowski M, O'Connor C, Petrovic E, Zhou X, et al. Critical role for leukocyte hypoxia inducible factor-1alpha expression in post-myocardial infarction left ventricular remodeling. *Circ Res* (2010) 106:601–10. doi:10.1161/CIRCRESAHA.109.208967
237. Dehn S, DeBerge M, Yeap XY, van-Charvet LY, Fang D, Eltzschig HK, et al. HIF-2alpha in resting macrophages tempers mitochondrial reactive oxygen species to selectively repress MARCO-dependent phagocytosis. *J Immunol* (2016) 197:3639–49. doi:10.4049/jimmunol.1600402
238. Imtiyaz HZ, Williams EP, Hickey MM, Patel SA, Durham AC, Yuan LJ, et al. Hypoxia-inducible factor 2alpha regulates macrophage function in mouse models of acute and tumor inflammation. *J Clin Invest* (2010) 120:2699–714. doi:10.1172/JCI39506
239. Takeda N, O'Dea EL, Doedens A, Kim JW, Weidemann A, Stockmann C, et al. Differential activation and antagonistic function of HIF-{alpha} isoforms in macrophages are essential for NO homeostasis. *Genes Dev* (2010) 24:491–501. doi:10.1101/gad.1881410
240. Choe SS, Shin KC, Ka S, Lee YK, Chun JS, Kim JB. Macrophage HIF-2alpha ameliorates adipose tissue inflammation and insulin resistance in obesity. *Diabetes* (2014) 63:3359–71. doi:10.2337/db13-1965
241. Eubank TD, Roda JM, Liu H, O'Neil T, Marsh CB. Opposing roles for HIF-1alpha and HIF-2alpha in the regulation of angiogenesis by mononuclear phagocytes. *Blood* (2011) 117:323–32. doi:10.1182/blood-2010-01-261792
242. Ridker PM, Everett BM, Thuren T, MacFadyen JG, Chang WH, Ballantyne C, et al. Antiinflammatory therapy with canakinumab for atherosclerotic disease. *N Engl J Med* (2017) 377:1119–31. doi:10.1056/NEJMoa1707914
243. Ridker PM, Thuren T, Zalewski A, Libby P. Interleukin-1beta inhibition and the prevention of recurrent cardiovascular events: rationale and design of the Canakinumab Anti-inflammatory Thrombosis Outcomes Study (CANTOS). *Am Heart J* (2011) 162:597–605. doi:10.1016/j.ahj.2011.06.012

244. Hasan AS, Luo L, Yan C, Zhang TX, Urata Y, Goto S, et al. Cardiosphere-derived cells facilitate heart repair by modulating M1/M2 macrophage polarization and neutrophil recruitment. *PLoS One* (2016) 11:e0165255. doi:10.1371/journal.pone.0165255
245. Harel-Adar T, Ben Mordechai T, Amsalem Y, Feinberg MS, Leor J, Cohen S. Modulation of cardiac macrophages by phosphatidylserine-presenting liposomes improves infarct repair. *Proc Natl Acad Sci U S A* (2011) 108:1827–32. doi:10.1073/pnas.1015623108
246. Ben-Mordechai T, Kain D, Holbova R, Landa N, Levin LP, Elron-Gross I, et al. Targeting and modulating infarct macrophages with hemin formulated in designed lipid-based particles improves cardiac remodeling and function. *J Control Release* (2017) 257:21–31. doi:10.1016/j.jconrel.2017.01.001

Conflict of Interest Statement: The authors declare that the research was conducted in the absence of any commercial or financial relationships that could be construed as a potential conflict of interest.

18

Proteinase 3 Interferes with C1q-Mediated Clearance of Apoptotic Cells

Pascale Tacnet-Delorme[1], Julie Gabillet[2,3,4], Simon Chatfield[2,3,4,5], Nathalie Thieblemont[2,3,4,5], Philippe Frachet[1] and Véronique Witko-Sarsat[2,3,4,5]*

[1] Université Grenoble Alpes, CEA, CNRS, IBS, Grenoble, France, [2] INSERM U1016, Cochin Institute, Paris, France, [3] CNRS UMR 8104, Paris, France, [4] Université Paris-Descartes, Sorbonne Paris Cité, Paris, France, [5] Center of Excellence, LABEX Inflamex, Paris, France

Edited by:
Kirsten Lauber, Ludwig-Maximilians-Universität München, Germany

Reviewed by:
Uday Kishore, Brunel University London, United Kingdom
Ian Dransfield, University of Edinburgh, United Kingdom

***Correspondence:**
Philippe Frachet
philippe.frachet@ibs.fr

Proteinase 3 (PR3) is the autoantigen in granulomatosis with polyangiitis, an autoimmune necrotizing vasculitis associated with anti-neutrophil cytoplasmic antibodies (ANCAs). Moreover, PR3 is a serine protease whose membrane expression can potentiate inflammatory diseases such as ANCA-associated vasculitis and rheumatoid arthritis. During apoptosis, PR3 is co-externalized with phosphatidylserine (PS) and is known to modulate the clearance of apoptotic cells through a calreticulin (CRT)-dependent mechanism. The complement protein C1q is one mediator of efferocytosis, the clearance of altered self-cells, particularly apoptotic cells. Since PR3 and C1q are both involved in the clearance of apoptotic cells and immune response modulation and share certain common ligands (i.e., CRT and PS), we examined their possible interaction. We demonstrated that C1q binding was increased on apoptotic rat basophilic leukemia (RBL) cells that expressed PR3, and we demonstrated the direct interaction between purified C1q and PR3 molecules as shown by surface plasmon resonance. To better understand the functional consequence of this partnership, we tested C1q-dependent phagocytosis of the RBL cell line expressing PR3 and showed that PR3 impaired C1q enhancement of apoptotic cell uptake. These findings shed new light on the respective roles of C1q and PR3 in the elimination of apoptotic cells and suggest a novel potential axis to explore in autoimmune diseases characterized by a defect in apoptotic cell clearance and in the resolution of inflammation.

Keywords: proteinase 3, C1q, apoptotic cells, efferocytosis, autoimmunity

INTRODUCTION

Proteinase 3 (PR3) is a neutrophil-derived serine protease located together with its homologs, human neutrophil elastase, and cathepsin G, in azurophilic granules as reviewed in Martin and Witko-Sarsat (1). One particular feature of PR3 is its affinity for membranes leading to its surface expression on viable and apoptotic neutrophils (2). Indeed, PR3 possesses a unique membrane insertion domain composed of four basic (R193, R194, K195, and R227) and four hydrophobic (F180, F181, L228, and F229) amino acids that allows it to anchor to the membrane (3, 4). During neutrophil activation

---

Abbreviations: ANCA, anti-neutrophil cytoplasmic antibody; CRT, calreticulin; cC1q, collagenous-like fragment of C1q; gC1q, globular region of C1q; GPA, granulomatosis with polyangiitis,; PS, phosphatidylserine; PLSCR1, phospholipid scramblase 1; PR3, proteinase 3; RBL, rat basophilic leukemia; SPR, surface plasmon resonance.

and apoptosis, membrane expression of PR3 increases, and soluble PR3 is also released into the extracellular environment during degranulation (4). PR3 is a pro-inflammatory factor whose membrane expression can potentiate chronic inflammatory diseases such as anti-neutrophil cytoplasmic antibodies (ANCAs) systemic vasculitis (AAV) and rheumatoid arthritis (5). PR3 has been characterized as the autoantigen in granulomatosis with polyangiitis (GPA) (6). During apoptosis, PR3 is co-externalized with phosphatidylserine (PS) *via* its association with phospholipid scramblase 1 (4). Furthermore, it has been proposed that PR3 can modulate apoptotic cell clearance (7) through a mechanism linked to the ability of PR3 to associate with calreticulin (CRT), a protein involved in apoptotic cell recognition and an important "eat-me" signal (8).

Apoptotic cells release "find-me" signals that recruit phagocytes that will recognize, engulf, and degrade them (9, 10) promoting a monocyte/macrophage program that promotes inflammation resolution, tissue repair, and wound healing (11). The function of the complement protein C1q, well known for its role in innate immunity, has been reconsidered over the past 15 years with evidence that it is one mediator of efferocytosis, the mechanism of clearance of altered self-cells and of apoptotic cells in particular (12, 13). C1q serves as a physical bridge between the phagocyte and its prey. Numerous C1q-binding molecules on both sides of the phagocytic synapse have been characterized (14). Among these, cell surface CRT and PS have also been characterized as PR3 partners (7). C1q is a hexamer of heterotrimers, which consists of two typical regions and a collagenous-like fragment of C1q (cC1q) from which six globular regions (GR) [globular region of C1q (gC1q)] emerge. gC1q is involved in the specific recognition of apoptotic cells, and cC1q has primarily been described in C1q recognition by phagocyte membranes (15). However, as the C1q collagenous tail (cC1q) is known to interact with several membrane receptors (14), widely distributed on various cell types, C1q can enter into a vast array of interactions by binding of its heads or/and its stalks depending on their accessibility in a particular situation. Of note, C1q deficiency is strongly associated with autoimmune diseases, such as systemic lupus erythematosus (SLE) and glomerulonephritis and may be associated with compromised removal of apoptotic cells (16).

One other major effect of C1q modulation concerns its function in regulating immune cells, independently from efferocytosis. This includes the role of C1q in neutrophil function. It has previously been shown that the C1q–CRT interaction modulates cytokine release by macrophages, and CRT is released from activated neutrophils (17). Thus, it might be hypothesized that C1q–CRT interaction could also interfere in neutrophil-mediated inflammatory processes.

Given the evidence that PR3 and C1q are involved in both immune response modulation and in clearance of apoptotic cells and share common ligands (i.e., CRT and PS) (7, 18, 19), this study was designed to examine their possible interaction. We investigated C1q binding to apoptotic neutrophils and showed the direct interaction between purified C1q and PR3. To better understand the functional consequence of this partnership, we tested the C1q-dependent phagocytosis of rat basophilic leukemia (RBL) cell line expressing PR3.

These findings shed new light on the respective role of C1q and PR3 in the elimination of apoptotic cells. A number of autoimmune diseases are characterized by defects in apoptotic cell clearance, and this novel potential axis may play a role in the appropriate resolution of inflammation.

MATERIALS AND METHODS

Proteins, Antibodies

C1q was purified from human serum, C1q GR, and the collagen-like region (CLF) were prepared and quantified as described previously (20). Rabbit polyclonal antibody directed against human C1q was from IRPAS group (IBS, Grenoble, France). Mouse monoclonal antibody against C1q (A201) was from Quidel (San Diego, CA, USA), and mouse monoclonal anti-PR3 (clone CLB12.8) was from Sanquin (Amsterdam, Holland). Ficolin 3 was obtained from Nicole Thielens's team (IBS, Grenoble, France). PR3 was from Athens Research and Technology.

Blood Cell Isolation, Cells Culture, and Apoptosis Induction

This study was carried out and approved in accordance with the recommendations of the INSERM Institutional Review Board and the Cochin Hospital Ethics Committee (Paris, France). Blood from healthy donors was provided by the Etablissement Français du Sang (Paris, France). Human neutrophils were isolated from EDTA-anticoagulated healthy donor blood using density-gradient centrifugation through polymorphoprep (Nycomed) as previously described (5). To induce physiologic apoptosis, neutrophils were resuspended at 2×10^6/ml in RPMI (Gibco) supplemented with 10% fetal calf serum (FCS) and kept for 16 h at 37°C in a humidified 5% CO_2 atmosphere. RBL and RBL-PR3 used in this study refers to a previously described cell line (21), transfected with pcDNA3 plasmid alone, or pcDNA3/PR3, respectively. To induce the differentiation of THP1 monocyte cells to macrophages, the cells were treated with 10 nM PMA for 72 h as previously described (20). Apoptosis of RBL cells was induced as follows: briefly, cells were grown in sterile dishes overnight to 60–80% confluence and exposed to 1,000 mJ/cm^2 UV-B irradiation at 312 nm in a fresh DMEM medium. Cells were then incubated for 17 h at 37°C under 5% CO_2. Measurement of apoptosis was performed by flow cytometry using the FITC-Annexin V Kit (MACS Miltenyi Biotec) according to the manufacturer's instructions.

PR3 and C1q Immunolabeling

RBL cells were washed with PBS and then incubated with C1q (80 μg/ml) during 40 min at 4°C. For all experiments on RBL cells, FcγR blockade was performed with the FcγR-blocking solution (Miltenyi Biotec). Then cells were incubated for 30 min with mouse anti-C1q Abs (A201, Quidel) diluted 1:100. PR3 was detected by a monoclonal anti-PR3 Abs (CLB12.8) at 2 μg/ml. Bound antibodies were visualized with Alexa 488-conjugated goat anti-mouse IgG or cyanine-3 (Cy3) rat anti-mouse IgG. Alexa488-labeled C1qGR and Alexa 647-labeled C1q were prepared using the AlexaFluor-488 and AlexaFluor-647 labeling kits (Invitrogen, ThermoFisher Scientific). For experiments on PMN by confocal microscopy, cells were induced to adhere to poly-L-lysine-precoated coverslip cells. Fixed PMN were incubated with FcγR-blocking solution and then with anti-PR3 Abs, then incubated with Cy3-conjugated anti-mouse IgG before labeling with Alexa488-C1qGR or with Alexa 488-conjugated rat anti-mouse IgG before labeling with Alexa647-C1q (80 μg/ml). Cell slides were mounted glass slides using Vectashield solution with 4′,6′-diamidino-2-phenylindole (Vector Laboratories) and were visualized under a laser spinning-disk confocal microscope (Olympus and Andor, M4D cell imaging platform, IBS). Data were evaluated with Volocity software.

Uptake of Apoptotic Cells

Rat basophilic leukemia cells were labeled with CFSE (CellTraceTM CFSE Cell Proliferation Kit, Life Technologies) as follows: cells were washed twice and then resuspended at 1×10^6 cells/ml in PBS and incubated with 1 μM CFSE at 37°C for 20 min. The remaining CFSE was quenched with the addition of DMEM-10% FCS for at least 10 min. Cells were then pelleted by centrifugation and resuspended in DMEM-10% FCS before the induction of apoptosis. THP1 cells were labeled with PKH26 dye (Sigma-Aldrich) before PMA induction. Apoptotic RBL cells were then added to THP1-derived macrophages that had been preincubated or not with C1q, at a ratio of 5:1 (RBL:THP1) for 1 h at 37°C, 5% CO_2 in RPMI medium supplemented or not with 10% of decomplemented FCS. After incubation, cells were washed and harvested with 0.25% trypsin/EDTA and analyzed by flow cytometry (MACSQuant VYB Cytometer, Miltenyi Biotec), and collected data were treated with MACSQuantify software. Phagocytosis was calculated as the percentage of the double CFSE and PKH26 labeled cells in the THP1 macrophage population. Phagocytosis negative controls performed at 4°C or at 37°C in the presence of 5 μM of cytochalasin were subtracted. Significance was tested using non-parametric Wilcoxon signed-rank test for paired samples.

Surface Plasmon Resonance (SPR) Spectroscopy

Analyses were carried out on a BIAcore 3000 instrument (BIAcore, GE Healthcare). 100 μg of purified PR3 was resuspended in H_2O to give a final concentration of 1 mg/ml. The running buffer for PR3 immobilization was 10 mM HEPES, 145 mM NaCl, 5 mM EDTA, pH 7.4. PR3 was diluted in acetate buffer (pH 4.5) to achieve a final concentration of 50 μg/ml and was immobilized onto a CM5 sensor chip (GE Healthcare) using the BIAcore amine coupling kit. Binding of C1q, GR, or CLF to immobilized PR3 was measured at a flow rate of 20 ml/min in the running buffer of 50 mM Tris, 150 mM NaCl, 2 mM Ca^{2+} containing 0.005% surfactant P20 (pH 7.4). Surfaces were regenerated by one injection of 5 μl of 20 mM NaOH. The specific binding signal shown was obtained by subtracting the background signal, obtained by injection of the sample over an activated-deactivated surface. Data were analyzed by global fitting to a 1:1 Langmuir binding model of the association and dissociation phases for several concentrations of PR3, using the BIAevaluation 3.2 software (GE Healthcare) and were obtained with a statistic χ^2 value < 2. The apparent equilibrium dissociation constants (K_D) were calculated from the ratio of the dissociation and association rate constants (k_{off}/k_{on}).

PR3 Proteolytic Activity

Proteinase 3 activity on fibronectin (FN) and C1q was analysis by the method described by Rao et al. (22). Briefly, digestion of purified FN and C1q by PR3 was performed in 0.15 M NaCl, 0.002 M $CaCl_2$, 0.01 M HEPES buffer, pH 7.4 at 37°C for 18 h, with an enzyme/substrate molecular ratio of 1:25. The samples were analyzed by SDS-PAGE under reducing conditions. The gel was stained with Coomassie Blue R-250.

RESULTS

C1q Binding to Apoptotic Cells Increases With Surface PR3 Exposure

We first examined the colocalization of C1q and PR3 at the surface of apoptotic neutrophils from healthy human donors. As shown in **Figure 1**, under physiologic apoptosis induction, apoptotic neutrophils were recognized by C1q and PR3 partially colocalized with C1q. Similar colocalization was also observed using the C1q globular head (not shown). As it has been shown that PR3 expression on the neutrophil population is non-homogenous and is characterized by important interindividual variability, further experiments were conducted using the previously established RBL cell line that expresses PR3 at the cell surface during apoptosis (7). Late apoptosis was induced in RBL-PR3 and RBL (transfected with the empty plasmid) cells using UVB irradiation and demonstrated by double Annexin V/PI labeling (**Figure 2A**). Immunolabeling experiments of RBL cells confirmed that wild-type RBL cells did not express PR3 (**Figure 2B**, blue curves). PR3 was significantly externalized at the surface of apoptotic RBL-PR3 cells whereas almost no PR3 was detected on untreated PR3-RBL cells (**Figure 2B**, PR3 labeling, red curves). In addition, we analyzed the C1q binding to these cell populations. As expected from its capacity to recognize apoptotic bodies, C1q bound to both RBL and PR3-RBL cell subsets which appeared after UV irradiation (**Figure 2C**, P3 subset). However, we observed an increase of the C1q binding to apoptotic PR3-RBL compared with RBL cells (**Figures 2C,D**). Together, these observations prompted us to investigate the direct PR3–C1q interaction.

Purified C1q Binds PR3

To further investigate whether there is a direct interaction between PR3 and C1q, SPR experiments were performed using purified PR3 immobilized on a sensor chip. Intact C1q, its globular heads (gC1q), or its collagenous tail (cC1q) were used as soluble ligands. As illustrated in **Figure 3**, intact C1q, its gC1q and cC1q regions, all bound to immobilized PR3 while no interaction was detected using the C1q-related protein Ficolin 3 (data not shown). The kinetic parameters of PR3–C1q interaction were determined by recording sensorgrams at varying ligand concentrations. For intact C1q and its globular heads, the kinetic (k_a, k_d) and dissociation (K_D) constants were calculated with a simple 1:1 Langmuir

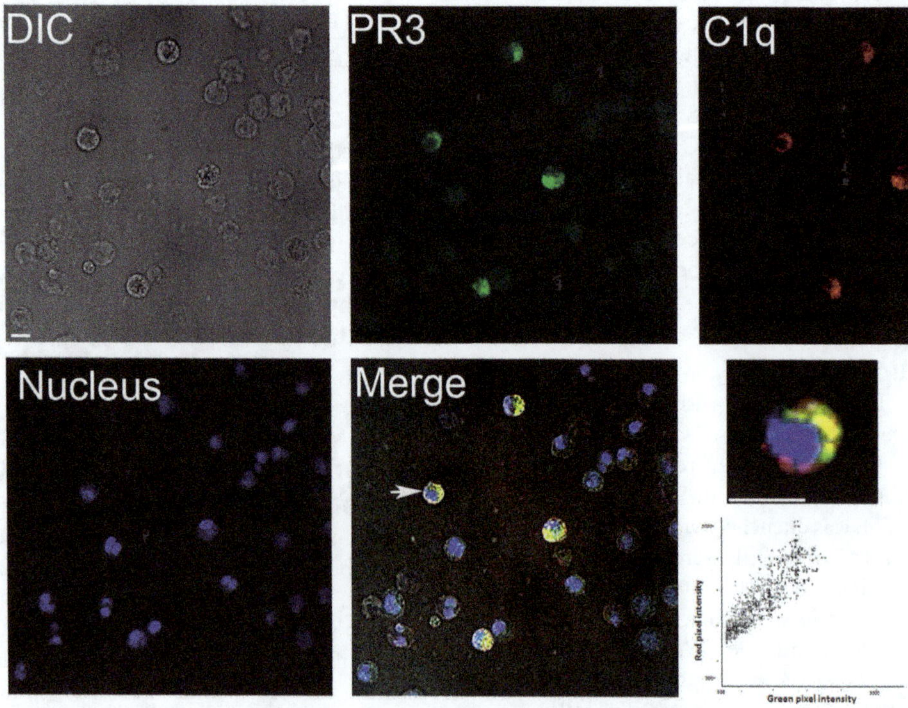

FIGURE 1 | Colocalization of proteinase 3 (PR3) and C1q in neutrophils. Apoptotic neutrophils (prepared as described in Section "Materials and Methods") were double labeled for membrane PR3 expression and C1q binding using an anti-PR3 mAb followed by an Alexa488-conjugated anti-mouse IgG and then incubating cells with Alexa647-C1q. Nuclei were labeled with DAPI. Samples were visualized by confocal microscopy under differential interference contrast (DIC), DAPI, A488, and A647 filters, and merge is shown (as indicated). Higher magnification is shown for one selected cell indicated by a white arrow. Scatterplot of red and green pixel intensities collected from the focal plane of the cell shown is represented. Scale bar 8 μm. Colocalization (yellow regions in merge) was evaluated by Pearson's correlation coefficient ≥0.9.

binding model (**Table 1**) with K_D values of 1.9×10^{-7} and 4.0×10^{-7} M, respectively. Thus, the K_D value of the interaction with PR3 was significantly lower for the full-length C1q compared with its isolated GR, accounting for an increased binding avidity of C1q arising from its hexameric structure. In addition, cC1q bound in a dose-dependent manner to PR3. Although the curve fitting was not adequate to determine accurately the kinetic value of this interaction, this result suggests that components of the collagen tail of C1q could also be involved in the interaction.

PR3 Impaired C1q Enhancement of Apoptotic Cell Uptake

To analyze the possible PR3/C1q-dependent effect on the uptake of apoptotic cells by macrophages, we measured the capacity of PMA-stimulated THP1 cells to phagocytose PR3-expressing RBL cells. Late apoptotic RBL cells or RBL-PR3 cells were added to THP1 macrophages, and their uptake was determined by flow cytometry. To assess the effect of C1q on phagocytosis specifically, whole C1q at a physiological concentration (80 μg/ml) was added to opsonize apoptotic cells before macrophage–target cell contact. RBL cells were efficiently engulfed by THP1 macrophages, and the phagocytosis increased significantly with C1q either in presence or absence of serum (**Figure 4**). Remarkably, this C1q enhancement of phagocytosis was abolished for PR3-expressing cells as we did not observe any effect of C1q on the uptake of RBL-PR3 cells. In addition, we observed that apoptotic RBL-PR3 cells were more readily phagocytosed by THP1 macrophages than the RBL cells in absence of serum ($P = 0.043$, $n = 5$). Interestingly, this difference is not apparent when experiments were done in presence of serum ($P = 0.108$, $n = 4$). This is possibly due to the presence of serum protein(s) that could interfere with PR3.

C1q Was Not Proteolytically Cleaved by PR3

As PR3 is able to cleave a broad range of matrix proteins such as elastin, FN, or laminin, we tested the hypothesis that PR3 could degrade C1q and thus impair its ability to enhance phagocytosis. After 18 h of incubation, PR3 did not degrade C1q but efficiently cleaved FN used as a control (**Figure 5**). This suggests that proteolytic cleavage of C1q was not the mechanism explaining the failure of C1q to enhance uptake of apoptotic RBL-PR3 cells.

DISCUSSION

This study provides the first experimental evidence that C1q, well known to be involved in the recognition of apoptotic cells and their uptake by phagocytes interacts with the neutrophil-specific serine protease PR3 with potentially relevant consequences for apoptotic cell uptake. Our conclusion is based on the following observations: (1) C1q and PR3 colocalize at the surface of

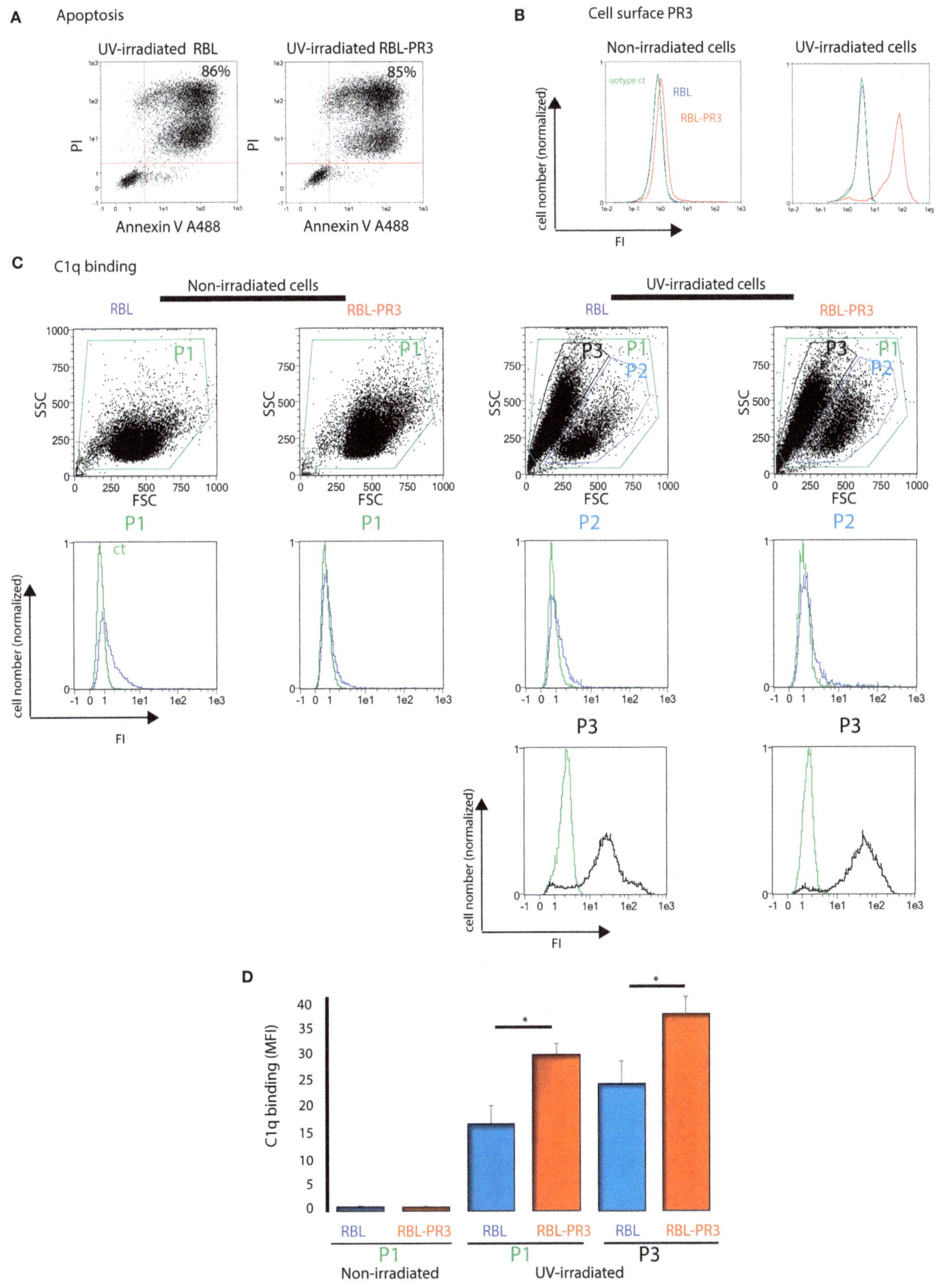

FIGURE 2 | Continued

FIGURE 2 | C1q binding increases on apoptotic cells that express proteinase 3 (PR3) on their surface. **(A)** Apoptosis of rat basophilic leukemia (RBL) and RBL-PR3 was analysed by double Annexin V/PI labeling. FSC/SSC dot plot of the irradiated population is shown, and gating strategy used for AV/PI analysis is indicated. Percentages correspond to the AV/PI positive population. **(B)** Untreated and UV-B irradiated RBL and RBL-PR3 cells were labeled with anti-PR3 antibody. Isotype controls in green for PR3-RBL cells are shown. **(C)** Cells were incubated with soluble C1q, and its binding was detected using a monoclonal antibody against C1q. Representative experiments are shown. Blue curves correspond to C1q binding on P1 gate (non-irradiated cells) or on P2 gate (UV-irradiated cells). Black curves represent the fluorescence of the P3 gate which appears with apoptosis; green curves represent the fluorescence of cells in the same gate in the absence of C1q. **(D)** Quantification of the C1q binding as shown in panel **(C)**, done on the P1 and P3 subsets (three independent experiments, *$P < 0.05$, Student's t-test). Analyses were monitored by flow cytometry as described in Section "Materials and Methods." Abbreviations: FI, fluorescence intensity; MFI, median fluorescence intensity; SSC, side scatter; FSC, forward scatter.

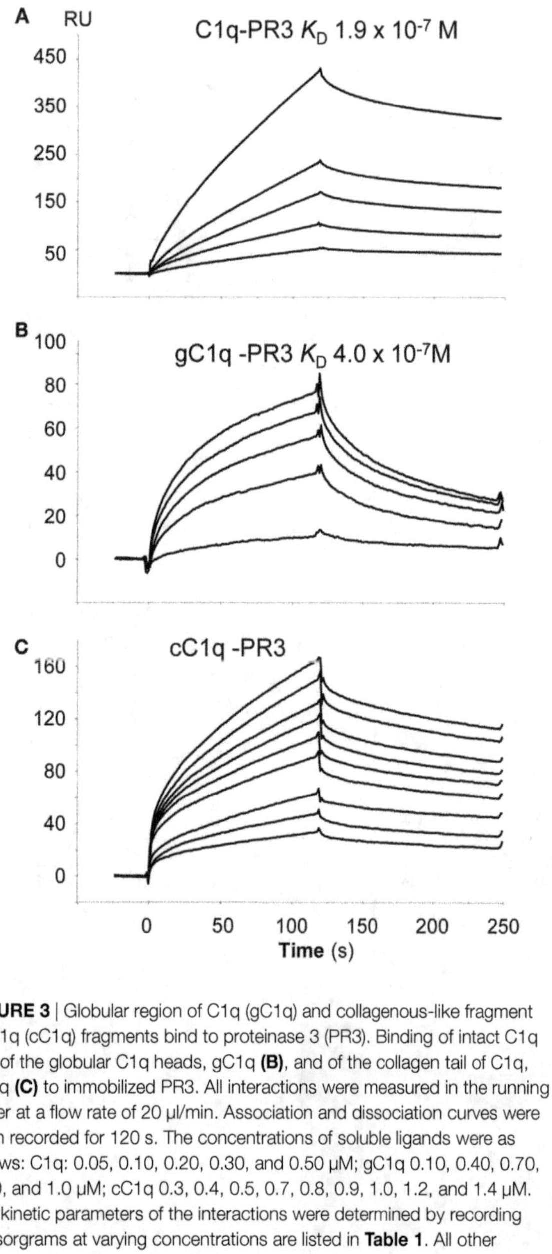

FIGURE 3 | Globular region of C1q (gC1q) and collagenous-like fragment of C1q (cC1q) fragments bind to proteinase 3 (PR3). Binding of intact C1q **(A)**, of the globular C1q heads, gC1q **(B)**, and of the collagen tail of C1q, cC1q **(C)** to immobilized PR3. All interactions were measured in the running buffer at a flow rate of 20 µl/min. Association and dissociation curves were each recorded for 120 s. The concentrations of soluble ligands were as follows: C1q: 0.05, 0.10, 0.20, 0.30, and 0.50 µM; gC1q 0.10, 0.40, 0.70, 0.80, and 1.0 µM; cC1q 0.3, 0.4, 0.5, 0.7, 0.8, 0.9, 1.0, 1.2, and 1.4 µM. The kinetic parameters of the interactions were determined by recording sensorgrams at varying concentrations are listed in **Table 1**. All other conditions are described in Section "Materials and Methods."

TABLE 1 | Kinetic constants for the binding of C1q to immobilized proteinase 3 (PR3).

Soluble analyte	PR3		
	k_{on} (M^{-1} s^{-1})	k_{off} (s^{-1})	K_D (M)
C1q	7.8×10^3	1.51×10^{-3}	1.9×10^{-7}
gC1q (globular domain)	1.8×10^4	7.2×10^{-3}	4.0×10^{-7}
cC1q (collagenous-like fragment)	nd	nd	nd

Binding of C1q, its globular region of C1q (gC1q), and collagenous-like fragment of C1q (cC1q) to PR3 were measured as described in Section "Materials and Methods." The association (k_{on}) and dissociation (k_{off}) rate constants were determined by global fitting of the data using a 1:1 Langmuir binding model (A + B ⇔ AB, BIAevaluation 3.2). The dissociation constants K_D were determined from the k_{off}/k_{on} ratios. The data presented were obtained with a statistic χ^2 value < 2.

ND, not determined (i.e., despite the dose-dependent response the K_D value cannot be determined satisfactorily).

neutrophils; (2) C1q binds more efficiently apoptotic RBL cells when they expose PR3 at their surface and; (3) C1q recognizes PR3 with a sub-micromolar affinity ($K_D = 1.9 \times 10^{-7}$ M) as shown by SPR analysis. The ability of C1q to bind PR3 on the cell surface is further substantiated by the effect of PR3 exposure on the phagocytosis of apoptotic cells in the presence of C1q. Remarkably, when apoptotic RBL cells express cell surface PR3, the C1q effect on their uptake is undetectable in contrast to the wild-type RBL cells and what has previously been observed with other apoptotic cell types (13, 23). The mechanism of inhibition does not appear to be simple cleavage of C1q by the protease activity of PR3. It should be noted that in this phagocytosis model, PR3 expression did not impair the uptake of apoptotic cells even in absence of C1q. However, together with the characterization of the PR3–C1q direct interaction, it might suggest that PR3 could affect C1q-mediated functions.

Interestingly, we have previously observed using an *in vivo* phagocytosis assay of RBL cells in mice, that PR3 membrane expression efficiency decreases the phagocytosis of apoptotic cells by peritoneal macrophages (7) in contrast to the increased phagocytosis *in vitro* using macrophages differentiated from the THP1 monocyte cell line in this study. We hypothesize that PR3 could bind to other mediators implicated in the phagocytosis process, e.g., serum proteins such as C1q. Our present study showing that C1q-mediated enhancement of phagocytosis is impaired when cells expose PR3 on their surface supports this hypothesis and strongly suggests that PR3 binding to C1q disables C1q-mediated phagocytic macrophage function. We have also observed differences between phagocytosis in the presence and absence of serum, indicating that other serum molecules could interfere with this process in the specific tissue environment. Of note, it has been demonstrated that C1q and PR3 share common binding partners involved in efferocytosis, the well-known "eat-me signals" and immune modulators CRT and PS (7, 18, 19, 24, 25).

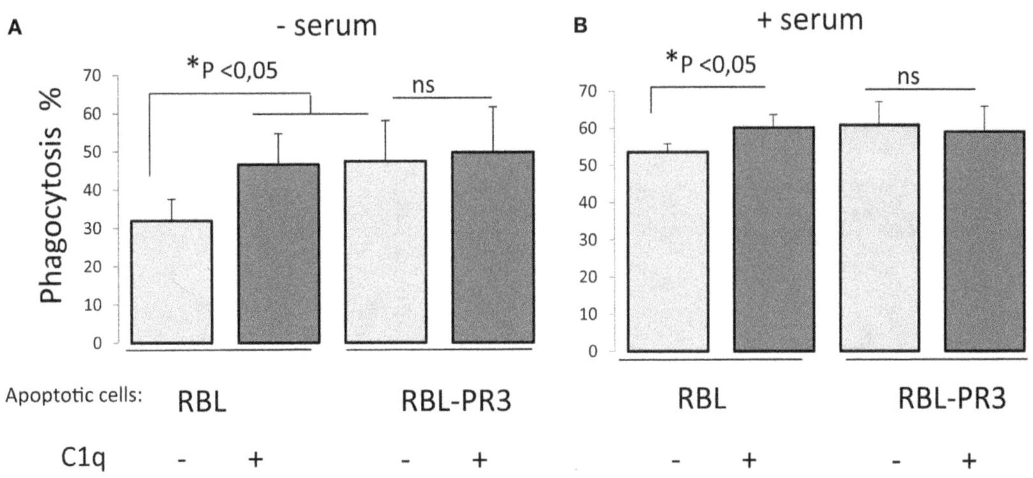

FIGURE 4 | Surface proteinase 3 (PR3) exposure supresses C1q enhancement of apoptotic cell phagocytosis. Late apoptotic rat basophilic leukemia (RBL) cells labeled with CFSE, opsonized or not with C1q (80 µg/ml), were incubated with PMA-treated THP1 cells labeled with PKH26 dye, 1 h at 37°C in absence **(A)** or in presence of serum **(B)**. Phagocytosis is expressed as the percentage of the double-labeled cells in the macrophage population (i.e., PKH26 and CFSE labeled cells). Data are the mean ± SD of independent experiments. *Significance was tested using non-parametric two tail Wilcoxon signed-rank test for paired samples, $n = 5$ (without serum) and $n = 4$ (with serum).

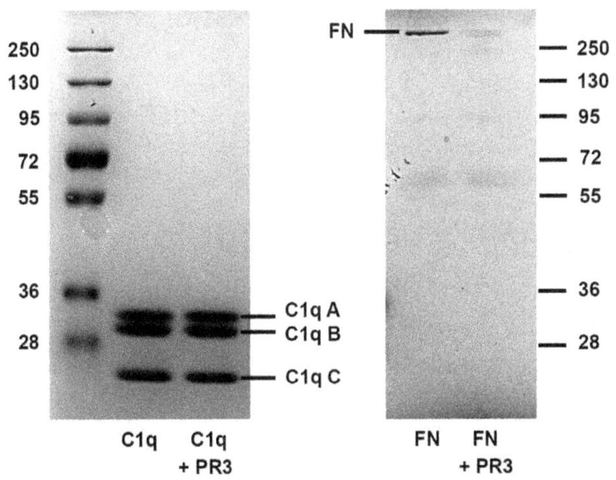

FIGURE 5 | C1q is not cleaved by proteinase 3 (PR3). Fibronectin (FN) and C1q were submitted to PR3 digestion at 37°C for 18 h. The digestion products were separated on a 10% SDS-PAGE gel under reducing conditions and visualized by Coomassie staining. The position of the three (A–C) subunits of C1q and of FN is shown.

Importantly, we have previously shown that the ectoCRT–C1q interaction modulates the efferocytosis process. As co-externalization of CRT and PR3 occurs during neutrophil apoptosis, this presumably could interfere with the C1q-dependent PR3 effect on phagocytosis. These findings should be taken into consideration when interpreting the role of PR3 under pathophysiological conditions.

Neutrophils play a key role in the molecular pathology of number of autoimmune diseases since these cells can generate neo-epitopes that have the potential to break immune tolerance resulting in the generation of autoantibodies (26). Of particular interest, PR3 is the major target of ANCA in GPA, and a high percentage of neutrophils bearing membrane PR3 is considered a risk factor for autoimmune vasculitis. Indeed, we have previously proposed that PR3 can inhibit the clearance of apoptotic neutrophils by phagocytes, acting as a "don't eat me" signal by interfering with the ability of CRT to promote phagocytosis (7) and perturbing the normally anti-inflammatory response following phagocytosis of apoptotic cells by macrophages (27). Furthermore, PR3 acts as an alarmin inducing pro-inflammatory cytokine and chemokine production by macrophages and dendritic cell activation leading to dysregulated T cell polarization and favoring autoimmunity (27). Another key protein of immune tolerance is C1q through its facilitation of apoptotic cell clearance. Indeed, C1q deficiency is linked to development of autoimmune diseases including SLE, probably through the alteration of efferocytosis (28). Genetic variants in the region of the C1q genes have also been associated with rheumatoid arthritis (29). Interestingly, the frequency of circulating PR3-high neutrophils also increases in patients with rheumatoid arthritis (5). Our current data reinforce the idea that the PR3 pro-inflammatory effect could be mediated by its ability to specifically bind molecules involved in the safe removal of apoptotic cells. On the other hand, the variability of PR3 expression on the neutrophil population raises the question of its physiological role. It is tempting to speculate that PR3 membrane exposure could be associated with subpopulations of neutrophils that exhibit specific regulatory functions linked to innate clearance or inflammatory response.

In summary, together with previously published data, our findings highlight the existence of interactions between key serum and tissue proteins involved in the efferocytosis process (recognition and/or phagocyte signaling pathways) including multivalent proteins (e.g., C1q) with a propensity to aggregate molecules (**Figure 6**). We hypothesize that regulation of the immune response and the initiation of pathological events may be dependent on

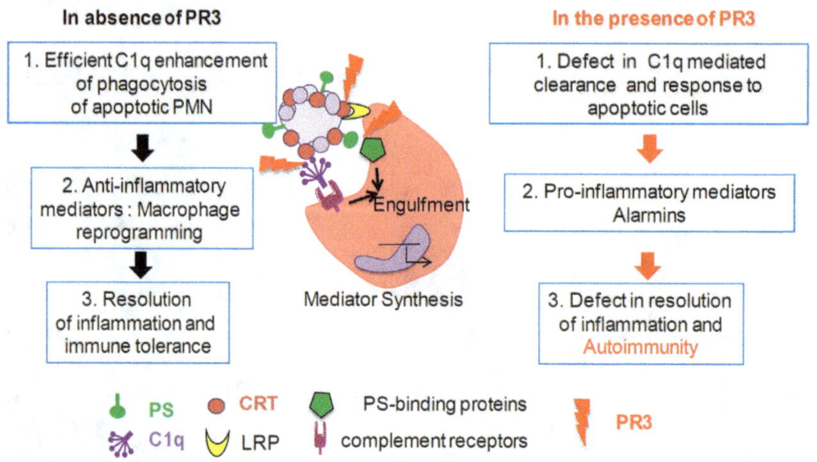

FIGURE 6 | Proteinase 3 (PR3) expression on apoptotic cells decreases C1q-mediated phagocytosis. *In absence of membrane PR3*, C1q opsonization of apoptotic cells through apoptotic cell-associated molecular pattern recognition [such as phosphatidylserine (PS), calreticulin (CRT), and others] enhances phagocytosis through macrophage receptors (such as LRP and others) and triggers anti-inflammatory responses leading to the resolution of inflammation. *In the presence of membrane PR3*, expression on apoptotic cells, PR3 that binds PS, CRT, and C1q, could interfere with the multimolecular complexes at the interface between apoptotic neutrophils and macrophage, delaying or affecting phagocytosis of apoptotic cells, activating a pro-inflammatory immune response favoring autoimmunity.

the relative abundance of these proteins. Efficient apoptotic body removal has important inflammatory consequences, and these molecules may be important to target in neutrophil-associated autoimmune disorders.

ETHICS STATEMENT

This study was carried out and approved in accordance with the recommendations of the INSERM Institutional Review Board and the Cochin Hospital Ethics Committee (Paris, France). Blood from healthy donors was provided by the Etablissement Français du Sang (Paris, France).

AUTHOR CONTRIBUTIONS

PF and VW-S designed the study and analyzed and interpreted data. JG, NT, and PT-D contributed to the study design and acquired, analyzed, and interpreted data. PF, NT, SC, and VW-S wrote the manuscript. All the authors approved the submitted version.

ACKNOWLEDGMENTS

The authors are grateful to Nicole Thielens for having provided them the Ficolin 3. The authors greatly acknowledge the Cochin Cytometry and Immunobiology Facility at the Cochin Institute and the Etablissement Français du Sang (Centre Crozatier, Saint-Antoine Hospital, Paris France).

FUNDING

This work was supported in part by the French National Research Agency (ANR-C1qEffero-11-16) (PF); the platforms of the Grenoble Instruct—ERIC Center: ISBG; UMS 3518 CNRS-CEA-UGA-EMBL (PF) with support from FRISBI: ANR-10-INSB-05-02 (PF) and GRAL: ANR-10-LABX-49-01 (PF) within the Grenoble Partnership for Structural Biology (PSB); and Investissements d'Avenir programme ANR-11-IDEX-0005-02, Sorbonne Paris Cite, Labex INFLAMEX (VW-S), the Chancellerie des Universités de Paris (Legs Poix; VW-S), and the Arthritis Foundation (VW-S).

REFERENCES

1. Martin KR, Witko-Sarsat V. Proteinase 3: the odd one out that became an autoantigen. *J Leukoc Biol* (2017) 102(3):689–98. doi:10.1189/jlb.3MR0217-069R
2. Hajjar E, Broemstrup T, Kantari C, Witko-Sarsat V, Reuter N. Structures of human proteinase 3 and neutrophil elastase – so similar yet so different. *FEBS J* (2010) 277(10):2238–54. doi:10.1111/j.1742-4658.2010.07659.x
3. Hajjar E, Mihajlovic M, Witko-Sarsat V, Lazaridis T, Reuter N. Computational prediction of the binding site of proteinase 3 to the plasma membrane. *Proteins* (2008) 71(4):1655–69. doi:10.1002/prot.21853
4. Kantari C, Pederzoli-Ribeil M, Amir-Moazami O, Gausson-Dorey V, Moura IC, Lecomte MC, et al. Proteinase 3, the Wegener autoantigen, is externalized during neutrophil apoptosis: evidence for a functional association with phospholipid scramblase 1 and interference with macrophage phagocytosis. *Blood* (2007) 110(12):4086–95. doi:10.1182/blood-2007-03-080457
5. Witko-Sarsat V, Lesavre P, Lopez S, Bessou G, Hieblot C, Prum B, et al. A large subset of neutrophils expressing membrane proteinase 3 is a risk factor for vasculitis and rheumatoid arthritis. *J Am Soc Nephrol* (1999) 10(6):1224–33.
6. Jennette JC, Falk RJ. Pathogenesis of antineutrophil cytoplasmic autoantibody-mediated disease. *Nat Rev Rheumatol* (2014) 10(8):463–73. doi:10.1038/nrrheum.2014.103
7. Gabillet J, Millet A, Pederzoli-Ribeil M, Tacnet-Delorme P, Guillevin L, Mouthon L, et al. Proteinase 3, the autoantigen in granulomatosis with polyangiitis, associates with calreticulin on apoptotic neutrophils, impairs macrophage phagocytosis, and promotes inflammation. *J Immunol* (2012) 189(5):2574–83. doi:10.4049/jimmunol.1200600
8. Gardai SJ, McPhillips KA, Frasch SC, Janssen WJ, Starefeldt A, Murphy-Ullrich JE, et al. Cell-surface calreticulin initiates clearance of viable or apoptotic cells through trans-activation of LRP on the phagocyte. *Cell* (2005) 123(2):321–34. doi:10.1016/j.cell.2005.08.032

9. Poon IK, Lucas CD, Rossi AG, Ravichandran KS. Apoptotic cell clearance: basic biology and therapeutic potential. *Nat Rev Immunol* (2014) 14(3):166–80. doi:10.1038/nri3607
10. Ariel A, Ravichandran KS. 'This way please': apoptotic cells regulate phagocyte migration before and after engulfment. *Eur J Immunol* (2016) 46(7):1583–6. doi:10.1002/eji.201646505
11. Ariel A, Serhan CN. New lives given by cell death: macrophage differentiation following their encounter with apoptotic leukocytes during the resolution of inflammation. *Front Immunol* (2012) 3:4. doi:10.3389/fimmu.2012.00004
12. Hochreiter-Hufford A, Ravichandran KS. Clearing the dead: apoptotic cell sensing, recognition, engulfment, and digestion. *Cold Spring Harb Perspect Biol* (2013) 5(1):a008748. doi:10.1101/cshperspect.a008748
13. Ghebrehiwet B, Hosszu KK, Valentino A, Peerschke EI. The C1q family of proteins: insights into the emerging non-traditional functions. *Front Immunol* (2012) 3:52. doi:10.3389/fimmu.2012.00052
14. Frachet P, Tacnet-Delorme P, Gaboriaud C, Thielens NM. Role of C1q in efferocytosis and self-tolerance – links with autoimmunity. In: Chatzidionysiou K, editor. *Autoimmunity – Pathogenesis, Clinical Aspects and Therapy of Specific Autoimmune Diseases*. Croatia: InTech (2015). p. 21–51.
15. Navratil JS, Watkins SC, Wisnieski JJ, Ahearn JM. The globular heads of C1q specifically recognize surface blebs of apoptotic vascular endothelial cells. *J Immunol* (2001) 166(5):3231–9. doi:10.4049/jimmunol.166.5.3231
16. Botto M, Walport MJ. C1q, autoimmunity and apoptosis. *Immunobiology* (2002) 205(4–5):395–406. doi:10.1078/0171-2985-00141
17. Kishore U, Sontheimer RD, Sastry KN, Zappi EG, Hughes GR, Khamashta MA, et al. The systemic lupus erythematosus (SLE) disease autoantigen-calreticulin can inhibit C1q association with immune complexes. *Clin Exp Immunol* (1997) 108(2):181–90. doi:10.1046/j.1365-2249.1997.3761273.x
18. Paidassi H, Tacnet-Delorme P, Verneret M, Gaboriaud C, Houen G, Duus K, et al. Investigations on the C1q-calreticulin-phosphatidylserine interactions yield new insights into apoptotic cell recognition. *J Mol Biol* (2011) 408(2):277–90. doi:10.1016/j.jmb.2011.02.029
19. Martin KR, Kantari-Mimoun C, Yin M, Pederzoli-Ribeil M, Angelot-Delettre F, Ceroi A, et al. Proteinase 3 is a phosphatidylserine-binding protein that affects the production and function of microvesicles. *J Biol Chem* (2016) 291(20):10476–89. doi:10.1074/jbc.M115.698639
20. Verneret M, Tacnet-Delorme P, Osman R, Awad R, Grichine A, Kleman JP, et al. Relative contribution of c1q and apoptotic cell-surface calreticulin to macrophage phagocytosis. *J Innate Immun* (2014) 6(4):426–34. doi:10.1159/000358834
21. Witko-Sarsat V, Canteloup S, Durant S, Desdouets C, Chabernaud R, Lemarchand P, et al. Cleavage of p21waf1 by proteinase-3, a myeloid-specific serine protease, potentiates cell proliferation. *J Biol Chem* (2002) 277(49):47338–47. doi:10.1074/jbc.M202789200
22. Rao NV, Wehner NG, Marshall BC, Gray WR, Gray BH, Hoidal JR. Characterization of proteinase-3 (PR-3), a neutrophil serine proteinase. Structural and functional properties. *J Biol Chem* (1991) 266(15):9540–8.
23. Fraser DA, Laust AK, Nelson EL, Tenner AJ. C1q differentially modulates phagocytosis and cytokine responses during ingestion of apoptotic cells by human monocytes, macrophages, and dendritic cells. *J Immunol* (2009) 183(10):6175–85. doi:10.4049/jimmunol.0902232
24. Paidassi H, Tacnet-Delorme P, Garlatti V, Darnault C, Ghebrehiwet B, Gaboriaud C, et al. C1q binds phosphatidylserine and likely acts as a multiligand-bridging molecule in apoptotic cell recognition. *J Immunol* (2008) 180(4):2329–38. doi:10.4049/jimmunol.180.4.2329
25. Osman R, Tacnet-Delorme P, Kleman JP, Millet A, Frachet P. Calreticulin release at an early stage of death modulates the clearance by macrophages of apoptotic cells. *Front Immunol* (2017) 8:1034. doi:10.3389/fimmu.2017.01034
26. Thieblemont N, Wright HL, Edwards SW, Witko-Sarsat V. Human neutrophils in auto-immunity. *Semin Immunol* (2016) 28(2):159–73. doi:10.1016/j.smim.2016.03.004
27. Millet A, Martin KR, Bonnefoy F, Saas P, Mocek J, Alkan M, et al. Proteinase 3 on apoptotic cells disrupts immune silencing in autoimmune vasculitis. *J Clin Invest* (2015) 125(11):4107–21. doi:10.1172/JCI78182
28. Scott D, Botto M. The paradoxical roles of C1q and C3 in autoimmunity. *Immunobiology* (2016) 221(6):719–25. doi:10.1016/j.imbio.2015.05.001
29. Trouw LA, Daha N, Kurreeman FA, Bohringer S, Goulielmos GN, Westra HJ, et al. Genetic variants in the region of the C1q genes are associated with rheumatoid arthritis. *Clin Exp Immunol* (2013) 173(1):76–83. doi:10.1111/cei.12097

Conflict of Interest Statement: The authors declare that the research was conducted in the absence of any commercial or financial relationship that could be considered a potential conflict of interest.

19

Induction of Apoptosis and Subsequent Phagocytosis of Virus-Infected Cells as an Antiviral Mechanism

Firzan Nainu[1,2], Akiko Shiratsuchi[2] and Yoshinobu Nakanishi[2]**

[1] Laboratory of Pharmacology and Toxicology, Faculty of Pharmacy, Hasanuddin University, Makassar, Indonesia,
[2] Laboratory of Host Defense and Responses, Graduate School of Medical Sciences, Kanazawa University, Kanazawa, Japan

Edited by:
Estee Kurant,
University of Haifa, Israel

Reviewed by:
Junji Xing,
Houston Methodist Research Institute, United States
Claudia Ida Brodskyn,
Centro de Pesquisa Gonçalo Moniz-FIOCRUZ/BA, Brazil

***Correspondence:**
Firzan Nainu
firzannainu@unhas.ac.id;
Yoshinobu Nakanishi
nakanaka@p.kanazawa-u.ac.jp

Viruses are infectious entities that hijack host replication machineries to produce their progeny, resulting, in most cases, in disease and, sometimes, in death in infected host organisms. Hosts are equipped with an array of defense mechanisms that span from innate to adaptive as well as from humoral to cellular immune responses. We previously demonstrated that mouse cells underwent apoptosis in response to influenza virus infection. These apoptotic, virus-infected cells were then targeted for engulfment by macrophages and neutrophils. We more recently reported similar findings in the fruit fly *Drosophila melanogaster*, which lacks adaptive immunity, after an infection with *Drosophila* C virus. In these experiments, the inhibition of phagocytosis led to severe influenza pathologies in mice and early death in *Drosophila*. Therefore, the induction of apoptosis and subsequent phagocytosis of virus-infected cells appear to be an antiviral innate immune mechanism that is conserved among multicellular organisms. We herein discuss the underlying mechanisms and significance of the apoptosis-dependent phagocytosis of virus-infected cells. Investigations on the molecular and cellular features responsible for this underrepresented virus–host interaction may provide a promising avenue for the discovery of novel substances that are targeted in medical treatments against virus-induced intractable diseases.

Keywords: viral infection, apoptosis, phagocytosis, innate immunity, antiviral mechanism

INTRODUCTION

Viruses are one of the most abundant entities present in the environment. All species, including microbial pathogens, such as bacteria and fungi, are subject to infections by viruses (1, 2). Greater susceptibility to viral infection has been reported in higher metazoans, such as humans, which live in a community system (3). In this system, close interactions exist between species, and, thus, infection easily spreads among members of the community (3), particularly under the condition of compromised immunity (4). Irrespective of the types of genomes and other structural and functional characteristics, viruses behave in a similar manner after invading host organisms. Most viruses, if not all, are obligate intracellular parasites and, thus, require immediate access to the cytosolic and/or nuclear compartments of host cells (2, 5). In the cytoplasm, viruses hijack the ribosomes of host cells to generate proteins encoded by their own genomes for the production of new infective virions (5–8). In the nucleus, viruses may utilize, when necessary, host enzymes to replicate their genomes and synthesize mRNA. On the other hand, host cells are equipped with an array of intracellular and extracellular immune responses to limit this viral proliferation process (9, 10). The final result of this race between the host and virus decides the outcome of infection, from which infected host organisms become ill or remain healthy. Although drugs have been developed to combat diseases caused by viral infections, their efficacy, unlike those against bacteria and fungi, is limited to certain types of viruses: targets for effective drugs are nucleoside kinases of herpes virus, protease and reverse transcriptase of human immunodeficiency virus, neuraminidase of influenza virus, and non-structural proteins of hepatitis C virus (11–13). Some infectious diseases, such as those caused by Ebola virus and highly pathogenic influenza virus, have been challenging to treat and often result in a large number of deaths (14–17). Therefore, medical treatments that are effective against different types of viruses are urgently required.

We herein highlight an underrepresented virus–host interaction, the apoptosis-dependent phagocytosis of virus-infected cells, which enables the elimination of viruses as an innate immune response. This mechanism may be effective against most types of viruses and appears to be conserved among multicellular organisms. Therefore, it may provide a better rationale for the development of novel medical treatments against virus-induced diseases (18, 19).

PHAGOCYTIC ELIMINATION OF CELLS UNDERGOING APOPTOTIC DEATH

As one of the host responses evoked upon viral infection, host cells are induced to undergo apoptotic death (20, 21). Apoptosis is an orchestrated process of self-demolition, which is observed across metazoan species and considered to be a major form of programmed cell death (22–24). The pathways for the induction of apoptosis have been documented for three model animals—the nematode *Caenorhabditis elegans*, the fruit fly *Drosophila melanogaster*, and the mouse *Mus musculus*—and are shown to be fundamentally equivalent (22, 25, 26), as illustrated in **Figure 1**. All cellular changes observed during the apoptotic process are generally attributed to the actions of cysteine-proteases, caspases,

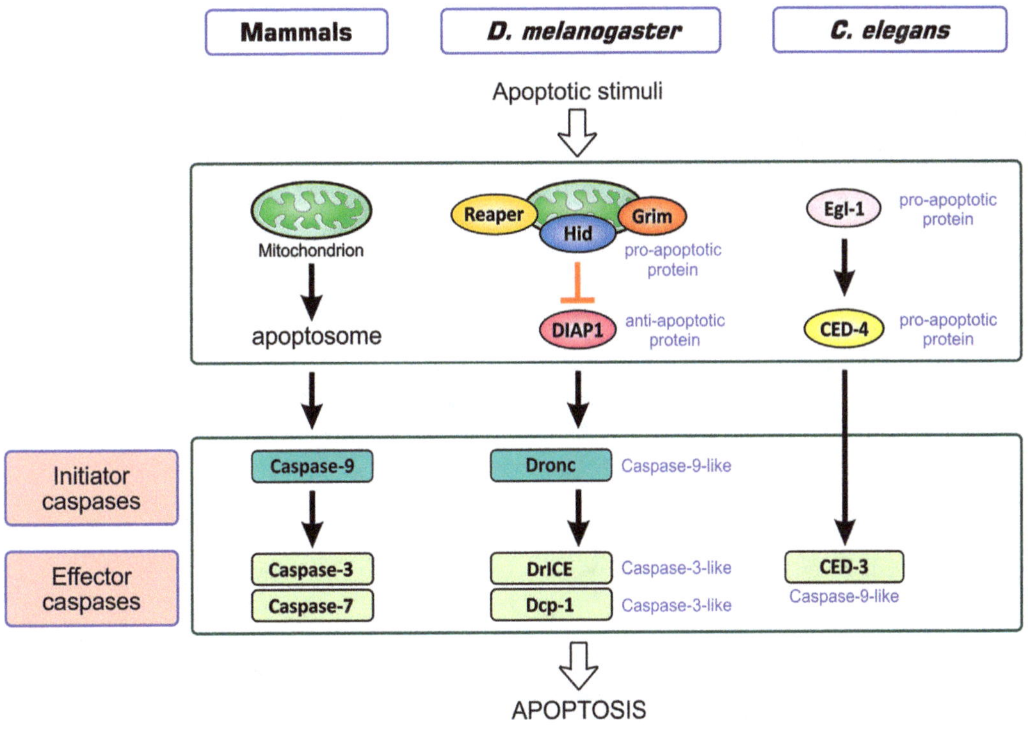

FIGURE 1 | Similarity in apoptosis-inducing pathways in three model animals. Pathways for the induction of apoptosis in mammals, *Drosophila melanogaster*, and *Caenorhabditis elegans* are shown. Not all signal mediators are indicated. In the activation of initiator caspases, the mitochondrion is involved in mammals, probably involved in *Drosophila*, and not involved in *C. elegans*. Initiator caspases are caspase-9 in mammals, caspase-9-like Dronc in *Drosophila*, and absent in *C. elegans*, while effector caspases are caspase-3 and caspase-7 in mammals, caspase-3-like DrICE and Dcp-1 in *Drosophila*, and caspase-9-like CED-3 in *C. elegans*. Effector caspases, once activated by initiator caspases, degrade a number of cellular proteins, leading to structural and functional features that are typical of apoptosis. In mammals, two different modes of apoptosis-inducing pathways exist, and the so-called intrinsic pathway that involves the action of mitochondria is shown. The other one, the extrinsic pathway, which is initiated by extracellular death factors and their receptors independent of mitochondrial actions, is not shown (see **Figure 4**). CED, cell death abnormal; Dcp-1, death caspase-1; DrICE, *Drosophila* interleukin-1 β-converting enzyme; Dronc, *Drosophila* Nedd2-like caspase; Egl-1, egg-laying defective-1.

and the onset of apoptosis involves the activation of initiator caspases that, in turn, partially cleave and activate another group of caspases, the effector caspase (24, 27, 28). Activated effector caspases then cleave a number of cellular proteins, resulting in the structural and biochemical features of apoptosis, such as the shrinkage of cells, fragmentation of DNA, and condensation of chromatin (22).

Apoptotic cells completely disappear: they are engulfed and digested by immune cells, a process-dubbed phagocytosis (29–31). Phagocytosis was described more than 100 years ago through the seminal studies of the late professor Elie Metchnikoff (32, 33). Researchers initially investigated the mechanisms underlying the phagocytosis of microbial pathogens that invade the human body and later identified apoptotic cells as another target. The phagocytosis of microbial pathogens is evident in innate and adaptive immune responses: phagocytes bind to surface structures specific to the target in the former response, while immunoglobulin, which binds antigens on the surface of pathogens and is often called an opsonin, functions as a ligand for an engulfment receptor, i.e., the Fc receptor, of phagocytes in the adaptive response. In contrast, antibodies are not involved in the phagocytosis of apoptotic cells, and the surface structures of the target that undergo modifications during the apoptotic process are recognized by the engulfment receptors of phagocytes. Under certain conditions, soluble proteins called bridging molecules connect apoptotic cells and phagocytes, similar to opsonins; however, these bridging molecules do not include immunoglobulin. Therefore, apoptotic cell clearance is categorized into an innate immune response to eliminate cells unwanted by the body. The phagocytic elimination of cells undergoing apoptosis is so rapidly accomplished that apoptotic cells are rarely detected in tissues and organs. Cells undergoing apoptosis maintain the integrity of plasma membrane permeability until engulfment by phagocytes, and thus the noxious components of cells do not leak out and damage surrounding tissues. Therefore, apoptosis is considered to be a physiological, silent mode of cell death (34, 35).

The entire process of the phagocytosis of apoptotic cells is shown in **Figure 2**. Apoptotic cells that are close to engulfment release substances, which are often referred to as find-me signals, to recruit phagocytes. A number of molecules have been reported to act as such signals, including proteins, lipids, and nucleotides, and their receptors as well as downstream signal transduction pathways have been mostly identified (36–38). Phagocytosis is initiated when apoptotic cells are in close proximity to phagocytes, which allows engulfment receptors on the surface of

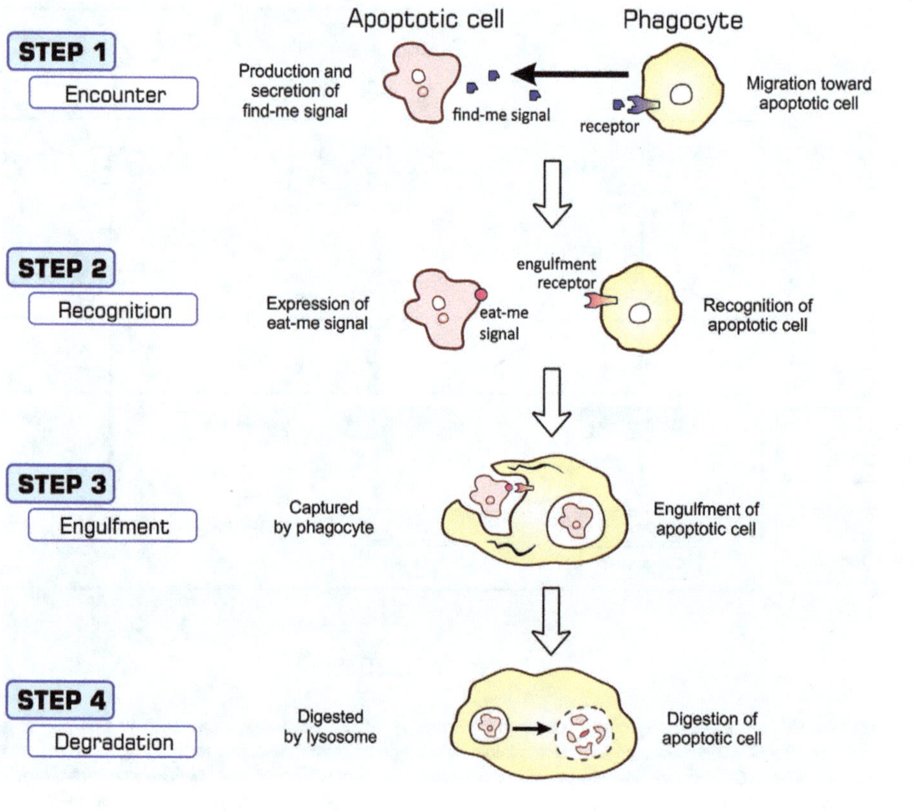

FIGURE 2 | Processes of apoptosis-dependent phagocytosis. Cells undergoing apoptosis secrete substances that attract phagocytes (STEP 1), often called find-me signals, and simultaneously express eat-me signals on their surface (STEP 2). Phagocytes that come into close proximity to apoptotic cells recognize and bind eat-me signals using engulfment receptors (STEP 2), and activate signaling pathways for the induction of phagocytosis. The culmination of this signal transduction is the generation of pseudopodia that help phagocytes surround and incorporate apoptotic cells (STEP 3). Materials engulfed exist as phagosomes, which subsequently fuse with the lysosomes for degradation (STEP 4).

phagocytes to recognize and bind to ligands on the surface of target apoptotic cells (34, 35). The ligands for engulfment receptors are called eat-me signals or markers for phagocytosis, which appear on the cell surface during the apoptotic process (37, 39, 40). The engagement of eat-me signals to the corresponding receptors activates signaling pathways that ultimately generate pseudopodia, extensions of plasma membranes that surround and engulf target cells (30, 37, 40). Apoptotic cells are then incorporated, forming specialized membrane vesicles called phagosomes. Phagosomes subsequently fuse with lysosomes, giving rise to phagolysosomes (31, 34), and the components of apoptotic cells are then subjected to digestion through the actions of lysosomal enzymes. There are two partly overlapping pathways for the induction of phagocytosis, which are conserved among the nematode, fruit fly, and mammals (22, 25, 31, 34), as are those for the induction of apoptosis, and are shown in **Figure 3**. In the figure, the names of eat-me signals, engulfment receptors, and intracellular signal mediators of *C. elegans*, *Drosophila*, and mammals are shown. Phosphatidylserine (PS) is an eat-me signal common among these animal species, and transthyretin-like protein 52 (TTR52) in *C. elegans* and milk fat globule epidermal growth factor protein 8 (MFG-E8) in mammals are PS-binding proteins that bridge apoptotic cells and phagocytes. The *C. elegans* cell death abnormal protein 1 (CED-1) and its counterparts, Draper in *Drosophila* and multiple epidermal growth factor-like domains 10 (MEGF10) in mammals, and integrin INA1-PAT3 of *C. elegans* and its counterparts, αPS3–βν of *Drosophila* and α_v–β_3 and α_v–β_5 of mammals, are engulfment receptors located at the furthest upstream of the two pathways. CED-6 in *C. elegans* and its counterparts, dCed-6 in *Drosophila* and engulfment adapter protein (GULP) in mammals, and CED-2 in *C. elegans* and its counterparts, CT10 regulator of kinase (Crk) in *Drosophila*, and CrkII in mammals, are adaptor proteins that directly bind the engulfment receptors upon activation by eat-me signals. The *C. elegans* CED-5 and its counterparts, myoblast city (Mbc) of *Drosophila* and dedicator of cytokinesis 180 (Dock180) of mammals, are guanine nucleotide exchange factors that activate small G proteins. CED-12 in *C. elegans* and engulfment and cell motility (ELMO) in mammals are another adaptor proteins constituting one pathway, but their counterpart in *Drosophila*, dElmo, seems to be dispensable (41). The two pathways converge on the small G proteins CED-10 in *C. elegans*, Rac1 and Rac2 in *Drosophila*, and Rac1 in mammals, which remodel the actin cytoskeleton for the generation of pseudopodia. CED-7 in *C. elegans* and its counterparts, CG1718 in *Drosophila* and ATP-binding cassette (ABC) protein A1 in mammals, are ABC transporters whose actions in the pathways remain to be solved. Some signal mediators remain missing in these pathways and need to be identified. Other eat-me signals, bridging molecules, engulfment receptors, and signal mediators have been reported, which could be incorporated into the pathways shown here or constitute additional pathways.

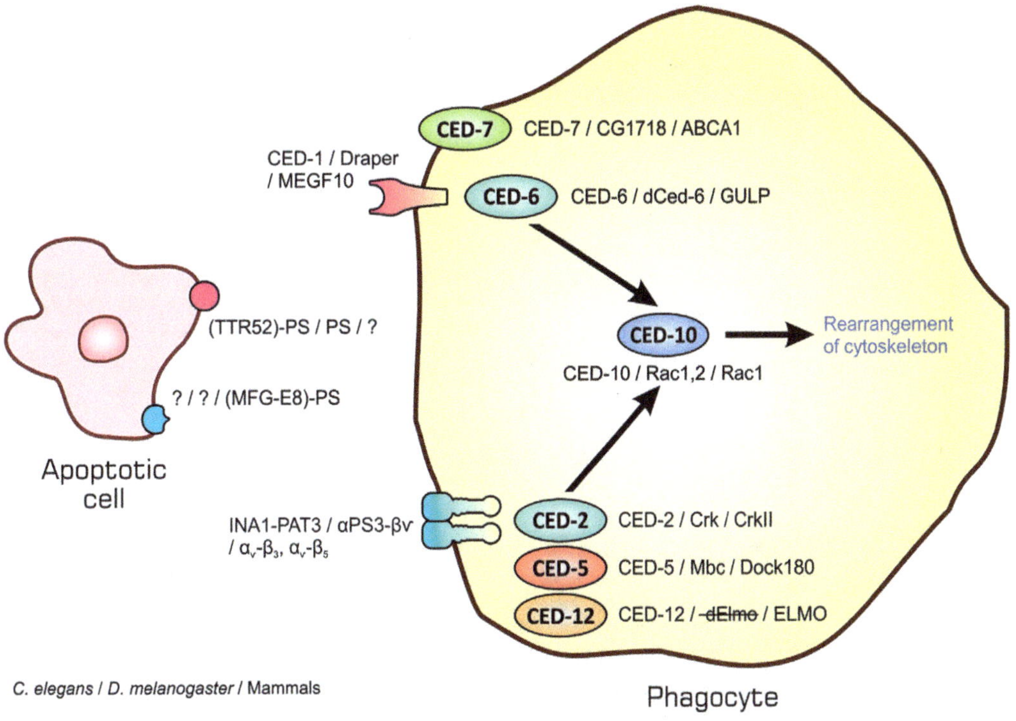

FIGURE 3 | Similarity in signaling pathways for the induction of apoptotic cell clearance in three model animals. Molecules constituting two partly overlapping pathways for the induction of phagocytosis in the nematodes, insects, and mammals are shown. The names of eat-me signals, bridging molecules, engulfment receptors, and intracellular signal mediators of *Caenorhabditis elegans*, *Drosophila melanogaster*, and mammals are shown from left to right with slashes in between. All molecules in each individual category are counterparts to each other in three animal species. Refer to the text for explanation.

INDUCTION OF APOPTOSIS AND SUBSEQUENT PHAGOCYTOSIS OF INFLUENZA VIRUS-INFECTED CELLS

Ranges of cell types, either primarily cultured cells or established cell lines, are susceptible to infection with influenza virus and are subsequently induced to undergo apoptosis (42–47). Previous studies demonstrated that HeLa cells and Madin–Darby canine kidney cells become apoptotic upon influenza A virus infection, characterized by the cleavage of host chromosomal DNA (48), condensation of chromatin (48), surface exposure of PS (49), and activation of initiator and effector caspases (50). Further studies demonstrated that the initiation of apoptosis in HeLa cells infected with influenza H3N2 virus may due to an elevated levels of Fas and the Fas ligand, a death receptor and its ligand (48, 49, 51). Upon infection, the activity of the transcription factor CCAAT/enhancer-binding protein β (C/EBPβ) increased, possibly through the action of double-stranded RNA-activated protein kinase (52), and this factor enhances the transcription of Fas- and Fas ligand-encoding genes (51, 53). Influenza virus-infected cells with elevated levels of Fas and the Fas ligand on the cell surface most likely interact with each other for the induction of apoptosis (54) (**Figure 4**). Besides the above-described study, influenza virus-induced cell death also appears to occur through the actions of apoptosis-inducing factor, another cell death-inducing ligand, in the human alveolar epithelial cell line A549, independent of caspases (55), and the upregulated expression of tumor necrosis factor-related apoptosis-inducing ligand (TRAIL) and tumor necrosis factor-α was observed in human monocyte-derived macrophages exposed to influenza H5N1 virus (56). Similarly, the induction of TRAIL was reported in natural killer cells, helper T cells, and cytotoxic T cells during infection with influenza H1N1 virus (57). Viral clearance was found to be markedly delayed in the presence of an anti-TRAIL monoclonal antibody, suggesting an important role for TRAIL in the antiviral immune response (57). A recent study demonstrated that the upregulated expression of B-cell lymphoma-2-associated X protein may be an alternative cause of the induction of apoptosis in influenza virus-infected cells (58). Nevertheless, all findings revealed that influenza virus-infected cells were induced to undergo apoptotic cell death (48, 59–61).

Although typical apoptosis mediated by caspases is evident in influenza virus-infected cells, virus replication in these cells did not appear to be impaired (48). This may be because this type of virus rapidly produces its progeny after entering host cells. In order to examine the role of apoptosis, we investigated whether influenza virus-infected cells are targeted for engulfment by phagocytes. The findings obtained from an assay for phagocytosis *in vitro* using mouse peritoneal macrophages as phagocytes showed that HeLa cells became susceptible to phagocytosis when infected with influenza A virus (62), and that this leads to the inhibition of viral propagation (63). The phagocytosis of influenza virus-infected cells appeared to be mediated by PS, the eat-me signal characterized in the most detail, and carbohydrate moieties on the surface of macrophages, which are modified by influenza virus neuraminidase expressed in virus-infected cells (62, 64) (**Figure 5**). Further investigations using an *in-vivo* model of infection in mice revealed the involvement of macrophages and neutrophils in the phagocytosis of cells infected with influenza A virus, and this contributed to the mitigation of influenza

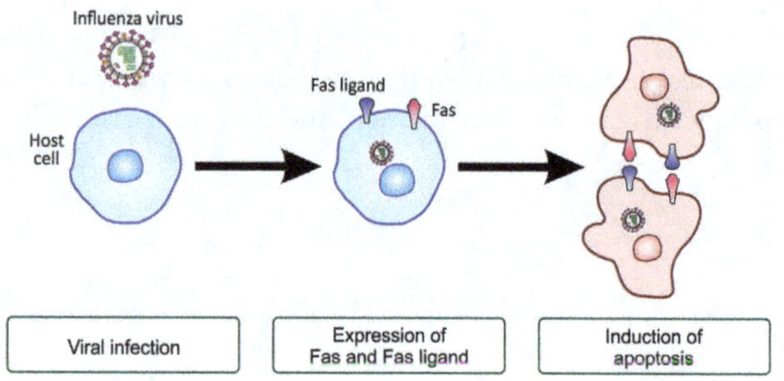

FIGURE 4 | Fas and Fas ligand-induced apoptosis in influenza virus-infected cells. Upon infection with influenza virus, the production of the death receptor Fas and its ligand, the Fas ligand, is enhanced at the level of gene transcription. As a result, virus-infected cells have higher levels of Fas and the Fas ligand on their surface. When these cells associate with each other, the ligand-engaged receptor activates an intracellular signaling pathway for the induction of apoptosis. This mode of apoptosis induction is called the extrinsic pathway and does not involve mitochondria (see **Figure 1**).

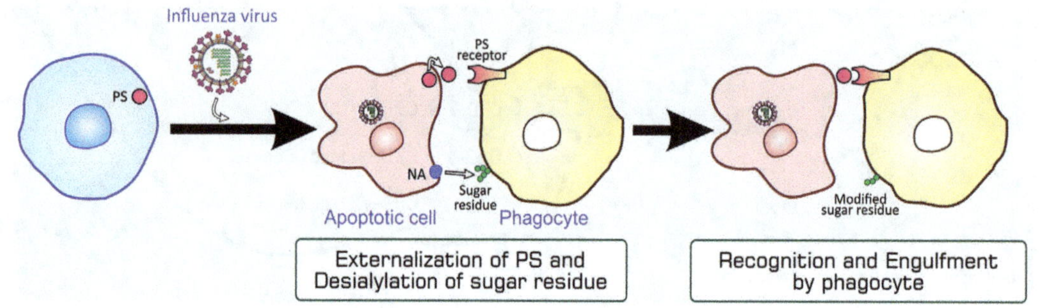

FIGURE 5 | PS-mediated and sugar residue-stimulated phagocytosis of influenza virus-infected cells. Influenza virus-infected cells are induced to undergo apoptosis and express PS and viral NA on their surfaces. When phagocytes, macrophages and neutrophils, bind to these cells through interactions between PS and a PS-recognizing engulfment receptor, NA enzymatically modifies sugar residues that exist on the surface of phagocytes. The PS-bound receptor activates a signaling pathway for the induction of phagocytosis, while modified sugar residues somehow stimulate phagocytosis. NA, neuraminidase; PS, phosphatidylserine.

pathologies in mice (65). Find-me signals responsible for the recruitment of these phagocytes to the site of virus-infected cells remain to be known. The phagocytic activity of alveolar macrophages prepared from influenza virus-infected mice was stronger than that of macrophages from uninfected counterparts (65). Furthermore, the increased mortality of Toll-like receptor 4-lacking mice infected with influenza virus suggested a role for this pattern recognition receptor in antiviral mechanisms (65). The rapid mobilization of neutrophils and macrophages to target sites soon after influenza virus infection may explain the importance of pattern recognition receptors (65). Collectively, apoptosis in influenza virus-infected cells makes them susceptible to phagocytosis, and this mechanism for the direct elimination of the virus serves as an antiviral immune response.

INDUCTION OF APOPTOSIS AND SUBSEQUENT PHAGOCYTOSIS OF *Drosophila* C VIRUS-INFECTED CELLS

We then examined whether a similar antiviral mechanism exists in insects with no adaptive immunity. We used *D. melanogaster* as a host for infection with *Drosophila* C virus (DCV), a natural pathogen of *Drosophila* (66, 67). DCV is a non-enveloped, positive-strand picorna-like RNA virus that belongs to the Dicitroviridae genus *Cripavirus* (66, 68). When S2 cells, a *Drosophila* culture cell line, were incubated with DCV, they underwent apoptosis, as was evident from chromatin condensation, DNA fragmentation, and caspase activation, accompanied by the propagation of the virus (69). We found that the amount of *Drosophila* inhibitor of apoptosis protein 1 (DIAP1), a *Drosophila* protein that inhibits caspases, decreased upon infection with DCV. All these changes in S2 cells after viral infection became undetectable when a synthetic inhibitor of caspase was present in cell cultures or the virus was pretreated with UV. The mechanisms underlying apoptosis in DCV-infected cells have not yet been elucidated; however, several studies suggested the involvement of a mechanism similar to that observed during the early developmental stages of *Drosophila* (69–71). Upon infection with Flock house virus or the DNA virus *Autographa californica* multicapsid nucleopolyhedrovirus, the expression of *reaper* and *head involution defective* (*hid*), the products of which antagonize DIAP1, was significantly increased in a manner mediated by the transcription control regions of the two genes, namely, a p53-bound sequence and sequence named the irradiation-responsive enhancer region (71). A similar mechanism appears to exist in mosquitoes when they are infected with *Culex nigripalpus* nucleopolyhedrovirus (72). We anticipate the following pathway for the induction of apoptosis in DCV-infected cells, as shown in **Figure 6**: the propagation of the virus enhances the transcription of *reaper* and *hid*; Reaper and/or Hid suppress DIAP1; the initiator caspase *Drosophila* Nedd2-like caspase (Dronc) is activated; Dronc cleaves and activates the effector caspases *Drosophila* interleukin-1 β-converting enzyme (DrICE) and death caspase-1 (Dcp-1); and the activated effector caspases degrade cellular proteins.

The presence of l(2)mbn cells, a *Drosophila* cell line derived from larval hemocytes, in cultures of DCV-infected S2 cells induced a decrease in the amount of the virus (69). Therefore, we examined whether DCV-infected cells are phagocytosed in a manner that is dependent on apoptosis and found that this was the case. Phagocytosis was partly inhibited in the presence of a PS-containing liposome that interferes with the action of this phospholipid as an eat-me signal. *Drosophila* phagocytes used the engulfment receptors Draper and integrin αPS3–βν for the phagocytosis of apoptotic cells (30), and the inhibited expression of either receptor by RNA interference (RNAi) caused a decrease in the level of phagocytosis. Phagocytosis was decreased further after the simultaneous RNAi of both receptors. These findings collectively suggested that DCV-infected cells are subjected to apoptosis-dependent phagocytosis by *Drosophila* phagocytes, depending on, at least partly, the eat-me signal PS and engulfment receptors Draper and integrin αPS3–βν. In order to assess

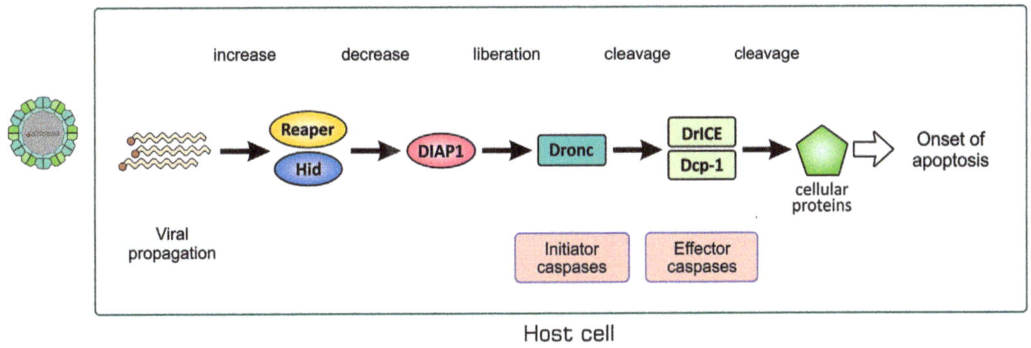

FIGURE 6 | Possible mechanism for the induction of apoptosis in *Drosophila* C virus-infected cells. The process for the propagation of the virus somehow increases pro-apoptotic proteins Reaper and Hid, which cause a decrease in the level of DIAP1. This leads to the liberation and, thus, activation of the initiator caspase Dronc that partially cleaves the effector caspases DrICE and Dcp-1 for activation. Activated DrICE and Dcp-1, in turn, cleave a number of cellular proteins for apoptotic changes. The brownish dot at the end of the viral genome indicates a viral protein called virion protein, genome-linked, which plays a role in the synthesis of viral RNA. Dcp-1, death caspase-1; DIAP1, *Drosophila* inhibitor of apoptosis protein 1; DrICE, *Drosophila* interleukin-1 β-converting enzyme; Dronc, *Drosophila* Nedd2-like caspase.

this *in vivo*, we established a fatal infection of *Drosophila* adults with DCV. The findings of an assay for survivorship revealed that Draper and integrin αPS3–βν were both involved in the protection of flies from viral infection. Measurements of the viral load during infection indicated that these engulfment receptors were responsible for reducing viral propagation in adult flies. The ectopic expression of a PS-binding protein made flies more severely succumb to DCV infection and increased the viral load, confirming the PS-mediated phagocytosis of DCV-infected cells in adult flies. Hemocytes contained in the adult hemocoel appeared to be responsible for the phagocytosis of virus-infected cells. These findings indicate that the PS-mediated, Draper and integrin αPS3–βν-dependent phagocytosis of DCV-infected, apoptotic cells by hemocytes plays a role in antiviral mechanisms in *Drosophila* (see **Figure 3**).

In *Drosophila*, RNAi-based antiviral innate immunity has been intensely investigated (73–81). Our study now adds another mechanism for the immune response, i.e., the apoptosis-dependent, phagocytosis-based elimination of virus-infected cells. A recent study demonstrated that *Drosophila* hemocytes spread double-stranded RNA, which induces virus-specific RNAi, in the entire body of adult flies upon viral infection; therefore, cells uninfected with the virus acquired the competence for RNAi (82). The phagocytosis of virus-infected cells may make hemocytes gain a source for the production of double-stranded RNA. More importantly, the apoptosis-dependent phagocytosis of virus-infected cells serves as an antiviral mechanism in *Drosophila*, which is only equipped with innate immunity, indicating that this mechanism is an innate immune response against viral infection and common among multicellular organisms.

CONTRIBUTION OF APOPTOSIS-DEPENDENT PHAGOCYTOSIS TO IMMUNITY CONTROL

Recent studies on the mechanisms and consequences of apoptosis-dependent phagocytosis have revealed that this type of phagocytosis achieves not only the elimination of unwanted cells but also endows additional effects that contribute to the maintenance of tissue homeostasis. These effects may cooperatively control, with the direct removal of virus-infected cells, immunity to fight against viral infection, as shown in **Figure 7**.

Antigen Presentation

The presentation of antigens by certain types of immune cells toward T lymphocytes is a prerequisite for the induction of adaptive immunity. Antigen presentation is mainly accomplished by dendritic cells and macrophages, antigen-presenting cells (APCs), which sample the peptides of foreign materials, such as microbial pathogens, and expose them on the cell surface together with the major histocompatibility complex (83–86). Although APCs may process the foreign peptides synthesized in these cells, the presentation of microbial antigens by APCs that are apparently not infected with the corresponding pathogens is often observed, and this is an immunological process called cross-presentation (84, 87–89). Apoptosis-dependent phagocytosis provides a mechanistic basis for this phenomenon; APCs that are not infected by the virus engulf virus-infected cells undergoing apoptosis, process viral proteins, and present viral antigens to cytotoxic T lymphocytes (CTLs) for their activation (90, 91). Activated CTLs induce apoptosis in cells that are infected with the same virus. Therefore, the phagocytosis of virus-infected cells may lead to the activation of adaptive immune responses.

Inflammation Resolution

The phagocytic elimination of cells undergoing apoptotic cell death is an immunologically silent reaction; inflammation is not evoked. Previous research by Fadok, Henson, and others demonstrated that this mode of phagocytosis more actively resolves inflammation (92–95). They showed that macrophages engulf apoptotic neutrophils to remove cells that produce and secrete pro-inflammatory cytokines, and that neutrophil-engulfing macrophages simultaneously secrete anti-inflammatory cytokines such as transforming growth factor-β. Therefore, phagocytes alter the repertoire of cytokines after the engulfment of virus-infected cells so that possible inflammation ceases in addition to the direct elimination of virus.

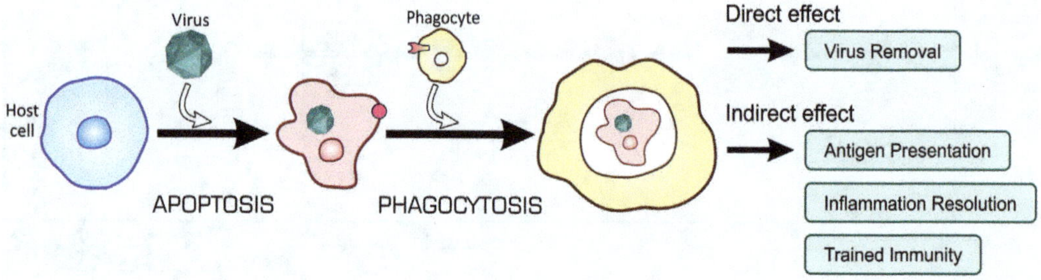

FIGURE 7 | Orchestrated antiviral mechanisms initiated by the phagocytosis of virus-infected cells. Virus-infected cells are subjected to apoptosis-dependent phagocytosis, which results in the digestion of viruses together with host cells, Virus Removal. At the same time, phagocytes process incorporated viral proteins for the presentation of antigens toward CTLs for their activation, Antigen Presentation. Moreover, phagocytes change the pattern of gene expression at a transcriptional level, so that the repertoire of cytokines shifts to the mitigation of inflammation, Inflammation Resolution; the phagocytic activity of phagocytes is enhanced, Trained Immunity; and virus-specific RNAi is systemically induced, Trained Immunity. CTLs, cytotoxic T lymphocytes.

Trained Immunity

In animals equipped with innate and adaptive immunity, the first encounter with foreign substances such as microbial pathogens makes the host organism prepare for a second encounter with the same substances (96, 97). In the first encounter, the major players to combat invaders are the components of innate immunity, and they take actions not only to eliminate the invaders but also setup the activation of adaptive immunity so that adaptive responses are evoked in a rapid and robust manner in the second encounter, a mechanism called trained immunity or immunological memory. The culmination of trained immunity that involves apoptotic cell clearance is expected to be antigen presentation as described above.

Until recently, trained immunity was generally considered to not exist in innate immunity, which does not involve antibodies and lymphocytes. However, recent studies using *Drosophila* cast doubts on this assumption. Hemocytes, *Drosophila* macrophages, exhibit enhanced phagocytic activity when they encounter targets, namely, bacteria (98) and apoptotic cells (41). This is regarded as preparation for the next encounter with the same targets in order to eliminate them. The enhancement of phagocytic activity observed in both studies may have been due to a higher level of engulfment receptors. An increase of engulfment receptors in phagocytes after phagocytosis was previously reported for mammalian macrophages (96, 97), and, thus, trained immunity that leads to the activation of phagocytes is most likely conserved among multicellular organisms. Our findings also demonstrated that gene expression patterns in phagocytes changed after the engulfment of apoptotic cells, including the enhanced transcription of genes coding for engulfment receptors (41). In addition, a change in the mode of cytokine production in macrophages appears to occur at the level of transcription (99). Therefore, the mode of gene transcription appears to change in phagocytes upon the engulfment of apoptotic cells in order to more effectively control tissue homeostasis.

CONCLUDING REMARKS

Most living organisms are always exposed to potentially fatal infections by viruses and, thus, have acquired several distinct mechanisms to prevent the invasion, proliferation, and release of viruses. The apoptosis-dependent phagocytosis of virus-infected cells is one such mechanism, through which organisms directly remove viruses from the body. Cells are induced to undergo apoptosis upon infection with a number of viruses, and this process inactivates the cellular machineries for gene expression and proliferation, which are required by invading viruses to produce their progeny in infected cells. Although apoptosis itself retards the growth of invaders, virus-infected cells appear to be equipped with a more active process, subsequent to apoptosis, for the direct elimination of viruses; virus-infected cells become susceptible to phagocytosis for degradation. Recent studies demonstrated that not only apoptotic cells, but also those undergoing other types of cell death are subjected to phagocytic elimination (100–102). Nevertheless, apoptosis remains a major biological process for the safe removal of cells unwanted by the body because it is the only mode of cell death during which the control of plasma membrane permeability is maintained. As opposed to viruses, *Leishmania major*, an intracellular parasite, appears to exploit the mechanisms described above for the establishment and dissemination of infection (103). Upon entry into animals, these types of protozoa are first captured by neutrophils that subsequently undergo apoptosis. Then, *L. major*-infected, apoptotic neutrophils are phagocytosed by macrophages, which are primary host cells for these protozoa. The pathogens survive and replicate in macrophages, and, at the same time, macrophages create an anti-inflammatory environment. As a result, *L. major* disseminates its infection. The mechanisms by which *L. major* evades killing and digestion in neutrophils and macrophages remain to be clarified.

Individual processes that constitute the apoptosis-dependent phagocytosis of virus-infected cells may be targeted by the development of novel medical treatments against virus-induced diseases. The enhancement of apoptosis in virus-infected cells may be one such treatment. Apoptosis is induced in cells that ideally need to be retained, and, thus, this treatment needs to be restricted to cells infected with viruses. In order to achieve this, the mechanisms underlying virus-induced apoptosis need to be elucidated in more detail. Another concern is the presence of proteins that inhibit apoptosis by antagonizing caspases in some types of viruses, particularly DNA viruses (20, 21, 104). The development of a method to repress apoptosis-inhibiting viral proteins may be an effective treatment. On the other hand, molecules involved in the process of the phagocytosis of virus-infected cells have been largely identified, and the stimulation of phagocytic activity is not always harmful to health. A substance that acts as an agonist for engulfment receptors is a promising candidate for an effective drug. Alternatively, the secondary effects of the phagocytosis of apoptotic cells may be targeted. The administration of apoptotic cells to patients may contribute to mitigating inflammation and stimulating phagocytes, and antiviral adaptive immunity is expected when apoptotic cells harboring viral antigens are used to treat patients. However, such efforts toward inventing novel medical treatments require great care, because the apoptosis-dependent phagocytosis of microbe-infected cells could favor the pathogens, as an example shown above.

AUTHOR CONTRIBUTIONS

FN mostly conducted the experiments in our study quoted in this review. FN, AS, and YN wrote the paper.

ACKNOWLEDGMENTS

A word of thanks goes to Rudi Arfiansyah, S.Si., Apt of Hasanuddin University, for his kind support in preparing the figures. We also thank the members of the Laboratory of Host Defense and Responses for discussion.

FUNDING

This work was partly supported by KAKENHI Grant from Japan Society for the Promotion of Science (to AS and YN) and a Directorate General of Higher Education scholarship from the Ministry of Research, Technology, and Higher Education of the Republic of Indonesia (to FN).

REFERENCES

1. Pedulla ML, Ford ME, Houtz JM, Karthikeyan T, Wadsworth C, Lewis JA, et al. Origins of highly mosaic mycobacteriophage genomes. *Cell* (2003) 113:171–82. doi:10.1016/S0092-8674(03)00233-2
2. Edwards RA, Rohwer F. Viral metagenomics. *Nat Rev Microbiol* (2005) 3:504–10. doi:10.1038/nrmicro1163
3. Kucharski AJ, Kwok KO, Wei VWI, Cowling BJ, Read JM, Lessler J, et al. The contribution of social behaviour to the transmission of influenza A in a human population. *PLoS Pathog* (2014) 10:e1004206. doi:10.1371/journal.ppat.1004206
4. Lindahl JF, Grace D. The consequences of human actions on risks for infectious diseases: a review. *Infect Ecol Epidemiol* (2015) 5:30048. doi:10.3402/iee.v5.30048
5. Breitbart M, Rohwer F. Here a virus, there a virus, everywhere the same virus? *Trends Microbiol* (2005) 13:278–84. doi:10.1016/j.tim.2005.04.003
6. Schmid M, Speiseder T, Dobner T, Gonzalez RA. DNA virus replication compartments. *J Virol* (2014) 88:1404–20. doi:10.1128/JVI.02046-13
7. Novoa RR, Calderita G, Arranz R, Fontana J, Granzow H, Risco C. Virus factories: associations of cell organelles for viral replication and morphogenesis. *Biol Cell* (2005) 97:147–72. doi:10.1042/BC20040058
8. Netherton CL, Wileman T. Virus factories, double membrane vesicles and viroplasm generated in animal cells. *Curr Opin Virol* (2011) 1:381–7. doi:10.1016/j.coviro.2011.09.008
9. Braciale TJ, Hahn YS. Immunity to viruses. *Immunol Rev* (2013) 255:5–12. doi:10.1111/imr.12109
10. Yan N, Chen ZJ. Intrinsic antiviral immunity. *Nat Immunol* (2012) 13:214–22. doi:10.1038/ni.2229
11. De Clercq E, Li G. Approved antiviral drugs over the past 50 years. *Clin Microbiol Rev* (2016) 29:695–747. doi:10.1128/cmr.00102-15
12. Wegzyn CM, Wyles DL. Antiviral drug advances in the treatment of human immunodeficiency virus (HIV) and chronic hepatitis C virus (HCV). *Curr Opin Pharmacol* (2012) 12:556–61. doi:10.1016/j.coph.2012.06.005
13. De Clercq E. Antiviral drugs in current clinical use. *J Clin Virol* (2004) 30:115–33. doi:10.1016/j.jcv.2004.02.009
14. Feldmann H, Geisbert TW. Ebola haemorrhagic fever. *Lancet* (2011) 377:849–62. doi:10.1016/S0140-6736(10)60667-8
15. Ströher U, Feldmann H. Progress towards the treatment of Ebola haemorrhagic fever. *Exp Opin Investig Drugs* (2006) 15:1523–35. doi:10.1517/13543784.15.12.1523
16. De Clercq E. Strategies in the design of antiviral drugs. *Nat Rev Drug Discov* (2002) 1:13–25. doi:10.1038/nrd703
17. Ferguson NM, Cummings DAT, Fraser C, Cajka JC, Cooley PC, Burke DS. Strategies for mitigating an influenza pandemic. *Nature* (2006) 442:448–52. doi:10.1038/nature04795
18. Martins N, Imler J-L, Meignin C. Discovery of novel targets for antivirals: learning from flies. *Curr Opin Virol* (2016) 20:64–70. doi:10.1016/j.coviro.2016.09.005
19. Majzoub K, Hafirassou Mohamed L, Meignin C, Goto A, Marzi S, Fedorova A, et al. RACK1 controls IRES-mediated translation of viruses. *Cell* (2014) 159:1086–95. doi:10.1016/j.cell.2014.10.041
20. Roulston A, Marcellus RC, Branton PE. Viruses and apoptosis. *Annu Rev Microbiol* (1999) 53:577–628. doi:10.1146/annurev.micro.53.1.577
21. Barber GN. Host defense, viruses and apoptosis. *Cell Death Differ* (2001) 8:113–26. doi:10.1038/sj.cdd.4400823
22. Fuchs Y, Steller H. Programmed cell death in animal development and disease. *Cell* (2011) 147:742–58. doi:10.1016/j.cell.2011.10.033
23. Fogarty CE, Bergmann A. The sound of silence: signaling by apoptotic cells. *Curr Top Dev Biol* (2015) 114:241–65. doi:10.1016/bs.ctdb.2015.07.013
24. Taylor RC, Cullen SP, Martin SJ. Apoptosis: controlled demolition at the cellular level. *Nat Rev Mol Cell Biol* (2008) 9:231–41. doi:10.1038/nrm2312
25. Hay BA, Huh JR, Guo M. The genetics of cell death: approaches, insights and opportunities in *Drosophila*. *Nat Rev Genet* (2004) 5:911–22. doi:10.1038/nrg1491
26. Stuart LM, Ezekowitz RA. Phagocytosis and comparative innate immunity: learning on the fly. *Nat Rev Immunol* (2008) 8:131–41. doi:10.1038/nri2240
27. Nagata S, Tanaka M. Programmed cell death and the immune system. *Nat Rev Immunol* (2017) 17:333–40. doi:10.1038/nri.2016.153
28. Bergmann A, Steller H. Apoptosis, stem cells, and tissue regeneration. *Sci Signal* (2010) 3:re8. doi:10.1126/scisignal.3145re8
29. Erwig LP, Henson PM. Clearance of apoptotic cells by phagocytes. *Cell Death Differ* (2007) 15:243–50. doi:10.1038/sj.cdd.4402184
30. Nakanishi Y, Nagaosa K, Shiratsuchi A. Phagocytic removal of cells that have become unwanted: implications for animal development and tissue homeostasis. *Dev Growth Differ* (2011) 53:149–60. doi:10.1111/j.1440-169X.2010.01224.x
31. Hochreiter-Hufford A, Ravichandran KS. Clearing the dead: apoptotic cell sensing, recognition, engulfment, and digestion. *Cold Spring Harb Perspect Biol* (2013) 5:a008748. doi:10.1101/cshperspect.a008748
32. Gordon S. Phagocytosis: an immunobiologic process. *Immunity* (2016) 44:463–75. doi:10.1016/j.immuni.2016.02.026
33. Tauber AI. Metchnikoff and the phagocytosis theory. *Nat Rev Mol Cell Biol* (2003) 4:897–901. doi:10.1038/nrm1244
34. Arandjelovic S, Ravichandran KS. Phagocytosis of apoptotic cells in homeostasis. *Nat Immunol* (2015) 16:907–17. doi:10.1038/ni.3253
35. Lauber K, Blumenthal SG, Waibel M, Wesselborg S. Clearance of apoptotic cells: getting rid of the corpses. *Mol Cell* (2004) 14:277–87. doi:10.1016/S1097-2765(04)00237-0
36. Peter C, Wesselborg S, Herrmann M, Lauber K. Dangerous attraction: phagocyte recruitment and danger signals of apoptotic and necrotice cells. *Apoptosis* (2010) 15:1007–28. doi:10.1007/s10495-010-0472-1
37. Ravichandran KS. Beginnings of a good apoptotic meal: the find-me and eat-me signaling pathways. *Immunity* (2011) 35:445–55. doi:10.1016/j.immuni.2011.09.004
38. Medina CB, Ravichandran KS. Do not let death do us part: 'find-me' signals in communication between dying cells and the phagocytes. *Cell Death Differ* (2016) 23:979–89. doi:10.1038/cdd.2016.13
39. Ravichandran KS. "Recruitment signals" from apoptotic cells: invitation to a quiet meal. *Cell* (2003) 113:817–20. doi:10.1016/s0092-8674(03)00471-9
40. Ravichandran KS, Lorenz U. Engulfment of apoptotic cells: signals for a good meal. *Nat Rev Immunol* (2007) 7:964–74. doi:10.1038/nri2214
41. Nonaka S, Ando Y, Kanetani T, Hoshi C, Nakai Y, Nainu F, et al. Signaling pathway for phagocyte priming upon encounter with apoptotic cells. *J Biol Chem* (2017) 292:8059–72. doi:10.1074/jbc.M116.769745
42. Nichols JE, Niles JA, Roberts NJ. Human lymphocyte apoptosis after exposure to influenza A virus. *J Virol* (2001) 75:5921–9. doi:10.1128/jvi.73.13.5921-5929.2001
43. Brydon EWA, Smith H, Sweet C. Influenza A virus induced apoptosis in bronchiolar epithelial (NCI-H292) cells limits pro-inflammatory cytokine release. *J Gen Virol* (2003) 84:2389–400. doi:10.1099/vir.0.18913-0
44. Mok CKP, Lee DCW, Cheung C-Y, Peiris M, Lau ASY. Differential onset of apoptosis in influenza A virus H5N1- and H1N1-infected human blood macrophages. *J Gen Virol* (2007) 88:1275–80. doi:10.1099/vir.0.82423-0
45. Thitithanyanont A, Engering A, Ekchariyawat P, Wiboon-ut S, Limsalakpetch A, Yongvanitchit K, et al. High susceptibility of human dendritic cells to avian influenza H5N1 virus infection and protection by IFN-α and TLR ligands. *J Immunol* (2007) 179:5220–7. doi:10.4049/jimmunol.179.8.5220
46. Daidoji T, Koma T, Du A, Yang C-S, Ueda M, Ikuta K, et al. H5N1 avian influenza virus induces apoptotic cell death in mammalian airway epithelial cells. *J Virol* (2008) 82:11294–307. doi:10.1128/jvi.01192-08
47. Gu J, Xie Z, Gao Z, Liu J, Korteweg C, Ye J, et al. H5N1 infection of the respiratory tract and beyond: a molecular pathology study. *Lancet* (2007) 370:1137–45. doi:10.1016/S0140-6736(07)61515-3
48. Takizawa T, Matsukawa S, Higuchi Y, Nakamura S, Nakanishi Y, Fukuda R. Induction of programmed cell death (apoptosis) by influenza virus infection in tissue culture cells. *J Gen Virol* (1993) 74:2347–55. doi:10.1099/0022-1317-74-11-2347
49. Fujimoto I, Takizawa T, Ohba Y, Nakanishi Y. Co-expression of Fas and Fas-ligand on the surface of influenza virus-infected cells. *Cell Death Differ* (1998) 5:426–31. doi:10.1038/sj.cdd.4400362
50. Takizawa T, Tatematsu C, Ohashi K, Nakanishi Y. Recruitment of apoptotic cysteine proteases (caspases) in influenza virus-induced cell death. *Microbiol Immunol* (1999) 43:245–52. doi:10.1111/j.1348-0421.1999.tb02400.x
51. Wada N, Matsumura M, Ohba Y, Kobayashi N, Takizawa T, Nakanishi Y. Transcription stimulation of the Fas-encoding gene by nuclear factor for interleukin-6 expression upon influenza virus infection. *J Biol Chem* (1995) 270:18007–12. doi:10.1074/jbc.270.30.18007

52. Takizawa T, Ohashi K, Nakanishi Y. Possible involvement of double-stranded RNA-activated protein kinase in cell death by influenza virus infection. *J Virol* (1996) 70:8128–32.
53. Takizawa T, Fukuda R, Miyawaki T, Ohashi K, Nakanishi Y. Activation of the apoptotic Fas antigen-encoding gene upon influenza virus infection involving spontaneously produced beta-interferon. *Virology* (1995) 209:288–96. doi:10.1006/viro.1995.1260
54. Nakanishi Y, Hashimoto Y, Takizawa T, Shiratsuchi A. Mechanisms and consequences of phagocytosis of influenza virus-infected cells. *Antiinflamm Antiallergy Agents Med Chem* (2008) 7:97–100. doi:10.2174/187152308784533122
55. Qu X, Ding X, Duan M, Yang J, Lin R, Zhou Z, et al. Influenza virus infection induces translocation of apoptosis-inducing factor (AIF) in A549 cells: role of AIF in apoptosis and viral propagation. *Arch Virol* (2017) 162:669–75. doi:10.1007/s00705-016-3151-x
56. Zhou J, Law HKW, Cheung CY, Ng IHY, Peiris JSM, Lau YL. Functional tumor necrosis factor-related apoptosis-inducing ligand production by avian influenza virus-infected macrophages. *J Infect Dis* (2006) 193:945–53. doi:10.1086/500954
57. Ishikawa E, Nakazawa M, Yoshinari M, Minami M. Role of tumor necrosis factor-related apoptosis-inducing ligand in immune response to influenza virus infection in mice. *J Virol* (2005) 79:7658–63. doi:10.1128/jvi.79.12.7658-7663.2005
58. Fan N, Wang J. MicroRNA 34a contributes to virus-mediated apoptosis through binding to its target gene Bax in influenza A virus infection. *Biomed Pharmacother* (2016) 83:1464–70. doi:10.1016/j.biopha.2016.08.049
59. Mori I, Komatsu T, Takeuchi K, Nakakuki K, Sudo M, Kimura Y. In vivo induction of apoptosis by influenza virus. *J Gen Virol* (1995) 76:2869–73. doi:10.1099/0022-1317-76-11-2869
60. Hinshaw VS, Olsen CW, Dybdahl-Sissoko N, Evans D. Apoptosis: a mechanism of cell killing by influenza A and B viruses. *J Virol* (1994) 68:3667–73.
61. Fesq H, Bacher M, Nain M, Gemsa D. Programmed cell death (apoptosis) in human monocytes infected by influenza A virus. *Immunobiology* (1994) 190:175–82. doi:10.1016/S0171-2985(11)80292-5
62. Shiratsuchi A, Kaido M, Takizawa T, Nakanishi Y. Phosphatidylserine-mediated phagocytosis of influenza A virus-infected cells by mouse peritoneal macrophages. *J Virol* (2000) 74:9240–4. doi:10.1128/JVI.74.19.9240-9244.2000
63. Fujimoto I, Pan J, Takizawa T, Nakanishi Y. Virus clearance through apoptosis-dependent phagocytosis of influenza A virus-infected cells by macrophages. *J Virol* (2000) 74:3399–403. doi:10.1128/JVI.74.7.3399-3403.2000
64. Watanabe Y, Shiratsuchi A, Shimizu K, Takizawa T, Nakanishi Y. Role of phosphatidylserine exposure and sugar chain desialylation at the surface of influenza virus-infected cells in efficient phagocytosis by macrophages. *J Biol Chem* (2002) 277:18222–8. doi:10.1074/jbc.M201074200
65. Hashimoto Y, Moki T, Takizawa T, Shiratsuchi A, Nakanishi Y. Evidence for phagocytosis of influenza virus-infected, apoptotic cells by neutrophils and macrophages in mice. *J Immunol* (2007) 178:2448–57. doi:10.4049/jimmunol.178.4.2448
66. Xu J, Cherry S. Viruses and antiviral immunity in *Drosophila*. *Dev Comp Immunol* (2014) 42:67–84. doi:10.1016/j.dci.2013.05.002
67. Kapun M, Nolte V, Flatt T, Schlötterer C. Host range and specificity of the *Drosophila* C virus. *PLoS One* (2010) 5:e12421. doi:10.1371/journal.pone.0012421
68. Marques JT, Imler J-L. The diversity of insect antiviral immunity: insights from viruses. *Curr Opin Microbiol* (2016) 32:71–6. doi:10.1016/j.mib.2016.05.002
69. Nainu F, Tanaka Y, Shiratsuchi A, Nakanishi Y. Protection of insects against viral infection by apoptosis-dependent phagocytosis. *J Immunol* (2015) 195:5696–706. doi:10.4049/jimmunol.1500613
70. Settles EW, Friesen PD. Flock house virus induces apoptosis by depletion of *Drosophila* inhibitor-of-apoptosis protein DIAP1. *J Virol* (2008) 82:1378–88. doi:10.1128/JVI.01941-07
71. Liu B, Behura SK, Clem RJ, Schneemann A, Becnel J, Severson DW, et al. P53-mediated rapid induction of apoptosis conveys resistance to viral infection in *Drosophila melanogaster*. *PLoS Pathog* (2013) 9:e1003137. doi:10.1371/journal.ppat.1003137
72. Liu B, Becnel JJ, Zhang Y, Zhou L. Induction of reaper ortholog mx in mosquito midgut cells following baculovirus infection. *Cell Death Differ* (2011) 18:1337–45. doi:10.1038/cdd.2011.8
73. Lamiable O, Imler J-L. Induced antiviral innate immunity in *Drosophila*. *Curr Opin Microbiol* (2014) 20:62–8. doi:10.1016/j.mib.2014.05.006
74. Sabin LR, Hanna SL, Cherry S. Innate antiviral immunity in *Drosophila*. *Curr Opin Immunol* (2010) 22:4–9. doi:10.1016/j.coi.2010.01.007
75. Karlikow M, Goic B, Saleh M-C. RNAi and antiviral defense in *Drosophila*: setting up a systemic immune response. *Dev Comp Immunol* (2014) 42:85–92. doi:10.1016/j.dci.2013.05.004
76. Kemp C, Imler J-L. Antiviral immunity in *Drosophila*. *Curr Opin Immunol* (2009) 21:3–9. doi:10.1016/j.coi.2009.01.007
77. Buchon N, Silverman N, Cherry S. Immunity in *Drosophila melanogaster*: from microbial recognition to whole-organism physiology. *Nat Rev Immunol* (2014) 14:796–810. doi:10.1038/nri3763
78. Wang X-H, Aliyari R, Li W-X, Li H-W, Kim K, Carthew R, et al. RNA interference directs innate immunity against viruses in adult *Drosophila*. *Science* (2006) 312:452–4. doi:10.1126/science.1125694
79. Mueller S, Gausson V, Vodovar N, Deddouche S, Troxler L, Perot J, et al. RNAi-mediated immunity provides strong protection against the negative-strand RNA vesicular stomatitis virus in *Drosophila*. *Proc Natl Acad Sci U S A* (2010) 107:19390–5. doi:10.1073/pnas.1014378107
80. Kemp C, Mueller S, Goto A, Barbier V, Paro S, Bonnay F, et al. Broad RNA interference-mediated antiviral immunity and virus-specific inducible responses in *Drosophila*. *J Immunol* (2013) 190:650–8. doi:10.4049/jimmunol.1102486
81. Spellberg MJ, Marr MT. FOXO regulates RNA interference in *Drosophila* and protects from RNA virus infection. *Proc Natl Acad Sci U S A* (2015) 112:14587–92. doi:10.1073/pnas.1517124112
82. Tassetto M, Kunitomi M, Andino R. Circulating immune cells mediate a systemic RNAi-based adaptive antiviral response in *Drosophila*. *Cell* (2017) 169:314–25.e13. doi:10.1016/j.cell.2017.03.033
83. Watts C, Amigorena S. Phagocytosis and antigen presentation. *Semin Immunol* (2001) 13:373–9. doi:10.1006/smim.2001.0334
84. Savina A, Amigorena S. Phagocytosis and antigen presentation in dendritic cells. *Immunol Rev* (2007) 219:143–56. doi:10.1111/j.1600-065X.2007.00552.x
85. Blander JM, Medzhitov R. On regulation of phagosome maturation and antigen presentation. *Nat Immunol* (2006) 7:1029–35. doi:10.1038/ni1006-1029
86. Lim JJ, Grinstein S, Roth Z. Diversity and versatility of phagocytosis: roles in innate immunity, tissue remodeling, and homeostasis. *Front Cell Infect Microbiol* (2017) 7:191. doi:10.3389/fcimb.2017.00191
87. Amigorena S, Savina A. Intracellular mechanisms of antigen cross presentation in dendritic cells. *Curr Opin Immunol* (2010) 22:109–17. doi:10.1016/j.coi.2010.01.022
88. Houde M, Bertholet S, Gagnon E, Brunet S, Goyette G, Laplante A, et al. Phagosomes are competent organelles for antigen cross-presentation. *Nature* (2003) 425:402–6. doi:10.1038/nature01912
89. Joffre OP, Segura E, Savina A, Amigorena S. Cross-presentation by dendritic cells. *Nat Rev Immunol* (2012) 12:557–69. doi:10.1038/nri3254
90. Guermonprez P, Valladeau J, Zitvogel L, Théry C, Amigorena S. Antigen presentation and T cell stimulation by dendritic cells. *Annu Rev Immunol* (2002) 20:621–67. doi:10.1146/annurev.immunol.20.100301.064828
91. Heath WR, Carbone FR. Cross-presentation in viral immunity and self-tolerance. *Nat Rev Immunol* (2001) 1:126–34. doi:10.1038/35100512
92. Fadok VA, Bratton DL, Konowal A, Freed PW, Westcott JY, Henson PM. Macrophages that have ingested apoptotic cells in vitro inhibit proinflammatory cytokine production through autocrine/paracrine mechanisms involving TGF-beta, PGE2, and PAF. *J Clin Invest* (1998) 101:890–8. doi:10.1172/JCI1112
93. Fadok VA, McDonald PP, Bratton DL, Henson PM. Regulation of macrophage cytokine production by phagocytosis of apoptotic and post-apoptotic cells. *Biochem Soc Trans* (1998) 26:653–6. doi:10.1042/bst0260653
94. Fadok VA, Chimini G. The phagocytosis of apoptotic cells. *Semin Immunol* (2001) 13:365–72. doi:10.1006/smim.2001.0333
95. Chung EY, Kim SJ, Ma XJ. Regulation of cytokine production during phagocytosis of apoptotic cells. *Cell Res* (2006) 16:154–61. doi:10.1038/sj.cr.7310021
96. Sarang Z, Joós G, Garabuczi É, Rühl R, Gregory CD, Szondy Z. Macrophages engulfing apoptotic cells produce nonclassical retinoids to enhance their phagocytic capacity. *J Immunol* (2014) 192:5730–8. doi:10.4049/jimmunol.1400284
97. A-Gonzalez N, Bensinger SJ, Hong C, Beceiro S, Bradley MN, Zelcer N, et al. Apoptotic cells promote their own clearance and immune tolerance through activation of the nuclear receptor LXR. *Immunity* (2009) 31:245–58. doi:10.1016/j.immuni.2009.06.018

98. Weavers H, Evans Iwan R, Martin P, Wood W. Corpse engulfment generates a molecular memory that primes the macrophage inflammatory response. *Cell* (2016) 165:1658–71. doi:10.1016/j.cell.2016.04.049
99. Xiao YQ, Freire-de-Lima CG, Schiemann WP, Bratton DL, Vandivier RW, Henson PM. Transcriptional and translational regulation of TGF-β production in response to apoptotic cells. *J Immunol* (2008) 181:3575–85. doi:10.4049/jimmunol.181.5.3575
100. Krysko DV, D'Herde K, Vandenabeele P. Clearance of apoptotic and necrotic cells and its immunological consequences. *Apoptosis* (2006) 11:1709–26. doi:10.1007/s10495-006-9527-8
101. Petrovski G, Ayna G, Májai G, Hodrea J, Benkő S, Mádi A, et al. Phagocytosis of cells dying through autophagy induces inflammasome activation and IL-1β release in human macrophages. *Autophagy* (2011) 7:321–30. doi:10.4161/auto.7.3.14583
102. Li Z, Venegas V, Nagaoka Y, Morino E, Raghavan P, Audhya A, et al. Necrotic cells actively attract phagocytes through the collaborative action of two distinct PS-exposure mechanisms. *PLoS Genet* (2015) 11:e1005285. doi:10.1371/journal.pgen.1005285
103. van Zandbergen G, Klinger M, Mueller A, Dannenberg S, Gebert A, Solbach W, et al. Cutting edge: neutrophil granulocyte serves as a vector for *Leishmania* entry into macrophages. *J Immunol* (2004) 173:6521–5. doi:10.4049/jimmunol.173.11.6521
104. Fuentes-González AM, Contreras-Paredes A, Manzo-Merino J, Lizano M. The modulation of apoptosis by oncogenic viruses. *Virol J* (2013) 10:182. doi:10.1186/1743-422x-10-182

Conflict of Interest Statement: The authors declare that the research was conducted in the absence of any commercial or financial relationships that could be construed as a potential conflict of interest.

20

Requirement of Gamma-Carboxyglutamic Acid Modification and Phosphatidylserine Binding for the Activation of Tyro3, Axl, and Mertk Receptors by Growth Arrest-Specific 6

Edited by:
Amiram Ariel,
University of Haifa, Israel

Reviewed by:
Carla Rothlin,
Yale University, United States
Ian Dransfield,
University of Edinburgh,
United Kingdom

***Correspondence:**
Raymond B. Birge
birgera@njms.rutgers.edu

Ke Geng[1‡], Sushil Kumar[1†‡], Stanley G. Kimani[1], Vladyslav Kholodovych[2,3], Canan Kasikara[1], Kensaku Mizuno[4], Oleta Sandiford[5], Pranela Rameshwar[5], Sergei V. Kotenko[1] and Raymond B. Birge[1]*

[1]Department of Microbiology, Biochemistry, and Molecular Genetics, New Jersey Medical School Cancer Center, Rutgers, State University of New Jersey, Newark, New Jersey, United States, [2]Office of Advanced Research Computing (OARC), Rutgers, State University of New Jersey, Newark, New Jersey, United States, [3]Department of Pharmacology, Robert Wood Johnson Medical School, Rutgers, State University of New Jersey, Piscataway, NJ, United States, [4]Department of Biomolecular Sciences, Graduate School of Life Sciences, Tohoku University, Sendai, Miyagi, Japan, [5]Department of Medicine, Rutgers, State University of New Jersey, Newark, New Jersey, United States

The Tyro3, Axl, and Mertk (TAM) receptors are homologous type I receptor tyrosine kinases that have critical functions in the clearance of apoptotic cells in multicellular organisms. TAMs are activated by their endogenous ligands, growth arrest-specific 6 (Gas6), and protein S (Pros1), that function as bridging molecules between externalized phosphatidylserine (PS) on apoptotic cells and the TAM ectodomains. However, the molecular mechanisms by which Gas6/Pros1 promote TAM activation remains elusive. Using TAM/IFNγR1 reporter cell lines to monitor functional TAM activity, we found that Gas6 activity was exquisitely dependent on vitamin K-mediated γ-carboxylation, whereby replacing vitamin K with anticoagulant warfarin, or by substituting glutamic acid residues involved in PS binding, completely abrogated Gas6 activity as a TAM ligand. Furthermore, using domain and point mutagenesis, Gas6 activity also required both an intact Gla domain and intact EGF-like domains, suggesting these domains function cooperatively in order to achieve TAM activation. Despite the requirement of γ-carboxylation and the functional Gla domain, non-γ-carboxylated Gas6 and Gla deletion/EGF-like domain deletion mutants still retained their ability to bind TAMs and acted as blocking decoy ligands. Finally, we found that distinct sources of PS-positive cells/vesicles (including apoptotic cells, calcium-induced stressed cells, and exosomes) bound Gas6 and acted as cell-derived or exosome-derived ligands to activate TAMs. Taken together, our findings indicate that PS is indispensable for TAM activation by Gas6, and by inference, provides new perspectives on how PS, regulates TAM receptors and efferocytosis.

Keywords: phosphatidylserine, Tyro3, Axl, and Mertk receptors, growth arrest-specific 6, vitamin K, γ-carboxylation, tumor exosomes

INTRODUCTION

Tyro3, Axl, and Mertk (TAMs) comprise homologous type I transmembrane receptor tyrosine kinases that are implicated as both oncogenic kinases that drive transformation and tumorigenicity of cancer cells, as well as receptors for the clearance of apoptotic cells and regulate innate immunity and dampen inflammation (1–3). Structurally, TAM receptors have a similar topology and domain organization, whereby their ecto-domains are comprised of dual tandem immunoglobulin-like domains (Igl and Ig2) that bind ligand, two fibronectin type III repeats, followed by a transmembrane and intracellular tyrosine kinase domain with a conserved sequence KW(I/L)A(I/L)ES (4–7).

The main ligands for TAMs are two homologous proteins, growth arrest-specific 6 (Gas6) and protein S (Pros1), that are γ-carboxylated by a process that depends on vitamin K. Gas6 and Pros1 also share similar spatial topological homology, including an N-terminus γ-carboxyglutamic acid (Gla) domain, four tandem EGF-like repeats followed by LG1 and LG2 domains, that bind to Igl and Ig2 domains of TAM receptors to initiate receptor activation (8, 9). However, despite a high degree of homology between TAMs and their ligands, the ligand-inducible TAM activation follows a biochemical hierarchy whereby (i) Axl is preferentially activated by Gas6 with 100–1,000× higher binding affinity over Mertk and Tyro3 (kd in the nM range), (ii) Tyro3 is preferentially activated by Pros1, and (iii) Merk displays lower sensitivity to both ligand proteins (in the μM range) (10–12). Notably however, both Tyro3 and Mertk become hyperactivated in the presence of phosphatidylserine (PS)-positive liposomes and apoptotic cells, implying that Mertk and Tyro3 may act as "PS sensors" for externalized PS on apoptotic cells and in the tumor microenvironment (13, 14). By utilizing this arrangement, the TAM receptors act as indirect receptors for apoptotic cells, and through their bridging molecules Gas6 and Pros1, are critical for the clearance of apoptotic cells under both homeostatic and stress conditions (15, 16). In mouse models, knockout of TAMs results in inefficient apoptotic cell clearance, and subsequently, in the development of a SLE-like autoimmune condition (17). Single knockout of Mertk partially phenocopies these defects on clearance, whereby resident and bone-marrow derived macrophages fail to engulf apoptotic cells leading to increased circulating inflammatory cytokines and subsequent anti-dsDNA antibodies (18). Therefore, at a functional level, TAMs are thought to have homeostatic roles in higher metazoans that mediate the tolerogenic clearance of apoptotic cells as well as the resolution of inflammation. In other physiological processes, TAMs are also expressed on retinal pigmented epithelial cells (RPEs), Sertoli cells, resident brain microglia where they are involved in the uptake and clearance of rod outer segments, immature spermatogonium, and apoptotic neurons/pruned synapses, respectively, processes that also appear to depend on externalized PS (19–23).

In addition to their homeostatic roles under physiological conditions and in the resolution phase of inflammation, in recent years TAMs have been implicated in human cancers where they have dual roles, first as direct oncogenes expressed on cancer cells to influence proliferation, metastasis, and chemoresistance (3, 24–28), and second as potential checkpoint inhibitors on myeloid cells that induce expression of immunosuppressive cytokines to drive immune escape (26, 29, 30). Therefore, from a targeting therapeutic standpoint, TAMs are intriguing receptor targets (31), since they utilize the same receptor ligand pairs to promote both oncogenic progression and immune escape, and as such, there is great excitement in the field to develop TAM therapeutics. However, despite the importance of TAMs and their ligands in signaling, presently the mechanisms by which Gas6 and Pros1 activate TAMs, and the role that PS plays in this process, is not well understood. Furthermore, no crystal structures or cryo-EM models for Gas6-induced TAM dimerization/activation have been reported. Therefore, how TAMs achieve ligand-dependent dimerization and activation is not clear.

In the present study, we used biochemical, molecular, and reporter cell-based models to study Gas6-mediated activation of TAMs as well as the requirements for PS. Using TAM/IFNγR1 reporter cell lines to monitor functional TAM activity, we found that Gas6 activity was exquisitely dependent on vitamin K-mediated γ-carboxylation, and that replacing vitamin K with anticoagulant warfarin abolished the γ-carboxylation of Gas6, and abrogated activity toward TAM receptors. Furthermore, using domain mutagenesis, we found that Gas6 activity required both an intact Gla domain and intact EGF-like repeats, as mutant Gas6 that carries intact Gla and LG1 and LG2 domains but only lacks EGF-like repeats was also inactive, suggesting these domains are used cooperatively in order to achieve TAM activation, possibly by facilitating dimerization. Using LC/MS/MS to map important γ-carboxylation sites predicted to bind PS, we found that E54/E55 mutants abrogated Gas6 activity, supporting a direct role for PS binding in Gas6-mediated activation of TAMs. In addition, we found that different forms of PS-positive cells and tumor exosomes (comprising important sources of PS in the tumor microenvironment) all recruited Gas6 to their surfaces and acted as cell-based or exosome-based ligands to activate TAMs. Taken together, our findings indicate that PS is indispensable for TAM activation by Gas6, and add new perspectives on how PS impinges on the activation of TAMs.

MATERIALS AND METHODS

Human Gas6-Containing Media (Gas6-CM) and Cell Culture

HEK293 cells were grown in DMEM media supplemented with 10% fetal bovine serum (FBS, Sigma) and 1% penicillin/streptomycin and incubated in 37°C and 5% CO_2 humidified incubator. To produce wild-type Gas6-CM, when cells reached approximately 60% confluency, pSecTaq-hGas6 (10) was transfected into the cells using LipoD293 transfection reagent (SignaGen). pUcD2SRα-rGas6-Myc plasmids that encode the domain-deleted mutant Gas6 were described previously (32). ΔG, ΔE, and ΔGE represent the mutant Gas6 proteins with Gla domain deleted, EGF-like repeats deleted, and both Gla domain and EGF-like repeats deleted, respectively. The expected molecular weight of these mutant Gas6 proteins is ΔG, 70 kDa; ΔE, 59 kDa; and ΔGE, 49 kDa. These plasmids were transfected into the HEK293 cells with the same method to produce mutant Gas6-CM. 18 h after transfection, the cells were washed twice with PBS and the growth media was replaced with serum-free DMEM media supplemented with 4.4 μM vitamin K (Hospira) or 2 μM warfarin (Sigma) to promote, or abrogate, γ-carboxylation, respectively. The Gas6-CM (Gas6-CM) was collected 72 h later and filtered through 0.22 μm filter. Gas6 concentrations were evaluated using a standard curve against purified recombinant Gas6 (R&D Systems) as previously reported, and unless otherwise stated, approximately 250 nM was used to stimulate TAM receptors in this study (13). Human TAM/IFNγR1 chimeric reporter cells were grown in HAM's F12 media supplemented with 10% FBS, 2 mM glutamine, and 400 μg/ml of G418 as previously described (10). H1299 and Jurkat cells were grown in RPMI1640 media supplemented with 10% FBS and 1% penicillin/streptomycin. MDA-MB231 and U118 cells were grown in DMEM media supplemented with 10% FBS and 1% penicillin/streptomycin.

Site-Directed Mutagenesis

Primers for mutagenizing E54 and E55 were designed from QuikChange and purchased from Sigma. The sequences of primers are: Forward 5′-gcgcctttcaggtcttcgacgacgccaagcagg-3′; Reverse 5′-cctgcttggcgtcgtcgaagacctgaaaggcgc-3′. Site-directed mutagenesis was performed by using QuikChange II XL site-directed mutagenesis kit and following the protocol provided within the kit.

Mass Spectrometric Analysis

The Gas6-CM collected from the transfected HEK293 cells was subjected to SDS-PAGE electrophoresis followed by Coomassie-blue staining. Gel band at approximately 72 kDa of Gas6 was excised and sent for mass spectrometric analysis. In-gel trypsin digestion was performed and the resulting peptides were analyzed by LC-MS/MS on Orbitrap Velos MS. The MS/MS spectra were searched against a Swissprot human database using a local MASCOT search engine (V.2.3).

Homology Modeling of Gas6/PS Association

A homology model of Gla domain of Gas6 was build based on the available X-ray structure of bovine prothrombin that contained a PS lipid and Ca^{2+} ions, PDB access code is 1NL2. An amino acid sequence of the Gla domain of hGAS6 was obtained from UniProt server (http://www.uniprot.org/), access code Q14393, residues 53–94. After a homology model was created in Molecular Operating Environment (MOE) 2016.08, all CGU (carboxylated glutamic acid residues), a PS lipid and calcium ions were transferred from 1NL2 into the homology model. The model refinement was performed by all atom minimization with Amber10EHT force field in MOE. The second PS lipid was manually docked into the refined model of Gas6 around residue 55 which followed by the second refinement by all atom minimization as described earlier. A structural analysis and a visualization of the interaction network of PS lipids with Gas6 were also completed in MOE (2016.08; Chemical Computing Group ULC, Montreal, QC, Canada, H3A 2R7, 2017).

Western Blotting

Western blotting was performed as described previously (10). Briefly, Gas6-CM or hTAM/IFNγR1 cell lysate was mixed with 6x sample buffer and subjected to SDS-PAGE gel for electrophoresis, and the antibodies used were as follows: anti-hGas6 monoclonal antibody (R&D Systems), anti-Myc antibody (Cell Signaling Technology), anti-γ-carboxyglutamyl residues monoclonal antibody (Sekisui Diagnostics), anti-STAT1 (pY701) antibody (BD Biosciences), and anti-β-actin antibody (Cell Signaling Technology).

TAM/IFNγR1 Reporter Cells Stimulation

Human TAM/IFNγR1 cells were seeded in 6-well plate one day prior to stimulation experiment. When the cells reached approximately 90% confluency on the day of experiment, growth media was replaced with serum-free HAM's F12 medium to starve the cells for 6 h. Gas6-CM or purified Pros1 (isolated from human plasma, purchased from Haematologic Technologies, Inc. HCPS-0090) was added to the reporter cells for 30 min in 37°C incubator. After removing the stimulant, hTAM/IFNγR1 cells were washed with ice-cold PBS, and cellular proteins were extracted by using 1% HNTG lysis buffer (20 mM HEPES, 150 mM NaCl; Triton X100; 10% glycerol; 1 mM phenylmethylsulphonyfluoride; 1 mM sodium vanadate; 0.1 mM sodium molybdate; and 20 μg/ml aprotinin). Protein concentration of cell lysate was determined by Bradford assay using Protein Assay Reagent (Bio-Rad). Same amount of protein from lysate was subjected to SDS-PAGE gel for electrophoresis, and the activation of hTAM/IFNγR1 chimeric receptors was determined by phosphorylated-STAT1 immunoblotting.

For TAM/IFNγR1 cells stimulation by PS positive cells equivalent numbers of apoptotic (UV-treated) or calcium-stressed cells (calcium ionophore A23187-treated) were mixed with Gas6-CM or Pros1 and incubated at room temperature for 30 min first, then the mixture was added to the starved reporter cells for another 30 min at 37°C to trigger receptor activation. To induce PS externalization on native hAxl/IFNγR1 cell lines, the cells were first treated with calcium ionophore A23187, at concentrations of 1, 5, or 10 μM, for 15 min at 37°C. Conversely, to block the externalized PS, PS targeting antibody PGN635 (a gift from Peregrine Pharmaceuticals) was added to the serum-free HAM's F12 medium at concentrations of 100 and 200 μg/ml and then incubated with the hAxl/IFNγR1 reporter cells for 15 min at 37°C. The antibody-containing solution was washed off by PBS, and Gas6-CM was added to activate the hAxl/IFNγR1 reporter cells.

PS Externalization

To induce PS externalization by UV-mediated apoptosis, H1299, Jurkat or MDA-MB231 cells were radiated by UV (CL-1000 Ultra Violet Crosslinker, UVP) for 5 min (25,000 $\mu J/cm^2$) to activate apoptosis pathways. Then the cells were kept in serum-free medium for 2–3 h at 37°C. To induce PS externalization under calcium stress, H1299, Jurkat, and MDA-MB231 cells were detached from plates, washed and treated with 10 μM calcium ionophore A23187 in Ca^{2+}/Mg^{2+} free PBS for 15 min at 37°C.

The apoptotic or stressed cells were stained with FITC-Annexin V apoptosis detection kit with Propidium Iodide (Biolegend) to assess the extent of PS exposure by flow cytometry following the product's protocol.

Exosome Purification

Cell-free media from MDA-MB-231 were collected for exosome isolation using two methods: ExoQuick–TC Exosome Precipitation Solution (System Biosciences) or serial ultracentrifugation at 100,000 *g* for 80 min as previously described (33, 34). The pellets from both methods were washed with PBS and then resuspended in 1 mL of PBS. The samples were diluted at 1/200 and studied by NanoSight Range (Malvern, Westborough, MA, USA). In six different analyses the average particle size was 90 nm with 5.35×10^{10} particles/ml.

PS Staining on the Surface of Exosomes

Exosomes purified from MDA-MB-231 cells were first isolated by using exosome-human CD63 isolation reagent following manufacturer's instruction (Invitrogen Thermofisher). Briefly, the exosomes were incubated with the Dynabeads overnight at 4°C. After this, the beads were washed twice with PBS, with the aid of a magnetic stand. The bound exosomes were stained with FITC-Annexin V using the Apoptosis Detection Kit (BioLegend) to assess PS distribution on the surface by flow cytometry.

Gas6 Binding by Flow Cytometry

Growth arrest-specific 6 binding assays were performed following the method described by Dransfield et al. (35). Briefly, calcium-stressed H1299 cells were firstly prepared as described above and then washed once and re-suspended in washing buffer (PBS with 1 mM $CaCl_2$ and 1 mM $MgCl_2$). 50 nM of recombinant Gas6 (R&D Systems) was added to the cell and incubated in rotation at room temperature for 30 min. The cells were then washed with washing buffer and re-suspended in washing buffer with diluted (1:100) primary Gas6 antibody (Abcam, Ab136249) and incubated in rotation at 4°C for 40 min. After this, the cells were washed twice using washing buffer to remove unbound antibody, and an Alexa Fluor 647-conjugated secondary antibody (Rabbit IgG, Thermo Fisher) diluted (1:500) in washing buffer was added to the cells for 30 min incubation at 4°C in dark. The cells were washed again to remove unbound antibody, and Gas6 binding to the cell surface was assessed by flow cytometry.

To assess Gas6 binding to Axl receptor, hAxl/IFNγR1 cells were detached from the plate and washed twice with PBS that contains 5 mM EDTA to eliminate natively-bound proteins. The cells were then centrifuged and resuspended in the Myc-tagged wild-type or mutant Gas6-CM for 30 min in rotation at room temperature. The cells were then washed once with washing buffer and resuspended in diluted (1:100) anti-Myc-PE antibody (Cell Signaling Technology) for 40 min at 4°C in dark. The cells were washed again to remove unbound antibody, and Gas6 binding to Axl receptors was assessed by flow cytometry.

Statistical Analysis

Student's *t*-tests or two-way ANOVA followed by post/hoc tests were performed to analyze statistical differences between groups using GraphPad Prism software. *p* Values lesser than 0.05 were considered significant ($*p < 0.05$; $**p < 0.005$; $***p < 0.001$; $****p < 0.0001$).

RESULTS

γ-Carboxylation of the Gas6 Gla Domain Is Required for Gas6 Activity

In a previous study, we developed TAM/IFNγR1 chimeric receptor CHO reporter cell lines, whereby the extracellular and trans-membrane domains of each human and mouse TAM receptor were fused in frame to the intracellular domain of IFNγR1 to investigate the molecular mechanisms of TAM activation by endogenous ligands Gas6 and Pros1 (10). Using these reporters, and subsequently validating their utility with native TAM receptors, we reported that TAMs have differential specificities for Gas6 and Pros1, as well as differential affinities as PS sensing receptors to stimulate TAM activity and induce efferocytosis (13). Here, we extended these studies to investigate the requisite role of PS with respect to Gas6-mediated TAM activation, as well as inquiring whether PS externalization on apoptotic cells, calcium-stressed cells, or tumor exosomes have different capabilities to activate TAMs.

To examine the requirement of γ-carboxylation and Gas6 activity toward TAM receptors, we generated recombinant non-γ-carboxylated Gas6 by culturing Gas6 expressing HEK293 cells in the presence of warfarin, an anticoagulant that blocks vitamin K epoxide reductase, an enzyme that reduces vitamin K to its active form, thereby blocking an essential step in vitamin K-dependent γ-carboxylation (36). As shown in **Figure 1A**, when we assessed non-γ-carboxylated versus fully γ-carboxylated Gas6 proteins (produced in the presence of exogenous vitamin K), the warfarin-treated Gas6 showed no detectable γ-carboxylation compared with its vitamin K-treated counterpart, and this was confirmed by LC/MS/MS (data not shown). Moreover, while overall production yields of non-γ-carboxylated and γ-carboxylated Gas6 were similar (**Figure 1A**. lower panel), only γ-carboxylated Gas6 activated hAxl/IFNγR1 reporter cells and the non-γ-carboxylated Gas6 was inactive under these conditions [compare warfarin (W) versus vitamin K (K) in **Figure 1B**]. Interestingly, however, while only γ-carboxylated Gas6 was effective to activate hAxl-IFNγR1 cells, both non-carboxylated (Gas6-W) and γ-carboxylated (Gas6-K) forms of Gas6 are capable of binding to the hAxl-IFNγR1 cells with similar efficacy (i.e., geometric mean intensity) (**Figure 1C**), suggesting that while γ-carboxylation is not required for Axl binding, it appears to be critical to induce a conformational change following Gas6 binding to Axl, presumably in order to achieve dimerization and postreceptor signaling.

Structure/Activity Analysis of Gas6 by Domain Mutagenesis

To better understand why N-terminal γ-carboxylation is required for TAM activation (which is mediated *via* the binding of the carboxyl-terminal LG1 and LG2 domains to TAMs), we performed structure activity experiments on Gas6 (for TAM activation) using a series of Gas6 domain deletion mutants (**Figure 1D**).

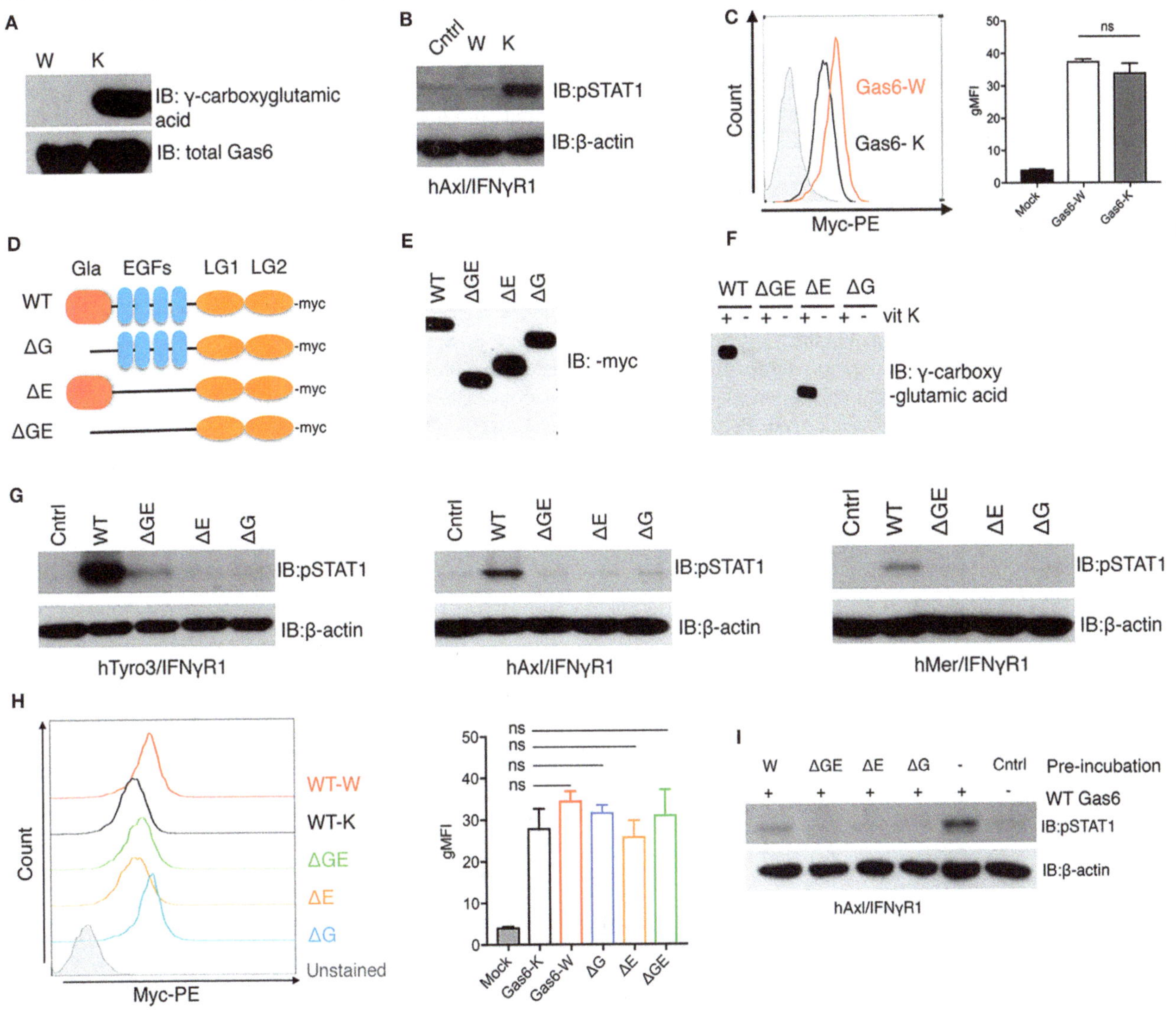

FIGURE 1 | γ-Carboxylation is required for growth arrest-specific 6 (Gas6) activity. **(A)** HEK293 cells were transfected with pSecTaq-hGas6 plasmid (10) to collect Gas6-containing media (CM). Serum-free culture media was supplemented with either 2 μM warfarin (W) or 4.4 μM vitamin K (K) to abolish or induce γ-carboxylation of Gas6. The conditioned media was subjected to SDS-PAGE gel to assess total Gas6 and γ-carboxylation level by Gas6 and γ-carboxylglutamic acid immunoblotting. **(B)** hAxl/IFNγR1 reporter cells were stimulated by warfarin-treated Gas6-CM (W) or vitamin K-treated Gas6-CM (K). The activation of Axl receptor was assessed by pSTAT1 immunoblotting. **(C)** hAxl/IFNγR1 cells were incubated with warfarin-treated Gas6 (Gas6-W, red) conditioned media or vitamin K-treated Gas6 (Gas6-K, black) conditioned media. The binding of Gas6 to hAxl/IFNγR1 receptors was assessed in flow cytometry by using anti-Myc-PE antibody (left), with geometric MFI quantification from independent experiments ($n = 3$) in the right panel. Error bar, SEM. n.s., not significant. **(D)** Schematic structure of Myc-tagged mutant Gas6 proteins. WT, wild-type Gas6; ΔG, Gas6 that lacks Gla domain; ΔE, Gas6 that lacks EGF repeats; ΔGE, Gas6 that lacks both Gla domain and EGF repeats. **(E)** HEK293 cells were transfected with the pUcD2SRα-rGas6-Myc plasmid (32) in the presence or absence of vitamin K. The expression level of the Gas6 proteins in the conditioned media was assessed by Myc-immunoblotting. **(F)** The γ-carboxylation level of the Gas6 proteins was assessed by γ-carboxylglutamic acid immunoblotting. **(G)** Human TAM/IFNγR1 reporter cells were stimulated by wild-type or mutant Gas6-CM, and the activations of TAM receptors were assessed by pSTAT1 immunoblotting. **(H)** The binding of different (warfarin or vitamin K-treated wild-type, or mutant) Gas6 proteins to hAxl/IFNγR1 cells was assessed by flow cytometry using anti-Myc-PE antibody (left), with MFI quantification from independent experiments ($n = 3$) in the right panel. Error bar, SEM, n.s., not significant. **(I)** Warfarin-treated (W) or mutant Gas6 was added to hAxl/IFNγR1 cells in culture dish for 30 min. The cells were then washed by PBS and wild-type Gas6 with fully γ-carboxylation was added to the cells. The activation of Axl was assessed by pSTAT1 immunoblotting. The Western blotting results are representative results from at least three independent experiments.

Previous studies, for example, predicted that non-γ-carboxylated Gas6 might form an intramolecular inhibitory structure to block LG domains, that becomes released by γ-carboxylation of the Gla domain to activate Gas6 (32). To explore this idea, we expressed Myc-tagged domain mutant recombinant Gas6 proteins (**Figure 1D**), either in the presence of warfarin or vitamin K to generate secreted recombinant mutant proteins expressed in similar amounts (**Figure 1E**). As expected, only WT Gas6 and

ΔE mutant Gas6 (that deletes the 4 tandem EGF-like domains) showed vitamin K-dependent γ-carboxylation (**Figure 1F**, lanes 1 and 5), which were subsequently blocked in Gas6-producing cells treated with warfarin (**Figure 1F**, lanes 2 and 6). Interestingly, however, when Gas6 domain mutants were tested on the TAM/IFNγR1 cell lines (Tyro3/IFNγR1, Axl/IFNγR1, and Mertk/IFNγR1 lines), none of the mutant Gas6 proteins (ΔG, ΔE, and ΔGE) showed detectable activity for TAM receptors compared to WT Gas6 (**Figure 1G**). Notably, the lack of activity of the ΔE mutant Gas6 (that contains Gla and LG1/LG2 domains) implies that the 4 tandem EGF-like repeats are also required for Gas6-mediated TAM activation, a domain region of Gas6 that has not previously been implicated in regulating the function of Gas6.

To explore whether Myc-Gas6 mutants retained ability to bind to the TAM receptors (and are functional proteins), we incubated each of the aforementioned proteins with hAxl/IFNγR1 reporter cells. All of the aforementioned proteins bound TAMs as evident by flow cytometry (**Figure 1H**) with similar efficacy (i.e., geometric mean intensity) but could not dimerize/activate the TAMs (**Figure 1G**). Indeed, when mutant proteins were first incubated with hAxl/IFNγR1 cells, and subsequently cells were treated with WT Gas6, prior exposure to inactive mutant proteins blocked subsequent Gas6 activity, suggesting they may act as dominant negative or ligand traps to block TAM function (**Figure 1I**). Indeed, these data are consistent with previous reports that warfarin can block TAM signaling in cancer models by producing inactive TAM ligands (29).

Mapping Gas6 γ-Carboxylation by LC-MS/MS

The above mentioned experiments employing a series of Gas6 domain mutants implicated γ-carboxylation as an essential feature for Gas6-mediated activation of TAMs. To investigate the relationship between γ-carboxylation and binding to PS in more detail, we performed LC-MS/MS analysis on hGas6, prepared from vitamin K-treated Gas6-producing cells (**Figure 2A**). After SDS-PAGE electrophoresis, the Gas6 band at ~72 kDa was excised, and peptide bands were assessed for γ-carboxylation (i.e., increased mass of 44 Da). Under these conditions, (whereby

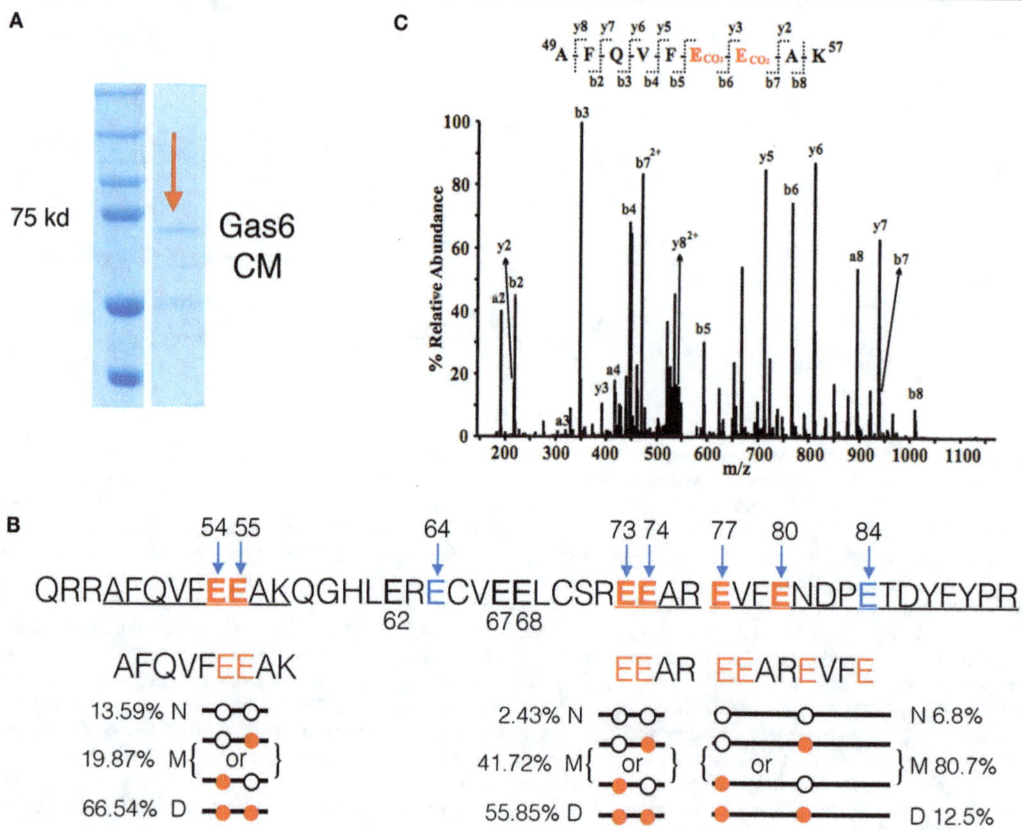

FIGURE 2 | LC-MS/MS analysis of γ-carboxylation sites in Gla domain of growth arrest-specific 6 (Gas6). **(A)** Gas6-containing media (Gas6-CM) was subjected to SDS-PAGE electrophoresis, and the band at 72 kDa was excised and subjected to LC-MS/MS analysis. **(B)** In the Gla domain of Gas6, glutamic acids 54, 55, 64, 73, 74, 77, 80, and 84 were identified as γ-carboxylated by LC-MS/MS. Among them, three independent carboxylated peptides, ^{49}AFQVFEEAK^{57}, ^{73}EEAR^{76}, and ^{77}EVFENDPETDYFYPR^{91}, were identified with glutamic acid residues shown in red. Each peptide was found in three forms: non-γ-carboxylated (N), single carboxylated at one of the E residues (M), or dually carboxylated at both E residues **(D)**. The frequencies of each form of the peptides were listed. **(C)** Representative MS/MS spectrum of a doubly-charged ion (*m/z* 578.76) is corresponding to the peptide sequence of ^{49}AFQVFEEAK^{57} with carboxylation modification at E54 and E55. The observed *y*- and *b*-ion series confirmed the peptide sequence and modification.

Gas6 is functionally active; **Figure 1**), we isolated three peptides from the trypsin-digested protein for stoichiometry analysis, and found a complex pattern of γ-carboxylated glutamic acids in the active species of Gas6. A representative MS/MS spectrum of the γ-carboxylated peptide ^{49}AFQVFEEAK^{57} was shown in **Figure 2C**. For example, in the peptide ^{49}AFQVFEEAK^{57}, E54 and E55 were found either dually γ-carboxylated at a frequency of 66.5%, or single γ-carboxlated at one or the other glutamic acid at a frequency of 19.87%, or non-carboxylated at a frequency of only 13.6%. Similarly, for peptide ^{73}EEAR^{76}, E73 and E74 were either dually γ-carboxylated at a frequency of 55.85%, or single γ-carboxylated at one or the other residue at 41.72%, or non-carboxylated at a frequency of 2.43%, while in a third γ-carboxylated peptide ^{77}EVFENDPETDYFYPR^{91}, the frequency of dually, single, and non-γ-carboxylation at E77 and E80 were found to be 12.5, 80.7, and 6.8%, respectively (**Figure 2B**). Other residues at lower stoichiometry included E64, E67, and E84. Together, these data suggest that γ-carboxylation of Gas6 is highly dynamic, likely to fine-tune Gas6 activity *via* the activity of enzymes that are required for this post-translational modification, such as gamma-glutamyl carboxylase (GGCX) and vitamin K epoxide reductase complex I (VCOR1).

Model for Interaction between Gas6-Gla Domain and PS Based on the PDB Structure of Bovine Prothrombin

Having identified a complex pattern in the γ-carboxylation of active Gas6, including residues E54, E55, E64, E73, E74, E77, E80, and E84 we aligned hGas6 across species, as well as to other γ-carboxylated clotting factors that belong to the vitamin K-dependent proteins family, that require γ-carboxylated in their Gla domains to bind PS exposed on activated platelets (37). As shown in **Figures 3A,B**, the arrangement of glutamic acid motifs in Gla domain of Gas6 is well-conserved across species, and pairwise analysis showed considerable conservation among members of Gla-containing proteins in human such as F7, F9, thrombin, Pros1, Gas6, and bovine prothrombin, that latter of which there is a derived crystal structure in the presence of PS (38). Indeed, based on the available structure and the high degree of homology between bovine Prothrombin and human Gas6 (**Figure 3C**), we generated homology models of hGas6 in complex with PS.

From these models and the experimental-derived constellation of γ-carboxylated glutamic residues identified in the Gla domain of Gas6 by LC-MS/MS, a synergy model is predicted whereby two or more PS head-groups can interact symmetrically with the Gla domain. In these models, one molecule of PS (PS1) is predicted to interact with E54 and E64 on one surface, while a second molecule of PS (PS2) is predicted to interact with E55, E73, and E77 on the opposite surface of the Gla domain (**Figures 3D,E**). Indeed, since the E54/E55 motif are located at the base of the Gla domain that makes a direct contact with phospholipids, and therefore predicted to bind PS directly, we generated a Gas6 E54/E55 double mutant (**Figure 3F**). As indicated, this mutant abrogated Gas6-mediated activation of hAxl/IFNγR1 cells, even though E54D/E55D retained almost complete net γ-carboxylation levels, as evident from Western blotting with an anti-γ-carboxyglutamic acid (Gla) specific mAb (**Figure 3F**). These data suggest that critical PS interacting glutamic acid residues in Gas6, when substituted, abrogate ligand-binding activity of Gas6, further supporting this idea that direct PS binding is indispensable for Gas6-mediated activation of TAMs.

PS-Positive Cells Opsonize Gas6 and Stimulate TAMs

Given the essential role for PS for the Gas6-mediated activation of TAMs, we next examined physiological sources of PS, and how they influence Gas6-mediated activation of TAMs (**Figure 4**). It is well known that PS is predominantly distributed on the inner leaflet of the plasma membrane, but can be externalized (i) during apoptosis, (ii) during cell stress (particularly calcium induced stress), and (iii) on exosomes derived from tumor cells (tumor exosomes) (39, 40). Over the past several years, molecular mechanisms for PS externalization have been described, including the identification of scramblases and mechanisms of PS externalization under the aforementioned conditions. For example, during apoptosis, PS is externalized *via* activation of a caspase 3/caspase 7-activated scramblase called Xkr8 (41), while PS is externalized in calcium-stress cells by a calcium-activated scramblase called TMEM16F (42).

In order to compare physiological sources of externalized PS, including PS externalized following calcium ionophore treatment (stress) versus PS externalized following UV irradiation (apoptosis), and how these cells impinge on Gas6-mediated TAM activation, we induced PS externalization on distinct cancer cell lines, including Jurkat T cell leukemia cells, MDA-MB-231 breast cancer cells, and H1299 lung cancer cells, and tested their ability to (i) recruit soluble recombinant Gas6 to their surface and (ii) to activate, as cell-based ligands, TAM/IFNγR1 reporter lines *via* cell to cell interactions (**Figure 4**).

As shown in **Figure 4A**, When Jurkat cells, MDA-MB-231 cells, or H1299 cells were transiently exposed to calcium ionophore A23187 for 15 min as an acute non-apoptotic stressor, all three cell lines displayed clear PS externalization (as evident from FITC-Annexin V positive/PI-negative staining). In an independent experiment, these cells also expose PS on their surface due to the UV radiation-induced apoptosis (**Figure 4B**). Quantifications using geometric mean staining of annexin V-positive cells revealed similar PS externalization under both conditions (**Figure 4C**).

Next, to assess whether PS positive cells induced by stress versus apoptosis recruited Gas6, cells were normalized for the extent of PS exposure, and subsequently incubated cells with 50 nM recombinant Gas6. Under these conditions, both cell populations recruited Gas6 to a similar amount (~30%) as evident by flow cytometric analysis (**Figure 4D**). Subsequently, aforementioned PS positive cells from **Figure 4A** were normalized for PS (after flow cytometry), incubated with Pros1 or Gas6-CM, and further added to the TAM/IFNγR1 reporter lines as opsonized cell-based ligands. As indicated, both apoptotic cells and calcium stressed cells profoundly enhanced Mertk and Tyro3 activation by Gas6 and Pros1, respectively, compared with receptor activation induced by Gas6 or Pros1 alone. Axl activation by Gas6, on the other hand, was not further enhanced in the presence of apoptotic cells and decreased by calcium stressed cells in some cases

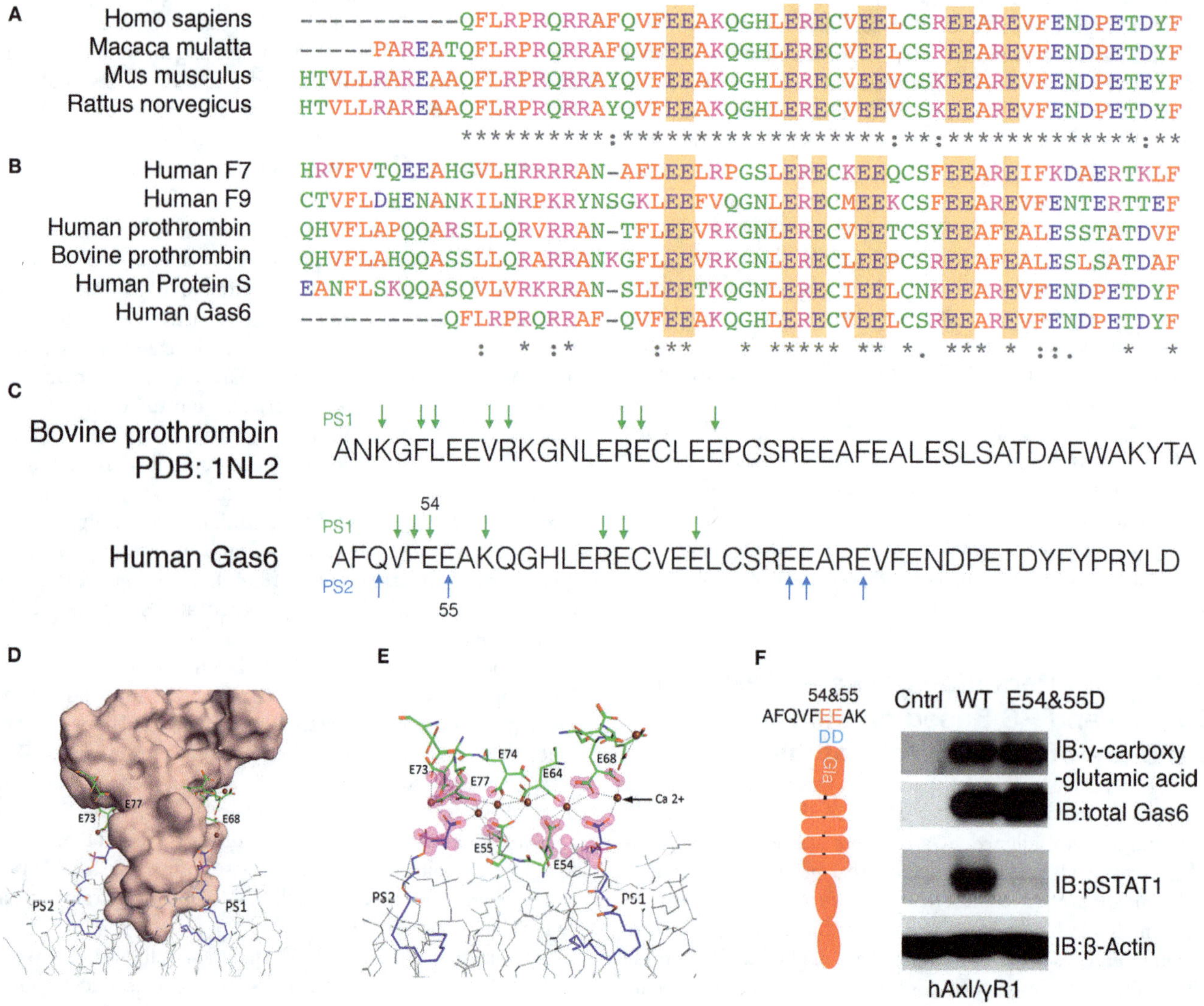

FIGURE 3 | Homology modeling for the association between Gla domain of growth arrest-specific 6 (Gas6) and phosphatidylserine (PS). **(A)** The sequences of Gla domain of Gas6 proteins between species and **(B)** the sequences of Gla domain of different human vitamin K-dependent proteins were pairwise aligned with conserved glutamic acids highlighted in orange. **(C)** Homology model of Gla domain of Gas6 is based on Gla domain of bovine prothrombin PDB: 1NL2. The amino acids that may provide binding sites for PS were pointed out by green arrows. In human Gas6, two PS molecules (PS1 and PS2) are predicted to be associated with Gla domain from two symmetrical interfaces of the protein; those amino acids were pointed out by green and blue arrows separately. **(D)** Molecular surface of Gla domain of Gas6 was shown in pink, and the γ-carboxylated E residues shown in green, associated with calcium ions as brown balls, two PS molecules are shown in blue and POC lipids as gray wire models. **(E)** Possible contacts between PS and E residues are highlighted with magenta clouds. **(F)** Point mutations were induced to switch 54 and 55 glutamic acids (E) to aspartic acids (D). HEK293 cells were transfected with the mutant Gas6 plasmid, and the γ-carboxylation and expression of wild-type or E54 and 55D Gas6 proteins were assessed by γ-carboxylglutamic acid or Gas6 immunoblotting (upper panel). hAxl/IFNγR1 reporter cells were stimulated by wild-type or E54 and 55D Gas6, and the activation of Axl was assessed by pSTAT1 immunoblotting (lower panel). The Western Blotting results are representative results from at least three independent experiments.

(**Figure 4E**). These data support our previous models and suggest that among TAMs, mainly Tyro3 and Mertk act as PS sensors to amplify PS signaling from stress-activated or apoptotic cells.

PS-Positive Tumor Exosomes Opsonize Gas6 and Stimulate TAMs

Having established that apoptotic and calcium-mediated stressed cells, *via* externalized PS, could act as cell-based ligands to activate TAMs in the presence of Gas6/Pros1, we next explored whether PS-positive tumor exosomes could also activate TAMs (**Figure 5**). To address this, tumor-derived exosomes were first isolated from either cultured MDA-MB-231 cells or U118 glioblastoma cells (data not shown) after which, nanoparticles with diameter in the range of 70–100 nm were purified by ultra-centrifugation (**Figure 5A**). To assess whether the tumor-derived exosomes were positive for PS, we subsequently incubated the vesicles from **Figure 5A** with Dynabeads conjugated with anti-CD63 antibody, followed FITC-Annexin V staining (**Figure 5B**). As shown in **Figure 5C**, only the "exosome + beads" were stained

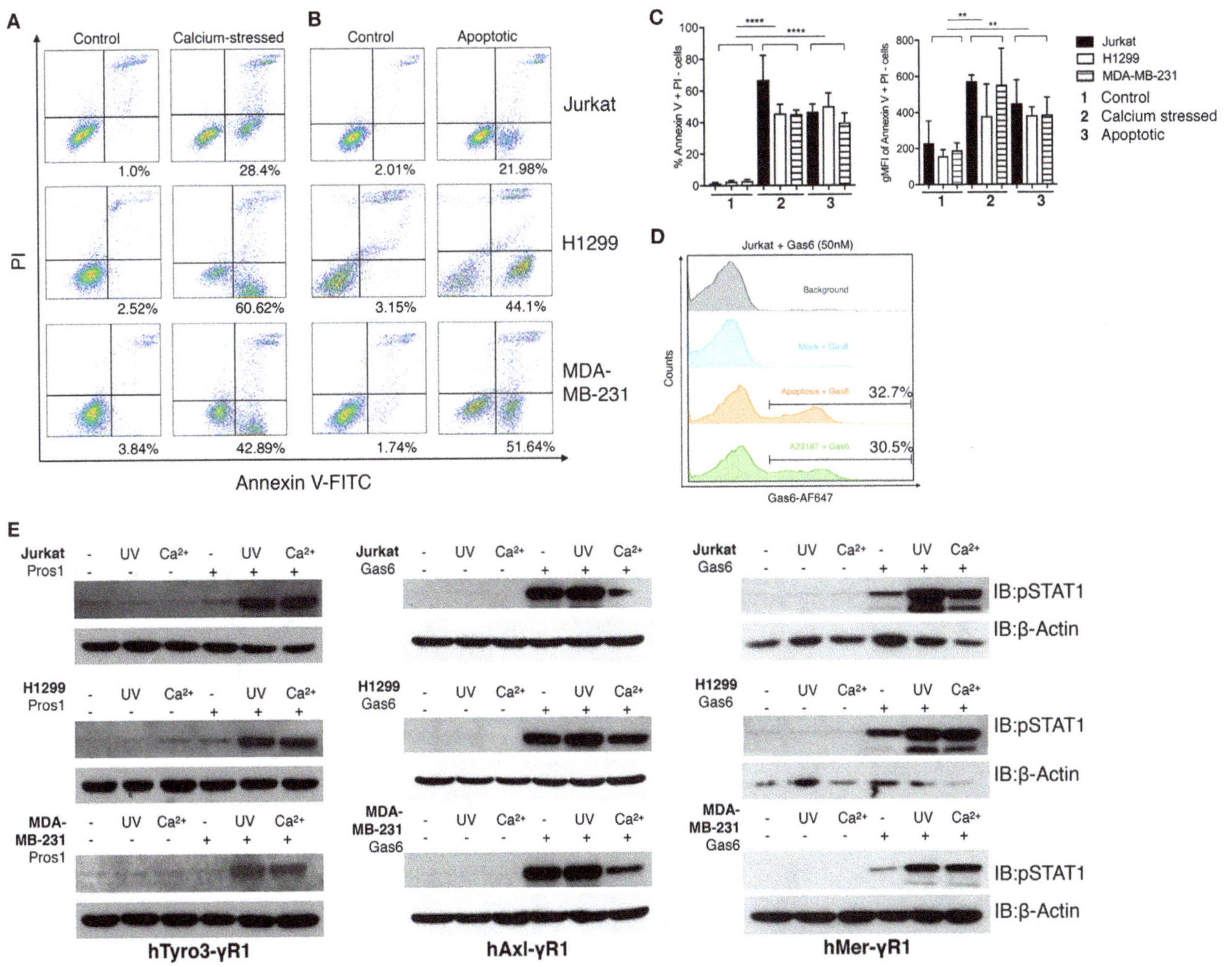

FIGURE 4 | Phosphatidylserine (PS)-positive apoptotic and calcium-stressed cells differentially affect Tyro3, Axl, and Mertk (TAMs) activation by growth arrest-specific 6 (Gas6) or protein S (Pros1). **(A)** Jurkat, H1299, and MDA-MB231 cells (top to bottom) were treated with calcium ionophore A23187 (10 μM) for 15 min in 37°C and then stained with FITC-Annexin V and PI to assess PS externalization. **(B)** Jurkat, H1299, and MDA-MB231 cells (top to bottom) were treated with UV radiation and then stained with FITC-Annexin V and PI to assess PS externalization. **(C)** Percentage (left) and MFI (right) quantification of the Annexin V$^+$/PI$^-$ cells that were induced by calcium stress or UV radiation from independent experiments ($n = 3$). Error bar, SEM, $^{**}p < 0.005$, $^{****}p < 0.0001$. **(D)** Recombinant Gas6 protein at concentration of 50 nM was added to 1.0×10^6 apoptotic or calcium-stressed Jurkat cells and subsequently stained with anti-hGas6 primary and antirabbit Alexa Fluor 647 secondary antibody to assess Gas6-cell surface association by flow cytometry. Background (gray): Jurkat cells without treatment were stained with primary and secondary antibodies; Mock + Gas6 (blue): Jurkat cells without treatment were incubated with 50 nM Gas6 and then stained with primary and secondary antibodies. **(E)** Three modes (resting, calcium-stressed and apoptotic) of Jurkat, H1299 and MDA-MB231 cells were mixed with Gas6-containing media (Gas6-CM, Gas6 concentration estimated as 250 nM) or Pros1 (100 nM) to stimulate hTAM/IFNγR1 reporter cells. The activation of TAM receptors was assessed by pSTAT1 immunoblotting. The Western blotting results are representative results from at least three independent experiments.

the FITC-Annexin V, indicating PS distribution on the surface of the exosomes. These results revealed that in addition to apoptotic and calcium-stressed cancer cells, tumor exosomes are also PS-positive, in agreement with recent published results of Schroit and colleagues (40). To assess whether Gas6/Pros1-opsonized exosomes could activate TAMs, they were subsequently incubated with the hTAM/IFNγR1 reporter cells. Under these conditions, PS-positive exosomes (opsonized with their TAM ligands, Gas6 or Pros1) reproducibly hyper-activated all three TAM receptors (**Figures 5D,E**). This pattern of activation of Axl by PS + exosomes is distinct to the Axl activation observed for PS + apoptotic cells and PS + calcium stressed cells (**Figure 4E**), possibly suggesting different mechanisms of Axl activation by exosomes.

Cell Intrinsic Role for Gas6 in the Activation of TAMs

The aforementioned findings that externalized PS from calcium-activated cells or apoptotic cells could act *in trans* as cell-derived opsonins to activate TAM reporter lines. However, to evaluate whether PS externalization on TAM expressing cells could also act intrinsically, i.e., *in cis, to* activate TAMs by an autocrine mechanism (**Figure 6**), we treated the hAxl/IFNγR1 cell line with 10 μM calcium ionophore A23187 to achieve PS externalization.

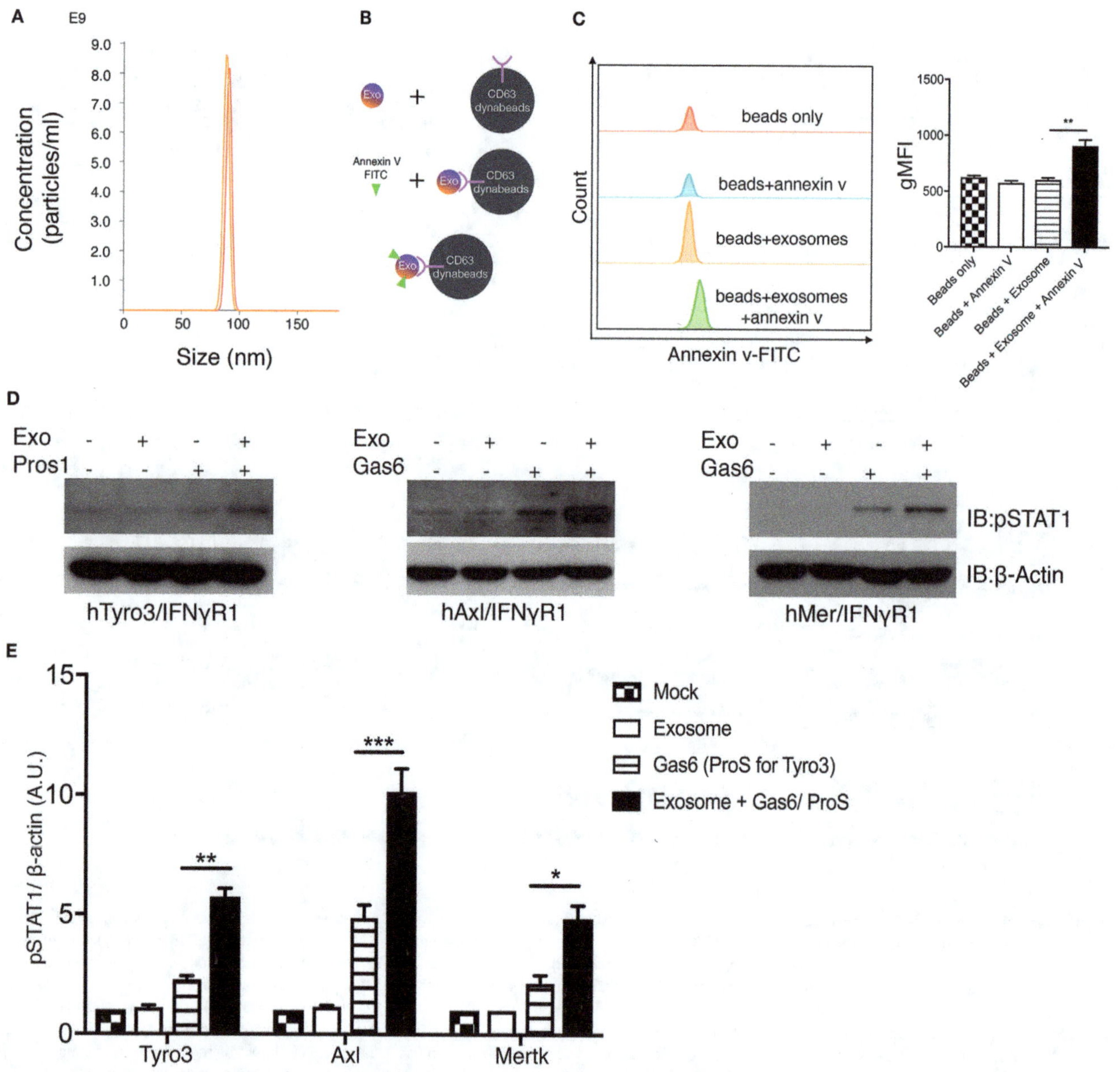

FIGURE 5 | Phosphatidylserine (PS)-positive tumor exosomes enhance Tyro3, Axl, and Mertk (TAMs) activation by growth arrest-specific 6 (Gas6) or protein S (Pros1). **(A)** Tumor-derived exosomes were purified from MDA-MB-231 cell's culture media, and the size and concentration of the exosomes were analyzed by NanoSight Tracking Analysis. **(B)** Schematic procedure of PS staining of exosomes using CD63-Dynabeads. Exosomes isolated from MDA-MB231 cells were incubated with CD63-Dynabeads and then stained with FITC-Annexin V to assess PS exposure on the surface of exosomes. The experiment results are shown in **(C)** in which the samples from top to bottom are: beads alone (red), beads + FITC-Annexin V (blue), beads + exosomes (orange), beads + exosomes + FITC-Annexin V (green). The MFI quantification is shown in the right panel from independent experiments ($n = 3$). Error bar, SEM, $^{**}p < 0.005$. **(D)** Exosomes that were isolated from MDA-MB231 cells were mixed with Gas6-containing media (Gas6-CM) or Pros1 (100 nM) to stimulate hTAM/IFNγR1 reporter cells. The activation of TAM receptors was assessed by pSTAT1 immunoblotting. **(E)** Densitometric quantification of hTAM/IFNγR1 reporter cells' activation by exosomes (pSTAT1) normalized to β-actin from independent experiments ($n = 3$). Error bar, SEM. $^{*}p < 0.05$, $^{**}p < 0.005$, $^{***}p < 0.001$.

As shown in **Figure 6A**, 16.1% of the cells were stained with FITC-Annexin V, indicating the native existence of PS-positive cells in cell culture. After 15 min of calcium ionophore treatment, 48.4% of the cells displayed PS exposure on their surface. When PS targeting antibody PGN632 was preincubated with the non-treated hAxl/IFNγR1 cells to mask the native PS-positive cells, the subsequently added Gas6 induced drastically reduced hAxl/IFNγR1 activation compared with the cells that were not treated with PS antibody (**Figure 6B**, lanes 2–4), demonstrating the requirement of PS for Axl activation. Further, to assess whether PS externalization is sufficient for TAM activation, we increased the concentration of calcium ionophore A23187 (of 1, 5, and 10 μM) and subsequently exposed cells to a subthreshold concentration of Gas6. As shown in **Figure 6C**, Axl activation

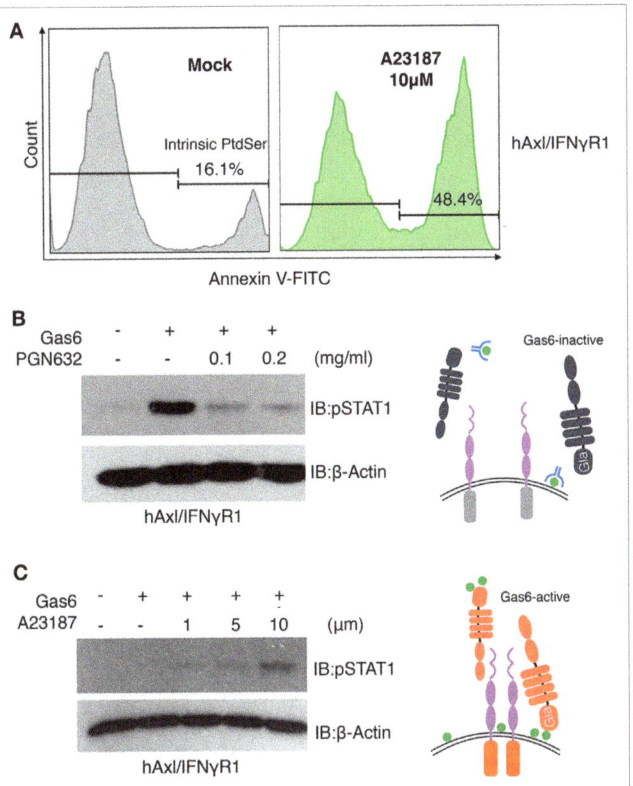

FIGURE 6 | Phosphatidylserine (PS) is involved in growth arrest-specific 6 (Gas6)-Axl activation after calcium-ionophore treatment. **(A)** hAxl/IFNγR1 cells that received no treatment (gray) or 10 μM calcium ionophore A23187 (green) at 37°C for 15 min were stained with FITC-Annexin V to assess PS externalization. **(B)** PGN632, a PS-targeting reagent, was added to hAxl/IFNγR1 reporter cells at two concentrations (0.1 and 0.2 mg/ml) for 20 min. The reporter cells were washed and stimulated by hGas6-CM subsequently. The activation of Axl receptor was assessed by pSTAT1 immunoblotting. Schematic illustration of this experiment is shown in the right panel. **(C)** hAxl/IFNγR1 reporter cells were first treated with calcium ionophore A23187 at concentrations of 1, 5, and 10 μM and then stimulated by hGas6-CM. The activation of Axl receptor was assessed by pSTAT1 immunoblotting. Schematic illustration of this experiment is shown in the right panel. The Western blotting results are representative results from at least three independent experiments.

by Gas6 was most profoundly enhanced when cells were treated with 10 μM calcium ionophore, indicating native (intrinsic) PS externalization could also induce TAM activation. These results suggest that PS-Gas6 can activate TAM-expressing cells as both cell-based ligands and possibly also in an autocrine fashion (see model in **Figure 6C**).

DISCUSSION

In the present study, we used structure-function mutagenesis and cell biological approaches to investigate molecular mechanisms for the Gas6-mediated activation of TAM receptors. Using a series of engineered TAM/IFNγR1 reporter lines, designed to detect ligand-inducible TAM receptor dimerization and postreceptor tyrosine phosphorylation (i.e., TAM receptor activation), our studies demonstrate that activation of TAM receptors by Gas6 requires γ-carboxylation of the N-terminal Gla domain. Supportive evidences include findings that; (i) non-γ-carboxylated Gas6 produced in the presence of warfarin fails to activate TAM receptors, as well as structure/activity analysis that (ii) Gla-less Gas6 fails to activate TAM receptors, and (iii) point mutations of key glutamic acid residues (E54/E55) that directly interact with PS abrogate Gas6-mediated TAM activation. By inference that γ-carboxylated Gla domains bind the surface of anionic phospholipids such as PS, our studies indicate that Gas6 requires interaction with PS and that the active ligand is a functional protein/lipid hybrid. These data are consistent with previous observations showing that Gas6 γ-carboxylation is required for the proliferation of vascular smooth muscle cells (43), for the antiapoptotic function of Gas6 on fibroblasts and endothelial cells (44), and most relevant to the present study, for the clearance of rod outer segments by RPEs (45, 46), as well as apoptotic cell clearance and synaptic pruning in the central nervous system that is mediated by TAM receptors expressed on resident microglia (23). Furthermore, elevated distribution of PS on the surface of many virus, including HIV, has been thought to function in concert with Gas6 and TAM receptors on the targeted cells to facilitate viral entry and infection effectively (47–50).

Although Gas6 requires γ-carboxylation for TAM receptor activation (dimerization), γ-carboxylation does not appear to be required for Gas6/TAM ligand/receptor interactions, as both non-carboxylated and γ-carboxylated forms of Gas6 bind to the surface of Axl/IFN-γR1 cells as shown by flow cytometry (**Figures 1C,H**). These data are consistent with previous reports of Lemke and colleagues showing that Gla-less Gas6 could still bind to TAM receptors (12), as well as our previous data that soluble TAM ecto-domains could also bind non-carboxylated Gas6 (10). Although it is not clear whether Gas6/TAM receptor complexes exist preformed in the absence of PS *in vivo*, or whether there is sufficient steady-state levels of extracellular PS vesicles or PS + cells to achieve the activation of Gas6, the high level of externalized PS dysregulation that occurs in stressed tissues, virally infected tissues, or in the tumor microenvironment is expected provide a strong activation platform for the Gas6/TAM axis, and a molecular rationale for why TAMs are active in the cancer microenvironment and in virally infected tissues. Similarly, it is also possible that non-γ-carboxylated Gas6 proteins, to possibly developed as cancer therapeutics, might act as ligand traps to bind and prevent subsequent TAM activation. This latter idea might offer a molecular explanation that the *in vivo* administration of warfarin could block the Gas6-Axl signaling complex in a pancreatic cancer model (29), as well as that warfarin exerts an antimetastatic action in mice *via* the Cbl-Gas6-TAM axis in NK cells (25).

The finding that Gas6 requires γ-carboxylation and PS binding for biological activity has important implications in cancer biology. This is because, unlike the case of native tissues under homeostatic conditions where externalized PS is undetectable, externalized PS is widely dysregulated and elevated in the tumor microenvironment *via* three interactive events that include (i) the high apoptotic index of highly proliferative cancers, (ii) the high metabolic and hypoxic stress of viable cancer cells, and (iii) tumor-derived exosomes (51).

Mechanistically, PS externalization following apoptosis and stress is mediated by distinct molecular processes; Apoptotic cells externalize PS *via* the combined activity of Xkr8 and ATP11C (41, 52), while stress-induced PS-externalization is regulated by TMEM16 family members, in response to (transient) rise in intracellular calcium (42). Recent studies suggest that tumor exosomes are also PS-positive, likely based on the constitutive externalization of PS on native cancer cells (40). Here, we show that all of the aforementioned sources of externalized PS are able to bind γ-carboxylated Gas6 and serve as cell-based or exosome-based scaffolds to activate TAMs, although they appear to do so with different biological characteristics; Apoptotic cells and stressed cells hyperactivate Mertk and Tyro3 as cell-based opsonins with Gas6, while cell-based sources of PS do not appear to be hyperactive for Axl. On the other hand, 70–100 nm exosomes, derived from MDA-MB-231 cells are also PS positive, and these small extracellular vesicles hyperactivate all three TAMs. Whether PS + exosomes reflect a functional difference in the activation of TAMs in the tumor microenvironment is an important consideration.

Finally, while these studies support an essential role for γ-carboxylation of Gas6, and subsequent interaction with a lipid source of PS, for its ability to activate TAM receptors, we also show by using LS-MS/MS that Gas6 γ-carboxylation has a complex and likely dynamic arrangement in this post-translational modification. Both of the enzymes required for γ-carboxylation, namely GGCX and VCOR1, appear to be broadly expressed in cells, including a variety of cells that express Gas6 and contribute to the tumor microenvironment, such as monocytic cells (macrophages) and cancer cells themselves. It is possible that overexpression of these enzymes, or upregulation in their activity, could enhance the γ-carboxylation of Gas6, and by inference from the results developed here, enhance the activity of TAM ligands (model in **Figure 6**). Further studies examining the levels of GGCX and VCOR1 in the tumor microenvironment are meritorious, as well as developing suitable mouse models to genetically manipulate these rate-limiting enzymes for γ-carboxylation.

CONCLUSION

In summary, we show here by using functional TAM reporter cell lines that both γ-carboxylation and PS binding are indispensable for Gas6 to be active as a TAM ligand. Since externalized PS is frequently dysregulated in the tumor microenvironment, the study has important implications for not only how TAMs function in efferocytosis but also how they skew immune response in cancer.

AUTHOR CONTRIBUTIONS

KG and SKumar designed, performed, and analyzed experiments and wrote the manuscript. SKimani, CK, and OS provided technical help in experiments. VK performed homology modeling experiments. KM provided reagents and intellectual and technical advice. PR and SVK provided intellectual and technical advice. RB conceived and coordinated the project, designed experiments, analyzed results, and wrote the manuscript. All authors reviewed the manuscript.

ACKNOWLEDGMENTS

We would like to thank David Calianese and Viralkumar Davra for helpful discussions, as well as Anita Antes for maintaining the clonally-derived TAM-IFN reporter lines, and Cyril Empig and Bruce Freimark at Peregrine Pharmaceuticals for providing the PS targeting antibody PGN632. We would like to also acknowledge the mass spectrometry and advanced research computing core facilities. The mass spectrometry data were obtained from a Orbitrap instrument funded in part by an NIH grant NS046593, for the support of the UMDNJ Neuroproteomics Core Facility. RBB was supported in part form NIH CA 165077 as well as a grant New Jersey Research Scholars award to Sushil Kumar.

REFERENCES

1. Davra V, Kimani SG, Calianese D, Birge RB. Ligand activation of TAM family receptors-implications for tumor biology and therapeutic response. *Cancers (Basel)* (2016) 8. doi:10.3390/cancers8120107
2. Graham DK, DeRyckere D, Davies KD, Earp HS. The TAM family: phosphatidylserine sensing receptor tyrosine kinases gone awry in cancer. *Nat Rev Cancer* (2014) 14:769–85. doi:10.1038/nrc3847
3. Wu G, Ma Z, Hu W, Wang D, Gong B, Fan C, et al. Molecular insights of Gas6/TAM in cancer development and therapy. *Cell Death Dis* (2017) 8:e2700. doi:10.1038/cddis.2017.113
4. O'Bryan JP, Frye RA, Cogswell PC, Neubauer A, Kitch B, Prokop C, et al. axl, a transforming gene isolated from primary human myeloid leukemia cells, encodes a novel receptor tyrosine kinase. *Mol Cell Biol* (1991) 11:5016–31. doi:10.1128/MCB.11.10.5016
5. Graham DK, Dawson TL, Mullaney DL, Snodgrass HR, Earp HS. Cloning and mRNA expression analysis of a novel human protooncogene, c-mer. *Cell Growth Differ* (1994) 5:647–57.
6. Lai C, Gore M, Lemke G. Structure, expression, and activity of Tyro 3, a neural adhesion-related receptor tyrosine kinase. *Oncogene* (1994) 9:2567–78.
7. Lemke G, Burstyn-Cohen T. TAM receptors and the clearance of apoptotic cells. *Ann N Y Acad Sci* (2010) 1209:23–9. doi:10.1111/j.1749-6632.2010.05744.x
8. Manfioletti G, Brancolini C, Avanzi G, Schneider C. The protein encoded by a growth arrest-specific gene (gas6) is a new member of the vitamin K-dependent proteins related to protein S, a negative coregulator in the blood coagulation cascade. *Mol Cell Biol* (1993) 13:4976–85. doi:10.1128/MCB.13.8.4976
9. Stitt TN, Conn G, Gore M, Lai C, Bruno J, Radziejewski C, et al. The anticoagulation factor protein S and its relative, Gas6, are ligands for the Tyro 3/Axl family of receptor tyrosine kinases. *Cell* (1995) 80:661–70. doi:10.1016/0092-8674(95)90520-0
10. Tsou WI, Nguyen KQ, Calarese DA, Garforth SJ, Antes AL, Smirnov SV, et al. Receptor tyrosine kinases, TYRO3, AXL, and MER, demonstrate distinct patterns and complex regulation of ligand-induced activation. *J Biol Chem* (2014) 289:25750–63. doi:10.1074/jbc.M114.569020
11. Zagorska A, Traves PG, Lew ED, Dransfield I, Lemke G. Diversification of TAM receptor tyrosine kinase function. *Nat Immunol* (2014) 15:920–8. doi:10.1038/ni.2986
12. Lew ED, Oh J, Burrola PG, Lax I, Zagórska A, Través PG, et al. Differential TAM receptor-ligand-phospholipid interactions delimit differential TAM bioactivities. *Elife* (2014) 3. doi:10.7554/eLife.03385
13. Kasikara C, Kumar S, Kimani S, Tsou WI, Geng K, Davra V, et al. Phosphatidylserine sensing by TAM receptors regulates AKT-dependent chemoresistance and PD-L1 expression. *Mol Cancer Res* (2017) 15:753–64. doi:10.1158/1541-7786.MCR-16-0350

14. Lemke G. Phosphatidylserine is the signal for TAM receptors and their ligands. *Trends Biochem Sci* (2017) 42:738–48. doi:10.1016/j.tibs.2017.06.004
15. Scott RS, McMahon EJ, Pop SM, Reap EA, Caricchio R, Cohen PL, et al. Phagocytosis and clearance of apoptotic cells is mediated by MER. *Nature* (2001) 411:207–11. doi:10.1038/35075603
16. Wu Y, Tibrewal N, Birge RB. Phosphatidylserine recognition by phagocytes: a view to a kill. *Trends Cell Biol* (2006) 16:189–97. doi:10.1016/j.tcb.2006.02.003
17. Lu Q, Lemke G. Homeostatic regulation of the immune system by receptor tyrosine kinases of the Tyro 3 family. *Science* (2001) 293:306–11. doi:10.1126/science.1061663
18. Cohen PL, Caricchio R, Abraham V, Camenisch TD, Jennette JC, Roubey RA, et al. Delayed apoptotic cell clearance and lupus-like autoimmunity in mice lacking the c-mer membrane tyrosine kinase. *J Exp Med* (2002) 196:135–40. doi:10.1084/jem.20012094
19. Burstyn-Cohen T, Lew ED, Través PG, Burrola PG, Hash JC, Lemke G. Genetic dissection of TAM receptor-ligand interaction in retinal pigment epithelial cell phagocytosis. *Neuron* (2012) 76:1123–32. doi:10.1016/j.neuron.2012.10.015
20. Chen Y, Wang H, Qi N, Wu H, Xiong W, Ma J, et al. Functions of TAM RTKs in regulating spermatogenesis and male fertility in mice. *Reproduction* (2009) 138:655–66. doi:10.1530/REP-09-0101
21. Lu Q, Gore M, Zhang Q, Camenisch T, Boast S, Casagranda F, et al. Tyro-3 family receptors are essential regulators of mammalian spermatogenesis. *Nature* (1999) 398:723–8. doi:10.1038/19554
22. Sun B, Qi N, Shang T, Wu H, Deng T, Han D. Sertoli cell-initiated testicular innate immune response through toll-like receptor-3 activation is negatively regulated by Tyro3, Axl, and mer receptors. *Endocrinology* (2010) 151:2886–97. doi:10.1210/en.2009-1498
23. Fourgeaud L, Través PG, Tufail Y, Leal-Bailey H, Lew ED, Burrola PG, et al. TAM receptors regulate multiple features of microglial physiology. *Nature* (2016) 532:240–4. doi:10.1038/nature17630
24. Nguyen KQ, Tsou WI, Calarese DA, Kimani SG, Singh S, Hsieh S, et al. Overexpression of MERTK receptor tyrosine kinase in epithelial cancer cells drives efferocytosis in a gain-of-function capacity. *J Biol Chem* (2014) 289:25737–49. doi:10.1074/jbc.M114.570838
25. Paolino M, Choidas A, Wallner S, Pranjic B, Uribesalgo I, Loeser S, et al. The E3 ligase Cbl-b and TAM receptors regulate cancer metastasis via natural killer cells. *Nature* (2014) 507:508–12. doi:10.1038/nature12998
26. Cook RS, Jacobsen KM, Wofford AM, DeRyckere D, Stanford J, Prieto AL, et al. MerTK inhibition in tumor leukocytes decreases tumor growth and metastasis. *J Clin Invest* (2013) 123:3231–42. doi:10.1172/JCI67655
27. Kimani SG, Kumar S, Davra V, Chang YJ, Kasikara C, Geng K, et al. Normalization of TAM post-receptor signaling reveals a cell invasive signature for Axl tyrosine kinase. *Cell Commun Signal* (2016) 14:19. doi:10.1186/s12964-016-0142-1
28. Vouri M, Hafizi S. TAM receptor tyrosine kinases in cancer drug resistance. *Cancer Res* (2017) 77:2775–8. doi:10.1158/0008-5472.CAN-16-2675
29. Kirane A, Ludwig KF, Sorrelle N, Haaland G, Sandal T, Ranaweera R, et al. Warfarin blocks Gas6-mediated Axl activation required for pancreatic cancer epithelial plasticity and metastasis. *Cancer Res* (2015) 75:3699–705. doi:10.1158/0008-5472.CAN-14-2887-T
30. Akalu YT, Rothlin CV, Ghosh S. TAM receptor tyrosine kinases as emerging targets of innate immune checkpoint blockade for cancer therapy. *Immunol Rev* (2017) 276:165–77. doi:10.1111/imr.12522
31. Verma A, Warner SL, Vankayalapati H, Bearss DJ, Sharma S. Targeting Axl and Mer kinases in cancer. *Mol Cancer Ther* (2011) 10:1763–73. doi:10.1158/1535-7163.MCT-11-0116
32. Tanabe K, Nagata K, Ohashi K, Nakano T, Arita H, Mizuno K. Roles of gamma-carboxylation and a sex hormone-binding globulin-like domain in receptor-binding and in biological activities of Gas6. *FEBS Lett* (1997) 408:306–10. doi:10.1016/S0014-5793(97)00448-1
33. Bliss SA, Sinha G, Sandiford OA, Williams LM, Engelberth DJ, Guiro K, et al. Mesenchymal stem cell-derived exosomes stimulate cycling quiescence and early breast cancer dormancy in bone marrow. *Cancer Res* (2016) 76:5832–44. doi:10.1158/0008-5472.CAN-16-1092
34. Thery C, Amigorena S, Raposo G, Clayton A. Isolation and characterization of exosomes from cell culture supernatants and biological fluids. *Curr Protoc Cell Biol* (2006) Chapter 3:Unit 3.22. doi:10.1002/0471143030.cb0322s30
35. Dransfield I, Zagorska A, Lew ED, Michail K, Lemke G. Mer receptor tyrosine kinase mediates both tethering and phagocytosis of apoptotic cells. *Cell Death Dis* (2015) 6:e1646. doi:10.1038/cddis.2015.18
36. Wallin R, Hutson SM. Warfarin and the vitamin K-dependent gamma-carboxylation system. *Trends Mol Med* (2004) 10:299–302. doi:10.1016/j.molmed.2004.05.003
37. Danziger J. Vitamin K-dependent proteins, warfarin, and vascular calcification. *Clin J Am Soc Nephrol* (2008) 3:1504–10. doi:10.2215/CJN.00770208
38. Huang M, Rigby AC, Morelli X, Grant MA, Huang G, Furie B, et al. Structural basis of membrane binding by Gla domains of vitamin K-dependent proteins. *Nat Struct Biol* (2003) 10:751–6. doi:10.1038/nsb971
39. Kimani SG, Geng K, Kasikara C, Kumar S, Sriram G, Wu Y, et al. Contribution of defective PS recognition and efferocytosis to chronic inflammation and autoimmunity. *Front Immunol* (2014) 5:566. doi:10.3389/fimmu.2014.00566
40. Lea J, Sharma R, Yang F, Zhu H, Ward ES, Schroit AJ. Detection of phosphatidylserine-positive exosomes as a diagnostic marker for ovarian malignancies: a proof of concept study. *Oncotarget* (2017) 8:14395–407. doi:10.18632/oncotarget.14795
41. Suzuki J, Denning DP, Imanishi E, Horvitz HR, Nagata S. Xk-related protein 8 and CED-8 promote phosphatidylserine exposure in apoptotic cells. *Science* (2013) 341:403–6. doi:10.1126/science.1236758
42. Suzuki J, Umeda M, Sims PJ, Nagata S. Calcium-dependent phospholipid scrambling by TMEM16F. *Nature* (2010) 468:834–8. doi:10.1038/nature09583
43. Nakano T, Kawamoto K, Kishino J, Nomura K, Higashino K, Arita H. Requirement of gamma-carboxyglutamic acid residues for the biological activity of Gas6: contribution of endogenous Gas6 to the proliferation of vascular smooth muscle cells. *Biochem J* (1997) 323(Pt 2):387–92. doi:10.1042/bj3230387
44. Hasanbasic I, Rajotte I, Blostein M. The role of gamma-carboxylation in the anti-apoptotic function of gas6. *J Thromb Haemost* (2005) 3:2790–7. doi:10.1111/j.1538-7836.2005.01662.x
45. Rajotte I, Hasanbasic I, Blostein M. Gas6-mediated signaling is dependent on the engagement of its gamma-carboxyglutamic acid domain with phosphatidylserine. *Biochem Biophys Res Commun* (2008) 376:70–3. doi:10.1016/j.bbrc.2008.08.083
46. Hall MO, Obin MS, Prieto AL, Burgess BL, Abrams TA. Gas6 binding to photoreceptor outer segments requires gamma-carboxyglutamic acid (Gla) and Ca(2+) and is required for OS phagocytosis by RPE cells in vitro. *Exp Eye Res* (2002) 75:391–400. doi:10.1006/exer.2002.2030
47. Callahan MK, Popernack PM, Tsutsui S, Truong L, Schlegel RA, Henderson AJ. Phosphatidylserine on HIV envelope is a cofactor for infection of monocytic cells. *J Immunol* (2003) 170:4840–5. doi:10.4049/jimmunol.170.9.4840
48. Mercer J, Helenius A. Vaccinia virus uses macropinocytosis and apoptotic mimicry to enter host cells. *Science* (2008) 320:531–5. doi:10.1126/science.1155164
49. Meertens L, Carnec X, Lecoin MP, Ramdasi R, Guivel-Benhassine F, Lew E, et al. The TIM and TAM families of phosphatidylserine receptors mediate dengue virus entry. *Cell Host Microbe* (2012) 12:544–57. doi:10.1016/j.chom.2012.08.009
50. Morizono K, Xie Y, Olafsen T, Lee B, Dasgupta A, Wu AM, et al. The soluble serum protein Gas6 bridges virion envelope phosphatidylserine to the TAM receptor tyrosine kinase Axl to mediate viral entry. *Cell Host Microbe* (2011) 9:286–98. doi:10.1016/j.chom.2011.03.012
51. Birge RB, Boeltz S, Kumar S, Carlson J, Wanderley J, Calianese D, et al. Phosphatidylserine is a global immunosuppressive signal in efferocytosis, infectious disease, and cancer. *Cell Death Differ* (2016) 23:962–78. doi:10.1038/cdd.2016.11
52. Segawa K, Kurata S, Yanagihashi Y, Brummelkamp TR, Matsuda F, Nagata S. Caspase-mediated cleavage of phospholipid flippase for apoptotic phosphatidylserine exposure. *Science* (2014) 344:1164–8. doi:10.1126/science.1252809

Conflict of Interest Statement: The authors declare that the research was conducted in the absence of any commercial or financial relationships that could be construed as a potential conflict of interest.

21

Soluble Mediators Produced by Pro-Resolving Macrophages Inhibit Angiogenesis

*Shira Michaeli[1], Vivian Dakwar[1], Keren Weidenfeld[1], Ortal Granski[1], Odelya Gilon[1], Sagie Schif-Zuck[1], Anatolii Mamchur[2], Imad Shams[2] and Dalit Barkan[1]**

[1] Department of Human Biology, University of Haifa, Haifa, Israel, [2] Department of Evolutionary and Environmental Biology, Institute of Evolution, University of Haifa, Haifa, Israel

Edited by:
Estee Kurant,
University of Haifa, Israel

Reviewed by:
Michal Amit Rahat,
Technion – Israel Institute of Technology, Israel
Hynda K. Kleinman,
George Washington University, United States

***Correspondence:**
Dalit Barkan
dalitbrk@gmail.com

Different subtypes of macrophages have been shown to participate in different stages of inflammation and tissue repair. In the late stage of tissue repair, the macrophages, following their engulfment of apoptotic neutrophils, acquire a new phenotype termed alternatively activated macrophages. These macrophages produce growth factors, such as vascular endothelial growth factor (VEGF), that facilitate the angiogenic response as part of tissue restoration. Then, in the later stages of tissue healing, capillary regression takes place. It is presently unknown whether macrophages play an antiangiogenic role in the final stages of tissue repair. Here, we examined whether soluble mediators secreted by pro-resolving CD11b^{low} macrophages (Mres) inhibit angiogenesis in the context of the resolution of tissue repair. Our findings indicate that soluble mediators produced by *ex vivo* generated Mres (CM-Mres) attenuate angiogenesis *in vitro* by inhibiting human umbilical vein endothelial cell (HUVEC) proliferation by lowering their cyclin D1 expression. In addition, CM-Mres lowered HUVEC survival by inducing caspase 3/7 activation, and also inhibited VEGFR2 activation *via* VEGF. HUVEC migration and differentiation to tubular-like structure was also inhibited by CM-Mres. Similarly, CM-Mres significantly inhibited neovascularization as depicted *ex vivo* by utilizing the rat aorta ring assay and *in vivo* by utilizing the chick chorioallantoic membrane assay. Notably endostatin, which was shown previously to exert its antiangiogenic effect by inhibiting proliferation, survival, motility, and morphogenesis of endothelial cells *via* inhibition of VEGFR2 activation, is produced by Mres. Taken together, our results suggest that a specialized subset of macrophages that appear during the resolution of inflammation can produce antiangiogenic mediators, such as endostatin. These mediators can halt angiogenesis, thereby restoring tissue structure.

Keywords: pro-resolving macrophages, tissue repair, angiogenesis, antiangiogenic factors, resolution of inflammation, endostatin

INTRODUCTION

Inflammation and tissue repair are adaptive responses to tissue damage induced by pathogen infiltration or mechanical or chemical injury. These responses involve sequential stages which are orchestrated by recruitment and activation of various non-hematopoietic and hematopoietic cells such as neutrophils, macrophages, fibroblasts, and endothelial cells (1). The return of the tissue to its homeostatic state is dependent on the tight regulation and final resolution of the inflammatory response and the wound healing processes. However, dysregulated and exaggerated tissue repair that fails to subside and resolve will result in fibrosis and consequently will lead to organ failure (2, 3). Therefore, it is important to understand the contribution of the different cellular mediators in the resolution of inflammation and the various stages of tissue repair.

Macrophages are highly versatile immune cells that can acquire functionally distinct phenotypes (4, 5). Indeed, recent reports suggest the role of specific subpopulations of macrophages in regulating the different stages of tissue repair (6) and resolution of inflammation (7). In the initial stage of the inflammatory response to injury, leukocyte infiltration is followed by recruitment of monocytes to the site of injury. The monocytes differentiate to classically active macrophages also known as M1-like macrophages (pro-inflammatory) (4, 5). The phenotype of M1-like macrophages that engulf apoptotic polymorphonuclear leukocytes cells (PMN) shifts to that of alternatively activated macrophages. The latter are also referred to as M2-like macrophages and are involved in attenuating inflammation and promoting tissue repair (8, 9). Specifically, M2-like macrophages promote tissue repair by secreting growth factors such as transforming growth factor beta-1 which induces myofibroblast differentiation and deposition of extracellular matrix, and vascular endothelial growth factor (VEGF) which promotes angiogenesis (2, 10). Hence, tight temporal regulation of macrophage phenotype is required to promote resolution of inflammation, tissue repair, and reinstatement of homeostasis. The potential role of macrophages in resolving tissue repair and inflammation has been recently described (7, 11). Schif-Zuck et al. characterized a novel subset of pro-resolving macrophages designated CD11b^{low} macrophages that appear during the resolution of zymosan-induced murine peritonitis. These macrophages secrete pro-resolving mediators and are generated *in vivo* and *ex vivo* from M2-like macrophages following the engulfment of apoptotic leukocytes (11). However, these macrophages display a distinct enzyme expression signature from either M1 or M2, are devoid of phagocytic potential, and are prone to migrate to lymphoid tissues. We recently demonstrated that these pro-resolving macrophages can secrete anti-fibrotic mediators, thus preventing the establishment of a fibrotic-like milieu by preventing the expression of type I collagen (Col-I) by activated myofibroblasts (Gilon et al., submitted for publication). Notably, Col-I remodeling and vasculature regression are evident in the late phase of resolution. Furthermore, intraperitoneal zymosan injection is a model of acute inflammation which self-resolves within 48–72 h (12). Hence, we hypothesized that the recently characterized, pro-resolving macrophages secrete antiangiogenic mediators in addition to anti-fibrotic mediators, thus finalizing tissue repair. Here, we demonstrate that *ex vivo* generated pro-resolving CD11b^{low} macrophages (Mres) secrete antiangiogenic mediators such as endostatin thereby inhibiting angiogenesis by endothelial cells.

MATERIALS AND METHODS

Cell Line Cultures

Human umbilical vein endothelial cells (HUVECs) (kindly provided by Prof. Gera Neufeld, Technion, Israel) were grown on 10 cm plates, coated with 0.2% gelatin in Dulbecco's phosphate-buffered saline (PBS; Biological Industries, Israel) and overlaid with growth medium comprised of Earle's salt base (M-199) medium supplemented with 20% fetal bovine serum (FBS), 1% antibiotics, 1% vitamins, and glutamine (Biological Industries, USA) and freshly added basic fibroblast growth factor (bFGF) (PeproTech, Israel) (5 ng/ml). Jurkat T cells (kindly provided by Prof. Debbie Yablonski, Technion, Israel) were maintained in RPMI-1640 (Gibco–Life Technology, USA) with high glucose, 10% heat inactivated FBS, and 1% antibiotics. All cells were incubated at 37°C, 5% CO_2 incubator.

Animals

7- to 8-week-old male C57BL/6 mice and 6-week-female Sprague Dawley rats were purchased from Harlan Biotech Israel. All animals were maintained under specific pathogen-free conditions. Care and handling of animals was in compliance with University of Haifa's experimental protocols. This study was carried out in accordance with the recommendations of University of Haifa Animal Ethics Committee guidelines. The protocol was approved by the University of Haifa Animal Ethics Committee.

Ex Vivo Generation of Pro-Resolving CD11b^{low} Macrophages

Male C57BL/6 mice were injected i.p. with zymosan A (1 mg) purchased from Sigma-Aldrich, Israel. After 66 h, peritoneal exudates were collected, and exudate cells were stained with PE-conjugated rat anti-F4/80 (BioLegend Inc., USA). Macrophages were isolated using EasySep PE selection magnetic beads following the manufacturer's instructions (StemCell Technologies, Israel). Isolated macrophages were co-stained with FITC-conjugated rat anti-Ly-6G and PerCP-conjugated rat anti-mouse CD11b (BioLegend Inc., USA) and analyzed by FACSCanto II (BD Biosciences, USA) and the FACSDiva software. Jurkat T cells were treated with 1 μM staurosporine (Sigma-Aldrich, Israel) to induce apoptosis and washed. Then, peritoneal macrophages were incubated in the presence or absence of apoptotic Jurkat T cells [1:5 macrophage to apoptotic cell (AC) ratio]. After 8 h of incubation, the cells were washed with PBS and overlaid with fresh media; RPMI-1640 with high glucose 10% FBS, and 1% antibiotics for additional 12 h of incubation. Next, conditioned media were collected, and the macrophages were further characterized for their conversion to the CD11b^{low} phenotype using flow cytometry.

Preparation of the Different Conditioned Media for the Experimental Assays

The following conditioned media, listed below, were diluted with HUVEC Assay Medium (M-199 medium supplemented with 10% FBS, 1% antibiotics, 1% vitamins, 1% glutamine, and freshly added bFGF 5 ng/ml) at a ratio of 1:1. This step was carried out to ensure viability of HUVEC in all the experimental assays described below.

Condition medium (CM): Baseline conditioned media comprise RPMI supplemented with 10% heat inactivated FBS and 1% antibiotics.

CM-Mϕ: Conditioned media collected after 12 h from culture enriched with CD11b^{high} macrophages (Mϕ).
CM-Mres: Conditioned media collected after 12 h from culture enriched with CD11b^{low} macrophages (Mres).
CM-AC: Conditioned media of un-engulfed ACs.

Proliferation Assay

Human umbilical vein endothelial cell grown in 10 cm plates were treated for 2 h in M-199, 5% FBS, 1% antibiotics, 1% vitamins, and 1% glutamine medium. Next, the above treated cells (3×10^3 cells/well) were cultured in 96-wells plate coated with Cultrex® growth factor-reduced basement membrane extract (BME) (Trevigen Inc., USA) and treated with the different conditioned media. After overnight incubation at 37°C, 5% CO_2 incubator Cell Titer 96 AqueousOne Solution cell proliferation assay kit (Promega, USA) was added to the wells for 2 h to measure cell proliferation according to the manufacturer's instructions. The absorbance was recorded at 490 nm.

Immunofluorescence Staining

Human umbilical vein endothelial cell cultured in 8-well chamber glass slides coated with BME (Trevigen Inc., USA), as described previously (13), were treated for 5 min with mixture containing 0.1% Triton X-100 and 4% PFA containing 5% sucrose, and fixed for an additional 25 min with 4% PFA containing 5% sucrose. The cells were washed for 10 min with PBS and an additional 15 min with PBS containing 0.05% Tween 20 (PBS-T). Next, fixed cells were blocked with IF buffer (130 mM NaCl, 7 mM Na_2HPO_4, 3.5 mM NaH_2PO_4, 7.7 mM NaN_3, 0.1% BSA, 0.2% Triton X-100, and 0.05% Tween 20) containing 10% donkey serum for 1 h followed by overnight incubation at 4°C with rabbit anti-active-caspase-3 (1:400) (Cat # 559565; BD Biosciences). The cells were washed three times with PBS for 15 min each and incubated for 1 h with donkey anti-rabbit conjugated to Alexa Fluor® 647 (Invitrogen, USA) at room temperature. Next, the cells were washed as mentioned earlier and mounted with VECTASHIELD mounting medium with 4′,6-diamidino-2-phenylindole (DAPI). Immunofluorescent images were captured by Zeiss LSM 700 confocal laser scanning microscope (magnification 40×).

For F-actin staining, cells were incubated overnight with Alexa Fluor 488 Phalloidin (1:40) (Molecular Probes, USA), washed three times with PBS for 15 min each and mounted with VECTASHIELD mounting medium with DAPI.

Caspase 3/7 Activity

Human umbilical vein endothelial cells grown in 10 cm plates were treated for 2 h in M-199, 5% FBS, 1% antibiotics, 1% vitamins, and 1% glutamine medium. Next, the aforementioned treated cells were cultured in 96 wells coated with 50 μl BME (3×10^3 cells/well) and were overlaid with the different conditioned media. After overnight incubation at 37°C, 5% CO_2 incubator, Caspase-Glo® reagent (Promega, USA) was added for each well according to the manufacturer's instructions, and plates were incubated at room temperature for 1 h. Luminescence of each sample was measured using a plate-reading infinite M200PRO, TECAN luminometer.

Wound Migration Assay

Wound migration assay was performed using a 12-well plate coated with 0.2% gelatin in PBS (10×10^4 HUVEC/well). 17 h post seeding, a wound was induced by mechanical application of a 1,000 μl sterile tip. Images of wound formation and healing were acquired at time 0 and at 2.5, 5.5, 7, and 8.5 h post-induction, using a light inverted microscope magnification 10× (Nikon Eclipse TS100). Analysis of the wound healing was carried using the Nikon A1R confocal laser scanning microscopy software (NIS Elements AR version 4.3, by Nikon). The area of the scratch was quantified and normalized to the area of the scratch at time 0.

Time-Lapse Microscopy

Human umbilical vein endothelial cells (6×10^4) were plated on top of gelatinized 15 mm glass-bottom cell culture dishes (Nest Scientific USA Inc.) and overlaid with either CM-Mϕ or CM-Mres. Cells were incubated at 37°C, 5% CO_2 incubator for 30 min to allow adherence of the cells. Thereafter, cell motility was followed by time-lapse video microscopy using Nikon A1R confocal laser scanning microscope (20× magnification). Differential interference contrast (DIC) microscopy images were acquired every 1 min for a period of 2 h. Motility of the cells for a period of 2 h was measured by determining the average velocity of 10 different cells for each treatment utilizing ImageJ software (with the "win 64" plug).

Tube-Formation Assay

Human umbilical vein endothelial cells grown in 10 cm plates were treated for 2 h in M-199, 5% FBS, 1% antibiotics, 1% vitamins, and 1% glutamine medium. These cells were then cultured in eight-chamber glass slides (Lab-TEK® II, Naperville, IL, USA) coated with BME and overlaid with the different conditioned media for 16 h. For positive control M-199, 20% FBS, 1% antibiotics, 1% vitamins, and 1% glutamine medium was applied. Pictures were acquired by Nikon Eclipse TS100 light microscopy (10× magnification), and the number of bifurcations per field was quantified using ImageJ software.

Aortic Ring Assay

Aorta ring assay was carried out as previously described (14) with slight modification. Briefly, thoracic aorta rings were prepared from female Sprague Dawley rats according to the protocol by Bellacen and Lewis (14) and placed in 48-well plates coated with 150 μl Cultrex® growth factor-reduced BME. Plates were incubated for 10 min at 37°C in a 5% CO_2 incubator. Following incubation, each well was overlaid with an additional 150 μl BME and incubated at 37°C, in a 5% CO_2 incubator for 20–30 min. Next, the aorta rings were subjected to the different treatments for a period of 6 days. For a negative control, vascular cell basal medium (ATCC® PCS-100-030™) was applied, and for positive control vascular cell basal medium supplemented with components listed in Table S1 in Supplementary Material was utilized. Images were acquired by Stemi SV 6, ZEISS light microscope (10× magnification). The micro-vessel sprouting area was analyzed by ImageJ software. Sprouting area for each treatment at time point $T = 0$ was subtracted from the sprouting area at 2, 4, and 6 days of treatment.

Chick Chorioallantoic Membrane (CAM) Assay

The CAM assay was carried out as previously described (15) with slight modification. Embryonated chicken eggs (~30 per treatment) were incubated at 38°C incubator. At day 3, ovalbumin was removed (3 ml per egg), a window was opened [according to the protocol of Ponce and Kleinmann (15)], and inserts treated with the different conditioned media were applied. The inserts were composed of autoclaved filtered paper (5.5 mm in diameter), which were treated either with CM-Mϕ or CM-Mres (10 µl/insert). Eggs were incubated for additional 48 h at 38°C incubator. Images were acquired by binocular (0.8× magnification) at time 0 and after 48 h incubation with the above inserts. Quantification of blood vessel density was carried out using ImageJ (by using "win 64" plug). For each treatment, an area of 15 cm^2 was analyzed with the filter paper at its center. Vascular density measured at 48 h posttreatment was normalized to the vascular density measured at time 0.

Mouse Angiogenesis Array

Mouse angiogenesis array kit (R&D Biosystems) was used according to the manufacture's instructions using either CM-Mϕ or CM-Mres. Membranes were analyzed using ImageQuant LAS-4000 analyzer (GE Healthcare Life Sciences, Pittsburgh, PA, USA) and "ImageQuant LAS-4000" software (GE Healthcare Life Sciences). Densitometry analysis was performed using ImageQuant total lab-7 (GE Healthcare Life Sciences) image analysis software.

VEGFR2 Phosphorylation

Confluent HUVEC cells (cultured on gelatin as described earlier) were treated in M-199, 5% FBS, 1% antibiotics, 1% vitamins, and 1% glutamine medium for 12 h. Next, the cells were overlaid with the different conditioned media for 30 min followed by VEGF supplementation (25 ng/ml) for 5 min. Cell pellets were prepared for western blot analysis.

Western Blot Analysis

Cell pellets were lysed in WCE (whole-cell extract) buffer [25 mM HEPES, pH 7.7, 0.3 M NaCl, 1.5 mM $MgCl_2$, 0.2 mM EDTA, 0.1% Triton X-100, 100 µg/ml PMSF, and 25 mM protease inhibitor cocktail (Roche)], and for p-VEGFR2 detection the cell pellets were lysed in WCE buffer supplemented with 20 mM NAF, 2 mM Na_3VO_4, and 0.5 mM DTT. The proteins from cell lysate or from the conditioned media were separated by SDS-PAGE followed by transfer on to a nitrocellulose membrane. The membranes were blocked with 5% (w/v) non-fat dried skimmed milk powder either in PBS supplemented with 0.05% Tween 20 (PBS-T) for protein detection or for phosphorylated VEGFR2 in Tris-buffered saline supplemented with 0.1% Tween 20 (TBS-T) for 1 h at room temperature. Membrane was then probed either with rabbit anti-GAPDH (1:500), rabbit anti-cyclin D1 (1:200), or rabbit anti-VEGFR2 (1:1,000) (Santa Cruz, Dallas, TX, USA). Rabbit anti-phospho-VEGFR2-Tyr951 (1:500) (Cell Signaling, Danvers, MA, USA), rabbit anti-VEGF (Abcam, Cambridge, UK), or monoclonal anti-endostatin (1:1,000) (Merck, Darmstadt, Germany) at 4°C overnight. Next, the membranes were incubated with the appropriate horseradish peroxidase-conjugated secondary antibodies (1:10,000; Jackson ImmunoResearch Laboratories, West Grove, PA, USA) for 1 h at room temperature and washed 15 min 3× with either PBS-T or TBS-T (for phosphor-protein detection). Western Bright ECL (Advansta, Menlo Park, CA, USA) was added to the membrane for 1 min and analyzed using ImageQuant LAS-4000 analyzer (GE Healthcare Life Sciences, Pittsburgh, PA, USA) and "ImageQuant LAS-4000" software (GE Healthcare Life Sciences). Densitometry analysis was performed using ImageQuant total lab-7 (GE Healthcare Life Sciences) image analysis software.

Statistical Analysis

Student's unpaired *t*-test was used accordingly. Two-tailed *p* values of 0.05 or less were considered to be statistically significant. Repeated measures ANOVA comparison test was used accordingly. Values of 0.05 or less were considered to be statistically significant.

RESULTS

Ex Vivo Generation of Secreted Factors of Pro-Resolving CD11b^{low} Macrophages

To this end, peritonitis was induced, and 66 h later peritoneal exudates were collected. The percentage of macrophages (Mϕ) was determined in peritoneal exudates, based on their size and granularity and positive staining for F4/80 as previously described (11). Next, CD11b^{high} Mϕ were collected and were either untreated or treated with apoptotic Jurkat cells {a common apoptotic leukocyte target for macrophages in experimental procedures [(11, 16) at a ratio of 1:5 respectively]}. Incubation with ACs resulted in 79% conversion of CD11b^{high}-Mϕ to CD11b^{low}-Mϕ compared with untreated macrophages where only 13% of CD11b^{high}-Mϕ were converted to CD11b^{low}-Mϕ, as determined by surface expression of CD11b by FACS analysis (Figure S1 in Supplementary Material). Conditioned media were collected after 12 h from cultures enriched with either CD11b^{high} macrophages (CM-Mϕ) or pro-resolving CD11b^{low} macrophages (CM-Mres) and from un-engulfed AC (CM-AC).

Proliferation of HUVECs is Modulated by Factors Secreted by Pro-Resolving CD11b^{low} Macrophages

Angiogenesis, characterized by sprouting of preexisting vasculature to form new vessels, requires several coordinated endothelial cell activities, such as proliferation, migration, and morphogenesis (17, 18). Therefore, initially we determined whether CM-Mres was able to inhibit the proliferation of endothelial cells. To this end, conditioned media (condition medium (CM), CM-Mϕ, CM-Mres, or CM-AC) were overlaid on HUVECs cultured on top of growth factor-reduced reconstituted BME. The proliferation was measured after overnight incubation. One-way repeated measures ANOVA analysis was conducted followed with repeated contrasts to probe the differences between the four groups of treatments. Significant difference was found between the different treatments [$F(3, 12) = 5.699$, $p = 0.012$]. Furthermore, significant difference was found between the treatments CM-Mϕ vs. CM-Mres ($p = 0.022$) (**Figure 1A**). By contrast, no significant difference was found between CM-Mϕ vs. CM-AC ($p > 0.05$). Given that we observed a reduction in cell number of HUVEC treated with CM-Mres, we next tested whether this inhibition is also due to an increase in apoptosis of HUVEC. HUVECs were cultured as described earlier and scored for apoptosis by (1) percentage of TUNEL positive cells (Figure S2 in Supplementary Material), (2) activated caspase 3 detected by immunofluorescence staining (**Figure 1B**), and caspase 3/7 activity (**Figure 1C**). Increase in the percentage of HUVEC positive for TUNEL staining was evident upon treatment with CM-Mres compared with CM-Mϕ (Figure S2 in Supplementary Material). Activated caspase 3 was apparent by immunofluorescence staining (red staining) in HUVEC treated with CM-Mres (**Figure 1B**), whereas no staining was evident in CM-Mϕ. In addition, a significant increase in caspase 3/7 activity was observed in HUVEC treated either with CM-Mres or CM-AC compared with CM-Mϕ as determined by one-way repeated measures ANOVA analysis followed with repeated contrasts. Specifically, significant difference was found between the different treatments [$F(4, 20) = 4.549$, $p = 0.009$]. Furthermore, significant difference was found between CM-Mres compared with CM-Mϕ ($p = 0.014$) (**Figure 1C**) and CM-AC compared with CM-Mϕ ($p = 0.030$). Hence, our results suggest that CM-Mres inhibits proliferation. Of note, activation of caspase 3/7 by CM-Mres may be also attributed to the presence of mediators secreted by un-engulfed ACs.

CM-Mres Inhibits the Motility of HUVEC

Migration of vascular endothelial cells plays an important role in angiogenesis (18). Therefore, we tested whether CM-Mres may impact the motility of HUVEC. To this end, wound migration assay was utilized to study the rate of the wound closure in the plate (cell migration toward the wound/scratch) as detailed below. Wounded monolayers of HUVECs were incubated with the different conditioned media (CM, CM-Mϕ, CM-Mres, or CM-AC). The filling of the "wound" was monitored in a period of 2.5–8.5 h by measuring the % of remaining clear surface, compared with the $T = 0$ (scratch initiation). Our results demonstrate that treatment with CM-Mres delayed overtime the closure of the wound (**Figures 2A,B**) compared with treatment with CM-Mϕ. Repeated measures ANOVA and repeated contrasts were conducted with treatment (CM, CM-Mϕ, CM-Mres, and CM-AC) as repeated measures, at time 8.5 h. A significant difference between all treatments was found [$F(3, 3) = 24.261$, $p = 0.013$].

In addition, at 8.5 h posttreatment, CM-Mres significantly inhibited the closure of the wound compared with CM-Mϕ as determined by repeated contrast [$F(1, 5) = 7.548$, $p = 0.040$].

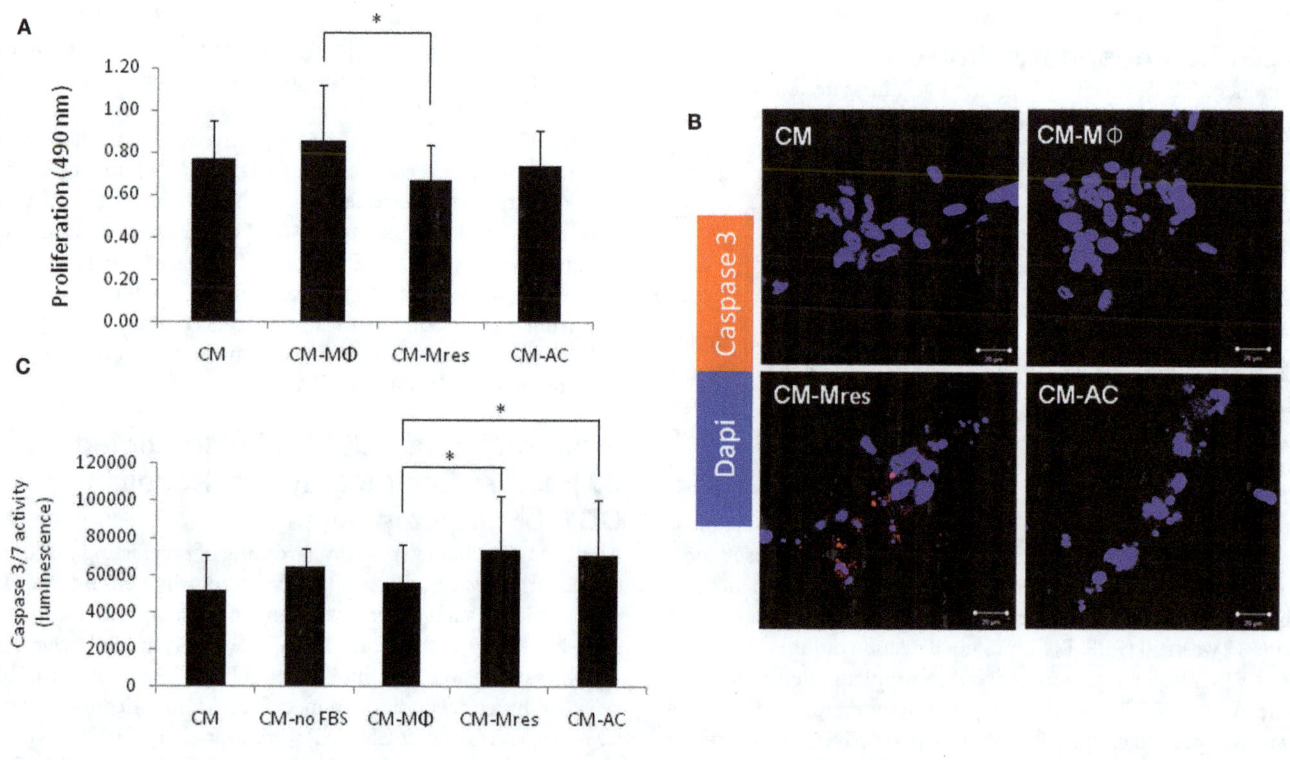

FIGURE 1 | CM-Mres inhibits proliferation and induces apoptosis of human umbilical vein endothelial cell (HUVEC). HUVEC were cultured on basement membrane extract and treated with condition medium (CM), CM-Mϕ, CM-Mres, or CM-AC. **(A)** Representative proliferation of HUVEC after overnight treatment. $n = 5$ with four to five replicates. **(B)** Representative confocal images of HUVEC stained for activated caspase 3 (red) and nuclei (DAPI, blue). Magnification 40×, bar = 50 µm. **(C)** Caspase 3/7 activity in HUVEC either starved [CM-no-fetal bovine serum (FBS); positive control] or treated with the indicated conditioned media overnight. Columns; mean, bars; STD, $n = 6$ with three replicates for each experiment. One-way repeated measures ANOVA analysis with repeated contrasts, $^{*}p \leq 0.05$.

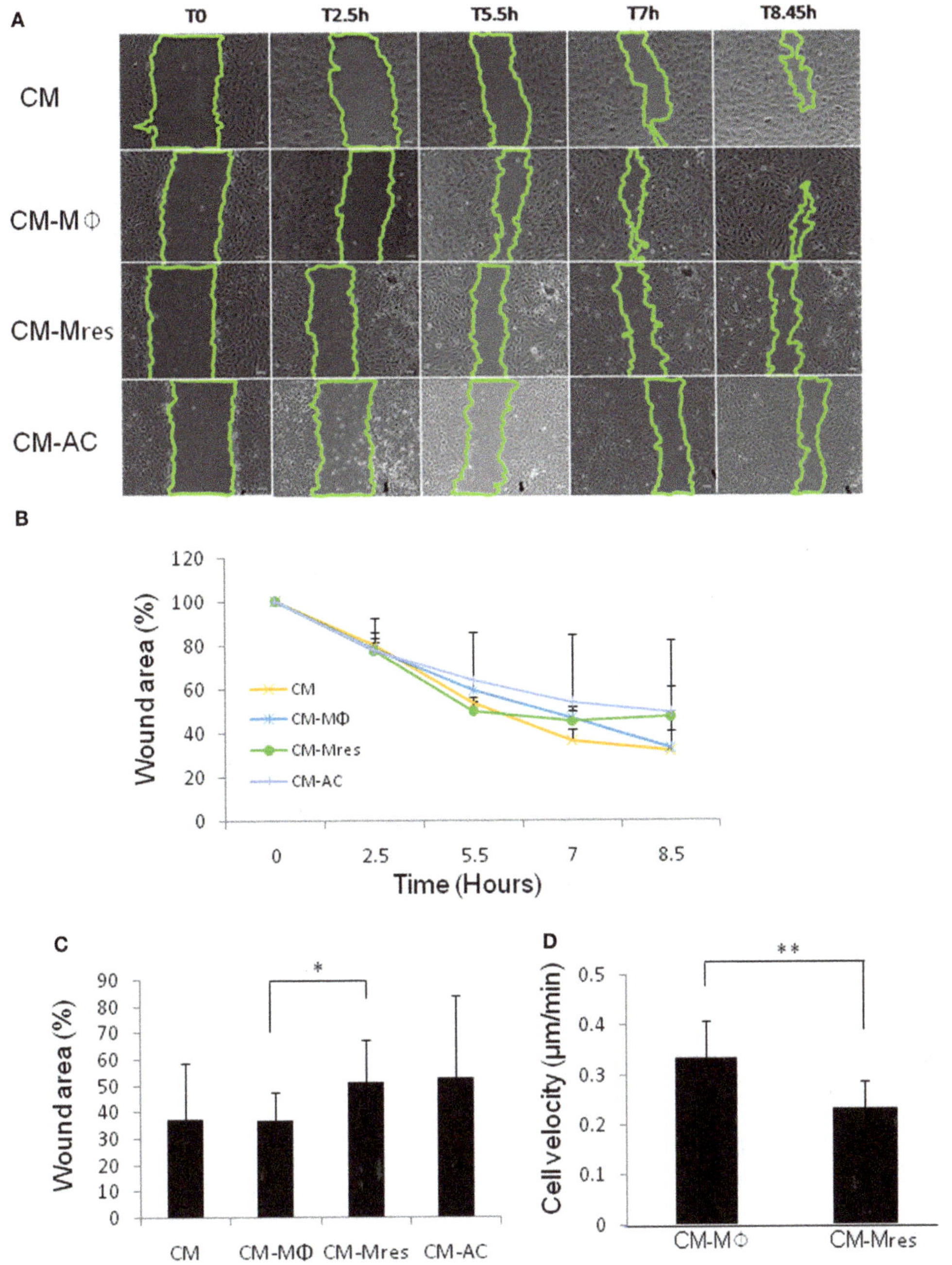

FIGURE 2 | Motility of human umbilical vein endothelial cell (HUVEC) is attenuated upon treatment with CM-Mres. **(A,B)** Representative results (n = 4) of wound migration assay of HUVEC treated with the indicated conditioned media. **(A)** Images taken at indicated time points demonstrate quicker wound closure upon treatment with either condition medium (CM) or CM-Mϕ compared with treatment with CM-Mres or CM-AC. (B) Quantification of the wound closure (motility of the cells) over time. Points; mean, bars; STD, n = 4. **(C)** Quantification of the wound closure at 8.5 h post wounding. Columns; mean, bars; STD, n = 6, one-way repeated measures ANOVA analysis with repeated contrasts, $^*p \leq 0.05$. **(D)** Motility of HUVEC cultured on gelatin for a period of 2 h upon treatment with either CM-Mϕ or CM-Mres was monitored by time-lapse microscopy. An average velocity of 10 different cells for each treatment was determined. Columns; mean, bars; STD, n = 10 cells for each treatment, t-test, $^*p \leq 0.05$, $^{**}p \leq 0.01$.

Specifically, 51% of the original area remained (**Figure 2C**) upon treatment with CM-Mres whereas, 36% of the original area remained upon treatment with CM-Mϕ (**Figure 2C**). Whereas, there was no significant difference between CM-Mϕ compared with CM-AC (p > 0.05).

To further validate that the delay in wound closure was due to inhibition of motility of HUVEC, we conducted time-lapse video microcopy for a period of 2 h (DIC images were captured every minute). Indeed, CM-Mres treatment significantly attenuated the motility of HUVEC cultured on gelatin (Video S1 in

Supplementary Material; **Figure 2D**) compared with treatment with CM-Mϕ (Video S2 in Supplementary Material; **Figure 2D**). Interestingly, CM-Mϕ altered HUVEC morphology and induced cell expansion and contraction often associated with membrane blebbing upon their movement, whereas treatment with CM-Mres induced a spindle shape morphology and no cell expansion or membrane blebbing was apparent in the migrating HUVEC.

In summary, our findings suggest that CM-Mres inhibits the motility of HUVEC.

Soluble Mediators Secreted by Pro-Resolving Macrophages Prevent HUVEC Morphogenesis to Tubular Structures

Next, we determined whether soluble mediators secreted by pro-resolving macrophages were able to inhibit the differentiation of endothelial cells to capillary-like networks. To this end, HUVECs were cultured on BME and treated with CM, CM-Mϕ, CM-Mres, CM-AC, or Assay Medium that promotes HUVEC differentiation to tubular structures. After 16–18 h, the extent of HUVEC differentiation to tubular structures was determined by light microscopy (**Figure 3A**). Quantification of the number of bifurcations of vessel-like tubular structures was carried out in three independent fields per each experimental condition using ImageJ software (**Figures 3A,C**). F-actin organization of the tubular-like structures was determined by phalloidin staining (**Figure 3B**).

HUVEC grown on BME with Assay Medium undergo spontaneous alignment into hollow tubes, forming capillary-like networks (19). However, we found that endothelial tubule formation and stability was impaired in the presence of CM-Mres compared with endothelial cells treated with CM, Assay Medium, or CM-Mϕ (**Figures 3A,B**). This was further supported by one-way repeated measures ANOVA analysis demonstrating significant differences between the different treatments [$F(4, 16) = 9.997$, $p < 0.001$]. Utilizing repeated contrasts analysis, we found a significant reduction in the number of bifurcations in the tubular network formed by the endothelial cell treated with CM-Mres in comparison to treatment with CM-Mϕ ($p = 0.012$) (**Figure 3C**). By contrast, there was no significant difference between CM-AC in comparison to CM-Mϕ ($p > 0.05$). The tube-formation assay is based on the ability of endothelial cells to form three-dimensional capillary-like tubular structures in the 3D BME system. In this system, endothelial cells proliferate, differentiate, directionally migrate to align, branch, and form the tubular polygonal networks of blood vessels. This is a well-established assay to study angiogenesis (19, 20). Given that a significant effect on formation of the capillary-like tubular structures was only evident upon treatment with CM-Mres, we proceeded to further validate the antiangiogenic effect of CM-Mres and compared it to CM-Mϕ, as described below.

Sprouting Angiogenesis *Ex Vivo* and *In Vivo* are Attenuated by CM-Mres

To further verify our *in vitro* findings, we utilized the rat aorta ring assay as an *ex vivo* model of angiogenesis. This organ culture assay scores for sprouting angiogenesis from the segmented aorta ring cultured on BME. Our results demonstrate that CM-Mϕ promoted neovascularization (see black arrow; **Figure 4A**; Figure S3 in Supplementary Material), whereas, CM-Mres restrained neovascularization (see white arrows; **Figure 4A**). We conducted repeated measures two-way ANOVA analysis, and we found significant difference between the different treatments [$F(4, 40) = 10.828$, $p = 0.003$]. Furthermore, we found significant interactions between time and treatment [$F(8, 40) = 14.127$, $p < 0.001$]. Planned comparisons were carried out on data from day 6 to probe the interaction. We found significant difference between CM-Mres compared with CM-Mϕ [$F(1, 5) = 55.574$, $p = 0.001$]. Furthermore, a significant difference was found between negative control vs. CM [$F(1, 5) = 6.663$, $p = 0.049$]. Whereas, no significant difference was found between CM-Mres vs. negative control [$F(1, 5) = 4.839$, $p = 0.079$] (**Figure 4B**). Hence, CM-Mres contain soluble mediators that can inhibit sprouting angiogenesis in the rat aorta ring assay. Next, we tested whether CM-Mres can inhibit angiogenesis *in vivo* by using the chick CAM assay. Similarly, to the *in vitro* and *ex vivo* results, exposure of the CAM to CM-Mres reduced vessel density (compare vessel density at time $T = 0$ vs. $T = 48$ h post CM-Mres treatment; **Figure 4C**, see white arrows and **Figure 4D**) whereas, treatment with CM-Mϕ for 48 h increased vessel density compared with $T = 0$ (**Figure 4C** see black arrows and **Figure 4D**). Altogether, these results suggest that CM-Mres contains antiangiogenic mediators.

Increased Levels of the Antiangiogenic Mediator Endostatin and Decrease in Angiogenic Factor VEGF in CM-Mres Compared with CM-Mϕ

Angiogenesis is regulated by a balance between angiogenic and antiangiogenic factors. Given that CM-Mres inhibited angiogenesis *in vitro*, *ex vivo*, and *in vivo*, has promoted us to determine the presence and predominance of antiangiogenic mediator/s over angiogenic factors in CM-Mres. To this end, mouse angiogenesis array was utilized to detect pro- and antiangiogenic factors in CM-Mres compared with CM-Mϕ (**Figures 5A–C**). Initial dot blot analysis revealed several pro-angiogenic factors with lower levels in CM-Mres compared with CM-Mϕ such as osteopoetin (21), HGF (22), CXCL16 (23), and CCL2 (24) (**Figure 5B**). Whereas, increase in the levels of the antiangiogenic factors endostatin, PEDF and thrombospondin-2 [reviewed in Nyberg et al. (25)] was observed in CM-Mres compared with their levels in CM-Mϕ (**Figure 5C**). Notably, the levels of the central pro-angiogenic mediator VEGF (26) decreased significantly in CM-Mres compared with its levels in CM-Mϕ (determined by western blot analysis; **Figures 5D,E**), whereas the levels of endostatin; an inhibitor of VEGF mediated signaling (25), was significantly higher in CM-Mres compared with CM-Mϕ (determined by western blot analysis; **Figures 5D,F**). Next, we tested VEGFR2 phosphorylation, given that endostatin was previously shown to block VEGF-induced tyrosine phosphorylation of VEGFR2 in HUVEC (see Materials and Methods) (27). Indeed, VEGFR2 phosphorylation on Y195 was reduced upon treatment with CM-Mres compared with treatment with CM-Mϕ (**Figures 5G–H**). Furthermore, CM-Mres inhibited

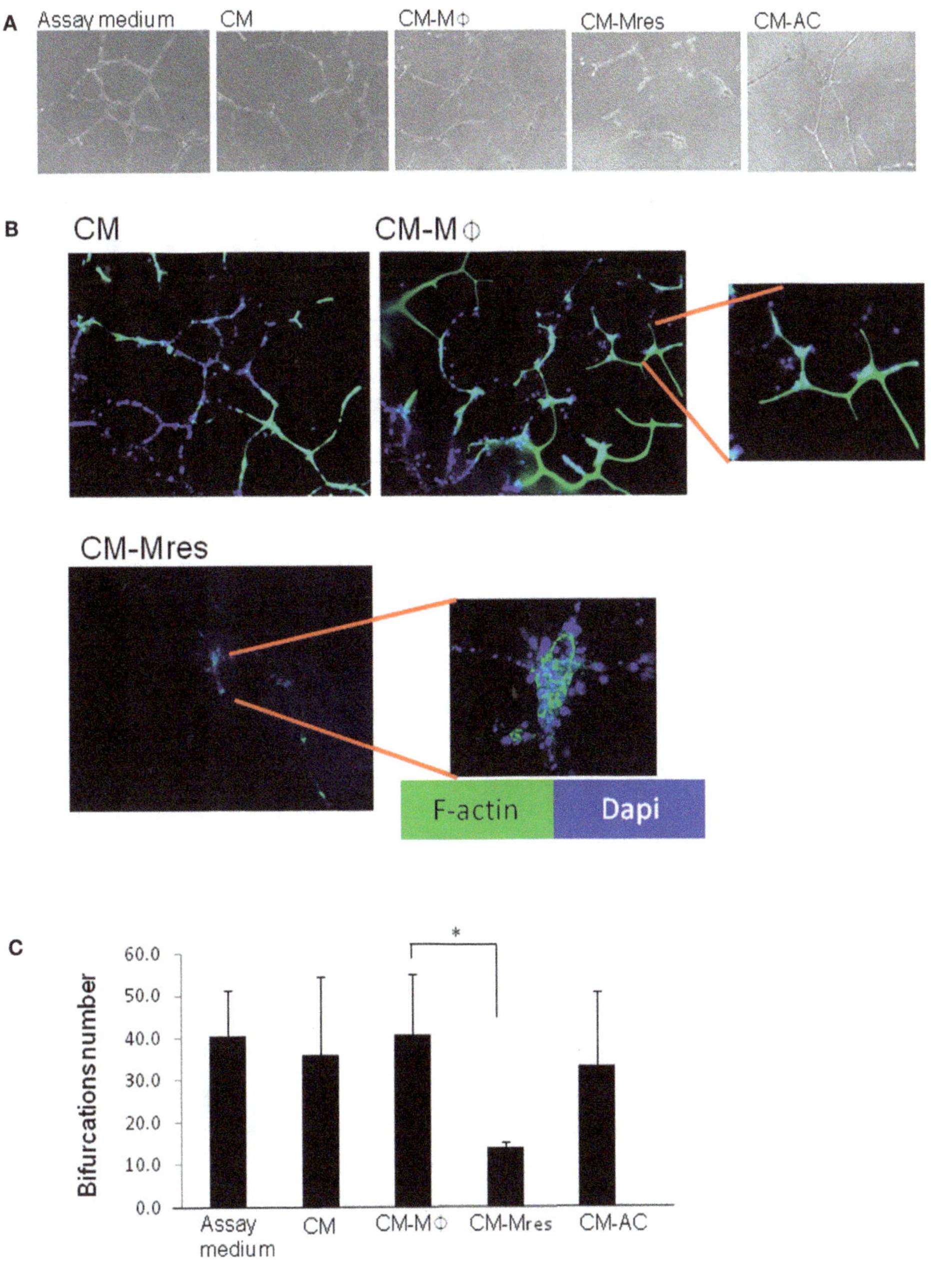

FIGURE 3 | CM-Mres hinders human umbilical vein endothelial cell (HUVEC) differentiation to tubular-like structures. HUVECs were cultured on basement membrane extract for 16–19 h with Assay Medium, condition medium (CM), CM-Mϕ, CM-Mres, or CM-AC. **(A)** Representative light microscopy images ($n = 5$). Magnification 10×. **(B)** Fluorescence staining of HUVEC cells for F-actin (phalloidin; green) and nuclei (DAPI; blue). Representative confocal images are shown. Magnification 40×. Bars = 50 µm. **(C)** Quantification of the bifurcation number of vessel-like tubular structures obtained from three to five microscopic fields. Columns; mean, bars; STD, $n = 5$, one-way repeated measures ANOVA analysis with repeated contrasts, $^{*}p < 0.05$.

significantly Cyclin D1 expression in HUVEC compared with treatment with CM-Mϕ (determined by western blot analysis; **Figures 5I,J**). These results are in accordance with previous studies demonstrating downregulation of Cyclin D1 expression upon endostatin treatment (28). Overall, our results suggest that reduction in the levels of VEGF and increase in endostatin levels in CM-Mres may mediate in part the antiangiogenic effect of CM-Mres by inhibiting VEGFR2 mediated signaling.

DISCUSSION

Macrophages play a fundamental role in wound healing by generating bioactive mediators that stimulate angiogenesis and fibroplasia (29, 30). However, the potential role of macrophages in resolving tissue repair by inhibiting angiogenesis is largely unknown (9).

Angiogenesis is a multifaceted process required to facilitate restoration of the damaged tissue during wound healing. This

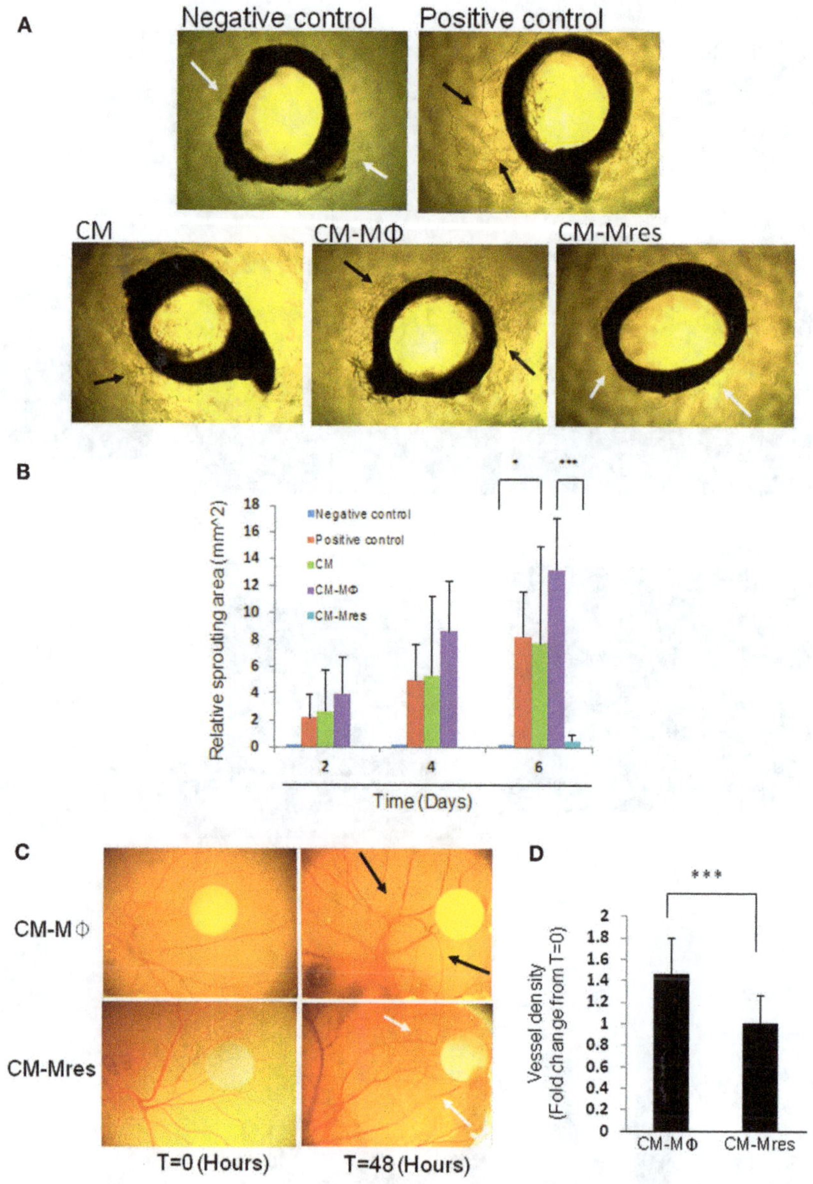

FIGURE 4 | Effect of CM-Mres on angiogenesis of rat aorta ring and chick chorioallantoic membrane (CAM). **(A)** Rat aortic ring assay. Representative photomicrographs of rat aortic ring sections from 6-week-old rat cultured on basement membrane extract and treated for 6 days with basal media (negative control), basal media with supplementations (positive control, see Materials and Methods), condition medium (CM), CM-Mϕ, or CM-Mres. Vascularized area is indicated by black arrow heads, and avascular area is indicated by white arrow heads. **(B)** Quantification of the sprouting area of endothelial cells relative to $T = 0$. Columns; mean, bars; STD, $n = 2$ with three replicates for each treatment, two-way repeated measures ANOVA analysis with repeated contrasts $^{*}p \leq 0.05$, $^{***}p \leq 0.001$. **(C)** Representative images of the chick CAM treated for 48 h with either CM-Mϕ ($n = 15$) or CM-Mres ($n = 12$). Vascularized area is indicated by black arrow heads, and avascular area is indicated by white arrow head. **(D)** Quantification of vessel density in panel **(C)**. Columns; mean of the fold change in vessel density from $t = 0$ for each treatment, bars; STD, *t*-test, $^{***}p \leq 0.001$.

process is orchestrated by: (1) remodeling of the extracellular matrix, (2) proliferation and migration/chemotaxis of endothelial cells, and (3) assembly of endothelial cells to vessel tube and its stabilization by pericytes and smooth muscle cells (17). In this study, we demonstrate for the first time to our knowledge that mediators generated by pro-resolving CD11b^{low} macrophages (Mres) that participate during resolution of acute murine peritonitis (11) inhibit angiogenesis *in vitro* and *in vivo*. The multiple parameters of angiogenesis, proliferation, viability, and motility, were modulated by CM-Mres thus culminating in overall significant and robust inhibition of the angiogenic process.

In vitro, we demonstrated that bioactive mediators produced by Mres inhibited HUVEC proliferation significantly and enhanced their apoptosis. This was depicted by increase in the percentage of cells positive for TUNEL and increase in caspase 3/7 activity. Furthermore, a significant reduction in the migration capacity

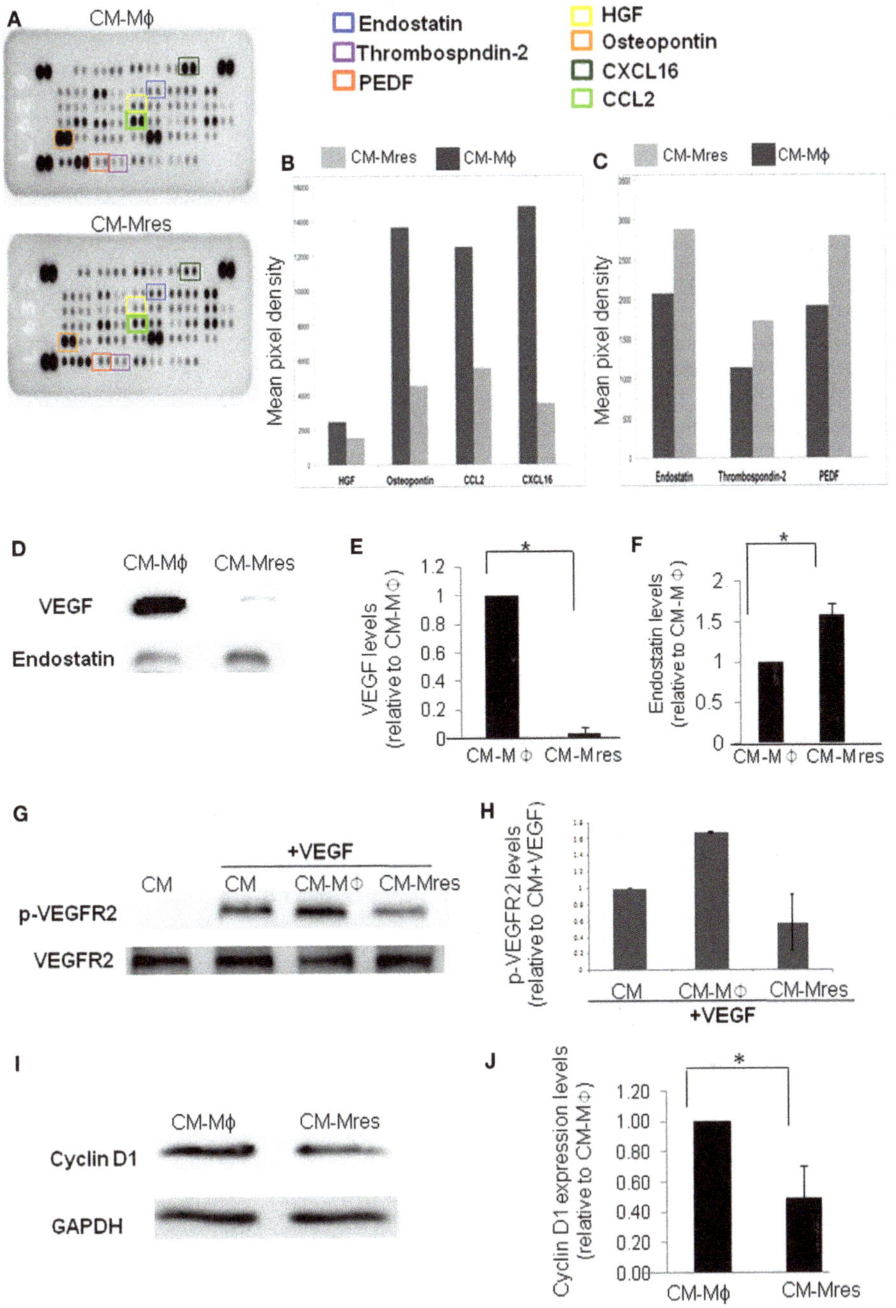

FIGURE 5 | Antiangiogenic factors levels are higher in CM-Mres compared with CM-Mϕ. **(A)** The Proteome Profiler Mouse Angiogenesis Array Kit (Catalog # ARY015) was used to simultaneously assess the relative levels of 53 mouse angiogenesis-related proteins in either CM-Mϕ (upper panel) or CM-Mres (lower panel). **(B)** Quantification of pro-angiogenic factors; HGF, osteopontin, CXCL16, and CCL2. **(C)** Quantification of antiangiogenic factors; endostatin, thrombospondin-2, and PEDF. The histograms **(B,C)** were generated by quantifying the mean spot pixel density from the arrays using image software analysis. **(D)** Representative western blot analysis of vascular endothelial growth factor (VEGF) and endostatin levels in CM-Mϕ compared with CM-Mres. **(E,F)** Quantification of VEGF **(E)** and endostatin levels **(F)**. Densitometry values in panels **(E,F)** were normalized to CM-Mϕ. $n = 3$, t-test, $^{*}p < 0.05$. **(G)** Representative western blot of VEGFR2 phosphorylation in human umbilical vein endothelial cell (HUVEC) induced by VEGF (25 ng/ml) in the presence of the different conditioned media and its quantification **(H)**. **(H)** Densitometry values of p-VEGFR2 were normalized to treatment with condition medium (CM) + VEGF ($n = 2$). Columns; mean, bars; STD. **(I)** Representative western blot of cyclin D1 expression levels in HUVEC treated with either CM-Mϕ or CM-Mres and its quantification **(J)**. **(J)** Densitometry values were normalized to treatment with CM-Mϕ. Columns; mean, bars; STD, $n = 3$, t-test, $^{*}p < 0.05$.

of the cells was observed upon exposing HUVEC to CM-Mres compared with treatment with CM-Mϕ as determined by the wound migration assay and time-lapse live video microscopy. Furthermore, tubular formation of HUVEC on BME was also significantly reduced upon exposure to CM-Mres in comparison to treatment with CM-Mϕ. This was further supported by a significant reduction in the number of bifurcations in the tubular network formed by the endothelial cells treated with CM-Mres in comparison to treatment with CM-Mϕ. Similarly, CM-Mres restrained vascular outgrowth, whereas CM-Mϕ promoted vascular outgrowth, in the rat aorta ring and CAM model systems. Notably, rat aorta ring assay and CAM assay exhibit multiple cell processes involved in angiogenesis as depicted *in vitro* and allow analysis of angiogenesis in an environment composed of multiple cell types and in the physiological context, respectively (31–33). Hence, our results demonstrate that bioactive mediators generated by Mres can inhibit different stages in the angiogenesis process.

Inhibition of angiogenesis is dependent on tilting the delicate balance between pro- and antiangiogenic factors. Therefore, if antiangiogenic factors predominate the angiogenic factors then angiogenesis will not occur (17, 34). Initial insight into the balance between pro- and antiangiogenic factors in CM-Mres vs. CM-Mϕ was obtained by performing a mouse angiogenesis array. Preliminary results demonstrate reduction in some of the pro-angiogenic growth factors and chemokines such as HGF (22), osteopontin (21), CCL2 (35), and CXCL16 (36) in CM-Mres compared with CM-Mϕ. Notably, these mediators were shown previously to modulate motility proliferation and/or survival of endothelial cells. Furthermore, an increase in some of the antiangiogenic mediators such as PEDF, thrombospondin-2 and endostatin [reviewed in Nyberg et al. (25)] was observed in CM-Mres in comparison to CM-Mϕ. Endostatin is a potent inhibitor of angiogenesis. It is a 20-kDa proteolytic fragment of collagen XVIII, which exerts its antiangiogenic effect by inhibiting VEGF binding to VEGFR2 and thus prevents VEGFR2 phosphorylation and activation (27). Notably, along with the increase in endostatin levels in CM-Mres, VEGF levels were significantly reduced in comparison to CM-Mϕ. Hence, this tilt in the balance between VEGF and endostatin in CM-Mres may have resulted in the antiangiogenic effect of the CM-Mres on HUVEC. To further explore this, we tested whether CM-Mres can inhibit VEGF-induced tyrosine phosphorylation of VEGFR2 in HUVEC (27). Indeed, VEGFR2 phosphorylation on Y195 was reduced upon treatment with CM-Mres compared with treatment with CM-Mϕ. Furthermore, we demonstrated reduction in cyclin D1 expression in HUVEC treated with CM-Mres compared with treatment with CM-Mϕ. Notably, in these experiments HUVECs were exposed to angiogenic factors that were added, for instance, bFGF (supplemented in all prepared conditioned media) or present in the CM-Mϕ, such as VEGF. This is in line with a previous report demonstrating endostatin-induced downregulation of cyclin D1 (28) resulting in G1 arrest of endothelial cells that were either treated with bFGF or VEGF. Similarly, the reduction in motility and increase in caspase 3 activity upon treatment with CM-Mres can also be attributed to endostatin angiostatic activity, as described previously (37). Overall, our results suggest that reduction in the levels of VEGF and increase in endostatin levels in CM-Mres may mediate in part the antiangiogenic effect of CM-Mres by inhibiting VEGFR2-mediated downstream signaling. VEGFR2 activation and induction of its downstream signaling by VEGF is one of the key pathways in the angiogenesis process and wound healing repair (38, 39). Hence, the potential antiangiogenic activity of CM-Mres *via* inhibition of VEGFR2 downstream signaling warrants further future experimentation. Overall, our results suggest that pro-resolving $CD11b^{low}$ macrophages can resolve tissue repair by secreting angiostatic mediators, such as endostatin, thus insuring tissue restoration to its homeostatic state.

ETHICS STATEMENT

This study was carried out in accordance with the recommendations of University of Haifa Animal Ethics Committee guidelines. The protocol was approved by the University of Haifa Animal Ethics Committee.

AUTHOR CONTRIBUTIONS

SM, VD, KW, and OGranski performed the experiments and analyzed the data. SM, VD, and KW prepared the figures. SS and OGilon prepared the conditioned media of the macrophages and performed FACS analysis. AM and IS performed the aorta ring assay and its analysis. IS reviewed the manuscript. DB conceived the project, designed the experiments, analyzed the data, and wrote the manuscript.

ACKNOWLEDGMENTS

We thank Dr. Amiram Ariel for the use of FACSCanto II (BD) equipment, Dr. Ram Reshef for assisting with the CAM assay and Dr. Haim Shapiro and Shany Barkan for reviewing the manuscript.

FUNDING

This research was supported by the British and Swiss friends of University of Haifa, University of Haifa seed money (DB), and by John Templeton Foundation (grant # 53057; AM and IS).

REFERENCES

1. Wynn TA. Cellular and molecular mechanisms of fibrosis. *J Pathol* (2008) 214:199–210. doi:10.1002/path.2277
2. Stramer BM, Mori R, Martin P. The inflammation-fibrosis link? A Jekyll and Hyde role for blood cells during wound repair. *J Invest Dermatol* (2007) 127:1009–17. doi:10.1038/sj.jid.5700811
3. Wynn TA, Ramalingam TR. Mechanisms of fibrosis: therapeutic translation for fibrotic disease. *Nat Med* (2012) 18:1028–40. doi:10.1038/nm.2807
4. Mantovani A, Sica A, Sozzani S, Allavena P, Vecchi A, Locati M. The chemokine system in diverse forms of macrophage activation and polarization. *Trends Immunol* (2004) 25:677–86. doi:10.1016/j.it.2004.09.015
5. Martinez FO, Helming L, Gordon S. Alternative activation of macrophages: an immunologic functional perspective. *Annu Rev Immunol* (2009) 27:451–83. doi:10.1146/annurev.immunol.021908.132532
6. Duffield JS, Lupher M, Thannickal VJ, Wynn TA. Host responses in tissue repair and fibrosis. *Annu Rev Pathol* (2013) 8:241–76. doi:10.1146/annurev-pathol-020712-163930
7. Ariel A, Serhan CN. New lives given by cell death: macrophage differentiation following their encounter with apoptotic leukocytes during the resolution of inflammation. *Front Immunol* (2012) 3:4. doi:10.3389/fimmu.2012.00004
8. Fadok VA, Bratton DL, Konowal A, Freed PW, Westcott JY, Henson PM. Macrophages that have ingested apoptotic cells in vitro inhibit proinflammatory cytokine production through autocrine/paracrine mechanisms involving TGF-beta, PGE2, and PAF. *J Clin Invest* (1998) 101:890–8. doi:10.1172/JCI1112
9. Koh TJ, DiPietro LA. Inflammation and wound healing: the role of the macrophage. *Expert Rev Mol Med* (2011) 13:e23. doi:10.1017/S1462399411001943
10. Wynn TA, Vannella KM. Macrophages in tissue repair, regeneration, and fibrosis. *Immunity* (2016) 44:450–62. doi:10.1016/j.immuni.2016.02.015
11. Schif-Zuck S, Gross N, Assi S, Rostoker R, Serhan CN, Ariel A. Saturated-efferocytosis generates pro-resolving CD11b low macrophages: modulation by resolvins and glucocorticoids. *Eur J Immunol* (2011) 41:366–79. doi:10.1002/eji.201040801
12. Cash JL, White GE, Greaves DR. Chapter 17. Zymosan-induced peritonitis as a simple experimental system for the study of inflammation. *Methods Enzymol* (2009) 461:379–96. doi:10.1016/S0076-6879(09)05417-2
13. Barkan D, Green JE. An in vitro system to study tumor dormancy and the switch to metastatic growth. *J Vis Exp* (2011) (54). doi:10.3791/2914
14. Bellacen K, Lewis EC. Aortic ring assay. *J Vis Exp* (2009) (33):1564. doi:10.3791/1564
15. Ponce ML, Kleinmann HK. The chick chorioallantoic membrane as an in vivo angiogenesis model. *Curr Protoc Cell Biol* (2003) 18(1):19.5.1–19.5.6. doi:10.1002/0471143030.cb1905s18
16. Lawrence T. Inflammation and cancer: a failure of resolution? *Trends Pharmacol Sci* (2007) 28:162–5. doi:10.1016/j.tips.2007.02.003
17. Distler JH, Hirth A, Kurowska-Stolarska M, Gay RE, Gay S, Distler O. Angiogenic and angiostatic factors in the molecular control of angiogenesis. *Q J Nucl Med* (2003) 47:149–61.
18. Herbert SP, Stainier DY. Molecular control of endothelial cell behaviour during blood vessel morphogenesis. *Nat Rev Mol Cell Biol* (2011) 12:551–64. doi:10.1038/nrm3176
19. Arnaoutova I, Kleinman HK. In vitro angiogenesis: endothelial cell tube formation on gelled basement membrane extract. *Nat Protoc* (2010) 5:628–35. doi:10.1038/nprot.2010.6
20. Kubota Y, Kleinman HK, Martin GR, Lawley TJ. Role of laminin and basement membrane in the morphological differentiation of human endothelial cells into capillary-like structures. *J Cell Biol* (1988) 107:1589–98. doi:10.1083/jcb.107.4.1589
21. Dai J, Peng L, Fan K, Wang H, Wei R, Ji G, et al. Osteopontin induces angiogenesis through activation of PI3K/AKT and ERK1/2 in endothelial cells. *Oncogene* (2009) 28:3412–22. doi:10.1038/onc.2009.189
22. Bussolino F, Di Renzo MF, Ziche M, Bocchietto E, Olivero M, Naldini L, et al. Hepatocyte growth factor is a potent angiogenic factor which stimulates endothelial cell motility and growth. *J Cell Biol* (1992) 119:629–41. doi:10.1083/jcb.119.3.629
23. Isozaki T, Arbab AS, Haas CS, Amin MA, Arendt MD, Koch AE, et al. Evidence that CXCL16 is a potent mediator of angiogenesis and is involved in endothelial progenitor cell chemotaxis: studies in mice with K/BxN serum-induced arthritis. *Arthritis Rheum* (2013) 65:1736–46. doi:10.1002/art.37981
24. Stamatovic SM, Keep RF, Mostarica-Stojkovic M, Andjelkovic AV. CCL2 regulates angiogenesis via activation of Ets-1 transcription factor. *J Immunol* (2006) 177:2651–61. doi:10.4049/jimmunol.177.4.2651
25. Nyberg P, Xie L, Kalluri R. Endogenous inhibitors of angiogenesis. *Cancer Res* (2005) 65:3967–79. doi:10.1158/0008-5472.CAN-04-2427
26. Hoeben A, Landuyt B, Highley MS, Wildiers H, Van Oosterom AT, De Bruijn EA. Vascular endothelial growth factor and angiogenesis. *Pharmacol Rev* (2004) 56:549–80. doi:10.1124/pr.56.4.3
27. Kim YM, Hwang S, Kim YM, Pyun BJ, Kim TY, Lee ST, et al. Endostatin blocks vascular endothelial growth factor-mediated signaling via direct interaction with KDR/Flk-1. *J Biol Chem* (2002) 277:27872–9. doi:10.1074/jbc.M202771200
28. Hanai J, Dhanabal M, Karumanchi SA, Albanese C, Waterman M, Chan B, et al. Endostatin causes G1 arrest of endothelial cells through inhibition of cyclin D1. *J Biol Chem* (2002) 277:16464–9. doi:10.1074/jbc.M112274200
29. Leibovich SJ, Ross R. The role of the macrophage in wound repair. A study with hydrocortisone and antimacrophage serum. *Am J Pathol* (1975) 78: 71–100.
30. Polverini PJ, Cotran PS, Gimbrone MA Jr, Unanue ER. Activated macrophages induce vascular proliferation. *Nature* (1977) 269:804–6. doi:10.1038/269804a0
31. Nicosia RF, Bonanno E, Villaschi S. Large-vessel endothelium switches to a microvascular phenotype during angiogenesis in collagen gel culture of rat aorta. *Atherosclerosis* (1992) 95:191–9. doi:10.1016/0021-9150(92)90022-9
32. Nicosia RF, Villaschi S. Rat aortic smooth muscle cells become pericytes during angiogenesis in vitro. *Lab Invest* (1995) 73:658–66.
33. Staton CA, Reed MW, Brown NJ. A critical analysis of current in vitro and in vivo angiogenesis assays. *Int J Exp Pathol* (2009) 90:195–221. doi:10.1111/j.1365-2613.2008.00633.x
34. Tabruyn SP, Griffioen AW. Molecular pathways of angiogenesis inhibition. *Biochem Biophys Res Commun* (2007) 355:1–5. doi:10.1016/j.bbrc.2007.01.123
35. Mehrad B, Keane MP, Strieter RM. Chemokines as mediators of angiogenesis. *Thromb Haemost* (2007) 97:755–62. doi:10.1160/TH07-01-0040
36. Zhuge X, Murayama T, Arai H, Yamauchi R, Tanaka M, Shimaoka T, et al. CXCL16 is a novel angiogenic factor for human umbilical vein endothelial cells. *Biochem Biophys Res Commun* (2005) 331:1295–300. doi:10.1016/j.bbrc.2005.03.200
37. Dhanabal M, Ramchandran R, Waterman MJ, Lu H, Knebelmann B, Segal M, et al. Endostatin induces endothelial cell apoptosis. *J Biol Chem* (1999) 274:11721–6. doi:10.1074/jbc.274.17.11721
38. Millauer B, Wizigmann-Voos S, Schnurch H, Martinez R, Moller NP, Risau W, et al. High affinity VEGF binding and developmental expression suggest Flk-1 as a major regulator of vasculogenesis and angiogenesis. *Cell* (1993) 72:835–46. doi:10.1016/0092-8674(93)90573-9
39. Shalaby F, Rossant J, Yamaguchi TP, Gertsenstein M, Wu XF, Breitman ML, et al. Failure of blood-island formation and vasculogenesis in Flk-1-deficient mice. *Nature* (1995) 376:62–6. doi:10.1038/376062a0

Conflict of Interest Statement: The authors declare that the research was conducted in the absence of any commercial or financial relationships that could be construed as a potential conflict of interest.

The handling Editor declared a shared affiliation, though no other collaboration, with the authors.

Permissions

All chapters in this book were first published in FI, by Frontiers Media SA; hereby published with permission under the Creative Commons Attribution License or equivalent. Every chapter published in this book has been scrutinized by our experts. Their significance has been extensively debated. The topics covered herein carry significant findings which will fuel the growth of the discipline. They may even be implemented as practical applications or may be referred to as a beginning point for another development.

The contributors of this book come from diverse backgrounds, making this book a truly international effort. This book will bring forth new frontiers with its revolutionizing research information and detailed analysis of the nascent developments around the world.

We would like to thank all the contributing authors for lending their expertise to make the book truly unique. They have played a crucial role in the development of this book. Without their invaluable contributions this book wouldn't have been possible. They have made vital efforts to compile up to date information on the varied aspects of this subject to make this book a valuable addition to the collection of many professionals and students.

This book was conceptualized with the vision of imparting up-to-date information and advanced data in this field. To ensure the same, a matchless editorial board was set up. Every individual on the board went through rigorous rounds of assessment to prove their worth. After which they invested a large part of their time researching and compiling the most relevant data for our readers.

The editorial board has been involved in producing this book since its inception. They have spent rigorous hours researching and exploring the diverse topics which have resulted in the successful publishing of this book. They have passed on their knowledge of decades through this book. To expedite this challenging task, the publisher supported the team at every step. A small team of assistant editors was also appointed to further simplify the editing procedure and attain best results for the readers.

Apart from the editorial board, the designing team has also invested a significant amount of their time in understanding the subject and creating the most relevant covers. They scrutinized every image to scout for the most suitable representation of the subject and create an appropriate cover for the book.

The publishing team has been an ardent support to the editorial, designing and production team. Their endless efforts to recruit the best for this project, has resulted in the accomplishment of this book. They are a veteran in the field of academics and their pool of knowledge is as vast as their experience in printing. Their expertise and guidance has proved useful at every step. Their uncompromising quality standards have made this book an exceptional effort. Their encouragement from time to time has been an inspiration for everyone.

The publisher and the editorial board hope that this book will prove to be a valuable piece of knowledge for researchers, students, practitioners and scholars across the globe.

List of Contributors

Qian Zheng, Lei Yuan, Ning Gao, Qi Feng and Hui Xiao
Key Laboratory of the Ministry of Education for Medicinal Plant Resources and Natural Pharmaceutical Chemistry, National Engineering Laboratory for Resource Development of Endangered Crude Drugs in the Northwest of China, College of Life Sciences, Shaanxi Normal University Xi'an, Xi'an, China

AiYing Ma
Key Laboratory of the Ministry of Education for Medicinal Plant Resources and Natural Pharmaceutical Chemistry, National Engineering Laboratory for Resource Development of Endangered Crude Drugs in the Northwest of China, College of Life Sciences, Shaanxi Normal University Xi'an, Xi'an, China
College of Biological Sciences and Engineering, Beifang University of Nationalities, Yinchuan, NingXia, China

Nathalie C. Franc
Department of Cell and Molecular Biology, The Scripps Research Institute, La Jolla, CA, United States

Zsuzsa Szondy and Krisztina Köröskényi
Department of Biochemistry and Molecular Biology of Medical Faculty, University of Debrecen, Debrecen, Hungary
Department of Basic Medical Sciences of Dental Faculty, University of Debrecen, Debrecen, Hungary

Zsolt Sarang, Beáta Kiss and Éva Garabuczi
Department of Biochemistry and Molecular Biology of Medical Faculty, University of Debrecen, Debrecen, Hungary

Catherine Lynch, Maria Panagopoulou and Christopher D. Gregory
MRC Centre for Inflammation Research, Queen's Medical Research Institute, University of Edinburgh, Edinburgh, United Kingdom

Siamon Gordon
College of Medicine, Graduate Institute of Biomedical Sciences, Chang Gung University, Taoyuan City, Taiwan
Sir William Dunn School of Pathology, University of Oxford, Oxford, United Kingdom

Annette Plüddemann
Nuffield Department of Primary Care Health Sciences, University of Oxford, Oxford, United Kingdom
Uriel Trahtemberg
General Intensive Care Unit, Hadassah-Hebrew University Medical Center, Jerusalem, Israel

Dror Mevorach
Rheumatology Research Center, Department of Medicine, Hadassah-Hebrew University Medical Center, Jerusalem, Israel

Delphine Lumbroso
Faculty of Dental Medicine, Institute for Dental Sciences, Hadassah Medical School, Hebrew University of Jerusalem, Jerusalem, Israel
Department of Biology, The Faculty of Natural Sciences, University of Haifa, Haifa, Israel

Soaad Soboh, Sagie Schif-Zuck and Amiram Ariel
Faculty of Dental Medicine, Institute for Dental Sciences, Hadassah Medical School, Hebrew University of Jerusalem, Jerusalem, Israel

Tal Burstyn-Cohen and Avi Maimon
Department of Biology, The Faculty of Natural Sciences, University of Haifa, Haifa, Israel

Michaela Jung, Andreas Weigert and Claudia Rehwald
Faculty of Medicine, Institute of Biochemistry I, Goethe University Frankfurt, Frankfurt, Germany

Christina Mertens
Faculty of Medicine, Institute of Biochemistry I, Goethe University Frankfurt, Frankfurt, Germany
Faculty 15, Biological Sciences, Goethe University Frankfurt, Frankfurt, Germany

Bernhard Brüne
Faculty of Medicine, Institute of Biochemistry I, Goethe University Frankfurt, Frankfurt, Germany
Project Group Translational Medicine and Pharmacology TMP, Fraunhofer Institute for Molecular Biology and Applied Ecology, IME, Frankfurt, Germany

Philippe Saas
INSERM, EFS BFC, UMR1098, Interactions Hôte-Greffon-Tumeur/Ingénierie Cellulaire et Génique, Fédération Hospitalo-Universitaire INCREASE, LabEx LipSTIC, Université Bourgogne Franche-Comté, Besançon, France
INSERM CIC-1431, University Hospital of Besançon, Clinical Investigation Center in Biotherapy, Fédération Hospitalo-Universitaire INCREASE, LabEx LipSTIC, Besançon, France

Francis Bonnefoy and Sylvain Perruche
INSERM, EFS BFC, UMR1098, Interactions Hôte-Greffon-Tumeur/Ingénierie Cellulaire et Génique, Fédération Hospitalo-Universitaire INCREASE, LabEx LipSTIC, Université Bourgogne Franche-Comté, Besançon, France

Eric Toussirot
INSERM CIC-1431, University Hospital of Besançon, Clinical Investigation Center in Biotherapy, Fédération Hospitalo-Universitaire INCREASE, LabEx LipSTIC, Besançon, France
Department of Rheumatology, University Hospital of Besançon, Besançon, France
Department of Therapeutics, Université Bourgogne Franche-Comté, UPRES EA 4266, Pathogenic Agents and Inflammation, Besancon, France

Jesmond Dalli
Lipid Mediator Unit, William Harvey Research Institute, Bart's and the London School of Medicine, Queen Mary University of London, London, United Kingdom

Charles N. Serhan
Center for Experimental Therapeutics and Reperfusion Injury, Harvard Institutes of Medicine, Brigham and Women's Hospital and Harvard Medical School, Boston, MA, United States

Nicole D. Barth, John A. Marwick, Marc Vendrell, Adriano G. Rossi and Ian Dransfield
MRC Centre for Inflammation Research, Queen's Medical Research Institute, University of Edinburgh, Edinburgh, United Kingdom

Aizhen Yang, Fengwu Chen and Chao He
Cyrus Tang Hematology Center, Collaborative Innovation Center of Hematology, Soochow University, Suzhou, China

Junsong Zhou and Yi Wu
Cyrus Tang Hematology Center, Collaborative Innovation Center of Hematology, Soochow University, Suzhou, China
The Sol Sherry Thrombosis Research Center, Temple University School of Medicine, Philadelphia, PA, United States

Yi Lu
Wuhan Thalys Medical Technology Inc., Wuhan, China

Jihong Dai
The Sol Sherry Thrombosis Research Center, Temple University School of Medicine, Philadelphia, PA, United States
Department of Pathology and Laboratory Medicine, Rutgers University-New Jersey Medical School, Newark, NJ, United States

Raymond B. Birge
Department of Microbiology, Biochemistry and Molecular Genetics, Rutgers University-New Jersey Medical School, Newark, NJ, United States

Gerhard E. Grossmayer, Sebastian Boeltz, Jürgen Rech, Martin Herrmann and Luis E. Muñoz
Department of Internal Medicine 3 - Rheumatology and Immunology, Universitätsklinikum Erlangen, Friedrich-Alexander- Universität Erlangen-Nürnberg, Erlangen, Germany

Hildegard Keppeler
Department of Internal Medicine II, University of Tübingen, Tübingen, Germany

Christina Janko
Department of Internal Medicine 3 - Rheumatology and Immunology, Universitätsklinikum Erlangen, Friedrich-Alexander- Universität Erlangen-Nürnberg, Erlangen, Germany
Department of Otorhinolaryngology, Head and Neck Surgery, Universitätsklinikum Erlangen, Else Kröner-Fresenius-Stiftung Professorship, Erlangen, Germany

Kirsten Lauber
Department of Radiation Oncology and Radiotherapy, University Hospital, Ludwig-Maximilians-University, Munich, Germany

Lindsey D. Hughes
Department of Immunobiology, School of Medicine, Yale University, New Haven, CT, United States

Lidia Bosurgi
I. Medizinische Klinik und Poliklinik, Universitätsklinikum Hamburg-Eppendorf, Hamburg, Germany
Bernard-Nocht-Institut für Tropenmedizin, Hamburg, Germany

Sourav Ghosh
Department of Pharmacology, School of Medicine, Yale University, New Haven, CT, United States
Department of Neurology, School of Medicine, Yale University, New Haven, CT, United States

Carla V. Rothlin
Department of Immunobiology, School of Medicine, Yale University, New Haven, CT, United States
Department of Pharmacology, School of Medicine, Yale University, New Haven, CT, United States

Scott P. Davies and Zania Stamataki
Centre for Liver Research, College of Medical and Dental Sciences, Institute for Immunology and Immunotherapy, University of Birmingham, Birmingham, United Kingdom

Gary M. Reynolds
Centre for Liver Research, College of Medical and Dental Sciences, Institute for Immunology and Immunotherapy, University of Birmingham, Birmingham, United Kingdom
Centre for Liver Research and National Institute for Health Research (NIHR) Birmingham Liver Biomedical Research Unit, College of Medical and Dental Sciences, Institute of Immunology and Immunotherapy, University of Birmingham, Birmingham, United Kingdom
University Hospitals Birmingham NHS Foundation Trust, Birmingham, United Kingdom

David S. Ucker
Department of Microbiology and Immunology, University of Illinois College of Medicine, Chicago, IL, United States

Jerrold S. Levine
Department of Medicine, Division of Nephrology, University of Illinois College of Medicine, Jesse Brown Veterans Affairs Medical Center, Chicago, IL, United States

Sandy B. Serizier
Department of Biology, Boston University, Boston, MA, United States
Graduate Program in Molecular Biology, Cell Biology and Biochemistry, Boston University, Boston, MA, United States

Kimberly McCall
Department of Biology, Boston University, Boston, MA, United States

Matthew DeBerge, Shuang Zhang, Kristofor Glinton, Luba Grigoryeva, Islam Hussein, Esther Vorovich, Karen Ho, Xunrong Luo and Edward B. Thorp
Department of Pathology, Feinberg Cardiovascular Research Institute, Feinberg School of Medicine, Northwestern University, Chicago, IL, United States

Pascale Tacnet-Delorme and Philippe Frachet
Université Grenoble Alpes, CEA, CNRS, IBS, Grenoble, France

Julie Gabillet
INSERM U1016, Cochin Institute, Paris, France
CNRS UMR 8104, Paris, France
Université Paris-Descartes, Sorbonne Paris Cité, Paris, France

Simon Chatfield, Nathalie Thieblemont and Véronique Witko-Sarsat
INSERM U1016, Cochin Institute, Paris, France
CNRS UMR 8104, Paris, France
Université Paris-Descartes, Sorbonne Paris Cité, Paris, France
Center of Excellence, LABEX Inflamex, Paris, France

Firzan Nainu
Laboratory of Pharmacology and Toxicology, Faculty of Pharmacy, Hasanuddin University, Makassar, Indonesia
Laboratory of Host Defense and Responses, Graduate School of Medical Sciences, Kanazawa University, Kanazawa, Japan

Akiko Shiratsuchi and Yoshinobu Nakanishi
Laboratory of Host Defense and Responses, Graduate School of Medical Sciences, Kanazawa University, Kanazawa, Japan

Ke Geng, Sushil Kumar, Stanley G. Kimani, Canan Kasikara and Sergei V. Kotenko
Department of Microbiology, Biochemistry, and Molecular Genetics, New Jersey Medical School Cancer Center, Rutgers, State University of New Jersey, Newark, New Jersey, United States

Vladyslav Kholodovych
Office of Advanced Research Computing (OARC), Rutgers, State University of New Jersey, Newark, New Jersey, United States
Department of Pharmacology, Robert Wood Johnson Medical School, Rutgers, State University of New Jersey, Piscataway, NJ, United States

Kensaku Mizuno
Department of Biomolecular Sciences, Graduate School of Life Sciences, Tohoku University, Sendai, Miyagi, Japan

Oleta Sandiford and Pranela Rameshwar
Department of Medicine, Rutgers, State University of New Jersey, Newark, New Jersey, United States

Shira Michaeli, Vivian Dakwar, Keren Weidenfeld, Ortal Granski, Odelya Gilon, Sagie Schif-Zuck and Dalit Barkan
Department of Human Biology, University of Haifa, Haifa, Israel

Anatolii Mamchur and Imad Shams
Department of Evolutionary and Environmental Biology, Institute of Evolution, University of Haifa, Haifa, Israel

Index

I

L

M

N

O

P

R

S

T

V

W

www.ingramcontent.com/pod-product-compliance
Lightning Source LLC
LaVergne TN
LVHW080952210926
848485LV00004B/827